Phenology: An Integrative Environmental Science

Mark D. Schwartz
Editor

Phenology: An Integrative Environmental Science

Second Edition

Editor
Mark D. Schwartz
Department of Geography
University of Wisconsin-Milwaukee
Milwaukee, USA

ISBN 978-94-007-6924-3 ISBN 978-94-007-6925-0 (eBook)
DOI 10.1007/978-94-007-6925-0
Springer Dordrecht Heidelberg New York London

Library of Congress Control Number: 2013943824

Printed on acid-free paper

Springer is part of Springer Science+Business Media (www.springer.com)

Foreword

I was both surprised and delighted when Mark D. Schwartz asked me to write the foreword for the 2nd Edition of *Phenology: An Integrative Environmental Science*. I came into phenological science and networks fairly recently and, frankly, through the back door. In August 2004, Pat Mulholland (Oak Ridge National Laboratories), David Breshears (University of Arizona), and I convened a 3-day workshop in Tucson, Arizona, on how existing and planned national networks in the U.S.A. might be used to understand, monitor and forecast ecological responses to climate variability and change (http://www.neoninc.org/documents/neon-climate-report.pdf).

David, Pat, and I gathered 30 prominent scientists, encompassing the fields of ecology, hydrology, and climatology. It fast became clear in the group that there were serious mismatches in assumptions and scales of investigation, not just mutually-unintelligible jargon. Some pressed to make comprehensive measurements at a few sites to get at complex responses in ecological process, while others advocated broad scale monitoring to reveal emergent and large-scale patterns driven by climatic variations and trends.

The tension at the 2004 workshop only served to reinforce my own prejudices about some of the interdisciplinary challenges facing global change biology. If ecologists are to be successful in distinguishing competing and interacting causes of large-scale ecological changes and associated feedbacks to the atmosphere and hydrosphere, they will need to match the spatial and temporal scales of analysis employed routinely by climatologists and hydrologists. A fundamental need are networks of routine, standardized, and integrated observations, on the ground and from space, strategically deployed to gage ecological variability and change across the mosaic of hydroclimatic areas, biomes, and anthromes (a term I learned from Chap. 26 by de Beurs and Henebry) that comprise the United States...and the world.

Phenology, the gateway to climatic effects on the biosphere and associated feedbacks to the atmosphere, seemed as good a place to start as any. Right after the 2004 workshop, Steve Running connected me with Mark D. Schwartz.

In collaboration with several colleagues and institutions, including base stable support from the U.S. Geological Survey, Mark and I helped launch the USA-National Phenology Network (USA-NPN).

Previously, I had studied phenology only from afar. To get up to speed, for the past few years I have lugged my copy of the 1st Edition of *Phenology: An Integrative Environmental Science* on planes and even in the field, loaning it repeatedly to students and colleagues, always anxious to get it back. To continue my ongoing education, I have now read each updated and new chapter in the 2nd Edition front to back. This new tome, and the exponential growth in primary literature since the 1st Edition, marks "the rebirth of phenology…as a critical element of global change research" (Richardson et al. 2012, p. 157).

Having toiled at it myself, I very much appreciate that the authors of each chapter pay homage to the history of network development in each country. I spent hours listening to the late Joseph M. Caprio talk about the beginnings of the U.S. Western States Phenological Network in the 1950s, and I can imagine similar stories told by the late Coching Chu, who pioneered phenological networks in China during the 1930s (Chap. 2 by Chen). In 2007, Kjell Bolmgren, then a postdoc at University of California-Berkeley, sat next to me in one of our USA-NPN planning workshops. There was then a gleam in his eye, and there is now something called the Swedish Phenological Network.

This proliferation in networks worldwide is exciting, and with it comes the responsibility to forge a meaningful and effective, global community of practice. Over the long term, some of this could be accomplished through the integration of phenology in higher education. *Phenology: An Integrative Environmental Science* would make a great textbook for a novel course that could be taught globally. Such a course would blend classroom and online learning of first principles, integrated systems, and quantitative and modeling skills with the generation and use of data products from both observational networks and remote sensing. The examples given in Chap. 31 (by de Beurs and others) provide an excellent start.

I will take license here and point to some future directions for a 3rd Edition of *Phenology: An Integrative Environmental Science*. As Helmut Lieth remarked in his Foreword to the 1st Edition, phenology arises from planet Earth tumbling around the sun. Moreover, atmospheric planetary-scale waves drive temporally and spatially averaged exchanges of heat, momentum, and water vapor that ultimately determine and synchronize large-scale patterns in phenology, growth, demography, disturbance, biogeochemical cycling, and atmospheric feedbacks. At its core, phenology is the biological expression of climatology. To make real progress, particularly when it comes to prediction, we must fully engage climatologists to focus on regional to global patterns and sources for seasonal timing variations and trends in the climate system. To date, this has not been a particular focus in the climate community, but it is fertile and essential ground for integrative environmental science and, specifically, phenology.

For example, the annual phasing of temperatures advanced about 1.5 days over the Northern Hemisphere, due in large part to changes in atmospheric circulation in the 1980s, but we are not totally sure why (Stine et al. 2009; Stine and Huybers 2012).

Additionally, so-called warming holes[1] extend from the southeastern U.S. across the Atlantic from Scandinavia to Siberia and northern China. Such warming holes have muted advances and even delayed spring onset in the southeastern U.S. and Eurasia, and have been explained as intrinsic decadal variability in the Pacific (Meehl et al. 2012) and poorly understood interactions between Eurasian snow cover and the Arctic Oscillation (Cohen et al. 2012), respectively. Not all regional trends in phenology can reliably be attributed to greenhouse warming.

Phenology is maturing as a global change science, and with this maturity comes an obligation to get it right. Discrepancies always arise from comparative approaches in global ecology. Recent meta-analyses and cross-method comparisons reveal poor agreement between: (1) seasonality of ecosystem-scale CO_2 exchange in terrestrial biosphere models and actual flux tower measurements (Richardson et al. 2013; but see Kovalskyy et al. 2012); (2) different remote sensing platforms and algorithms used to define start of season (White et al. 2009); and (3) temperature-sensitivities for timing of leaf-out and flowering identified from warming experiments versus historical observations across the Northern Hemisphere (Wolkovich et al. 2012). So how do we choose which numbers are right to use in assessments and models to identify vulnerabilities and predict the future?

The 1st and the 2nd Editions of *Phenology: An Integrative Environmental Science* laid the necessary groundwork for a critical component of global change research. The phenological community should strive to resolve these seminal questions, and I very much look forward to reading the solutions in the 3rd Edition.

Senior Scientist, U.S. Geological Survey
Reston, VA, USA, March 2013
Julio L. Betancourt

References

Cohen JL, Furtado JC, Barlow MA, Alexev VA, Cherry JE (2012) Arctic warming, increasing snow cover and widespread boreal winter cooling. Environ Res Lett 26:345–348

Kovalskyy V, Roy DP, Zhang X, Ju J (2012) The suitability of multi-temporal web-enabled Landsat data NDVI for phenological monitoring – a comparison with flux tower and MODIS NDVI. Remote Sens Lett 3:325–334

Meehl GA, Arblaster JM, Branstator G (2012) Mechanisms contributing to the warming hole and the consequent U.S. east–west differential of heat extremes. J Clim 25:6394–6408

Richardson AD, Anderson RS, Arain MA, Barr AG, Bohrer G, Chen G, Chen JM, Ciais P, Davis KJ, Desai AR, Dietze MC, Dragoni D, Garrity SR, Gough CM, Grant R, Hollinger DY, Margolis HA, McCaughey H, Migliavacca M, Monson RK, Munger JW, Poulter B, Raczka BM, Ricciuto DM, Sahoo AK, Schaefer K, Tian H, Vargas R, Verbeeck H, Xiao J, Xue Y (2012) Terrestrial biosphere models need better representation of vegetation phenology: results from the North American Carbon Program Site Synthesis. Global Change Biol 18:566–584

[1] Large regions in the world where either seasonal cooling has occurred or the rate of warming has been slower than elsewhere.

Richardson AD, Keenan TF, Migliavacca M, Sonnentag O, Ryu Y, Toomey M (2013) Climate change, phenology, and phenological control of vegetation feedbacks to the climate system. Agric For Meteorol 169:156–173

Stine AR, Huybers P (2012) Changes in the seasonal cycle of temperature and atmospheric circulation. J Clim 25:7362–7380

Stine AR, Huybers P, Fung IY (2009) Changes in the phase of the annual cycle of surface temperature. Nature 457:435–440

White MA, de Beurs KM, Didan K, Inouye DW, Richardson AD, Jensen OP, O'Keefe J, Zhang G, Nemani RR, van Leeuwen WJD, Brown JF, de Wit A, Schaepman M, Lin X, Dettinger M, Bailey AS, Kimball J, Schwartz MD, Baldocchi DD, Lee JT, Lauenroth WK (2009) Intercomparison, interpretation, and assessment of spring phenology in North America estimated from remote sensing for 1982 to 2006. Global Change Biol 15(10):2335–2359

Wolkovich EM, Cook BI, Allen JM, Crimmins TM, Travers S, Pau S, Regetz J, Davies TJ, Betancourt JL, Kraft NJB, Ault TR, Bolmgren K, Mazer SJ, McCabe GJ, McGill BJ, Parmesan C, Salamin N, Schwartz MD, Cleland EE (2012) Warming experiments underpredict plant phenological responses to climate change. Nature 485(7399):494–497

Preface

In the preface to the first edition I described my personal phenological research "journey", and reviewed the conditions which enabled the successful creation of a general phenological reference volume in 2003. Those conditions were: (1) sufficient interest in the topic by the general scientific community; and (2) an interconnected community of phenological researchers with the diversity of research expertise necessary to cover the range of required topics.

Looking back over the last 10 years, it is clear that interest in phenological research has grown significantly while an increase in the number and range of scientific publications indicates an expansion in the diversification of the subject. The validity of phenological research is evident by inclusion, in the IPCC's 4th Assessment Report (2007), of a range of phenological records to demonstrate that climate change was having a detectable impact on the natural environment. In addition, a number of national phenological observation networks have been initiated in several countries, including Australia, Sweden, Turkey, and the United States (I am co-founder of the USA National Phenology Network, USA-NPN). Furthermore, two successful interdisciplinary international phenology conferences have also been held: "Phenology 2010" (Dublin, Ireland) and "Phenology 2012" (Milwaukee, Wisconsin, USA).

Thus, over the last decade a dynamic international and interdisciplinary phenological research community has matured, which this second edition of *Phenology: An Integrative Environmental Science* is designed to nurture and serve as we move forward over the coming decade.

Milwaukee, WI, USA, March 2013 Mark D. Schwartz

Acknowledgements

As with the first edition, I am grateful for the help I received from many individuals as this second edition of the book was constructed. I would especially like to thank my wife, Ann Lessner Schwartz, for her support and patience. My graduate student assistant, Isaac Park, provided invaluable support, including many long hours of tedious work reformatting and checking references. The following individuals generously took the time to review one or more of the chapter manuscripts: Gregory Carbone, Lynda Chambers, Xiaoqiu Chen, Frank-M. Chmielewski, Theresa Crimmins, Ellen Denny, Ankur Desai, Emanuele Eccel, Jonathan Hanes, Sandra Henderson, Geoffrey Henebry, Stein-Rune Karlsen, Marie Keatley, Elisabeth Koch, Koen Kramer, Liang Liang, Cary Mock, Anders Møller, David Moore, Eric Post, Jacques Régnière, Thomas Rötzer, Patricia Selkirk, and Rudi Stickler. In particular, I want to recognize the exceptional assistance I received from Alison Donnelly in not only reviewing numerous chapters, but also providing logistical support and advice. I also appreciate the assistance and patience of all the editors and staff at Springer.

Contents

Part II Phenologies of Selected Bioclimatic Zones

Part III Phenological Models and Techniques

Part IV Sensor-Derived Phenology

Contributing Authors

Pablo Arroyo-Mora Department of Ecology and Evolutionary Biology, University of Connecticut, Storrs, CT, USA

LoriAnne Barnett Education Program, USA National Phenology Network, National Coordinating Office, Tucson, AZ, USA

Elisabeth G. Beaubien Department of Renewable Resources, University of Alberta, Edmonton, AB, Canada

Dana M. Bergstrom Australian Antarctic Division, Kingston, TAS, Australia

Ekko Bruns Department of Networks and Data, German Meteorological Service, Offenbach, Germany

Maria Gabriela G. Camargo Departamento de Botânica, Laboratório de Fenologia, Plant Phenology and Seed Dispersal Research Group, Instituto de Biociências, Universidade Estadual Paulista UNESP, São Paulo, Brazil

Carla Cesaraccio Institute of Biometeorology, National Research Council, Sassari, Italy

Lynda E. Chambers Centre for Australian Weather and Climate Research, Bureau of Meteorology, Melbourne, VIC, Australia

Xiaoqiu Chen College of Urban and Environmental Sciences, Peking University, Beijing, China

Frank-M. Chmielewski Agricultural Climatology, Department of Crop and Animal Sciences, Faculty of Agriculture and Horticulture, Humboldt-University of Berlin, Berlin, Germany

Isabelle Chuine Centre d'Ecologie Fonctionnelle et Evolutive, CNRS, Montpellier, France

Robert B. Cook Environmental Sciences Division, Oak Ridge National Laboratory, Oak Ridge, TN, USA

Humphrey Q.P. Crick Natural England, Peterborough, UK

Theresa M. Crimmins National Coordinating Office, USA National Phenology Network, Tucson, AZ, USA

Kirsten M. de Beurs Department of Geography and Environmental Sustainability, The University of Oklahoma, Norman, OK, USA

Iñaki Garcia de Cortazar-Atauri AGROCLIM INRA, Avignon, France

Pierpaolo Duce Institute of Biometeorology, National Research Council, Sassari, Italy

Peter O. Dunn Department of Biological Sciences, University of Wisconsin-Milwaukee, Milwaukee, WI, USA

Yuling Fu Institute of Geographical Sciences and Natural Resources Research, Chinese Academy of Sciences, Beijing, China

Eliana Gressler Departamento de Botânica, Laboratório de Fenologia, Plant Phenology and Seed Dispersal Research Group, Instituto de Biociências, Universidade Estadual Paulista UNESP, São Paulo, Brazil

Wulf Greve Senckenberg Research Institute, German Center for Marine Biodiversity Research, Hamburg, Germany

Lianhong Gu Environmental Sciences Division, Oak Ridge National Laboratory, Oak Ridge, TN, USA

Brian Haggerty Department of Ecology, Evolution, and Marine Biology, University of California, Santa Barbara, CA, USA

Jonathan M. Hanes Department of Geography, University of Wisconsin-Milwaukee, Milwaukee, WI, USA

Heikki Hänninen Department of Biosciences, University of Helsinki, Helsinki, Finland

Stefan Heider Agricultural Climatology, Department of Crop and Animal Sciences, Faculty of Agriculture and Horticulture, Humboldt-University of Berlin, Berlin, Germany

Geoffrey M. Henebry Geographic Information Science Center of Excellence, South Dakota State University, Brookings, SD, USA

Alisa Hove Biology Department, Warren Wilson College, Asheville, NC, USA

David W. Inouye Department of Biology and Rocky Mountain Biological Laboratory, University of Maryland, College Park, MD, USA

Gregory V. Jones Department of Environmental Studies, Southern Oregon University, Ashland, OR, USA

Margaret E. Kalacska Earth and Atmospheric Sciences Department, University of Alberta, Edmonton, AB, Canada

Marie R. Keatley Department of Forest and Ecosystem Science, University of Melbourne, Creswick, VIC, Australia

Jeffrey Kerby Department of Biology, The Pennsylvania State University, University Park, PA, USA

Stephen Klosterman Department of Organismic and Evolutionary Biology, Harvard University, Cambridge, MA, USA

Koen Kramer Alterra - Green World Research, Wageningen University and Research Centre, Wageningen, The Netherlands

Liang Liang Department of Geography, University of Kentucky, Lexington, KY, USA

Jorge A. Lobo Biology Department, Universidad de Costa Rica, San Jose, Costa Rica

Yiqi Luo Department of Microbiology and Plant Biology, University of Oklahoma, Norman, OK, USA

Susan Mazer Department of Ecology, Evolution, and Marine Biology, University of California, Santa Barbara, CA, USA

Annette Menzel Department of Ecology and Ecosystem Management, Technische Universität München, Freising, Germany

L. Patrícia C. Morellato Departamento de Botânica, Laboratório de Fenologia, Plant Phenology and Seed Dispersal Research Group, Instituto de Biociências, Universidade Estadual Paulista UNESP, São Paulo, Brazil

Susanne Moryson Agricultural Climatology, Department of Crop and Animal Sciences, Faculty of Agriculture and Horticulture, Humboldt-University of Berlin, Berlin, Germany

Shuli Niu Institute of Geographical Sciences and Natural Resources Research, Chinese Academy of Sciences, Beijing, China

Department of Microbiology and Plant Biology, University of Oklahoma, Norman, OK, USA

Rebecca Phillips Traditional Ecological Knowledge Coordinator, Parks Victoria, Melbourne, VIC, Australia

Bob R. Pohlad Ferrum College, School of Natural Sciences and Mathematics, Ferrum, VA, USA

Eric Post Department of Biology, The Pennsylvania State University, University Park, PA, USA

James A. Powell Department of Mathematics and Statistics, Utah State University, Logan, UT, USA

Mauricio Quesada Centro de Investigaciones en Ecosistemas, Universidad Nacional Autónoma de México, Morelia, Mexico

Jacques Régnière Natural Resources Canada, Canadian Forest Service, Quebec City, QC, Canada

Andrew D. Richardson Department of Organismic and Evolutionary Biology, Harvard University, Cambridge, MA, USA

Arturo Sanchez-Azofeifa Earth and Atmospheric Sciences Department, University of Alberta, Edmonton, AB, Canada

Mark D. Schwartz Department of Geography, University of Wisconsin-Milwaukee, Milwaukee, WI, USA

Richard L. Snyder Department of Land, Air, and Water Resources, University of California, Davis, CA, USA

Leonid V. Sokolov Biological Station Rybachy, Zoological Institute, Russian Academy of Sciences, St. Petersburg, Russia

Donatella Spano Department of Science for Nature and Environmental Resources (DipNet), University of Sassari, Sassari, Italy

Euro-Mediterranean Centre for Climate Change (CMCC), Sassari, Italy

Tim H. Sparks Institute of Zoology, Poznań University of Life Sciences, Poznań, Poland

Fachgebiet für ökoklimatologie and Institute for Advanced Study, Technische Universität München, Munich, Germany

Sigma, Coventry University, Coventry, UK

Kathryn E. Stoner Centro de Investigaciones en Ecosistemas, Universidad Nacional Autónoma de México, Morelia, Mexico

Carolyn L. Thomas Ferrum College, School of Natural Sciences and Mathematics, Ferrum, VA, USA

Michael Toomey Department of Organismic and Evolutionary Biology, Harvard University, Cambridge, MA, USA

Jake F. Weltzin National Coordinating Office, USA National Phenology Network, Tucson, AZ, USA

Frans E. Wielgolaski Department of Bioscience, University of Oslo, Oslo, Norway

Eric J. Woehler Institute of Marine and Antarctic Studies, University of Tasmania, Sandy Bay, TAS, Australia

Chapter 1
Introduction

Mark D. Schwartz

Abstract Phenology has been used as a proxy for climate and weather throughout human history particularly in relation to agriculture, but only within the last two centuries has it emerged as a science in its own right. Moreover, during the last half of the twentieth century the value of phenological science has been recognized as an integrative measure of plant and animal responses to climate and other environmental change that can be scaled from a local to a global level. Multiple examples, concepts, and applications of phenology have been systematically compiled to create this book. Together, they serve to reemphasize the valuable contribution of phenological research, in particular related to environmental change, to-date, and highlight the urgent need for more data collection, networks, and global collaborations in the future.

1.1 Basic Concepts and Background

Phenology, which is derived from the Greek word *phaino* meaning to show or to appear, is the study of recurring plant and animal life cycle stages, especially their timing and relationships with weather and climate. Sprouting and flowering of plants in the spring, color changes of leaves in the fall, bird migration and nesting, insect hatching, and animal hibernation are all examples of phenological events (Dubé et al. 1984). Seasonality is a related term, referring to similar non-biological events, such as timing of the fall formation and spring break-up of ice on fresh water lakes.

Human knowledge and activities connected to what is now called phenology are probably as old as civilization itself. Surely, soon after farmers began to continuously dwell in one place—planting seeds, observing crop growth, and carrying out the harvest year after year—they quickly became aware of the connection of

---

M.D. Schwartz (✉)
Department of Geography, University of Wisconsin-Milwaukee, Milwaukee, WI 53211, USA
e-mail: mds@uwm.edu

M.D. Schwartz (ed.), *Phenology: An Integrative Environmental Science*,
DOI 10.1007/978-94-007-6925-0_1, © Springer Science+Business Media B.V. 2013

changes in their environment to plant development. Ancient records and literature, such as observations taken up to 3,000 years ago in China (see Chap. 2), and references in the Christian Bible, testify to a common level of understanding about phenology among early peoples:

> Learn a lesson from the fig tree. Once the sap of its branches runs high and it begins to sprout leaves, you know that summer is near. *Gospel of Mark 13:28*

Unfortunately, these ancient "roots" did not translate into systematic data collection across large areas over the centuries, nor did they provide impetus for the early development of phenology as a scientific endeavor and discipline. For a long time the field remained tied almost exclusively to agricultural applications, and even those were only deemed practical at the local scale (i.e., every place was different, and generalizations difficult or impossible). With the establishment of continuous and continental-scale observation networks by the mid-1900s (though still largely confined to Europe, see Chap. 4), and contributions of early researchers such as Schnelle (1955), phenology began to emerge as an environmental science. Lieth's (1974) book was the first modern synthesis to chart the interdisciplinary extent of the field, and demonstrate its potential for addressing a variety of ecological system and management issues. These foundations prepared the way for the first edition of the current book (in 2003) and subsequently this second edition.

1.2 Organization and Use

Phenological research has traditionally been identified with studies of mid-latitude plants (mostly trees and shrubs) in seasonal climates, but other areas of the field are also progressing. Thus, a principal goal in organizing this book was to overcome this mid-latitude plant bias with a structure that would facilitate a thorough examination of wider aspects of plant and animal phenology.

The first section, "Phenological Data, Networks, and Research," adopts a regional approach to assess the state and scope of phenological research around the world with chapters on "East Asia" (2), "Australia and New Zealand" (3), "Europe" (4), "North America" (5, excluding Mexico), "...South and Central America" (6), and "Antarctica" (7). Several major regions, most notably Africa and central Asia were not included due to an inability to identify researchers working in these geographical areas. While some efforts were made in Chaps. 2, 3, 4, 5, 6 and 7 to survey the history of regional data collection and research, more emphasis was given to an assessment of recent developments. My assumption was that since Lieth's (1974) book had make an extensive survey of the history of phenological research up to the early 1970s, there was no great need to reproduce all that historical information in this volume. The final chapter in this section explores plans for developing global phenological networks, "International Phenological Observation Networks" (8).

Part II, "Phenology of Selected Bioclimatic Zones," examines phenological research in areas outside of mid-latitudes, with chapters on "Tropical Dry Climates" (9) and "Phenology at High Latitudes" (13). Other chapters in this section document phenology in drier mid-latitude biomes, including "Mediterranean Phenology" (10), and "North American Grassland Phenology" (11). A new chapter was added to this edition examining "Mesic Temperate Deciduous Forest Phenology" (12) which has been a traditional region of intensive phenological research. Lastly, the particular responses of "Phenology at High Altitudes" are explored in Chap. 14.

Part III, "Phenological Models and Techniques" presents a survey of phenological research methodologies and strategies. Model building and development is outlined in chapters addressing plants (15), and animal life cycles (16, concentrating on poikilotherms). The challenges of spatial and temporal modeling are explored in Chap. 17, and other chapters address the issues of temperature measurement (20), methods to detect climate change (18) and comparing high-resolution ground and moderate-resolution satellite-derived phenology (19). The next section (Part IV) is devoted entirely to the important area of "Sensor-Derived Phenology", but now includes both a chapter on "Satellite-Sensor Derived Phenology (21) and the recently emerging "Near-Surface Sensor-Derived Phenology" (22).

Part V, "Phenology of Selected Lifeforms" looks at research and developments in animal phenology, including chapters on "Aquatic Plants and Animals" (23), "Birds" (24), and "Reproductive Phenology of Large Mammals" (25). The final section of the book (Part VI) details "Applications of Phenology" to a variety of topics. Chapter 26 looks specifically at "Vegetation Phenology in Global Change Studies," Chap. 27 explores frontiers related to "...Photosynthesis Phenology in Northern Ecosystems," and Chap. 28 examines "Phenology and Evapotranspiration." Several remaining chapters in this section explore applications in traditional field agriculture and horticulture (29) and winegrape growth and care (30). Lastly, selected phenological applications in higher education are examined in the final Chap. (31).

Therefore, this volume's structure is primarily designed to serve the basic reference needs of phenological researchers and students interested in learning more about specific aspects of the field, or evaluating the feasibility of new ideas and projects. However, it is also an ideal primer for ecologists, climatologists, remote sensing specialists, global change scientists, and motivated members of the public who wish to gain a deeper understanding of phenology and its potential uses.

1.3 Future Directions and Challenges

When I chose the name for this book, I deliberately selected the word "integrative" because of its implication of a process. Phenology is an interdisciplinary environmental science, and as such brings together individuals from many different scientific backgrounds, but the full benefits of their combined disciplinary perspectives to enrich phenological research have yet to be realized. Thus, the term "integrative" as in moving together, rather than "integrated" implying already being together.

The nearly 10 years which have passed since publication of the first edition (in 2003) have seen steady progress in the transmission of "phenological perspectives" into the mainstream of science, especially related to the needs of global change research, but considerable work remains to be done. While other parts of phenological research are still important and need to progress, I still contend that it is global change science that will stimulate, challenge, and transform the discipline of phenology most in the coming decades. In order to maximize the benefits of phenology for global change research as rapidly as possible, commitments to integrative thinking and large-scale data collection must continue. First of all, the limitations of the primary forms of data collection (remote sensing derived, native species, cloned indicator species, and model output) must be accepted. None of these data sources can meet the needs of all research questions, and an "integrative approach" that combines data types provides synergistic benefits (Schwartz 1994, 1999).

The most needed data are traditional native and cloned plant species observations. Networks that select a small number of common and cloned plants for coordinated observation among national and global scale networks will prove the most useful. I am deeply gratified by the role that I have played in the development of the USA National Phenology Network (USA-NPN), using that basic framework. Over the last 10 years USA-NPN has progressed from being little more than an idea in my head, to a real operational network, thanks to efforts of my many dedicated colleagues. Such networks should continue to be created, embraced and (where possible) integrated into the missions of national and global data collection services around the world. A little more than 100 years ago, the countries of the world began to cooperate in a global-scale network of weather and climate monitoring stations. The results of this long-term investment are the considerable progress that has been made in understanding the workings of the earth's climate systems. I continue to believe that we have a similar opportunity with phenological data, and that small investments in national and global-scale observation networks are crucial to global change science, and will yield an impressive return in the years ahead.

References

Dubé PA, Perry LP, Vittum MT (1984) Instructions for phenological observations: Lilac and honeysuckle. Vermont Agricultural Experiment Station Bulletin 692. University of Vermont, Burlington

Lieth H (ed) (1974) Phenology and seasonality modeling. Ecological studies, vol 8. Springer, New York

Schnelle F (1955) Pflanzen-Phänologie. Probleme Der Bioklimatologie, vol 3. Akademische Verlagsgesellschaft, Leipzig

Schwartz MD (1994) Monitoring global change with phenology: the case of the spring green wave. Int J Biometeorol 38(1):18–22

Schwartz MD (1999) Advancing to full bloom: planning phenological research for the 21st century. Int J Biometeorol 42(3):113–118

Part I
Phenological Data, Networks, and Research

Chapter 2
East Asia

Xiaoqiu Chen

Abstract Phenological observations and research have a long history in East Asia. Countrywide phenological networks have been established mostly by national meteorological administrations or agencies during 1950s to 1980s. Since 2000, phenological research has made significant progress in China, Japan, and South Korea. The recent network-related research focuses mainly on three aspects: first, the temporal and spatial variation of plant phenology and its responses to climate change at individual and community levels by means of statistical methods; second, the effect of genetic diversity on phenological responses to climate change; and third, identification and extrapolation of the vegetation growing season on the basis of plant community phenology and satellite data.

2.1 Phenological Observation and Research in China

2.1.1 Historical Background

Modern phenological observation and research in China started in the 1920s with Dr. Coching Chu (1890–1974). As early as 1921 he observed spring phenophases of several trees and birds in Nanjing. In 1931, he summarized phenological knowledge from the past 3,000 years in China. He also introduced phenological principles (e.g. species selection, criteria of phenological observations and phenological laws) developed in Europe and the United States from the middle of the eighteenth to the early twentieth century (Chu 1931). According to his literature survey, phenological observation can be traced back to the eleventh century B.C. in China. The earliest phenological calendar, Xia Xiao Zheng, stems from this period and recorded (on a monthly basis) phenological events, weather,

X. Chen (✉)
College of Urban and Environmental Sciences, Peking University, Beijing, China
e-mail: cxq@pku.edu.cn

M.D. Schwartz (ed.), *Phenology: An Integrative Environmental Science*,
DOI 10.1007/978-94-007-6925-0_2,

astronomical phenomena, and farming activities in the region between the Huai River drainage area and the lower reaches of Yangtze River. In addition, extensive phenological data were recorded in other ancient literatures over the past 3,000 years. These data could to some extent reflect past climate. Using ancient phenological data and other data, he reconstructed a temperature series of the past 5,000 years in China (Chu 1973).

2.1.2 Networks and Data

In 1934, Dr. Coching Chu established the first phenological network in China. Observations covered some 21 species of plants, nine species of animals, some crops, and several hydro-meteorological events, and ceased in 1937 because of the War of Resistance Against Japan (1937–1945). Twenty-five years later the Chinese Academy of Sciences (CAS) established a countrywide phenological network under the guidance of Dr. Chu. The observations began in 1963 and continued until 1996. Observations resumed in 2003, but with a reduced number of stations, species, and phenophases.

The observation program of the CAS network included a total of 173 observed species. Of these, 127 species of woody and herbaceous plants had a localized distribution. Table 2.1 lists the 33 species of woody plants, two species of herbaceous plants, and 11 species of animals that were observed across the network (Institute of Geography at the Chinese Academy of Sciences 1965). During 1973–1986, several stations added phenological observation of major crops, such as rice, winter wheat, spring wheat, corn, grain sorghum, millet, cotton, soybean, potato, buckwheat, rape, etc. The observations were carried out mainly by botanical gardens, research institutes, universities and middle schools according to unified observation criteria (Institute of Geography at the Chinese Academy of Sciences 1965; Wan and Liu 1979). The phenophases of woody plants included bud swelling, budburst, first leaf unfolding, 50 % leaf unfolding, flower bud or inflorescence appearance, first flowering, 50 % flowering, the end of flowering, fruit or seed maturing, first fruit or seed shedding, the end of fruit or seed shedding, first leaf coloration, full leaf coloration, first defoliation, and the end of defoliation. Changes to the stations and in observers over the years resulted in data that were spatially and temporally inhomogeneous. The number of active stations varied over time. The largest number of stations operating was 69 in 1964 and the lowest number occurred between 1969 and 1972 with only 4–6 stations active. The phenological data from 1963 to 1988 were published in form of Yearbooks of Chinese Animal and Plant Phenological Observation (Volume 1–11). Since then, the data have not been published.

In 1980 the China Meteorological Administration (CMA) established another countrywide phenological network. The CMA phenological network is affiliated with the national-level agrometeorological monitoring network and came into operation in 1981. The phenological observation criteria for woody and herbaceous plants, and animals were adopted from the CAS network. There are 28 common species of woody plants, one common species of herbaceous plant and 11 common species of animals. The main phenophases are the same as those of the CAS network. In addition to the natural phenological observations, the network also carries out professional

Table 2.1 Common observation species of the CAS phenological network in China

Latin names
Woody plants
Ginkgo biloba
Metasequoia glyptostroboides
Platycladus orientalis
Sabina chinensis
Populus simonii
Populus canadensis
Salix babylonica
Juglans regia
Castanea mollissima
Quercus variabilis
Ulmus pumila
Morus alba
Broussonetia papyrifera
Paeonia suffruticosa
Magnolia denudata
Firmiana simplex
Malus pumila
Prunus armeniaca
Prunus persica
Prunus davidiana
Albizia julibrissin
Cercis chinensis
Sophora japonica
Robinia pseudoacacia
Wisteria sinensis
Melia azedarach
Koelreuteria paniculata
Zizyphus jujuba
Hibiscus syriacus
Lagerstroemia indica
Osmanthus fragrans
Syringa oblata
Fraxinus chinensis
Herbaceous plants
Paeonia lactiflora
Dendranthema indicum
Animals
Apis mellifera
Apus apus pekinensis
Hirundo rustica gutturalis
Hirundo daurica japonica
Cuculus canorus canorus
Cuculus micropterus micropterus
Cryptotympana atrata
Gryllulus chinensis
Anser fabalis serrirostris
Oriolus chinensis diffusus
Rana nigromaculata

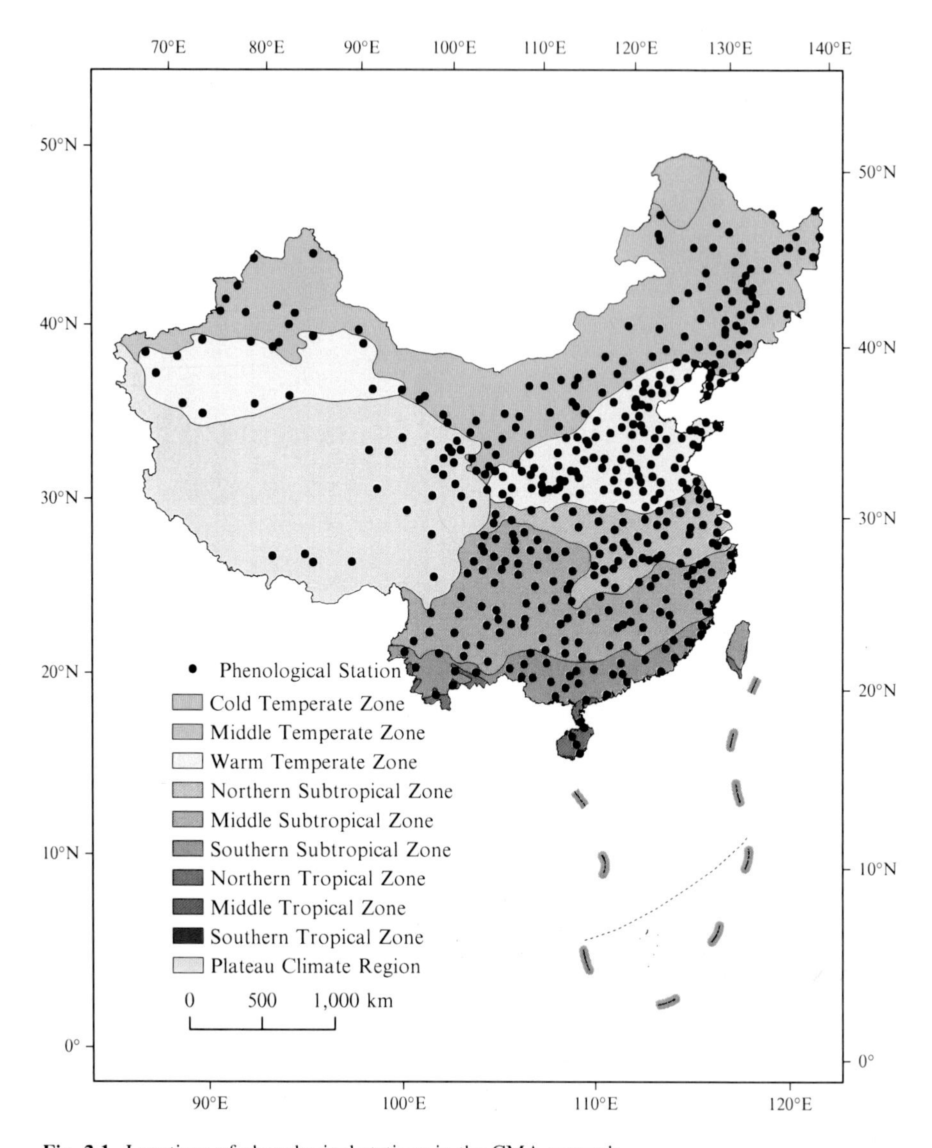

Fig. 2.1 Locations of phenological stations in the CMA network

phenological observation of crops on the basis of a specific observation criterion (National Meteorological Administration 1993). The main crop varieties include rice, wheat, corn, grain sorghum, millet, sweet potato, potato, cotton, soybean, rape, peanut, sesame, sunflower, sugarcane, sugar beet, and tobacco. In grassland areas, phenophases of dominant grass species are also observed.

The CMA network is the largest phenological observation system in China. There were 587 agrometeorological measurement stations in 1990. At present, 446 stations are undertaking phenological observations. These phenological stations are distributed in different climate zones and regions throughout the country, especially in eastern China (Fig. 2.1). The CMA-archive keeps the original phenological observation records from 1981 to the present and provides the data freely to research institutes and universities. As the phenological and meteorological observations are mostly parallel at the same location in this network, the data are especially valuable for understanding phenology-climate relationships. Moreover, these data can also be used to provide an agrometeorological service and prediction on crop yield, soil moisture and irrigation amounts, plant diseases and insect pests, and forest fire danger (Cheng et al. 1993). Recently, Xiaoqiu Chen and his students have digitized the phenological data set with permission from the China Meteorological Administration.

In addition, there were also some regional phenological networks. One example was the network established by Guodong Yang and Xiaoqiu Chen during the period 1979–1990. The network consisted of approximately 30 stations in the Beijing area (about 1,6410.54 km^2). Based on the observed data of this network, they worked out and published 16 phenological calendars, about one phenological calendar per 1,026 km^2 (Yang and Chen 1995).

2.1.3 Recent Network-Related Phenological Research

2.1.3.1 Measuring Plant Community Seasonality in Eastern China

Determining phenological seasons of plant communities is important for identifying the vegetation growing season combining surface phenology and satellite data at regional scales (Chen et al. 2000, 2001, 2005). Other than the empirical method for identifying phenological seasons (Chen 2003), a simulating method of phenological cumulative frequency has been developed to measure plant community seasonality using phenological data of the CAS network. The basic idea of the method was to establish a mixed data set composed of the occurrence dates of all phenophases of observed deciduous trees and shrubs (Chen 2003) at each station and for each year. The phenological data were acquired from seven stations in the temperate zone (Chen and Han 2008) and four stations in the northern subtropical zone (Chen et al. 2011) of eastern China. The study period was from 1982 to 1996. Based on the data set, the cumulative frequency of the occurrence dates of phenophases in every 5-day period (pentad) throughout each year and at each station was calculated. In order to simulate the phenological cumulative frequency using a logistic function, the empirical phenological cumulative frequency curve was divided into two parts, namely, the spring cumulative frequency curve and the autumn cumulative frequency curve (Chen 2003). Further, corresponding dates with the maximum changing rate of the curvature were computed on the spring and autumn simulating curves as onset dates of phenological stages (Fig. 2.2). Because four turning points with the maximum changing rate of the curvature are detectable on the spring and autumn simulating curves, four phenological stages were identified in each year and at each station, namely, green-up, active photosynthesis, senescence, and dormancy stages.

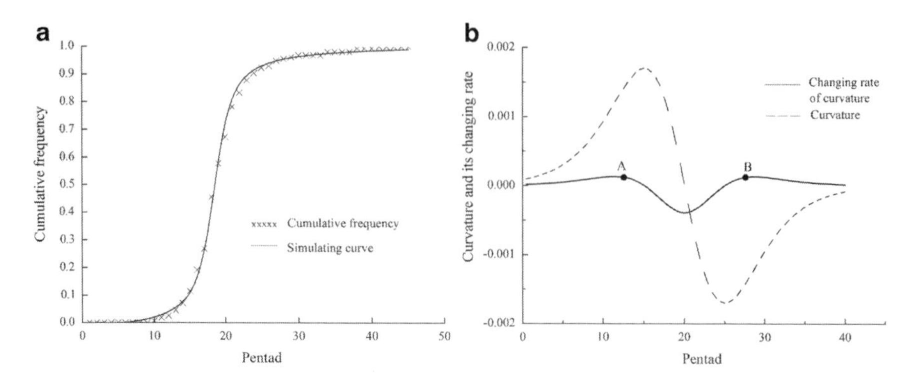

Fig. 2.2 Simulating cumulative frequencies of spring or autumn phenophases (**a**) and determining turning points of phenological stages (**b**) A: onset date of green-up or senescence stage, B: onset date of active photosynthesis or dormancy stage

Table 2.2 Comparison in phenological stage onset dates (month, day) and growing season durations (days) between the temperate zone and the northern subtropical zone

Climate zone	Green-up onset date	Active photosynthesis onset date	Senescence onset date	Dormancy onset date	Growing season duration
Temperate zone	3.19	5.23	9.9	11.9	235
Northern subtropical zone	3.1	5.8	9.7	11.24	268

The comparison in multiyear mean onset dates of phenological stages between temperate stations and northern subtropical stations indicates that the onset dates of green-up and active photosynthesis stages are obviously earlier in the northern subtropical zone than in the temperate zone. In contrast, the onset dates of dormancy stage are obviously later in the northern subtropical zone than in the temperate zone. The growing season duration (from the onset date of the green-up stage to the end date of the senescence stage) in the northern subtropical zone is 33 days longer than that in the temperate zone (Table 2.2). From each northern subtropical station (located between 29 °50′N and 33 °21′N) to the northernmost temperate station (located at 45 °45′N), the multiyear mean onset dates of green-up and active photosynthesis stages show a significant delay at a rate of 2.7–4.0 days and 1.8–2.8 days per latitudinal degree northwards, respectively, whereas the multiyear mean onset dates of senescence and dormancy stages indicate a non-significant advancement and a significant advancement at a rate of 2.9–3.3 days per latitudinal degree northwards, respectively.

2.1.3.2 Measuring Plant Community Growing Season in Eastern China

Since phenological stations and conventional phenological data of the CAS network are comparatively scarce in eastern China, the only option for detecting growing season trends at regional scales is to estimate the growing season of land vegetation using the limited station phenology data and satellite data (Chen and Pan 2002). Because metrics and thresholds of vegetation indices may not directly correspond to conventional, ground-based phenological events, but rather provide indicators of vegetation dynamics, a detailed comparison of these satellite measures with ground-based phenological events is needed. In recent years, some studies have been carried out to compare satellite sensor-derived onset and offset of greenness with surface phenological stages of individual plant species, mono-specific forests, and mixed forests for selected biomes (White et al. 1997; Duchemin et al. 1999; Schwartz et al. 2002; Badeck et al. 2004). Other than the above top-down method, namely determining the satellite-sensor-derived growing season at a regional scale first, and then validating it using conventional phenological data at local scales, Chen et al. (2000, 2001) developed a bottom-up method, namely determining the phenological growing season at sample stations first and then finding out the corresponding threshold values of normalized difference vegetation index (NDVI) at pixels overlaying the sample stations in order to extrapolate the phenological growing season at a regional scale. Using phenological and NDVI data from 1982 to 1993 at seven sample stations in temperate eastern China, Chen et al. (2005) calculated the cumulative frequency of leaf unfolding and leaf coloration dates for deciduous species every 5 days throughout the study period. Then, they determined the growing season beginning and end dates by computing times when 50 % of the species had undergone leaf unfolding and leaf coloration for each station in every year. Next, they used these beginning and end dates of the growing season as time markers to determine corresponding threshold NDVI values on NDVI curves for the pixels overlaying phenological stations. Based on a cluster analysis of the annual NDVI curves, they determined extrapolation areas for each phenological station in every year, and then, implemented the spatial extrapolation of growing season parameters from the seven sample stations to 87 meteorological stations in the study area.

Results show that spatial patterns of growing season beginning and end dates correlate significantly with spatial patterns of mean air temperatures in spring and autumn, respectively. Contrasting with results from similar studies in Europe and North America, this study suggests that there is a significant delay in leaf coloration dates, along with a less pronounced advance of leaf unfolding dates in different latitudinal zones and the whole area from 1982 to 1993. The growing season has been extended by 1.4–3.6 days per year in the northern zones and by 1.4 days per year across the entire study area on average (Table 2.3). The apparent delay in growing season end dates is associated with regional cooling from late spring to summer, while the insignificant advancement in beginning dates corresponds to inconsistent temperature trend changes from late winter to spring. On an interannual basis, growing season beginning and end dates correlate negatively with mean air temperatures from February to April and from May to June, respectively (Chen et al. 2005).

Table 2.3 Linear trends (days per year) of growing season beginning (BGS) and end dates (EGS) and lengths (LGS) in different latitudinal zones and the whole area during 1982 to 1993

Region	BGS	EGS	LGS
Zone 1 (32 °N–34.99 °N)	0.7	0.0	−0.7
Zone 2 (35 °N–37.99 °N)	−0.7	0.9**	1.6**
Zone 3 (38 °N–40.99 °N)	−0.7	1.8***	2.5**
Zone 4 (41 °N–43.99 °N)	−1.7*	1.9****	3.6***
Zone 5 (≥44 °N)	−0.7	0.7	1.4**
Whole area	−0.4	1.0***	1.4***

$*P < 0.1$, $**P < 0.05$, $***P < 0.01$, $****P < 0.001$

2.1.3.3 Assessing the Phenology of Individual Plant Species

So far, almost all conventional phenological studies in China have been based on discontinuous time series from a few stations in a data set of the CAS network. Since the only continuous phenological records exist in Beijing, several studies focused on plant phenological responses to urban climate change. The trends detected are more or less similar to those observed for a variety of tree species in Europe and North America.

Lu et al. (2006) analyzed spring flowering dates of four species during 1950–2004 in Beijing, and found that flowering date of early-blossom species advanced much quicker than other late-blossom species, which tend to stretch the flowering interval among species. With regard to phenological response to temperature, the flowering sensitivity of four tree species to daily maximum, minimum and average temperature is ‘species-specific’. On the basis of spring and autumn phenological data of three species during 1962–2004 in Beijing, Luo et al. (2007) assessed differences in phenological responses of plant to urban climate change for the period 1962–1977 and 1978–2004, and found that the urban heat island effect from 1978 onwards is the dominant cause of the observed phenological changes.

With respect to phenological variations across a broad area, Zheng et al. (2006) used the discontinuous time series of 32 spring phenophases during the period 1963–1996 at 16 stations of eastern China to analyze statistical relationships between plant phenology and temperature. Only seven phenophases at five stations show a significant advancing trend ($P < 0.1$), and six phenophases at five stations indicate a significant delay trend ($P < 0.1$). In general, the advancing trends in both early and late spring phenophases occurred mainly in the northern regions. This is in line with the observed warming temperature trend. Opposite changes, that is a delay, were detected in the southwestern regions consistent with the cooling temperature trend. In addition, the individual phenophases were significantly correlated either with the mean temperature of that month in which the mean phenophase occurred or preceding months, and sometimes with both. Similar results were also obtained by another study using first flowering data of 23 plant species at 22 stations of eastern China during 1963–2006 (Ge et al. 2011).

2.2 Phenological Observation and Research in Japan and South Korea

2.2.1 Networks and Data

In 1953, the Japan Meteorological Agency (JMA) established a national phenological observation network consisting of 102 stations. The aims were to monitor local climate via phenological phenomena of some specific plants and animals. The Observation Division at the Headquarters of the Japan Meteorological Agency in Tokyo is responsible for phenological observations in Japan. The observation program consists of 12 species of plants and 11 species of animals, and related phenophases (Table 2.4). The observation criteria are defined in "Guidelines for the Observation of Phenology" (Japan Meteorological Agency 1985). The phenological data were published monthly in the "Geophysical Review" under categories of "Agrometeorological Summary" or "Applied Meteorology", and are available from

Table 2.4 Phenological observation program in Japan

Plants	Budding	First flowering	Full flowering	Color change	Leaf fall
Prunus yedoensis		X	X		
Prunus mume		X			
Camellia japonica		X			
Taraxacum (*T. platycarpum*, *T. albidum*, and *T. japonicum*)		X			
Rhododendron kaempferi		X			
Wisteria floribunda		X			
Lespedeza bicolor var. *japonica*		X			
Hydrangea macrophylla var. *otaksa*		X			
Lagerstroemia indica		X			
Miscanthus sinensis		X			
Ginkgo biloba	X			X	X
Acer palmatum				X	X

Animals	First heard	First seen	Last seen
Alauda arvensis	X		
Cettia diphone	X		
Lanius bucephalus	X		
Graptopsaltria nigrofuscata	X		
Tanna japonensis	X		
Hirundo rustica		X	X
Pieris rapae crucivora		X	
Papilio machaon hippocrates		X	
Orthetrum albistylum speciosum		X	
Lampyridae (*Luciola cruciata* and *L. lateralis*)		X	
Rana nigromaculata		X	

Japan Meteorological Agency (http://www.jmbsc.or.jp/english/index-e.html). The phenological network in Japan was encountering difficulties in continuing reliable observations because of the effects of urbanization, and the function of phenological observation in monitoring local climate has been weakened with the modernization of surface weather observation network (from personal correspondence with Dr. Mitsuhiko Hatori, Director of Observations Division, Observations Department, JMA).

There is less information on the phenological observation network in South Korea. As reported by Primack et al. (2009a), the Weather Service of South Korea has been gathering data on 20 phenological events at 74 weather stations with some observations dating from 1921. The data are available from Korea Meteorological Administration: (http://web.kma.go.kr/edu/unv/agricultural/seasonob/1173374_1389.html).

2.2.2 Recent Network-Related Phenological Research

In 2001, Ministry of the Environment of Japan published a report addressing the effects of global warming on Japan, in which influences of climate change on plant and animal phenology were summarized (Ministry of the Environment 2001). Since then, phenological research in Japan has made great progress.

The recent research undertaken on phenology in Japan and South Korea includes a wide range of issues, such as phenological data-based temperature reconstruction of past centuries (Aono and Kazui 2008; Aono and Saito 2010), plant phenology and its relation to the local environment based on specified field observations (Yoshie 2010; Ohashi et al. 2011), pollen fertility and flowering phenology (Ishida and Hiura 1998), application of digital camera for phenological observation (Ide and Oguma 2010) and phenological responses to climate change (Matsumoto et al. 2003; Ho et al. 2006; Doi and Katano 2008; Doi and Takahashi 2008; Chung et al. 2009; Primack et al. 2009b; Doi et al. 2010; Fujisawa and Kobayashi 2010; Matsumoto 2010; Doi 2011) etc. It is worth noting that the network-related phenological research focuses mainly on phenological responses to climate change.

Matsumoto et al. (2003) systematically studied the extension of the growing season between 1953 and 2000 in *Ginkgo biloba* in relation to climate change using a data set of the annual budding and leaf fall dates from 67 stations in Japan. In contrast to the traditional method for detecting phenological responses to temperature by computing monthly mean temperature (Chen 1994; Sparks et al. 2000; Chmielewski and Rötzer 2001; Menzel 2003; Gordo and Sanz 2010), they used a daily mean temperature-based method proposed by Shinohara (1951). In this method, the length of the period (*LP*) during which a particular daily mean temperature might influence a phenological event is defined as:

$$LP = EP - BP \tag{2.1}$$

where *LP* is the period length (number of days), *EP* is the end date of the period (day of year, DOY), and *BP* is the beginning date of the period (DOY). The *EP* is defined as average budding or leaf fall dates between 1953 and 2000. Using this equation, they calculated the correlation coefficient between the date of budding or leaf fall and the average daily temperature during the period $EP - BP$. The highest correlation appears within the optimum *LP* during which air temperature affected phenological events most markedly. Using the average optimum *LP* at all station, they analyzed the phenology-temperature relationship.

The results show that the advancement rate (−0.9 days per decade) in the beginning date of the *Ginkgo biloba* growing season was smaller than the delay rate (1.6 days per decade) of the end date. With regard to the phenological response to temperature, variation in the growing season beginning date (BGS) is closely related to air temperatures in the 45-day period before the average budding date, whereas variation in the growing season end date (EGS) is affected mostly by air temperatures in the 85-day period before the average leaf fall date. On average, an increase in the average air temperature of 1 °C in spring may advance BGS by about 3 days. If the average autumn air temperature increases by 1 °C, EGS may be delayed by about 4 days. Furthermore, LGS may be extended by about 10 days when the mean annual air temperature increases by 1 °C.

Further, Matsumoto (2010) examined spatial patterns in long-term phenological trends (1961–2000) and their causal factors using a data set of the annual budding and leaf fall dates of *Ginkgo biloba* from 60 stations in Japan. The results show that there was no significant relationship between phenological trends and geographical variables: latitude, longitude, and altitude, with the exception of a negative relationship between the trend of leaf fall date and latitude. Namely, the linear trends of leaf fall date at lower latitudes were larger than those at higher latitudes. With respect to relationship between the air temperature trend and the phenological trend, a negative relationship was found with the budding trend, but there was no obvious relationship with the leaf fall trend. By contrast, the spatial variability of the phenological sensitivity to temperature displayed a significant linear relationship with trends in budding and leaf fall. That is, where trees had higher sensitivity to temperature, they showed earlier budding and delayed leaf fall. Therefore, the spatial variations in phenological trends were dependent more on phenological sensitivity to air temperature than temperature trends.

Other than the above work, Doi and Takahashi (2008) examined the latitudinal pattern of phenological response to temperature in Japan using a data set of two species during 1953–2005. Negative relationships were found between the phenological response of leaves to temperature and latitude based on leaf coloring and leaf fall data of *Ginkgo biloba* and *Acer palmatum* at 63 and 64 stations, respectively. Single regression slopes of the phenological responses at lower latitudes were larger than those at higher latitudes. Similar results were also obtained by another study using leaf budburst and leaf fall data of *Morus bombycis* during 1953–2005 at 25 stations in Japan (Doi 2011).

During the last several decades, many studies have estimated the phenological response of plants to temperature. However, the effect of genetic diversity on phenological responses to climate change has less been considered. Doi et al. (2010) tested whether variations in phenological responses to temperature depend on genetic diversity based on flowering dates of ten species and leaf budburst dates of one species across Japan from 1953 to 2005. The results show that the within-species variations of phenological response to temperature as well as regional variations were less in the plant populations with lower genetic diversity. Thus, genetic diversity influences the variation in phenological responses of plant populations. Under increased temperatures, low variation in phenological responses may allow drastic changes in the phenology of plant populations with synchronized phenological timings. Maintaining genetic diversity may alleviate the drastic changes in phenology due to future climate change.

References

Aono Y, Kazui K (2008) Phenological data series of cherry tree flowering in Kyoto, Japan, and its application to reconstruction of springtime temperatures since the 9th century. Int J Climatol 28(7):905–914. doi:10.1002/Joc.1594

Aono Y, Saito S (2010) Clarifying springtime temperature reconstructions of the medieval period by gap-filling the cherry blossom phenological data series at Kyoto, Japan. Int J Biometeorol 54(2):211–219

Badeck FW, Bondeau A, Bottcher K, Doktor D, Lucht W, Schaber J, Sitch S (2004) Responses of spring phenology to climate change. New Phytol 162(2):295–309. doi:10.1111/j.1469-8137.2004.01059.x

Chen XQ (1994) Untersuchung zur zeitlich-raeumlichen Aehnlichkeit von phaenologischen und klimatologischen Parametern in Westdeutschland und zum Einfluss geooekologischer Faktoren auf die phaenologische Entwicklung im Gebiet des Taunus, Selbstverlag des Deutschen Wetterdienstes, Offenbach am Main

Chen XQ (2003) Assessing phenology at the biome level. In: Schwartz MD (ed) Phenology: an integrative environmental science. Kluwer Academic Publishers, Dordrecht

Chen XQ, Han JW (2008) Seasonal aspection stage of plant communities and its spatial-temporal variation in temperate Eastern China. J Plant Ecol (Chinese Version) 32:336–346

Chen XQ, Pan WF (2002) Relationships among phenological growing season, time-integrated normalized difference vegetation index and climate forcing in the temperate region of Eastern China. Int J Climatol 22(14):1781–1792. doi:10.1002/Joc.823

Chen XQ, Tan ZJ, Schwartz MD, Xu CX (2000) Determining the growing season of land vegetation on the basis of plant phenology and satellite data in Northern China. Int J Biometeorol 44(2):97–101

Chen XQ, Xu CX, Tan ZJ (2001) An analysis of relationships among plant community phenology and seasonal metrics of normalized difference vegetation index in the Northern part of the monsoon region of China. Int J Biometeorol 45(4):170–177

Chen XQ, Hu B, Yu R (2005) Spatial and temporal variation of phenological growing season and climate change impacts in temperate Eastern China. Glob Chang Biol 11(7):1118–1130. doi:10.1111/j.1365-2486.2005.00974.x

Chen XQ, Qi XR, A S, Xu L (2011) Spatiotemporal variation of plant community aspections in the northern subtropical zone of Eastern China. Acta Ecol Sin (Chinese Version) 31:3559–3568

Cheng C, Feng X, Gao L, Shen G (1993) Climate and agriculture in China. China meteorological press, Beijing

Chmielewski FM, Rotzer T (2001) Response of tree phenology to climate change across Europe. Agric For Meteorol 108(2):101–112
Chu C (1931) New monthly calendar (in Chinese). Bull Chin Meteorol Soc 6:1–14
Chu C (1973) A preliminary study on the climate fluctuation during the last 5,000 years in China (in Chinese). Scientia Sinica 16:226–256
Chung U, Jung JE, Seo HC, Yun JI (2009) Using urban effect corrected temperature data and a tree phenology model to project geographical shift of cherry flowering date in South Korea. Clim Chang 93(3–4):447–463. doi:10.1007/s10584-008-9504-z
Doi H (2011) Response of the *Morus bombycis* growing season to temperature and its latitudinal pattern in Japan. Int J Biometeorol. doi:10.1007/s00484-011-0495-5
Doi H, Katano I (2008) Phenological timings of leaf budburst with climate change in Japan. Agric For Meteorol 148(3):512–516. doi:10.1016/j.agrformet.2007.10.002
Doi H, Takahashi M (2008) Latitudinal patterns in the phenological responses of leaf colouring and leaf fall to climate change in Japan. Glob Ecol Biogeogr 17(4):556–561. doi:10.1111/j.1466-8238.2008.00398.x
Doi H, Takahashi M, Katano I (2010) Genetic diversity increases regional variation in phenological dates in response to climate change. Glob Chang Biol 16(1):373–379. doi:10.1111/j.1365-2486.2009.01993.x
Duchemin B, Goubier J, Courrier G (1999) Monitoring phenological key stages and cycle duration of temperate deciduous forest ecosystems with NOAA/AVHRR data. Remote Sens Environ 67 (1):68–82
Fujisawa M, Kobayashi K (2010) Apple (*Malus pumila var. domestica*) phenology is advancing due to rising air temperature in Northern Japan. Glob Chang Biol 16(10):2651–2660. doi:10.1111/j.1365-2486.2009.02126.x
Ge QS, Dai JH, Zheng JY, Bai J, Zhong SY, Wang HJ, Wang WC (2011) Advances in first bloom dates and increased occurrences of yearly second blooms in Eastern China since the 1960s: further phenological evidence of climate warming. Ecol Res 26(4):713–723. doi:10.1007/s11284-011-0830-7
Gordo O, Sanz JJ (2010) Impact of climate change on plant phenology in Mediterranean ecosystems. Glob Chang Biol 16(3):1082–1106. doi:10.1111/j.1365-2486.2009.02084.x
Ho CH, Lee EJ, Lee I, Jeong SJ (2006) Earlier spring in Seoul, Korea. Int J Climatol 26 (14):2117–2127. doi:10.1002/Joc.1356
Ide R, Oguma H (2010) Use of digital cameras for phenological observations. Ecol Inform 5(5):339–347. doi:10.1016/j.ecoinf.2010.07.002
Institute of Geography at the Chinese Academy of Sciences (1965) Yearbook of Chinese animal and plant phenological observation, no. 1. Science Press, Beijing (in Chinese)
Ishida K, Hiura T (1998) Pollen fertility and flowering phenology in an androdioecious tree, *Fraxinus lanuginosa* (Oleaceae), in Hokkaido, Japan. Int J Plant Sci 159(6):941–947
Japan Meteorological Agency (1985) Guidelines for the Observation of Phenology, 3rd edn. Japan Meteorological Agency, Tokyo (in Japanese)
Lu PL, Yu Q, Liu JD, Lee XH (2006) Advance of tree-flowering dates in response to urban climate change. Agric For Meteorol 138(1–4):120–131. doi:10.1016/j.agrformet.2006.04.002
Luo ZK, Sun OJ, Ge QS, Xu WT, Zheng JY (2007) Phenological responses of plants to climate change in an urban environment. Ecol Res 22(3):507–514. doi:10.1007/s11284-006-0044-6
Matsumoto K (2010) Causal factors for spatial variation in long-term phenological trends in *Ginkgo biloba L.* in Japan. Int J Climatol 30(9):1280–1288. doi:10.1002/joc.1969
Matsumoto K, Ohta T, Irasawa M, Nakamura T (2003) Climate change and extension of the *Ginkgo biloba L.* growing season in Japan. Glob Chang Biol 9(11):1634–1642. doi:10.1046/j.1529-8817.2003.00688.x
Menzel A (2003) Plant phenological anomalies in Germany and their relation to air temperature and NAO. Clim Chang 57(3):243–263
Ministry of the Environment, Investigation Committee of Global Warming Problems (2001) Impact of global warming on Japan 2001. Ministry of the Environment, Tokyo (in Japanese)

National Meteorological Administration (1993) Agrometeorological observation criterion, vol 1. Meteorological Press, Beijing (in Chinese)

Ohashi Y, Kawakami H, Shigeta Y, Ikeda H, Yamamoto N (2011) The phenology of cherry blossom (*Prunus yedoensis* "Somei-yoshino") and the geographic features contributing to its flowering. Int J Biometeorol 56(5):903–914. doi:10.1007/s00484-011-0496-4

Primack RB, Ibáñez I, Higuchi H, Lee SD, Miller-Rushing AJ, Wilson AM, Silander JA (2009a) Spatial and interspecific variability in phenological responses to warming temperatures. Biol Conserv 142(11):2569–2577

Primack RB, Higuchi H, Miller-Rushing AJ (2009b) The impact of climate change on cherry trees and other species in Japan. Biol Conserv 142(9):1943–1949. doi:10.1016/j.biocon.2009.03.016

Schwartz MD, Reed BC, White MA (2002) Assessing satellite-derived start-of-season measures in the conterminous USA. Int J Climatol 22(14):1793–1805. doi:10.1002/Joc.819

Shinohara H (1951) On the period, the temperature of which mostly influences the flowering date of the cherry (*P. yedoensis*). J Agric Meteorol 7:19–20 (in Japanese)

Sparks TH, Jeffree EP, Jeffree CE (2000) An examination of the relationship between flowering times and temperature at the national scale using long-term phenological records from the UK. Int J Biometeorol 44(2):82–87

Wan M, Liu X (1979) Method of Chinese phenological observation. Science Press, Beijing (in Chinese)

White MA, Thornton PE, Running SW (1997) A continental phenology model for monitoring vegetation responses to interannual climatic variability. Global Biogeochem Cycles 11(2):217–234

Yang G, Chen XQ (1995) Phenological calendars and their applications in the Beijing area. Capital Normal University Press, Beijing (in Chinese)

Yoshie F (2010) Vegetative phenology of alpine plants at Tateyama Murodo-daira in central Japan. J Plant Res 123(5):675–688. doi:10.1007/s10265-010-0320-y

Zheng JY, Ge QS, Hao ZX, Wang WC (2006) Spring phenophases in recent decades over eastern China and its possible link to climate changes. Clim Chang 77(3–4):449–462. doi:10.1007/s10584-005-9038-6

Chapter 3
Australia and New Zealand

Marie R. Keatley, Lynda E. Chambers, and Rebecca Phillips

Abstract This chapter outlines the historical context of phenological observation and study in Australia and New Zealand. Details of early records are given as they provide a valuable baseline against which current phenology may be assessed. It also summarizes the results of phenological studies undertaken in recent years and identifies further long-term phenological data yet to be analysed. The information presented here begins to address the acknowledged lack of phenological studies undertaken in both countries. Community-based phenological networks and their contribution to the collection of phenological data are also described.

3.1 Historical Context

The Northern Hemisphere, particularly Europe and parts of Asia, have a long history of recording phenological events (Nekovář et al. 2008; Sakurai et al. 2011). In the last 20 years or so these records have been used to contribute to the understanding of the impacts of climate change on natural and managed systems (Rosenzweig et al. 2007). In the Southern Hemisphere, phenological records are sparse by comparison. However, this chapter highlights recent advances that have been made in documenting phenology in Australia and New Zealand and that significant historical information exists in the form of traditional knowledge.

M.R. Keatley (✉)
Department of Forest and Ecosystem Science,
University of Melbourne, Creswick, VIC, Australia
e-mail: mrk@unimelb.edu.au

L.E. Chambers
Centre for Australian Weather and Climate Research,
Bureau of Meteorology, Melbourne, VIC, Australia

R. Phillips
Traditional Ecological Knowledge Coordinator, Parks Victoria, Melbourne, VIC, Australia

M.D. Schwartz (ed.), *Phenology: An Integrative Environmental Science*,
DOI 10.1007/978-94-007-6925-0_3, © Springer Science+Business Media B.V. 2013

3.1.1 *Traditional Knowledge*

Traditional Ecological Knowledge (TEK) is increasingly being studied because of the insights it can offer into climate forecasting and climate change (Riseth et al. 2011).

Aboriginal people have occupied the Australian landscape for over 50,000 years making their culture the oldest living culture in the world (Head 1993). They have endured major climatic changes such as earthquakes, severe drought and flood, ice ages and the rise and fall of oceans, which have produced long-term vegetation changes, yet they managed to adapt, survive and acquire knowledge of these occurrences (Gott 2005).

3.1.1.1 Traditional Ecological Knowledge

There are more than 500 different Countries[1] in Australia and each Aboriginal group has detailed knowledge pertaining to the ecosystems within their traditional boundary and how they fit into it. Castellano (2000) explains,

> Indigenous Knowledge is gained by three processes: observation, traditional teachings and revelation. Indigenous observation is undertaken over long time scales ... traditional teachings encompass knowledge that has been passed down through generations, for example creation stories. Knowledge acquired through revelation, such as dreams, visions and intuition, is sometimes regarded as spiritual knowledge.

Country is understood on many levels (social, emotional, spiritual and physical) with various weather conditions and signs of particular culturally significant species and the interaction with them. It is these species that signify the cues for ceremonies, hunting, gathering, breeding times and movements, with the transition of seasons interpreted through cultural values and beliefs. TEK is a holistic form of many types of environmental knowledge and practice now studied within a variety of scientific disciplines, including phenology.

3.1.1.2 Traditional Ecological Knowledge and Phenology

TEK and the study of phenology share three common factors; they are both dependent and built upon observations of ecological timing, in a specific area, and utilise key species or events of interest triggered or influenced by climate. One observable example of this is documented using Aboriginal seasonal calendars. There are also differences between phenology and TEK, one obvious difference is the timescale of recorded knowledge. For example, the D'harawal from Sydney

---

[1] Country is a place that gives and receives life. It encapsulates everything from flora and fauna, topographical features, dreaming stories, values, totems and the ancestral spirits within the land (Parks Victoria 2010 Healthy Country, Healthy People Digital Story. http://www.youtube.com/watch?v=2UmVNOpC1zU. Accessed February 2012).

area have two cycles that run considerably longer than the yearly cycle, the Mudong, or life cycle which covers about 11 or 12 years, and the Garuwanga, or Dreaming, which is a cycle of about 12,000–20,000 years (Kingsley 2003). The second difference is how environmental observations are interpreted through language, cultural values and beliefs systems that are shaped by lore and society. A third difference is the way of accumulating that knowledge such as oral histories, songs, dance and art rather than written data and modern technologies, such as carbon dating. In Aboriginal communities, specialised knowledge holders have varying responsibilities with imparting knowledge that is collective. Much was remembered and shared through stories, songs and art. Other knowledge is earned through demonstration of trust to ensure the knowledge will not be exploited to the detriment of Country or community.

3.1.1.3 Seasonal Calendars

There have been several studies of Australian seasonal calendars (e.g. Hoogenraad and Robertson 1997; Rose 2005). A commonly used western method of recording this information is the calendar wheel, to show the ongoing cycles of life. There are mixed views on whether this captures Aboriginal knowledge in an appropriate way (Rose 2005). However, some acknowledged benefits are that seasonal calendars prompt discussion about TEK and the use of local language by the traditional owners, where other languages are dominant, and allow the information to be passed on to younger generations (Hoogenraad and Robertson 1997). Some Aboriginal calendars are more detailed than others, reflecting contrasting weather patterns across the Australian continent.

3.1.1.4 Ecological Timing

Known seasonal calendars range from the two seasons (wet – Wantangka and dry – Yurluurrp) of the Wallabunnba people north of Alice Springs, to the 13 seasons in the Ngan'gi Seasons Calendar of the Nauiyu – Daly River people (http://www.bom.gov.au/iwk). This highlights a very complex and detailed understanding of climatic variations, events and accumulation of ecological knowledge. Within Victoria, the numbers of seasons recorded are fairly consistent (e.g. the six seasons of Gariwerd, the Gunditjmara six seasons and the seven seasons of the Wurundjeri) probably reflecting a more similar climate. Each season has particular cues for selected resource use, ceremony, management practice or movement to another area. To explain this further, Uncle Banjo Clarke (Gunditjmara people) describes the custom of *keeping the bird families strong* (Clarke and Chance 2003): they would visit a particular lake at a certain time of year (*Flowering Time* season) when thousands of swans would gather to nest and breed. It was traditional practice to collect swan eggs specifically from the nests in excess of four eggs. Only the excess would be taken carefully without leaving human scent. This prevented any food from being wasted on the weakest cygnets that would eventually be kicked out of the nest.

The European phenology calendar for the Middle Yarra (Victoria), defines six seasons of the Melbourne area as Autumn, Winter, Pre-spring, True Spring, Early Summer and Late Summer (Jameson 2001). Some of these align with observations recorded by Wurundjeri, but the descriptive language of the seasons imposed from the Northern Hemisphere is still evident.

In a holistic approach, one link that is not captured by phenology is the use of the stars. Aboriginal people believe that what happens on the land is reflected in the night sky. The appearance and disappearance of certain constellations were linked with regulating hunting regimes, so breeding times would be undisturbed. During various times of the year, the brightness and movements of the stars and planets signaled the arrival of particular birds or other species and the blooming of certain vegetation.

> For the Boonwurrung (people), the coastal cliffs and beaches along the bay were the focus of Old-Man Sun activity. The seasons were not mapped by a calendar but by the movement of the stars and the blooming of the plants. In the bay area the beginning of summer and the return of the snapper in the bay were signalled by the flowering in early November of the coast tee-tree and the late black wattle (Briggs 2008).

The connection Aboriginal people have to their Country is built upon evolutionary interrelationships and life cycles. This is demonstrated by the tracts of collective knowledge that has been shared and recorded as well as the specialized knowledge that is earned. It is evident the landscape has shaped their culture and their culture has shaped the land. This is why Traditional Ecological Knowledge is so important in understanding the story of that Country and for advancing the study of phenology and improving the management of that land.

3.1.2 Early European Phenological Observations

Having an understanding of why phenological observations were recorded can assist in locating data sources for examination which in turn can add to our understanding of the impacts in the climate change in the Southern Hemisphere.

The following section summarizes some of the known early European phenological observations undertaken in Australia and New Zealand.

3.1.2.1 Australia

There is no confirmed date as to when systematic European phenological observation or monitoring commenced in Australia. Table 3.1 summarizes early (prior to 1970) Australian phenological history, focusing on plants, and reflects extant records.

The earliest series of phenological observations undertaken by an individual are those attributed to von Mueller in 1856 (Prince 1891) – 21 years after the official European settlement of Victoria (Carron 1985). In the same year Hannaford (1856)

Table 3.1 Early phenological studies undertaken in Australia

Date	Comments	References
1856–?	List of plants with their month/s of flowering; Victorian focus	Hannaford (1856)
1856–?	Phenological recordings undertaken by Baron von Mueller, Victorian government botanist	Prince (1891)
1857–1895 (?)	Irregular collection of flowering observations by Maplestone	Maplestone (1895b)
1856–1885	Leafing, flowering, and fruiting of standard plants recorded in the Royal Society of Tasmania's Garden	Anon (1856) and Chambers and Keatley (2010b)
1886–1887	Monthly listing of plants in flower around Sydney	e.g. Haviland (1886, 1888)
1891	Call for establishment of phenological network in Victoria	Prince (1891)
1895	Flowering phenology of orchids, method outlined for recording phenological observations	French (1895) and Maplestone (1895a, b)
1903	Nature Study Calendar for Victoria	Gillies and Hall (1903)
1905, 1908	New South Wales Undersecretary for Lands requests that foresters record flowering within their district	Maiden (1909)
1906	Call for South Australian Royal Society's field naturalist's section to commence a list of flowering times	Anon (1906)
1907–1954	Climate observers requested to undertake phenological observations	Bureau of Meteorology (1925, 1954) and Commonwealth Meteorology (1907)
1909–1921	Broad information of flowering times of honey flora published in apiarists' journals	e.g. Beuhne (1914), McLachlan (1921) and Penglase and Armour (1909)
1909– 1924	Books published on general flora and fauna observations taken throughout the year	Mack (1909, 1924)
	A plea for the study of Australian phenological phenomena by New South Wales government botanist	Maiden (1909, 1924)
1925	Positive response to the Royal Meteorological Society's Phenological Committee 1924 call for the establishment of an International Phenological Network	Clark (1924, 1925) and Ploughshare (1926)
1929–1949	Flowering dates of 7 orchid species at 3 locations in Western Australia	Erickson (1950)
1925–1981	Monitoring undertaken by the Forest Commission of the various States	Keatley et al. (1999), Loneragan (1979), Steane (1931) and Tout (1935)
1934–1949	Arrival dates of Pallid Cuckoo and Nankeen Kestrel, breeding dates of Willie Wagtail in Western Australia	Sedgwick (1947, 1949, 1950)
1940–1962	Records of eucalypt flowering undertaken by the Victorian Forest Commission.	Keatley et al. (2002)

(continued)

Table 3.1 (continued)

Date	Comments	References
1949	Another call for phenological studies to be undertaken.	Gentilli (1949)
	"Meteorological Service" established a program for phenological observations.	Anon (1949) and Wang (1967)
1949–1954 (?)	Australia-wide ornithological program	Jarman (1950)

published a list of plants, primarily focused on Victoria, with months of flowering provided for most species within the list. Whilst lists without specific dates of flowering are now considered limited, it is highly likely that the range and months of flowering of many of the species were unknown at the time.

Other early observations were primarily undertaken by individuals (e.g. Haviland 1886–1888, Maplestone 1861–1895) associated with scientific organisations. It was not until the late eighteenth century that the general public in Australia were encouraged to be involved in science (Newell and Sutherland 1997).

The Royal Society of Tasmania appears to have instigated the earliest set of phenological recordings undertaken by an Australian organisation (1856–1885). The majority of the observations, leaf break, leaf colouration and falling, flowering, fruit ripening and harvesting, are of exotic plant species (Chambers and Keatley 2010b).

Between 1905 and 1924, Joseph Maiden, the New South Wales' government botanist and director of the Sydney Botanical Gardens (1896–1924) (Hall 1978), made requests of the Bureau of Meteorology and the Forests Department of New South Wales, to undertake phenological observations (Maiden 1909, 1924).

In the early 1920s the English Royal Meteorology Society put out a call for the establishment of an international phenological network (Clark 1924) and apparently received a positive response from Australia (Clark 1925; Ploughshare 1926).

The Bureau of Meteorology recognized the value of undertaking phenological observations and requests were made of meteorological observers to record the flowering of native plants, the arrival of migratory birds and butterflies in their weather notes (Commonwealth Meteorology 1907; Bureau of Meteorology 1925, 1954). The Bureau did not, however, insist on these being taken.

There is also a specific mention of a phenological network being established in 1949 (Wang 1967). It seems that the Bureau of Meteorology endeavored to establish a network after the 1947 Conference of the International Meteorological Organisation (International Meteorological Committee 1949), but again apparently relied on volunteers (Anon 1949).

In 1948, the Annual Congress of the Royal Australian Ornithologists' Union (now Birdlife Australia) adopted a proposal that an Australian-wide community network of observers be established to provide data on bird distributions and movements and the influence of climate on these. The program had commenced by 1949 with 200 observers. It seems that Tasmania and South Australia were not included as they already had similar networks operating (Jarman 1950).

Table 3.2 Early phenological studies, prior to 1972, undertaken in New Zealand

Date	Comments	References
1883	List of flowering months for 23 orchid species	Adams (1883)
1900	Records of first flowering dates for nine species (1893–1900)	Cockayne (1899)
1914–1955	Weekly sightings of Humpback whales in Cook Strait	Dawbin (1956)
1936–1954	Laying dates of Yellow-eyed Penguins on Otago Peninsula	Richdale (1957)
1938–1946	Departure dates of the New Zealand Bronze Cuckoo from Dunedin as well as detailed arrival dates from across New Zealand (>120 locations) for 1945	Fell (1947)
1938–1947	Arrival (8 seasons) and breeding (6 seasons) records of a single Erect-crested Penguin on Otago Peninsula.	Richdale (1950)
1938–1953	Arrival dates of Shining Cuckoo at various locations	Cunningham (1953, 1955)
1964–1969	Laying dates of Southern Royal Albatross at Campbell Island later than those reported by Richdale on Otago Peninsula	Waugh (1997)
1954–1964	Breeding timing of California Quail in Central Otago and Central North Island	Williams (1967)
1964–1972	Laying dates (3 seasons) of Red-billed Gulls at Kaikoura Peninsula	Mills (1979)

3.1.2.2 New Zealand

As with Australian records, it is not known when European phenological recording commenced in New Zealand and long-term studies are also limited (McGlone and Walker 2011). Table 3.2 lists the early New Zealand records.

In 1883, Adams (1883) listed the months of flowering for 23 orchid species and, as with the Haviland and Hannaford records, it is highly likely that the months of flowering of many of the species were unknown at the time. The earliest plant phenological exact dates located so far for New Zealand are first flowering dates for nine species, encompassing the years 1893–1899 (Cockayne 1899).

Using historical sources (e.g. logbooks, diaries, catch numbers) and his own research, Dawbin (1956) defined the north and south migration routes of Humpback whales around New Zealand. The number of whales sighted per week for Cook Strait is provided for the period 1914–1955, providing early phenological information. For example, the mean length of migration over that period was 86 days (range of 64–110 days). The longest seasons were in 1920 and 1937 when the first whales were sighted in the first week of May and last sighted in the third week of August (Dawbin 1956).

Prominent in the early avian phenology literature is Lance Richdale. A teacher in Otago and an amateur ornithologist, Richdale undertook a number of major research projects on seabirds (including penguins, albatross and petrels) from the 1930s to 1950s, mainly during weekends and evenings (Fleming and Warham 1985). Early records indicated very little variability from year to year in arrival and breeding timing in Royal Albatross (Richdale 1942) or in breeding timing of Erect-crested Penguins (Richdale 1950).

3.2 Forest Agency Data and Research

Australian forest agencies have collected phenological data in the majority of states (Fig. 3.1). The early history of which has been detailed by Keatley and Fletcher (2003). These observations cover various durations of flowering and/or budding of commercial forest and/or 'honey' trees. The original aim seems to have been the enumeration of seed crops for silvicultural management. Currently, phenological studies are much reduced but continues in major commercial species (e.g. *Eucalyptus regnans*) and now includes the forecasting of the size of flowering and seed crops (Bassett in prep).

The New Zealand Forest Service was formed out of its predecessor the State Forest Service in 1949. In the same year the Forest Experiment Station in Rotorua became the Forest Research Institute (FRI). One example of early phenological research undertaken by FRI is the study of the seed abundance and crop periodicity of Rimu (*Dacrydium cupressinum*), Kahikatea (*Dacrycarpus dacrydioides*), Matai (*Prumnopitys taxifolia*), Miro (*P. ferruginea*) and Totara (*Podocarpus totara*) between 1958 and 1970 (Beveridge 1964, 1973).

The Department of Conservation (DOC) was launched in 1987, encompassing the New Zealand Forest Service as well as other previous land management agencies, Department of Land and Survey and Wildlife Service. DOC, often in association with volunteers, monitors many aspects of native and introduced species in New Zealand, including phenology (e.g. Mander et al. 1998). Monks (2007)

Fig. 3.1 Overview of Australian forest agencies' phenological data collection (1925–1981)

used seed data collected by DOC (and its predecessors) as well as other researchers to predict seedfall in Beech species (*Nothofagus* spp.), snow tussock (*Chionchloa* spp.) and Rimu.

3.3 Data and Research by Other Organisations

As shown by the Royal Society of Tasmania records there was, and remains, an interest in determining what crops would be suited to particular areas. Annual crops may be considered to have "false" phenophases in that their timing is influenced by management (Menzel and Sparks 2006). However, historical agricultural (annual and perennial) and horticultural records have been shown to be useful in determining the response of crops to climate (Sparks et al. 2005) and are required in developing adaptation measures to climate change and variability (Craufurd and Wheeler 2009). Table 3.3 highlights that long-term phenological data related to horticulture are available in Australia (Darbyshire et al. 2013); though much of data still needs to be uncovered and analysed in relation to climate (though this work is currently underway, personal communication, R. Darbyshire, 2011).

Table 3.3 Examples of horticultural data (provided by R. Darbyshire, University of Melbourne)

Location	State	Variety	Phenology phase(s)	Years	Source
Lenswood	SA	Jonathan apple	Green tip	1963–ongoing	SARDI
Tatura	Vic	Granny Smith apple	Full bloom	1982–ongoing	Grower
Tatura	Vic	Josephine pear	Full bloom	1983–2007	Grower
Tatura	Vic	Packham's Triumph pear	Full bloom	1982–ongoing	Grower
Tatura	Vic	Williams' Bon Chretien pear	Full bloom	1982–ongoing	Grower
Tatura	Vic	Golden Queen peach	Full bloom	1945–ongoing	DPI
Yarra Valley	Vic	Golden Delicious apple	Full bloom	1977–2005	Grower
Yarra Valley	Vic	Red Delicious apple	Full bloom	1977–2001	Grower
Yarra Valley	Vic	Granny Smith apple	Full bloom	1976–2005	Grower
Batlow	NSW	Fuji apple	Full bloom & first pick	1992–ongoing	Grower
Batlow	NSW	Pink Lady apple	Full bloom & first pick	1995–ongoing	Grower
Batlow	NSW	Royal Gala apple	Full bloom & first pick	1991–ongoing	Grower
Donnybrook	WA	Gala apple	Full bloom & first pick	1996–ongoing	Grower
Donnybrook	WA	Golden Delicious apple	Full bloom & first pick	1996–ongoing	Grower
Donnybrook	WA	Pink Lady apple	Full bloom & first pick	1996–ongoing	Grower

In the 1890s New South Wales established agricultural research stations. "Farm cards" (covering the period 1927–1969) provide a synopsis of experiments on wheat and oat varieties (e.g. seeding rates, fertilizer applied) and list the dates of several phenostages (e.g. planting and harvesting commencement dates) along with rainfall and yield per plot (Keatley et al. 2009).

The South Australian Research and Development Institute (SARDI) focuses on primary industry research (e.g. from understanding the effects of fishing on wild fisheries to the development of new horticultural varieties). SARDI has been recording the phenological phase "greentip" in Jonathan apples since 1963 (Table 3.3). They are also investigating the impacts of climate on maturity dates of wine grapes (Sadras and Petrie 2011), as part of their climate applications and crop physiology program. SARDI also monitors Australian Sea Lions *Neophoca cinerea* in South Australia, including breeding season timing from 2002 at eight colonies (Goldsworthy et al. 2009) and for Seal Bay, Kangaroo Island, from 1985. In conjunction with the South Australian Museum, SARDI also monitors New Zealand fur seals *Arctocephalus forsteri* on Kangaroo Island (since 1989; personal communication, Simon Goldsworthy, 2011).

In Victoria, the Dept. of Primary Industries (DPI) is responsible for agriculture, fisheries, earth resources, energy and forestry. Full bloom of Golden Queen peach has been recorded since 1945 at its Tatura research centre (Table 3.3).

3.3.1 Waite Arboretum

Recording of eucalypt flowering began at the Waite Arboretum (formerly the Waite Agricultural Research Institute) around 1951 (Boomsma 1972). The arboretum contains 998 individual trees made up of 432 species. The trees were planted between 1911 and 2011 and approximately 42 % of the trees have had their flowering observed. Observations of the timing and intensity of flowering were undertaken weekly between approximately 1958 and 1993. These data contributed to determining the flowering period of 37 of the eucalypts in the early 1970s (Boomsma 1972).

In addition to the eucalypt flowering records the arboretum also has an extensive collection of ornamental pears (approximately 90 specimens) on which they record various phenological phases including leafing, leaf drop, leaf colour, budding, the beginning, full and ending of flowering.

3.3.2 Pollen Studies

Short-term aerobiological studies were undertaken in Australia initially to determine which species are present and likely to cause allergic respiratory symptoms such as hay fever (e.g. Sharwood 1935; Stevenson et al. 2007). A long running (since 1984) data set of pollen grains counts is held by the University of Melbourne www.botany.unimelb.edu.au/botany/pollencount/counts_pollen.html.

The daily count of pollen grains usually commences at the beginning of September and finishes at the end of December (austral spring through to early summer). These data have been used to predict hourly grass pollen counts in Melbourne and to determine the influence of climate on grass pollen (de Morton et al. 2011).

3.3.3 New Zealand

New Zealand has a long history of agricultural crop research (since 1928) with research stations established across the country (e.g., Havelock North, Te Puke, Clyde, Motueka).

Apple breeding research commenced at Havelock North in 1969 (White 1988) with leafing and flowering dates being two of the commonly assessed traits (Kumar et al. 2010). Peach breeding research began in 1976 (Malone 1994). Using 20 years of flowering data from the Havelock North research centre for calibration, Atkins and Morgan (1990) modelled the impacts of climate change on pip and stone fruit. Flowering dates of Delicious apples from Havelock North (1987–1997) and Nelson (1969–1987) were used to examine the changes in bloom and maturity dates as well as apple size under three different greenhouse gas emission scenarios (Austin and Hall 2001). Unfortunately, the flowering data are not included in these papers.

3.4 Community-Based Phenological Networks

The need for volunteers to be involved in phenological monitoring has gone hand-in-hand with the call for the establishment of phenological networks (e.g. Prince 1891; Kanangra 1949; Keatley and Fletcher 2003) and some of the major community-based groups collecting phenological data are listed below.

3.4.1 ClimateWatch

In the first edition of this chapter the authors (Keatley and Fletcher 2003) highlighted the need for a national Australian phenological network. They recognized that in order to be successful in a country the size of Australia a website would be required as a focus for data collection. ClimateWatch (http://www.climatewatch.org.au), launched in 2009, is now meeting this need. By December 2011 ClimateWatch had around 11,300 observations from over 2,800 registered participants at over 1,200 locations. Participants can record phenological information on over 100 species of plants, birds, mammals, insects, etc.

A recent addition to the ClimateWatch project is the development of ClimateWatch trails. The use of trails, where observers record phenological observations along an established route, can be an effective means to introducing people to ClimateWatch and encourages repeat visits to sites (increasing data reliability and usefulness).

3.4.2 *Timelines*

The main philosophy behind Timelines is that European seasons are inappropriate for Australia (Reid and Beckett 1995). Timelines aims to develop appropriate Australian seasonal calendars similar to the aboriginal calendars of Northern Australia (Jameson 2001).

Participants are encouraged to record anything of interest to them as long as the reason for recording the data is also listed. Hence, people may concentrate on birds, insects, flowers or any one particular species of these. They are also asked to record the month, species, activity (e.g. preening), number and location.

A national Timelines program was launched in 1997, although individual programs operated at a local level from 1994 (Jameson 2001). Timelines is coordinated by Alan Reid (personal communication, Alan Reid, 2012), with earlier sponsorship by The Gould League of Victoria, an environmental education organization, who published a recording diary called "Banksias and Bilbies" (Reid and Beckett 1995) and a CD called "Timelines" (Gould League of Victoria 1998).

The project has developed 64 bioregions for Australia (http://www.timelines.org.au/australias-bioregions) which in time will be populated with the characteristics of each of their seasons.

3.4.3 *Birdlife Australia*

Birds Australia (BA; http://www.birdlife.org.au/) and Bird Observation and Conservation Australia (BOCA) merged in 2012 to become Birdlife Australia. BA brings to Birdlife Australia about 8,000 members, 25,000 supporters and two observatories. The organisation's journal, The Emu, is one of Australia's oldest scientific journals and the source of many historical phenological observations. BOCA has 61 branches, affiliates and Special Interest Groups around Australia with activities including education and bird surveys, including the long-running Melbourne City Bird Watch (1959–1996).

Key activities of Birdlife Australia, of relevance to phenology, include bird atlasing, Birds in Backyards and the Nest Record Scheme. Birds Australia had two main atlas periods, 1977–1981 and 1998 onwards. Although the main aim of the atlas project is to collect information on bird distributions, the information collected can also be used to investigate migration and breeding timing (though only in a limited sense, see Gibbs et al. 2011). Birds in Backyards (http://www.birdsinbackyards.net) started in 1998 and activities include online surveys (e.g. recording the arrival dates of the Common Koel and Channel-billed Cuckoo). The Nest Record Scheme is Australia's longest-running bird survey, with the database containing breeding information (including timing) for hundreds of species.

3.4.4 New Zealand Plant Phenology Websites

The New Zealand Plant Conservation Network hosts a Phenology Recording System (http://nzpcn.org.nz/page.asp?flora_phenology) where phenological observations on any vascular plant (native or exotic species) in New Zealand can be submitted. Officially launched in June 2010 (Crisp 2010), by February 2011 the network had more than 3,000 records (Anon 2011).

Landcare Research and Lincoln University have developed the New Zealand Biodiversity Recording System (http://www.nzbrn.org.nz/index.aspx). The system covers birds, plants, fungi, mammals, invertebrates, frogs and lizards. Observations are wide ranging from feeding resting, mating and egg laying in invertebrates, regeneration under exotic canopy for plants to records of road kill for animals. The Biodiversity Recording System therefore has a wider focus than the phenological recording system of the Plant Conservation Network. As of February 2012 there were more than 370,000 records covering the period 1882–2012 contributed by over 7,800 individuals.

3.4.5 Ornithological Society of New Zealand

The Ornithological Society of New Zealand (OSNZ; http://osnz.org.nz) was founded in 1939. As at 2005, OSNZ had ~1,000 members. The Society's aims include encouraging the recording and archiving of observations and studies of birds, particularly for the New Zealand region. OSNZ also runs a number of projects which are of particular relevance to phenology:

The Moult Recording Scheme, started in 1981, collates information on the timing and pattern of moult in New Zealand birds, particularly wing and tail moult.

The OSNZ Nest Record Scheme began in 1950 and has over 26,000 cards for 144 species (as of January 2012). Information contained within this scheme was used to show that Welcome Swallows in New Zealand have advanced their breeding timing (Evans et al. 2003).

eBird New Zealand was launched in May 2008 and provides a real-time on-line checklist for bird observations. Information from this project can provide regional information on migration timing and so far has been used to map arrival timing in Shining Cuckoos throughout New Zealand.

3.5 Recent Phenological Research

The last two IPCC assessments (IPCC 2001, 2007) highlighted the lack of phenological studies in both Australia and New Zealand. However, a comprehensive survey of the literature reveals that many additional phenological studies have appeared in recent years (Tables 3.4 and 3.5, Fig. 3.2) and that further long-term phenological data are available for analysis (Sect. 3.3). Most of the studies to date assessing temporal trends in phenology have come from Australia (722 of 732 data sets analysed).

Overall, most species studied in Australia and New Zealand have not shown any tendency to shift their phenology in response to climate change, with ~70 % having no significant trend towards either earlier or later life-cycle events (Fig. 3.2). Where a shift was observed, it was generally towards earlier events over time, around 20–25 % of data series, though some later events have been observed. Birds were the most commonly studied group (318 data series), followed by plants (252), invertebrates (160) and reptiles (2). The average rate of change for species with significant advances in life-cycles was 1.55 d/y earlier (range 0.09–6.94), while the later events averaged 2.12 d/y later (range 0.19–13.09) (Fig. 3.3).

3.5.1 Australia

A summary of recent long-term phenological studies undertaken in Australia is given in Table 3.4 and is discussed below.

3.5.1.1 Plants

Two studies (Gallagher et al. 2009; Green 2010) have examined flowering focus on the alpine region of New South Wales. Of the 20 species Gallagher et al. (2009) examined, only Alpine groundsel (*Senecio pectinatus*), showed a significant advance in its first flowering date (0.69 days per year (d/y)). Green (2010) found that the first flowering of Marsh marigold (*Psychrophila introloba*) and Mueller's snow gentian (*Chionogentias muelleriana*) was significantly correlated with the date of snow melt. Over the observation period (1954–2008), snow melt has advanced significantly, by 0.3 d/y.

Along the Victorian coastline the first flowering of four species: Marsh saltbush (*Atriplex paludosa*), Mistletoe (*Dendrophthoe vittellina*), Leafy peppercress (*Lepidium foliosum*) and Oval-leaf logania (*Logania ovata*) advanced by an average of 0.86 d/y (Rumpff et al. 2010). In South Australia peak flowering of the wall-flower orchid (*Diuris orientis*) shifted by 0.17 days earlier per year between 1897 and 2005 (MacGillivray et al. 2010).

Table 3.4 Recent long-term (>10 years) phenological studies undertaken in Australia

Region	Phenological change	Years of study	
South-eastern Australia	Arrival of cherries into Adelaide and Sydney metropolitan markets became significantly earlier	Longest individual series Sydney: 1836–2009	Keatley (2010)
	Generally earlier events over time; no evidence of later phenology, based on date of designated maturity & harvest date	Longest individual series 1895–2009	Webb et al. (2011)
	2 of 6 bird species shifted migration phenology. Rainfall and/or temperature influenced phenology of 4 species	Longest individual series 1958–2004	Chambers and Keatley (2010a)
	Earlier avian arrival & later departure, differences between short and long distance migrants. Earlier arrival linked to increasing minimum temperature	1960–2004	Beaumont et al. (2006)
	Earlier arrival (7 of 15 species) & later departure (2 of 13 species) of migrants at Blaxland, near Sydney	1981–2010	Smith and Smith (2012)
South-eastern Australia, Alpine	5 of 20 species earlier flowering when warmer; 1 species earlier flowering over time	1950–2007	Gallagher et al. (2009)
	Earlier flowering linked to earlier snow melt. Arrival of Richard's Pipit earlier when snowmelt earlier, Flame Robin migration timing linked to low-altitude temperatures	1979–2008	Green (2010)
Victoria	Flowering dates of 65 native species, over 24 years; 8 species flowered significantly earlier, 5 later	1983–2006	Keatley and Hudson (2007)
	8 of 101 species showed trend towards earlier flowering over time	Longest individual series 1854–2007	Rumpff et al. (2010)
	Daily pollen count significantly influence by spring precipitation over 16 seasons in Melbourne	1991–2008	de Morton et al. (2011)
	Earlier breeding in Helmeted Honeyeater related to reduction in rainfall and mild warming	1989–2006	Chambers et al. (2008b)
	Little Penguin: warmer oceans leads to earlier laying & more successful breeding seasons	1968–2009	Cullen et al. (2009)
	Shifts in arrival (3 of 12 species), departure (4 of 11 species) & timing of peak abundance (2 of 13 species). 68 % of all seasonal movement associated with climate drivers	1976–1997	Chambers (2010)
	Earlier emergence of the Common Brown Butterfly	1941–2005	Kearney et al. (2010)

(continued)

Table 3.4 (continued)

Region	Phenological change	Years of study	
South Australia	Earlier peak flowering in 2 Australian orchids.	1893–2005	MacGillivray et al. (2010)
	Pairing dates in Sleepy Lizards earlier when winters warmer & drier	1983–1997	Bull and Burzacott (2002)
	Earlier maturity (3 grape varieties; 18 locations). Advance in maturity associated with higher temperatures in Chardonnay and Cabernet Sauvignon	Longest individual series 1993–2006	Petrie and Sadras (2008)
	Earlier maturity (3 grape varieties; 3 climate regions) associated with higher temperatures	1995–2009	Sadras and Petrie (2011)
Western Australia	Avian migration timing shifts in south-west Australia; stronger relationship to rainfall than temperature	1973–2000	Chambers (2008)
	Arrival & departure dates of birds in semi-arid region; overall trend for earlier arrival &departure; generally earlier arrivals with increasing maximum and minimum temperatures	1984–2003	Chambers (2005)
	Stronger Leeuwin Current & warmer ocean temperatures extend the egg laying period in Little Penguins on Penguin Island	1986–2009	Cannell et al. (2012)
	Later breeding of 3 tern species at Pelsaert Island	1991–2007	Surman and Nicholson (2009)
Continental Australia	6 of 68 butterfly species delayed first flight dates, while 6 advanced their date	1950–2010	McClellan (2011)
	Shifts in breeding timing in Masked Lapwings varied by region (later in south-east; earlier in north-east) – rainfall effect in some regions	1957–2002	Chambers et al. (2008a)
	Shifts in breeding timing in 16 species of Australian birds – climate variables: latitude, altitude, South Oscillation Index	1963–1999	Gibbs et al. (2011)
	Shifts in breeding timing in Australian magpie influenced by latitude, altitude, South Oscillation Index	1967–2001	Gibbs (2007)

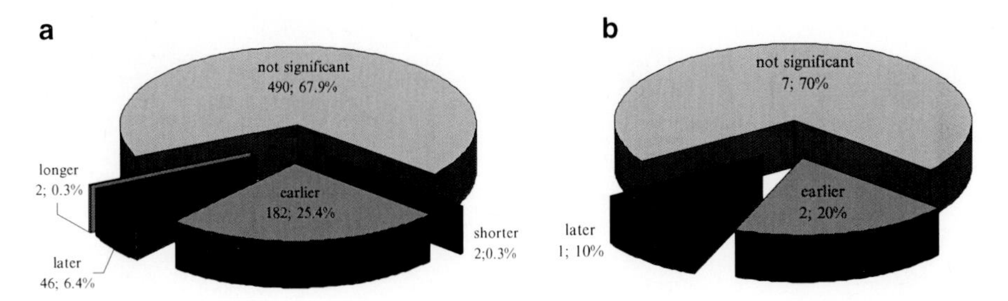

Fig. 3.2 Summary of observed direction of trend for long-term phenological studies (>10 years) in Australia (**a**) and New Zealand (**b**); confined to studies including data post 1970. Total number of data series is 732. Also shown are the number and percentage of data series in each category. Note that the New Zealand trend results are only for avian studies

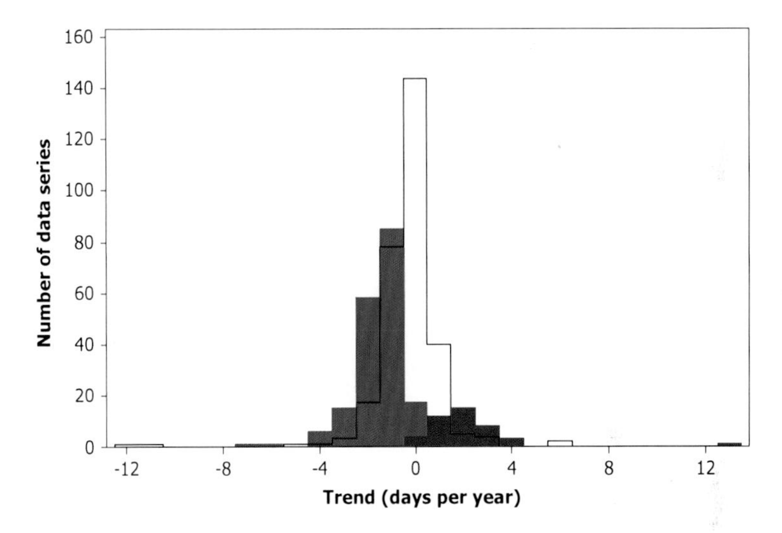

Fig. 3.3 Magnitude of trends observed for long-term phenological studies. Trends significant at the 5 % level are colored

Examination of the first flowering dates of 65 species at a single location in Victoria found that the first flowering dates in 13 species had changed significantly: eight species flowered on average 1.7 d/y earlier and 5 species 1.8 d/y later (Keatley and Hudson 2007).

Three studies (Petrie and Sadras 2008; Sadras and Petrie 2011; Webb et al. 2011) have examined the changes in wine grape (*Vitis vinifera*) phenology. Each study found an overall shift toward earlier maturity.

Of the phenological studies presented in this chapter, plants have the highest percentage of data series with significant trends over time. Most were towards earlier events (113 of 252 data series), particularly for harvest and maturity dates (Fig. 3.4), though some plants were observed to flower later (Keatley and Hudson 2007; Gallagher et al. 2009). The average rate of change for plants with significantly

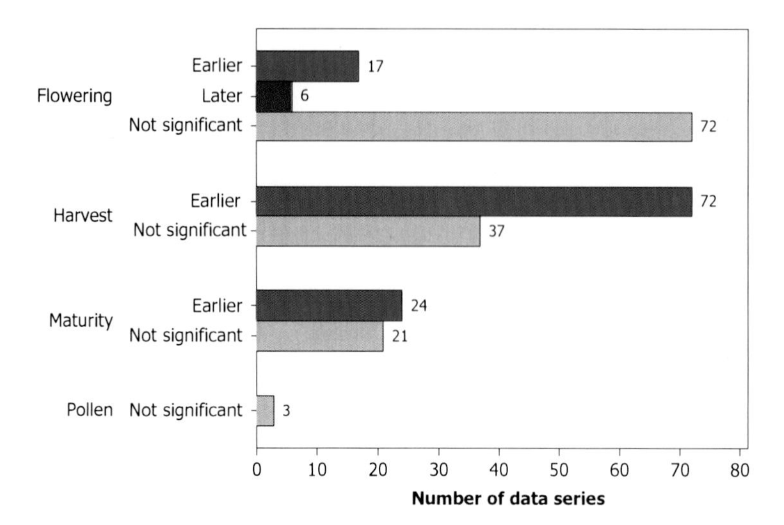

Fig. 3.4 Summary of observed direction of trend for long-term phenological studies in Australia plants. There were no records for New Zealand plants

earlier events over time was 1.6 d/y, while later events averaged 1.9 d/y. Although based on a small number of studies, the percentage of species with significantly changed phenology (45 %) is greater than that determined via meta-analysis for European plant phenology (30 %) (Menzel et al. 2006). However, the percentage of species with significantly delayed phenology is similar (2 % this study; 3 % Menzel et al. 2006).

3.5.1.2 Birds

The majority of the bird data series (218 of 318 data series) did not show any significant trend over time. When a trend was observed, it was more often towards earlier (63 data series) than later events (33 data series). This was particularly true for migration (Fig. 3.5). The average rate of change for birds was similar to that observed for plants, with significantly earlier events occurring around 1.6 d/y earlier (range 0.1 – 6.9), and while later events averaged 2.5 d/y later (range 0.2–13.1).

Three studies assessed shifts in the timing of breeding for terrestrial Australian birds over wide-spatial scales (Gibbs 2007; Chambers et al. 2008a; Gibbs et al. 2011). Importantly, these studies highlighted both regional and species differences in response to climate variability and change. For example, the timing of breeding in most Australian bird species, as is the case in many other countries, varies with both altitude and latitude, with later breeding occurring at higher elevations and at more southerly latitudes (Gibbs et al. 2011) (i.e. in generally cooler locations). Trends towards earlier breeding over time were only consistently observed in south-eastern Australia, with most other regions showing either few significant trends over time or mixes of both earlier and later breeding, depending on the species. For

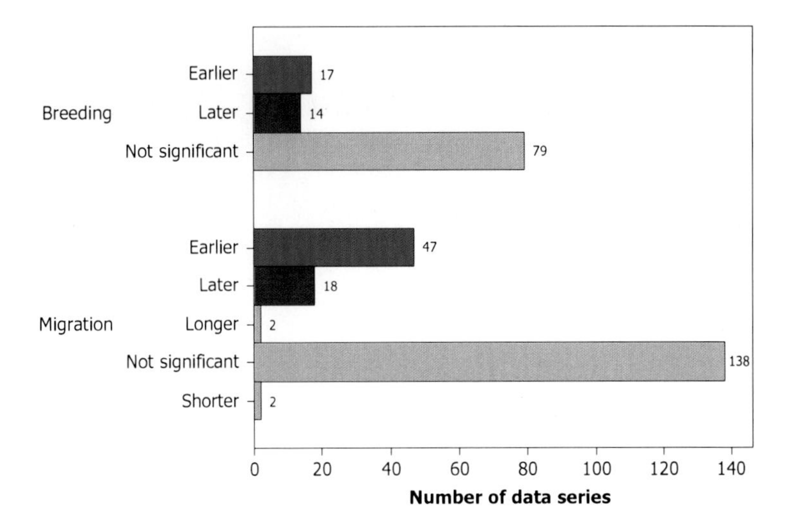

Fig. 3.5 Summary of observed direction of trend for long-term phenological studies in Australia and New Zealand for birds

individual species, differences in the sign and magnitude of trends in timing were observed, according to where in Australia breeding took place; a result also found by Chambers et al. (2008a), Gibbs (2007), and in general (Fig. 3.5).

For marine species, breeding timing has been linked to oceanographic conditions. Warmer ocean temperatures in south-eastern Australia have been linked to an earlier start to breeding in the Little Penguins *Eudyptula minor* of Phillip Island, Victoria (Cullen et al. 2009). In seasons where breeding commences earlier, more and heavier chicks are produced, resulting in a more successful breeding season. Similarly, in south-western Australia, warmer ocean temperatures and a stronger Leeuwin Current correspond to longer breeding seasons, again resulting in improved likelihood of a successful breeding season (Cannell et al. 2012). Further north, in the Houtman Abrolhos, the timing of breeding in tern species is becoming later while, for the long-distance migrant, the Wedge-tailed Shearwater *Ardenna pacifica*, breeding timing remains unchanged. The shearwater has a relatively long breeding season length (120 days) and this, together with its migratory strategy, is likely to constrain its ability to alter its breeding timing in response to changes in environmental conditions (Surman and Nicholson 2009).

A number of studies have investigated shifts in migration timing in Australian birds (Table 3.4), including in relation to climate variability and change. Although many of the species studied have no noticeable trend in their migration timing (Fig. 3.5), when changes were observed they were much more likely to be towards earlier events. Arrival dates, generally in spring, were more likely to occur earlier over time (32 of 105 earlier, compared to 7 later), while there was a fairly even split towards both earlier and later departure dates (13 and 11 of 74, respectively) and the date of peak abundance (1 of 19 for both earlier and later). Many studies (e.g. Chambers 2005, 2008, 2010; Chambers and Keatley 2010a; Green 2010) found significant associations between migration timing and climate variables, particularly temperature, the number of raindays or rainfall totals. Rainfall changes appeared to be particularly important for the timing of movement in waterbirds and those associated with littoral zones, as well as for regions where rainfall has declined (e.g. south-western Australia, Chambers 2008).

3.5.1.3 Mammals

Improved census methods have enhanced our knowledge of the timing of breeding in marine mammals. In South Australia, Goldsworthy et al. (2009) found that both environmental and physiological factors affect the timing and duration of the Australian Sea Lion reproductive cycle. For this species a seasonal drift in timing of pupping was observed (over the period 2002–2006); thought to be due to a breeding cycle of slightly less than 18 months. This species does not have synchronous breeding between colonies, which may indicate the ability of the species to adjust its breeding timing in response to local prey availability.

Environmental variability also plays an important role in the timing and success of breeding in Australian fur seals *Arctocephalus pusillus doriferus*. Over the period 2003–2007, at Kanowna Island in Bass Strait, median birth dates varied from 21 to 25 November (Gibbens and Arnould 2009). Earlier pupping dates corresponded to more pups being produced.

We were unable to locate any long-term phenological studies of Australian terrestrial mammals.

3.5.1.4 Other Vertebrates

The only known long-term phenological study of an Australian reptile, the Sleepy Lizard, is that of Bull and Burzacott (2002). Over a 15-year period, near Mt Mary in the mid-north of South Australia, reproductive timing, or pairing timing, became earlier. The start date of pairing was earlier in years when temperatures during the austral winter were warmer. There was no observed trend in the date when pairing ended, though pairing tended to end earlier in years with warmer spring temperatures. Years with higher spring rainfall corresponded to an increased likelihood of lizard pairings being observed.

3.5.1.5 Invertebrates

The results from McClellan's (2011) butterfly study dominated the observed trends seen in Australian invertebrate studies. There was a fairly even split between studies showing earlier events over time ($n = 7$; mean 7.3 d/y) and those that were later ($n = 8$; mean 1.0 d/y). However, for most species and regions no trend in the timing of migration or emergence was observed ($n = 145$).

One of the first long-term Australian phenological-climate change studies of an invertebrate was that of Kearney et al. (2010), who studied changes in emergence dates of the Common brown butterfly *Heteronympha merope* in relation to climate. Based on 65 years of data, emergence dates have advance by 1.6 days per decade (d/d), which were consistent with a modelled rate of advance of 1.3 d/d, adding strength to the argument that the earlier emergence is driven by the effects of warmer air temperatures on the butterflies' development rate.

Using data from 68 butterfly species from around Australia, McClellan (2011) found that, over the period 1950–2010, 12 of species had significant temporal trends in the date of first record. Six species were seen on average 0.8 d/y earlier, while 6 delayed their first flight date by 1.0 d/y. Trends towards earlier flight were more common in inland regions of south-eastern Australia and in south-western Australia, while species in the coastal regions of south-eastern Australia and those commencing flight later in the season were more likely to have delayed first flight dates over time.

Over a 30 year period in the Snowy Mountains, Green (2010) investigated the phenology of alpine species, including three invertebrates. In this region snow melt has advanced by 2.8 d/d, but was unrelated to the arrival or first emergence dates of the Bogong moth *Agrotis infusa*, Macleay's swallowtail *Graphium macleayanum* or march flies *Scaptia* spp., which showed varied responses over time. The Bogong moth arrived later over time and there was no change in march fly emergence dates. Green's (2010) study is important as it highlights that potential mismatches may be occurring in the Australia alpine region. In short, he found that the flowering season has advanced but that this was coupled with no change in pollinator timing (a key pollinator in this region being march flies). This is likely to affect plant reproductive success. In addition, the later arrival of Bogong moths may impact on many species dependent on them as a food source (e.g. Richard's Pipit *Anthus novaeseelandiae*, Flame robin *Petroica phoenicea*, Mountain pygmy possum and Dusky antechinus). Mismatches in timing are more likely if these dependent species change their phenology in response to changes in snow melt or temperature, as has been seen for the Flame robin and Richard's Pipit (Green 2010).

3.5.2 New Zealand

A summary of recent long-term phenological studies undertaken in New Zealand is shown in Table 3.5. Compared to Australia, few long-term phenological datasets have been compiled and analysed for change in this region.

3.5.2.1 Plants

A number of studies have examined the mast flowering and seeding in Snow tussocks (*Chionochloa* spp.), Beech (*Nothofagus* spp.) and other species (e.g.

Table 3.5 Recent long-term (>10 years) phenological studies undertaken in New Zealand

Region	Phenological change	Years of study	References
New Zealand	Earlier breeding in Welcome Swallow	1962–1996	Evans et al. (2003)
Lower Hutt	Later breeding in Common Starling; non-linear relationship with El Niño – Southern Oscillation	1970–2003	Tryjanonwski et al. (2006)
Kaikoura Peninsula	Laying date in Red-billed Gulls linked to food availability and population size	1983–2003	Mills et al. (2008)
New Zealand	Some indication New Zealand Dotterels nest earlier in years of warmer winter or early spring	1937–2000	Pye and Dowding (2002)
Invercargill	Date of first (14 years) and last egg (9 years) of Caspian Tern. There was no significant trend over time in either egg date[a]	1964–1993	Barlow and Dowding (2002)
Flea Bay, Banks Peninsula	Later relative laying date, increased hatching and breeding success in White-flippered Penguin. No significant temporal trend in laying date[b]	1996–2009	Allen et al. (2011)
Near Harihari, South Island	No significant correlation between seed-fall in Rimu and rainfall. Seedfall negatively correlated with temperature in summer and autumn two seasons prior to seedfall and positively with summer and autumn temperature of seedfall season. No assessment of trends over time	1954–1986	Norton and Kelly (1988)
Canterbury	Mountain Beech: Prolific flowering appears to follow warm temperatures at the time of floral primordia formation. Lack of seed after prolific flowering attributed to extreme frosts or wet conditions	1965–1988	Allen and Platt (1999)
Mt Hutt, Canterbury	Flowering in snow tussocks highly variable with heavier flowering following warm Januarys the year before flowering	1986–2008	Kelly et al. (2008)
Canterbury	Mountain Beech: 7 Year periodicity in total and viable seed counts at each elevation	1965–2007	Allen et al. (2012)
Takahe Valley, Fiordland	16 datasets from 11 species of snow tussock. Within and, to a lesser extent, between sites highly synchronous flowering. Heavier flowering following warm summers	1973–1998	Kelly et al. (2000)
New Zealand	25 datasets of 4 species Beech; 23 datasets from 10 species of Snow tussock; Rimu 4 datasets. Some of the datasets listed (e.g. Rimu 1954–1986 were used in other studies (Norton and Kelly 1988))	Longest individual series Beech: 1965–2002 Snow tussock: 1973–2003 Rimu: 1954–1986	Monks (2007)

[a] Calculated from values provided in Table 2 of Barlow and Dowding (2002)

[b] Calculated from raw data provided by WJ Allen

Norton and Kelly 1988; Monks 2007; Kelly et al. 2008; Allen et al. 2012). Some of these studies have also investigated the influence of climate on the mast flowering and seeding processes, along with the synchrony within and between species and the periodicity of flowering and seeding. Such studies form the basis of predictions of seedfall for the management of feral animals (e.g. Mice *Mus musculus* and stoats *Mustela erminea*) and endangered fauna (e.g. Kakapo *Strigops habroptila* and Kaka *Nestor meridionalis*) However, the methods in which the data have been collected, whilst appropriate for the particular study, preclude detection of trends over time. For example, Kelly et al. (1992) counted inflorescences per tussock in late January or early February for *Chionochloa* spp.

3.5.2.2 Birds

Only a small number of recent studies have assessed trends over time in breeding dates of New Zealand birds (Table 3.5). As was the case in Australia, a similar number of studies found either earlier or later breeding over time. Not all long-term phenological studies considered trends over time, with many instead concerned with identifying drivers of variability, both between years and between individuals. For example, the timing of laying in Red-billed Gulls is influenced by both environmental variability as well as the size of the population (Mills et al. 2008). However, Mills et al. (2008) found that these relationships are not always static. When the population was at its maximum, as prey increased in availability, laying dates became earlier and productivity increased, whereas later, when the population began to decline, even when prey availability increased, laying dates became later.

Laying date has also been shown to vary with the age of the individual (Mills 1973; Low et al. 2007). Generally, older females tend to lay earlier, possibly due to enhanced foraging skills allowing them to reach breeding condition earlier. For example, female Red-billed Gulls who retained mates from previous seasons also bred earlier, suggesting that establishing a new pair-bond may decrease the amount of time available for foraging prior to laying (Mills 1973). However, the relationship is not always linear, with a delay in breeding timing also evident after birds reach peak breeding age (Low et al. 2007) or after 3 years of age in the case of the Yellow-eyed Penguin (Richdale 1957).

Timing of laying has been described as a key variable influencing breeding success (e.g. Low et al. 2007), with earlier laying in many species providing an opportunity for multiple clutches in a season, thus increasing reproductive output.

Two studies examined the timing of breeding in the introduced Common Starling *Sturnus vulgaris* (Bull and Flux 2006; Tryjanonwski et al. 2006). Although laying in starlings, and many other birds, is generally delayed at higher latitudes, Bull and Flux (2006) found this was not the case in their study, with more birds at more southerly locations commencing breeding earlier and producing more young. Using a longer dataset from a single location, Tryjanowski et al. (2006) found a delay in laying date over time and observed that laying was earlier in both El Niño and La Niña years. The timing of egg-laying also varied spatially in the Welcome

Swallow *Hirundo tahitica*, with birds in eastern regions laying earlier than those in the west and there was a non-linear relationship between breeding timing and altitude (Evans et al. 2003). Over time egg-laying has become earlier in this species.

3.6 Conclusions

Although our current knowledge of the drivers of phenological change in Australia and New Zealand is limited (temporally, spatially and by taxa), significant advances have been made in recent years in understanding these drivers of change. Further advances in knowledge are expected as phenological data becomes more readily available, through the collation of historical information, continued monitoring programs and improved observation networks. Advances in knowledge are also expected through the further joint exploration of Traditional Ecological Knowledge. The increasing involvement of community-based science in phenology has benefits not only for expanding the network of data collection, but also for increasing community awareness and thus support for the importance of phenological studies.

Acknowledgments We thank Rebecca Darbyshire, University of Melbourne for supplying the details of the apple, pear and peach data, Jennifer Gardner, Waite Arboretum, University of Adelaide for information on the flowering of eucalypts and ornamental plums at the Waite Arboretum, Fran MacGillivray, University of Adelaide for additional information on the ornamental plums, Peter Smith and Judy Smith of P. & J. Smith Ecological Consultants, for providing a copy of their paper prior to publication, Owen Bassett, Forest Solutions for an early copy of his report. We would also like to thank Amy Winnard, Bureau of Meteorology for assistance with sourcing many of the papers on New Zealand phenology as well as the Australian National Meteorological Library for their assistance in sourcing historical information. We would also like to thank Tim Fletcher, University of Melbourne, and Lesley Hughes, Macquarie University, and to the anonymous reviewer for providing helpful comments on the manuscript.

References

Adams J (1883) On the botany of the Thames Goldfield. Trans N Z Inst XVI:385–393

Allen RB, Platt KH (1999) Annual seedfall variation in *Nothofagus solandri* (Fagaceae), Canterbury, New Zealand. Oikos 57:199–206

Allen WJ, Helps FW, Molles LE (2011) Factors affecting breeding success of the Flea Bay white-flippered penguin (*Eudyptula minor albosignata*) colony. N Z J Ecol 35:199–208

Allen RB, Mason NWH, Richardson SJ, Platt KH (2012) Synchronicity, periodicity, and bimodality in inter-annual tree seed production along an elevation gradient. Oikos 121(2):367–376

Anon (1856) Royal Society of Tasmania: annual general meeting. Colonial Times, p. 2

Anon (1906) Field naturalists. The Advertiser

Anon (1949) Tasmanian naturalists co-operating in international survey. The Mercury, p. 8

Anon (2011) Phenology records top 3000. Trilepidea: newsletter of the New Zealand Plant Conservation Network, vol 87. New Zealand Plant Conservation Network, Wellington

Atkins TA, Morgan ER (1989) Modelling the effects of possible climate change scenarios on the phenology of New Zealand fruit crops. In: Anderson JL (ed) Second international symposium on computer modelling in fruit research and Orchard Management, 5–8 Sept 1989. Logan, Utah, 1990. ISHS, Wageningen

Austin PT, Hall AJ (2001) Temperature impacts on development of apple fruits. In: Warrick RA, Kenny GJ, Harman JJ (eds) The effects of climate change and variation in New Zealand: an assessment Using the CLIMPACTS system. International Global Change Institute, The University of Waikato, Hamilton

Barlow ML, Dowding JE (2002) Breeding biology of Caspian terns (*sterna caspia*) at a colony near Invercargill, New Zealand. Notornis 49(2):78–90

Bassett OD (in prep) Seed crop monitoring and assessment. Native forest silvicultural guidelines No 1. 2nd edn

Beaumont LJ, McAllan IAW, Hughes L (2006) A matter of timing: changes in the first date of arrival and last date of departure of Australian migratory birds. Glob Chang Biol 12:1–16

Beuhne FR (1914) The Honey Flora of Victoria. J Dept Agric Vic XII(10):610–618

Beveridge AE (1964) Dispersal and destruction of seed in central North Island Podocarp forests. Proc N Z Ecol Soc 11:48–56

Beveridge AE (1973) Regeneration of Podocarps in a Central North Island Forest. N Z J For 18:23–35

Boomsma CD (1972) Native trees of South Australia. Bulletin No 19. 1st edn. Woods and Forests Department, South Australia

Briggs C (2008) The journey cycles of the Boonwurrung: stories with Boonwurrung language. Victorian Aboriginal Corporation for Languages, Melbourne

Bull CM, Burzacott D (2002) Changes in climate and in the timing of pairing of the Australian lizard, *Tiliqua rugosa*. J Zool Lond 256:383–387

Bull PC, Flux JEC (2006) Breeding dates and productivity of starlings (*Sturnus vulgaris*) in Northern, central and Southern New Zealand. Notornis 53:208–214

Bureau of Meteorology (1925) Australian meteorological observer's handbook. H.J. Green, Government Printer, Melbourne

Bureau of Meteorology (1954) Australian meteorological observers' handbook. Commonwealth of Australia, Melbourne

Cannell B, Chambers LE, Wooller RD, Bradley JS (2012) Poorer breeding by Little Penguins near Perth, Western Australia is correlated with above average sea surface temperatures and a stronger Leeuwin Current. Mar Freshwater Res 63(10):914–915

Carron LT (1985) A history of forestry in Australia. Australian National University Press, Canberra

Castellano MB (2000) Updating aboriginal traditions of knowledge. In: Sefa Dei GJ, Hall BL, Rosenburg DG (eds) Indigenous knowledges in global contexts. University of Toronto Press, Toronto

Chambers LE (2005) Migration dates at Eyre bird observatory: links with climate change? Clim Res 29:157–165

Chambers LE (2008) Trends in timing of migration of south-western Australian birds and their relationship to climate. Emu 108:1–14

Chambers LE (2010) Altered timing of avian movements in a peri-urban environment and its relationship to climate. Emu 110:48–53

Chambers LE, Keatley MR (2010a) Australian bird phenology – a search for climate signals. Aust Ecol 35(8):969–979

Chambers LE, Keatley MR (2010b) Phenology and climate – early Australian botanical records. Aust J Bot 58(6):473–484. doi:10.1071/BT10105

Chambers LE, Gibbs H, Weston MA, Ehmke GC (2008a) Spatial and temporal variation in the breeding of masked lapwings (*Vanellus miles*) in Australia. Emu 108:115–124

Chambers LE, Quin BR, Menkhorst P, Franklin DC, Smales I (2008b) The effects of climate on breeding in the Helmeted Honeyeater. Emu 108:15–22

Clark JE (1924) International co-operation in phenology. Nature 114:607–608
Clark JE (1925) International co-operation in phenological research. Nature 115(2895):602–603
Clarke B, Chance C (2003) Wisdom Man. Viking, Camberwell
Cockayne L (1899) A sketch of the plant geography of the Waimakariri river basin, considered chiefly from an œcological point of view. Trans N Z Inst 32:95–136
Commonwealth Meteorology (1907) Instructions to country observers. William Applegate Gullick, Government Printer, Sydney
Craufurd PQ, Wheeler TR (2009) Climate change and the flowering time of annual crops. J Exp Bot 60:2529–2539
Crisp P (2010) Message from the President. Trilepidea: newsletter of the New Zealand plant conservation network, vol 79. New Zealand Plant Conservation Network, Wellington
Cullen JM, Chambers LE, Coutin PC, Dann P (2009) Predicting the onset and success of breeding of Little Penguins, *Eudyptula minor*, on Phillip Island from ocean temperatures off south east Australia. Mar Ecol Prog Ser 378:269–278
Cunningham JM (1953) The dates of arrival of the Shining Cuckoo in New Zealand in 1952. Notornis 5(6):192–195
Cunningham JM (1955) The dates of arrival of the Shining Cuckoo in New Zealand in 1953. Notornis 6(4):121–130
Darbyshire R, Webb L, Goodwin L, Barlow EWR (2013) Evaluation of recent trends in Australian pome fruit spring phenology. Int J Biometeorol 57:409–421
Dawbin DH (1956) The migrations of the Humpback whale which pass the New Zealand coast. Trans R Soc N Z 84(1):147–196
de Morton J, Bye J, Pezza A, Newbigin E (2011) On the causes of variability in amounts of airborne grass pollen in Melbourne, Australia. Int J Biometeorol 55:613–622
Erickson R (1950) Flowering dates of orchids. West Aust Nat 2(3):72
Evans KL, Tyler C, Blackburn TM, Duncan RP (2003) Changes in the breeding biology of the Welcome Swallow in New Zealand since colonisation. Emu 103:215–220
Fell HB (1947) The migration of the New Zealand Bronze Cuckoo, *Chalcites lucides lucide*s (Gmelin). Trans R Soc N Z 76(4):504–515
Fleming C, Warham J (1985) Obituary: Launcelot Eric Richdale, O.B.E. (1900–1983). Emu 85 (1):53–54
French CJ (1895) Observations on the flowering times and habitats of some Victorian orchids. Vic Nat 12:31–34
Gallagher RV, Hughes L, Leishman MR (2009) Phenological trends among Australian alpine species: using herbarium records to identify climate-change indicators. Aust J Bot 57:1–9
Gentilli J (1949) Phenology – a new field for Australian naturalists. West Aust Nat 2(1):15–20
Gibbens J, Arnould JPY (2009) Interannual variation in pup production and the timing of breeding in benthic foraging Australian fur seals. Mar Mamm Sci 25:573–587
Gibbs H (2007) Climatic variation and breeding in the Australian Magpie (*Gymnorhina tibicen*): a case study using existing data. Emu 107:284–293
Gibbs HM, Chambers LE, Bennett AF (2011) Temporal and spatial variability of breeding in Australian birds and the potential implications of climate change. Emu 111:283–291
Gillies AM, Hall R (1903) Nature studies in Australia with a natural history calendar, summaries of the chapters and complete index. Whitcombe & Tombs, Melbourne
Goldsworthy SD, McKenzie J, Shaughnessy PD, McIntosh RR, Page B, Campbell R (2009) An update of the report: understanding the impediments to the growth of Australian sea lion populations. Report to the Department for Environment, Water, Heritage and the Arts. SARDI research report series no. 356, West Beach, South Australia
Gott B (2005) Aboriginal fire management in south-eastern Australia: aims and frequency. J Biogeogr 32(7):1203–1208
Gould League of Victoria (1998) Timelines. Viridans Biological Databases, Brighton East
Green K (2010) Alpine taxa exhibit differing responses to climate warming in the Snowy Mountains of Australia. J Mt Sci 7(2):167–175

Hall N (1978) Botanists of the eucalypts. CSIRO, Melbourne
Hannaford S (1856) Jottings in Australia: or, notes on the flora and fauna of Victoria. With a catalogue of the more common plants, their habitats and dates of flowering. James J Blundell & Co, Melbourne
Haviland E (1886) Flowering seasons of Australian plants No 1. Proc Linn Soc N S W 1(11):1049
Haviland E (1888) Flowering seasons of Australian plants No 8. Proc Linn Soc NSW 3:267–268
Head L (1993) The value of long-term perspective: environmental history and traditional ecological knowledge. In: Williams NM, Baines G (eds) Traditional ecological knowledge: wisdom for sustainable development. CRES, Canberra
Hoogenraad R, Robertson GJ (1997) Seasonal calendars from central Australia. In: Webb EK (ed) Windows on meteorology: Australian perspective. CSIRO Publishing, Melbourne
International Meteorological Committee (1949) CAgM Toronto 1947: II phenological networks. In: Conference of directors: final report, Washington, DC, 1947. Lausanne
IPCC (2001) Climate Change 2001: impacts, adaptation, and vulnerability. Contribution of working group II to the third assessment report of the Intergovernmental Panel on Climate Change (IPCC). Cambridge University Press, Cambridge
IPCC (2007) Climate change 2007 – impacts, adaptation and vulnerability. Contribution of Working Group II to the Fourth Assessment Report of the IPCC. Cambridge University Press, Cambridge
Jameson G (2001) Timelines calendars: entering the landscape. In: Interpretation Australia Association (ed) Getting to the heart of it: connecting people with heritage. The ninth annual conference interpretation Australia Association, Alice Springs, 2001. Interpretation Australia Association, Collingwood
Jarman HEA (1950) Proceedings of the annual congress of the Royal Australian Ornithological Union. Emu 49:238
Kanangra (1949) New science would have us all watching birds. The Sydney Morning Herald, 10 November 1949
Kearney MR, Briscoe NJ, Karoly DJ, Porter WP, Norgate M, Sunnucks P (2010) Early emergence in a butterfly causally linked to anthropogenic warming. Biol Lett – UK 6:674–677
Keatley MR (2010) The first box of cherries. Paper presented at the Phenology 2010, Trinity College, Dublin
Keatley MR, Fletcher TD (2003) Phenological data, networks, and research: Australia. In: Schwartz MD (ed) Phenology: an integrative environmental science. Kluwer Academic Publishers, Dordrecht
Keatley MR, Hudson IL (2007) Shift in flowering dates of Australian plants related to climate: 1983–2006. In: Oxley L, Kulasiri D (eds) MODSIM 2007 international congress on modelling and simulation. Land, water and environmental management: integrated systems for Sustainability Modelling and Simulation Society of Australia and New Zealand. Christchurch, New Zealand
Keatley MR, Hudson IL, Fletcher TD (1999) The use of long-term records for describing flowering behaviour: a case-study in Victorian Box-ironbark forests. In: Dargavel J, Wasser B (eds) Australia's ever-changing forests IV. Australian University Press, Canberra
Keatley MR, Fletcher TD, Hudson IL, Ades PK (2002) Phenological studies in Australia: potential application in historical and future climate analysis. Int J Clim 22(14):1769–1780
Keatley MR, Chambers LE, Martin RAU (2009) PhenoARC: an Australia-wide phenological data archive. Paper presented at the Greenhouse 2009: climate change and resources, Burswood convention centre, Perth, Western Australia, 23–26th March
Kelly D, Harrison AL, Lee WG, Payton IJ, Wilson PR, Schauber EM (2000) Predator satiation and extreme mast seeding in 11 species of *Chionochloa* (Poaceae). Oikos 90:477–488
Kelly D, Turnbull MH, Pharis RP, Sarfati MS (2008) Mast seeding, predator satiation, and temperature cues in *Chionochloa* (Poaceae). Popul Ecol 50:343–355
Kingsley D (2003) The lost seasons. http://www.abc.net.au/science/features/indigenous/. Accessed 10 Feb 2012

Kumar S, Volz RK, Alspach PA, Bus VGM (2010) Development of a recurrent apple breeding programme in New Zealand: a synthesis of results, and a proposed revised breeding strategy. Euphytica 173:207–222

Loneragan OW (1979) Karri (*Eucalyptus diversicolor* F. Muell.) phenological studies in relation to reforestation, vol Bulletin 90. Forest Department of Western Australia, Perth

Low M, Pärt T, Forslund P (2007) Age-specific variation in reproduction is largely explained by the timing of territory establishment in the New Zealand Stitchbird *Notiomystis cincta*. J Anim Ecol 76:459–470

MacGillivray F, Hudson IL, Lowe AJ (2010) Herbarium collections and photographic images: alternative data sources for phenological research. In: Hudson IL, Keatley MR (eds) Phenological research: methods for environmental and climate change analysis. Springer, Dordrecht

Mack AE (1909) A bush calendar. Angus and Robertson, Sydney

Mack AE (1924) A bush calendar, 3rd edn. Cornstalk Publishing Company, Sydney

Maiden JH (1909) A plea for the study of phenological phenomena in Australia. Proc R Soc NSW 43:157–170

Maiden JH (1924) Phenology. A form of nature study, with very practical applications. Aust For J 7:4–7

Malone MT (1994) Peach and nectarine breeding in New Zealand. Orchardist 67(10):16

Mander C, Hay R, Powlesland R (1998) Monitoring and management of kereru (*Hemiphaga novaeseelandiae*). Department of conservation technical series; No. 15. Department of Conservation, Wellington

Maplestone C (1895a) Calendars and the indexing of natural history observations. Vic Nat 12 (10):120–122

Maplestone C (1895b) Flowering times of orchids. Vic Nat 12(7):82–83

McClellan KE (2011) The responses of Australian butterflies to climate change. Macquarie University, Sydney

McGlone M, Walker S (2011) Potential effects of climate change on New Zealand's terrestrial biodiversity and policy recommendations for mitigation, adaptation and research. Dept. of Conservation, Wellington

McLachlan RG (1921) Victoria Valley experiences. Vic Bee J 2(6):64–65

Menzel A, Sparks T (2006) Temperature and plant development: phenology and seasonality. In: Morison JIL, Morecroft MD (eds) Plant growth and climate change. Blackwell Publishing Ltd, Ames

Menzel A, Sparks TH, Estrella N, Koch E, Aasa A, Ahas R, Alm-Kubler K, Bissolli P, Brasavská O, Briede A, Chmielewski F-M, Crepinsek Z, Curnel Y, Dahl A, Defila C, Donnelly A, Filella Y, Jatczak K, Mage F, Mestre A, Nordli Ø, Peñuelas J, Pirinen P, Remišová V, Scheifinger H, Striz M, Susnik A, van vliet AJH, Wielgolaski F-E, Zach S, Zust A (2006) European phenological response to climate change matches the warming pattern. Glob Chang Biol 12(10):1969–1976

Mills JA (1973) The influence of age and pair-bond on the breeding biology of the red-billed gull *Larus novaehollandiae scopulinus*. J Anim Ecol 42:147–162

Mills AM (1979) Factors affecting the egg size of Red-billed gulls *Larus novaehollandiae scopulinus*. Ibis 121:53–67

Mills JA, Yarrall JW, Bradford-Grieve JM, Uddstrom MJ, Renwick JA, Merilä J (2008) The impact of climate fluctuation on food availability and reproductive performance on the planktivorous red-billed gull *Larus novaehollandiae scopulinus*. J Anim Ecol 77:1129–1142

Monks A (2007) Climatic prediction of seedfall in *Nothofagus*, *Chionochloa* and *Dacrydium cupressinum*. DOC research and development series 276. Dept. of Conservation, Wellington

Nekovář J, Koch E, Kubin E, Nejedlik P, Sparks T, Wielgolaski F-E (eds) (2008) COST Action 725: the history and current status of plant phenology in Europe. Finnish forest research institute Muhos Research unit and COST office, Vammalan Kirjapaino Oy, Finland

Newell J, Sutherland D (1997) Scientists and Colonists. Australas Sci 18(4):56

Norton DA, Kelly D (1988) Mast seeding over 33 years by *Dacrydium cupressinum* Lamb. (Rimu) (Podocarpaceae) in New Zealand: the importance of economies of scale. Func Ecol 2:399–408

Penglase, Armour J (1909) Victorian honeys and where they come from. Federal independent Beekeeper (March 1):2–4

Petrie PR, Sadras VO (2008) Advancement of grapevine maturity in Australia between 1993 and 2006: putative causes, magnitude of trends and viticultural consequences. Aust J Grape Wine R 14:33–45

Ploughshare (1926) International phenology. The Mercury, 25 October 1926, p. 8

Prince JE (1891) Phenology and rural biology. Vic Nat 8:119–127

Pye DA, Dowding JE (2002) Nesting period of the northern New Zealand dotterel (*Charadrius obscurus aquilonius*). Notornis 49(4):259–260

Reid AJ, Beckett A (1995) Banksias and Bilbies: seasons of Australia. Gould League of Victoria, Moorabbin

Richdale LE (1942) Supplementary notes on the Royal Albatross. Emu 41:169–184

Richdale LE (1950) Further notes on the erect-crested Penguin. Emu 49:153–166

Richdale LE (1957) A population study of penguins. Oxford University Press, Oxford

Riseth JA, Tømmervik H, Helander-Renvall E, Labba N, Johansson C, Malnes E, Bjerke JW, Jonsson C, Pohjola V, Sarri L-E, Schanche A, Callaghan TV (2011) Sámi traditional ecological knowledge as a guide to science: snow, ice and reindeer pasture facing climate change. Polar Rec 47:202–217

Rose D (2005) Rhythms, patterns, connectivities: indigenous concepts of seasons and change. In: Sherratt T, Griffiths T, Robin L (eds) A change in the weather: climate and culture in Australia. National Museum of Australia Press, Canberra

Rosenzweig C, Casassa G, Karoly DJ, Imeson A, Liu C, Menzel A, Rawlins S, Root TL, Seguin B, Tryjanowski P (2007) Assessment of observed changes and responses in natural and managed systems. Climate Change 2007: impacts, adaptation and vulnerability. Contribution of working group II to the fourth assessment report of the Intergovernmental Panel on Climate Change. Cambridge University Press, Cambridge

Rumpff L, Coates F, Morgan J (2010) Biological indicators of climate change: evidence from long-term flowering records of plants along the Victorian coast, Australia. Aust J Bot 58:428–439

Sadras VO, Petrie PR (2011) Climate shifts in south-eastern Australia: early maturity of Chardonnay, Shiraz and Cabernet Sauvignon is associated with early onset rather than faster ripening. Aust J Grape Wine R 17:199–205

Sakurai R, Jacobson SK, Koboric H, Primack R, Oka K, Komatsu N, Machida R (2011) Culture and climate change: Japanese cherry blossom festivals and stakeholders' knowledge and attitudes about global climate change. Biol Conserv 144(1):654–658

Sedgwick EH (1947) Breeding of the Black and White Fantail. West Aust Nat 1(1):14–17

Sedgwick EH (1949) Proceedings of the annual congress of the Royal Australian Ornithological Union, Perth 1948. Emu 48:177–211

Sedgwick EH (1950) The Pallid Cuckoo in the south-west. West Aust Nat 2(5):119

Sharwood (1935) The pollen content of the Melbourne air during the hay fever season of August 1933-March 1934. Med J Aust 1:326–332

Smith P, Smith J (2012) Climate change and bird migration in south-eastern Australia. Emu 112(4):333–342

Sparks TH, Croxton PJ, Collinson N, Taylor PW (2005) Examples of phenological change, past and present, in UK farming. Ann Appl Biol 146:531–537

Steane SW (1931) Report of the forestry department for the year ended 30th June, 1930. Forestry Department, Hobart

Stevenson J, Haberle SG, Johnston FH, Bowman DMJS (2007) Seasonal distribution of pollen in the atmosphere of Darwin, tropical Australia: preliminary results. Grana 46:34–42

Surman CA, Nicholson LW (2009) The good, the bad and the ugly: ENSO driven oceanographic variability and its influence on seabird diet and reproductive performance at the Houtman Abrolhos, Eastern Indian Ocean. Mar Ornithol 37:129–138

Tout SM (1935) Enquires on method of collection of data in regard to flowering and fruiting of native trees from Forestry Commission of New South Wales. Forest Commission of Victoria, Melbourne, Unpublished Correspondence in VPRS 11563/P/0001, File FCV 35/3123 HONEY, Location L/AZ/068/01/08, Public Records Office

Tryjanonwski P, Flux JEC, Sparks TH (2006) Date of breeding of the starling *Sturnus vulgaris* in New Zealand is related to El Niño Southern Oscillation. Aust Ecol 31:634–637

Wang JY (1967) Agricultural meteorology. Agriculture Weather Information Service, San Jose

Waugh SM (1997) Laying dates, breeding success and annual breeding of Southern Royal Albatrosses *Diomedea epomorphora epomorphora* at Campbell Island during 1964–1969. Emu 97:194–199

Webb LH, Whetton PH, Barlow EWR (2011) Observed trends in winegrape maturity in Australia. Glob Chang Biol 17(8):2707–2719

White A (1988) Apple breeding in New Zealand. ISHS Acta Horticulturae 224:119–121

Williams GR (1967) The breeding biology of California quail in New Zealand. Proc N Z Eco Soc 14:88–99

Chapter 4
Europe

Annette Menzel

Abstract Europe has a long tradition of systematic phenological data collection from a range of different environments. In recent times this data has proved invaluable in demonstrating the impact of climate warming on our natural environment together with providing a means by which to ground truth remotely sensed information. However, since the networks evolved in different countries with different traditions of data observation and collection aggregation at a continental scale is challenging. Here we provide a snapshot of some of the professional and citizen science-based national phenology networks, and describe a number of recently established pan-European initiatives to explore ways to establish a standardized framework for plant monitoring, data collection, quality control and transfer. Finally, we highlight areas such as species or groups of high value for nature conservation which require further research.

4.1 Introduction

While the longest written phenological record originates in Japan at the Royal court of Kyoto (the beginning of cherry flowering since 705 AD), the most vital and broadest tradition of phenological monitoring is found in Europe (Menzel 2002). In many countries long-term data sets exist, thus Europe is a particularly suitable region for investigating phenological changes or providing "ground truth" to satellite data. However, phenological information exists in numerous countries at a local, regional, or national level with quite different histories and traditions of observation. Thus, we are far away from a homogenous plant phenological data set at a continental level necessary for the applications indicated above, although new

---

A. Menzel (✉)
Department of Ecology and Ecosystem Management, Technische Universität München, Freising, Germany
e-mail: amenzel@wzw.tum.de

M.D. Schwartz (ed.), *Phenology: An Integrative Environmental Science*,
DOI 10.1007/978-94-007-6925-0_4,

initiatives, such as COST725 and the resulting PEP725 European phenological database, now offer structured, broad datasets. Unfortunately, it is impossible to provide a complete overview of European phenology here; however, current important national networks are compared to new schemes, such as Nature's Calendar, and to international initiatives (e.g. International Phenological Gardens, ICP Forests).

One of the oldest European phenological records is the famous Marsham family record in Norfolk 1736–1947 (Sparks and Carey 1995). Following Schnelle's (1955) historical overview, the first phenological network was then established by Linné (1750–1752) in Sweden. The first international (European) phenological network was run by the Societas Meteorologica Palatina at Mannheim (1781–1792), and a second famous one by Hoffmann and Ihne (1883–1941). Schnelle (1955) summarized the development of phenological observations in Germany as well as in other countries, such as (considering only European ones) Austria, Poland, Czech Republic and Slovakia, Russia, Finland, Sweden, Norway, United Kingdom, The Netherlands, Belgium, France, Switzerland, Spain, Italy, Greece and the former Yugoslav Republic through the middle of the last century. However, their subsequent history was very patchy.

4.2 International Networks

4.2.1 International Phenological Gardens

The International Phenological Gardens (IPG) is a unique phenological network in Europe, which was founded in 1957 by F. Schnelle and E. Volkert. Since then, it has been maintained on a voluntary basis, coordinated by a chairman (Chmielewski 1996). Manuals and annual observations have been published in the journal of the IPGs, the *Arboreta Phaenologica*. At present the network is coordinated by the Humboldt University Berlin (http://ipg.hu-berlin.de/). The core idea of this network was to obtain comparable phenological data across Europe by observing genetically identical plants (clones), which permanently remained at one site. Thus, the records are not influenced by different genetic codes of the plants and the variability and potential inaccuracy of the observations is reduced compared to data from the national phenological networks (Baumgartner and Schnelle 1976). In 1959, the first IPG started its observations in Offenbach (near Frankfurt am Main), Germany. Subsequently, additional IPGs were established with vegetatively propagated species of trees and shrubs at different sites across Europe. In 2002 about 50 IPGs record up to 7 phases of 23 plants species (~50 clones), in 2012 ~20 of them were still active and almost 50 new stations have been established since 2002. The network in Europe covers a large area from 38 to 69°N (Portugal to Scandinavia) and from 10°W to 27°E (Ireland to Finland), comprising different climate regions in Europe. Recent studies comprehensively analyzing the data of the IPGs revealed a lengthening of the growing season across Europe, provided the necessary ground truth to satellite data and CO_2 records, and linked the changed onset of spring to spring temperature and the North Atlantic Oscillation Index (NAO) (Menzel and Fabian 1999; Menzel 1997, 2000; Chmielewski and Rötzer 2001, 2002).

Most recently, following the example of the IPG, a new phenological network of arctic-alpine botanical gardens has been initiated by the Alpine Botanical Garden Schachen (Germany) including a new observation key for selected and propagated cloned alpine plant species (Schuster et al. 2011).

4.2.2 ICP Forests

Phenological observations are also made at Level II plots of ICP Forests. ICP Forests is the International Co-operative Programme on Assessment and Monitoring of Air Pollution Effects on Forests, which was launched in 1985 under the convention on long-range transboundary air pollution of the United Nations Economic Commission for Europe. ICP Forests monitors the forest condition in Europe using two monitoring intensity levels. The second level (so called Level II) has been operating since 1994 in selected forest ecosystems. On these plots, soil and soil solution chemistry, foliar nutrient status, increment, meteorological condition, ground vegetation and deposition of air pollutants are measured in addition to the annual crown condition assessments. On an optional basis, phenological observations are made to provide supplementary information on the status and development of forest tree condition during the year. Since 1999, additional phenological phases are recorded to determine the course of the annual development of forest trees, to explain possible changes in relation to environmental factors, and to utilize this knowledge in interpreting observed changes in tree condition. The ICP Expert Panel on Meteorology, Phenology and Leaf Area Index gives an additional focus on seasonal variations in phenology and LAI. Information about this network including the complete manual for phenological observations is available at http://www.icp-forests.org/.

4.3 National Networks

In Europe two major types of national (countrywide) phenological networks can be distinguished. In several countries, such as Albania, Austria, Czech Republic, Estonia, Germany, Poland, Russia, Slovak Republic, Slovenia, Spain, and Switzerland, the National Weather Services have been running (plant) phenological schemes during the second half of the twentieth century, and some networks already existed at the beginning of the twentieth century (see Schnelle 1955). In contrast to these "traditional" networks, "younger" networks have been (re) established recently.

Table 4.1 Basic identification information about the selected European phenological networks portrayed

Network contact	Internet	Current number of stations	Manual
Central Institute for Meteorology and Geodynamics, Austria	www.zamg.ac.at	~100	ZAMG (2000)
Czech Hydrometeorological Institute	www.chmi.cz	158 (46 forests, 84 crops, 28 fruit)	Guidebooks available
Slovak Hydrometeorological Institute	www.shmu.sk	221 (61 forests, 53 crops, 15 fruit, 92 com. phenology)	SHMÚ Bratislava (1988a, b, 1996a, b)
German Meteorological Service	www.dwd.de/phaenologie	~1,250 in the basic network plus approx. 400 immediate reporters	DWD (1991)
Estonian Hydrometeorological Institute	www.emhi.ee	21	EMHI (1987)
Environmental Agency of Slovenia	www.arso.gov.si/en/	61	Observation guidelines
MeteoSchweiz	www.meteoswiss.ch/	160	Meteo-Schweiz (2003)

4.3.1 Traditional Networks of the National Weather Services

Basic information concerning the phenological networks of selected countries, such as web site and network contact, number of observers, and species and phases, are listed in Tables 4.1 and 4.2. The recorded phenological information is used by the monitoring networks themselves or externally, primarily for research in agri-, horti-, and viniculture, forestry, ecology, human health, as well as for climatic evaluations and evaluation of potential global change impacts. In most cases, National Weather Services' phenological networks were intended to gather additional (integrated) climate information. Thus, observations of wild plants were used to monitor phenological seasons; agricultural observations of different crops and fruit trees mainly served to predict growing success, delivered data for modeling, and facilitated agro-meteorological consulting. Other purposes include the prognosis of onset dates, pollen forecasts, frost risk management, and monitoring of biotic damage.

The recent history (since 1950) and special characteristic of selected networks are quite similar: In **Germany**, the Deutscher Wetterdienst in 1949 and the Hydrometeorologische Dienst der DDR in 1951 took over the phenological network, which has been founded by the Reichswetterdienst in 1936 by combining different regional phenological networks. Since 1991, these networks are unified again, and the phenological network is managed by the Deutscher Wetterdienst (1991). At the same time, the observational program was adjusted. Similarly, in **Austria** and **Switzerland** phenological phases have been recorded continuously since 1951 in networks run

Table 4.2 Basic observation program information from selected European phenological networks portrayed (S species, P phases) (as 2002)

Network Contact	First (current) network	Wild plant + trees	Agricultural crops + farmer's activities	Fruit trees + grape vines
Central Institute for Meteorology and Geodynamics, Austria	(1928) 1951–today	15 + 9 S 7 P (68 P_{tot})	11 S 12 P (78 P_{tot})	5 + 1 S 9 P (26 P_{tot})
Czech Hydrometeorological Institute	(1923) 1986–today	45 S 26 P	19 S 33 P	14 + 1 S 16 / 2 P
Slovak Hydrometeorological Institute	(1923) 1986–today			
German Meteorological Service	(1936) 1951–today	30 S 59 P_{tot}	10 S 66 P_{tot}	10 S 32 P_{tot}
Russian Geographical Society		>102 S (wild plants, trees, crops, animals, AGROMET. EVENTS), >32 P		
Estonian Hydrometeorological Institute	1948–today	68[a] S 7 P	9 S 16 P	8 S 11 P

[a] The selection of native species is voluntary for the observers

by their National Weather Services. In Austria after the World War II, a new phenological network was established by the Zentralanstalt für Meteorologie und Geodynamik in Vienna, based on an older network started in 1928.

Unfortunately, the number of stations decreased from around 500 in the 1970s to 80 currently. The Swiss phenological observation network was founded in 1951 and initially consisted of 70 observation posts; the phenological observation program was slightly modified in 1996. The first phenological record in **Slovenia** is Scopoli's work *Calendarium Florae Carniolicae* from 1761. Modern phenology data collection started in 1950/1951 with the establishment of a phenological network within the Agrometeorological service, thus data are mainly used for research and applicable agriculture purposes. The recent network consists of 61 phenological stations, which are evenly distributed by a regional climatic key over the entire territory of Slovenia. The observations are carried out on species of non-cultivated plants (herbaceous plants, forest trees and bushes, clover and grasses) and of cultivated plants, such as field crops and fruit trees. In some portions of **the Slovak and the Czech Republic,** phenological observation was also conducted for a short time in the last half of the nineteenth century, but regular and managed phenological observation did not start until the twentieth century. From 1923–1955, the observational program comprised more than 80 plant species (crops, fruit trees, native plants), but also some migratory birds, insects, as well as agro-technical data. From 1956 to 1985 observations were made following the first instruction guide edited by the Hydrometeorological Institute, with an enlarged program including, for example, agro-meteorological observations and crop diseases. In 1985/86 a new system of phenological observation (including new guides) was instituted with three special sub-networks for field crops, fruit trees, and forest plants, and respective stations in regions with intensive agricultural production, in orchards and vineyard regions, and in forest regions. Three developmental stages (10, 50 and 100 %) are now observed.

In **Slovakia**, some historical stations were maintained in the (so called) "common" phenological network. In this network, general phenological observations (crops, fruit trees and grapevine, forest plants, migrating birds some agro-meteorological and agro-technical data) are made by volunteers, in contrast to the "special" networks, where experts (e.g., with agronomic education) do the recording. In 1996 the guides for the common and special observation of forest plants were modified. The species and scales of phenophases are now very close to those in use before 1985.

The former **USSR** area has many phenological observation programs supervised mainly by Russian organizations. The Russian Geographical Society started phenological studies in 1850s with more than 600 observers, mostly in European Russia. Today, the archive of this voluntary network in St. Petersburg is one of the most important phenological centers in Russia, with more than 2000 observation sites all over the former USSR area, and regional subprograms which have different observation manuals and species lists (Hydrometeorological Printing House 1965; Schultz 1981). The second important network is the Hydrometeorological Service's agrimeteorological observation program, organized by Schigolev in 1930, with a unique and strict methodology and very detailed observations of agricultural crops and some natural tree species, including climate parameters, such as soil temperature and moisture, and snow and precipitation, at the same site (Davitaja 1958; Hydrometeorological Printing House 1973). Today, the database at the central archive in Obninsk is not actively used, because the data is not digitized. Several other phenological observation programs exist that are run by the plant protection service, agricultural selection service, forestry department (Schultz 1982) or in Nature Conservation areas that use their materials for study and educational purposes (Kokorin et al. 1997).

In Estonia, the first scientific phenological observation program (with more than 30 observed species) was set up in the botanical garden of University of Tartu in 1869, by the Estonian Naturalists Society (Oettingen 1882). The Estonian Naturalist's Society started organized phenological studies in the 1920s/1930s and a broad observation program in 1951 (Eilart 1959). Today, the society is the most active voluntary observer of plant, bird, fish, phenology and seasonal phenomena in the country (Eilart 1968; Ahas 1999). The agri-phenological network of the Estonian Meteorological and Hydrometeorological Institute (started in 1948) used standard observation methods similar to those used in the former USSR (Hydrometeorological Printing House 1973; EMHI 1987). Their observation list consisted of agricultural plants, selected tree species, and main characteristics of the physical environment. Until the 1990s, 21 stations were still in operation (Ahas 2001), but that number diminished to 10 in 2001, and 6 in 2002.

In **Poland**, the Hydrometeorological Institute ran a phenological network from 1951 to 1990 composed of around 70 stations. In the manual by Sokolowska (1980), phenological observations are described, and the main results are reported by Tomsazewska and Rutkowski (1999). In **Spain**, the phenological network organized by the Spanish Meteorological Institute is characterized by an enormous number of stations, species, and phases, but less continuity of observations at single sites.

In general, the traditional phenological networks of the national hydrometeorological services can be characterized by their long-term continuous records which are ideal for climate change impact monitoring.

4.3.2 New Citizen Science Based Networks

The three examples of the British "Nature's Calendar", the Dutch "De Natuurkalender" and the Swedish National Phenological Network (SWE-NPN) stand as examples of phenological networks which have been set up recently, mostly run on the Internet, and organized by non-governmental organizations (NGOs), media, and research institutions. They include a lot of observations on animals (e.g., birds and butterflies).

A national phenological network in the **United Kingdom** was established by the Royal Meteorological Society in 1857. However, the subsequent development of British phenology was quite different from the continental central European countries, as annual reports were published only up until 1948. In 1998 a pilot scheme to revive a phenology network in the UK was started by Tim Sparks, research biologist at the Centre for Ecology & Hydrology in Cambridge, comprising both plant and animal phases. In autumn 2000 the Woodland Trust forces joined with the Centre for Ecology & Hydrology to promote phenology to a far wider and larger audience. In 2001 the number of registered recorders across the UK rose to over 11,700, and by August 2002 it was 16,809 and still growing, with around half of these being online observers. The Nature's calendar's website (http://www.phenology.org.uk) provides information about the species observed, an online list of observations, and graphic presentations of trends.

In February 2001, Wageningen University and the national radio program VARA Vroege Vogels (Early Birds) started a phenological monitoring network in **the Netherlands**, called De Natuurkalender. This network aimed to increase understanding of changes in the onset of phenological phases, also due to climate change, for human health, agriculture, and forestry. Other aims were to strengthen the engagement of the public in their natural surroundings and to develop interactive educational programs for school children and adults. The observation program includes over 100 species of plants, birds, and butterflies with at least one phenophase per species. The phenophases are clearly defined in an observation manual. Over 2000 volunteers subscribed to the program, and send their observations via the Internet, a paper form, or a special telephone line (Fenolijn) to the coordinators of the network. The observers and other potentially interested people are informed about the results of the observations by a weekly report during the radio program, which is followed by 500,000 people every Sunday morning. Furthermore, the network uses an interactive website to provide direct feedback to the observers.

The **Swedish National Phenological Network** (SWE-NPN) officially started in Jan 2010 to set up a nationwide, geographically dense phenology database (see http://www.swe-npn.se/). Equally to all other networks, its aim is to observe phenological shifts and to understand and predict how they will feedback on the climate system, ecosystem productivity and processes, and human health (pollen forecasts). The observation network is based on both professional field stations and community-based volunteers. Phenological observations are reported to the http://www.blommar.nu/ website, which offers ample feedback and outreach as well as data access tools. The SWE-NPN has meanwhile digitized all historical observational data (1873–1926) which were recorded on around 50 plants and 25 animals at more than 300 sites. It is intended to use them as a reference point for changing climatological and phenological conditions.

4.4 Other Networks

This rough overview of phenological networks leaves many regions in Europe blank, due to the lack of current phenological networks in those places (e.g., Portugal, Greece). Other countries only have local networks due to regional organized research structure (e.g., Italy), current networks that are run by other institutions (e.g. Finland METLA, Norway), or they mainly have historical networks (e.g. Norway, Finland). Thus, this overview does not claim to be exhaustive, and it is fairly certain that in many other countries, national or regional networks existed, or are still running.

An evaluation of the World Meteorological Organization (WMO) RA VI agro-meteorological questionnaire on phenological observations and networks revealed that from 28 replying countries only 6 countries (Belgium, Bosnia and Herzegovina, Denmark, Luxembourg, Portugal and the United Kingdom) had no regular (agro-meteorological) phenological network, whereas 22 countries (Armenia, Austria, Croatia, Czech Republic, Estonia, France, Germany, Hungary, Ireland, Italy, Israel, Latvia, Lithuania, Macedonia, Moldavia, Romania, Russia, Slovakia, Slovenia, Spain, Switzerland, Syrian Arabic Republic) have regular phenological networks (WMO 2000). However, following this WMO evaluation, the phenological observations, the applied observations methods, the structure of the networks, the coding systems, and the practical usage of data are highly diverse.

The French phenological network that started in 1880 under the care of Meteo France may represent observations “fallen into oblivion.” Phenological observations have been reported continuously up to 1960 for most stations adjacent to meteorological stations, but only three of them continued their observations after that date, with the last one stopping in 2002. The inception of this network was similar to other central European ones still in operation, as the observational program comprised perennial wild species including trees (25 species), crops (10), and fruit trees (8), and an instruction booklet was provided to observers to standardize observations. Observations are contained in archives, but have not been digitized. These data have not been analyzed, except at the very beginning of the network by C.A. Angot from Meteo France (in a few Annales du Bureau Central Météorologique) who mapped isolines of the onset of phenophases for the decade 1881–1891. Meanwhile France has set up a common system on all phenological information available at the national level (http://www.gdr2968.cnrs.fr/) including a so called "season observatory" unifying more than 27 laboratories and research organizations working on phenology (http://www.obs-saisons.fr/), such as INRA which started the Phénoclim network on fruit trees and grapevine.

The PhenoAlp project (www.phenoalp.eu) may serve as an example for recent transnational collaboration, here co-funded by the EU, to establish specialized networks in those areas currently not well covered.

The picture won't be complete without a hint to other environmental networks in which vegetative or reproductive phenology is monitored as auxiliary data. For example, at around 100 European and over 400 so called fluxnet research sites, carbon dioxide and water vapor fluxes are measured by eddy covariance method along with other environmental data such as leaf phenology (www.fluxdata.org).

4.5 Towards A Pan European Phenology

In order to overcome the diversity in phenological monitoring at the European scale, the COST725 action (Establishing a European data platform for climatological applications, 2004–2009) aimed to standardize and harmonize phenological observation schemes and guidelines and to unify data stored in different locations and in different formats (Koch et al. 2001). Thus, the succeeded in defining species and phases of common interest, to develop recommendations for monitoring and quality control, and to assign the BBCH code (Meier 1997) to all phases observed. Furthermore, in cooperation with WMO, WCDMP and WCP, COST725 developed guidelines for plant phenological observations. A common COST725 database was established and maintained comprising main parts of the historical data in a common format and with BBCH coding. Moreover, a complete overview of national and international European phenological networks including 30 contributions from all COST725 member states plus Croatia, Bosnia and Herzegovina, Montenegro and the International Phenological Gardens was published by Necovář et al. (2008).

The most important scientific achievement of COST725 was the first Pan European study of observed phenological changes comprising more than 125,000 time series of 542 plant and 19 animal species in 21 countries for the period 1971–2000 (see Fig. 4.1). The corresponding paper entitled 'European phenological response to

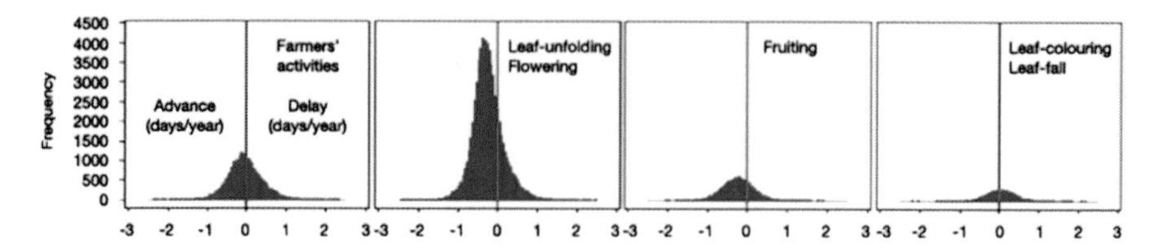

Fig. 4.1 Frequency distributions of trends in phenology (in days/year) over 1971 to 2000 for 542 plant species in 21 European countries (From Menzel et al. 2006; Rosenzweig et al. 2007, figure 1.6. on page 113)

climate change matches the warming pattern' (Menzel et al. 2006) received worldwide attention and was one of the backbones of the 4th Assessment Report of IPCC in 2007, demonstrating an attributable fingerprint of climate change in nature (Rosenzweig et al. 2007, 2008; see also Chapter 18 Plant Phenological Fingerprints). Out of the 29,000 time series of significant changes in natural and managed systems collated by IPCC, more than 28,000 were contributed by Europe, i.e. the COST725 publication of 2006 (AR4, IPCC, Summary for Policy Makers).

The successor of the COST725 action is the PEP725 database (www.pep725.eu) where 16 European meteorological network members (Austria, Belgium, Croatia, Czech Republic, Finland, Germany, Hungary, Ireland, Norway, Poland, Romania, Serbia, Slovak Republic, Slovenia, Spain, Switzerland) and 6 other partners have agreed to promote and facilitate phenological research by its database with unrestricted access for science, research and education. The database (2012) comprises 8.6 million records from almost 19,000 stations across 29 European countries, mainly since 1951.

4.6 Conclusions

The scientific community has a long list of requirements for monitoring. Long-term continuous data records of high quality are needed with good documentation, many auxiliary data, and often much more. Thus, special characteristics of the networks may be of interest to them. In general, observations are made by volunteers interested in nature, in special networks (Slovakia) or special stations (Germany, IPGs), or by experts. In the IPGs, observations are made on three specimens of each clone, whereas national network's rules describe how the observing area is defined and how the specimens have to be chosen in the surroundings of a phenological station. Constant specimens and locations are desired, however only in Slovenia are forest trees, shrubs, fruit trees, and vines permanently marked. All networks possess paper forms to note observations (even the new established networks in the UK and in the Netherlands do not want to exclude "offline" recorders and developed forms). The frequency of submitting forms varies from once a year to weekly to event-based data.

Data consistency and quality is difficult to evaluate. Most of the networks analyzed have monitoring guidelines, however they are very different, ranging from brand-new instructions, also available on the Internet (such as complete manual of ICP Forests, species and / or phenophase information of the Nature's Calendar or the German Weather Service) or substantial printed manuals (Germany Weather Service), to descriptions used since the beginning of the network (IPG). Quality control of the data mostly consists of only simple plausibility control. Accompanying disaster information does not exist at the moment, but could be available in the future (e.g., ICP Forests). The general data release policy varies as well, and in most cases data are open on an individual decision basis only.

Data formats are also quite different. In some countries older records still need to be digitized from paper, but most networks do have their data in files or databases. Nature's Calendar (with almost 50 % online observers) offers quick data access for registered observers and allows different kinds of data comparisons. In the National Weather Services' networks, observations on cultivated species are generally accompanied by information about varieties. However, associated data about the site (such as meteorology, soil, relief, and slope) are not available and (due to the coarse information about the station location) it is nearly impossible to gather exact auxiliary data.

Coding of phenophases following the BBCH code (Chapter 4.4, Biologische Bundesanstalt 1997) is a huge step forward in understanding phenophase definitions in different languages and making observations comparable.

Most recently, there is a clear tendency to complete long-term phenological monitoring in national networks by regional and/or experimental approaches in order to assess not only impacts of climate change on phenology, but to understand more about their triggers also under future conditions characterized by new, non-analog climates. In addition, it turned out that more species groups, e.g. of high value for nature conservation (endemic, invasive, protected) or community phenology (e.g. in grasslands), need further attention.

Acknowledgements I kindly thank my phenology colleagues from the COST 725 action, especially the chair Elisabeth Koch (Austria) as well as Rein Ahas (Estonia), Olga Braslavska (Slovak Republic), Isabelle Chuine (France), Zoltan Dunkel (Hungary), Elisabeth Koch (Austria), Jiri Nekovar (Czech Republic), Tim Sparks (United Kingdom), Andreja Susnik (Slovenia), and Arnold VanVliet (The Netherlands), for their valuable contributions.

References

Ahas R (1999) Spatial and temporal variability of phenological phases in Estonia. Dissertation, Universitatis Tartuensis, Tartu

Ahas R (ed) (2001) Estonian phenological calendar. Publicationes Instituti Geographici Universitatis Tartuensis 90:206

Baumgartner A, Schnelle F (1976) International phenological gardens (purpose, results, and development). In: 16th IUFRO World Congress Oslo subject group S1.03

Chmielewski FM (1996) The International Phenological Gardens across Europe. Present State Perspect Phenol Seas 1(1):19–23

Chmielewski FM, Rötzer T (2001) Response of tree phenology to climate change across Europe. Agric For Meteorol 108:101–112

Chmielewski FM, Rötzer T (2002) Annual and spatial variability of the beginning of growing season in Europe in relation to air temperature changes. Clim Res 19:257–264

Davitaja FF (1958) Agrometeorological problems, Moscow (in Russian with English Contents). Hydrometeorological Publishing House, Moscow

Deutscher Wetterdienst (ed) (1991) Anleitung für die phänologischen Beobachter des Deutschen Wetterdienstes (BAPH), Offenbach am Main

Eilart J (1959) Phytophenological observation manual (in Estonian). Estonian Naturalists Society, Tartu

Eilart J (1968) Teaduse ajaloo ehekülgi Eestis. In: Lumiste Ü (ed) Some aspects of history of phenology in Estonia (in Estonian with summary in German and Russian). Tallinn Acad Sci 1:169–176

EMHI (1987) Manual for hydrometeorological observation stations and points. Estonian Department of Hydrometeorology and Environmental Monitoring 11, I-II, Tallinn

Biologische Bundesanstalt für Land- und Forstwirtschaft (ed) (1997) Growth stages of plants – BBCH monograph. Blackwell Wissenschafts-Verlag, Berlin/Wien

Hydrometeorological Printing House (1965) Natural calendars of North Western USSR (in Russian). Geographical Society of USSR, Hydrometeorological Printing House, Leningrad

Hydrometeorological Printing House (1973) Methodology for hydrometeorological stations and observation points 11, Agri-meteorological observations in stations and observation points, 3rd edition (in Russian). Hydrometeorological Printing House, Leningrad

Koch E, Donnely A, Lipa W, Menzel A, Nekovář J (2001) EUR 23922 – COST Action 725 – final scientific report – establishing a European dataplatform for climatological applications. Cost Off Brux. doi:10.2831/10279

Kokorin AO, Kozharinov AV, Minin AA (eds) (1997) Climate change impact on ecosystems, nature protected areas in Russia, analyses of long-term observations, WWF Russia policy book No. 4 (in Russian with English Summaries). World Wildlife Foundation 174

Meier U (ed) (1997) Growth stages of mono- and dicotyledonous plants BBCH Monograph. Blackwell Wissenschaftsverlag Berlin, Wien

Menzel A (1997) Phänologie von Waldbäumen unter sich ändernden Klimabedingungen. Dissertation at the Forest Faculty of the LMU Munich, Forstlicher Forschungsbericht

Menzel A (2000) Trends in phenological phases in Europe between 1951 and 1996. Int J Biometeorol 44:76–81

Menzel A (2002) Phenology, its importance to the global change community. Clim Chang 54:379–385

Menzel A, Fabian P (1999) Growing season extended in Europe. Nature 397:659

Menzel A, Sparks T, Estrella N, Koch E, Aasa A, Ahas R, Alm-Kübler K, Bissolli P, Braslavská O, Briede A, Chmielewski FM, Crepinsek Z, Curnel Y, Dahl Å, Defila C, Donnelly A, Filella Y, Jatczak K, Måge F, Mestre A, Nordli Ø, Peñuelas J, Pirinen P, Remišová V, Scheifinger H, Striz M, Susnik A, van Vliet AJH, Wielgolaski FE, Zach S, Zust A (2006) European phenological response to climate change matches the warming pattern. Glob Change Biol 12 (10):1969–1976. doi:10.1111/j.1365-2486.2006.01193.x

MeteoSchweiz (ed) (2003) Pflanzen im Wandel der Jahreszeiten – Anleitung für phänologische Beobachtungen. Geographica Bernensia

Nekovář J, Koch E, Kubin E, Nejedlik P, Sparks T, Wielgolaski FE (2008) The history and current status of plant phenology in Europe. COST, Brussels

Oettingen AJ (1882) Phänologie der Dorpater Lignosen, In: Archiv Naturk. Liv-, Est- u. Kurlands, II Serie. Bd. 8, Dorpat

Rosenzweig C, Casassa G, Karoly DJ, Imeson A, Liu C, Menzel A, Rawlins S, Root T, Seguin B, Tryjanowski P (2007) Assessment of observed changes and responses in natural and managed systems. In: Parry ML, Canziani OF, Palutikof JP, van der Linden PJ, Hanson CE (eds) Climate Change 2007: Impacts, Adaptation and Vulnerability. Contribution of Working Group II to the

Fourth Assessment Report of the Intergovernmental Panel on Climate Change. Cambridge University Press, Cambridge

Rosenzweig C, Karoly D, Vicarelli M, Neofotis P, Wu Q, Casassa G, Menzel A, Root TL, Estrella N, Seguin B, Tryjanowski P, Liu C, Rawlins S, Imeson A (2008) Attributing physical and biological impacts to anthropogenic climate change. Nature 453:353–357

Schnelle F (1955) Pflanzen-Phänologie, Akademische Verlagsgesellschaft. Geest and Portig, Leipzig

Schultz GE (1981) General phenology (in Russian). Nauka, Leningrad

Schultz GE (1982) Geographische Phänologie in der USSR (in German). Wetter und Leben 34:160–168

Schuster C, Gröger A, Breier M, Menzel A (2011) High elevation phenology in mountain forests and alpine botanical gardens. Geophys Res Abstr 13:EGU2011–EGU11034

SHMÚ Bratislava:Ed (1988a) Metodický predpis 2 Návod na činnosť fenologických staníc Poľné plodiny (Manual for special crop station)

SHMÚ Bratislava:Ed (1988b) Metodický predpis 3 Návod na činnosť fenologických staníc Ovocné plodiny (Manual for special fruit and grapevine station)

SHMÚ Bratislava:Ed (1996a) Fenologické pozorovanie všeobecnej fenológie, Metodický predpis (Manual for general stations)

SHMÚ Bratislava:Ed (1996b) Fenologické pozorovanie lesných rastlín, Metodický predpis (Manual for forest stations)

Sokolowska J (1980) Przewodnik fenologiczny. Instytut Meteorologii i Gospodarki Wodnej. Wydawnictwa Komunikacji i Lacznosci, Warszawa

Sparks TH, Carey PD (1995) The responses of species to climate over two centuries: an analysis of the Marsham phenological record, 1736–1947. J Ecol 83:321–329

Tomsazewska T, Rutkowski Z (1999) Fenologiczne pory roku u uch zmiennosc w wieloleciu 1951–1990. Materialy Badawcze, Seris Meteorlogia – 28, Instytut Meteorologii i Gospodarki Wodnej, Warszawa

World Meteorological Organization, Commission for Agricultural Meteorology:Ed (2000) Report of the RA VI Working Group on Agricultural Meteorology. CAgM report No. 82, WMO/TD No. 1022

Zentralanstalt für Meteorologie und Geodynamik, ZAMG (2000) Anleitung zur phänologischen Beobachtung in Österreich. Anleitungen und Betriebsunterlagen Nr 1 der ZAMG, Austria

Chapter 5
North America

Mark D. Schwartz, Elisabeth G. Beaubien, Theresa M. Crimmins, and Jake F. Weltzin

Abstract Plant phenological observations and networks in North America have been largely local and regional in extent until recent decades. In the USA, cloned plant monitoring networks were the exception to this pattern, with data collection spanning the late 1950s until approximately the early 1990s. Animal observation networks, especially for birds have been more extensive. The USA National Phenology Network (USA-NPN), established in the mid-2000s is a recent effort to operate a comprehensive national-scale network in the United States. In Canada, PlantWatch, as part of Nature Watch, is the current national-scale plant phenology program.

5.1 United States

5.1.1 Early Observations and Research

Throughout the early history of the United States, extending into the first years of the twentieth century, there were few attempts to create organized phenological networks. One of the most noteworthy in this period was started by the Smithsonian Institution in 1851, and included observations on 86 plant species, birds, and insects in 33 states, but only lasted till 1859 (Hough 1864; Hopp 1974). A few individuals who were part of this and subsequent weather/climate observation networks did

M.D. Schwartz (✉)
Department of Geography, University of Wisconsin-Milwaukee, Milwaukee, WI 53211, USA
e-mail: mds@uwm.edu

E.G. Beaubien
Department of Renewable Resources, University of Alberta, Edmonton, AB, Canada

T.M. Crimmins • J.F. Weltzin
National Coordinating Office, USA National Phenology Network, Tucson, AZ 85721, USA

M.D. Schwartz (ed.), *Phenology: An Integrative Environmental Science*,
DOI 10.1007/978-94-007-6925-0_5,

record phenological data at selected sites during other periods. For example, Dr. Samuel D. Martin's April 1865 report (from Pine Grove, Kentucky) contains the dates of numerous phenological events (Martin 1865). Thomas Mikesell at Wauseon, Ohio, compiled another important local record over the period 1873–1912 (Smith 1915). Later instructions to Weather Bureau observers included lists of phenological phenomena to record (concentrating on agricultural crops, but also including timing of leaf opening and fall in deciduous forests, Weather Bureau 1899). However, there is little evidence to suggest that large numbers of observations were taken based on these instructions. Hopp (1974) notes that the Weather Bureau made a final limited attempt to start a phenological network at 20 cooperative sites in the state of Indiana, during 1904–1908, and lists several other extensive local records taken in Indiana, Kansas, and Minnesota.

An important phenological research contribution from the United States during the first half of the twentieth century was Hopkins' (1938) "Bioclimatic Law." The most well-known part of this law states that (other conditions being equal) the south to north progression of spring phenological events in temperate portions of North America is delayed by 4 days for each degree of latitude northward, for each 5° of longitude eastward, and for each 400-foot increase in elevation. This model was developed from data available around the northern hemisphere at the time. Hopp (1974) observed that the law is highly generalized, has geographical limitations, and is difficult to apply to individual plant species in any one season. Despite these limitations, Hopkins' Bioclimatic Law became one of the best-understood concepts of phenology for other scientists and the public. Paradoxically, its simplicity could have made phenology seem too easily predictable, which may have hindered and delayed efforts to develop new data collection networks, especially in the United States.

5.1.2 Agricultural Experiment Station Regional Networks

The first extensive U.S. phenological observation networks began in the 1950s with a series of regional agricultural experiment station projects, designed to employ phenology to characterize seasonal weather patterns and improve predictions of crop yield (Schwartz 1994). J. M. Caprio at Montana State University began the first of these projects, W-48 "Climate and Phenological Patterns for Agriculture in the Western Region" in 1957. This network contained up to 2,500 volunteer observers distributed throughout 12 Western states (Caprio 1957, Fig. 5.1). Common purple lilac plants (*Syringa vulgaris*) were observed initially, with two honeysuckle cultivars (*Lonicera tatarica* 'Arnold Red' and *L. korolkowii* 'Zabeli' added later. Observations ended in 1994, however, a few observers have again reported data since the later 1990s (Cayan et al. 2001).

Encouraged by Caprio's program, similar projects were started in the central U.S. (NC-26 "Weather Information for Agriculture") by W. L. Colville at the University of Nebraska in 1961, and in the northeastern U.S. with the renewal of

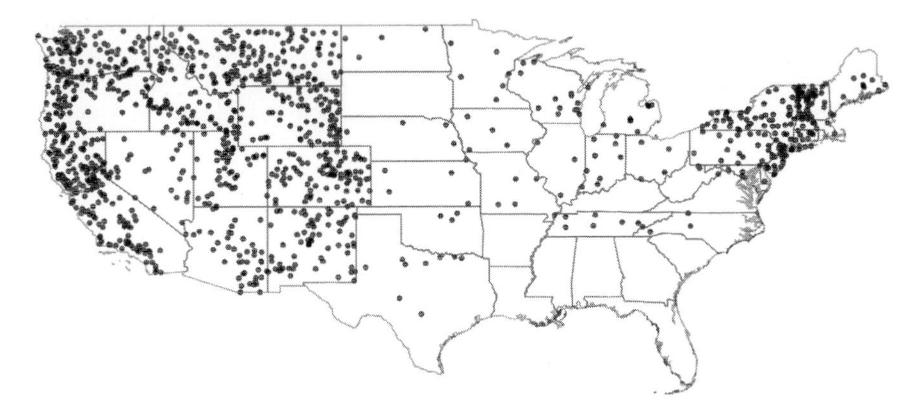

Fig. 5.1 Locations in the USA with 5 years or more of lilac phenology data, 1956–2008

NE-35 ("Climate of the Northeast – Analysis and Relationships to Crop Response") by R. Hopp at the University of Vermont in 1965. Both of these networks observed cloned plants of the lilac cultivar Syringa chinensis 'Red Rothomagensis' and the two honeysuckle cultivars from W-48. In 1970, NC-26 and NE-35 were combined as part of a new regional project, NE-69 "Atmospheric Influences on Ecosystems and Satellite Sensing." The program expanded to about 300 observation sites (Fig. 5.1), with three individuals as unofficial leaders: B. Blair (Purdue University), R. Hopp (University of Vermont), and P. Dubé (Laval University). In 1975 NE-69 was replaced by another new project, NE-95, "Phenology, Weather and Crop Yields," which was replaced by still another, NE-135, "Impacts of Climatic Variability on Agriculture" in 1980. This project was coordinated by M. T. Vittum (Cornell University) until 1985, when responsibility was turned over to R. C. Wakefield (University of Rhode Island). The phenology portion of NE-135 was briefly supervised by W. Kennard (University of Connecticut) until the eastern U.S. network lost funding and was terminated at the end of 1986. The eastern USA agricultural experiment station networks also included observations from a limited number of stations in eastern Canada, primarily in Ontario.

After the "decommissioning" of eastern U. S. lilac-honeysuckle phenology network operations by the Agricultural Experiment Stations in 1986, M. D. Schwartz corresponded with the most recent network supervisors (Schwartz 1994). They granted him permission to contact the observers and invite them to continue participating in an "interim" network, pending new funding. Approximately 75 observers responded to a renewed survey form sent out in March 1988, returning data for 1988 and in many cases 1987 as well. From that time until it became part of the USA National Phenology Network (see Sect. 5.1.4) in 2005, Schwartz continued to operate this interim "Eastern North American Phenology Network" with approximately 50 observers reporting lilac or honeysuckle event dates each year.

As an extension of the lilac/honeysuckle regional networks, a statewide phenological garden system of 12 stations operated in Indiana during the 1960s and

1970s. Numerous protocols were developed, and observations were taken on up to 14 species at each site (Blair et al. 1974). Also, an extensive phenology network observing redbud (*Cercis canadensis*), dogwood (*Cornus florida*), and red maple (*Acer rubrum*) operated in North Carolina during the 1970s (Reader et al. 1974).

5.1.3 Early Network-Related Phenological Research

A large number of horticultural and physiological studies have reported results regarding site-specific phenological characteristics of (commercially important) fruit tree species, and general phenological responses of woody plants. These are adequately summarized elsewhere, and will not be addressed here (e.g., Flint 1974; Schwartz 1985; Schwartz et al. 1997). Relatively few researchers have taken advantage of the Agricultural Experiment Station Regional Network data to examine phenological relationships on the continental scale. Caprio (1974) was the first, developing the Solar Thermal Unit Concept from lilac phenological data recorded in the western U. S. These data were also used in a recent study to examine the relationship between lilac-honeysuckle phenology and the timing of spring snowmelt-runoff pulses, in the context of global change. Earlier spring onsets since the late 1970s are reported throughout most of the region (Cayan et al. 2001).

Schwartz (1985) began an extensive phenological research program in the mid-1980s that has made intense use of lilac-honeysuckle network data from the eastern U. S. Areas explored include modeling; resulting in the Spring Indices (e.g., Schwartz 1998; Schwartz and Reiter 2000; Schwartz et al. 2006, 2013), spring plant growth impacts on the lower atmosphere (e.g., Schwartz 1992; Schwartz and Crawford 2001), and analyses and comparisons to remote sensing measurements.

Another past network of note used in research was run by the Wisconsin Phenological Society (www.wps.uwm.edu). Data for a large number of native and cultivated flowers and shrubs extend from early 1960s to early 2000s. Unfortunately, there is considerable variation in the number and types of plants observed at each site (selected from a standard form), and most individual station records are less than 10 years long. However, these data are now largely in digital form, and have contributed to an innovative methodological study exploring ways to “fill-in” the gaps in such incomplete records (Zhao and Schwartz 2003).

5.1.4 Current Networks

Over the last decade, there has been an explosion of interest in phenology across the continent, particularly in the United States. This interest stems from the increasing realization that phenology is not only an integrative science, but that it is critical to decision-support and management of managed and natural ecological systems, and that it is a leading indicator of ecological and climatological impacts in a rapidly changing environment. In addition, phenology is being used as a platform for education, outreach, science translation, and science literacy, as the inclusion of "citizen scientists" in phenological monitoring can engage the public in the process of science, from discovery to analysis to application.

The bellwether event in the United States over the last decade has been the conceptualization and implementation of the USA National Phenology Network (USA-NPN; discussed in more detail below), which stemmed from recommendations of a working group charged with assisting in the conceptual development of the NSF-funded National Ecological Observatory Network (Schwartz et al. 2012). The USA-NPN, established in 2007, has served as a platform for the development of a variety of models for phenological monitoring, from national networks to local organized programs to non-organized projects to individual observers. A number of other programs and projects have developed, and are operated, independently of the USA-NPN. The following sections first describe the USA-NPN, and then highlight representative contemporary national, regional and local programs and projects that operate in collaboration with, or independently of, the USA-NPN.

5.1.4.1 National-Scale Programs

USA National Phenology Network

The USA-NPN was established in 2007 with support from the U.S. National Science Foundation, the U.S. Geological Survey, and several other agencies and organizations (Schwartz et al. 2012). The USA-NPN is defined as a consortium of individuals and organizations that collect, share, and use phenological data, models, and related information, with a mission to serve science and society by promoting a broad understanding of plant and animal phenology and its relationship with environmental change. Professional and volunteer observers collect data on hundreds of species of plants and animals across the nation, including the common and cloned lilac species observed by Network precursors. In turn, the Network makes phenology data, models, and related information freely available to empower scientists, resource managers, and the public in decision-making and adapting to variable and changing climates and environments.

The National Coordinating Office (NCO) of the Network maintains a website (www.usanpn.org) and provides data management services on behalf of the broader Network, promotes the use of standardized approaches to monitoring phenology, encourages the widespread collection of phenological data, and facilitates communication within and beyond the network. The NCO also facilitates basic and applied research on phenology and promotes the development and dissemination of decision-support tools, educational materials, and other information or activities related to phenology.

One of the functions of the USA-NPN is to provide infrastructure, including scientifically-vetted monitoring protocols, species information pages, education materials, a data management system, and data visualization tools for other organizations to utilize. This structure facilitates partner organizations, which do not need to build and create their own infrastructure but which can contribute to the national database of standardized observations. The USA-NPN recognizes three types of partners who leverage on the Network in different ways:

- Geographic Affiliates – These groups are geographically organized, ranging in scope from a town or university to several states, and have been organized for the purpose of monitoring phenology. Examples of USA-NPN Geographic Affiliates include the California Phenology Project, the Wisconsin Phenological Society, PennPhen (at Pennsylvania State University), and Signs of the Seasons: A Maine Phenology Program (described below).
- Collaborating Organizations – Collaborating organizations are groups that exist independent of the USA-NPN and have established a relationship with the USA-NPN to accomplish a wide variety of goals, including, but not limited to, engaging members in phenology monitoring, using the USA-NPN as an archive for phenology data, and pursuing joint funding proposals. The USA-NPN recognizes many collaborating organizations including the National Park Service, The Wildlife Society, and the Appalachian Trail Conservancy.
- Collaborating Projects – USA-NPN Collaborating Projects are variable and can address topics ranging from specific data collection efforts to the development of decision-support tools. These efforts are typically short in duration (lasting a few years), have specific goals, and are grant-funded. The Juniper Pollen Project, a collaborative effort among the USA-NPN, NASA, several universities, and state health departments to develop a pollen early warning system for allergy and asthma sufferers in the southwestern US, is an example of a collaborating project.

There are a number of other national and international programs that operate independently within the United States; many of these were developed in the last decade, though others have been operational for decades. The following paragraphs provide a few examples of the diversity of programs, including those that focus on science, education, outreach or public engagement at the national scale. Additional national-scale programs are listed in Table 5.1 (this list is not comprehensive).

Project BudBurst

The National Ecological Observatory Network's Project BudBurst (www.budburst.org) is a national field campaign with the primary goals of education and outreach. Characterized by uncomplicated protocols for observing the timing of plant leafing, flowering, and fruiting, this program is appropriate for all ages and skill levels. Data submitted by observers ultimately reside in the National Phenology Database maintained by the USA National Phenology Network and have been used by scientists across the country to better understand plants' relationship with climate.

Table 5.1 Phenology networks and programs in North America

Program name	Taxa	Geog. scope	Web address
GLOBE Phenology Network	Plants, animals	Internat.	www.globe.gov/web/phenology-and-climate/overview
Journey North	Animals, tulip gardens	Internat.	www.learner.org/jnorth/
Operation RubyThroat	Hummingbirds	Internat.	www.rubythroat.org/
Priority Migrant eBird	Birds	Internat.	ebird.org/content/primig/
The Great Sunflower Project	Plants, animals	Internat.	www.greatsunflower.org/
The Hummingbird Monitoring Network	Hummingbirds	Internat.	www.hummonnet.org/
USA National Phenology Network	Plants, animals	Internat.	www.usanpn.org
Bee Hunt!	Bees	National	www.discoverlife.org/bee/
Breeding Bird Atlas (Cornell Lab of Ornithology)	Birds	National	bird.atlasing.org/
FrogWatch USA	Frogs	National	www.aza.org/frogwatch/
Honey Bee Net	Honeybees	National	honeybeenet.gsfc.nasa.gov/
NestWatch	Birds	National	watch.birds.cornell.edu/nest/home/index
Plantwatch	Plants	National	www.naturewatch.ca/english/plantwatch/intro.html
Frogwatch	Frogs	National	www.frogwatch.ca
Pollinators.info bumble bee photo group	Bumble bees	National	www.pollinators.info/archives/tag/bumble-bee-pollinators
Project BudBurst	Plants	National	neoninc.org/budburst/
Project Feeder Watch	Birds	National	www.birds.cornell.edu/pfw/
Spring Watch USA (Animal Planet)	Animals	National	animal.discovery.com/tv/spring-watch/spring-watch.html
The Foliage Network	Deciduous trees	National	www.foliagenetwork.com/
Watch the Wild	Plants, animals	National	www.natureabounds.org/Watch_the_Wild.html
Birds in Forested Landscapes	Birds	N. Amer.	www.birds.cornell.edu/bfl/
Monarch Larva Monitoring Project	Monarch butterfly	N. Amer.	www.mlmp.org/
Monarch Watch	Monarch butterfly	N. Amer.	www.monarchwatch.org/
North American Amphibian Monitoring Program	Amphibians	N. Amer.	www.pwrc.usgs.gov/naamp/
North American Breeding Bird Survey	Birds	N. Amer.	www.pwrc.usgs.gov/BBS/
The Goldenrod Challenge	Plants, animals	N. Amer.	www.discoverlife.org/goldenrod/
Grunion Greeters	Grunion (Fish)	Regional	grunion.pepperdine.edu/
Mountain Watch (Appalachian Mountain Club)	Plants	Regional	www.outdoors.org/conservation/mountainwatch/mtplant.cfm
Ohio State University (OSU) Phenology Network	Plants, animals	Regional	phenology.osu.edu/

(continued)

Table 5.1 (continued)

Program name	Taxa	Geog. scope	Web address
Pika Monitoring	Pika	Regional	www.adventureandscience.org/pika.html
PikaNet	Pika	Regional	www.mountainstudies.org/index.php?q=content/pikanet-citizen-science-monitoring-program-american-pika
Prairie Chicken Project	Birds	Regional	www.iowadnr.gov/Environment/WildlifeStewardship/NonGameWildlife/DiversityProjects/PrairieChickenProject.aspx
Signs of the Seasons	Plants/animals	Regional	umaine.edu/signs-of-the-seasons

FrogWatch USA

The Association of Zoos and Aquariums' FrogWatch USA program (www.aza.org/frogwatch) allows individuals and families to learn about the wetlands in their communities and help conserve amphibians by reporting the calls of local frogs and toads. Data collected since the late 1990s have proven valuable to state wildlife agencies and for informing management strategies of these important animals.

GLOBE

The Global Learning and Observations to Benefit the Environment (GLOBE) program is a worldwide hands-on, primary and secondary school-based science and education program that includes a suite of phenology observation protocols (www.globe.gov). Documented protocols include generalized budburst, and green-up and green-down, as well as protocols for individual species or genera including lilac, ruby-throated hummingbirds, seaweed, and arctic bird migration. GLOBE data have been used to investigate urban heat island effects on phenology of leaf budburst within urban environments in cities around the world; results suggest that while vegetation phenology is consistently different between urban and rural areas, a uniform paradigm based on the explanatory variables in this study did not emerge (Gazal et al. 2008).

eBird

eBird is a real-time, online checklist program launched in 2002 by the Cornell Lab of Ornithology and National Audubon Society (www.ebird.org). This avian observation and reporting system is designed to maximize the utility and accessibility of the vast numbers of bird observations made each year by recreational and professional bird watchers, and provides a rich data source for basic information on bird abundance and distribution at a variety of spatial and temporal scales. The program is amassing one of the largest and fastest growing biodiversity data resources in existence: for example, in January 2010, participants reported more than 1.5 million bird observations across North America. The observations of each participant join those of others in an international network of *eBird* users. *eBird* shares these observations with a global community of educators, land managers, ornithologists, and conservation biologists.

5.1.4.2 Regional and Local Programs

A large number of more regional or local programs track phenology of plants and/or animals at various levels of detail (Table 5.1). Some programs focus primarily on tracking phenology; others are organized around specific species or taxa and collect many types of observations including phenology. Geographic scope for these programs can range from a neighborhood or community group to a region or broader; programmatic foci range from science to education and outreach.

Signs of the Seasons: A Maine Phenology Program

Participants in the *Signs of the Seasons*: *A Maine Phenology Program*, a geographic affiliate of USA-NPN established in 2010, help scientists document the local effects of global climate change by observing and recording the phenology of common plants and animals living in their own backyards and communities. UMaine Extension and Maine Sea Grant coordinate the program in partnership with USA-NPN, Acadia National Park, Schoodic Education and Research Center, US Fish and Wildlife Service, Maine Maritime Academy, Maine Audubon, and climate scientists and educators at the University of Maine (www.umaine.edu/signs-of-the-seasons).

Ohio State University Phenology Garden Network

The OSU Phenology Garden Network, initiated in 2004, consists of 44 gardens established across the state of Ohio (www.phenology.osu.edu). Cooperators at each site track first and full bloom of woody and herbaceous perennials, many of them cultivars, with the goal of establishing a biological calendar for the state and region. Results are correlated with insect emergence data and used in pest monitoring and management. In addition, data are being used to test the hypothesis that phenological events occur in the same order throughout the state, and that a comprehensive biological calendar that was developed in Wooster, Ohio is relevant state-wide.

Journey North

Journey North is a national, internet-based program that explores the interrelated aspects of animal migrations and seasonal change, with a focus on K-12 students and the general public (www.learner.org/jnorth). Participants record and share observations of migration patterns of monarch butterflies, robins, hummingbirds, whooping cranes, gray whales, and bald eagles among other animals, as well as the phenological development of plants and changes in natural environments.

5.2 Canada

Canada has a long and rich history of phenological observations. Since deglaciation some eight to ten thousand years ago, First Nations and Inuit have perfected their oral knowledge of "nature's calendar" to maximize their survival and to find resources efficiently across a wide landscape. The earliest recorded observations will likely be found in the journals of fur traders and missionaries. Today phenological data continue to "serve as a check of season against season, and region against region" (Minshall 1947, p. 56).

Phenological studies vary in the size of area surveyed and in the duration of observations, but they can basically be divided into three types:

- the "snapshot" study, in which many observers survey phenology over a large area at one point in time.
- the intensive study, in which one or a small number of people survey a small area over a period of one or more growing seasons.
- the extensive study, in which a network of observers surveys a large area over a period of years. This chapter section concentrates mainly on the involvement of Canadians in such networks as of the time of writing, 2012.

The studies described here focus mainly on plant phenology and are divided into two major sections; national and regional networks. First, the national networks described include extensive studies such as the Royal Society of Canada survey launched in 1881, participation by eastern Canadian observers in the North East Agricultural Project in the United States in the 1950s, and in recent decades: PlantWatch, a program engaging Canadians with coordinators in each Province and Territory. Second, the regional and localized networks and research are described by region, from east to west, and in the north.

5.2.1 National Networks in Canada

5.2.1.1 First Nations and Inuit: Traditional Phenological Knowledge

For Canada's First Nations, phenology was a well-honed tool. The Blackfoot in Alberta used the flowering time of *Thermopsis rhombifolia* (golden bean or buffalo bean) to indicate the best time in spring to hunt bison bulls (Johnston 1987). In

British Columbia more than 20 cultural/linguistic groups used over 140 indicators (Lantz and Turner 2003). These authors note that phenological indicators permitted the most efficient use of human resources in acquiring food or materials from the land. One example is the Stl'atl'imx peoples, who used the blooming of wild rose (Rosa spp.) to indicate the best time to collect cedar roots and basket grass. Another is the west coast Nuu-Chah-Nulth peoples who used the ripening of salmonberries (*Rubus spectabilis*) as an indicator that adult sockeye salmon were starting to run in freshwater streams. Thirdly, the west coast Comox peoples used the bloom time of oceanspray (*Holodiscus discolor*) to alert them to dig for butter clams. In the Okanagan area, First Nations observed that female black bears generally headed to dens when the western larch needles turned gold in the fall. Later denning often meant these bears would not produce cubs the following spring, perhaps due to poor berry crops and thus insufficient weight gain. Across North America accurate timing was the key to survival, and phenology was common sense to those who lived so close to the land.

5.2.1.2 Royal Society of Canada Survey

About 20 years after Canadian Confederation in 1867, a countrywide phenology survey was initiated. In 1890 the Royal Society of Canada passed a resolution requesting affiliated natural history and scientific societies to:

> obtain accurate records in their individual localities of meteorological phenomena, dates of the first appearance of birds, of the leafing and flowering of certain plants, and of any events of scientific interest for collation and publication in the 'Transactions of the Society'. (Royal Society of Canada 1893, p. 54)

In the following year, 1891, the Botanical Club of Canada was formed in affiliation with the Royal Society (MacKay 1899). By 1897 this nation-wide phenology survey included observations of 100 events including first bloom dates of many species of native plants, arrivals of spring birds, days with thunderstorms, and timing of ice melt on rivers. By 1895 reports were received from nine Provinces, and the numbers of events and observation of locations increased over the duration of this extensive survey. The secretary of the Botanical Club, Dr. A. H. MacKay, coordinated the survey until 1910 when the club was dissolved, and F.F. Payne of the Meteorological Service then coordinated the survey until 1922. Observations for 1892–1922 were published annually in the *Proceedings and Transactions of the Royal Society of Canada*.

5.2.1.3 Participation of Atlantic and Central Provinces in U.S. Agricultural Experiment Networks

The United States Regional Agricultural Experiment Station Northeast Project, NE-69, added stations in several Canadian provinces in 1970 (see also Sect. 5.1.2). The next year Quebec initiated a large observer network to track the same plant species. Observations were made until 1977 at over 300 locations, of which 51 were adjacent to meteorological stations (Dubé and Chevrette 1978). Three indices (earliness index, summer index and growing season index) were derived from the data, which were used to define bioclimatic zones. By 1977 observers in the six provinces east of Manitoba were involved. Quebec had the largest number of observers, with 268 active observation sites in 1977, versus New York State, the next largest at 84 sites (Vittum and Hopp 1978). Pierre André Dubé of Laval University coordinated the Quebec participants and also computerized results for the whole project. Analysis of the phenological and meteorological data confirmed the existence of significant differences among phenological zones in Quebec (Castonguay and Dubé 1985). The resulting maps were used to modify agricultural taxation zones.

5.2.1.4 Plantwatch: National Network

PlantWatch began in 1995 based at the University of Alberta's Devonian Botanic Garden. Reporting was via the Internet with data tables and maps updated regularly. The focus of this extensive survey was initially on Canadian students, ages 8–11 years, reporting spring bloom dates for 3 plant species across the Prairie Provinces. By 1997 the survey had expanded to a Canada-wide program for both adults and youth, with 7 indicator plants. International data were also gathered for one species, *Syringa vulgaris* (common purple lilac). Beginning in the early 1990s Elisabeth Beaubien gave talks across Canada to encourage the formation of provincial plant phenology programs. A teacher guide was posted on the Alberta PlantWatch website in 2001, providing curriculum applications in science, mathematics, social studies, etc. for students from elementary to high school level.

Beginning in 2000 PlantWatch (www.plantwatch.ca) expanded with assistance from Environment Canada NatureWatch coordinators Elizabeth Kilvert, followed by Heather Andrachuk and most recently Marlene Doyle. The author, as national science advisor for PlantWatch, found coordinators for each of the provinces and territories. The regional coordinators met in Ottawa in May 2000 and in Winnipeg in November 2001. E. Beaubien completed a science review and selected plant species for observation in this cross-Canada spring phenology survey. Based on phenology protocols used by the Deutscher Wetterdienst (the German weather service) and other European networks a simplified and standardized description of key phenophases was developed: first bloom, mid bloom, and leafing. In 2002 a guide booklet "PlantWatch: Canada in Bloom" was produced through Nature Canada. An updated version was completed in 2009.

This Canada PlantWatch program was developed in partnership with Environment Canada and Nature Canada (formerly the Canadian Nature Federation) as part of the NatureWatch (www.naturewatch.ca) suite of Canadian Internet-based citizen science monitoring programs. These programs also included an ice seasonality monitoring program (IceWatch) and programs to monitor frog (FrogWatch) and worm (WormWatch) distribution, diversity and abundance. Website materials were available in English and French.

Each Province and Territory selected plant species from the initial list, with some coordinators adding more species suitable for their particular ecozones. The number of species observed for Plantwatch expanded over the years, reaching 40 in 2009.

A national PlantWatch Educators Guide geared toward grades 4–6 (students aged 10–12) was posted on the website in 2009 with curriculum links for each province and territory. A DandelionWatch website (www.dandelionwatch.ca) was developed as an introduction to plant phenology monitoring for grades 4–6.

As of February 2012 the PlantWatch database contained over 12,000 observations, while the IceWatch seasonality database contained over 9,400 observations. All NatureWatch data was freely available, as of 2012, for download on the website www.naturewatch.ca/english/download.html. The data has been used to conduct local and national assessments of change (Canadian Council of Ministers of the Environment 2003). However in 2011 Environment Canada cut most of its support for the NatureWatch programs, though Marlene Doyle continues to coordinate conference calls two or three times a year. At the time of writing, Dr. Robert McLeman and Dr. Andre Viau of the University of Ottawa's Geography Department planned to establish a citizen science lab to house the NatureWatch citizen science web servers and analyze the data. The data will remain in the public domain and the program will continue to be offered in English and French. Program goals include developing a smart phone NatureWatch application to allow observers to enter data in the field and to engage more young Canadians.

Research publications based on the Plantwatch data include those in remote sensing (Beaubien and Hall-Beyer 2003; Kross et al. 2011; Pouliot et al. 2011). An analysis of the IceWatch data appears in Futter (2003).

5.2.2 Atlantic Region

5.2.2.1 Nova Scotia

MacKay Network 1897–1923

Dr. A.H. MacKay (see Sect. 5.2.1.2) was not only secretary for the Botanical Club of Canada but also Superintendent of Education for Nova Scotia. He promoted phenology very successfully, such that in 1898 eight hundred sets of observations on up to one hundred events were submitted by school classes (MacKay 1899). Observations by Nova Scotia schools continued as part of the Nature Studies curriculum until at least 1923 (MacKay 1927).

Recent Networks

In Nova Scotia this interest in tracking nature's calendar was rekindled through a number of programs. A "Peeper Program", based at the Nova Scotia Museum of Natural History started in 1994, for the public to report calling dates of spring peeper frogs. Nova Scotia PlantWatch began in the spring of 1996, tracking bloom times for 12 plant species at about 200 sites. Liette Vasseur, Peta Mudie, Robert Guscott and others formed the initial team to promote and coordinate PlantWatch. They produced an observer's guide, a webpage on the Environment Canada website, and a colorful newsprint poster of the 12 plant species. In 2001 Edward Reekie of Acadia University took over Liette Vasseur's functions, summarizing the data and sending out annual newsletters. Sixteen plant species were then tracked. A comparison was made of the PlantWatch results from the first 3 years 1996–1998 with the MacKay data for the same species. For most species no significant differences in bloom times were found (Vasseur et al. 2001). It is interesting to note that climate records for Atlantic Canada show a cooling trend for 1948–1995. The "Thousand Eyes" project (www.thousandeyes.ca) based at the Nova Scotia Museum of Natural History began in 2000 and was first coordinated by Elizabeth Kilvert and later by Chris Majka. It gathered records via the Internet from students in Nova Scotia on the timing of 50 events selected from the MacKay program.

In 2004 the coordination of PlantWatch Nova Scotia moved to Acadia University's Harriet Irving Botanical Gardens, under the direction of Conservation Horticulturist Melanie Priesnitz botanicalgardens.acadiau.ca/plantwatch.html). Each season the Gardens integrated PlantWatch into their environmental education and public outreach programs, holding public education seminars and group walks in order to promote PlantWatch to potential citizen scientists across Nova Scotia. One of the long-terms goals of PlantWatch Nova Scotia is to follow in MacKay's footsteps by adding the recording of phenological events to the public school curriculum. The re-introduction of PlantWatch to schools was done one school at a time by encouraging the planting of Red Maple trees through existing schoolyard greening projects. As of 2012 Nova Scotia continued to have a solid base of volunteer observers, some of whom had been making observations since the start of the program in 1996.

5.2.2.2 New Brunswick

Dr. Liette Vasseur moved to the University of New Brunswick in 2001 and started a New Brunswick PlantWatch that tracks 12 species. More recently Vanessa Roy-McDougall acted as coordinator through the organization Nature New Brunswick. The New Brunswick Naturalists have a rich history of participating in phenology programs, specifically with birds. This organization was formed in 1972 to represent nature clubs but now reaches over 1,000 youth each year through school presentations and has eight Young Naturalists Clubs throughout the province. Additionally, they have issued a quarterly publication (the New Brunswick

Naturalist) for close to 40 years (naturenb.ca/nbnaturalistnewsletter.html) which includes phenology information. Finally, naturalists in the Province are also very active on the *naturenb* listserv where they can post sightings etc. on a daily basis.

5.2.2.3 Prince Edward Island

The Bedeque Bay Environmental Management Association started a PEI PlantWatch in 2000 with Ilana Kunelius promoting the project to students and volunteers. In 2002 Charmaine Noonan took Ilana Kunelius' place as PlantWatch coordinator, promoting the tracking of 12 plant species. A PlantWatch video as well as a PEI PlantWatch Guide were produced. Tracy Brown has been coordinator for almost a decade to the time of writing. The PEI Natural History Society gathers phenology data from members and publishes it annually in the *Island Naturalist* newsletter.

5.2.2.4 Newfoundland and Labrador

Luise Hermanutz and Madonna Bishop of Memorial University started Newfoundland and Labrador PlantWatch in 1998. The number of plants observed has grown from 5 species in 1998 to 18 species in 2012. A teachers' resource kit, a web page (www.mun.ca/botgarden/plant_bio/PW) and annual newsletters have been produced with the support of Memorial University of Newfoundland Botanical Garden.

5.2.3 Central Canada

5.2.3.1 Quebec

Dr. P. A. Dubé coordinated a large phenology network in the 1970s (see Sect. 5.2.1.3). Intensive studies of tree phenology have been done by Dr. Martin Lechowicz and his students at McGill University (Hunter and Lechowicz 1992; Lechowicz 1995), and Dr. Lechowitz promoted phenology to monitor ecosystems for the effects of climate change (Lechowicz 2001). Starting in 2001 Opération Floraison or PlantWatch in Quebec was based at the Montréal Botanical Garden, coordinated by botanist Stéphane Bailleul. Thirty-two species of plants were tracked and blooming times for a number of species were recorded at the Garden. Public participation in the program in Québec has been limited.

5.2.3.2 Ontario

Starting in 1932 dates for the Ottawa district of flowering and fruiting for weeds and native plants were gathered by the Division of Botany and Plant Pathology of the federal Department of Agriculture (Minshall 1947). Minshall provides a brief review of early Canadian phenology research, and notes that in 1939–1940 an interdepartmental federal committee presented recommendations to coordinate all federal projects in phenology. Unfortunately World War II prevented action on these recommendations. Bassett et al. (1961) of the federal Department of Agriculture analyzed this Ottawa data for selected trees, shrubs, herbs and grasses for 1936–1960 and calculated the effective base temperatures for spring development of ten early-blooming tree species. The Royal Botanical Garden (RBG) in Hamilton gathered lilac data in the 1970s as part of the North East Network described in Sect. 5.2.1.3. This was the base for the PlantWatch program for Ontario starting in 2002.

Natalie Iwanycki coordinated Ontario PlantWatch at the time of writing. In order to promote the program and introduce visitors to some of the species on the Ontario PlantWatch list, interpretive signage was designed and installed in 2008 around certain species on the Ontario PlantWatch list that were planted in the Royal Botanical Garden's (RBG) Helen M. Kippax Wild Plant Garden. In 2009, a large permanent PlantWatch poster was installed at the RBG Centre as part of the Steadman Exploration Hall Exhibit in the main entrance. This bilingual poster introduced RBG's visitors to the program, summarizing how to participate and collect phenology data, and presenting photos of all 22 species on the Ontario list. Between 2009 and 2011 RBG held over a dozen educational programs on PlantWatch for school groups and teachers visiting the Gardens from local school boards, and also via video-conference for schools elsewhere.

In the early spring of 2009 RBG's science and horticulture staff launched a new initiative to record annual bloom dates for approximately 175 cultivated taxa and 100 wild species present within its gardens and nature sanctuaries. Many of these species were selected because they are tracked by the USA's National Phenology Network. Weekly bloom lists were prepared for garden visitors and were shared through RBG's website and also through social networking sites such as Facebook and Twitter.

5.2.4 *Western Canada*

5.2.4.1 Manitoba

Criddle (1927) observed 400 prairie species over 20 years in southern Manitoba and published flowering and seed-ripening times valuable for present-day use in reclamation work. Mitchener (1948) presented data on flowering times and pollen availability from beekeepers. From 2001 to 2011 Kim Monson coordinated Manitoba PlantWatch. This was cosponsored by the University of Winnipeg Geography Department and Nature Manitoba (previously the Manitoba Naturalists Society). Sixteen plants were tracked and products included a promotional pamphlet, an observer guide, a teacher guide, and annual newsletters. Observers received a certificate with space for an annual sticker to thank them for data submitted. Long-term observers received 5 and 10 years volunteer pins to commemorate their commitment to the program. PlantWatch Manitoba had 120 citizen scientist observers collecting spring flowering data which was used in a report on trends in spring flowering dates from northern Canada (Monson 2008).

5.2.4.2 Saskatchewan

Budd and Campbell (1959) reported first bloom dates for 145 prairie native plants recorded at Agriculture Canada's Experimental Farm in Swift Current. They recommended using the bloom of Wood's rose (*Rosa woodsii*) as the best indicator of "range readiness" (i.e., pasture plants can now withstand grazing). PlantWatch Saskatchewan began in 2001 based at Nature Saskatchewan and coordinated by Kerry Hecker. From 2005 Deanna Trowsdale-Mutafov was the Saskatchewan PlantWatch Coordinator and 20 plant species were observed. As of 2012, there were over 200 interested participants and close to 30 schools in the program. From 2005 to 2012, approximately 75 presentations were given to over 2,000 people, both youth and adults, on native plant identification, plant phenology, how to participate in the PlantWatch program, and on climate change information and awareness. An average of 20 observers per year submitted blooming data. PlantWatch also distributed hundreds of seedlings from SaskPower's Shand Greenhouse to interested schools to plant in their school yards.

Promotional items were produced including a brochure, poster, wall chart, yearly spring newsletters with datasheet and fall e-bulletins. A news release was sent to over 100 media sources each year to promote the program, and articles appeared in many newsletters, newspapers, magazines and in the Saskatchewan Science textbook for Grades 7–8.

The long-term objectives of PlantWatch Saskatchewan are to foster appreciation of native plants, maintain phenological studies throughout Saskatchewan through citizen science observations, and to measure impacts of climate change on ecosystems through analysis of plant phenological data.

5.2.4.3 Alberta

Moss (1960) recorded "height of bloom" dates for 25 spring-blooming shrubs and trees near the University of Alberta in Edmonton from 1926 to 1958. These flowering data were correlated with degree-days to determine the average amount of warmth the plants were exposed to before flowering.

Starting in 1976 an annual May Species Count, a "snapshot phenology" study, was coordinated by Nature Alberta on the last weekend of May. Naturalists participate in this count of wildlife including plants (species in bloom), birds, mammals, butterflies, etc. Numbers of plant species found in bloom indicate the relative earliness or lateness of the spring season.

Dr. Charles Bird, Professor of Botany at the University of Calgary, established a volunteer network to record the flowering of native plants, and results were published annually in *Alberta Naturalist*, the journal of the Federation of Alberta Naturalists, from 1973 to 1982 (Bird 1974).

This survey was revived and revised in 1987 by the author (Beaubien and Freeland 2000) as the Alberta Wildflower Survey, based at the University of Alberta. Between 150 and 200 volunteers per year reported dates of first (10 %), mid (50 %), and full (90 %) flowering for up to 15 native plant species from 1987–2001. Training was provided using printed program information with tips on site selection, protocols, and species identification, including color photographs and sketches. In 2002 the program was renamed Alberta PlantWatch and by 2012 provided a choice of 25 plant species to observe (www.plantwatch.fanweb.ca).

To encourage teachers, a PlantWatch Teacher Guide was posted on the Alberta website in 2001 and then updated in 2009 in English and French (www.plantwatch.ca). A wallchart helped maintain program visibility in schools and parks during the busy spring season (see 'educational materials' at the 'fanweb' website above).

Over the two decades beginning in 1987, over 600 Alberta PlantWatch volunteers reported 47,000 records, the majority contributed by observers who participated for more than 9 years (Beaubien and Hamann 2011a). This article makes recommendations on phenology program protocols.

An analysis of rural PlantWatch data from central Alberta found that bloom dates for the earliest-blooming species (*Populus tremuloides* and *Anemone patens*) advanced by 2 weeks over the 71 years 1936–2006. Later-blooming species' bloom dates advanced between 0 and 6 days. Winter temperatures also increased: February mean monthly temperatures went up by 5.3 °C over that period (Beaubien and Hamann 2011b).

5.2.4.4 British Columbia

The entomologist R. Glendenning (1943) summarized the methods, history and uses of phenology and recommended certain species as suitable across Canada. He suggested phenological events to observe in each month of the year for the British Columbia coast, based on his observations over 34 years. In 1984 Bill Merilees of the Federation of British Columbia Naturalists launched a study of the flowering of vascular plants, requesting 16 phenophases for up to 50 native species of the observer's choice. In 2000 he modified the survey, requesting bloom times for a shorter list of plants found on southeastern Vancouver Island. In 2001 Dave Williams of the University College of the Cariboo in Kamloops took on the task of coordinating PlantWatch in British Columbia, tracking the phenology of 15 plant species. Starting in 2010 Dawn Hanna, president of the Native Plant Society of British Columbia, coordinated BC PlantWatch with help from Josie Osborne of the Tofino Botanic Garden as well as Patrick Wilson.

5.2.5 Northern Canada

Erskine (1985) presented data for first bloom of native plants and bird arrival for five boreal sites across Canada, at a different site for each year from 1971 to 1975. He also noted the appeal of phenology observations: "such visible events as the flowering of plants and the arrival of birds have an appeal that is lacking in the cold statistics of the meteorological record" (p. 188).

Climate warming is predicted to show the biggest effects in arctic regions. In 1990 the International Tundra Experiment (ITEX) www.geog.ubc.ca/itex/index.php began, linking arctic and alpine scientists to study the effects of climate change on northern ecosystems (Henry and Molau 1997). In 1999 the Canadian researchers in this group started CANTTEX, the Canadian Tundra and Taiga Experiment (www.emannorth.ca/canttex) to develop a strategy for studying climatic and ecological change in Canada's north. The 2010 CANNTEX field manual includes the PlantWatch phenology protocols (www.ec.gc.ca/Publications/default.asp?lang=En&xml=5B22FA0C-A0C1-4AE4-9C11-D3900215EEBD.

The three northern territories, Yukon, Northwest Territories and Nunavut, all started PlantWatch programs in 2001. In Yukon, Lori Schroeder coordinated PlantWatch for many years through the Yukon Conservation Society, promoting tracking of 16 species. Brian Charles was acting as coordinator in 2012, as a member of the Yukon Wildlife Preserve which runs several phenology programs. In 2012 a new recruitment drive was started to bring in a new cohort of PlantWatch volunteers. The Yukon Wildlife Preserve (YWP) Operating Society promotes research and fosters appreciation of arctic and boreal ecology. The phenology programs of YWP included PlantWatch, Butterfly watch and Frog watch, as well as the monitoring of arrival dates and species types of migratory game birds. Other phenology-related programs included Yukon-wide surveys conducted by Environment Canada such as the arrival of migratory game birds and breeding bird surveys. At Herschel Island, on the north shore of Yukon in the Beaufort Sea, research on the timing of nesting song birds on the island was underway in 2012.

In the Northwest Territories PlantWatch was coordinated by the non-profit organization Ecology North, tracking 17 plant species. Jen Morin coordinated for many years and later, Shannon Ripley took this position. In 2009 volunteer Mike Mitchell and Ecology North staff worked with national PlantWatch coordinators to add *Betula papyrifera* (paper birch) as a species that is tracked in the NWT. The PlantWatch program was highlighted in a new Northwest Territories Experiential Science grade 10 program that focuses on terrestrial systems. In Nunavut, PlantWatch was seeking a coordinator as of 2012.

In 2001–2002 the coordinators from the 3 territories (Lori Shroeder, Jen Morin, Paula Hughson) and northern Manitoba (Kim Monson) with assistance from Elisabeth Beaubien and Leslie Wakelyn of Environment Canada's Ecological Monitoring and Assessment Network North (EMAN-North) applied as "PlantWatch North" and received funds from Environment Canada's Northern Ecosystems. They produced a PlantWatch North poster in 2002 with versions in English, French and Inuktitut, as well as recognition pins for observers. The funding also permitted several workshop meetings, and in 2003 an illustrated booklet describing the PlantWatch North program was published.

5.2.6 Conclusions

Canada enjoys a wealth of phenological studies, starting with early applications by First Nations and continued today by a variety of observers including scientists, naturalists, gardeners and students. Starting in 2000 Environment Canada embraced phenology through its "NatureWatch" programs, involving the public in finding out what is changing in the environment and why. PlantWatch continues to engage the public as "eyes of science", tracking nature's plant calendar and boosting awareness of the biotic response to climate change. Historically, many Canadians have enjoyed noting down the timing of seasonal events such as the appearance of wildflower blooms, spring birds, calling of frogs, and melt or freeze-up of lakes and rivers. These data are important as a baseline against which to compare current timing in our increasingly variable climate.

To attract and retain volunteers in Canada PlantWatch, this citizen science program needs regional coordination that identifies and meets the needs and interests of observers. It must provide to volunteers appropriate training, frequent feedback, and rewards. As this support of volunteers requires considerable financial and other resources, government support is essential and has been the backbone of many long-term phenology networks elsewhere. As Bonney et al. (2009) p. 983 notes: "An effective citizen science program requires staff dedicated to direct and manage project development; participant support; and data collection, analysis, and curation. Such a program can be costly; the Cornell Laboratory of Ornithology's current citizen science budget exceeds $1 million each year ... Considering the quantity of high-quality data that citizen science projects are able to collect once the infrastructure for a project is created, the citizen science model is cost effective over the long term."

Acknowledgments United States section: Thanks to Glen Conner for information on Dr. Samuel D. Martin and other early phenological observers and networks. We appreciate careful reviews provided by E. Denny, S. Newman, S. Phillips, and E. Stancioff. Erin Posthumus provided content for Table 5.1.

Canada section: Thanks to these regional coordinators for their contributions: S. Bailleul, M. Bishop, B. Charles, N. Iwanycki, K. Monson, M. Priesnitz, S. Ripley, V. Roy McDougall, D. Trowsdale-Mutafov. M. Doyle and R. McLeman contributed information from Environment Canada and the University of Ottawa respectively.

L. Seale and M. Hall-Beyer kindly edited the article. Thanks to Environment Canada and Nature Canada for their help in coordinating and promoting the PlantWatch program. A bouquet of flowers to each of the almost 700 Albertans who observed and reported over the years starting in 1987!

This chapter complies with US Geological Survey Fundamental Science Practice standards. It has undergone peer and policy review and approval.

References

Bassett IJ, Holmes RM, MacKay KH (1961) Phenology of several plant species at Ottawa, Ontario, and an examination of the influence of air temperatures. Can J Plant Sci 41:643–652

Beaubien EG, Freeland HJ (2000) Spring phenology trends in Alberta, Canada: links to ocean temperature. Int J Biometeorol 44(2):53–59

Beaubien EG, Hall-Beyer M (2003) Plant phenology in western Canada: trends and links to the view from space. Environ Monit Assess 88(1–3):419–429

Beaubien E, Hamann A (2011a) Plant phenology networks of citizen scientists: recommendations from two decades of experience in Canada. Int J Biometeorol 55(6):833–841. doi:10.1007/s00484-011-0457-y

Beaubien E, Hamann A (2011b) Spring flowering response to climate change between 1936 and 2006 in Alberta, Canada. BioScience 61(7):514–524

Bird CD (1974) 1973 flowering dates. Alta Nat 4(1):7–14

Blair RJ, Newman JE, Fenwick JR (1974) Phenology Gardens in Indiana. In: Lieth H (ed) Phenology and seasonality modeling. Springer, New York

Bonney R, Cooper C, Dickinson J, Kelling S, Phillips T, Shirk JL (2009) Citizen science: a developing tool for expanding science knowledge and scientific literacy. Bioscience 59:977–984

Budd AC, Campbell JB (1959) Flowering sequence of a local flora. J Range Manage 12:127–132

Caprio JM (1957) Phenology of lilac bloom in Montana. Science 126:1344–1345

Caprio JM (1974) Solar thermal unit concept in problems related to plant development and potential evapotranspiration. In: Lieth H (ed) Phenology and seasonality modeling. Springer, New York

Castonguay Y, Dubé PA (1985) Climatic analysis of a phenological zonation: a mutivariate approach. Agr Forest Meteorol 35:31–45

Cayan DR, Kammerdiener SA, Dettinger MD, Caprio JM, Peterson DH (2001) Changes in the Onset of Spring in the Western United States. Bull Am Met Soc 82:399–415

Criddle N (1927) A calendar of flowers. Can Field Nat 41:48–55

Dubé PA, Chevrette JE (1978) Phenology applied to bioclimatic zonation in Québec. In: Hopp RK (ed) Phenology: an aid to agricultural technology. Vt Agric Exper Sta Bull 684. Vermont Agricultural Experiment Station, Burlington

Erskine AJ (1985) Some phenological observations across Canada's boreal regions. Can Field Nat 99(2):185–195

Flint HL (1974) Phenology and Genecology of Woody Plants. In: Lieth H (ed) Phenology and Seasonality Modeling. Springer, New York

Futter M (2003) Patterns and trends in Southern Ontario Lake ice Phenology. Environ Monit Assess 88(1–3):431–444

Gazal R, White MA, Gillies R, Rodemaker E, Sparrow E, Gordon L (2008) GLOBE students, teachers, and scientists demonstrate variable differences between urban and rural leaf phenology. Global Change Biol 14:1–13

Glendenning R (1943) Phenology, the most natural of sciences. Can Field Nat 57:75–78

Henry GHR, Molau U (1997) Tundra plants and climate change: the International Tundra Experiment (ITEX). Global Change Biol 3(suppl 1):1–9

Hopkins AD (1938) Bioclimatics – a science of life and climate relations. US Dept Agr Misc Publ 280

Hopp RJ (1974) Plant phenology observation networks. In: Lieth H (ed) Phenology and seasonality modeling. Springer, New York

Hough FB (1864) Observations upon periodical phenomena in plants and animals from 1851 to 1859, with tables of the dates of opening and closing of lakes, rivers, harbors, etc. In: Results of meteorological observations, made under the direction of the United States Patent Office and the Smithsonian Institution, from the year 1854 to 1859, inclusive, report of the Commissioner of Patents, vol 2, part 1, Exec. Doc. 55, 36th congress, 1st session, U.S. Government Printing Office, Washington, DC

Hunter AF, Lechowicz MJ (1992) Predicting the timing of budburst in temperate trees. J Appl Ecol 29:597–604
Johnston A (1987) Plants and the Blackfoot, occasional paper no. 15, Lethbridge Historical Society. Historical Society of Alberta, Lethbridge
Kross A, Fernandes R, Seaquist J, Beaubien E (2011) The effect of the temporal resolution of NDVI data on season onset dates and trends across Canadian broadleaf forests. Remote Sens Environ 115:1564–1575
Lantz TC, Turner NJ (2003) Traditional phenological knowledge (TPK) of aboriginal peoples in British Columbia. J Ethnobiol 23:263–28
Lechowicz MJ (1995) Seasonality of flowering and fruiting in temperate forest trees. Can J Bot 73:175–182
Lechowicz MJ (2001) Phenology, in encyclopedia of global environmental change. In: Canadell JG (ed) Biological and ecological dimensions of global environmental change, vol 2. Wiley, London
MacKay AH (1899) Phenological observations in Canada. Can Rec Sci 8(2):71–84
MacKay AH (1927) The phenology of Nova Scotia, 1923. Trans Nova Scotia Inst Sci 16 (2):104–111
Martin SD (1865) Register of meteorological observations, Pine Grove, Kentucky. Smithsonian Institution, Washington, DC
Minshall WH (1947) First dates of anthesis for four trees at Ottawa, Ontario, for the period of 1936 to 1945. Can Field Nat 61:56–59
Mitchener AV (1948) Nectar & pollen producing plants of Manitoba. Sci Agric 28:475–480
Monson (2008) Trends in spring flowering dates from Churchill, Manitoba and Northern Labrador: an assessment of PlantWatch North Phenological Data. www.ec.gc.ca/Publications/9C760386-CD1C-4B5C-87A3-044BECEF17BD%5CTrendsInSpringFloweringDatesFromChurchill-MaintobaAndNorthernLaborador.pdf
Moss EH (1960) Spring phenological records at Edmonton, Alberta. Can Field Nat 74(13):118
Pouliot D, Latifovic R, Fernandes R, Olthof I (2011) Evaluation of compositing period and AVHRR and MERIS combination for improvement of spring phenology detection in deciduous forests. Remote Sens Environ 115:158–166
Royal Society of Canada (1893) Proceedings for 1892. Proceedings and Transactions of the Royal Society of Canada 10(3):53–55
Reader R, Radford JS, Lieth HL (1974) Modeling important phytophenological events in Eastern North America. In: Lieth HL (ed) Phenology and seasonality modeling. Springer, New York
Schwartz M D (1985) The advance of phenological spring across Eastern and Central North America. Dissertation, University of Kansas
Schwartz MD (1992) Phenology and springtime surface layer change. Mon Wea Rev 120 (11):2570–2578
Schwartz MD (1994) Monitoring global change with phenology: the case of the spring green wave. Int J Biometeorol 38:18–22
Schwartz MD (1998) Green-wave phenology. Nature 394(6696):839–840
Schwartz MD, Crawford TM (2001) Detecting energy-balance modifications at the onset of spring. Phys Geogr 21(5):394–409
Schwartz MD, Reiter BE (2000) Changes in North American Spring. Int J Climatol 20(8):929–932
Schwartz MD, Carbone GJ, Reighard GL, Okie WR (1997) Models to predict peach phenology from meteorological variables. HortSci 32(2):213–216
Schwartz MD, Ahas R, Aasa A (2006) Onset of spring starting earlier across the northern hemisphere. Glob Change Biol 12(2):343–351
Schwartz MD, Betancourt JL, Weltzin JF (2012) From Caprio's Lilacs to the USA National Phenology Network. Front Ecol Environ 10(6):324–327
Schwartz MD, Ault TR, Betancourt JL (2013) Spring onset variations and trends in the continental USA: past and regional assessment using temperature-based indices. Int J Climatol. doi:10.1002/joc.3625

Smith JW (1915) Phenological dates and meteorological data recorded by Thomas Mikesell between 1873 and 1912 at Wauseon, Ohio. Mon Wea Rev Suppl 2:23–93

Vasseur L, Guscott RL, Mudie PJ (2001) Monitoring of spring flower phenology in Nova Scotia: comparison over the last century. Northeast Nat 8(4):393–402

Vittum MT, Hopp RJ (1978) The N.E. 95 lilac phenology network. In: Hopp RJ (ed) Phenology, an aid to agricultural technology. Vt Agric Exper Sta Bull, 684. Vermont Agricultural Station, Burlington

Weather Bureau, Instructions for Voluntary Observers, U.S. Department of Agriculture, Washington, DC (1899)

Zhao T, Schwartz MD (2003) Examining the onset of spring in Wisconsin. Clim Res 24:59–70

Chapter 6
A Review of Plant Phenology in South and Central America

L. Patrícia C. Morellato, Maria Gabriela G. Camargo, and Eliana Gressler

Abstract Phenology has established its status as an integrative environmental science and as a key component for climate change research. We aim to present a review of phenological research in South and Central America. We describe the flowering and fruiting patterns of the main vegetation types studied to date, and highlight areas where phenological information is lacking. Phenological research is still required even though the number of papers published has increased, especially in the twenty-first century. The distribution of phenological studies on South American vegetation was very uneven over the different vegetation types and life forms. Tropical moist forest continues to be, by far, the most widely studied ecosystem while trees were the life-form observed in almost all papers surveyed. Currently, long-term phenological datasets are rare and few long-term monitoring systems are known for South and Central America. Few papers reviewed were concerned with the effects of climatic change and its evaluation using plant phenology, which differed greatly from Northern Hemisphere research which had a strong focus on phenology and climate change. Among new developments we mention phenology and edge effects and fragmentation, and remote monitoring of phenology with digital cameras. Finally, building phenology networks is the greatest challenge for South and Central American phenologists, demanding an effort from and cooperation among universities, research institutions, governmental, and non-governmental agencies.

L.P.C. Morellato (✉) • M.G.G. Camargo • E. Gressler
Departamento de Botânica, Laboratório de Fenologia, Plant Phenology and Seed Dispersal Research Group, Instituto de Biociências, Universidade Estadual Paulista UNESP, São Paulo, Brazil
e-mail: pmorella@rc.unesp.br

M.D. Schwartz (ed.), *Phenology: An Integrative Environmental Science*,
DOI 10.1007/978-94-007-6925-0_6,

6.1 Introduction

During the first decade of this new century phenology has established its status as an integrative environmental science and as a key component for climate change research (Peñuelas et al. 2009; Cleland et al. 2012). At the same time, we have witnessed a significant advancement in scientific research in Latin America, mainly in South America (Regalado 2010). Consequently, producing an overview of phenological research for South America was more challenging than 10 years ago primarily due to an increase in the number of data sources, and in the range of tools available by which to mine the data. Therefore, we now present an improved and updated review of South American phenology patterns and perspectives and expanded the previous review (Morellato 2003) to incorporate Central American studies, given their relevance to tropical ecology and phenological research.

Comprising about one eighth of the earth's land surface, the South American continent is situated between 12 °N–55 °S latitude and 80–35 °W longitude. It covers an area of approximately 17,850,000 km^2 divided among 13 countries. Eighty percent of its land is within the tropical zone, yet it extends into the sub-Antarctic region (Davis et al. 1997). Essentially, most life zones and vegetation formations are represented. The principal vegetation types are tropical evergreen and semi-evergreen moist forest, dry forest and woodland (cerrado or woody savanna), open grassy savanna, desert and arid steppe, Mediterranean-climate communities, temperate evergreen forest, and several montane assemblages (e. g. páramo, rupestrian fields or campos rupestres, puna). Within the large array of vegetation types there are some of the most diverse in the world including the upper Amazon forest and Atlantic forest, as well as vegetation types with high concentrations of endemic species. In addition, the Andean montane forests and the Mediterranean-climate region of central Chile are also included (Davis et al. 1997). At least 46 sites distributed over eight regions have been recognized as centers of plant diversity (Davis et al. 1997), and several are considered biodiversity hotspots for conservation priorities (Myers et al. 2000; Peñuelas et al. 2009; Murray-Smith et al. 2009). Despite its much smaller geographical extent (31 °35′0″N; 77 °2′0″W), Central America covers about 742,266 km^2 and 20 countries. There are roughly 20 life zones exhibiting a high diversity of species and vegetation types including coastal swamp and mangrove, semi-arid scrub, pine (*Pinus*) savanna, coniferous forest, tropical deciduous forest, semi-deciduous, and evergreen rain forest, cloud forest, and paramo (Davis et al. 1997).

This diversity of species and vegetation types has, as yet, not been comprehensively studied in respect of its floristic diversity, especially in South America. Consequently, only a small percentage of the species and vegetation types have been examined from the point of view of their seasonal changes. Phenological research is still required even though the number of papers published has increased over the last 20 years, especially in the twenty-first century. Currently, long-term phenological datasets are rare and few long-term monitoring systems are known for South and Central America (see Morellato 2003 and the present review).

Therefore, a comprehensive search of the major electronic databases (Web of Science, Scielo), and some smaller alternative ones (Periodicos Capes, NAL Catalog, Binabitrop) was undertaken in order to identify any reproductive phenological studies of native species from South and Central America. Vegetative phenophases (leaf budding, leafing, leaf fall) were not included in our keywords search. Phenological data from studies on agricultural or introduced species were also excluded as were those studies on herbivory, pollination, frugivory and seed dispersal where the main focus of the paper was the examination of animal feeding behavior. Comparisons of phenology studies across different vegetation types were based largely on community studies obtained from the Web of Science and Scielo, which focused on phenology and which also included information on flowering and/or fruiting. We considered any study that included four or more years of observations as long-term. In order to attain a comprehensive historical perspective, we searched, using electronic and internet search tools, databases of the main research agencies and institutions in South and Central America undertaking phenological research. In addition, we also searched for historical phenological data and manuscripts and for groups, universities, or institutions undertaking phenological research. However, the information was not easy to find and may be biased towards Brazil for two reasons. Firstly, the author's more than 15 years of phenological research in Brazil, and secondly, because the Brazilian National Counsel for Scientific and Technological Research (CNPq) maintains a very well organized database of all active researchers and research groups in Brazil (http://www.cnpq.br), making it easy to collect the necessary information.

The main goal of this chapter is to provide a comprehensive update and overview of phenological work carried out on South and Central American vegetation. We describe the flowering and fruiting patterns of the main vegetation types studied to date, and highlight areas where phenological information is lacking. In addition, we identified current research and research groups involved in phenology and propose a vision of the future of phenological research and networks in South and Central America. This is the first time the phenology of South and Central American vegetation has been summarized from this perspective.

6.2 Brief History of Phenological Data Collection

This historical view is based on all surveyed papers, regardless of whether the paper appeared in the main electronic data base or not. The oldest phenological information surveyed for South America was the descriptions of the seasonal periodicity of a tropical rain forest in Guyana (Davis and Richards 1933) and of the annual cycle of plants and animals in two Atlantic forests in Rio de Janeiro, Southeast Brazil (Davis 1945). Excluding the databases surveyed, other phenological data uncovered included information in comprehensive papers describing plant communities (Veloso 1945; Andrade 1967) or phenological notes, where the authors recorded the dates of flowering or fruiting for several tropical species from a particular site or botanical garden (Silveira 1935; Lima 1957; Santos 1979). We also found some phenological notes in Warming (1892) a book about Southeastern Brazil savanna species. Alvim (1964) was one of the first researchers to describe and analyze the phenology of native tropical forest trees from Bahia, Brazil (Alvim 1964; Alvim and Alvim 1978), although most of his work focused on the flowering of coffee and cocoa trees. We also uncovered studies carried out by Ramia (1977, 1978) in the Venezuelan savannas and, the influential comparative study by Monasterio and Sarmiento (1976), which includes two vegetation types and several plant life-forms, as well as contributing to the definitions of general concepts in phenology such as phenological strategy and phenodynamics.

The work of Araujo (1970) on the phenology of 36 species of Amazon lowland forest trees marks the beginning of contemporary studies on phenology in South America. This paper is especially important because it represents the primary report of the first and oldest as well as possibly unique long-term phenological data collection for South American tropical forest trees. The INPA (Instituto Nacional de Pesquisas da Amazônia) phenological work started in 1965, at Reserva Florestal Ducke (Manaus, Amazonas State, Brazil). Trees were systematically selected and a total of 300 trees (three per species) were marked over an area of 140.5 ha of native Amazon lowland tropical forest. In 1970 the sample size was extended to 500 trees (five per species), which are still monitored today (Pinto et al. 2008). In 1974, INPA researchers replicated the phenology study. They marked 500 trees of another 100 species from an Amazon lowland forest at Estação Experimental de Silvicultura Tropical, about 30 km from Reserva Florestal Ducke. Both studies performed monthly observations for changes in reproductive and vegetative phenology of ten defined phenophases (Araujo 1970; Alencar et al. 1979; Pinto et al. 2008).

A similar program of long-term phenological data collection was established by Companhia Vale do Rio Doce (CVRD) at Reserva Natural de Linhares (Northeast Espírito Santo State, Brazil), a lowland evergreen forest reserve locally known as "Tabuleiro forest" (Engel and Martins 2005). They employed the same methodology proposed by Araujo (1970), selecting 41 species and marking 205 trees (5 per species). The project started in 1982 and we are currently unaware if it is still active. Another interesting South American long-term database, although not active, is the one analyzed by Ter Steege and Persaud (1991). The authors compiled data on the flowering and fruiting of Guyanese forest trees collected over about 100 non-consecutive years.

For Central America the oldest paper was Allen's (1956) long-term phenological study comprising 673 species of a rainforest in Puntarenas Province, Costa Rica. The studies by Fournier and Salas (1966) in Costa Rica tropical moist forest trees and the Melastomataceae phenology described in the classic study by Snow (1966) are seminal papers. Together with Croat's (1969) comprehensive study on the phenology of Barro Colorado Island (BCI) vegetation as well as Janzen's (1967) paper on reproductive phenology of Costa Rican dry forest trees, they mark the contemporary phenology research in Central America, followed by Nevling's (1971) paper on Puerto Rico elfin forest (tropical moist forest).

The studies by Croat (1975, 1969) were the first comprehensive long-term phenological studies in Central America, including a compilation of direct observations and herbarium data from more than 1,100 species in the BCI, Panama. These studies included various plant life forms, probably the first that made this kind of comparative analysis in the tropics. In the same decade, three long-term works were conducted in Costa Rica: Daubenmire (1972), in a tropical semi-deciduous forest (36 species); Fournier and Salas (1966) in a tropical wet forest (100 woody species) and Stiles (1977) in the tropical rainforest at La Selva Biological Station (11 hummingbird-pollinated species).

We found three active phenological long-term data collection for Central America. The oldest one is run by the Smithsonian Tropical Research Institute at the BCI Tropical Moist Forest (Wright et al. 1999; Wright and Calderón 2006). The phenological data is based on the weekly collection of flowers and fruits in 200 traps distributed in the forest (Wright et al. 1999), differing from the usual direct observation of tree crowns. The same experimental design was set up in two other sites at Luquillo, Puerto Rico in 1992 and Parque Nacional San Lorenzo, Panama in 1997 (http://www.ctfs.si.edu/floss/).

We detected four recent long-term projects undertaking systematic, direct phenological observations on South American vegetation. One in cerrado-savanna vegetation, which started in 2003 see (Camargo et al. 2011, for study site description) and another observing native tree species in an urban area at the University campus of UNESP at Rio Claro (Morellato et al. 2010), both in Southeastern Brazil. Two long-term phenological study using a network of seed traps were established in 2000 in South America and now has around 12 years of data: in French Guiana in an area of ca. 40 ha of tropical rainforest (Norden et al. 2007) on flowering and fruiting, and 200 seed traps in Estacion Cientifica Yasuni, Ecuador (http://www.ctfs.si.edu/floss/).

Several Institutions and Universities conducting agronomic-related research have phenology programs for crops and some economically valuable species. For example, in Argentina, many of the National Universities have a discipline on climatology and agricultural phenology, and several plant species are investigated for phenological changes. Brazilian research institutions such as INPA (Instituto Nacional de Pesquisas da Amazônia), EMBRABA (Empresa Brasileira de Agropecuária) and IAC (Instituto Agronomico de Campinas) perform phenological observation of crop plants and native or exotic fruit trees. We uncovered one research network on phenological observation for 3–5 years in tropical dry forests across South and Central America, the Tropidry initiative (http://tropi-dry.eas.ualberta.ca/). They perform monthly direct observations of phenology, but we do not know if data is still collected nor the duration. Finally, we were unable to find any kind of citizen science phenology network, besides the GLOBE educational project, involving eight South American and seven Central American countries (for more information, see http://www.globe.gov). All other information surveyed refers to the collection and analyses of phenological data during a defined time span, usually between 1 and 3 years. Out of the 153 articles surveyed in this review addressing community reproductive phenology, 83 % cover less than 4 years of observations, while just 11 % span more than 4 years and around 6 % did not mention the duration of the observations. Therefore, there is an urgent need for long-term phenological studies and networks in South and Central America.

After the works of Araujo's (1970), Fournier and Salas (1966), and Snow (1966), additional papers were published in the 1970s (11), and the number increased during the 1980s (52 papers). There was an increase in the production of phenological information for South and Central American vegetation in the 1990s, with at least 111 papers in all the electronic databases surveyed. These papers are not evenly distributed over the 13 South American (85 papers) and 20 Central American (29 papers) countries. About half of the papers produced in the 90s were from Brazil (45), followed by Chile (13), Argentina (13), Venezuela (10) and with some papers from Colombia (4), Peru (2), Paraguay (2), French Guiana (2), Ecuator (2), Bolivia (1) and Guyana (1). Costa Rica (15) and Panama (11) lead the Central America production, followed by Nicaragua and Belize (2 each) and Honduras and Cuba (1). The number of published papers stays high in the first decade of the new century, with about 326 papers surveyed in 12 years, if we consider all phenological kinds of research, not only the community studies addressing flowering or fruiting. Of that total, 41 papers are from Central America and 288 from South America, mainly from Brazil (196), followed by Argentina (35), Colombia (19), Venezuela (13), Chile (10), Peru (8), Ecuador (8), Bolivia (5), French Guiana (2), and Uruguay and Guyana with one paper each. Central America is still dominated by studies from Costa Rica (26), Panama (10), and Cuba, Jamaica and Bahamas, having one paper each. Therefore, there is a general increasing trend in publication of phenology studies, particularly in South America. If we considered another sources of information (books, journals not indexed, etc.), the number of papers would be much higher (about twice) but the main countries producing papers remain the same. A complete list of all papers surveyed can be obtained from the corresponding author.

6.2.1 Actual State of Phenological Research

Among the 583 papers surveyed, approximately 30 % were phenology studies at the community level, describing the phenological patterns of several species belonging to one or different life-forms, from a defined geographic region and vegetation type. These papers are a good source of information for a large number of native species from different South and Central American vegetation types (Table 6.1). The other papers are mostly studies focused on one or few species, or on a group of species belonging to the same plant family. However, some recent studies have addressed remote phenology using satellite-derived information, as well as seasonal patterns of CO_2 assimilation by plants and gas exchange derived from flux towers (e.g. Asner et al. 2000; Parolin 2000; Newell et al. 2002; Hutyra et al. 2007; Myneni et al. 2007).

Specifically for Brazil, we were able to locate at least 40 Universities or Institutions carrying out phenological research on almost every type of native vegetation across the country, from the Amazon forest around the equator to the subtropical Araucaria forest of South Brazil. Some vegetation types have been the

Table 6.1 Number of published phenology studies at the community level in the six large lowland vegetation groups of South and Central America (Adapted from Davis et al. 1997), plus the montane formations

	Vegetation group	Plant formations included	Number of papers
I.	Tropical moist forest (evergreen or semi-evergreen rain forest)	Amazon forest, Rainforests, Brazilian Atlantic rainforest, Semi-deciduous forest, Gallery forest and Swamp forest, Pre-montane/montane Pacific rain forest, Chocó and lower Magdalena valley	86
II.	Dry forest (integrating into woodland)	Chaco, Cerrado, Caatinga, Woody Savanna Northern Colombia and Venezuela, coastal Equator and Peru, and the Deciduous (dry) forests	27
III.	Open grassy savanna	Pampas region, the Llanos, Cerrado grassland, Pantanal and Gran Sabana and Sipaliwini savanna in the Guyana region	12
IV.	Desert and arid steppe	Sechura and Atacama regions along west coast and in the Monte and Patagonian Steppes in the Southern Cone of South America	8
V.	Mediterranean-climate region	Central Chile	1
VI.	Temperate evergreen forest	Chile and Argentina, South Brazil	5
VII.	Montane formations	Complex montane formations along the Andean Cordillera, the Tepuis, in the coastal cordillera of Brazil and rupestrian fields in the Espinhaço cordillera	6

focus of research groups, such as the Amazon lowland forest or terra-firme forest, Atlantic forest, semi-deciduous forest and Cerrado. Over the last 10 years we found papers and ongoing projects focused on seasonal changes of plant species from Caatinga, Dry Forest, Gallery forest, Swamp forest, Amazon-inundated and seasonally-inundated forest, coastal plain vegetation, dunes, and mangroves, among others.

However, an interesting point is that almost any paper or project was concerned with the effects of climatic change and its evaluation using plant phenology (but see Wright and Calderón 2006), strongly differing from the marked trend of phenological research in the Northern Hemisphere focused on phenology and climate change. Most research surveyed in the present review focus on evolutionary questions or actual environmental drivers of tropical phenology.

6.3 Overview of Phenological Patterns for South and Central American Vegetation

Summarizing the phenology of South and Central America is difficult, due to the diversity of vegetation types and species. To see how the phenological information is distributed over the different vegetation types and to compare the phenological patterns observed, the land vegetation of South and Central America was subdivided into seven large vegetation groups (Table 6.1), including the more complex montane formations occurring along the Andean Cordillera, the Tepuis and in the coastal cordillera of Brazil (Davis et al. 1997). We only considered community studies including information on flowering and fruiting patterns, preferably for a large number of species. Patterns were described based on the number of species flowering and fruiting per month unless otherwise noted. Forest phenological patterns were also described based on tree phenology unless we referred to another life form.

I. Tropical moist forest – Despite the apparent lack of marked seasonal climate and the fact that the vegetation type includes evergreen and semi-evergreen (semi-deciduous) forests, the tropical moist forest presented a distinctive seasonal phenological pattern for more than 50 % of the studies surveyed, with a flowering peak in the dry season, wet season or dry-to-wet season transition (Table 6.2), was observed for more than 50 % of the moist forests surveyed. A dry season flowering pattern is described for most Central America rainforest and semi-deciduous forests (Frankie et al. 1974; Wright and Calderón 2006, and see historical papers by Croat 1969; Janzen 1967, among others). The same dry season flowering is observed for South American forests, including the Amazon lowland evergreen forests (Croat 1969; Araujo 1970; Alencar et al. 1979; Sabatier and Puig 1986; Ter Steege and Persaud 1991; Peres 1994b; Wallace and Painter 2002; Haugaasen and Peres 2005; Cotton 2007; Fig. 6.1), as well as for the Amazon "Campina" forest (Alencar 1990). However, for the more seasonal moist forests south of the Equator such as the semi-deciduous forests, most species flower at the end of the dry season or in the transition from the dry-to-wet season (Morellato et al. 1989; Stranghetti and Ranga 1997; Mikich and Silva 2001; Rubim et al. 2010). The Venezuelan semi-deciduous forest has species flowering in both seasons (Monasterio and Sarmiento 1976) and the same is described for climbers from Brazilian semi-deciduous forests (Morellato and Leitão-Filho 1996). Brazilian evergreen Atlantic rainforest the flowering occurs mainly in the wet season (Davis 1945; Morellato et al. 2000; Talora and Morellato 2000). Gallery forest shows a flowering pattern similar to Atlantic forest, with the flower peak occurring in the dry-to-wet season (Funch et al. 2002). Two papers addressing swampy forest phenology were surveyed (Wallace and Painter 2002; Spina et al. 2001), and both report flowering patterns similar to those observed for semi-deciduous forests. In the South America, the Bolivian Sartanejal forest, a vegetation type influenced by forest streams, has a flowering peak in the dry season (Wallace and Painter 2002). Only the The Alto Yunda Colombian premontane rain forest was recorded as being aseasonal (Hilty 1980, but see Zimmerman et al. 2007), and cloud forests may present different patterns and definitions, mainly peaking in the transition from the dry-to-wet season (Wheelwright 1985; Koptur et al. 1988).

Fruiting patterns were more variable than flowering patterns across forest types and locations (Table 6.2, Fig. 6.1). For most of Central American seasonal or semideciduous forests the timing of the fruiting peak was in the dry season. Across South America, nearly all Amazon lowland forests showed seasonal fruiting patterns

Table 6.2 Main phenological patterns and peak season for South and Central American vegetation types

Vegetation type[a]	Group[a]	Main phenological pattern	
		Flowering pattern/peak season	Fruiting pattern/peak season
Tropical seasonal	I	Seasonal/dry or dry-to wet	Dry
Tropical rainforest	I	Seasonal/wet	Dry
Amazon lowland forest	I	Seasonal/dry	Seasonal/wet to non-seasonal
Amazon floodplain forest	I	Seasonal/flooding	Seasonal/flooding
Semi-deciduous forest	I	Seasonal/dry or dry-to-wet	Seasonal/dry or dry-to-wet
Atlantic rain forest	I	Seasonal/wet	Non-seasonal/wet
Gallery forest	I	Seasonal/wet-to dry	Seasonal/wet
Swamp forest	I	Seasonal/dry-to-wet or wet	Non-seasonal
Sartanejal forest	I	Seasonal/dry	Seasonal/wet
Pre-montane rain forest	I	Non-seasonal	Non-seasonal
Savanna-forest mosaic	II	Seasonal/wet	Seasonal/wet or dry to wet
Cerrado (woody flora)	II	Seasonal/dry-to-wet	Seasonal to non-seasonal
Dry forest	II	Seasonal/dry or dry-to-wet	Seasonal/dry or wet
Caatinga	II	Seasonal/rain	Seasonal/rain
Campo cerrado	III	Seasonal/wet or wet-to-dry	Seasonal/wet
Desert	IV	Seasonal/after rain	Seasonal/after rain
Mediterranean-climate region	V	Seasonal/after rain	Seasonal/after rain
Temperate Valdivian rain forest	VI	Seasonal/early summer (dry)	Seasonal/late summer
Temperate Araucaria forest	VI	Wet	Wet or aseasonal
Chile Andean Zone (2,000–3,600 m altitude)	VII	Seasonal/summer	Seasonal/late summer-to-fall
Montane grassland	VII	Seasonal/summer	Seasonal/summer
Pre-montane sub-tropical forest	VII	Seasonal/dry	Seasonal/wet

Patterns are ranked as seasonal or non-seasonal. The time of flowering and fruiting peak is indicated as dry season and wet season for tropical climates, and as spring, fall, winter or summer for temperate climates

[a]See Table 6.1 and text for more detailed description of vegetation and source data

peaking during the wet season (Alencar et al. 1979; Sabatier 1985; Peres 1994b; Barlow et al. 2007; Bentos et al. 2008), or in the transition from the wet-to-dry season (Schöngart et al. 2002; Muniz 2008). Guianese forest may present an almost bimodal fruiting pattern, with both peaks occurring in the dry season (Ter Steege and Persaud 1991). For Colombian lowland forests and Campina forest fruiting was non-seasonal (Alencar 1990; Wallace and Painter 2002). The fruiting of semi-deciduous forests

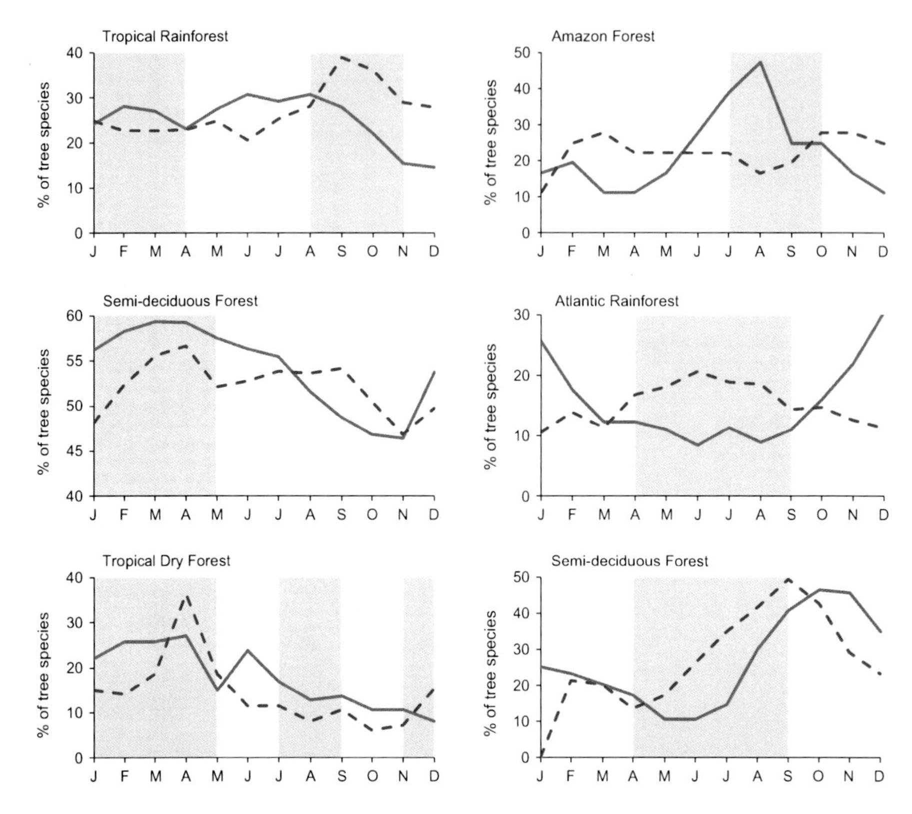

Fig. 6.1 Flowering (*continuous lines*) and fruiting (*dashed lines*) phenology of Central (*left*) and South (*right*) America forest sites. Tropical Moist Forests: Tropical Rainforest at La Selva, Costa Rica (n = 185 species; data source: Frankie et al. 1974), Tropical Semi-deciduous forest at Barro Colorado Island (*BCI*), Panama (n = 81; data source: Wright and Calderón 2006, using seed traps); Amazon lowland forest at Manaus, Brazil (n = 36, data source: Araujo 1970); Atlantic Rainforest at Picinguaba, Brazil (n = 214, data source: Morellato et al. 2000). Semi-deciduous forest at Campinas, Brazil (n = 103, data source: Morellato and Leitão-Filho 1995). Tropical Dry Forest at Comelco, Costa Rica (n = 113, data source: Frankie et al. 1974). *Shaded area*: dry season

were not found to be as seasonal as the flowering pattern, even though most species bear ripe fruits in the dry season or in the transition from the dry-to-wet season (Morellato et al. 1989; Stranghetti and Ranga 1997; Mikich and Silva 2001; Rubim et al. 2010). Fruiting is not seasonal for most of the Atlantic forest, increasing fruit production occurs in the wet season, but a seasonal pattern was observed for Coastal Plain forest and Restinga forest, peaking in the less wet season (Davis 1945; Morellato et al. 2000; Talora and Morellato 2000; Staggemeier and Morellato 2011). Gallery and Sartanejal forest have a fruiting peak in the wet season (Funch et al. 2002; Wallace and Painter 2002), and swamp forest fruiting was almost always non-seasonal (Spina et al. 2001). For the Amazon floodplain forests surveyed the peak of flowering and fruiting occurs during the aquatic (flooded) phase (Kubitzki and Ziburski 1994; Schöngart et al. 2002; Haugaasen and Peres 2005). Finally,

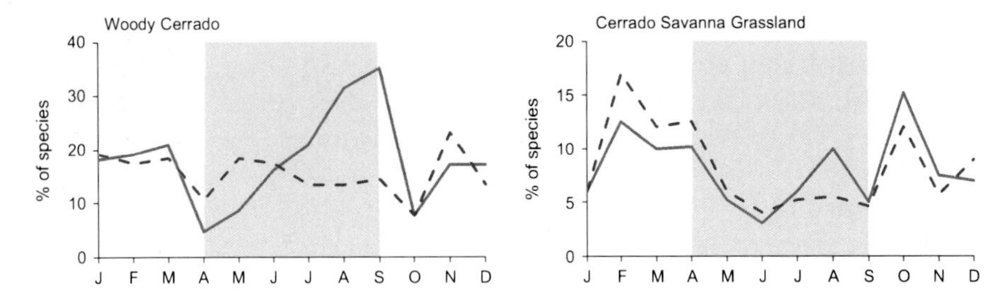

Fig. 6.2 Flowering (*continuous lines*) and fruiting (*dashed lines*) phenology of a Tropical Dry Forest (*left side*): Woody Cerrado Savanna (n = 105, data source: Camargo 2008), and of an Open Grassy Savanna (*right side*): Cerrado Savanna Grassland (n = 105, data source: Tannus et al. 2006), both at Itirapina, Southeastern Brazil. *Shaded area*: dry season

Colombian Premontane rain forest presented a relatively constant number of species in flower or fruit throughout the year (Hilty 1980).

II. Dry forest (integrated into woodland) – Venezuelan Savanna-forest mosaic and dry forest generally shows a marked seasonality for all habitats, with flowering peaking during the rainy season, and a fruiting peak towards the end of the rainy season (Table 6.2, Ramírez and Brito 1987; Ramírez 2002, 2009). Fruiting records are variable across the studies, peaking during the dry, transitional or wet seasons (Ramírez and Brito 1987; Ramírez 2002, 2009; Berg et al. 2007; Jara-Guerrero et al. 2011). Woody savannas or cerrado vegetation comprise trees, shrubs and herbs. It includes a range of structural vegetation types and we included two types in the dry forest: the cerrado *sensu stricto* or the typical woody savanna, an open vegetation dominated by shrubs and trees at different densities, and the Cerradão, a well-developed forest form of cerrado, with close canopy up to 20 m tall (Ribeiro and Tabarelli 2002). Generally, the woody flora of cerrado presents a seasonal phenology, with a flowering peak at the end of the dry season or in the beginning of the wet season and no fruiting peak in any particular season (Fig. 6.2, e.g. Batalha et al. 1997; Batalha and Mantovani 2000; Silberbauer-Gottsberger 2001; Costa et al. 2004; Salazar et al. 2012). Fruiting is more widespread over the year in the savannas, however, in some studies a middle-to-late rainy or dry season fruit peak is detected (Batalha and Mantovani 2000; Ribeiro and Tabarelli 2002; Wallace and Painter 2002). Herbaceous plants show a different pattern, the flowering peak occurring at the end of the rainy season and fruiting peak in the dry season (Batalha and Mantovani 2000; Ribeiro and Tabarelli 2002; Wallace and Painter 2002).

Flowering is clearly associated with the end of the dry season and the beginning of the rainy season for most dry forests from Jamaica and Costa Rica (Frankie et al. 1974; Opler et al. 1980; McLaren and McDonald 2005) to South America (e.g. Monasterio and Sarmiento 1976; Wallace and Painter 2002; Ragusa-Netto and Silva 2007). Flowering peak in the Bolivian dry forest occurs at the transition between the dry and the rainy seasons (Justiniano and Fredericksen 2000). There is a major peak of fruiting in the dry season and a minor one during the rainy season (Justiniano and Fredericksen 2000).

The Caatinga, a deciduous tree-shrub vegetation type common in Northeastern Brazil, experiences a low rainfall climate, which is very seasonal and variable among years (Machado et al. 1997). Reproductive events are concentrated in the rainy season. The flowering peak occurs early and the fruiting peak occurs late in the rainy period (e.g. Machado et al. 1997; Griz and Machado 2001; Amorim et al. 2009; Lima and Rodal 2010).

III. Open grassy savanna – The majority of the studies surveyed describe the phenology of Brazilian open savanna. The dominant perennial grasses from the Brazilian campo cerrado (open savanna grassland, Fig. 6.2) exhibited a flowering peak during the rainy season or in the transition from the wet-to-dry season while a fruiting peak occurred in the wet season (Almeida 1995; Munhoz and Felfili 2005, 2007; Tannus et al. 2006). Associated wet grassland may flower and fruit during the dry-to-wet and wet-to-dry season transitions, respectively (Tannus et al. 2006). The Venezuelan savanna is more variable with peaks in flowering occurring in both dry and wet seasons whereas fruiting is more widespread over the year, but in some areas peaking in the dry season (Ramírez 2002; Ramírez and Briceño 2011). Some studies in the Brazilian and Argentinean pampas also report plant reproduction phenology in the wet season or otherwise no peak is reported (e.g. Latorre and Caccavari 2009; Martini et al. 2010).

IV. Desert and arid step – Phenology studies were undertaken in a range of different places, such as arid Patagonia (Bertiller et al. 1991) and Monte Phytogeographical Province (Giorgetti et al. 2000) in Argentina, and the Southern Atacama Desert in Chile (Vidiella et al. 1999). Phenology is highly constrained by rainfall, which determines the onset of reproduction for most of the species.

V. Mediterranean-climate region – The phenology of shrub species from the coastal desert in North-Central Chile indicates the existence of at least two groups of species, with phenological patterns more or less dependent on precipitation (Olivares and Squeo 1999).

VI. Temperate evergreen forest – The timing of reproductive events and their ecological and climatic constraints in a Valdivian rain forest of Chiloé, Chile, one of the most widespread and species-rich forest types in austral South America, is discussed by Smith-Ramírez and Armesto (1994). Peak flowering for most species occurs in the dry season (late spring to early summer, Fig. 6.3). Ripe fruits are available all year round, but the number of species is lowest in early spring with the maximum in late summer (Smith-Ramírez and Armesto 1994). The mixed Araucaria forest from South Brazil show a flowering peak in the wet season whereas fruiting patterns are either aseasonal or tend to increase in the wet season (Marques et al. 2004; Paise and Vieira 2005; Liebsch and Mikich 2009).

VII. Montane formations – Includes a wide range of austral vegetation types, from different vegetation belts in the Andean zone to pre-montane subtropical forest. Phenological changes in flowering and fruiting of distinct vegetation belts in the Chile Andean zone (Andean scrub, cushion communities and fell-field) ranging from 2,000 to 3,600 m, have been described (Arroyo et al. 1979, 1981; Riveros 1983). The average growing season lasts 5–8 months, peak flowering coincides with the period of maximum temperature at lower altitudes and after

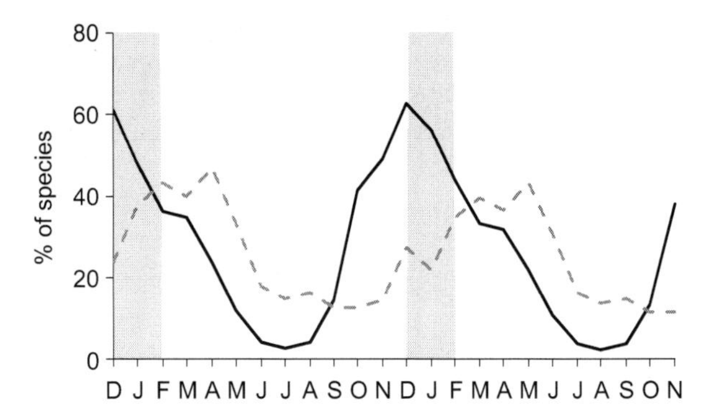

Fig. 6.3 Percent of species (trees, shrubs, and miscellaneous – epiphytes, vines, and hemi parasites) flowering (*continuous line*) and fruiting (*dashed line*) in the temperate rain forest of Chiloé, Chile (Data source Smith-Ramírez and Armesto 1994). *Shaded area*: Summer

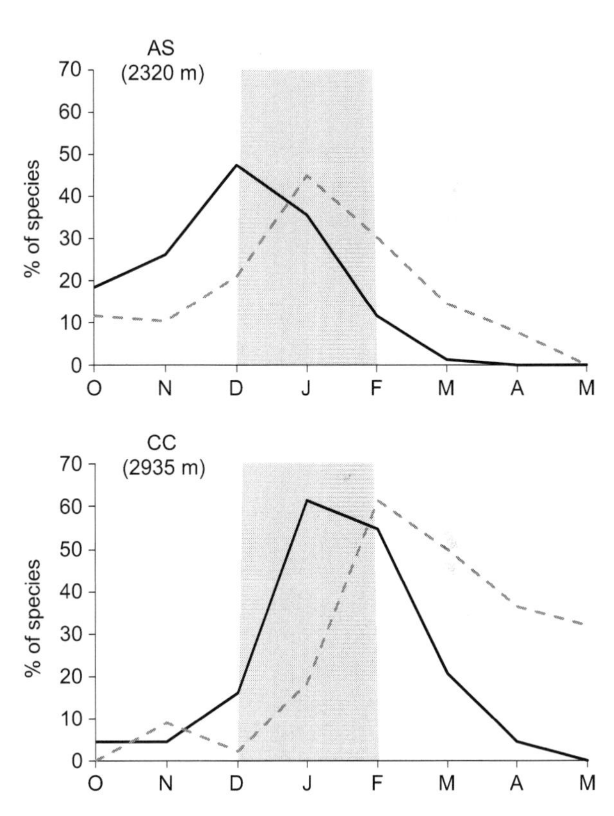

Fig. 6.4 Percent of species flowering (*continuous lines*) and fruiting (*dashed lines*) at Andean scrub (*AS*), and Cushion communities (*CC*) sites (altitudinal Andean vegetation) (Data source Arroyo et al. 1981). *Shaded area*: Summer

this period at higher altitudes, whilst fruiting peak takes place in the late summer and fall (Fig. 6.4, Arroyo et al. 1981).

Pre-montane subtropical forests in Argentina exhibit a seasonal reproductive phenology (Malizia 2001). Flowers are present throughout the dry season and the number of species with ripe fruit peaks during the wet season (Malizia 2001). In Argentina flowering and fruiting in montane grassland are concentrated in a short period during the summer months (Díaz et al. 1994).

Additional studies – One paper from the Caribbean region describes flowering and fruiting phenology in tropical semi-arid vegetation of Northeastern Venezuela (de Lampe et al. 1992). Venezuelan thorn woodland and thorn scrub desert formation show seasonality in their flowering and fruiting phenology (de Lampe et al. 1992). Flowering activity is concentrated in the rainy season. The mature fruit index peaks in the dry season for trees and tall shrubs, and is concentrated in the wet season for low shrubs and herbs (de Lampe et al. 1992). Phenological patterns of Brazilian coastal dune vegetation are described by Cordazzo and Seeliger (1988) and Castellani et al. (1999). Most species of the Southern Brazil coastal dune vegetation, under a warm temperate climate, flower during spring, summer and fall, while fruiting is concentrated in fall and winter. Mangroves present a general bimodal flowering pattern peaking in the wet and dry seasons and one fruiting peak in the wet season (Nadia et al. 2012).

6.4 Future Developments and Concluding Remarks

The distribution of phenological studies on South American vegetation was very uneven over the different vegetation types and life forms. Tropical moist forest continues to be, by far, the most studied ecosystem while trees were the life-form observed in almost all papers surveyed. By contrast a few studies surveyed temperate forest (Smith-Ramírez 1993) and only one addressed flowering and fruiting for Brazilian Araucaria evergreen forest (Marques et al. 2004). Studies focusing on climbers, epiphytes and especially on understory herbs were particularly uncommon (Seres and Ramírez 1993; Putz et al. 1995; Morellato and Leitao 1996; Marques et al. 2004), although some studies did include life forms other than trees (Peres 1994b). Several papers focused only on fruiting patterns (e.g. Zhang and Wang 1995; Stevenson et al. 1998; Develey and Peres 2000; Grombone-Guaratini and Rodrigues 2002). A number of studies on tropical and temperate forests were focused on one family only (e.g. Sist 1989; Peres 1994a; Riveros et al. 1995; Smith-Ramírez et al. 1998; Listabarth 1999; Henderson et al. 2000; Ruiz et al. 2000; Staggemeier et al. 2010). Dry forests, savannas and cerrados were the second most studied vegetation group, followed by open grasslands. Most of the studies were developed in savannas, and again the phenology of woody life-forms dominates, but more studies included other life-forms compared to tropical forests. Very few studies were undertaken on deserts, the Mediterranean-climate region, or montane formations. A number of studies focused on one phenophase (Rozzi et al. 1989), or a small number of species (Silva and Ataroff 1985; Rusch 1993; Fedorenko et al. 1996; Löewe et al. 1996; Jaramillo and Cavelier 1998; Velez et al. 1998; Rosello and Belmonte 1999; Rossi et al. 1999; Damascos and Prado 2001).

Therefore, there is still a need to expand phenological studies across South and Central America's vegetation. The condition is quite critical if we consider that some vegetation types or regions have high species diversity and a large number of endemic species. For example, the Mediterranean climatic region, Andean montane forest and rupestrian fields are basically unknown with respect to their seasonal

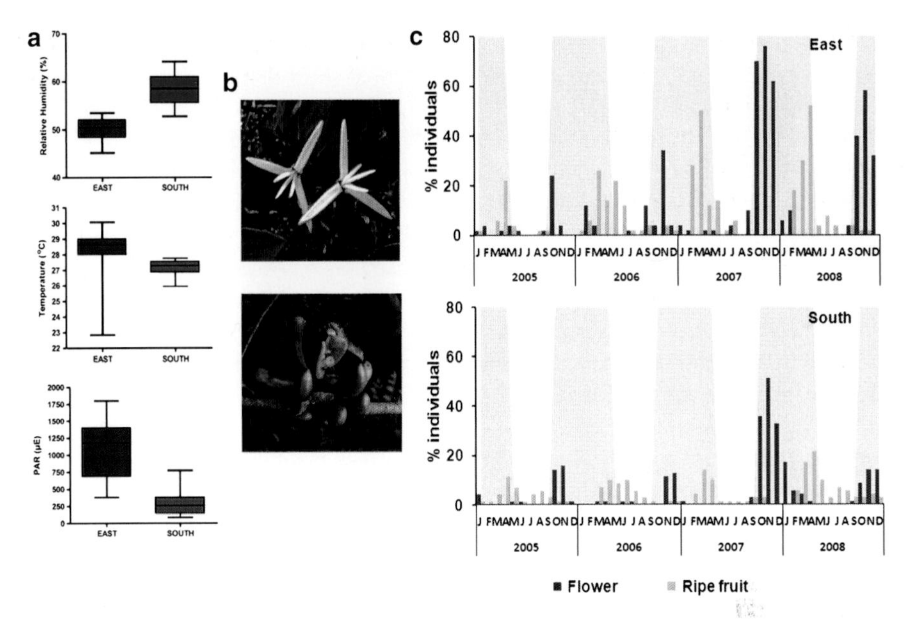

Fig. 6.5 Relationship between the reproductive phenology of *Xylopia aromatica* (Lam.) Mart. (Annonaceae) and microenvironmental conditions between the east (*light*) and south (*dark*) faces of a cerrado savanna on Southeastern Brazil (after Camargo et al. 2011). The box-plots (**a**) show the median and standard deviation of the relative humidity, temperature, and light on the east and south faces. The east face presents a significant higher light incidence and temperature and a lower relative humidity than the south face. Similar microenvironmental differences are found between savanna and forest edges and interiors. *Xylopia aromatica* (**b**) reproductive phenology (**c**) falls mainly in the rainy warm season (*shaded areas*). The flower and fruit patterns during the four reproductive seasons illustrates the consistent reduced proportion of trees flowering and fruiting (c) on the south face compared to the east face over the years. The proportion of trees contributing for a given reproductive season was two to six-folders higher on the east face, emphasizing the role of microenvironment on phenology (Photos: M.G.G. Camargo)

patterns. Community studies should be undertaken, making it possible to understand the seasonal changes in these vegetation types. For tropical forests, investigations exploring different life-forms are necessary. Long-term phenological observations are required in order to gain a better understanding of the effects of climatic change on plant phenology in South and Central America.

Recent new perspectives on phenological studies have been addressed or are of great potential in the context of fragmentation and climate changes. Studies addressing edge effects in plant phenology may offer a unique opportunity to understand how natural and human-induced environmental changes influence plant reproduction and its consequences for plant-animal interactions (Hagen et al. 2012). In this context, studies focusing on one or a few species may be more effective than whole community research (see Ramos and Santos 2005). For instance, Camargo et al. (2011) demonstrated the influence of face exposure in the phenology of *Xylopia aromatica*, an important cerrado-savanna tree (Fig. 6.5). D'Eça Neves and Morellato

(2012) also describe the edge effects in flower and fruit production of several tropical moist forest trees, while Alberti and Morellato (2008) show that even open trails may change the phenology of forest trees.

A new and innovative approach is to remotely monitor phenology using digital cameras (Richardson et al. 2007, 2009). This near-remote monitoring system have been successfully used as multichannel imaging sensors, and provide measures of leaf color changes or leafing patterns extracted from digital images in temperate forest plants (Richardson et al. 2007). A network of cameras has been established mainly in the Northern Hemisphere (http://phenocam.sr.unh.edu/webcam/). However, no near-remote information was found for the neotropical region. We know about two monitoring system: the Tropidry network (http://tropi-dry.eas.ualberta.ca/), with digital cameras in the dry forests in Brazil (Minas Gerais) and Costa Rica and one in the e-phenology project (http://www.recod.ic.unicamp.br/ephenology/) with digital cameras in Brazilian cerrado savanna (São Paulo). The main goal of the e-phenology is to monitor several vegetation types in a gradient of seasonality across Brazil, including tropical rainforest, seasonal forest, cerrados, caatinga, rupestrian fields and grasslands.

Finally, there is a growing number of scientists interested in plant phenology and its applicability. Building phenology networks is the greatest challenge for South and Central American phenologists, demanding an effort from and cooperation among universities, research institutions, governmental, and non-governmental agencies.

Acknowledgments We are grateful to Irene Mendoza for help in many ways during the literature and data survey. The authors were supported by research grants from FAPESP (Fundação de Amparo à Pesquisa do Estado de São Paulo); L.P.C.M. receives a research productivity fellowship from the Brazilian Research Council (CNPq); M.G.G.C. and E.G. received fellowships from FAPESP. We are also thankful to Linda Chambers, Alison Donnelly and Irene Mendoza for the chapter review and suggestions.

References

Alberti LF, Morellato LPC (2008) Influence of natural gaps and anthropic trails in the reproductive phenology of *Gymnanthes concolor* (Spreng.) Müll. Arg. (Euphorbiaceae). Revista Brasileira de Botânica 31(1):53–59

Alencar JC (1990) Interpretação fenológica de espécies lenhosas de campina na Reserva Biológica de Campina do Inpa ao norte de Manaus. Acta Amazonica 20:145–183

Alencar JC, Almeida RA, Fernandes NP (1979) Fenologia de espécies florestais em floresta tropical úmida de terra firme na Amazônia Central. Acta Amazonica 1:63–97

Allen PH (1956) The rainforests of Golfo Dulce. Univ. Florida Press, Gainesville

Almeida SP (1995) Phenological groups of perennial grass community on "campo-cerrado" area in the Federal District of Brazil. Pesquisa Agropecuária Brasileira 30(8):1067–1073

Alvim PT (1964) Periodicidade do crescimento das árvores em climas tropicais. In: Anais do XV Congresso Nacional de Botânica do Brasil, UFRG, Porto Alegre

Alvim PT, Alvim R (1978) Relation of climate to growth periodicity in tropical trees. In: Tomilson PB, Zimmermann MH (eds) Tropical trees as living systems. Cambridge University Press, London

Amorim IL, Sampaio EVSB, Araújo EL (2009) Phenology of woody species in the caatinga of Seridó, RN, Brazil. Revista Árvore 33(3):491–499

Andrade MAB (1967) Contribuição ao conhecimento da ecologia das plantas das dunas do litoral do Estado de São Paulo, vol 22, Boletim da Faculdade de Filosofia, Ciências e Letras da USP Série Botânica., pp 3–170

Araujo VC (1970) Fenologia de essências florestais amazônicas I, vol 4, Boletim do INPA – Série Pesquisas florestais., pp 1–25

Arroyo MTK, Armesto J, Villagran C, Uslar P (1979) High Andean plant phenology in Central Chile. Archivos De Biologia Y Medicina Experimentales 12(4):497–497

Arroyo MTK, Armesto JJ, Villagran C (1981) Plant phenological patterns in the high Andean Cordillera of Central Chile. J Ecol 69(1):205–223. doi:10.2307/2259826

Asner GP, Townsend AR, Braswell BH (2000) Satellite observation of El Nino effects on Amazon forest phenology and productivity. Geophys Res Lett 27(7):981–984

Barlow J, Gardner TA, Ferreira LV, Peres CA (2007) Litter fall and decomposition in primary, secondary and plantation forests in the Brazilian Amazon. For Ecol Manag 247(1–3):91–97. doi:10.1016/j.foreco.2007.04.017

Batalha MA, Mantovani W (2000) Reproductive phenological patterns of cerrado plant species at Pé-de-Gigante Reserve (Santa Rita do Passa Quatro, SP, Brazil): a comparison between the herbaceous and woody floras. Rev Bras Biol 60(1):129–145

Batalha MA, Aragaki S, Mantovani W (1997) Phenological variations of the cerrado species in Emas – Pirassununga, SP. Acta Botanica Brasilica 1(1):61–78

Bentos TV, Mesquita RCG, Williamson GB (2008) Reproductive phenology of Central Amazon pioneer trees. Trop Conserv Sci 1(3):186–203

Berg KS, Socola J, Angel RR (2007) Great Green Macaws and the annual cycle of their food plants in Ecuador. J Field Ornithol 78(1):1–10. doi:10.1111/j.1557-9263.2006.00080.x

Bertiller MB, Beeskow AM, Coronato F (1991) Seasonal environmental variation and plant phenology in arid Patagonia (Argentina). J Arid Environ 21(1):1–11

Camargo MG (2008) Influência da borda na frutificação e nas síndromes de dispersão de sementes em uma área de cerrado sensu stricto. UNIV Universidade Estadual Paulista, Rio Claro

Camargo MGG, Souza RM, Reys P, Morellato LPC (2011) Effects of environmental conditions associated to the cardinal orientation on the reproductive phenology of the cerrado savanna tree *Xylopia aromatica* (Annonaceae). An Acad Bras Cienc 83(3):1007–1019

Castellani TT, Caus CA, Vieira S (1999) Phenology of a foredune plant community in southern Brazil. Acta Botanica Brasilica 13(1):99–114

Cleland EE, Allen JM, Crimmins TM, Dunne JA, Pau S, Travers SE, Zavaleta ES, Wolkovich EM (2012) Phenological tracking enables positive species responses to climate change. Ecology 93 (8):1765–1771. doi:10.1890/11-1912.1

Cordazzo CV, Seeliger U (1988) Phenological and biogeographical aspects of coastal dune plant-communities in southern Brazil. Vegetatio 75(3):169–173

Costa IR, Araújo FS, Lima-Verde LW (2004) Flora and autecology's aspects of a disjunction cerrado at Araripe plateau, Northeastern Brazil. Acta Botanica Brasilica 18(4):759–770

Cotton PA (2007) Seasonal resource tracking by Amazonian hummingbirds. Ibis 149(1):135–142

Croat TB (1969) Seasonal flowering behavior in Central Panama. Ann Mo Bot Gard 56 (3):295–307

Croat TB (1975) Phenological behavior of habit and habitat classes on Barro Colorado Island (Panama Canal Zone). Biotropica 7(4):270–277

D'Eça Neves FF, Morellato LPC (2012) Efeito de borda na fenologia de árvores em floresta semidecídua de altitude na Serra do Japi. In: Vasconcellos-Neto J, Polli PR, Penteado-Dias AM (eds) Novos Olhares, Novos Saberes Sobre a Serra do Japi: Ecos de sua biodiversidade. CRV, Curitiba

Damascos MA, Prado C (2001) Leaf phenology and its associated traits in the wintergreen species Aristotelia chilensis (Mol.) Stuntz (Elaeocarpaceae). Rev Chil Hist Nat 74(4):805–815
Daubenmire R (1972) Phenology and other characteristics of tropical semi-deciduous forest in north-western Costa Rica. J Ecol 60(1):147–170. doi:10.2307/2258048
Davis DE (1945) The annual cycle of plants, mosquitoes, birds and mammals in two Brazilian forests. Ecol Monogr 15(3):245–295
Davis TAW, Richards PW (1933) The vegetation of Moraballi Creek, British Guiana: an ecological study of a limited area of tropical rain forest. Part I. J Ecol 21:350–384
Davis SD, Heywood VH, Herrera-MacBride O, Villa-Lobos J, Hamilton AC (1997) Centres of plant diversity: a guide and strategy for their conservation. The Americas, vol 3. IUCN Publications Unit, Cambridge
de Lampe MG, Bergeron Y, McNeil R, Leduc A (1992) Seasonal flowering and fruiting patterns in tropical semiarid vegetation of northeastern Venezuela. Biotropica 24(1):64–76. doi:10.2307/2388474
Develey PF, Peres CA (2000) Resource seasonality and the structure of mixed species bird flocks in a coastal Atlantic forest of southeastern Brazil. J Trop Ecol 16:33–53. doi:10.1017/s0266467400001255
Díaz S, Acosta A, Cabido M (1994) Grazing and the phenology of flowering and fruiting in a montane grassland in Argentina – a niche approach. Oikos 70(2):287–295. doi:10.2307/3545640
Engel VL, Martins FR (2005) Reproductive phenology of Atlantic forest tree species in Brazil: an eleven year study. J Trop Ecol 46(1):1–16
Fedorenko DEF, Fernandez OA, Busso CA, Elia IE (1996) Phenology of Medicago minima and Erodium cicutarium in semi- arid Argentina. J Arid Environ 33(4):409–416
Fournier LA, Salas S (1966) Algunas observaciones sobre la dinámica de la floración en el bosque tropical húmedo de Villa Colón. Rev Biol Trop 14(1):75–85
Frankie GW, Baker HG, Opler PA (1974) Comparative phenological studies of trees in tropical wet and dry forests in the lowlands of Costa Rica. J Ecol 62(3):881–919
Funch LS, Funch R, Barroso GM (2002) Phenology of gallery and montane forest in the Chapada Diamantina, Bahia, Brazil. Biotropica 34(1):40–50. doi:10.1111/j.1744-7429.2002.tb00240.x
Giorgetti HD, Manuel Z, Montenegro OA, Rodriguez GD, Busso CA (2000) Phenology of some herbaceous and woody species in central, semiarid Argentina. Phyton-Int J Exp Bot 69:91–108
Griz LMS, Machado ICS (2001) Fruiting phenology and seed dispersal syndromes in caatinga, a tropical dry forest in the northeast of Brazil. J Trop Ecol 17:303–321
Grombone-Guaratini MT, Rodrigues RR (2002) Seed bank and seed rain in a seasonal semi-deciduous forest in south-eastern Brazil. J Trop Ecol 18:759–774
Hagen M, Kissling WD, Rasmussen C, Aguiar MAM, Brown L, Carstensen DW, Santos IA, Dupont YL, Edwards FK, Genini J, Guimarães PR Jr, Jenkins GB, Jordano P, Kaiser-Bunbury CN, Ledger M, Maia KP, Marquitti FMD, Mclaughlin O, Morellato LPC, O'Gorman EJ, Trøjelsgaard K, Tylianakis JM, Vidal MM, Woodward G, Olesen JM (2012) Biodiversity, species interactions and ecological networks in a fragmented world. Adv Ecol Res 46:89–210
Haugaasen T, Peres CA (2005) Tree phenology in adjacent Amazonian flooded and unflooded forests. Biotropica 37(4):620–630. doi:10.1111/j.1744-7429.2005.00079.x
Henderson A, Fischer B, Scariot A, Pacheco MAW, Pardini R (2000) Flowering phenology of a palm community in a Central Amazon forest. Brittonia 52(2):149–159. doi:10.2307/2666506
Hilty SL (1980) Flowering and fruiting periodicity in a premontane rain-forest in pacific Colombia. Biotropica 12(4):292–306. doi:10.2307/2387701
Hutyra LR, Munger JW, Saleska SR, Gottlieb E, Daube BC, Dunn AL, Amaral DF, Camargo PB, Wofsy SC (2007) Seasonal controls on the exchange of carbon and water in an Amazonian rain forest. J Geophys Res-Biogeosci 112(G3):16. doi:G0300810.1029/2006jg000365
Janzen DH (1967) Synchronization of sexual reproduction of trees within dry season in Central America. Evolution 21(3):620–637

Jara-Guerrero A, De la Cruz M, Méndez M (2011) Seed dispersal spectrum of woody species in south Ecuadorian dry forests: environmental correlates and the effect of considering species abundance. Biotropica 43(6):722–730. doi:10.1111/j.1744-7429.2011.00754.x

Jaramillo MA, Cavelier J (1998) Fenologia de dos especies de Tillandsia (Bromeliaceae) en un bosque montano alto de la Cordillera Oriental Colombiana. Selbyana 19(1):44–51

Justiniano MJ, Fredericksen TS (2000) Phenology of tree species in Bolivian dry forests. Biotropica 32(2):276–281. doi:10.1111/j.1744-7429.2000.tb00470.x

Koptur S, Haber WA, Frankie GW, Baker HG (1988) Phenological studies of shrub and treelet species in tropical cloud forest of Costa Rica. J Trop Ecol 4:323–346

Kubitzki K, Ziburski A (1994) Seed dispersal in flood-plain forests of Amazonia. Biotropica 26 (1):30–43. doi:10.2307/2389108

Latorre F, Caccavari MA (2009) Airborne pollen patterns in Mar del Plata atmosphere (Argentina) and its relationship with meteorological conditions. Aerobiologia 25(4):297–312. doi:10.1007/s10453-009-9134-6

Liebsch D, Mikich SB (2009) Reproductive phenology of plant species of Mixed Ombrophilous Forest in Paraná, Brazil. Revista Brasileira de Botânica 32(2):375–391

Lima DA (1957) Notas para fenologia da zona da Mata de Pernambuco. Revista de Biologia Lisboa 1(2):125–135

Lima ALA, Rodal MJN (2010) Phenology and wood density of plants growing in the semi-arid region of northeastern Brazil. J Arid Environ 74(11):1363–1373. doi:10.1016/j.jaridenv.2010.05.009

Listabarth C (1999) The palms of the Surumoni area (Amazonas, Venezuela). II. Phenology and pollination of two flooded forest palms, Mauritiella aculeata and Leopoldinia pulchra. Acta Botanica Venezuelica 22(1):153–165

Löewe V, Alvear C, Salinas F (1996) Fenología de E. globulus, E. nitens y E. camaldulensis en la Zona Central de Chile: Estudio preliminar. Ciencia e Investigación Forestal 10(1):73–98

Machado ICS, Barros LM, Sampaio EVSB (1997) Phenology of caatinga species at Serra Talhada, PE, northeastern Brazil. Biotropica 29(1):57–68. doi:10.1111/j.1744-7429.1997.tb00006.x

Malizia LR (2001) Seasonal fluctuations of birds, fruits, and flowers in a subtropical forest of Argentina. Condor 103(1):45–61. doi:10.1650/0010-5422(2001)103[0045:sfobfa]2.0.co;2

Marques MCM, Roper JJ, Salvalaggio APB (2004) Phenological patterns among plant life-forms in a subtropical forest in southern Brazil. Plant Ecol 173(2):203–213

Martini A, Biondi D, Batista AC, Natal CM (2010) Phenology of native species with landscaping potential. Semina-Ciências Agrárias 31(1):75–84

McLaren KP, McDonald MA (2005) Seasonal patterns of flowering and fruiting in a dry tropical forest in Jamaica. Biotropica 37(4):584–590. doi:10.1111/j.1744-7429.2005.00075.x

Mikich SB, Silva SM (2001) Floristic and phenological study of zoochoric species of Semideciduous Seasonal Forest remnants in the mid-west region of Parana State, Brazil. Acta Botanica Brasilica 15(1):89–113

Monasterio M, Sarmiento G (1976) Phenological strategies of plant species in the tropical savanna and the semi-deciduous forest of the Venezuelan Llanos. J Biogeogr 3(4):325–355

Morellato LPC (2003) South America. In: Schwartz MD (ed) Phenology: an integrative environmental science. Kluwer Academic Publishers, Netherlands

Morellato PC, Leitao HF (1996) Reproductive phenology of climbers in a southeastern Brazilian forest. Biotropica 28(2):180–191. doi:10.2307/2389073

Morellato LPC, Leitão-Filho HF (1995) Ecologia e Preservação de Uma Floresta Tropical Urbana, Reserva de Santa Genebra, Editora da Unicamp, Campinas

Morellato LPC, Rodrigues RR, Leitão-Filho HF, Joly AC (1989) Estudo comparativo da fenologia de espécies arbóreas de floresta de altitude e floresta mesófila semidecídua na Serra do Japi, Jundiaí, São Paulo. Revista Brasileira de Botânica 12:85–98

Morellato LPC, Talora DC, Takahasi A, Bencke CC, Romera EC, Zipparro VB (2000) Phenology of Atlantic rain forest trees: a comparative study. Biotropica 32(4B):811–823. doi:10.1111/j.1744-7429.2000.tb00620.x

Morellato LPC, Camargo MGG, D'Eca Neves FF, Luize BG, Mantovani A, Hudson IL (2010) The influence of sampling method, sample size, and frequency of observations on plant phenological patterns and interpretation in tropical forest trees. In: Hudson IL, Keatley MR (eds) Phenological research: methods for environmental and climate change analysis. doi:10.1007/978-90-481-3335-2_5

Munhoz CBR, Felfili JM (2005) Phenology of the herbaceous layer in a campo sujo community in the Fazenda Água Limpa, Federal District, Brazil. Acta Botanica Brasilica 19(4):979–988

Munhoz CBR, Felfili JM (2007) Reproductive phenology of an herbaceous-subshrub layer of a Savannah (Campo Sujo) in the Cerrado Biosphere Reserve I, Brazil. Brazil J Biol 67 (2):299–307

Muniz FH (2008) Flowering and fruiting patterns of the Maranhense Amazon rainforest trees. Acta Amazonica 38(4):617–626

Murray-Smith C, Brummitt NA, Oliveira-Filho AT, Bachman S, Moat J, Lughadha EMN, Lucas EJ (2009) Plant diversity hotspots in the Atlantic coastal forests of Brazil. Conserv Biol 23 (1):151–163. doi:10.1111/j.1523-1739.2008.01075.x

Myers N, Mittermeier RA, Mittermeier CG, Fonseca GAB, Kent J (2000) Biodiversity hotspots for conservation priorities. Nature 403(6772):853–858

Myneni RB, Yang WZ, Nemani RR, Huete AR, Dickinson RE, Knyazikhin Y, Didan K, Fu R, Juarez RIN, Saatchi SS, Hashimoto H, Ichii K, Shabanov NV, Tan B, Ratana P, Privette JL, Morisette JT, Vermote EF, Roy DP, Wolfe RE, Friedl MA, Running SW, Votava P, El-Saleous N, Devadiga S, Su Y, Salomonson VV (2007) Large seasonal swings in leaf area of Amazon rainforests. Proceedings Nat Acad Sci USA 104(12):4820–4823. doi:10.1073/pnas.0611338104|ISSN 0027–8424

Nadia TL, Morellato LPC, Machado IC (2012) Reproductive phenology of a northeast Brazilian mangrove community: environmental and biotic constraints. Flora - Morphology, Distribution, Functional Ecology of Plants 207(9):682–692

Nevling LI (1971) The ecology of an elfin forest in Puerto Rico. 16. Flowering cycle and an interpretation of its seasonality. J Arnold Arboretum 52(4):586–613

Newell EA, Mulkey SS, Wright SJ (2002) Seasonal patterns of carbohydrate storage in four tropical tree species. Oecologia 131(3):333–342. doi:10.1007/s00442-002-0888-6

Norden N, Chave J, Belbenoit P, Caubère A, Châtelet P, Forget P, Thébaud C (2007) Mast fruiting is a frequent strategy in woody species of eastern South America. PLoS One 2(10):e1079

Olivares SP, Squeo FA (1999) Phenological patterns in shrubs species from coastal desert in north-central Chile. Rev Chil Hist Nat 72(3):353–370

Opler PA, Frankie GW, Baker HG (1980) Comparative phenological studies of treelet and shrub species in tropical wet and dry forests in the lowlands of Costa Rica. J Ecol 68(1):167–188. doi:10.2307/2259250

Paise G, Vieira EM (2005) Fruit production and spatial distribution of animal-dispersed angiosperms in a Mixed Ombrophilous Forest in State of Rio Grande do Sul, Brazil. Revista Brasileira de Botânica 28(3):615–625

Parolin P (2000) Phenology and CO2-assimilation of trees in Central Amazonian floodplains. J Trop Ecol 16:465–473

Peñuelas J, Rutishauser T, Filella I (2009) Phenology feedbacks on climate change. Science 324 (5929):887–888. doi:10.1126/science.1173004

Peres CA (1994a) Composition, density, and fruiting phenology of arborescent palms in an Amazonian terra-firme forest. Biotropica 26(3):285–294. doi:10.2307/2388849

Peres CA (1994b) Primate responses to phenological changes in an Amazonian terra-firme forest. Biotropica 26(1):98–112. doi:10.2307/2389114

Pinto AM, Morellato LPC, Barbosa AP (2008) Reproductive phenology of *Dipteryx odorata* (Aubl.) Willd (Fabaceae) in two forest areas in the Central Amazon. Acta Amazonica 38 (4):643–650

Putz FE, Romano GB, Holbrook NM (1995) Comparative phenology of epiphytic and tree-phase strangler figs in a Venezuelan palm savanna. Biotropica 27(2):183–189. doi:10.2307/2388994

Ragusa-Netto J, Silva RR (2007) Canopy phenology of a dry forest in western Brazil. Brazil J Biol 67(3):569–575

Ramia M (1977) Phenological observations on savannas of Middle Apure region (Venezuela). Acta Botanica Venezuelica 12(1/4):171–206

Ramia M (1978) Phenological pbservations in the savannas of Alto Apure region (Venezuela). Boletín de la Sociedad Venezolana de Ciencias Naturales 33:149–198

Ramírez N (2002) Reproductive phenology, life-forms and habitats of the Venezuelan Central Plain. Am J Bot 89(5):836–842. doi:10.3732/ajb.89.5.836

Ramírez N (2009) Correlations between the reproductive phenology of vegetation and climate variables in the Venezuelan Central Plain. Acta Botanica Venezuelica 32(2):333–362

Ramírez N, Briceño H (2011) Reproductive phenology of 233 species from four herbaceous–shrubby communities in the Gran Sabana Plateau of Venezuela. AoB Plants 2011:plr014. doi:010.1093/aobpla/plr1014. doi:10.1093/aobpla/plr014

Ramos FN, Santos FAM (2005) Phenology of Psychotria tenuinervis (Rubiaceae) in Atlantic forest fragments: fragment and habitat scales. Canadian Journal of Botany-Revue Canadienne De Botanique 83(10):1305–1316

Ramírez N, Brito Y (1987) Patterns of flowering and frutification in a swampy community, “Morichal” type (Calabozo, Edo Guarico, Venezuela). Acta Cient Venez 38(3):376–381

Regalado A (2010) Brazilian science: riding a gusher. Science 330(6009):1306–1312. doi:10.1126/science.330.6009.1306

Ribeiro LF, Tabarelli M (2002) A structural gradient in cerrado vegetation of Brazil: changes in woody plant density, species richness, life history and plant composition. J Trop Ecol 18:775–794

Richardson AD, Jenkins JP, Braswell BH, Hollinger DY, Ollinger SV, Smith ML (2007) Use of digital webcam images to track spring green-up in a deciduous broadleaf forest. Oecologia 152:323–334

Richardson AD, Braswell BH, Hollinger DY, Jenkins JP, Ollinger SV (2009) Near-surface remote sensing of spatial and temporal variation in canopy phenology. Ecol Appl 19:1417–1428

Riveros M (1983) Andean plant phenology, Volcan Casablanca, 40-degrees-S, X- region, Chile. Archivos De Biologia Y Medicina Experimentales 16(2):R181–R181

Riveros M, Palma B, Erazo S, Oreilly S (1995) Phenology and pollination in species of the genus Nothofagus. Phyton-Int J Exp Bot 57(1):45–54

Rosello NE, Belmonte SE (1999) Fenologia de Browningia candelaris (Meyen) Britt. et Rose en la Quebrada de Cardones, Norte de Chile. Idesia 17:47–55

Rossi BE, Debandi GO, Peralta IE, Palle EM (1999) Comparative phenology and floral patterns in *Larrea* species (Zygophyllaceae) in the Monte desert (Mendoza, Argentina). J Arid Environ 43 (3):213–226. doi:10.1006/jare.1999.0525

Rozzi R, Molina JD, Miranda P (1989) Microclimate and flowering periods on equatorial and polar- facing slopes in the Central Chilean Andes. Rev Chil Hist Nat 62(1):75–84

Rubim P, Nascimento HEM, Morellato LPC (2010) Interannual variation in the phenology of a tree community in a semideciduous seasonal forest in southeast Brazil. Acta Botanica Brasilica 24(3):756–764

Ruiz A, Santos M, Cavelier J, Soriano PJ (2000) Phenological study of Cactaceae in the dry enclave of Tatacoa, Colombia. Biotropica 32(3):397–407. doi:10.1111/j.1744-7429.2000.tb00486.x

Rusch VE (1993) Altitudinal variation in the phenology of *Nothofagus pumilio* in Argentina. Rev Chil Hist Nat 66(2):131–141

Sabatier D (1985) Fruiting periodicity and its determinants in a lowland rain-forest of French-Guiana. Revue D Ecologie-La Terre Et La Vie 40(3):289–320

Sabatier D, Puig H (1986) Phénologie et saisonnalité de la floraison et de la fructification en forêt dense guyanaise. Mémoires du Muséum Nacional d’Histoire Naturelle Série A: Zoologie. 132:173–184

Salazar A, Goldstein G, Franco AC, Miralles-Wilhelm F (2012) Seed limitation of woody plants in neotropical savannas. Plant Ecol 213(2):273–287. doi:10.1007/s11258-011-9973-4
Santos N (1979) Fenologia. Rodriguésia 31(50):223–226
Schöngart J, Piedade MTF, Ludwigshausen S, Horna V, Worbes M (2002) Phenology and stem-growth periodicity of tree species in Amazonian floodplain forests. J Trop Ecol 18:581–597. doi:10.1017/s0266467402002389
Seres A, Ramírez N (1993) Flowering and fructification of monocotyledons in a Venezuelan cloud forest. Rev Biol Trop 41(1):27–36
Silberbauer-Gottsberger I (2001) A hectare of cerrado. II. Flowering and fruiting of thick-stemmed woody species. Phyton-Annales Rei Botanicae 41(1):129–158
Silva JF, Ataroff M (1985) Phenology, seed crop and germination of coexisting grass species from a tropical savanna in Western Venezuela. Acta Oecologica-Oecologia Plantarum 6(1):41–51
Silveira FR (1935) Queda de folhas. Rodriguesia 1:1–6
Sist P (1989) Structure and phenology of the palm community of a French Guiana rain forest (Piste de St-Elie). Revue D Ecologie-La Terre Et La Vie 44(2):113–151
Smith-Ramírez C (1993) Hummingbirds and their floral resources in temperate forests of Chiloe Island, Chile. Rev Chil Hist Nat 66(1):65–73
Smith-Ramírez C, Armesto JJ (1994) Flowering and fruiting patterns in the temperate rain-forest of Chiloe, Chile – ecologies and climatic constraints. J Ecol 82(2):353–365. doi:10.2307/2261303
Smith-Ramirez C, Armesto JJ, Figueroa J (1998) Flowering, fruiting and seed germination in Chilean rain forest myrtaceae: ecological and phylogenetic constraints. Plant Ecol 136 (2):119–131
Snow DW (1966) A possible selective factor in evolution of fruiting seasons in tropical forest. Oikos 15(2):274–281
Spina AP, Ferreira WM, Leitão-Filho HF (2001) Flowering, fruiting and dispersal syndromes of a wet forest community. Acta Botanica Brasilica 15(3):349–368
Staggemeier VG, Morellato LPC (2011) Reproductive phenology of coastal plain Atlantic forest vegetation: comparisons from seashore to foothills. Int J Biometeorol 55(6):843–854. doi:10.1007/s00484-011-0482-x
Staggemeier VG, Diniz JAF, Morellato LPC (2010) The shared influence of phylogeny and ecology on the reproductive patterns of Myrteae (Myrtaceae). J Ecol 98(6):1409–1421. doi:10.1111/j.1365-2745.2010.01717.x
Stevenson PR, Quinones MJ, Ahumada JA (1998) Annual variation in fruiting pattern using two different methods in a lowland tropical forest, Tinigua National Park, Colombia. Biotropica 30 (1):129–134. doi:10.1111/j.1744-7429.1998.tb00376.x
Stiles FG (1977) Coadapted competitors: the flowering seasons of hummingbird-pollinated plants in a tropical forest. Science 198(4322):1177–1178
Stranghetti V, Ranga NT (1997) Phenological aspects of flowering and fruiting at the Ecological Station of Paulo de Faria-SP-Brazil. J Trop Ecol 38(2):323–327
Talora DC, Morellato PC (2000) Phenology of coastal-plain forest tree species from Southeastern Brazil. Revista Brasileira de Botânica 23(1):13–26
Tannus JLS, Assis MA, Morellato LPC (2006) Reproductive phenology in dry and wet grassland in an area of Cerrado at southeastern Brazil, Itirapina – SP. Biota Neotropica 6(3)
Ter Steege H, Persaud CA (1991) The phenology of Guyanese timber species – a compilation of a century of observations. Vegetatio 95(2):177–198
Velez V, Cavelier J, Devia B (1998) Ecological traits of the tropical treeline species *Polylepis quadrijuga* (Rosaceae) in the Andes of Colombia. J Trop Ecol 14:771–787. doi:10.1017/s026646749800056x
Veloso HP (1945) As comunidades e as estações botânicas de Teresópolis, ERJ. Boletim do Museu Nacional do Rio de Janeiro 3:3–95
Vidiella PE, Armesto JJ, Gutierrez JR (1999) Vegetation changes and sequential flowering after rain in the southern Atacama desert. J Arid Environ 43(4):449–458

Wallace RB, Painter RLE (2002) Phenological patterns in a southern Amazonian tropical forest: implications for sustainable management. Forest Ecol Manag 160(1–3):19–33. doi:10.1016/s0378-1127(00)00723-4

Warming E (1892) Lagoa Santa, contribuição para a geografia fitobiológica. In: Warming E, Ferri MG (eds) Lagoa Santa e a vegetação dos cerrados brasileiros. EDUSP, São Paulo.

Wheelwright NT (1985) Competition for dispersers and the timing of flowering and fruiting in a guild of tropical trees. Oikos 44(3):465–477. doi:10.2307/3565788

Wright SJ, Calderón O (2006) Seasonal, El Niño and longer term changes in flower and seed production in a moist tropical forest. Ecol Lett 9(1):35–44. doi:10.1111/j.1461-0248.2005.00851.x

Wright SJ, Carrasco C, Calderon O, Paton S (1999) The El Niño Southern Oscillation variable fruit production, and famine in a tropical forest. Ecology 80(5):1632–1647. doi:10.1890/0012-9658(1999)080[1632:tenoso]2.0.co;2

Zhang SY, Wang LX (1995) Comparison of three fruit census methods in French Guiana. J Trop Ecol 11:281–294

Zimmerman JK, Wright SJ, Calderón O, Pagan MA, Paton S (2007) Flowering and fruiting phenologies of seasonal and aseasonal neotropical forests: the role of annual changes in irradiance. J Trop Ecol 23(02):231–251. doi:10.1017/S0266467406003890

Chapter 7
Antarctica

Lynda E. Chambers, Marie R. Keatley, Eric J. Woehler, and Dana M. Bergstrom

Abstract Antarctica was the last continent to be discovered and colonized by people, and this has resulted in generally sparse meteorological, oceanographic and biological data for the Antarctic and much of the Southern Ocean. Within the Antarctic region, here defined to include all regions south of the Antarctic Polar Front, much of the land-based biological research occurs at or near international scientific stations, leading to some regions, such as the Amundsen Sea, being poorly researched. In the last decade, evidence has emerged of significant differences, but also some similarities, in species' responses to changing environmental conditions, including climate change. However, most of the studies have been confined to larger organisms, such as seabirds and marine mammals, with few long-term studies on the phenology of plants, invertebrates and other species. This highlights the need for greater spatial and species coverage in the southern regions of the globe to assess and quantify regional and ecosystem-scale processes and patterns.

L.E. Chambers (✉)
Centre for Australian Weather and Climate Research,
Bureau of Meteorology, Melbourne, VIC, Australia
e-mail: l.chambers@bom.gov.au

M.R. Keatley
Department of Forest and Ecosystem Science,
University of Melbourne, Creswick, VIC, Australia

E.J. Woehler
Institute of Marine and Antarctic Studies, University of Tasmania, Sandy Bay, TAS, Australia

D.M. Bergstrom
Australian Antarctic Division, Kingston, TAS, Australia

M.D. Schwartz (ed.), *Phenology: An Integrative Environmental Science*,
DOI 10.1007/978-94-007-6925-0_7, © Springer Science+Business Media B.V. 2013

7.1 Introduction

Antarctica is defined within this chapter as including all regions within the biological Antarctic boundary, which extends to the oceanic Antarctic Polar Front (formerly known as the Antarctic Convergence) and includes the islands of the Southern Ocean that lie on or close to the Front, and covers ~34.8 million km^2 (Young 1991; Griffiths 2010, Table 7.1). This biological boundary is not static, shifting with time and longitude, with an average position around 58°S, but extending as far north as 48°S.

The coldest continent on Earth, Antarctica is isolated from more northerly landmasses due to the perpetually circulating Antarctic Circumpolar Current and associated strong westerly wind field. Seasonality is extreme, with the thermal energy and light from the sun absent during winter at higher latitudes. Antarctic sea ice is also highly seasonal, more so than in the Arctic (Cattle 1991), and it is this variability that is an important climatic feature of the Southern Hemisphere. During winter, sea ice can reach the Polar Front, essentially doubling the continental surface area, but in summer only covers parts of the continental shelf (DASET 1992). The growth and decay of the pack ice, in conjunction with latitude and water mass movements, influence productivity of the system (Young 1991; Massom and Stammerjohn 2010) with the most productive regions of the Southern Ocean generally occurring within the sea ice zone, also known as the marginal ice zone (Griffiths 2010).

According to Young (1991, and others listed within) the Antarctic region can be divided into three biogeographical zones (Table 7.1). However, ecosystems on individual islands, and potentially their phenologies, may vary depending on the extent of sea ice or open water surrounding the islands during the summer period.

The northeast and western parts of the Antarctic Peninsula, and the surrounding seas, have seen significant warming trends in recent decades (Turner et al. 2005, 2007; Griffiths 2010; Trathan and Agnew 2010). Rapid changes in the marine environment include reduced sea ice formation and a shortening of the sea ice season in the Amundsen and Bellingshausen Seas, increased sea ice extent in the Ross Sea region, and altered coastal habitats associated with the collapse of ice shelves (Griffiths 2010; Trathan and Agnew 2010). Warming oceans have also been associated with a decrease in Antarctic Krill, an important food source for several species of fish, seabirds and marine mammals (Trathan and Agnew 2010). In terrestrial systems, changes have been observed in vascular plant populations, but are generally restricted to local population expansion (Convey et al. 2011).

Most coastal stations around Antarctica have recorded an increase in surface wind speeds over the last 50 years (Turner et al. 2005, 2007). At Casey, East Antarctica, increased winds have resulted in lower growth rates in mosses due to reduced water availability at some sites (Clarke et al. 2012).

Table 7.1 Antarctic bio-geographical zones

Zone	Maximum mean monthly temp	Comprises	Vegetation characteristics	Other biological characteristics
Cool (sea-ice free zone)	3–7 °C	Subantarctic islands close to or immediately south of the Polar Front, i.e. South Georgia, Marion and Prince Edward Is., Crozet Archipelago, Kerguelen Archipelago, Heard and McDonald Islands and Macquarie Island	Polar deserts and open fellfields at high altitudes developing to closed herbfield, mire and grasslands at lower altitudes. Dominated by bryophytes, graminoids, herbs and megaherbs	Terrestrial systems: high in marine-derived nutrients from seals and seabirds, high primary productivity at low altitudes. Low rates of decomposition Marine systems: nutrient concentrations variable reflecting complex oceanic regime. Influences from intrusions of Antarctic and subantarctic surface waters, and local off island enrichment from terrestrial (ex-marine) runoff. Antarctic Krill (*Euphausia superba*) is virtually absent Islands with permanent ice areas are exhibiting retreat of glaciers and decreases in the extents of permanent ice areas, opening areas for plant and animal colonization. In some cases, these colonisation events are facilitating animal population increases (e.g., fur seals and penguins)
Cold (seasonal sea-ice zone)	0–2 °C	Antarctic Peninsula to about 70°S, South Sandwich, South Shetland and South Orkney Island groups	Closed stands of bryophytes, lichens and algae extensive only in wet or mesic areas. Only two flowering species present.	Terrestrial systems: Low productivity, stark seasonality Marine systems: zone of highest productivity moves north and south with sea-ice migration; important zone for phytoplankton blooms; complex food web with krill at the centre
Extreme (permanent packice zone)	<0 °C	Continental Antarctica and islands close to continent, such as Peter I Is, Scott Is, Balleny Islands	Vegetation largely restricted to scattered colonies of mosses, lichens or algae and endolithic communities. Vascular plants absent, liverworts very rare	Terrestrial systems: low productivity, stark seasonality, low summer temperatures Marine systems: intense short period of high productivity for most species Strong seasonal pulse of biological activity for ~6 months (October–March inclusive) with high marine productivity manifested by bird and mammal breeding. Only vertebrate species that breeds on the continent in winter is the Emperor Penguin *Aptenodytes forsteri*

Based on Young (1991) and DASET (1992)

7.2 Agency Data and Research

Due to its extreme climate and isolation, Antarctica was the last continent to be discovered and colonized by people, though contemporary human residents are transitory. This has resulted in generally sparse meteorological, oceanographic and biological data for the Antarctic and much of the Southern Ocean (but see Laws 1984; Cattle 1991; El-Sayed 1994; and Knox 1994 for reviews).

Nations claiming territories in the Antarctic are: Argentina, Australia, Chile, France, New Zealand, Norway and the United Kingdom. Australia has the largest claim, its two sectors covering approximately 42 % of the continent. The Antarctic Treaty, enacted in 1961, is concerned with the governance of Antarctica and involves two tiers of membership: (i) nations which conduct significant science programs in the region, and (ii) those who subscribe to the provisions of the Treaty but are yet to establish scientific programs (DASET 1992).

Much of the land-based biological research in the Antarctic occurs at or near the scientific stations (37 of 64 are open year-round) (Griffiths 2010). Due to the locations of these stations, and their proximity to their home countries, this has meant that Australian, New Zealand and Asian (Japan, China, South Korea) research is concentrated in East Antarctica while European and South American research is conducted mainly in West Antarctica, the Scotia Sea and Weddell Sea areas. The United States and Russia conduct research in both regions (Griffiths 2010). As a result, the regions of Antarctica with no neighboring continent, such as the Amundsen Sea, have been and continue to be poorly researched. A summary of continuous long-term (commencing pre-1970) biological research programs is given below. Many other nations operate, or participate, in short-term biological monitoring programs in the Antarctic (http://www.scar.org), but these are not discussed here. Further details of logistic, operational and research efforts for all stations and Treaty partners are available at http://www.ats.aq/devAS/ie_annual.aspx?lang=e and http://www.ats.aq/devAS/ie_permanent.aspx?lang=e.

7.2.1 *Scientific Committee on Antarctic Research*

The Scientific Committee on Antarctic Research (SCAR) is composed of members from national scientific academies or research councils interested in Antarctic research, and the relevant scientific unions of the International Council for Science. Full membership is reserved for countries (31 as of 2011) with active Antarctic research programs (http://www.scar.org). SCAR's mission is to provide science leadership and to provide independent scientific advice to the Antarctic Treaty System. The SCAR scientific research program of most relevance to phenology is "*Evolution and Biodiversity in the Antarctic*", which includes the study of the 'response of life to change'.

7.2.2 Argentine Antarctic Institute

Argentina has six permanent research stations and three summer-only stations in the Antarctic, including Orcadas Base in the South Orkney Islands, which opened in 1904, making it the earliest Antarctic base (Griffiths 2010).

The Argentine Antarctic Institute (Instituto Antártico Argentino) was created in 1951. The Institute conducts a number of long-term biological studies including population dynamics and reproductive biology of seal and seabird species (http://www.scar.org/about/nationalreports/argentina/99to00/).

7.2.3 Australian Antarctic Program

The Australian Antarctic Program is a joint agency and university program, run from the Australian Antarctic Division (AAD: http://www.antarctica.gov.au) within the Federal Department of Sustainability, Environment, Water, Population and Communities. The basis of the science program is a 10 year Science Strategic Plan (http://www.antarctica.gov.au/science/australian-antarctic-science-strategic-plan-201112-202021).

The AAD is the lead Antarctic scientific agency in Australia and also administers the Australian Antarctic Territory (AAT) and Heard and McDonald Islands. Australia has three coastal Antarctic continent stations in East Antarctica (Mawson, Davis and Casey) and one subantarctic station on Macquarie Island. Visits to Heard Island are intermittent and at decadal scales.

Two of the four current priority research programs are relevant to phenology: *Southern Ocean Ecosystems: Environmental Change and Conservation* and *Terrestrial and Near Shore Ecosystems: Environmental Change and Conservation*. Long-term Adélie penguin (*Pygoscelis adeliae*) population and phenological studies occur near Mawson Station (Béchervaise Island) as a contribution to the Commission for the Conservation of Antarctic Marine Living Resources Ecosystem Monitoring Program (CCAMLR-CEMP). A number of other long-term population studies were undertaken in the AAT, some extending from the 1950s to the early 2000s, in addition to on-going long-term glaciological and meteorological research. Long-term albatross and fur seal population and phenological studies occur on Macquarie Island with intermittent penguin studies. No long-term terrestrial phenological studies are in progress at the time of writing.

7.2.4 British Antarctic Survey

The British Antarctic Survey (BAS, http://www.antarctica.ac.uk) has been responsible for the majority of the United Kingdom's Antarctic research for more than 60 years.

BAS has five research stations in the Antarctic (summer-only since 1996 at Signy, South Orkney Islands), two on South Georgia: Bird Island (continuously since 1982) and King Edward Point and Halley and Rothera, Antarctica.

Phenology fits within one of BAS's key challenges, "*understanding how Southern Ocean species respond to climate change*". Several long-term monitoring programs have been established and contribute to CCAMLR-CEMP, including seabird and seal research at Bird Island and penguin, seabird and seal biology at Signy research station. Signy research station also hosts long-term collaborative studies with the Japanese National Institute for Polar Research and the Netherlands Polar Programme.

7.2.5 Chilean Antarctic Institute

The Chilean Antarctic Institute (INACH, http://www.inach.cl) was established in 1963. It operates scientific stations on King George Island, Greenwich Island and Cape Shirreff (South Shetland Islands) and on the Palmer and Antarctic Peninsulas. Long-term monitoring programs investigate Antarctic flora, benthic communities, fur seals and seabirds. Recent initiatives include research on the impacts of regional climate change on marine and terrestrial systems.

7.2.6 French Polar Institute

France has had research bases in the Antarctic since 1950. The French Polar Institute Paul-Emile Victor is responsible for the implementation of research programs in polar and subpolar regions in both hemispheres.

The territory of the French Southern and Antarctic Lands (Terres Australes et Antarctiques Françaises) administers several island groups in the southern Indian Ocean and part of Antarctica (Adélie Land). The subantarctic territories include the Crozet and Kerguelen Archipelagos, St. Paul and Amsterdam Islands.

There are two continental stations (Dumont d'Urville and Concordia, which is managed in conjunction with Italy) and three subantarctic stations: Alfred Faure on the Crozet Archipelago, Port aux Français on the Kerguelen Archipelago and Martin-de-Viviès on Amsterdam Island.

Scientific research of relevance to phenology includes the project, "*penguins as indicators of climate anomalies in the Southern Ocean*". One of the longest phenological Antarctic and subantarctic projects (extending for more than 50 years), this study monitors the first arrival and first egg laying dates in Adélie Land, Crozet and Kerguelen Archipelagos and on Amsterdam Island (Table 7.2).

Table 7.2 Published long-term (at least 10 years) phenological studies from the Antarctic many of these studies are on-going

Location	Species	Phenological variable	Time period	Summary	References
Dumont d'Urville Station, Adélie Land, East Antarctica	Emperor Penguin *Aptenodytes forsteri*	First arrival date	1950–2004	No trend over time or relationship with sea ice extent	Barbraud and Weimerskirch (2006)
	Adélie Penguin *Pygoscelis adeliae*				
	Southern Giant Petrel *Macronectes giganteus*				
	Snow Petrel *Pagodroma nivea*		1970–2004		
	Southern Fulmar *Fulmarus glacialoides*	First arrival date	1950–2004	Later arrival over time; negative relationship with sea ice extent	Barbraud and Weimerskirch (2006)
	Cape Petrel *Daption capense*				
	South Polar Skua *Stercorarius maccormicki*				
	Wilson's Storm Petrel *Oceanites oceanicus*		1960–2004		
	Antarctic Petrel *Thalassoica antarctica*	First arrival date	1980–2004	Later arrival over time; no relationship with sea ice extent	Barbraud and Weimerskirch (2006)
	Adélie Penguin	Laying date: first eggs	1950–2004	Later laying over time; negative relationship with sea ice extent	Barbraud and Weimerskirch (2006)
	Cape Petrel				
	Emperor Penguin	Laying date: first eggs	1950–2004	No trend over time; negative relationship with sea ice extent	Barbraud and Weimerskirch (2006)
	Snow Petrel	Laying date: first eggs	1950–2004	No trend over time or relationship with sea ice extent	Barbraud and Weimerskirch (2006)
	South Polar Skua	Laying date: first eggs	1960–2004	Earlier laying over time; no relationship with sea ice extent	Barbraud and Weimerskirch (2006)

(continued)

Table 7.2 (continued)

Location	Species	Phenological variable	Time period	Summary	References
Béchervaise Island, East Antarctica	Adélie Penguin	Arrival date: first and average	1995–2008	No trend over time; later arrival when near-shore ice extensive and maximum temperatures lower	Emmerson et al. (2011)
	Adélie Penguin	Laying date	1990–2003	No trend over time; earlier laying when Southern Angular Mode positive and later when winds more southerly	Emmerson et al. (2011)
	Adélie Penguin	Hatching date	1990–2005	No trend over time	Emmerson et al. (2011)
	Adélie Penguin	Créching date	1992–2006	No trend over time	Emmerson et al. (2011)
	Adélie Penguin	Fledging date	1991–2004	No trend over time	Emmerson et al. (2011)
Casey, East Antarctica	Snow Petrel	Breeding success (7 years of laying and hatching dates)	1984–2003	Breeding success linked to sea ice extent, no relationship with Southern Oscillation Index (SOI) or Antarctic Circumpolar Wave	Olivier et al. (2005)
Cape Crozier, Ross Island, Ross Sea, Antarctica	Adélie Penguin	Dates of first arrival and first eggs	1962–1975	No long-term trend, inter-annual variability from sea-ice extent	Ainley et al. (1983)
Admiralty Bay, King George Island	Adélie Penguin Gentoo Penguin *Pygoscelis papua*	Laying date	1991–2006	Earlier egg laying when mean October temperatures warmer; overall trend in egg laying shared by all 3 species 0.15 days year^{-1} earlier	Lynch et al. (2009); Lynch et al. (2012a)
Cape Shirreff, Livingston Island	Chinstrap Penguin *Pygoscelis antarcticus*	Laying date	1997–2006	Earlier egg laying when mean October temperatures warmer; overall trend in egg laying shared by all 3 species 0.15 days year^{-1} earlier	Lynch et al. (2009); Lynch et al. (2012a)

Possession Island, Crozet Archipelago	King Penguin *Aptenodytes patagonicus*	Arrival date	1998–2008	Did not measure trend over time; difference in timing between banded and non-banded birds; banded birds arrived later at colonies	Saraux et al. (2001), Gauthier-Clerc et al. (2004), Le Bohec et al. (2008)
	King Penguin	Laying date	1998–2008	Did not measure trend over time; difference in timing between banded and non-banded birds	Saraux et al. (2001)
Marion Island	Macaroni Penguin *Eudyptes chrysolophus* Rockhopper Penguin *Eudyptes chrysocome*	Arrival date	1994–2005	Did not measure trend over time; mean and range vary by gender	Crawford et al. (2006)
	Macaroni Penguin Rockhopper Penguin	Fledging date	1994–2005	Did not measure trend over time	Crawford et al. (2006)
Macquarie Island	Royal Penguin *Eudyptes schlegeli*	Laying date	1960s (6 years), 1990s (5 years)	Earlier laying over time (-0.108 days^{-1}); lay later in years of high SOI (lower productivity)	McMahon and Hindell (2009)
	Southern Elephant Seal *Mirounga leonina*	Peak haul out date	1950–1959	Did not measure trend over time; mean and range according to age and breeding status	Hindell and Burton (1988)
	Leopard Seal *Hydrurga leptonyx*	First and last sighting dates	1949–1979	Irruption of seals onshore, cyclic haul out numbers. Cycles considered to be related to sea-ice proximity	Rounsevell and Eberhard (1980), Rounsevell (1988)

(continued)

Table 7.2 (continued)

Location	Species	Phenological variable	Time period	Summary	References
Bird Island, South Georgia	Antarctic Fur Seal *Arctocephalus gazella*	Birthing date	1984–2004	Later timing with extreme positive SST anomalies	Forcada et al. (2005)
	Gentoo Penguin	Laying date: first egg	Not provided	First egg ~10 days earlier than 18 years ago; advance in date when 75 % of eggs laid	Trathan and Agnew (2010)

7.2.7 New Zealand

Antarctica New Zealand is responsible for New Zealand government activities in the Southern Ocean and Antarctica. New Zealand manages the research station Scott Base, in Victoria Land, and has an operational presence in the Ross Dependency (http://www.antarcticanz.govt.nz/).

Research of relevance to phenology includes a long-term project investigating the population dynamics of the Adélie Penguin of the Ross Sea and its use as a biological indicator of local and broad-scale changes.

The Department of Conservation has long-term population and phenological studies established for New Zealand Sea Lions (*Phocarctos hookeri*) on Enderby Island. Long-term population and demographic studies of albatrosses at Campbell Island commenced in the 1940s (Waugh et al. 1999).

7.2.8 Norwegian Polar Institute

Norway operates a year round base in Dronning Maud Land and has conducted regular expeditions to the Antarctic since 1976 (http://www.npolar.no). Biological and climate-change related studies are conducted at their field station, Tor, and opportunistic studies on the island of Bouvetøya in the South Atlantic Ocean.

7.2.9 Russian Antarctic Expedition

Russia has had a research presence in the Antarctic since 1956 when Mirnyy station was established. The Russian Antarctic Expedition (http://www.aari.aq) currently maintains four other active stations: Vostok, Novolazarevskaya, Bellingshausen and Progress (summer only). Long-term monitoring of seabird populations has been undertaken at Mirnyy since the 1950s (Barbraud et al. 2011).

7.2.10 South African Antarctic Programme

The South African National Antarctic Programme (http://www.sanap.org.za/) maintains bases on Antarctica and on Marion and Gough Islands. The first Antarctic Expedition in 1959 established a permanent South African presence on Antarctica. Research themes include the response of biodiversity to earth system variability, and programs of phenological relevance include seabirds at Marion Island and individual variation in albatross reproduction.

7.2.11 United States Antarctic Program

The United States has had a research presence (including 3 year-round research stations) in Antarctica since 1956, with an aim to understand the region's ecosystems and its responses to global processes, including climate (http://www.nsf.gov/od/opp/antarct/usap.jsp).

Long-term biological studies are undertaken at Palmer Station on Anvers Island (Antarctic Peninsula), Admiralty Bay and King George Island, in the South Shetland Islands. This includes the breeding biology and demography of the Pygoscelid penguins. Long-term studies of Adélie Penguins are also conducted in the Ross Sea region on Beaufort and Ross Islands (Cape Crozier, Cape Bird and Cape Royds), and a long-term population study of Weddell Seals (*Leptonychotes weddellii*) in Erebus Bay, Ross Island.

7.2.12 National Institute of Polar Research (Japan)

Japan has had a research presence in the Antarctic since the 1950s, with Syowa Station, East Antarctica, opening in 1957 (http://www.nipr.ac.jp). Long-term biological studies are undertaken at Syowa Station and, until 1991, at Asuka Station. Since 1997, there has been an emphasis on monitoring for biological change.

7.3 Recent Phenological Research

Although low in number compared to many Northern Hemisphere regions, a number of long-term (at least 10 years of data) studies have been conducted in the Antarctic (Fig. 7.1, Table 7.2). The most notable being the 50+ year study of all seabird species in the East Antarctic (Barbraud and Weimerskirch 2006).

Most long-term phenological research has been conducted on birds (84 % of 45 data series); the balance has been conducted on mammals (16 %). Studies of breeding and migration phenology were equally represented.

7.3.1 Plants

Limited phenological studies have been conducted on terrestrial plants in the Antarctic and phenology (spelling was incorrect on line 221) on the subantarctic islands and most are short-term studies. Bryophyte and lichen phenology in particular, is significantly neglected as a topic of study. Many mosses and some lichens do not produce reproductive structures such as capsules or apothecia in the Antarctic or

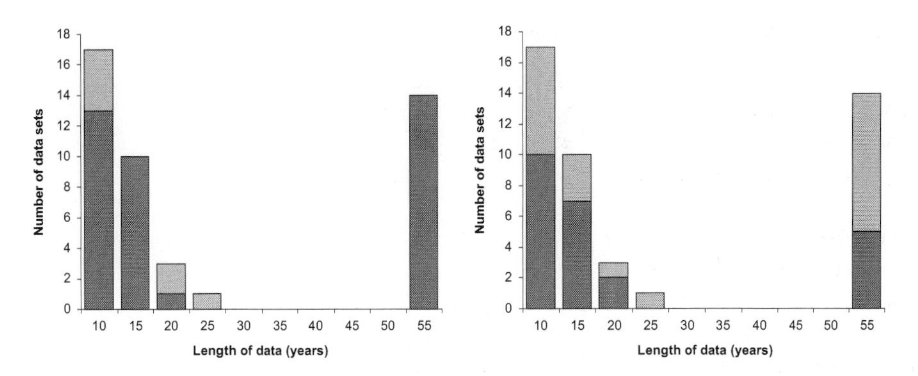

Fig. 7.1 Typical duration of long-term phenological data sets in the Antarctic. Broken down into (*left*) bird data sets (*dark grey*, n = 38) and mammal data sets (*light grey*, n = 7); (*right*) according to breeding data (*dark grey*, n = 24) and migration data (*light grey*, n = 21)

subantarctic (see Seppelt 2004; Olech 2004; Ochyra et al. 2008). In terrestrial areas of the extreme zone (continental Antarctica), only one third of the bryophyte flora (5 of 15) species have been observed fruiting. In the southern maritime Antarctic (68–72°S), this increases to 47 % at Marguerite Bay, 51 % at the South Shetlands, but decreases again on subantarctic islands (37 % South Georgia, Smith and Convey 2002; 33 % Macquarie Island [mosses data only], Bergstrom and Selkirk 1987). Reasons for failure to sexually reproduce include disjunct patterns in male and female plants, lack of one gender at a locality and failure for gametangial initiation (Smith and Convey 2002).

Phenology in flowering plants in the region is cued to seasonality in the light regime. Two species, the small cushion plant Antarctic Pearlwort *Colobanthus quitensis* and the grass Antarctic hair grass *Deschampsia antarctica* can be found along a latitudinal gradient from South Georgia extending to the cold Antarctic zone along the Antarctic Peninsula, to as far south as Alexander Island (69°22′0S, 71°50′7W). They have both been noted to flower most years and produce mature seed along their entire latitudinal range (Holtom and Greene 1967; Convey et al. 2011). Flower initiation for the species is in early to late summer and both species can produce seed within a single year but, with increased (southward) latitude, rates of development were observed to become lower and less predictable. Flower primordia for *D. antarctica* develop late in the previous summer season (Holtom and Greene 1967; Edwards 1974). In many years, seeds failed to reach maturity. With climate warming, however, seed maturation appears to be increasing (Convey 1996; Convey and Smith 2006).

In the subantarctic, flowering onset generally occurs in spring and is triggered by day length, but variation in the timing of stage development (e.g., flower bud opening, fruit ripening, seed shedding) can be found across the floras and is often associated with life history strategies. Small herbs generally have rapid development and some long-lived perennial stayers often do not complete a sexual reproductive cycle in a single year, with both preformation of flower buds the season previous and post winter ripening (see: Dorne 1977; Walton 1982; Bergstrom et al. 1997;

Table 7.3 Key plant phenological studies across the subantarctic region

Island	Author	Notes (number of species)
Kerguelen	Dorne (1977)	Floral phenology and germination (7)
	Frenot and Gloaguen (1994)	Reproductive biology (7)
	Hennion and Walton (1997)	Germination (8)
	Chapius et al. (2000)	Reproductive biology (1)
Macquarie	Taylor (1955)	Autecology including phenological notes (39)
	Bergstrom et al. (1997)	Phenology and reproductive behaviour (10)
	Tweedie (2000) and Shaw (2005)	Phenology and germination (6, 26)
Marion	Huntley (1970)	Phenology and distribution (12)
	Mukhadi (2010)	Vegetative and reproductive phenology (15)
South Georgia	Callaghan and Lewis (1971)	Reproductive performance (1)
	Tallowin (1977)	Phenology and reproductive biology (1)
	Walton (1977, 1979, 1982)	Phenology, germination, and hybridization (19)

Hennion and Walton 1997; Shaw 2005). Some non-native species, such as *Poa annua* and *Cerastium fontanum* exhibit longer reproductive seasons (Bergstrom et al. 1997; Mukhadi 2010). Table 7.3 summarises major studies on subantarctic vascular plant phenology.

Phenological patterns have been noted in some marine algae, with development finely tuned to seasonal light regimes. "Seasonal anticipators" become fertile in late autumn to winter, in contrast "seasonal responders" reproduce in summer under suitable irradiance, nutrient, and temperature conditions (see Wiencke et al. 2011 for review).

7.3.2 Birds

On a whole, more research has been conducted into seabirds than other taxa of the Antarctic region (Korczak-Abshire 2010). As these seabirds are long-lived (in some cases for more than 50 years), and thereby integrate environmental signals over multiple time and space scales, this research may provide useful insights into ecosystem status and processes (Trathan et al. 2007).

Many Antarctic seabirds have high degrees of breeding synchrony (e.g. Black-browed Albatross (Croxall et al. 1988), Cape Petrel (Weidinger 1997), Snow Petrel (Amundsen 1995; Barbraud et al. 2000)). For species that require snow-free areas for breeding, the amount of time available to breed can be relatively short and may lead to more breeding synchrony (and is broadly correlated with latitude); sea ice seasonality and photoperiod may also reduce the time available for breeding. Synchronous breeding may also result in predator swamping, thereby increasing the likelihood of individuals surviving the breeding period (but see Kharitonov and Siegel-Causey 1988; Siegel-Causey and Kharitonov 1990 for reviews of seabird coloniality).

Many Antarctic seabird species show strong latitudinal gradients in time of breeding, for example Lynch et al. (2009) found that *Pygoscelis* penguins breed approximately 4.77 days later per degree latitude south, later revised to 5.2 days per degree south (Lynch et al. 2012a). But despite this, not all species show a high degree of flexibility in phenology (e.g., Adélie and Emperor Penguins, Forcada and Trathan 2009).

Although species that are reliant on snow-free areas for breeding may advance their breeding in response to higher temperatures (e.g., Lynch et al. 2009), the rate of change is not always consistent among species, e.g., Gentoo Penguins advanced their breeding at more than twice the rate of either Adélie or Chinstrap Penguins (Lynch et al. 2009). Gentoo Penguins forage close to the breeding colonies during winter, potentially giving them the competitive advantage of local knowledge of prey distribution, compared to the dispersive Adélie and Chinstrap penguins (Lynch et al. 2012b).

Since the early 1950s, the community of nine species of Antarctic seabirds breeding at Dumont d'Urville on the East Antarctica coast have been monitored for their arrival dates and breeding chronologies (Barbraud and Weimerskirch 2006). Over the period of the study, this region saw no major warming or cooling events, but did experience a large reduction in sea ice extent. Although all nine species delayed their arrival over time (c. 9.1 days later than in 1950s), the only species with significantly later ($p < 0.05$) arrival dates were Southern Fulmars, Antarctic and Cape Petrels. The mean egg-laying dates for two species were also significantly delayed (Adélie Penguin and Cape Petrel). The combined delay in both arrival and egg-laying resulted in a c. 7 day compression of the pre-laying period (Barbraud and Weimerskirch 2006), possibly due to the birds requiring additional time to feed before returning to their colonies to commence breeding.

The extent of sea ice has been associated with the arrival timing of several species of Antarctic seabirds. For some penguin species at some colonies, years of extensive sea ice are associated with generally later arrival at breeding colonies, either due to longer travel times over the ice, penguins assessing the ice extent and delaying travel and/or the effect of the sea ice on prey availability and therefore the penguin's body condition (e.g., Ainley et al. 1983; Ainley 2002; Emmerson et al. 2011). For other seabird species, sea ice extent was either unrelated to arrival timing or resulted in earlier arrival when sea ice was more extensive (Olivier et al. 2005; Barbraud and Weimerskirch 2006). Changes in sea ice may not only affect the timing and success of breeding but, consequently, can have flow on effects to survival and population abundance (e.g., Fraser et al. 1992; Crawford et al. 2006; Korczak-Abshire 2010). In fact, changes in sea ice seasonality may explain many of the shifts observed in key phenological relationships in the Antarctic (Massom and Stammerjohn 2010).

Variation in body size has also been proposed as a means of explaining individual variation in laying date, at least in the case of the nest fidelity of Snow Petrels; with larger females having later laying dates (Barbraud et al. 2000). A higher proportion of larger females in larger colonies is also suggested as the main reason for laying occurring approximately 2 days later in larger breeding colonies compared to small colonies or isolated nests of the same species. The presence of flipper bands on birds has also been shown to delay arrival and breeding of penguins which had negative implications for breeding success and survival (Froget et al. 1998; Saraux et al. 2001; Gauthier-Clerc et al. 2004; Le Bohec et al. 2008).

The breeding cycle of the King Penguin is unique among penguins in that it takes more than 1 year to complete. As the timing of initiation of breeding depends on the success of the previous season there can be considerable asynchrony in laying dates over the colony (du Plessis et al. 1984). Early breeding leads to generally higher breeding success with more chicks surviving to fledging compared with later laying (du Plessis et al. 1984; Weimerskirch et al. 1992; Jouventin and Lagarde 1995). In the Crozet Archipelago, late breeding almost always leads to breeding failure (Weimerskirch et al. 1992). Successful breeders tended to return to the colony earlier for either moult or displaying (Jouventin and Lagarde 1995). Moult duration in King Penguins was shorter for individuals that commenced moult later in the season (Jouventin and Lagarde 1995).

7.3.3 Mammals

Although the peak haul out date for Southern Elephant Seals tends to vary with latitude, Authier et al. (2011) found no corresponding latitudinal pattern in the level of breeding synchrony. However, within sites, breeding synchrony tended to be lower in the harems that formed earlier in the breeding season than those forming later and may be age related (breeding experience); older, more experienced, females tended to haul out later (Authier et al. 2011). The level of haul out synchrony in Antarctic Fur Seals has been shown to vary between genders, and also at the individual and population levels (Duck 1990; Slip and Burton 1999). Both genders vary the timing of haul out in relation to environmental conditions, but females appear to be more constrained in timing to ensure that offspring are born and raised during the brief summer period.

A long-term study of Antarctic Fur Seals in South Georgia (Forcada et al. 2005, 2008) found that they are more likely to “optimize” individual fitness by altering their phenology (particularly birthing dates) when faced with short-term changes in their environment. Although later mean birthing dates tended to correspond to periods of higher sea surface temperatures, the overall long-term variability in birthing dates was small (range of 4–15 days), indicating that their ability to alter their phenology may be limited (Forcada et al. 2005, 2008). A similar result was found for all major Southern Elephant Seal populations (Hindell and Burton 1988). Antarctic and Subantarctic Fur Seals *A. tropicalis* at Marion Island also show little year-to-year variation in pupping dates (Hofmeyr et al. 2007).

There has been some debate in the past as to whether Leopard Seals are migratory or not (see Rounsevell and Eberhard 1980) and hence exhibit migration phenology. Long-term sighting dates of Leopard Seal haul outs on Macquarie Island indicate that non-breeding seals will over-winter in subantarctic and temperate oceans. The number of seals undertaking this northward movement varies from year to year, suggesting dispersion rather than migration, which may be driven by food shortages (Rounsevell and Eberhard 1980).

7.3.4 *Other Species*

Low thermal energy budgets in all three Antarctic zones results in reduced or zero growth rates in terrestrial invertebrates over winter (Convey 1996). This can lead to life-cycle events (such as moulting and oviposition) being synchronized following the winter period.

The onset of maximum production of Antarctic phytoplankton varies with latitude, from early spring in the northern zones (~50°S) to late summer or early autumn in the south (El-Sayed 1988). There is an inverse relationship between latitude and the length of the period of maximum production. The timing and magnitude of peak marine production, and the phytoplankton species associated with it, also varies inter-annually in association with oceanographic changes (El-Sayed 1988).

The timing, and intensity, of Antarctic Krill spawning is affected by a variety of environmental conditions including variation in the timing of sea ice formation and retreat, and by the amount and duration of sea-ice cover (Massom and Stammerjohn 2010). Spawning tends to be earlier when there is a longer duration of winter sea ice and delayed seasonal pack ice opening in spring. When spawning is delayed, larval production and or survival are generally poor, presumably due to larvae being less developed, and therefore less able to survive the food-limited winter sea ice conditions (Siegel and Loeb 1995).

Water temperature also influences moult in the sub-Antarctic dytiscid water beetle *Lancetes claussi* which delays moulting from its larval stage IV into pupa until surface water temperatures exceed 7.3 °C (Convey 1996). This delayed moulting results in early summer peaks of adult emergence and oviposition.

At subantarctic Heard and Kerguelen Islands, site, gender, and species differences have been observed in the timing of seasonal activity of Dipterans and, based on these data, earlier and longer lasting reproductive phases are expected in the future (Greenslade et al. 2011).

7.4 Conclusions

There are few long-term studies that collect phenological data for the Southern Hemisphere and our review draws heavily on information published on a limited number of species from a limited number of sites. For the few species with sufficient data to detect phenological changes, both over time and in response to climate variables, the responses varied by location and species. This highlights the need for greater spatial coverage to assess and quantify regional and ecosystem-scale processes and patterns. The proposed Southern Ocean Observing System (http://www.scar.org/soos), based on sustained multi-disciplinary observations for the detection, interpretation and responses to change may assist this goal. The ideal monitoring system comprises long-term observations over multiple trophic levels (Griffiths 2010), enabling investigators to detect potential mismatches in temporal scales between response from populations/communities/species and rates of change in environmental parameters.

For species with less flexible life histories (e.g., many Antarctic mammals and seabirds) changes to critical components of their environment (e.g., sea ice) may result in population changes. These may take the form of range shifts or demographic change. Some of these changes have already been observed (e.g., Forcada et al. 2008).

Acknowledgments We thank Lily Gao and the staff at the Australian (National Meteorological Library). Patricia Selkirk for advice and references. We thank Valeria Ruoppolo for assistance with details on the South American research programs and Amy Winnard, Steve Pendlebury, Scott Carpentier and Phillip Reid for helpful comments on earlier drafts.

References

Ainley DG (2002) The Adélie penguin: bellwether of climate change. Columbia University Press, New York

Ainley DG, LeResche RE, Sladen WJL (1983) Breeding biology of the Adélie penguin. University of California Press, Berkeley

Amundsen T (1995) Egg size and early nestling growth in the snow petrel. Condor 97:345–351

Authier M, Delord K, Guinet C (2011) Population trends of female Elephant Seals breeding on the Courbet Peninsula, îles Kerguelen. Polar Biol 34:319–328

Barbraud C, Weimerskirch H (2006) Antarctic birds breed later in response to climate change. PNAS 103:6248–6251

Barbraud C, Lormée H, LeNevé A (2000) Body size and determinants of laying date variation in the Snow Petrel *Pagodroma nivea*. J Avian Biol 31:295–302

Barbraud C, Gavrilo M, Mizin Y, Weimerskirch H (2011) Comparison of emperor penguin declines between Pointe Géologie and Haswell Island over the past 50 years. Antarct Sci 23:461–468

Bergstrom DM, Selkirk PM (1987) Reproduction and dispersal of mosses on Macquarie Island. Symposia Acta Biol Hung 35:247–257

Bergstrom DM, Selkirk PM, Keenan HM, Wilson ME (1997) Reproductive behaviour of ten flowering plant species on subantarctic Macquarie Island. Opera Bot 30:1–12

Callaghan TV, Lewis MC (1971) The growth of *Phleum alpinum* L. in contrasting habitats at a sub-Antarctic station. New Phytol 70:1143–1154

Cattle H (1991) Global climate models and Antarctic climate change. In: Harris C, Stonehouse B (eds) Antarctica and global climate change. Belhaven Press, London, pp. 21–34

Chapius JL, Hennion F, Roux VL, Cuziat JL (2000) Growth and reproduction of the endemic cruciferous species *Pringlea antiscorbutica* in Kerguelen Islands. Polar Biol 23:196–204

Clarke JL, Robinson AS, Hua Q, Ayre J, Fink D (2012) Radiocarbon bomb spike reveals biological effects of Antarctic climate change. Glob Chang Biol 18:301–310

Convey P (1996) Overwintering strategies of terrestrial invertebrates in Antarctica – the significance of flexibility in extremely seasonal environments. Eur J Entomol 93:489–505

Convey P, Smith RIL (2006) Reponses of terrestrial Antarctic ecosystems to climate change. Plant Ecol 182:1–10

Convey P, Hopkins DW, Roberts SJ, Tyler AN (2011) Global southern limit of flowering plants and moss peat accumulation. Polar Res 30:8929. doi:10.3402/polar.v30i0.8929

Crawford RJM, Dyer BM, Cooper J, Underhill LG (2006) Breeding numbers and success of *Eudyptes* penguins at Marion Island, and the influence of mass and time of arrival of adults. CCAMLR Sci 13:175–190

Croxall JP, McCann TS, Prince PA, Rothery P (1988) Reproductive performance of seabirds and seals at South Georgia and Signy Island, South Orkney Islands, 1976–1987: implications for Southern Ocean monitoring studies. In: Sahrhage D (ed) Antarctic Ocean and resource variability. Springer, Berlin, pp. 261–285

DASET (1992) Impact of climate change on Antarctica – Australia. Department of the Arts, Sport, the Environment and Territories, Canberra

Dorne AJ (1977) Analysis of the germination under laboratory and field conditions of seeds collected in the Kerguelen Archipelago. In: Llano GA (ed) Adaptations within Antarctic ecosystems – proceedings of the third SCAR symposium on Antarctic Biology. Smithsonian Institute, Washington, DC, pp. 1003–1113

du Plessis CJ, van Heezik YM, Seddon PJ (1984) Timing of king penguin breeding at Marion Island. Emu 94:216–219

Duck CD (1990) Annual variation in the timing of reproduction in Antarctic fur seals, *Arctocephalus gazelle*, at Bird Island, South Georgia. J Zool (Lond) 222:103–116

Edwards JA (1974) Studies in *Colobanthus quitensis* (Kunth) Bartl. and *Deschampsia antarctica* Desv. VI. Reproductive performance on Signy Island. Br Antarct Surv B 39:67–86

El-Sayed SZ (1988) Seasonal and interannual variabilities in Antarctic phytoplankton with reference to krill distribution. In: Sahrhage D (ed) Antarctic Ocean and resource variability. Springer, Berlin, pp. 101–119

El-Sayed SZ (1994) Southern Ocean ecology: the BIOMASS perspective. Cambridge University Press, Cambridge

Emmerson L, Pike R, Southwell C (2011) Reproductive consequences of environment-driven variation in Adélie penguin breeding phenology. MEPS 440:203–216

Forcada J, Trathan PN (2009) Penguin responses to climate change in the Southern Ocean. Glob Chang Biol 15:1618–1630

Forcada J, Trathan PN, Reid K, Murphy EJ (2005) The effects of global climate variability in pup production of Antarctic fur seals. Ecology 86:2408–2417

Forcada J, Trathan PN, Murphy EJ (2008) Life history buffering in Antarctic mammals and birds against changing patterns of climate and environmental variation. Glob Chang Biol 14:2473–2488

Fraser W, Trivelpiece WZ, Ainley DG, Trivelpiece SG (1992) Increases in Antarctic penguin populations: reduced competition with whales or a loss of sea ice due to environmental warming? Polar Biol 11:525–531

Frenot Y, Gloaguen J-C (1994) Reproductive performance of native and alien colonising phanerogams on a glacier foreland, Iles Kerguelen. Polar Biol 14:473–481

Froget G, Gauthier-Clerc M, LeMaho Y, Handrich Y (1998) Is penguin banding harmless? Polar Biol 20:409–413

Gauthier-Clerc M, Gendner J-P, Ribic CA, Fraser WR, Woehler EJ, Descamps S, Gilly C, Le Bohec C, Le Maho Y (2004) Long-term effects of flipper bands on penguins. Proc R Soc Lond B (Suppl) 271:S423–S426

Greenslade P, Vernon P, Smith D (2011) Ecology of Heard Island Diptera. Polar Biol 35 (6):841–850. doi:10.1007/s00300-011-1128-5

Griffiths HJ (2010) Antarctic marine biodiversity – what do we know about the distribution of life in the Southern Ocean. PLoS One 5(8):e11683

Hennion F, Walton DWH (1997) Seed germination of endemic species from Kerguelen phytogeographic zone. Polar Biol 17:180–187
Hindell M, Burton HR (1988) Seasonal haul-out patterns of the southern elephant seal (*Mirounga leonina* L.) at Macquarie Island. J Mammal 69:81–88
Hofmeyr GJG, Bester MN, Pistorius PA, Mulaudzi TW, de Bruyn PJN, Ramunasi JA, Tshithabane HN, McIntyre T, Radzilani PM (2007) Median pupping date, pup mortality and sex ratio of fur seals at Marion Island. S Afr J Wildl Res 37:1–8
Holtom A, Greene SW (1967) The growth and reproduction of Antarctic flowering plants. Philos Trans R Soc B 252:323–337
Huntley BJ (1970) Altitudinal distribution and phenology of Marion Island vascular plants. Tydskrif vir Natuurwetenskap 10:255–262
Jouventin P, Lagarde F (1995) Evolutionary ecology of the King Penguin *Aptenodytes patagonicus*: the self-regulation of the breeding cycle. In: Dann P, Norman I, Reilly P (eds) The penguins. Surrey Beatty, Sydney, pp. 80–95
Kharitonov SP, Siegel-Causey D (1988) Colony formations in seabirds. Curr Ornithol 5:223–269
Knox GA (1994) The biology of the Southern Ocean. Cambridge University Press, Cambridge
Korczak-Abshire M (2010) Climate change influences on Antarctic bird populations. Pap Glob Chang 17:53–66
Laws RM (1984) Antarctic ecology (2 vols). Academic, London
Le Bohec C, Durant JM, Gauthier-Clerc M, Stenseth NC, Park Y-H, Pradel R, Grémillet D, Gendner J-P, Le Maho Y (2008) King penguin population threatened by Southern Ocean warming. PNAS 105:2493–2497
Lynch HJ, Fagan WF, Naveen R, Trivelpiece SG, Trivelpiece WZ (2009) Timing of clutch initiation in *Pygoscelis penguins* on the Antarctic Peninsula: towards an improved understanding of off-peak census correction factors. CCAMLR Sci 16:149–165
Lynch HJ, Fagan WF, Naveen R, Trivelpiece SG, Trivelpiece WZ (2012a) Differential advancement of breeding phenology in response to climate may alter staggered breeding among sympatric pygoscelid penguins. MEPS 454:135–145. doi:10.3354/meps09252
Lynch HJ, Naveen R, Trathan PN, Fagan WF (2012b) Spatially integrated assessment reveals widespread changes in penguin populations on the Antarctic Peninsula. Ecology 93:1367–1377
Massom RA, Stammerjohn SE (2010) Antarctic sea ice change and variability – physical and ecological implications. Polar Sci 4:149–186
McMahon CR, Hindell MA (2009) Royal penguin phenology: changes in the timing of egglaying of a sub-Antarctic predator in response to a changing marine environment. In: Stienen E, Ratcliffe N, Seys J, Tack J, Mees J, Dobbelaere I (eds) Seabird Group 10th international conference VLIZ Special Publication 42. Communications of the research institute for nature and forest – INBO.M.2009.1. Provincial Court, Brugge, 27–30 March, 45p
Mukhadi FL (2010) Phenology of indigenous and alien vascular flowering plants on sub-Antarctic Marion Island. MSc thesis, University of Stellenbosch, Stellenbosch
Ochyra R, Bednarek-Ochyra H, Smith RIL (2008) Illustrated moss flora of Antarctica. Cambridge University Press, Cambridge
Olech M (2004) Lichens of King George Island, Antarctica. The Institute of Botany of the Jagiellonian University, Kraków
Olivier F, van Franeker JA, Creuwels JCS, Woehler EJ (2005) Variations of snow petrel breeding success in relation to sea-ice extent: detecting local responses to large-scale processes? Polar Biol 28:687–699
Rounsevell D (1988) Periodic irruptions of itinerant leopard seals within the Australasian sector of the Southern Ocean, 1976–86. Pap Proc R Soc Tasman 122:189–191
Rounsevell D, Eberhard I (1980) Leopard seals, *Hydrurga leptonyx* (Pinnipedia), at Macquarie Island from 1949 to 1979. Aust Wildl Res 7:403–415
Saraux C, Le Bohec C, Durant JM, Viblanc VA, Gauthier-Clerc M, Beaune D, Park Y-H, Yoccoz NG, Stenseth NC, La Maho Y (2001) Reliability of flipper-banded penguins as indicators of climate change. Nature 469:203–208

Seppelt R (2004) The moss flora of Macquarie Island. Australian Antarctic Division, Kingston
Shaw JD (2005) Reproductive ecology of vascular flora of sub-Antarctic Macquarie Island. Dissertation, University of Tasmania, Hobart
Siegel V, Loeb V (1995) Recruitment of Antarctic Krill Euphausia superba and possible causes for its variability. MEPS 123:45–56
Siegel-Causey D, Kharitonov SP (1990) The evolution of coloniality. Curr Ornithol 7:285–330
Slip DJ, Burton HR (1999) Population status and seasonal haulout patterns of the southern elephant seals (*Mirounga leonina*) at Heard Island. Antarct Sci 11:38–47
Smith RIL, Convey P (2002) Enhanced sexual reproduction in bryophytes at high latitudes in the maritime Antarctic. J Bryol 24:107–117
Tallowin JRB (1977) Studies in the reproductive biology of *Festuca contracta* T. Kirk on South Georgia: II. The reproductive performance. Br Antarct Surv B 45:117–129
Taylor BW (1955) Botany – the flora, vegetation and soils of Macquarie Island, vol II, ANARE reports, series B. Australian Antarctic Division, Melbourne
Trathan PN, Agnew D (2010) Climate change and the Antarctic marine ecosystem: an essay on management implications. Antarct Sci 22:387–398
Trathan PN, Forcada J, Murphy EJ (2007) Environmental forcing and Southern Ocean marine predator populations: effects of climate change and variability. Philos Trans R Soc B 362:2351–2365
Turner J, Colwell SR, Marshall GJ, Lachlan-Cope TA, Carleton AM, Jones PD, Lagun V, Reid PA, Iagovkina S (2005) Antarctic climate change during the last 50 years. Int J Climatol 25:279–294
Turner J, Overland JE, Walsh JE (2007) An Arctic and Antarctic perspective on recent climate change. Int J Climatol 27:277–293. doi:10.1002/joc.1406
Tweedie CE (2000) Climate change and the autecology of six plant species along an altitudinal gradient on subantarctic Macquarie Island. Dissertation, University of Queensland, Brisbane
Walton DWH (1977) Studies on *Acaena* (Roseaceae): I. Seed germination, growth and establishment in *A. magellanica* (Lam.) Vahl and *A. tenera* Alboff. Br Antarct Surv B 45:29–40
Walton DWH (1979) Studies on Acaena (Roseaceae) III. Flowering and hybridization on South Georgia. Br Antarct Surv B 55:11–25
Walton DWH (1982) Floral phenology in the South Georgian vascular flora. Br Antarct Surv B 55:11–25
Waugh SM, Weimerskirch H, Moore PJ, Sagar PM (1999) Population dynamics of Black-browed and Greyheaded Albatrosses *Diomedea melanophrys* and *D. chrysostoma* at Campbell Island, New Zealand, 1942–96. Ibis 141:216–225
Weidinger K (1997) Breeding cycle of the Cape petrel *Daption capense* at Nelson Island, Antarctica. Polar Biol 17:469–472
Weimerskirch H, Stahl JC, Jouventin P (1992) The breeding biology and population dynamics of King Penguins *Aptenodytes patagonicus* on the Crozet Islands. Ibis 134:107–117
Wiencke C, Gómez I, Dunton K (2011) Phenology and seasonal physiological performance of polar seaweed. In: Wiencke C (ed) Biology of polar benthic algae. De Gruyter, Berlin, pp. 181–194
Young EC (1991) Critical ecosystems and nature conservation in Antarctica. In: Harris C, Stonehouse B (eds) Antarctica and global climate change. Belhaven Press, London, pp. 117–146

Chapter 8
International Phenological Observation Networks: Concept of IPG and GPM

Frank-M. Chmielewski, Stefan Heider, Susanne Moryson, and Ekko Bruns

Abstract International phenological observation networks are of great importance for many applications in phenology. Data from these networks generally have a high quality standard (genetically identical plants, standardized observation guidelines, etc.) and cover different climatic regions. In this chapter we introduce two networks, the International Phenological Gardens in Europe (IPG) and the Global Phenological Monitoring Programme (GPM). Both observation networks are coordinated by the Humboldt-University of Berlin (Germany). These networks allow a phenological monitoring across lager geographical areas. At the end of each paragraph, we show some examples how these data can be used for scientific applications. They are of great importance to describe relationships between observed climate variability/change and plant development and they can be used to develop or validate phenological models which are able to project possible future shifts in plant development.

8.1 Introduction

In the 1990s interest in phenological research and thus demand for phenological observations has increased, substantially. Mainly rising air temperatures in the end of 1980 and the clear phenological response of plants and animals to this increase have caused a growing interest in phenology by researchers and in public (e.g. EEA 2004; Parmesan 2006; Rosenzweig et al. 2007). Furthermore, the potential use of these data in other fields (see the relevant chapters in this book) like remote

F.-M. Chmielewski (✉) • S. Heider • S. Moryson
Agricultural Climatology, Department of Crop and Animal Sciences,
Faculty of Agriculture and Horticulture, Humboldt-University of Berlin, Berlin, Germany
e-mail: chmielew@agrar.hu-berlin.de

E. Bruns
Department of Networks and Data, German Meteorological Service, Offenbach, Germany

M.D. Schwartz (ed.), *Phenology: An Integrative Environmental Science*,
DOI 10.1007/978-94-007-6925-0_8, © Springer Science+Business Media B.V. 2013

sensing, phenological modelling, agriculture and horticulture (Chap. 29), forestry, biodiversity, human health, and tourism has added value to phenological data and observations. Thus, climate and climate impact researchers have accepted the value of phenological data, and this renewed interest has increased demand for international cooperation in this area.

While the previous chapters mainly reported on different local, national, and regional phenological networks and cross-national analyses of these data, this chapter will introduce the only two *International Phenological Observation Networks* in the world.

The general advantage of these phenological networks is that the stations have identical standards. They require the observation of standardized, vegetatively propagated (genetically identical) plants. This is a great advantage if one considers that differences in the timing of the same phenological event (e.g. beginning of leafing or flowering) between two individuals may last several days or even some weeks. Differences in the maturity date of field crops (e.g. maize) or fruit trees can be even larger. Second, the definition of phenological stages is identical for all stations within the network and do not change over the years. Finally, the phenological stations are coordinated centrally, so that usually a high data quality is guaranteed.

The coordination of such networks is labour intensive and requires much dedication. Both networks, which are described below, have been coordinated by the Humboldt-University of Berlin for several years or decades. The central data collection and data management has been significantly improved in the recent years. General information about the monitoring network and real-time information on the current plant development can be found on the official webpages.

8.2 The International Phenological Gardens of Europe

The International Phenological Gardens (IPG) of Europe (http://ipg.hu-berlin.de) are a unique network for long-term phenological observations of plants representing the natural vegetation in Europe.

8.2.1 Historical Aspects

The idea to establish such an international monitoring network was initiated on the first meeting of the *Agrometeorological Commission* of the World Meteorological Organisation (WMO) in 1953. The aim of the new phenological network was to carry out large-scale and standardized phenological observations across Europe which are not influenced by the hereditary variability of the plants.

In the following years *Fritz Schnelle*, head of the Agrometeorological Division in the German Meteorological Service and *Erik Volkert*, University Professor of

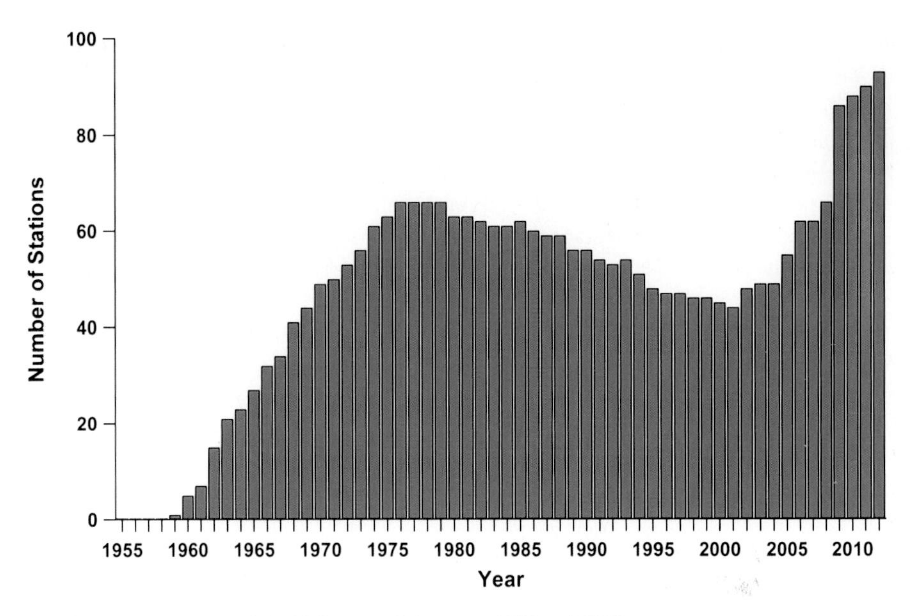

Fig. 8.1 Development of the IPG network between 1959 and 2012

Forestry, started to develop the concept for the International Phenological Gardens (Schnelle und Volkert 1957). At the beginning, they had to gather trees and shrubs from several places and institutions across Europe for the later propagation, find suitable places and institutions to establish phenological gardens, and draft an observation guide for the phenological observations. Simultaneously, they founded a parent garden at the *Federal Research Centre for Forestry and Forest Products* in Großhansdorf near Ahrensburg (Germany). This garden carried out the vegetative propagation and the plant dispatch for a long time.

After several years of preparation, in 1957 *Schnelle* and *Volkert* officially founded the International Phenological Gardens of Europe. Only 2 years later, the first IPG at Offenbach (Germany) reported phenological observations from six plants (*Picea abies*: early/late, *Populus canencens*, *Robinia pseudoacacia*, *Salix aurita*, *Salix smithiania*). In 1960 three IPG in Vienna (Austria) and another IPG in Germany (Trier) were added to the network. In the following years the number of IPG increased rapidly to 66 stations in the mid-late 1970s (Fig. 8.1).

After the start of the network the coordinating institutions changed three times. From 1973 to 1977 the network was co-ordinated by the *Institute of Biometeorology* of the University at Munich. Between 1978 and 1995 the *German Meteorological Service* (DWD) was responsible for the phenological gardens. Unfortunately, during this time the number of stations steadily decreased. There are different reasons for this trend. Probably, some stations lost their interest in long-term phenological observations. However, the main reason was the increasing shortage in young plants to establish new gardens, because in the 1990s the original parent garden was no longer able to manage the plant propagation and plant dispatch. When in 1996 the *Humboldt-University* of Berlin took over the coordination and the management, only 47 IPG belonged to the network. One of the first tasks of the new management was to find a new parent garden. In 1997, the young plants were first moved from the old parent garden in Großhansdorf to Ahrensburg (Germany), where a new site was established at the *JORDSAND* association. Finally in 2001, when the observation network comprised only 44 stations, the *Bavarian State Institute of Forestry* carried out the propagation of selected species from the IPG programme, so that by this the survival of the network was ensured. Some years later a further extension of the network was possible. The number of stations increased again. In 2009, when the Irish phenologists at the *Trinity College* Dublin decided to establish a national phenological network in Ireland, which was based on the IPG species, more than ten new stations joined the network.

The original IPG observation programme consisted of standard and expanded programmes. Both programmes comprise coniferous and deciduous trees. The *standard observation programme* includes 26 plants from the natural vegetation including different provenances (Table 8.1). In 2001 three further plant species (*Corylus avellana*, *Forsthia suspensa*, *Syringa chinensis*) were added to this programme. Two of them are plants from the Global Phenological Monitoring Programme (see Sect. 8.3), in order to have a link between these two networks. Over and above this, some IPG also observe the *expanded programme*. This programme includes 23 other plants, but only three new varieties (*Picea omorica*, *Betula pendula*, *Fagus orientalis*). The remaining 20 plants are further provenances of the species in the standard programme.

Currently, the plant propagation is only focussed on 21 species which now belong to the IPG kernel programme (bold numbers in Table 8.1). New established IPG will be supplied only with these plants. Sometimes, due to climatic restrictions, it is not possible to grow all species of the programme at one site. In this case, the gardens have only a few plants in stock.

The phenological phases are recorded according to the BBCH-code (Meier 1997), which classifies plant growth stages of many of species according to a standardized system. The extended BBCH scale is an internationally recognized standard and can be applied to most of the plants.

In Germany, the phenological stages of fruit crops were described initially by the BBCH-code (Bruns 1995). Since the start of the EPN-project (European Phenology Network) in 2003, the acceptance of the BBCH-code increased in Europe and later worldwide. In this European project the BBCH-code was adapted to the observation programme of 11 national phenological networks in Europe and to the IPG programme, as well (Bruns and van Vliet 2003).

The BBCH system is an excellent illustration for the standardization of phenological observations and the IPG programme an outstanding example for the standardization of plant species which are observed.

Table 8.1 Plants in the IPG Standard programme, including the observed phenological stages 1 May shoot (BBCH 10), 2 leaf unfolding (BBCH 11), 3 St. John's sprout, 4 beginning of flowering (BBCH 60), 5 general flowering (BBCH 65), 6 first ripe fruits (BBCH 86), 7 autumn colouring (BBCH 94), 8 leaf fall (BBCH 95)

Plant no.	Botanical name	English name	1	2	3	4	5	6	7	8
111	Larix decidua	European larch		x		x	x		x	x
121	Picea abies (early)	Norway spruce	x			x	x			
122	Picea abies (late)	Norway spruce	x			x	x			
123	Picea abies (northern)	Norway spruce	x			x	x			
131	Pinus silvestris	Scotch pine, Fir	x			x	x			
211	Betula pubescens	White birch		x		x	x		x	x
221	Fagus sylvatica 'Har'	Common beech		x		x	x	x	x	x
222	Fagus sylvatica 'Düd'	Common beech		x		x	x	x	x	x
223	Fagus sylvatica 'Tri'	Common beech		x		x	x	x	x	x
231	Populus canescens	Grey poplar		x		x	x	x	x	x
235	Populus tremula	Trembling poplar		x		x	x	x	x	x
241	Prunus avium 'Bov'	Wild cherry		x		x	x	x	x	x
242	Prunus avium 'Lut'	Wild cherry		x		x	x	x	x	x
251	Qercus petraea 'Zell'	Sessile oak		x	x	x	x	x	x	x
256	Qercus robur 'Wol'	Common oak		x	x	x	x	x	x	x
257	Qercus robur 'Bar'	Common oak		x	x	x	x	x	x	x
261	Robinia pseudoacacia	Common robinia		x		x	x	x	x	x
271	Sorbus aucuparia	Mountain ash		x	x	x	x	x	x	x
281	Tilia cordata	Small-leafed lime		x		x	x	x	x	x
311	Ribes alpinum	Alpine currant		x		x	x	x	x	x
321	Salix aurita	Roundear willow		x		x	x		x	x
323	Salix acutifolia	Pussy willow		x		x	x		x	x
324	Salix smithiana	Smith's willow		x		x	x		x	x
325	Salix glauca	Grey leafed willow		x		x	x		x	x
326	Salix viminalis	Basket willow		x		x	x		x	x
331	Sambucus nigra	Common elder		x		x	x	x	x	x
411	Corylus avellana	Common hazel		x		x	x	x	x	x
421	Forsythia suspensa	Forsythia		x		x	x		x	x
431	Syringa vulgaris	Common lilac		x		x	x		x	x

Bold plant numbers indicate the 21 plants from the current kernel program

8.2.2 The Network Today

Today, the network ranges across 28 latitudes from Scandinavia to Macedonia and across 37 longitudes from Ireland to Finland in the north and from Portugal to Macedonia in the south (Fig. 8.2). The IPG network now consists of 93 stations in 19 countries of Europe (stand 2012). All IPG are placed in similar surroundings, usually mainly plain surfaces with meadows and some trees. They are hosted by institutions such as universities, botanical gardens, meteorological services, forest research centres, etc. In the vicinity of each garden is usually an official meteorological station.

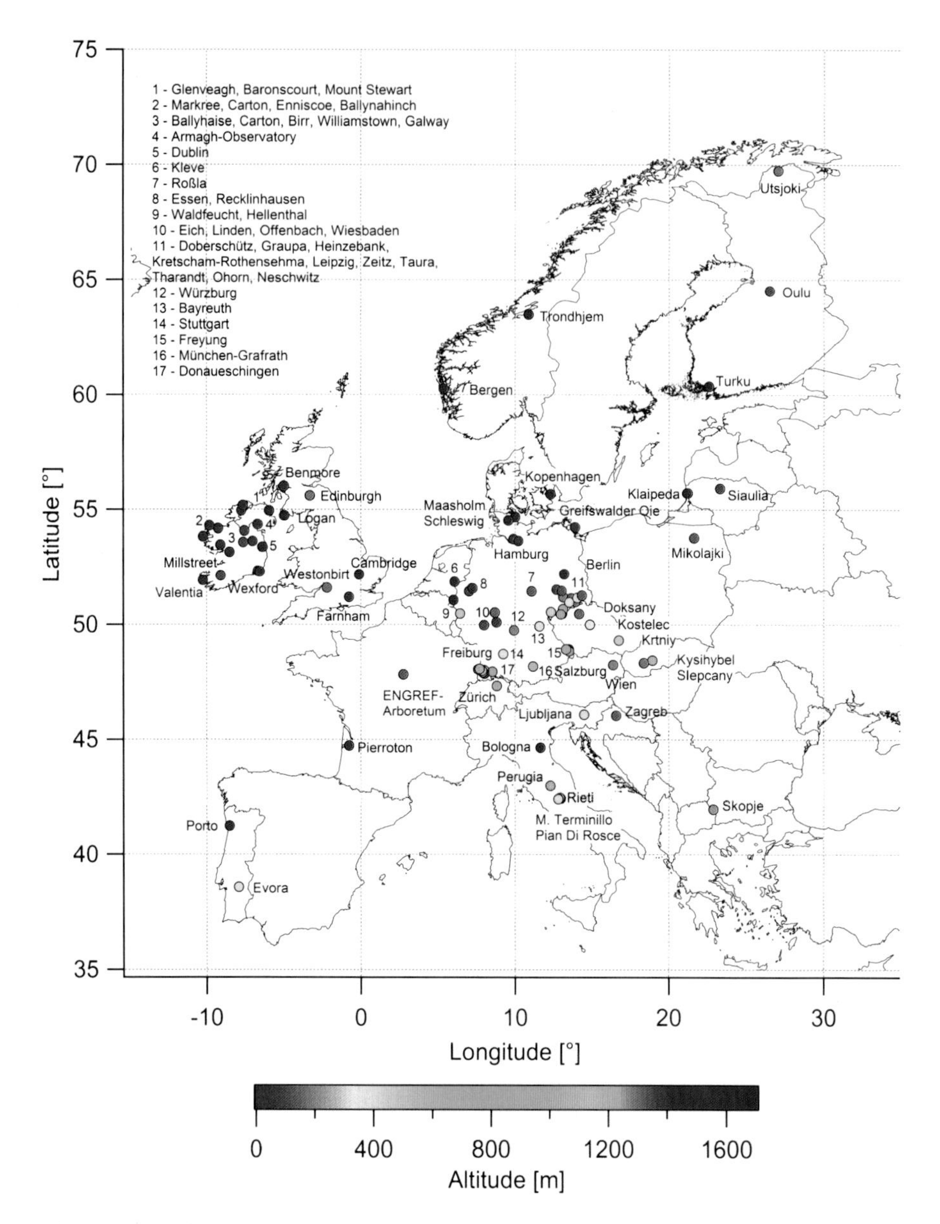

Fig. 8.2 Sites of the International Phenological Gardens of Europe (IPG), 2012

All observations are gathered and stored in the IPG database. The observations can be transmitted over the internet. In the members section of the IPG homepage the user can manage their station, add site information for the public (climate conditions, soil types, etc.) and can edit their observations. It is possible that a "super user" can serve several stations in a region. Additionally, the homepage gives everyone detailed information on the observed plants, the exact definition of the phenological

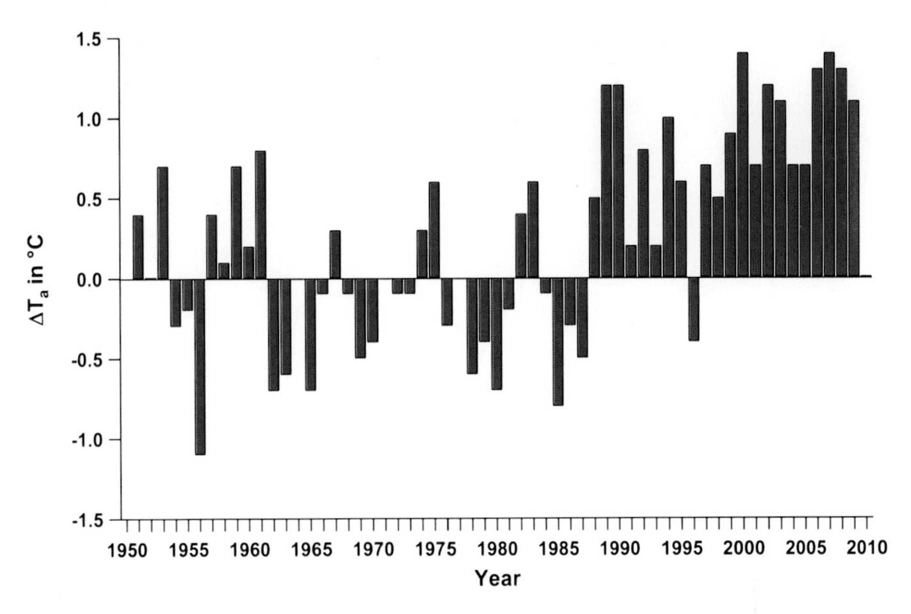

Fig. 8.3 Deviations of the mean annual air temperature (ΔT_a) in Europe (70°N–40°N, 10°W–25°E), 1950–2010 to the reference period 1961–1990 (E-OBS Data, Haylock et al. 2008)

stages (BBCH-code), and the number of active stations, including their climate conditions and average phenological data. For each year, beginning in 1959, the timing of phenological phases across Europe can be animated. Shown are deviations among all stations in timing of a phenological event from very early to very late.

8.2.3 Applications

The observational data of the International Phenological Gardens were used in many studies to detect regional or local differences in the timing of phenological events (e.g. Schnelle 1977, 1986; Schmittnägel 1983; Lauscher 1985; Sandig 1992; Seidler 1995) or trends (e.g. Menzel and Fabian 1999; Menzel 2000; Menzel et al. 2006), to present relationships between temperature variations and the beginning, end or length of growing season (e.g. Chmielewski and Rötzer 2001, 2002; Atkinson 2002; Donelly 2002; Köstner et al. 2005; Rödiger 2012), for different modelling purposes (e.g. Kramer 1996; Menzel 1997; Rötzer et al. 2004) or to verify phenological models (e.g. Chuine 2001; Caffarra and Donnelly 2011), and to calculate phenological maps for Europe (Rötzer and Chmielewski 2001).

The strong relationships between air temperature and plant development in mid- and high-latitudes make phenological observations to sensitive indicators that can be used to evaluate possible biological impacts of climate change. Since the end of the 1980s, clear changes in air temperature have been observed in Europe (Fig. 8.3)

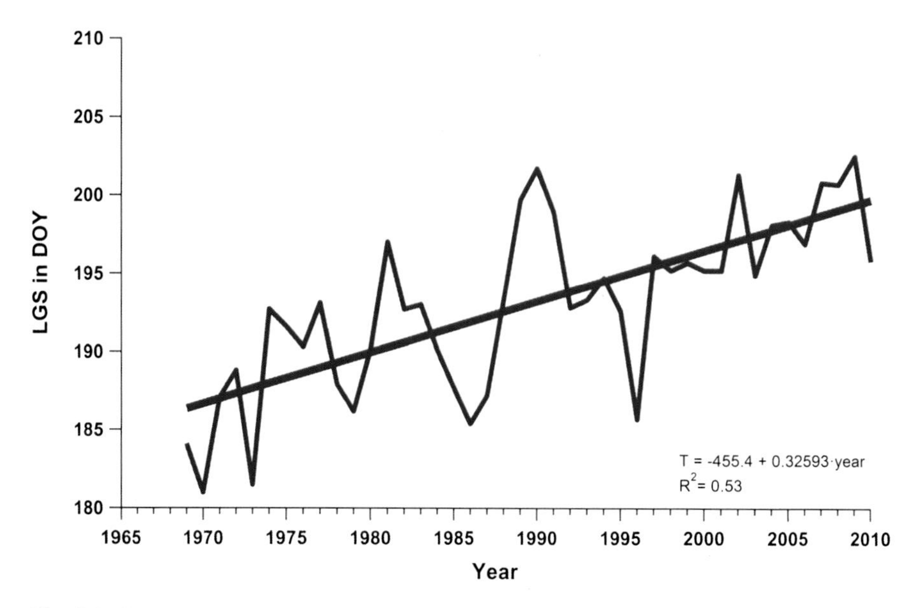

Fig. 8.4 Observed changes in the length of growing season across Europe between 1969 and 2010. The growing season length here is the average time between leaf unfolding and leaf colouring of selected IPG species (see description in text)

and in many other parts of the world. Mainly the temperatures in winter and in early spring – which are decisive for the spring plant development – changed distinctly. Most recent years were warmer than the long-term average.

These observed changes in temperature correspond well to changes in the circulation pattern over Europe (Chmielewski and Rötzer 2001). The increased frequency of positive phases in the North Atlantic Oscillation (NAO) since 1989 led to milder temperatures in winter and in spring, because of the prevailing westerly winds from the Atlantic Ocean during this time of the year. Cold spells were predominant only in the last two winters (2009/10, 2010/11).

These climate changes led to clear reactions in flora and fauna. Between 1969 and 2010, the average length of growing season in Europe, here defined as the average time between leaf unfolding and leaf fall of the IPG species 211, 241, 242, 271 or 324, and 311 (see Table 8.1), has extended by 14 days. This corresponds to a significant trend of 3.26 days/decade (Fig. 8.4). In accordance with climatic changes, mainly since the end of the 1980s, longer periods now occur. Between 1989 and 2010, 19 out of 22 years had an extended growing season, compared to the long-term average from 1969–2010.

In Europe the growing season lasts on average 193 days (s = 5.5 days) with large regional differences (Rötzer and Chmielewski 2001). The longest period can be found in the southern part of France and in the coastal regions of southern

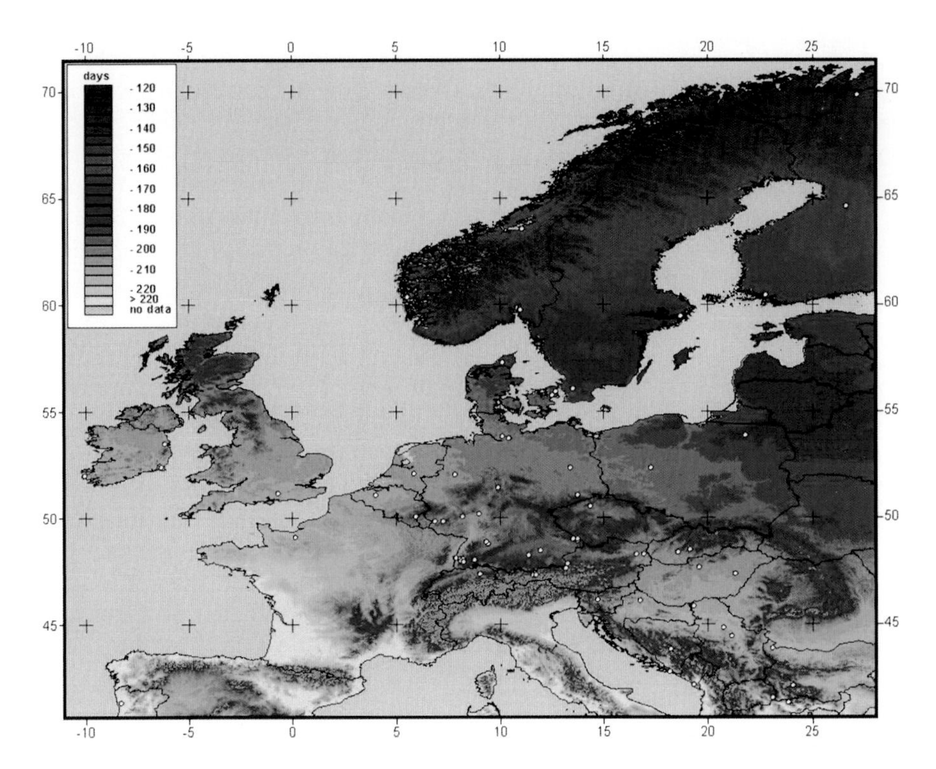

Fig. 8.5 Average length of growing season across Europe (Rötzer and Chmielewski 2001)

Europe (Fig 8.5). These findings were confirmed by Lebourgeois et al. 2010. They calculated for France a growing season length between 180 and 190 days in the east and more than 210 days in the west and south-west.

In large parts of Ireland, southern England, the Netherlands and in Belgium, and most parts of France, Hungary and southern Europe (expect in the mountainous regions) the growing season lasts between 200 and 220 days. In Scotland, Denmark, Germany, Switzerland, Austria, the Czech Republic, Slovenia and Poland, and in the southern part of Sweden, the growing season lasts more than 180 days, but less than 200 days. Shorter growing seasons, with less than 180 days, are calculated for high altitudes and for nearly all of Scandinavia. High regions in Scandinavia as well as the areas north of the Arctic Circle show growing seasons under 150 days. These growing season lengths correspond to the average duration of snow cover (180–222 days) in most northern regions of Scandinavia.

8.3 The Global Phenological Monitoring Program

The Global Phenological Monitoring (GPM) Programme (http://gpm.hu-berlin.de) was originally an initiative of the ISB Phenology Group and is still in development. A further successful establishment of this network requires the efforts of phenologists from different parts of the world. Currently, the aim of the study group is to establish a *Global Phenology Programme* by using, analysing and linking data from existing monitoring networks worldwide. First examples of this initiative are the European Phenology Network (van Vliet et al. 2003) and the Pan European Phenological database PEP725 (Koch et al. 2010). This is a welcome development, but can never substitute for a standardized global monitoring programme such as GPM or IPG. Some advantages of this network are highlighted at the end of this chapter.

8.3.1 Historical Aspects

The plan for establishing a new Global Phenological Monitoring network were started by the '*Phenology Group*' of the International Society for Biometeorology (ISB) at the 1993 meeting in Canada. At a second meeting in May 1995 (hosted by the DWD in Offenbach), the Phenology Group drew up concrete benchmarks that facilitated the network implementation. In 1996, the preparations of a Global Phenological Monitoring programme were completed at the 14th ISB Congress in Ljubljana, Slovenia. Phenologists from all over the world discussed the set-up of GPM stations. They agreed that the establishment a Global Phenological Monitoring programme was an important tool to meet the objectives of the ISB Phenology Group.

In 1998 and 1999 three GPM gardens were established in Germany (Deuselbach, Blumberg, Tharandt). In 2000, they reported the first observation results. Two new international gardens were established in 2002 (one garden in Beijing/China and another one in Milwaukee/USA). Unfortunately, the GPM in China had to be already abandoned in 2006. Despite of intensive efforts, it was not possible to set up a new garden in this region. Further gardens in Germany, the Czech Republic, Estonia, Italy, Slovakia, and Turkey quickly followed (Fig. 8.6).

The data from the GPM-stations are gathered at the Humboldt-University of Berlin. For the GPM gardens a similar database was established, such as for the IPG network. The annual observations can be also submitted via internet. In the members and in the public section of the GPM homepage the features are the same as those available for the IPG network.

GPM focuses mainly on temperature impacts on the timing of life cycle events. Since in arid and semiarid tropics or subtropics phenology is mainly driven by precipitation, the global network will be restricted to mid-latitudes of about 35° north to the Arctic Circle, and from the Tropic of Capricorn to 50° south.

The selection of plants was an important factor in determining the orientation of the monitoring program. The focus in GPM should not be the same (natural vegetation) as in the IPG programme. A number of criteria were used to choose species:

- plants should be *economically important*,
- species should have a broad geographic distribution and/or ecological amplitude,
- plants should be easy to propagate and vegetative propagation of these species should be common practice,

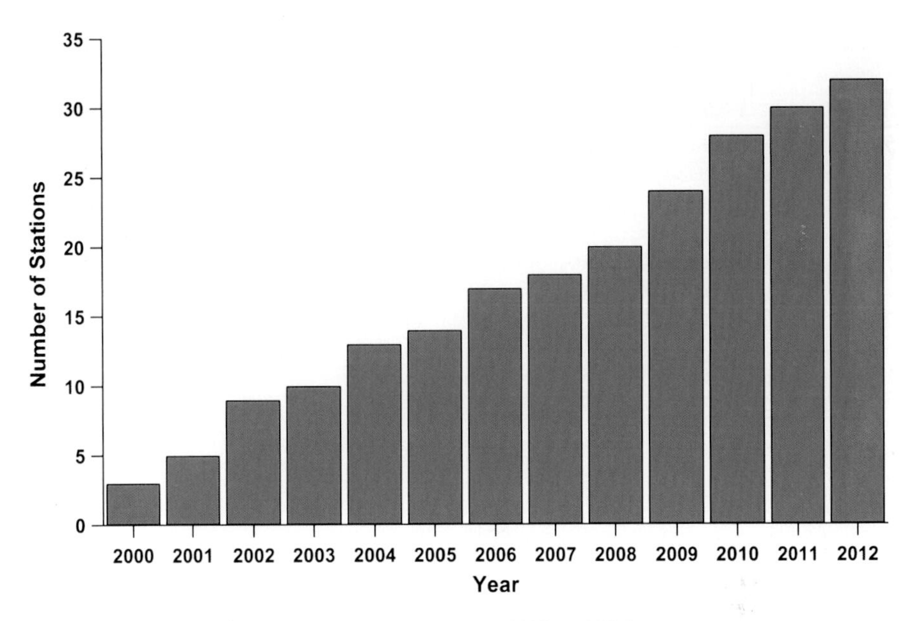

Fig. 8.6 Development of the GPM network between 2000 and 2012

- the start of the phases should be sensitive to air temperature, and
- plants should have phenological stages which are easy to recognize and to observe.

Based on these criteria, 16 plants were selected for the GPM observation programme (Tables 8.2 and 8.3). These species consist mainly of fruit trees (specified varieties), some park bushes, and spring flowers.

The fruit species represent the so-called '*Standard Programme*', which is required for each GPM-garden that will be established. The Standard Programme can be supplemented by a '*Flowering Phase Programme*' (ornamental shrubs and snow-drops). Due to different environmental conditions it is not possible to have all plants of the programme at all stations in middle and high latitudes. Thus, there are some gardens which only observe the standard programme and gardens which have a reduced number of species on-site. The minimum distance between trees shall be used for the same species. Larger distances are desirable and consequently not an issue.

Although temperature is the main factor affecting plant development, other environmental factors are also important. Therefore, to improve data analysis, a number of requirements for the phenological garden were specified to standardize the monitoring programme.

With the focus on temperature, precipitation impacts were excluded by allowing irrigation in case of extreme water shortage. Another requirement was that the location should be characteristic of the larger region around the observation area. Sites are to be avoided which, due to specific sun exposure (e.g. southern slope), shady side, topographical conditions, (e.g. frost hollow), or urban development, are

Table 8.2 Standard GPM-Observation programme and minimum distances (D) between the plants

Species	Variety	Latin name	Rootstock	D (m)
Almond	Perle der Weinstraße	Prunus dulcis	St. Julien A	3.0
Red currant	Werdavia, white cultivar	Ribus rubrum	own-rooted	1.5
Sweet cherry	Hedelfinger, type Diemitz	Prunus avium	GiSelA 5	3.0
Sour cherry	Vladimirskaja	Prunus cerasus	own-rooted	3.0
Pear	Doyenne de Merode	Pyrus communis	OHF 333	3.0
Apple	Yellow Transparent	Malus x domestica	Malus transitoria	2.5
Apple	Golden Delicious, type Golden Reinders	Malus x domestica	M26	3.0
European chestnut	Dore de Lyon	Castanea sativa	seedling	detached

Table 8.3 Flowering phase GPM-Observation programme and minimum distances (D) between the plants

Species	Variety	Latin name	D (m)
Witch hazel	Jelena	Hamamelis × intermedia	2.5
Snowdrop	Scharlokii	Galanthus nivalis	–
Forsythia	Fortunei	Forsythia suspensa	1.5
Lilac	Red Rothomagensis	Syringa chinensis	2.5
Mock-orange	(genuine)	Philadelphus coronarius	3.0
Heather	Allegro	Calluna vulgaris	0.5
Heather	Long white	Calluna vulgaris	0.5
Witch hazel	(genuine)	Hamamelis virginiana	2.5

All plants are own-rooted

known to have climatic anomalies, or where deviations from characteristic conditions can be expected. The plants should be planted on level ground (slopes of up to 3° in all directions are still acceptable). The trees and shrubs do not have to be planted in a specified order. The optimum growing site is open ground without obstacles, traffic routes, detrimental (for example, herbivores) or favourable influences (for example, artificial light), or other factors affecting the plants (shading).

If the observed plants are located near obstacles the following issues apply. The minimum distance from the base of any obstacle (building, tree, wall, etc.) should be at least 1.5 times the height of the obstacle (more, two times, from the edge of forested areas). The distance from a two-lane road should be at least 8 m, and from any larger (eight-lane) highway, at least 25 m. All plants must be protected against herbivores (consumption by wild or domestic animals) by a wire-netting fence or individually by an anti-game protective agent. So-called "plant protection covers" (e.g. tube protection and growth covers) are unsuitable, as they can accelerate growth considerably (heat congestion). Thus, preference should be given to wire-netting systems.

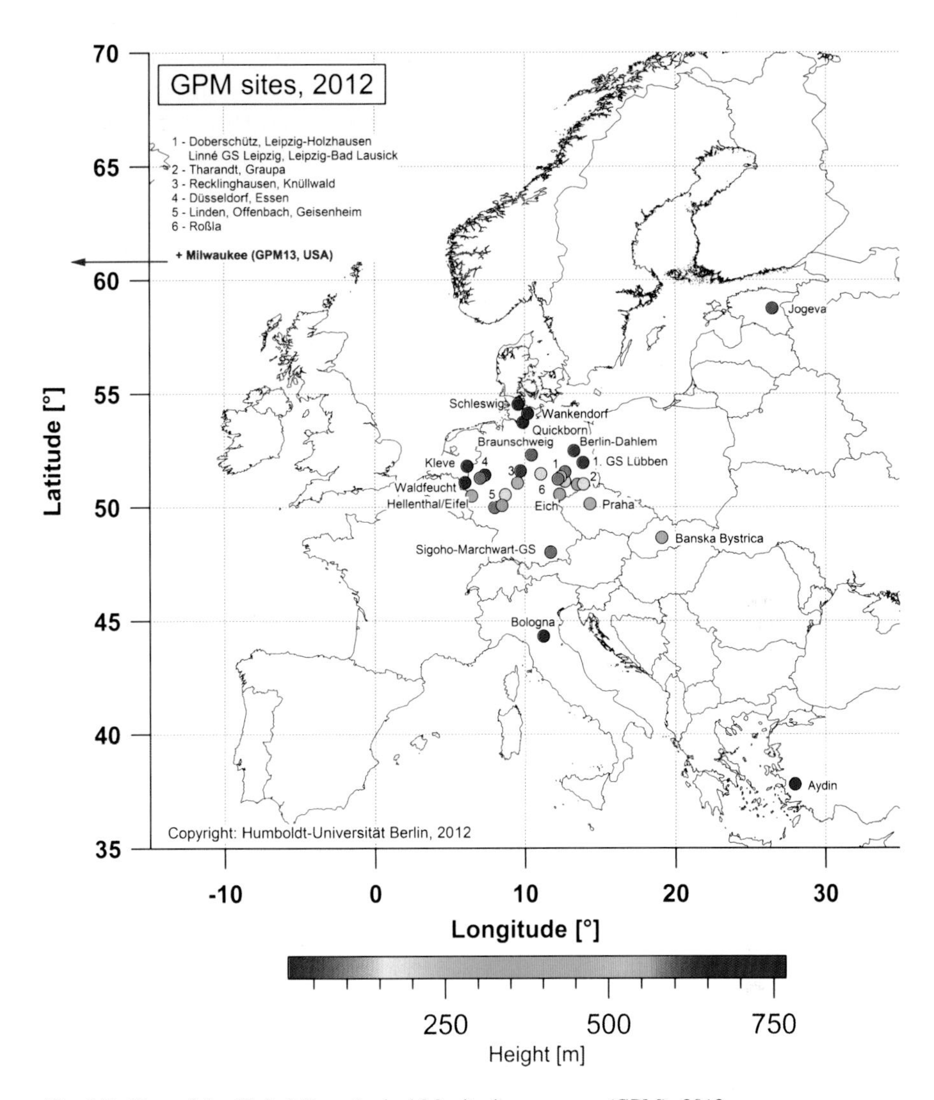

Fig. 8.7 Sites of the Global Phenological Monitoring program (GPM), 2012

8.3.2 *The Network Today*

Figure 8.7 shows the current distribution of the GPM-gardens in Europe. The only garden in the USA (Milwaukee) is not presented on this map. The current network includes 32 phenological gardens in 7 countries. One sees immediately that most gardens are located still in Germany. Additionally, three schools are involved in this network (*Linné Elementary School* in Leipzig, *First Elementary School* in Lübben, and the *Sigoho-Marchwart Elementary School* in Siegertsbrunn). This is absolutely desirable and a possibility to introduce students to science. Phenology is well-suited for this purpose, which meets the original goal of the ISB Phenology Group: '*To stimulate public interest in science, especially among pupils and students*'.

A future expansion of the network would be desirable. On the international scale import-requirements for plants are usually restricted. Within the European Union (EU), the export of plants without restrictions is possible. However, non-EU countries in Europe are already subject to more restrictions. For this reason it would be reasonable to establish *GPM parent gardens* at least in Asia, Australia, North and South America, which are specialized in growing plants, and are able to propagate and distribute the plant material. For Europe such a parent garden was established in Abstatt (Germany). The tree nursery 'Krauß' propagates the GPM plants successfully for several years. Plants can be ordered from anywhere in Europe. Since the fruit trees of the GPM are exactly defined (species, variety) it could be also possible to buy the plants on-site if they are really genetically identical. However, great care should be exercised here. This could be an option for counties with strong restrictions on imports of plants to at least partially join the monitoring programme.

In recent years, the Global Phenological Monitoring Programme has steadily increased in size. Set-up issues have been thoroughly explored, and new sites successfully implemented. GPM will continue to contribute to the further expansion of phenological gardens, to improve the use of phenological information and to improve cooperation and communication between the actors involved in phenology. The programme is now poised for future expansion into other parts of the world. However, this is only possible if there is an active support of phenologists outside of Europe.

8.3.3 Applications

The oldest GPM stations now have phenological records of more than 10 years. This allows for the first statistical analysis. However, the shorter time-series at many other stations limits use of the data.

Observations from GPM can contribute to investigations of possible impacts of climate variability or climate change on fruit crops. This is a very important task, because impact studies for perennial crops are limited compared to field crops (see also Chap. 29). For instance, the data can be used to analyse regional differences in the timing of phenological stages. It should be possible to explain these differences as a result of different climatic conditions. For example, the beginning of sour cherry blossom in Bologna (Italy) starts already on 12 April. Here, we have the warmest climate with an average annual air temperature (T_a) of 12.9 °C. In Jõgeva (Estonia) the annual mean is only 5.1 °C and the beginning of blossom only starts on 23 May (difference of 41 days). In Germany the beginning of sour cherry blossom starts on 24 April ($T_a = 9.2$ °C) and in Milwaukee (USA) on 10 May

Table 8.4 Validation of a phenological model for the beginning of sour cherry blossom (BB) on phenological observations of the GPM programme

GPM station	BB (number of years)	$RMSE_{val}$ (days)
Braunschweig (D)	113.9 (7)	2.68
Berlin-Dahlem (D)	108.2 (5)	2.53
Geisenheim (D)	108.7 (7)	1.37
Offenbach (D)	103.0 (4)	2.43
Schleswig (D)	119.0 (8)	3.95
Tharandt (D)	121.3 (7)	4.36
Linden (D)	115.0 (8)	2.91
Graupa (D)	111.5 (5)	3.32
Praha (CZ)	111.8 (6)	2.87
Banska Bystrica (SK)	116.3 (8)	2.14
Milwaukee (USA)	130.0 (9)	3.89
Average across stations	114.4 (6.7)	2.95

$RMSE_{val}$ root mean square error between calculated and observed data, variety 'Vladimirskaja', Model parameter: $t_0 = 244$ DOY, $C^* = 74.1$ CP, $T_{BF} = 1.0$ °C, $F^* = 567.4$ PTU, $RMSE_{opt} =$ 2.19 days

($T_a = 7.8$ °C). This very rough estimation shows that there is indeed a clear response between climate and the beginning of cherry blossom.

Climate change will probably shift climate and growing zones worldwide. Phenological observations can help to forecast how the plants could respond to warmer environmental conditions, if phenological gardens already exist in warm climates such as in Turkey. GPM also has a great potential for modelling tasks. Plants in the international networks are standardized, so that the phenological observations are well-suited to develop models for the beginning of fruit tree blossom or picking ripeness. These blossoming models can later be used to evaluate the late frost hazard due to climate change in different parts of the world. Additionally, phenological observations are necessary to develop yield and water budget models for fruit crops.

Currently, most of the records are not long enough for modelling purposes, but they can already be used to verify phenological models. Matzneller et al. (2013) developed a phenological model for the beginning of sour cherry blossom for Germany based on phenological observations from the DWD. The model was developed for one important cherry growing region in Germany (Rhineland-Palatinate). For this modelling task a combined 'Chilling Portion/Growing Degree Day Model' with daylength factor was used (see Chap. 29). The model was verified at several *GPM stations*. The results show that at nearly all stations the model performed very well (Table 8.4), even when the model is used outside of Germany. This example clearly shows how important standardized phenological observations are for model development and validation.

The later validation of the same model on data from *experimental stations* in Germany, Hungary, Poland, and Michigan/USA lead also to reasonable results, but the average RMSE were higher (4.66 days), compared to the validation at the GPM programme (2.95 days). The phenological observations at these experimental stations are certainly very precise, but the definition of the phenological stages varies between the sites. Variable definitions contributed to higher model error. Additionally, the cultivars differ from site to site, so that the calculations by the model are not easily comparable with the observations. The treatment and management of the fruit trees also differs from site to site, which led to larger differences between the modelled and observed values.

References

Atkinson MD (2002) Phenological studies of Sambucus nigra. Arbor Phaenol 45:20–25

Bruns E (1995) News for the phenological observers of the German Meteorological Service. Phenol J 3:2

Bruns E, van Vliet AJH (2003) Standardization of phenological monitoring in Europe. Final Project Report, Wageningen University and Deutscher Wetterdienst, 79p

Caffarra A, Donnelly A (2011) The ecological significance of phenology in four different tree species: effects of light and temperature on bud burs. Int J Biometeorol 55:711–721

Chmielewski FM, Rötzer T (2001) Response of tree phenology to climate change across Europe. Agric For Meteorol 108:101–112

Chmielewski FM, Rötzer T (2002) Annual and spatial variability of the beginning of growing season in Europe in relation to air temperature changes. Clim Res 19(1):257–264

Chuine I (2001) Using the IPG observations to test for model estimates stability across environments. Arbor Phaenol 44:6–8

Donelly A (2002) Trees as climate change indicators for Ireland. Arbor Phaenol 45:7–19

EEA (2004) Impacts of Europe's changing climate: an indicator-based assessment. European Environment Agency Copenhagen Report 2. Office for Official Publications of the European Communities, Luxembourg

Haylock MR, Hofstra N, Klein Tank AMG, Klok EJ, Jones PD, New M (2008) A European daily high-resolution gridded dataset of surface temperature and precipitation. J Geophys Res Atmos 113:D20119. doi:10.1029/2008JD10201

Koch E, Adler S, Lipa W, Ungersböck M, Zach-Hermann S (2010) The pan European phenological database PEP725. In: Berichte des Meteorologischen Institutes der Albrecht-Ludwigs-Universität Freiburg Nr. 20. Proceedings of the 7th conference on Biometeorology. http://www.mif.uni-freiburg.de/biomet/bm7/report20.pdf

Köstner B, Niemand C, Prasse H (2005) A 40-year study period of tree phenology at Tharandt international phenological garden. Arbor Phaenol 48:197–202

Kramer K (1996) Phenology and growth of European trees in relation to climate change. Master's thesis, Landbouw Universiteit, Wageningen

Lauscher F (1985) Zur Phänologie vegetative vermehrter Pflanzen einheitlicher Herkunft – Beobachtungen in phänologischen Pflanzgärten in Norwegen 1963–1982. Phyton 25:253–272

Lebourgeois F, Pierrat JC, Perez V, Piedallu C, Ceccini S, Ulrich E (2010) Simulating phenological shifts in French temperate forests under two climate change scenarios and for driving global circulation models. Int J Biometeorol 54:563–581

Matzneller P, Blümel K, Chmielewski FM (2013) Models for the beginning of sour cherry blossom. Int J Biometeorol. http://dx.doi.org/10.1007/s00484-013-0651-1

Meier U (1997) Growth stages of mono- and dicotyledonous plants. BBCH-Monograph Blackwell, Berlin

Menzel A (1997) Phänologie von Waldbäumen unter sich ändernden Klimabedingungen – Auswertung der Beobachtungen der Internationalen Phänologischen Gärten in Deutschland und Möglichkeiten der Modellierung von Phänodaten. Forstliche Forschungsberichte München 164:147

Menzel A (2000) Trends in phenological phases in Europe between 1951 and 1996. Int J Biometeorol 44:76–81

Menzel A, Fabian P (1999) Growing season extended in Europe. Nature 397:659

Menzel A, Sparks TH, Estrella N, Koch E, Aasa A, Ahas R, Alm-Kübler K, Bissolli P, Braslavská O, Briede A, Chmielewsk FM, Crepinsek Z, Curnel Y, Dahl Å, Defila C, Donnelly A, Filella Y, Jatczak K, Måge F, Mestre A, Nordli Ø, Peñuelas J, Pirinen P, Remišová V, Scheifinger H, Striz M, Susnik A, van Vliet AJH, Wielgolaski FE, Zach S, Zust A (2006) European phenological response to climate change matches the warming pattern. Glob Chang Biol 12:1969–1976

Parmesan C (2006) Ecological and evolutionary responses to recent climate change. Annu Rev Ecol Evol Syst 37:637–669

Rödiger MS (2012) Temperature changes in Europe 1969–2010 and impacts on the beginning and end of growing season. Master's thesis, Humboldt University Berlin

Rosenzweig C, Casassa G, Karoly DJ, Imeson A, Liu C, Menzel A, Rawlins S, Seguin B, Tryjanowski P (2007) Assessment of observed changes and responses in natural and managed systems. Climate change 2007: impacts, adaptation and vulnerability. Contribution of working group II to the fourth assessment report of the intergovernmental panel on climate change. Cambridge University Press, Cambridge

Rötzer T, Chmielewski FM (2001) Phenological maps of Europe. Clim Res 18(3):249–257

Rötzer T, Grote R, Pretzsch H (2004) The timing of bud burst and its effect on tree growth. Int J Biometeorol 48:109–118

Sandig S (1992) Der Internationale Phänologische Garten Grafrath – langjährige Beobachtungen und Ergebnisse. Diplomarbeit Universität München

Schmittnägel HR (1983) Vergleich phänologischer Phasen verschiedener Baumarten und -herkünfte in Europa anhand der Beobachtungsdaten der Internationalen Phänologischen Gärten. Diplomarbeit Universität Göttingen

Schnelle F (1977) Beiträge zur Phänologie Europas III. Ergebnisse aus den Internationalen Phänologischen Gärten (Regionale und Jahresunterschiede 1966–1975). Berichte des Deutschen Wetterdienstes 144:1–31

Schnelle F (1986) Ergebnisse aus den Internationalen Phänologischen Gärten in Europa – Mittel 1973–1982. Wetter und Leben 38:5–17

Schnelle F, Volkert E (1957) Vorschläge zur Einrichtung „Internationaler Phänologischer Gärten" als Stationen eines Grundnetzes für internationale phänologische Beobachtungen. Meteorologische Rundschau 10:130–133

Seidler C (1995) Langjährige phänologische Beobachtungen in Hartha (Tharandter Wald). Wissenschaftliche Zeitschrift der TU Dresden 44:69–72

van Vliet AJH, de Groot RS, Bellens Y, Braun P, Bruegger R, Bruns E, Clevers J, Estreguil C, Flechsig M, Jeanneret F, Maggi M, Marten P, Menne B, Menzel A, Sparks T (2003) The European phenology network. Int J Biometeorol 47:202–212

Part II
Phenologies of Selected Bioclimatic Zones

Chapter 9
Tropical Dry Climates

Arturo Sanchez-Azofeifa, Margaret E. Kalacska, Mauricio Quesada, Kathryn E. Stoner, Jorge A. Lobo, and Pablo Arroyo-Mora

Abstract Tropical dry climates are home to unique forest ecosystems, many of which are affected by strong phenological patterns. In the Americas, tropical dry forest ecosystems account for 40 % of their original extension, and are highly affected by deforestation patterns given the fact that they are located on high fertile forests. Moreover, forest presented in these ecosystems can be considered natural barometers to understand the impact of climate change. In this chapter we first presented an overview of the ecological characteristics of forests present in tropical dry climates. We explore also linkages between environmental change, conservation biology and land-use/cover change (including the role of remote sensing) in tropical dry environments. We conclude this chapter with an overview of future research avenues to improve our knowledge of the ecological mechanisms associated to forests in tropical dry climates.

9.1 Introduction

Based on the Holdridge life zone system (Holdridge 1967) approximately 111,599,269 km^2 around the world have a climate favorable for dry forest (Leemans 1992, Fig. 9.1). Of that area, 94 % is located in the tropics. Tropical

A. Sanchez-Azofeifa (✉) • M.E. Kalacska
Earth and Atmospheric Sciences Department, University of Alberta, Edmonton, AB, Canada
e-mail: arturo.sanchez@ualberta.ca

M. Quesada • K.E. Stoner
Centro de Investigaciones en Ecosistemas, Universidad Nacional Autónoma de México, Morelia, Mexico

J.A. Lobo
Biology Department, Universidad de Costa Rica, San Jose, Costa Rica

P. Arroyo-Mora
Department of Ecology and Evolutionary Biology, University of Connecticut, Storrs, CT 06269, USA

M.D. Schwartz (ed.), *Phenology: An Integrative Environmental Science*,
DOI 10.1007/978-94-007-6925-0_9, © Springer Science+Business Media B.V. 2013

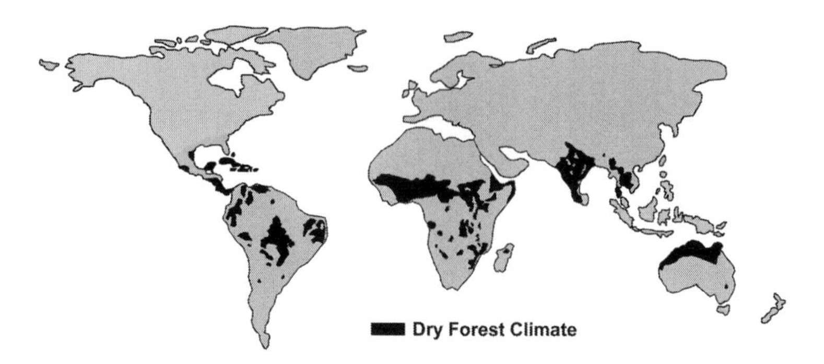

Fig. 9.1 Areas around the world with a climate favorable for supporting a dry forest ecosystem. Spatial resolution: 0.5° latitude by 0.5° longitude (Modified after Leemans 1992)

dry forests are found between the two parallels of latitude, the Tropics of Cancer and Capricorn (23°27′ N and S) where there are several months of little or no precipitation (Holdridge 1967). In general, tropical dry forests present two well-defined seasons: dry and wet season. During the dry season, usually with duration between 3 and 7 months, these forests drop between 85 and 100 % of their leaves. The dominant factor controlling leaf onset and fall is soil moisture (Janzen 1983). The degree of deciduousness is in general controlled by ecosystem composition, topography and forest age (Janzen 1983; Murphy and Lugo 1986; Lüttge 1997; Piperno and Pearsall 1998).

In general, Neotropical dry forests are less species rich than moist or wet forests. For example, 430 species of woody plants have been recorded in the wet forest of La Selva Biological Station, Costa Rica (Hartshorn and Hammel 1994), while in the dry forest of the Santa Rosa National Park, Costa Rica, 160 species (51 families) have been inventoried (Kalácska and Sánchez-Azofeifa, unpublished data). In addition, Gentry (1995) reports a range of 21–121 species (9 and 41 families) from various 0.1 ha plots around the Neotropics. However, there is more structural (e.g., wood specific gravity) and physiological (e.g., growth seasonality) diversity in the plant life forms of dry forests than in wet forests (Medina 1995).

Tropical forests that once formed a continuous habitat across Mesoamerica and some regions of the Pacific and Atlantic regions of South-America, are now found in fragmented patches (Whitmore and Sayer 1992; Heywood et al. 1994; Trejo and Dirzo 2000). Tropical deforestation is likely to affect both biotic and abiotic factors that control the phenological expression of plant communities with severe consequences to plant populations and the communities that interact or depend on them (Cascante et al. 2002; Fuchs et al. 2003). However, fortunately in certain regions of the Neotropics such as in Costa Rica, the secondary forests are in a state of regeneration through which the dry forests are also starting to recuperate (Arroyo-Mora 2002).

Both savannahs and dry forests (T-df) can co-occur in areas with the same climate, but the dry deciduous forests have a tendency to be found in areas with greater soil fertility (Ratter et al. 1973; Mooney et al. 1995). In many areas however, the occurrence of either savannah or dry forest is principally controlled by human disturbance (Maass 1995; Menaut et al. 1995).

The tropical dry forest ecosystem is one of the most fragile and least protected ecosystems in the world. Due to the favorable climatic conditions in which they are found, tropical dry forests have been heavily exploited for agriculture (Ewel 1999). Piperno and Pearsall (1998) argue that historically, tropical wet and dry forests had completely different associations with human activities. They state that the deciduous and semi-evergreen forests (especially the T-df) were the locations of the majority of the early human settlements in addition to being the home of the wild ancestors of many crop plants as well as the origin of animal husbandry. Even today, in most tropical countries, the majority of the agriculture and pasturelands are located in areas that used to be dry and moist forest. This pattern of both higher population density as well as higher intensity food production in the T-df as compared to wetter life zones may be a reflection of the historical tendency for humans to settle these areas (Piperno and Pearsall 1998). It appears that the tuberous plants that are rich in starch for human consumption seem to be more common in the seasonal forests, since the tuber is developed in part for energy storage during the dry season. The long dry season aided in burning of the vegetative cover in order to prepare the fields for agriculture (Piperno and Pearsall 1998). In addition, weeds and pests are less aggressive in the drier environments (Murphy and Lugo 1986).

Tropical dry forest phenology is an area that is still in its early stages of academic discovery, since historically more emphasis has been placed on tropical evergreen forests, especially the Amazon Basin (Lüttge 1997). Therefore, there is a need for efforts to understand tropical dry forest's phenological patterns and integrate its mechanisms at two levels: (1) In the context of conservation biology and (2) the context of land use and land cover change that are taking place on this rich agricultural frontier. In this chapter we document different aspects related to leaf phenology in the tropical dry forest ecosystem and its implications for satellite remote sensing. Emphasis is placed on presenting a description of the causes of leaf phenological change in this threatened ecosystem, and how these can be linked with conservation biology and land use/land cover change at the regional level.

9.2 Causes of Phenological Change

Several studies have indicated that the phenological expression of leaves, flowers and fruits are affected by biotic and abiotic factors. Abiotic factors such as changes in water level stored by plants (Reich and Borchert 1984; Borchert 1994, but also see Wright and Cornejo 1990; Wright 1991), seasonal variations in rainfall (Opler et al. 1976), changes in temperature (Ashton et al. 1988; Williams-Linera 1997), photoperiod (Leopold 1951; Tallak et al. 1981; van Schaik 1986), irradiance (Wright and Vanschaik 1994) or sporadic climatic events (Sakai et al. 1999), have been proposed as the main causes of leaf production or leaf abscission in tropical dry forest plants. In contrast, biotic factors, such as competition for pollinators or pollinator attraction (Robertson 1895; Janzen 1967; Gentry 1974; Stiles 1975; Appanah 1985; Murray et al. 1987; Sakai et al. 1999; Lobo et al. 2003), competition for seed dispersers, and avoidance of herbivory (Marquis 1988; Aide 1993; van Schaik et al. 1993) have been considered as the factors regulating the intensity and duration of leaf and flower production. The abiotic and biotic factors are not mutually exclusive, and it is likely that several are interacting to regulate the expression of each phenological phase.

In tropical dry forests, apart from foliage seasonality, relationships between water availability and structural and physiological characteristics such as hydraulic architecture or sensitivity to water stress produce a variety of phenological behaviors (Murphy and Lugo 1986; Bullock 1995; Holbrook et al. 1995; Lüttge 1997, among others). One of the major causes of the leaf phenological patterns (as mentioned above) in all tropical dry forest is the length of the dry season. This difference may be partly responsible for the differences in physical characteristics such as canopy height or biomass. Apart from leaf phenology, the length of the dry season and the seasonality of precipitation are also important for evolutionary adaptations of gene and seed dispersal, which are distinct in dry forests from the wet forests. In general, in dry forests most trees have conspicuous flowers and wind-dispersed seeds. Dry forests also have a lower biomass and a smaller stature than wet forests (Gentry 1995). Two other main factors that influence leaf phenological patterns are edaphic associations and topography since they determine the spatial heterogeneity of the available water (Murphy and Lugo 1995).

Water stress can vary at both regional and local scales. This variability induces a multitude of tree life forms with different leaf phenological patterns (Mooney et al. 1995). At the regional scale, the structure of the forest is greatly affected. It has been shown that as water availability decreases, so does the number of canopy stories as well as the horizontal continuity of the canopy (Murphy and Lugo 1995). Figure 9.2 compares the climate diagrams (Walter 1971) for fourteen sites from different life zones, ranging from wet to dry forests in Costa Rica as well as the dry forest in Chamela, Mexico. The mean monthly temperature (°C) and the monthly precipitation (mm) are scaled to represent the potential evapotranspiration. Dry months are represented by dotted areas, humid months by the vertical lines, and months with rain in excess of 100 mm are in solid black. Differences in the severity of the dry season as well as the pattern of the rainfall can cause the different leaf phenological patterns observed at various sites. For example, in Guanacaste, Costa Rica, Gentry (1995) estimates that 40–60 % of the tree species are deciduous whereas over 70 % are deciduous in Chamela, Mexico, where the severity of the dry season is more pronounced (Fig. 9.2).

The general response of tropical dry forest trees to the dry season is drought deciduousness where the woody plants lose between 85 and 100 % of their leaves. Some exceptions (wet season deciduous defined as inverse phenology, Fanjul and Barradas 1987) are present on dry evergreen forests and evergreen succulent plants in dry forests (Gentry 1995; Holbrook et al. 1995). Occasional anomalous rains in

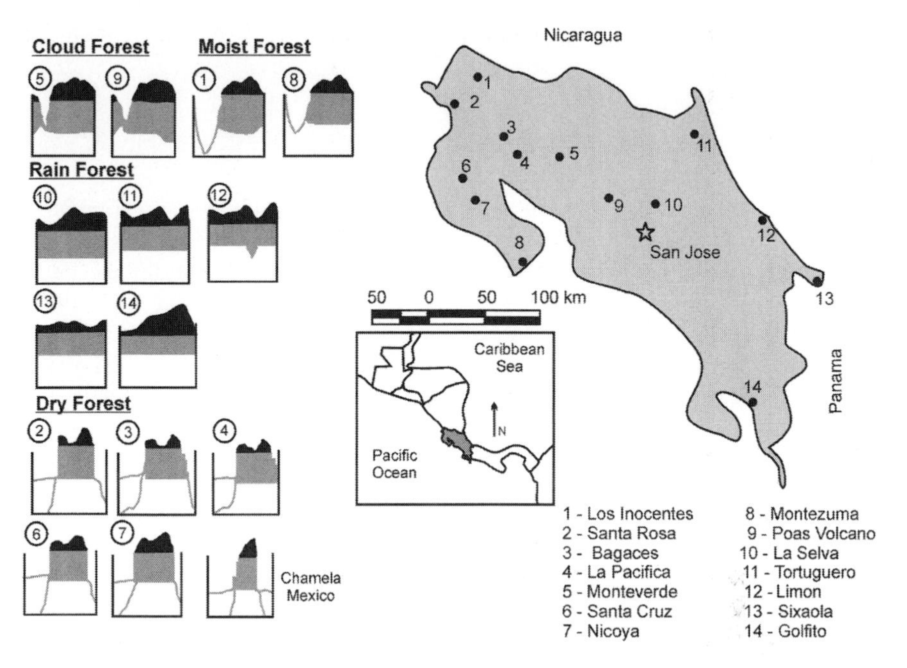

Fig. 9.2 Climate diagrams for 14 representative sites in Costa Rica and Chamela, Mexico

the dry season and drought spells in the wet season complicate this variation in resource availability. Growing periods are thus affected by the variability in flushing as it occurs in response to anomalous rains in the dry season or variation in the drying out process (Murphy and Lugo 1995). In a comparison between wet (La Selva) and dry (Comelco) sites in Costa Rica, Frankie et al. (1974) found that the forest at La Selva maintained its evergreen appearance throughout the year. However, even this wet forest experienced increased leaf flushing with the onset of the wet season. In Comelco they found that while leaf fall began as early as October, the majority of the trees lost their leaves in the dry season, with the peak in leaf fall occurring in March. Of the 113 species they inventoried at Comelco, 83 partially or completely lost their leaves and 19 were evergreen (ex. *Clusia rosea*, *Styrax argenteus*). The trees in the Riparian zones lost their leaves, but were simultaneously replaced. One species, *Lysiloma seemannii* had an unusual leaf-flushing pattern in that after it lost its leaves in the dry season, the new leaves did not appear until 1 month after the rainy season began. Certain species also brought new leaves in January and March but most of these species (for example, *Anacardium excelsum, Coccoloba padiformis*) were from the Riparian zones. In total, Frankie et al. (1974) found that 75 % of the species are affected by the seasonality of the precipitation in the dry forest, compared to 17 % in the wet forest. The timing of leaf flushing was also found to be very different: in the wet forest, most of the leaves were produced in the dry season, whereas in the dry forest the leaves were produced at the beginning of the wet season.

9.3 Phenology and Conservation Biology

In this section, we review the literature and present some of the main consequences that change or disruption of plant phenology may have on the viability of plant populations and animal communities that interact with them.

Biotic factors, such as competition for pollinators or pollinator attraction have been interpreted as important adaptive forces responsible for phenological patterns in tropical plants (Robertson 1895; Janzen 1967; Gentry 1974; Stiles 1975; Appanah 1985; Murray et al. 1987; Zimmerman et al. 1989; Sakai et al. 1999; Lobo et al. 2003). A disruption of the flowering phenological patterns (delayed or early flowering) caused by disturbance or fragmentation is likely to affect the behavior and visitation rate of pollinators. Fragmented landscapes reduce the amount of resources available, as well as appropriate areas for roosting and perching for nectarivorous bats and birds that serve as important pollinators for many tropical plant species (Feinsinger et al. 1982; Andren and Angelstam 1988; Bierregaard and Lovejoy 1989; Rolstad 1991; Saunders et al. 1991; Helversen 1993; Quesada et al. 2003). If the flowering pattern of plants that share pollinators of the same guild is displaced over time (Frankie et al. 1974; Stiles 1975; Fleming 1988), competition for the same pollinators will occur, resulting in negative consequences for the reproductive success of the plants and the ability of the pollinators to obtain resources over time. For example, in the tropical dry forest of the Chamela-Cuixmala Biosphere Reserve, Mexico, trees of the family Bombacaceae provided the main resource to the nectarivorous bats *Leptonycteris curasoae* during 8 months and *Glossophaga soricina* during 6 months. Both bat species concentrated on one bombacaceous species each month (Stoner et al. 2003). The sequential use of bombacaceuos species by these bats coincided with the flowering phenology of the tree species. These data suggest that changes in the flower phenology (e.g. reduction on the overall flower production) caused by habitat disruption may result in competition between these bat species and ultimately may result in local extinction, especially of endemic species that are common in this dry forest. A rare endemic nectarivorous bat that is only found in 4 states in Mexico, *Musonycteris harrisoni*, foraged on the bombacaceous tree *Ceiba grandiflora* during 3 months of the year (Stoner et al. 2002). Since this species has such a restricted distribution and is a specialist nectarivore, changes in flower phenology could be catastrophic for populations of this bat.

Timing of leaf flushing directly affects insect herbivores that depend upon flushing species to complete part of their life cycle (Janzen 1970, 1983; Dirzo and Dominguez 1995). Phenological changes caused by habitat loss will also disrupt the pollination patterns of many long-distance pollinators and trap-liners such as some large bees, hawkmoths, nectarivorous bats, and hummingbirds that have been shown to follow the flowering phenology of plants (Stiles 1977; Haber and Frankie 1989; Fleming et al. 1993; Frankie et al. 1997; Haber and Stevenson 2003). For example, in Costa Rica, hawkmoths regularly move from the lowland tropical dry forest to surrounding areas at higher elevations, following patterns of flowering resources (Haber and Stevenson 2003). Similarly, in México and the southwestern U.S. some nectarivorous bats have been shown to migrate following the availability of flower resources, mainly from the family Cactaceae and Agavaceae (Fleming et al. 1993).

Intra-specific variation in the frequency, duration, amplitude and synchrony of flowering by individuals also has been proposed as an important factor that affects the reproduction and the genetic structure of tropical plant populations in disturbed habitats (Murawski et al. 1990; Murawski and Hamrick 1992; Newstrom et al. 1994; Doligez and Joly 1997; Nason and Hamrick 1997). Flowering phenology directly determines the effective number of pollen donors and the density of flowering individuals, both of which affect the patterns of pollen flow between trees (Stephenson 1982; Murawski and Hamrick 1992). Plants with asynchronous flowering may experience a decrease in reproductive output, the amount of pollen, the number of pollen donors and the levels of outcrossing compared to individuals blooming during the same period. Fuchs et al. (2003) suggested that pollinator behavior is likely to change the mating patterns of *P. quinata*. This study showed that in disturbed fragmented habitats or in trees with early or late peak flowering, bat pollinators are more likely to promote selfing within trees (i.e., geitonogamy) and they have a tendency to produce singly sired fruits, whereas in undisturbed natural forests outcrossing is higher and multiple paternity is more common. The long-tongued bat (*Glossophaga soricina*), one of the main pollinators of *P. quinata*, has been shown to adopt a territorial behavior within a single plant in disturbed isolated environments with limited resources (Lemke 1984, 1985).

The timing of fruiting during the year (e.g. early or delayed fruiting), which may be altered as a result of environmental changes (e.g. edge effects on micro-meteorological variables such as air temperature and relative humidity, soil temperature, and solar radiation) associated with habitat disturbance, may affect potential vertebrate seed dispersers that, in turn, may affect the reproductive success of the plants they disperse (Fleming and Sosa 1994). Frugivorous Old World and New World bats are known to migrate or change habitats depending on the availability of fruit resources (Eby 1991; Stoner 2001). Similarly, the abundance of temperate and altitudinal migrant birds in tropical forests is closely associated with fruit abundance (Levey et al. 1994). Furthermore, displacement of fruiting phenology of tree species that are keystone resources because they provide fruits when resources are relatively scarce, could have negative consequences on populations of birds and mammals that disperse their seeds and ultimately negative effects on recruitment of the species they disperse (Howe 1984). Seed dispersal by animals is negatively affected by deforestation and results in lower recruitment in forest fragments.

Another factor affected by forest fragmentation is seed predation. In a tropical dry forest seed predation by bruchid beetles on the tree *Samanea saman* was higher in populations of trees found in continuous forest and found to be much less in isolated trees (Janzen 1978). The bruchid beetles, *Merobruchus columbinus* and *Stator limbatus* (Bruchidae) are specific seed predators of *S. saman*. It is likely that the populations of these bruchid species are affected by density dependent factors related to the availability and fluctuation of food resources within fragments, including seeds and flowers. Another explanation is that adult bruchids have to fly greater distances to find isolated trees than trees in continuous populations. This pattern of higher seed predation in populations from continuous forest also has been observed in the dry forest tree, *Bahuinia pauletia*.

Finally, the ultimate consequences of habitat reduction and phenological disruption is a decrease in reproductive plants, increasing the negative effects of endogamy, reducing the quantity and quality of pollen, and lowering the genetic variability of the progeny (Cascante et al. 2002). This likely will affect the viability and establishment of plant populations over time.

9.4 Phenology and Land Use/Cover Change

Remote sensing data provides the possibility for an instantaneous look at a large area with the opportunity of acquiring frequent repeat imagery for the same area. This is important for phenological studies because the temporal variability of the ecosystem can be captured at large scales. In particular, it is essential to consider leaf phenology in order to correctly characterize areas of deciduous forest. Since they measure surface reflectance, optical sensors have been widely used for land cover classification and characterization. However, it must be taken into consideration that one of the greatest limitations to optical sensors is cloud cover. And in the tropics, cloud cover is especially prevalent in the wet season where leaves are on making it difficult to obtained appropriate images for remote sensing applications. In contrast, cloud free imagery is more easily acquired during the dry season, with the drawback that the majority of the trees are leafless (Arroyo-Mora 2002). In addition, vegetation studies using reflectance data have generally focused on green leaves, with both dry vegetation and non-green components being neglected in comparison (van der Meer 1999). However, in areas of deciduous forest green leaves will not always dominate the spectral signature of the forest. In the dry season, only a small fraction of the spectra will be representative of green foliage. The majority of the pixels will be representing leaf litter, bark, branches and soil in various combinations. Therefore, this temporal variability of the spectral signatures that can be extracted from imagery must be taken into consideration in such environments.

As an example, two false color composite images of the same area of dry forest surrounding the Chamela Biological Station, Mexico, were acquired during the dry (March) and wet (August) seasons from the Landsat 7 ETM + sensor (not shown). While the two images visually look completely different, more importantly, the spectral signature of the forest also changes with the seasons. This is key because many algorithms rely on spectral signatures to classify areas. If the same unsupervised classification algorithm (Isodata) is run on the two images, 180 km^2 of forest cover is extracted from the wet season image, while only 26 km^2 of land cover

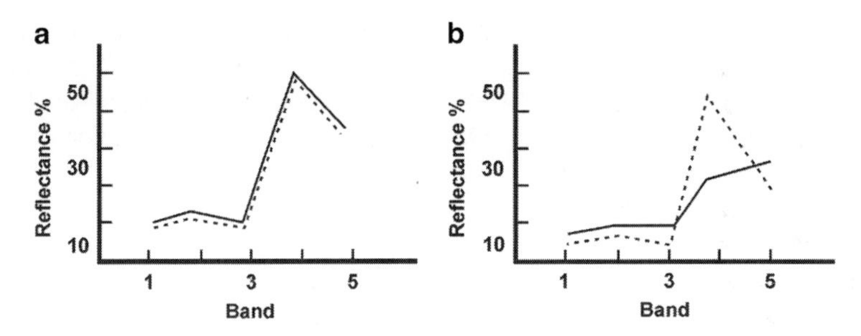

Fig. 9.3 Spectral signatures of the dry forest at the Santa Rosa National Park from 5 TM images. (**a**) Wet season (October) and (**b**) dry season (April). *Solid line* deciduous forest, *dashed line* evergreen forest

exhibits the spectral signature of forest in the dry season (Kalacska et al. 2001). In the dry season image, only the Riparian areas appear to have forest cover.

In a similar case study from the Santa Rosa National Park, Costa Rica, two images (dry season – April and wet season – October) of Landsat 5 TM were classified using an unsupervised classification into forest and non-forest classes. From the wet season image, 61 km^2 of forest were extracted, whereas from the dry season image only 18 km^2 were classified as forest (Kalacska et al. 2001). The discrepancy in the amount of forest extracted from the images in the two seasons is because dry deciduous forests (where trees lose their leaves), may seem to have the spectral signature of pasturelands or agricultural fields in the dry season (Fig. 9.3).

In the wet season, (Fig. 9.3a) the spectra for both the evergreen and deciduous components of the forest are similar. However, in the dry season (Fig. 9.3b), the spectral signature of the deciduous forest no longer resembles that of the evergreen forest. In fact, there is more than a 20 % difference in the near infrared band (band 4) between the two forest classes in the dry season. While these results are important at a local scale, their implications become more profound if regional or global scales are considered. For example, Sader and Joyce (1988) reported the total forest cover for Costa Rica as 17 %. If their map of forest distribution is examined, it can be seen that the province of Guanacaste and the Nicoya Peninsula, both with large extents of deciduous forest, are shown as almost completely non-forest. In a more recent classification of Guanacaste and the Nicoya Peninsula, using Landsat 7 ETM + imagery, Arroyo-Mora (2002) shows that the forest cover is actually 45 %. At the national scale, in the most recent remotely sensed forest cover inventory to date of the entire country of Costa Rica, Sanchez-Azofeifa and Calvo (2002) report a total forest extent 58 % greater than the other previous studies (Castro-Salazar and Arias-Murillo 1998). Seasonal changes in leaf phenology in the deciduous forest are part of the reason for those differences. Even at the spatial resolution of most global monitoring systems (1 km) significant areas of forest can be missed if only dry season images are used or if the phenological changes in leaf cover are not taken into consideration. This forest, which has been ignored by previous remote sensing analysis, is not uniform and includes different stages of succession with different levels of deciduousness

(Arroyo-Mora 2002). For example, in the recent global land cover classification from the MODIS Land Cover Classification Program, neither the area encompassing the Chamela Biological Station, Mexico nor the Santa Rosa National Park, Costa Rica is classified as forest. These complications are important not only for classification purposes, but also in many cases outputs from such data sets are used in global models like CENTURY. The calculations from such models are then further used to calculate baselines and benefits of a given policy for carbon sequestration, for example.

9.5 Final Remarks

Since so many organisms depend upon phenological patterns in tropical forests, it is crucial to document how these phenological patterns may be changed by deforestation and the resulting habitat fragmentation. In addition, studies aimed to understand shifts on phenological patterns (e.g. long terms of duration of growing season, length of dry season, and overall ecosystems productivity) are necessary to quantify the level of stress that tropical dry forests are under both climate and land-use/cover change. Because of their strong phenological patterns, these forests should be considered the number one barometer in tropical environments to quantify many important ecosystems responses to environmental change.

Future studies on phenological patterns of tropical plants should attempt to document intra-specific variation within distinct habitat types and under different levels of disturbance, in order to provide a clear understanding of ecosystem phenological response to different levels and types of disturbance. This information will be important in quantifying the effects of forest fragmentation on phenological patterns and ultimately on tropical ecosystems.

A wealth of information is available on studies conducted with remotely sensed data in both the temperate and tropical regions. And while the image processing techniques may be similar, the ground validation techniques are very different in certain aspects. The complexity (structural and temporal) of the tropical deciduous forests also requires special consideration when field data are being collected. In certain cases, for example when collecting Leaf Area Index (LAI), new sampling techniques need to be developed to account for the spatial and temporal heterogeneity of the forest. This is also the case if there are certain specific phenological patterns of interest. Both the scale of the sampling, as well as the technique should be determined by the required data. For example, biophysical parameters of the canopy such as LAI, vegetation fraction (VF) and the fraction of photosynthetically active radiation (f_{PAR}) have been successfully linked to remotely sensed data in many studies in conifer stands, temperate broad leaf forests and agricultural fields (Chen and Black 1991; Price and Bausch 1995; Chen and Cihlar 1996; Chen et al. 1997). However, similar techniques have not been as thoroughly explored in tropical dry forest environments, nor is there a clear understanding of the impact of phenology in these important biophysical variables. In addition, with the exception of a few studies such as Arroyo-Mora (2002) or Clark (2002) optical remote sensing studies in tropical environments have been predominantly conducted with either the Landsat (TM and ETM+) or AVHRR sensors. However, high spatial resolution multispectral sensors such as IKONOS (4 and 1 m spatial resolution and 4 spectral bands) and Quickbird (2 m and 60 cm spatial resolution, 4 spectral bands) have begun acquiring substantial worldwide archives and can play a key role in monitoring phenological processes in tropical dry forest environments. Also, with the introduction of ASTER (15 m spatial resolution, 14 spectral bands) data can be obtained quite economically. All three of these sensors may be used to capture detailed temporal changes in the dry deciduous forest. In addition, ALI (Advanced Land Imager) a new sensor from the EO-1 platform provides a more cost effective alternative for acquiring Landsat-type data.

Increased spectral resolution may also be an option to characterize deciduous forests from a remote sensing point of view. Hyperspectral sensors such as Hyperion (30 m spatial resolution and 220 spectral bands) or the air-borne sensor HYDICE (1 m spatial resolution and 220 bands) offer new possibilities for describing the phenological changes in the deciduous forest, but their application will be limited to the short life span of this sensor type. More small changes at the canopy level can be observed with these sensors than can be captured by multispectral sensors. These changes can be correlated to ground measurements such as chlorophyll concentrations as a function of age and complexity in order to begin modeling the seasonal changes in the ecosystem in greater detail. Hyperspectral data sets will provide a greater range of possibilities for deriving indices that may be more sensitive to the vegetation characteristics, as well as to phenological changes in dynamic environments.

References

Aide TM (1993) Patterns of leaf development and herbivory in a tropical understory community. Ecology 74:455–466

Andren H, Angelstam P (1988) Elevated predation rates as an edge effect in habitat islands – experimental-evidence. Ecology 69:544–547

Appanah S (1985) General flowering in the climax rain forests of Southeast Asia. J Trop Ecol 1:225–240

Arroyo-Mora P (2002) Forest cover assessment, Chorotega region, Costa Rica. Master's thesis, University of Alberta, Edmonton

Ashton PS, Givnish TJ, Appanah S (1988) Staggered flowering in the dipterocarpaceae – new insights into floral induction and the evolution of mast fruiting in the aseasonal tropics. Am Nat 132:44–66

Bierregaard ROJ, Lovejoy TE (1989) Effects of forest fragmentation on Amazonian understory bird communities. Acta Amaz 19:215–242

Borchert R (1994) Soil and stem water storage determine phenology and distribution of tropical dry forest trees. Ecology 75:1437–1449

Bullock SH (1995) Plant reproduction in neotropical dry forests. In: Mooney HA, Bullock SH, Medina E (eds) Seasonally dry tropical forests. Cambridge University Press, Cambridge

Cascante A, Quesada M, Lobo JJ, Fuchs EA (2002) Effects of dry tropical forest fragmentation on the reproductive success and genetic structure of the tree *Samanea saman*. Conserv Biol 16:137–147

Castro-Salazar R, Arias-Murillo G (1998) Costa Rica: toward the sustainability of its forest resources. FONAFIFO, San Jose

Chen JM, Black TA (1991) Measuring leaf-area index of plant canopies with branch architecture. Agric For Meteorol 57:1–12

Chen JM, Cihlar J (1996) Retrieving leaf area index of boreal conifer forests using Landsat TM images. Remote Sens Environ 55:153–162

Chen JM, Rich PM, Gower ST, Norman JM, Plummer S (1997) Leaf area index of boreal forests: theory, techniques, and measurements. J Geophys Res-Atmos 102:29429–29443

Clark D (2002) Applications of 1m and 4m resolution satellite imagery to studies of tropical forest ecology, management and secondary forest detection. In: Tropical forests: past, present, future. The Association of Tropical Biology Annual Meeting, Smithsonian Tropical Research Institute

Dirzo R, Dominguez CA (1995) Plant-herbivore interactions in Mesoamerican tropical dry forests. In: Bullock SH, Mooney HA, Medina E (eds) Seasonally dry tropical forests. Cambridge University Press, New York

Doligez A, Joly HI (1997) Genetic diversity and spatial structure within a natural stand of a tropical forest tree species, Carapa procera (Meliaceae), in French Guiana. Heredity 79:72–82

Eby P (1991) Seasonal movements of gray-headed flying-foxes, *Pteropus poliocephalus* (chiroptera, pteropodidae), from 2 maternity camps in northern New South Wales. Wildl Res 18:547–559

Ewel JJ (1999) Natural systems as models for the design of sustainable systems of land use. Agrofor Syst 45:1–21

Fanjul L, Barradas VL (1987) Diurnal and seasonal-variation in the water relations of some deciduous and evergreen trees of a deciduous dry forest of the western coast of Mexico. J Appl Ecol 24:289–303

Feinsinger P, Wolfe JA, Swarm LA (1982) Island ecology – reduced hummingbird diversity and the pollination biology of plants, Trinidad and Tobago, West-Indies. Ecology 63:494–506

Fleming TH (1988) The short-tailed fruit bat: a study in plant-animal interactions. University of Chicago Press, Chicago

Fleming TH, Sosa VJ (1994) Effects of nectarivorous and frugivorous mammals on reproductive success of plants. J Mammal 75:845–851

Fleming TH, Nunez RA, Sternberg LS (1993) Seasonal changes in the diets of migrant and non-migrant nectarivorous bats as revealed by carbon stable isotope analysis. Oecologia 94:72–75

Frankie GW, Baker HG, Opler PA (1974) Comparative phenological studies of trees in tropical wet and dry forests in lowlands of Costa Rica. J Ecol 62:881–919

Frankie GW, Vinson SB, Rizzardi MA, Griswold TL, O'Keefe S, Snelling RR (1997) Diversity and abundance of bees visiting a mass flowering tree species in disturbed seasonal dry forest, Costa Rica. J Kans Entomol Soc 70:281–296

Fuchs EJ, Lobo JA, Quesada M (2003) Effects of forest fragmentation and flowering phenology on the reproductive success and mating patterns of the tropical dry forest tree *Pachira quinata*. Conserv Biol 17:149–157

Gentry AH (1974) Flowering phenology and diversity in tropical Bignoniaceae. Biotropica 6:5

Gentry AH (1995) Diversity and floristic composition of neotropical dry forests. In: Bullock SH, Mooney HA, Medina F (eds) Seasonally dry tropical forests. Cambridge University Press, Cambridge

Haber WA, Frankie GW (1989) A tropical hawkmoth community: Costa Rican dry forest Sphingidae. Biotropica 21:17

Haber WA, Stevenson R (2003) Diversity, migration, and conservation of butterflies in northern Costa Rica. University of California Press, Berkeley

Hartshorn GS, Hammel BE (1994) Vegetation types and floristic patterns. In: McDade LA, Bawa KS, Hespenheide HA, Hartshorn GS (eds) La Selva: ecology and natural history of a neotropical rain forest. The University of Chicago Press, Chicago

Helversen OV (1993) Adaptations of flowers to the pollination by glossophaginae bats. In: Barthlott W (ed) Animal-plant interactions in tropical environments. Alexander Koenig Museum, Bonn

Heywood VH, Mace GM, May RM, Stuart SN (1994) Uncertainties in extinction rates. Nature 368:105–105

Holbrook NM, Witbeck JL, Mooney HA (1995) Drought responses of neotropical dry forest trees. In: Mooney HA, Bullock SH, Medina E (eds) Seasonally dry tropical forests. Cambridge University Press, Cambridge

Holdridge L (1967) Life zone ecology. Tropical Science Centre, San Jose

Howe HF (1984) Implications of seed dispersal by animals for tropical reserve management. Biol Conserv 30:261–281

Janzen DH (1967) Synchronization of sexual reproduction of trees within dry season in Central America. Evolution 21:620–637

Janzen DH (1970) Herbivores and the number of tree species in tropical forests. Am Nat 104:501–528

Janzen DH (1978) Reduction of seed predation on *Bauhinia pauletia* (Leguminosae) through habitat destruction in a Costa Rican deciduous forest. Brenesia 14–15:325–335

Janzen DH (1983) Costa Rica natural history. University of Chicago Press, Chicago

Kalacska M, SánchezAzofeifa GA, Rivard B, Arroyo-Mora P, Hall R, Zhang J, Dutchak K (2001) Implications of phenological changes in the extraction of Mesoamerican Tropical Dry Forests through remote sensing: a case study from Costa Rica and Mexico. Proceedings of the international conference, Wageningin University, Wageningin

Leemans R (1992) Global Holdridge life zone classifications. Digital Raster Data on a 0.5 degree Cartesian orthonormal geodetic (lat/long) 360x720 grid. In Global ecosystems database version 2.0, NOAA National Geophysical Data Center, Boulder

Lemke TO (1984) Foraging ecology of the long-nosed bat, *Glossophaga soricina*, with respect to resource availability. Ecology 65:538–548

Lemke TO (1985) Pollen carrying by the nectar-feeding bat Glossophaga Soricina in a suburban environment. Biotropica 17:107–111

Leopold AC (1951) Photoperiodism in plants. Q Rev Biol 26:247–263

Levey DJ, Moermond TC, Denslow JS (1994) Frugivory: an overview. In: McDade L, Bawa KS, Hespenheide HA, Hartshorn G (eds) La Selva: ecology and natural history of a neotropical rain forest. University of Chicago Press, Chicago

Lobo JA, Quesada M, Stoner KE, Fuchs EJ, Herrerias-Diego Y, Rojas J, Saborio G (2003) Factors affecting phenological patterns of bombacaceous trees in seasonal forests in Costa Rica and Mexico. Am J Bot 90:1054–1063

Lüttge U (1997) Physiological ecology of tropical plants. Springer, Heidelberg

Maass JM (1995) Conversion of tropical dry forest to pasture and agriculture. In: Mooney HA, Bullock SH, Medina E (eds) Seasonally dry tropical forests. Cambridge University Press, Cambridge

Marquis RJ (1988) Phenological variation in the neotropical understory shrub *Piper arieianum* – causes and consequences. Ecology 69:1552–1565

Medina E (1995) Diversity of life forms of higher plants in neotropical dry forests. In: Mooney HA, Bullock SH, Medina E (eds) Seasonally dry tropical forests. Cambridge University Press, Cambridge

Menaut J, Lepage M, Abbadie L (1995) Savannas, woodlands and dry forests in Africa. In: Mooney HA, Bullock SH, Medina E (eds) Seasonally dry tropical forests. Cambridge University Press, Cambridge

Mooney HA, Bullock SH, Medina E (1995) Introduction. In: Bullock SH, Mooney HA, Medina E (eds) Seasonally dry tropical forests. Cambridge University Press, New York, pp 1–8

Murawski DA, Hamrick JL (1992) Mating system and phenology of *Ceiba pentandra* (Bombacaceae) in central Panama. J Hered 83:401–404
Murawski DA, Hamrick JL, Hubbell SP, Foster RB (1990) Mating systems of 2 Bombacaceous trees of a neotropical moist forest. Oecologia 82:501–506
Murphy PG, Lugo AE (1986) Ecology of tropical dry forest. Annu Rev Ecol Syst 17:67–88
Murphy PG, Lugo AE (1995) Dry forests of Central America and the Caribbean. In: Mooney HA, Bullock SH, Medina E (eds) Seasonally dry tropical forests. Cambridge University Press, Cambridge
Murray KG, Feinsinger P, Busby WH, Linhart YB, Beach JH, Kinsman S (1987) Evaluation of character displacement among plants in 2 tropical pollination guilds. Ecology 68:1283–1293
Nason JD, Hamrick JL (1997) Reproductive and genetic consequences of forest fragmentation: two case studies of neotropical canopy trees. J Hered 88:264–276
Newstrom LE, Frankie GW, Baker HG, Colwell RK (1994) Diversity of long-term flowering patterns. In: McDade LA, Bawa KS, Hespenheide HA, Hartshorn GS (eds) La Selva: ecology and natural history of a neotropical rain forest. The University of Chicago Press, Chicago
Opler PA, Frankie GW, Baker HG (1976) Rainfall as a factor in the release, timing, and synchronization of anthesis by tropical trees and shrubs. J Biogeogr 3:231–236
Piperno DR, Pearsall DM (1998) Origins of agriculture in the lowland neotropics. Academic, New York
Price JC, Bausch WC (1995) Leaf-area index estimation from visible and near-infrared reflectance data. Remote Sens Environ 52:55–65
Quesada M, Stoner KE, Rosas-Guerrero V, Palacios-Guevara C, Lobo JA (2003) Effects of habitat disruption on the activity of nectarivorous bats (Chiroptera: Phyllostomidae) in a dry tropical forest: implications for the reproductive success of the neotropical tree *Ceiba grandiflora*. Oecologia 135:400–406
Ratter JA, Richards PW, Argent G, Gifford DR (1973) Observations on vegetation of northeastern Mato Grosso. Woody vegetation types of Xavantina Cachimbo expedition area. Philos Trans R Soc B 266:449–492
Reich PB, Borchert R (1984) Water-stress and tree phenology in a tropical dry forest in the lowlands of Costa Rica. J Ecol 72:61–74
Robertson C (1895) The philosophy of flower seasons and the phaenological relations of the entomophilous flora and the anthophilous insect fauna. Am Nat 29:97–117
Rolstad J (1991) Consequences of forest fragmentation for the dynamics of bird populations – conceptual issues and the evidence. Biol J Linn Soc 42:149–163
Sader SA, Joyce AT (1988) Deforestation rates and trends in Costa Rica, 1940 to 1983. Biotropica 20:11–19
Sakai S, Momose K, Yumoto T, Nagamitsu T, Nagamasu H, Hamid AA, Nakashizuka T (1999) Plant reproductive phenology over four years including an episode of general flowering in a lowland dipterocarp forest, Sarawak, Malaysia. Am J Bot 86:1414–1436
Sanchez-Azofeifa GA, Calvo J (2002) Final report on the extent to Costa Rica forest cover: year 2002:20
Saunders DA, Hobbs RJ, Margules CR (1991) Biological consequences of ecosystem fragmentation – a review. Conserv Biol 5:18–32
Stephenson AG (1982) The role of the extrafloral nectaries of *Catalpa speciosa* in limiting herbivory and increasing fruit production. Ecology 63:663–669
Stiles FG (1975) Ecology, flowering phenology, and hummingbird pollination of some Costa Rican Heliconia species. Ecology 56:285–301
Stiles FG (1977) Coadapted competitors – flowering seasons of hummingbird-pollinated plants in a tropical forest. Science 198:1177–1178
Stoner KE (2001) Differential habitat use and reproductive patterns of frugivorous bats in tropical dry forest of northwestern Costa Rica. Can J Zool 79:1626–1633

Stoner KE, Quesada M, Rosas-Guerrero V, Lobo JA (2002) Effects of forest fragmentation on the Colima long-nosed bat (*Musonycteris harrisoni*) foraging in tropical dry forest of Jalisco, Mexico. Biotropica 34:462–467

Stoner KE, Salazar KAO, Fernandez RCR, Quesada M (2003) Population dynamics, reproduction, and diet of the lesser long-nosed bat (*Leptonycteris curasoae*) in Jalisco, Mexico: implications for conservation. Biodivers Conserv 12:357–373

Trejo I, Dirzo R (2000) Deforestation of seasonally dry tropical forest: a national and local analysis in Mexico. Biol Conserv 94:133–142

Tallak NE, Muller WH (1981) Phenology of the drought deciduous shrub Lotus scoparius: climatic controls and adaptive significance. Ecol Monogr 51:323–341

van der Meer F (1999) Physical principals of optical remote sensing. In: Stein A, van der Meer F, Gorte B (eds) Spatial statistics for remote sensing – Remote sensing and digital image processing, vol 1. Kluwer Academic Publishers, New York

van Schaik CP (1986) Phenological changes in a Sumatran rain forest Indonesia. J Trop Ecol 2:327–348

van Schaik CP, Terborgh JW, Wright SJ (1993) The phenology of tropical forests – adaptive significance and consequences for primary consumers. Annu Rev Ecol Syst 24:353–377

Walter H (1971) Ecology of tropical and subtropical vegetation. Oliver and Boyd, Edinburgh

Whitmore TC, Sayer JA (1992) Deforestation and species extinction in tropical moist forests. In: Whitmore TC, Sayer JA (eds) Tropical deforestation and species extinction. Chapman & Hall, London

Williams-Linera G (1997) Phenology of deciduous and broadleaved-evergreen tree species in a Mexican tropical lower montane forest. Glob Ecol Biogeogr Lett 6:115–127

Wright SJ (1991) Seasonal drought and the phenology of understory shrubs in a tropical moist forest. Ecology 72:1643–1657

Wright SJ, Cornejo FH (1990) Seasonal drought and leaf fall in a tropical forest. Ecology 71:1165–1175

Wright SJ, Vanschaik CP (1994) Light and the phenology of tropical trees. Am Nat 143:192–199

Zimmerman JK, Roubik DW, Ackerman JD (1989) Asynchronous phenologies of a neotropical orchid and its euglossine bee pollinator. Ecology 70:1192–1195

Chapter 10
Mediterranean Phenology

Donatella Spano, Richard L. Snyder, and Carla Cesaraccio

Abstract This chapter describes the five Mediterranean zones around the world and discusses vegetation and environmental factors, including climate, that make the Mediterranean Climate zones unique. Several key reports on the role of climate and climate change on phenological development of Mediterranean ecosystems are presented and discussed. The chapter talks about the impact of current and projected temperature and precipitation on phenology and emphasizes the importance of precipitation patterns on response to higher temperature. One conclusion is that more studies are needed on drought impact on phenology since water stress can increase plant temperature and result in even faster phenological development. Drought can speed up phenological development, but it can also impede growth and lead to reduced productivity.

10.1 Mediterranean Characteristics

Mediterranean-type ecosystems are found in the far west regions of continents between 30° and 40° north and south latitude (Fig. 10.1). They cover about 2.73 million km^2 (IUCN 1999), with the majority (i.e., 73 %) of the ecosystem in the Mediterranean Basin including parts of Spain, Turkey, Morocco, and Italy

D. Spano (✉)
Department of Science for Nature and Environmental Resources (DipNet), University of Sassari, Sassari, Italy

Euro-Mediterranean Centre for Climate Change (CMCC), Sassari, Italy
e-mail: spano@uniss.it

R.L. Snyder
Department of Land, Air, and Water Resources, University of California, Davis, CA 95616, USA

C. Cesaraccio
Institute of Biometeorology, National Research Council, Sassari, Italy

M.D. Schwartz (ed.), *Phenology: An Integrative Environmental Science*,
DOI 10.1007/978-94-007-6925-0_10, © Springer Science+Business Media B.V. 2013

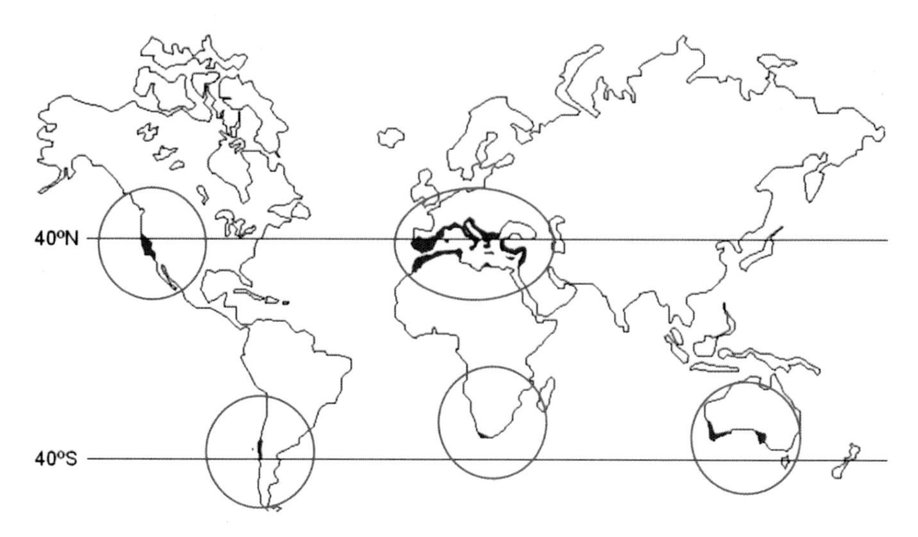

Fig. 10.1 Geographical distribution of Mediterranean-type ecosystems

(Rundel 1998). Areas are also found in California, Chile, Southwest and Southern Australia, and South Africa. In response to the climate, similar woody, shrubby plants, with evergreen sclerophyll leaves, have developed in communities of varying density. The names for the shrub vegetation vary by region because of language and plant structure. Common names for the vegetation include: *maquis* and *garrigue* in the Mediterranean Basin, *chaparral* in California, *matorral* in Chile, *fynbos* or *renosterveld* in South Africa, and *mallee* (*kwongan* or *heathlands*) in Australia.

Mediterranean ecosystems formed as a result of the unique climate, which falls in a transition between dry, tropical and temperate zones (Fig. 10.1).

The main characteristics are (1) variable winter rainfall, (2) summer droughts of variable length, (3) intensive summer sunshine, (4) mild to hot summers, and (5) cool to cold winters. Commonly, there is a cold ocean current off the West coast of regions with a Mediterranean climate that strongly influences the weather. The range of summer and winter temperatures mainly depends on proximity to the ocean (or sea) with higher temperatures near the coast during cooler periods and higher temperatures inland during warmer periods. Temperatures also vary with elevation having consistently cooler temperature in the mountains. Excluding mountains, the annual precipitation range at lower elevations typically varies between 250 and 900 mm with most falling in the winter and spring (i.e., November–April in the Northern Hemisphere and May–October in the Southern Hemisphere). Outside of the Mediterranean Sea region, westerly winds over cold ocean currents often lead to heavy marine fog that maintain low temperatures on the coast during summers. In the winter, the coastal areas tend to be fog free, whereas inland valleys that receive winter rainfall are prone to high-inversion radiation fog. Differences in relative humidity are mainly related to temperature variations over the zone rather than absolute humidity. Because the Mediterranean Sea has variable and warmer surface temperatures, the dew point temperatures are more variable over the Mediterranean Sea region.

Table 10.1 Climate classification based on length of summer drought period

Classification	Drought period (months)
Perarid	11–12
Arid	9–10
Semiarid	7–8
Subhumid	5–6
Humid	3–4
Perhumid	1–2

The five Mediterranean zones have similar characteristics, but there are important differences within each of the regions. Differences within a region are mainly related to the length of the summer drought period, which generally decreases as one moves poleward. For example, di Castri (1973, 1981) described a six-zone climate classification based on the length of drought period after Emberger (1962), as shown in Table 10.1.

Soil and climate both influence the development of natural vegetation, so a short discussion of soils is included here. More extensive discussions are presented by Thrower and Bradbury (1973), Zinke (1973), di Castri et al. (1981), Davis and Richardson (1995), Joffre et al. (1999), and Joffre and Rambal (2002). Most Mediterranean soils exhibit (1) considerable erosion, (2) alluvial deposition, (3) limited profile development, and (4) decreased soil development with increasing elevation. Because limestone is deficient in some areas, those soils often have water infiltration problems. Due to the lower precipitation, parent materials weather slower in Mediterranean zones than in more humid regions. Because of seasonal drying, some soils are dominated by shrinking and swelling processes and produce Vertisols. The soils tend to vary from reddish to brownish with increasing elevation. Higher precipitation and cooler temperatures at higher elevations have led to the development of predominant brownish podzolic soils with higher organic matter and moderate lime accumulations at middle elevations (500–1,000 m). At low elevations (0–500 m) with less precipitation and higher temperature, older *terra rossa* soils, which have lower organic matter and a reddish color due to iron oxidation, developed from limestone. In the river valleys, alluvial soils are found as highly weathered soils in terraces, light and well-drained in alluvial fans, and heavy and poorly drained in the valley floors. In some valley basins, fine textured soils have greatly inhibited drainage. In many areas within Mediterranean zones, older paleosoils, which were formed under different climate conditions, are prevalent.

10.2 Vegetation Types

Although the climate developed relatively recently in geologic time, distinctive flora with similar characteristics has evolved in the five Mediterranean zones. While the climate is similar within the five Mediterranean zones, high heterogeneity in plant communities is common. This heterogeneity developed because of large variations in landforms, microclimate, soils, phylogenetic origin, evolutionary strategy, ecological tolerance, and land use within the ecosystem.

The appearance of natural vegetation and landscape forms is strikingly similar between the five Mediterranean zones. The plants are woody, shrubby, and evergreen. The plant leaves tend to be small, broad, stiff, thick, and waxy or oily. In some locations, there are small trees with or without an understory of annual and herbaceous perennials. The vegetation represents different successional stages in relation to climate, topographical features, and human impact (di Castri et al. 1981), and it is prone to wildfires.

di Castri et al. (1981) presented a classification of six Mediterranean climate types (Table 10.1), based on the length of summer drought, and provided information on the structure of vegetation in each of the climate types. He noted that there were several overlapping clusters of characteristics between the five regions. However, the similarities between vegetation structures were most apparent between California and Chile, and between Australia and South Africa.

Mediterranean ecosystems have a large plant diversity including about 48,250 species, which is approximately 20 % of the world total (Cowling et al. 1996). The Mediterranean Basin, South Africa, Southwestern Australia, and California have about 25,000; 8,550; 8,000; and 900 species, respectively (Archibold 1995; Rundel 1998).

The Mediterranean Basin is mainly covered by scrub, sparse grass, or bare rock. However, there are scattered evergreen trees that suggest earlier presence of mixed forests. Several species of *Quercus* including the holm oak (*Quercus ilex*) prevail in the west with cork oak (*Q. suber*) dominant on non-calcareous soils. *Arbutus unedo* and other shrubs are found in the same plant communities. As aridity increased in the east, Kermes oak (*Q. coccifera*) became more prevalent than holm oak. Stone pine (*Pinus pinea*), cluster pine (*P. pinaster*), and Aleppo pine (*P. halepensis*) are common at higher elevations in the west. In the drier eastern region (e.g., Syria, Lebanon, and Israel), *Q. calliprinos,* which is an evergreen oak, and deciduous oaks are common. Corsican pine (*P. nigra*) and *P. brutia* often dominate in locations where wildfires occurred. *Q. ilex* is also found on the Atlas Mountains of North Africa at the elevation of 2,000 m. Shrublands are divided into *maquis*, which comprises evergreen shrubs and small trees about 2.0 m tall, *garrigue* on calcareous soils, and *jaral* on siliceous soils. All communities have representative species and the size depends on local conditions.

South African sclerophyll plant communities include mountain and coastal types (Moll et al. 1984). The mountain *fynbos* mainly consists of broad-leaved proteoid shrubs, which are found at elevations up to about 1,000 m and grow to heights between 1.5 and 2.5 m. At higher elevations, 0.2–1.5 m tall ericoid shrubs are dominant. In addition, 0.2–0.4 m tall shrubs and tussocky hemicryptophytes are present in the high elevation communities. Tussocky restioid shrubs, which reach 0.3 m, dominate communities at higher elevations. In high-elevation, arid regions, abundant succulent forms of karoo are the most common vegetation. The west coast is dominated by open ericoid cover with shrubs growing to 1.0 m tall. Small shrubs, grasses, and annuals form an open heath with 1–2 m tall proteoids along the south coast.

Western Australia is dominated by forests of karri (*Eucalyptus diversicolor*) and jarrah (*E. marginata*). Karri is restricted to regions with acidic soils (Rossiter and Ozanne 1970) and it grows in association with other tall eucalyptus. *Casuarina decussata* and species of *Banksia* are common in the understory of these forests. Jarrah forests occur on lateritic soils in areas with lower precipitation. These forests change to wandoo (*E. rudunca*) woodland as the annual precipitation decreases. The western region is separated from South Australia by the acacia shrubland. *Mallee* is the dominant cover in the southeastern Mediterranean zone. The prevalent species are *E. diversifolia* and *E. incrassata*. In more favorable sites, species such *E. behriana* grow with ground cover of herbs and grasses with few sclerophyllous shrubs (Specht 1981). These communities integrate with sclerophyll forests of stryngbark (*E. baxteri*) and messmate (*E. obliqua*).

The Chilean *matorral* communities occur in the coastal lowlands and on the west facing slopes of the Andes. Most *matorral* species are 1–3 m tall, evergreen shrubs with small sclerophyllous leaves. Many spinescent species and drought-deciduous shrubs are also important in these regions (Rundel 1981). *Salix chilensis*, *Cryptocarya alba*, and other trees are found in wetter regions with shrubs forming a cover. *Matorral* evergreen shrubs (e.g., *Lithaea caustica* and *Quillaja saponaria*) dominate coastal regions. In more arid locations, succulent species and *Fluorensia thurifera* are common. The central valley of Chile is dominated by *Acacia caven* (Ovalle et al. 1990, 1996).

California *chaparral* typically consists of a dense cover of 1–4 m tall, evergreen shrubs. In California, and particularly in the south, chamise (*Adenostoma fasciculatum*) is common and California lilac (*Ceanothus cuneatus*) is sometimes associated. In the Sierra Nevada foothills, *chaparral* occurs above 500 m elevation. Pure stands of California lilac are considered a fire-successional form in Southern California, but it is a dominant species of *chaparral* in Northern California (Hanes 1981). Manzanita (*Arctostaphylos spp.*) occurs throughout California, especially where there is snow and temperatures drop below freezing in winter. Various *Quercus* species may be present on lower hillsides. Coastal sage scrub (e.g., *Artemisia californica*) is the main vegetation along the coast.

Common characteristics of Mediterranean zones are summer drought, fire, tectonic instability, and variable floods and erosion during winter. Perhaps the most important of these is summer drought; however, drought tends to be more severe in California, Chile, and the subarid region of the Mediterranean Basin (Rundel 1995, 1998). In fact, the Mediterranean climate exhibits extreme year-to-year variability. In the last century, the rainfall trends were relatively consistent showing a general decrease. Mediterranean ecosystems are likely to be highly affected by climate change (Cubasch et al. 2001; IPCC 2001, 2007) with a higher variability of precipitation in many areas (Rodrigo 2002; Gao et al. 2006; Beniston et al. 2007; Giorgi and Lionello 2008; Somot et al. 2008). In the Mediterranean Basin, rainfall is projected to decrease by approximately 15 % for March–May, 42 % for June–August and 10 % for September–November (Somot et al. 2008). Concurrently, inter-annual variability is expected to increase (Gao and Giorgi 2008), and the frequency of long drought periods (4–6 months) to be multiplied by 3 at the end of this century (Sheffield and Wood 2008). In addition, warmer conditions may increase evapotranspiration demand by 200–300 mm in the south, which will intensify the characteristic summer drought of the Mediterranean region (Valladares et al. 2004). The drier summers could seriously impact on plant activity (Christensen et al. 2007) and ecosystem productivity (Valladares et al. 2004; Ogaya and Peñuelas 2007). Warming will also affect the other seasons and, although less intense, drought will probably extend farther into the spring and autumn (Giorgi et al. 2004). Since spring is the main vegetation growth season, changes in temperature and precipitation could strongly affect the structure and functioning of Mediterranean ecosystems effects on plant phenology and growth (Bernal et al. 2011).

Although less well documented, it is likely that more aridity will not eliminate the intermittent rainy years (Beniston et al. 2007) that occur in some regions. These sporadic rainy years have a strong impact for regeneration (Castro et al. 2005; Holmgren et al. 2006; Mendoza et al. 2009). Despite its importance, the role of intermittent rainy years in maintenance of ecosystem structure needs more study (Castro et al. 2005; Holmgren et al. 2006).

Dense cover and high woody biomass of shrublands of Mediterranean ecosystems make them prone to wildfire, which is an important disturbance regime in Mediterranean climates. Frequency of natural wildfire differs greatly between and within Mediterranean zones depending on many factors (Mooney and Conrad 1977; Rundel 1981, 1983; Trabaud and Prodon 1993; Oechel and Moreno 1994).

Although fire is a natural disturbance in Mediterranean ecosystems, the frequency and intensity of wildfires has increased dramatically in recent decades (Rundel 1998). This has led to changes in forest vigor, structure and soil stability (Kuzucuoglu 1989; Naveh 1990). Climate change is likely to increase fire frequency and fire extent (Fischlin et al. 2007). Greater fire frequencies are noted in Mediterranean Basin regions (Pausas and Abdel Malak 2004) with some exceptions (Mouillot et al. 2003). Double CO_2 climate scenarios increased projected wildfire events by 40–50 % in California (Fried et al. 2004), and doubled the fire risk in Cape Fynbos, South Africa (Midgley et al. 2005), favoring re-sprouting plants in Fynbos (Bond and Midgley 2003), fire-tolerant shrub dominance in the Mediterranean Basin (Mouillot et al. 2002), and vegetation structural change in California (e.g., from needle-leaved to broad-leaved trees and from trees to grasses) and reducing productivity and carbon sequestration (Lenihan et al. 2003). Studies by Viegas et al. (1992) helped to identify critical periods of high potential fire risk of Mediterranean shrubland ecosystems.

Pellizzaro et al. (2007) in Italy and Viegas et al. (2001) in Portugal and Spain showed that knowledge of both the mean moisture content and the phenology of plants are useful for fire risk assessment. Two groups of Mediterranean species were identified for different ranges of leaf fuel moisture content (LFMC) values throughout the year and different relationships between LFMC, seasonal changes of meteorological conditions and phenological stages. The experimental data reveal the different physiological and morphological responses by vegetation to cope with the summer drought season typical of the Mediterranean climate. Species such as *Cistus* and *Rosmarinus* avoid water stress by adjusting the growing period or by limiting water loss by reducing their transpiring surface. These species generally grow as shallow rooted shrubs and, therefore, are particularly affected by variations in moisture content of the surface soil layers (Correia et al. 1992; Munné-Bosch et al. 1999; Gratani and Varone 2004). *Pistacia lentiscus* and *Phillyrea angustifolia* are evergreen deep-rooted sclerophyllous species, tolerant to water stress and affected by drought conditions only when particularly severe (Kummerow 1981; Correia et al. 1992; Manes et al. 2002; Alessio et al. 2004). These species showed the highest values of LFMC in spring during sprouting and flowering phases. In summer, they partially reduced their vegetative activity that again increased in autumn. Consequently, *Pistacia* and *Phillyrea* showed a seasonal pattern of LFMC, although it was characterised by a range of values narrower than others such as *Cistus monspeliensis* and *Rosmarinus officinalis*.

Deforestation, grazing, agriculture, fire events, and fire suppression have changed vegetation community structure especially in recent decades. Increased urbanization and land abandonment has led to uneven management more frequent and larger wildfire disturbances (Rundel 1998).

Livestock grazing has greatly influenced Mediterranean ecosystems. A good example is in California, where livestock grazing converted much of the grassland from native perennials to exotic annuals from the Mediterranean Basin even prior to immigration by large numbers of people of European ancestry (Rundel 1998). In the late 1800s, agricultural expansion into the Central Valley and Southern California caused extensive changes in natural communities. Later, agricultural and urban expansion led to large changes in vegetation along the coast. Human activities influenced grassland and oak woodlands of the State mainly by replacing native perennial grasses with introduced annual grasses from Europe. Native Americans purposely set fires to control vegetation, but European immigrants introduced fire suppression as a management strategy in the late 1800s. This change in management has led to fewer but more intense wildfires (Minnich 1983; Rundel and Vankat 1989).

When Spanish settlers arrived in Chile in the mid-1500s, they introduced grazing and agriculture that greatly changed the natural ecosystems. The impact is most obvious in the semi-arid transition region where over-grazing has caused devegetation and desertification (Ovalle et al. 1990, 1996). Also, much of the Central Valley now is covered with exotic annual grasses rather than the native grasses (Gulmon 1977). Recently, Chile has become more urban having a plethora of abandoned farms and ranches as the population leaves rural areas. This has led to a big increase in mainly anthropogenic wildfires that have grown in size and intensity. Even more recently, the planting of winegrape vineyards has expanded dramatically in Chile and in California at the expense of native woodlands (Rundel 1998).

Agricultural development in Southwest Australia has resulted in widespread fragmentation of *mallee* ecosystems mixed in with agricultural lands (Rundel 1998). The fragmented habitats tend to be too small to maintain viable plant populations, which are also impacting on animal diversity. Deforestation is a big problem in native eucalypt forests, and the resulting rise in water tables has led to problems with saline paleosoil profiles (Rundel 1998), which threatens agriculture as well as the replanting of forests. The introduction of exotic species has resulted in problems with biological diversity in the Mediterranean climate zones (Thuiller et al. 2005; Gritti et al. 2006; García-de-Lomas et al. 2010).

Anthropogenic impacts on the Mediterranean ecosystems in South Africa are less obvious than in the other regions to a large extent because the soils of the region are not conducive to support cereal and vegetable production (Rundel 1998). However, large animal hunting and deforestation have impacted on the vegetation. A large introduction of non-native trees, especially Australian acacias, along rivers and streams, has occurred.

10.3 Phenology in the Mediterranean Climate

Mediterranean regions show seasonal changes in resource availability, which affect growth and reproductive activities of vegetation. Resource fluctuations have a strong influence not only on the structure and composition of the vegetation but also on the seasonal behavior pattern of the species. For example, the sclerophyllous forest can remain active throughout the year, but there is a distinct annual growth rhythm because photosynthesis is limited by drought and nutrients. However, several other species shed leaves during summer drought period.

Over recent decades, the economic, ecological, and cultural value of Mediterranean vegetation was increasingly recognized (Quezel 1977; Joffre and Rambal 2002; Rundel 2007), and many studies were devoted to improving management and protection of Mediterranean areas. In particular, comparative research on the structure of Mediterranean region ecosystems, which included a detailed assessment of phenological species behavior in the different areas, was performed. The first systematic study on Mediterranean vegetation was presented by Mooney et al. (1977) within the International Biological Program (IBP), which started in 1970. The authors summarized the results of the comparison of the structural, functional, and evolutionary features of California and Chile ecosystems. At the plant community level, there is a longer protraction of each phenological event in Chile than in California due to both the greater diversity of growth form and more moderate climate in Chile (Mooney et al. 1977). In addition, di Castri et al. (1981) pointed out that there were more species with non-overlapping phenological activities in Chile.

As more information on the phenology of ecosystems in the Mediterranean Basin, South Africa and Australia became available, it was noted that there is a pronounced seasonal rhythm in the vegetative growth throughout the year in Mediterranean regions. However, less similarity in phenological pattern was found when comparing Chile, California, and Mediterranean Basin with South Africa and Australia. In South Africa and Australia, shrubs grow in the summer as well as in the spring (Cody and Mooney 1978) because of differences in origin of the biota (Specht 1973) and nutrient availability in the soils (Specht 1979, 1981).

Comparative analysis of Mediterranean species development was intensified during the 1980s with more emphasis on the interactions between temperature and water as limiting factors. Tenhunen et al. (1987) summarized the results of years of cooperative work between several scientists on functional analysis in Mediterranean ecosystems. The work included studies on plant growth and development. Montenegro (1987) discussed the difficulty in comparing these ecosystems because of different methodologies used to quantifying phenology and growth. In Portugal, phenological observations conducted on different species (*Quercus coccifera* and *Q. suber*, *Arbutus unedo*, and *Cistus salvifolius*) showed that the flowering stage occurred during all times of the year except the driest months in late summer and the coldest months in winter. Shoot growth was intense in the absence of water stress, and leaf drop was possibly more intense during drought (Pereira et al. 1987). Similar results were obtained on *Q. coccifera* and *Arbutus unedo* in Greece (Arianoutsou and Mardilis 1987), although the responses to the physical environment were not synchronous for the two species. Moll (1987) observed that the differences between vegetation in South Africa and in other Mediterranean regions reported by Mooney and Kummerow (1981) were mostly due to the fact that they compared non-heath shrubland in Chile, California, and the Mediterranean Basin with heath shrubland in South Africa.

The occurrence of vegetative primary growth in spring is observed in the Mediterranean climate regions of the Northern hemisphere, and in Chile. In the South African *fynbos*, however, this phenophase is observed throughout the year, mainly due to the milder winters (Orshan 1989). The protraction of stem vegetative growth towards sub-optimal periods, like the end of winter or the beginning of summer, seems difficult to avoid for species with long phenological cycles, such as *Lonicera implexa*, *Buxus fruticosum* or *B. sempervirens* (Milla et al. 2010).

In recent decades, more attention was directed to the relationship between phenological events and seasonal fluctuations in nutrient and water uptake. A phenological survey conducted in central Italy (de Lillis and Fontanella 1992) showed the effect of increasing water stress and nutrient limitations on several species (*Cistus monspeliensis*, *Pistacia lentiscus*, *Calicotoma villosa*, *Quercus ilex*, *Erica arborea*, *Arbutus unedo*, *Phillyrea media*, *Smilax aspera*, and *Ruscus aculeatus*). Phenological rhythm of the community was closely correlated with changes in environmental conditions, and large variation occurred among species. In all species, peak growth was reached between March and early July, flowering occurred before July except for *A. unedo* and *S. aspera*, which flowered in autumn and winter, and fructification was unrelated to summer aridity. An analysis of water availability and growth modulation allowed for division into drought-tolerant species (*Pistacia lentiscus*, *Phillyrea media*, *Arbutus unedo*, and *Ruscus aculeatus*), drought-deciduous species (*Calicotoma villosa*), and semi-deciduous species (*Cistus monspeliensis*). Carbon leaf concentration peaked and nitrogen decreased when growth stopped. Correia et al. (1992) compared the phenological characteristics of four summer semi-deciduous (species of *Cistus*) and evergreen (*Pistacia lentiscus*) shrubs in Portugal, corresponding to earlier and later successional stages of vegetation. The *Cistus* species were similar in growth, flowering, and fruiting phenology, showing a long period of leaf emergence relative to *P. lentiscus*, which had a flush-type leaf emergence and an almost simultaneous leaf fall. In general, *Pistacia* showed lower leaf nitrogen contents than the *Cistus* species, with minimum value in winter, when the *Cistus* species had the highest concentrations of nitrogen. However, increased drought frequency and intensity is likely to greatly affect phenology of these species in the future. Little information is known about the relationship between phenological stage occurrence and duration and intensity of drought period.

Spano et al. (1999) recorded weekly phenology observations for a period of 11 years on the common species *Pistacia lentiscus*, *Olea europea*, *Myrtus communis*, *Quercus ilex*, *Spartium junceum*, and *Cercis siliquastrum*, and on the exotic species *Robinia pseudoacacia*, *Salix chrysocoma*, and *Tilia cordata* in Sardinia to investigate the impact of drought on phenology. The range of phenological event dates for the nine species varied widely, especially for flowering of the exotic species. The authors showed that difference in accumulated degree-days could not explain the variations in observed phenological development. During the winter and spring, there seemed to be little difference in the flowering dates of common species. However, the non-native species *Salix chrysocoma* and *Tilia cordata* showed more inter-annual variability and both exhibited later flowering when there was more rainfall during March (i.e., prior to flowering). There was no relationship with rainfall recorded two or more months prior to flowering.

Duce et al. (2000) conducted phenological observations on three *maquis* species and oak trees over the period 1997–1999 at *Giara di Gesturi*, a nature reserve located in Southern Sardinia, Italy. About 46 % oak trees (*Quercus suber*) and about 32 % successional Mediterranean *maquis* with four dominant species (*Arbutus unedo*, *Pistacia lentiscus*, *Phillyrea angustifolia*, and *Myrtus communis*) cover the reserve. Flowering and full ripe fruit stages occurred about 1 month later in 1997 for *Quercus suber* and *Pistacia lentiscus* and the response was related to rainfall distribution and water deficit. In 1997, both species were affected by the lack of spring rainfall, which led to a longer and more intense drought period. In 2002, Duce et al. showed a large species variation in terms of observed flowering dates and cumulative degree-day values, indicating that other factors in addition to heat units affected plant development (Duce et al. 2002). In general, the flowering date was postponed when the soil water was not limiting, so flowering occurred earlier during drought years.

Simões et al. (2008) analyzed the phenological patterns, growth and internal nutrient cycling of the Mediterranean shrubs *Cistus salvifolius* and *Cistus ladanifer* during 2 years of contrasted precipitation to compare their life responses and their competitive potential to cope with future climate change and drought. The two species exhibited different responses to summer drought. *C. salvifolius* showed high seasonal dimorphism in plant structure, with greater leaf shedding before summer drought, while the structure and biomass of *C. ladanifer* showed little change throughout the year. The increase in length and intensity of drought also caused greater variation on growth rates and leaf duration and shedding in *C. salvifolius* than in *C. ladanifer*. The results suggest that *C. ladanifer* has greater stress-tolerance ability against drought. The phenological pattern of *Halimium atriplicifolium* and *Thymus vulgaris* were analyzed by Castro-Díez et al. (2005) to provide information on their response to unfavorable periods of Mediterranean climate (winter and summer). The two species arrested all phenological activities, during colder months, probably due to a cold-induced decrease of meristem activity (Kozlowski and Pallardy 1997). In contrast, a species-dependent response to summer drought was found, as *T. vulgaris* ended all phenophases in June, while *H. atriplicifolium* extended most of them into a period with virtually no rainfall (July and August). *T. vulgaris* seems suffer from more severe water stress than *H. atriplicifilium* due to its shallower root system and arrested phenological activity earlier in the summer. The different morphological and phenological traits of long and short shoots in the two species suggest a specialization in carbon gain along different time periods of the year.

10.4 Phenology and Climate Change

The last IPCC AR4 report (Christensen et al. 2007) stated that the Mediterranean ecosystems may be one of the most impacted by global change drivers (Sala et al. 2000). Diverse Californian vegetation types may show substantial cover change for temperature increases greater than about 2 °C. For example, mixed deciduous forest may expand at the expense of evergreen conifer forest (Hayhoe et al. 2004). The bioclimatic zone of the Cape Fynbos biome could lose 65 % of its area under warming of 1.8 °C relative to 1961–1990 (2.3 °C, pre-industrial) with species extinction of 23 % in the long term (Thomas et al. 2004). For Europe, only minor biome-level shifts are projected for Mediterranean vegetation types (Parry 2000), contrasting with between 60 and 80 % of current species projected not to persist in the southern European Mediterranean region (global mean temperature increase of 1.8 °C) (Bakkenes et al. 2002). Inclusion of hypothetical and uncertain CO_2-fertilisation effects in biome-level modeling may partly explain this contrast. Land abandonment trends, however, facilitate ongoing forest recovery (Mouillot et al. 2003) in the Mediterranean Basin, complicating projections.

In Southwestern Australia, substantial vegetation shifts are projected under double CO_2 scenarios (Malcolm et al. 2002). Knowledge of the vegetation behavior under extreme climatic events is important for understanding the response and evolution of ecosystems in future climatic scenarios. This is particularly true for areas such as those in the Mediterranean regions that are currently subjected to a high degree of water stress (Peñuelas and Boada 2003) or to a progressive aridification (Peñuelas et al. 2002; Peñuelas and Boada 2003), that currently exhibit a great geographical and temporal variability in precipitation and water availability (Peñuelas 2001).

Several papers have presented the possible effects of changing temperature and water availability on the growth of forests. Kramer et al. (2000) presented models simulating physiological features of the annual cycle for boreal coniferous, temperate-zone deciduous, and Mediterranean forest ecosystems. In Spain, Peñuelas et al. (2002) compared phenological data from 1952 to 2000 providing a complete record of common plants, migratory birds and a common butterfly.

A conservative linear treatment of data showed that, in 2000, leaves unfolded on average 16 days earlier, leaf fall occurred about 13 days later, and plants flowered an average of 6 days earlier than in 1952. In addition, fruiting occurred about 9 days earlier in 2002 than in 1974. Butterflies appeared 11 days earlier and spring migratory birds arrived 15 days later than 1952. The biggest change in both temperature and phenophase timing occurred in the last 25 years.

García-Mozo et al. (2010) present the phenological trend of several species in response to climate change at six sites in southern Spain from 1986 to the present. They focused on vegetative and overall reproductive phenology in *Olea europaea* L. and *Vitis vinifera* L., as well as in various species of *Quercus* spp. and *Poaceae*. A trend towards earlier foliation, flowering and fruit ripening was observed for the trees, and temperature increase was identified as the cause. Herbaceous species were more affected than trees by changes in precipitation.

Morin et al. (2010) analyzed the phenological response to artificial climate change, obtained through experimental warming and reduced precipitation on several populations of three European oaks in a Mediterranean site. Experimental warming advanced the seedlings vegetative phenology, which caused a longer growing season, and advanced the leaf unfolding date. Conversely, soil water content did not affect the phenology of the seedlings or their survival. Thus, the phenological response of trees to climate change may be nonlinear, which suggests that predictions of phenological changes in the future should not be built on extrapolations of current observations.

Pinto et al. (2011) showed that air temperature was the main environmental driver of *Q. suber* budburst timing. This was also reported for other oak species of the Iberian Peninsula (Morin et al. 2010; Peñuelas et al. 2002; Sanz-Pérez et al. 2009). High mean and maximum daily temperatures in periods close to budburst accelerate more effectively bud development than in January and February. In the period with the best fit between budburst date and temperature (late-March to budburst) minimum daily temperature had no influence on budburst. The current differences in the timing of the budburst (earlier in *Q. faginea* than in *Q. ilex*) would be reduced in a global warming scenario, which could modify the competitive relationships between seedlings of these two species in the regeneration phase of mixed forests.

In many locations and species, chilling temperature accumulation is necessary to break bud dormancy (Cannell and Smith 1983; Hänninen 1990; Kramer 1994). Results from studies on Mediterranean species (García-Mozo et al. 2008), and even considering longer periods of temperature averaging (Pinto et al. 2011), however, failed to show any evidence of the chilling effect requirement. Conversely, other authors noted the importance of chilling even under Mediterranean conditions (Cesaraccio et al. 2004; Jato et al. 2007; Morin et al. 2010).

The relationship between rainfall and budburst date is a controversial topic. Pinto et al. (2011) found no relationship for the budburst triggering mechanisms in *Q. suber*, which seem species specific regardless of local soil and water conditions. Spano et al. (1999) found little effect of rainfall on budburst of Mediterranean species. Peñuelas et al. (2004), however, reported an overall relationship between

October to February rainfall and budburst date for a range of Mediterranean species.

Miranda et al. (2002) found that budburst in Mediterranean evergreen oaks is likely to occur earlier although probably within a range of species-specific photoperiod limits. Moreover, with respect to shoot elongation, two main situations may arise: (a) in water-limited areas, a drier spring and summer will lengthen the tree water stress period and restrict shoot elongation; (b) in fully watered places, shoot growth is limited by nutrient availability prior to budburst. Commonly, future phenology trends of Mediterranean evergreen oaks likely exhibit and earlier budburst and reduced shoot elongation (Pinto et al. 2011).

Temperature is the major factor responsible for phenological changes affecting flowering, fruit ripening, and leaf unfolding and shedding in plants (Peñuelas and Filella 2001); however, a delay in water supply is also of great importance (Dios Miranda et al. 2009). A long delay of the rainy season results in later flowering, as was shown for mesic Mediterranean environments (Peñuelas et al. 2002; Gordo and Sanz 2005). These papers reported significant correlations between precipitation and length of the life cycle. Importantly, changes in flowering date led to a reduction in the number of fruits, number of seeds, seed size, and seedling recruitment, affecting plant communities in the long term (Peñuelas et al. 2002).

The comprehensive analysis reported by Gordo and Sanz (2010) provides an essential tool to understand why flowering and leaf unfolding (spring phenophases) showed some of the largest phenological responses to climate change reported in plants (Menzel et al. 2006; Gordo and Sanz 2009). They used a dataset of more than 200,000 records for six phenological events of 29 perennial plant species monitored from 1943 to 2003. A comparison of sensitivity coefficients to temperature reported in literature for the same species in other parts of Europe suggests a higher sensitivity of populations in the Mediterranean. This fact would agree with the higher sensitivity found in plant populations from warmer regions (Menzel et al. 2005; Tryjanowski et al. 2006; Doi and Takahashi 2008), which could be a result of the lower probability of late frost damage (Askeyev et al. 2005). Differences in temporal responses of plant phenology to recent climate change are due to differences in the sensitivity to climate among events and species. Spring events are changing more than autumn events as they are more sensitive to climate and they are also undergoing the greatest alterations of climate relative to other seasons.

The phenology of Mediterranean plants is as responsive as the phenology of plants in colder biomes (Osborne et al. 2000; García-Mozo et al. 2002; Peñuelas et al. 2002; Mutke et al. 2003; Gordo and Sanz 2005). Prieto et al. (2009) analyzed the changes in the onset of spring growth in shrubland species in response to experimental warming along a north–south gradient in Europe. 'Bud break' was monitored in eight shrub and grass species in six European sites under control and experimentally warmer conditions generated by automatic roofs covering vegetation during the night. This study showed that warmer temperatures projected for coming decades have substantial to advance the spring growth of dominant species in different European shrublands. It also demonstrated the overall difficulties of applying simple predictive relationships to extrapolate the effects of global change on phenology. Various combinations of environmental factors occur concurrently at different European sites and the interactions between different drivers (e.g. water and chilling) can alter phenology significantly.

Results from Prieto et al. (2009) underscore the species-specific nature of the responsiveness of spring growth to temperature (Peñuelas et al. 2002, 2004; Hollister et al. 2005). The acceleration of the 'bud break' dates of the Spanish species is particularly noticeable. In general, the 'bud break' is related to the period when water first becomes available (Peñuelas et al. 2004). In a Mediterranean forest, the influence of water availability and temperature in the control of leaf development and spring flowering varies depending on the species (Ogaya and Peñuelas 2004). The spring growth for both tree species was associated with the mean temperature of the previous months, although only the 'bud break' of *Erica multiflora* was accelerated by warming treatment. The lack of significant acceleration in the 'bud break' of *Globularia alypum* in warming plots can be a consequence of its stronger dependence on the soil water status described for some ecophysiological parameters (Llorens et al. 2003) as well as for growth phenology. For *Erica multiflora*, the relationship with water availability was not significant, although the dry period between late winter and early spring in 2005 accelerated the onset of growth in *Erica multiflora* in control plots compared with 2003 and 2004. *Erica multiflora* is a species with a conservative strategy regarding water use (Llorens et al. 2003) and, in the light of the warming effects described in this study, the earlier growth in 2005 might be a consequence of an increased leaf temperature resulting from reduced stomatal conductance under lower water availability. The lower stomatal conductance reached in *Erica multiflora* in 2005 (winter and spring) relative to the rates in 2003 and 2004 support this hypothesis (Prieto 2007).

Plant responses to warming also depended on specific combinations of environmental drivers in particular years. For example, plant response depends on the temperature or the amount and distribution of rainfall throughout the season and preceding years. In *Erica multiflora*, in spite of the clear acceleration of 'bud break' dates in warming plots in 2003 and 2004, no significant change was observed in 2005, which was the year with the driest late winter and spring during of the 7 years. Moreover, the earlier 'bud break' date in 2003 and 2004 was only accompanied by greater spring shoot elongation in 2004 (Prieto 2007). This was partly due to the high temperature reached during the European heat wave in 2003, which enhanced evapotranspiration and reduced water availability for shoot growth.

Different phenological patterns of the various species partially help to explain their various productivity responses to warming reported by Peñuelas et al. (2007). Rainfall frequency reductions projected for some Mediterranean regions (Cheddadi et al. 2001) will exacerbate drought conditions, and these conditions have now been observed in the eastern Mediterranean (Körner et al. 2005). Soil water content controls ecosystem water and CO_2 flux in the Mediterranean Basin system (Rambal et al. 2003), and reductions are very likely to reduce ecosystem carbon and water flux (Reichstein et al. 2002). Many studies on the behavior of Mediterranean species in response to drought are reported in the IPCC (2007) report. Established *Pinus halepensis* (Borghetti et al. 1998) showed high drought resistance, but

Ponderosa pine forests had reduced productivity and evapotranspiration during a 1997 heat wave. The Ponderosa pine did not recover for the rest of the season, indicating threshold responses to extreme events (Goldstein et al. 2000). Mediterranean Basin pines (Martinez-Vilalta and Pinol 2002) and other woody species (Peñuelas et al. 2001) showed species-specific drought tolerance under field conditions. Experimental drying differentially reduced productivity of Mediterranean Basin shrub species (Llorens et al. 2003, 2004; Ogaya and Peñuelas 2004) and tree species (Ogaya and Peñuelas 2007), but delayed flowering and reduced flower production of Mediterranean Basin shrub species (Llorens and Peñuelas 2005), suggests complex changes in species relative success under drying scenarios. Drought may also act indirectly on plants by reducing the availability of soil phosphorus (Sardans and Peñuelas 2004).

Seasonal and inter-annual variation in climatic patterns (e.g., rainfall regimes) impacts on the pollination pattern in some anemophilous sub-desert plants. Alba-Sanchez et al. (2010) explored the effect that seasonal and inter-annual variation of rainfall regimes on pollination patterns in six anemophilous taxa located in the semiarid area of Almería (SE Spain), which is one of the most arid locations in Europe. The sampling from 1998 to 2005 showed that the pulsed and discrete rainfall events interspersed with drought periods are closely related to the alteration of the pollination in certain species. This is manifested in: (i) delayed onset of flowering until reaching the minimum threshold of soil water, in the case of some annual plants (Plantago, Rumex, and Poaceae), or (ii) scant variability both in the flowering period in plants with drought tolerance (Chenopodiaceae and Artemisia) or plants often linked to soil-moisture availability (Urticaceae).

As cited by Matias et al. (2011), under a global-change scenario where habitat as well as climatic conditions are altered (Houghton et al. 2001), the effect on dynamics of soil nutrients and its interaction with the plant community are not well known (Jensen et al. 2003; Andresen et al. 2010).

It is increasingly clear that changes in temperature or precipitation provoked by climate change will alter nutrient cycles (Sardans and Peñuelas 2007), and therefore nutrient availability for plants. Differences in carbon (C), nitrogen (N), and phosphorus (P) availability have severe effects for plant communities as these are fundamental nutrients for plant growth. Because P has strong implications in the water-use efficiency (Graciano et al. 2005), this modulates plant vulnerability to drought stress.

A dryer climate reduces microbial nutrient uptake, but increases soil nutrient availability. Higher nutrient availability in dry soil, however, cannot be exploited by plants due to the water deficit. This higher nutrient pool in soil, together with the higher torrential rainfall predicted for the coming decades (Houghton et al. 2001) may increase the risk of nutrient loss by leaching or erosion (De Luis et al. 2003; Ramos and Martinez-Casasnovas 2004), leading to a short to middle-term nutrient loss and soil impoverishment.

Matias et al. (2011) investigated the effect of three contrasting climatic scenarios on different carbon (C), nitrogen (N), and phosphorus (P) fractions in soil and microbial compartments among three characteristic habitats in a Mediterranean-type ecosystem:

forest, shrubland, and open areas. The climatic scenarios were (1) using a 30 % summer rainfall reduction, (2) simulating summer storms to reach the maximum historical records and (3) current climatic conditions. The results support the idea that higher rainfall boosts microbial and plant-nutrient uptake, and hence nutrient cycling. The rainfall reduction led to an accumulation of nutrients in the soil, increasing the risk of nutrient loss by leaching or erosion.

10.5 Conclusions

There are five Mediterranean zones around the world that are located near the west coasts of continents between 30° and 40° latitude. The climate represents a unique transition between arid zones towards the equator and temperate zones poleward. It is characterized by cold to cool, wet winters and warm to hot summers with varying periods of drought. The vegetation is similar in each region with woody, shrubby and evergreen shrubland plants, sparse grass, scattered evergreen trees, and many species of oak trees. In all zones, anthropogenic disturbances including deforestation, grazing, agricultural development, and fire starting and suppression have changed the vegetation community structure. In general, phenology in the five Mediterranean zones presents a pronounced seasonal rhythm related to vegetation and environmental characteristics, with large variation among species. Whereas heat unit accumulation is the main factor affecting phenology of well-watered plants, phenology of natural Mediterranean vegetation is influenced by drought and plant nutrition in addition to heat units. Climatic fluctuations and drought in particular, directly influence resources availability and indirectly phenology. Like other climate regions, more research is needed to better understand the interaction between weather factors and phenology.

The Gordo and Sanz (2009) analysis is a keystone to determine the role of recent climate change in the observed phenological shifts and to understand why plants are changing their phenology in Mediterranean ecosystems and how responses vary among species and events. Differences in temporal responses of plant phenology to recent climate change are due to differences in the sensitivity to climate among events and species. Spring events are changing more than autumn events as they are more sensitive to climate and are also undergoing the greatest alterations of climate relative to other seasons.

In Mediterranean climate regions, water availability and temperature are both key factors determining plant performance. For instance, water can restrict the length of growing season and affect flowering phenology. However, there are few studies on drought and phenology, so more research is needed to better characterized climate change effects on vegetation.

A drier climate will affect growth but also spring phenology of some Mediterranean species. A reduction in the cooling effect of transpiration could have the same effect as atmospheric warming and it could advance the initiation of growth in sensitive plants. Lengthening of the growing season, due to earlier phenological development may not result in higher productivity because drought can impede growth.

References

Alba-Sanchez F, Sabariego-Ruiz S, De La Guardia CD, Nieto-Lugilde D, De Linares C (2010) Aerobiological behaviour of six anemophilous taxa in semi-arid environments of southern Europe (Almeria, SE Spain). J Arid Environ 74(11):1381–1391

Alessio GA, de Lillis M, Brugnoli E, Lauteri M (2004) Water source and water use efficiency in Mediterranean coastal dune vegetation. Plant Biol 6:350–357

Andresen LC, Michelsen A, Jonasson S, Schmidt IK, Mikkelsen TN, Ambus P, Beier C (2010) Plant nutrient mobilization in temperate heathland responds to elevated CO2, temperature and drought. Plant Soil 328:381–396

Archibold OW (1995) Ecology of world vegetation. Chapman & Hall, London

Arianoutsou M, Mardilis TA (1987) Observations on the phenology of two dominant plants of the Greek maquis. In: Tenhunen JD, Catarino FM, Lange OL, Oechel WC (eds) Plant response to stress: functional analysis in Mediterranean ecosystems, vol 15, NATO Adv Sci Inst Ser G Ecol Sci. Springer, Berlin/Heidelberg

Askeyev OV, Tischin D, Sparks TH, Askeyev IV (2005) The effect of climate on the phe-nology, acorn crop and radial increment of pedunculate oak (Quercus robur) in the middle Volga region, Tatarstan, Russia. Int J Biometeorol 49:262–266

Bakkenes M, Alkemade JRM, Ihle F, Leemans R, Latour JB (2002) Assessing effects of forecasted climate change on the diversity and distribution of European higher plants for 2050. Glob Chang Biol 8:390–407

Beniston M, Stephenson DB, Christensen OB et al (2007) Future extreme events in European climate: an exploration of regional climate model projections. Clim Chang 81:71–95

Bernal M, Estiarte M, Peñuelas J (2011) Drought advances spring growth phenology of the Mediterranean shrub Erica multiflora. Plant Biol 13:252–257

Bond WJ, Midgley JJ (2003) The evolutionary ecology of sprouting in woody plants. Int J Plant Sci 164:S103–S114

Borghetti M, Cinnirella S, Magnani F, Saracino A (1998) Impact of long-term drought on xylem embolism and growth in Pinus halepensis Mill. Trees 12:187–195

Cannell MGR, Smith RI (1983) Thermal time, chill days and prediction of budburst in Picea sitchencis. J Appl Ecol 20:951–963

Castro J, Zamora R, Hodar JA, Gomez JM (2005) Alleviation of summer drought boosts establishment success of Pinus sylvestris in a Mediterranean mountain: an experimental approach. Plant Ecol 181:191–202

Castro-Díez P, Milla R, Virginia Sanz V (2005) Phenological comparison between two co-occurring Mediterranean woody species differing in growth form. Flora 200:88–95

Cesaraccio C, Spano D, Snyder RL, Duce P (2004) Chilling and forcing model to predict bud-burst of crop and forest species. Agric Forest Meteorol 126:1–13

Cheddadi R, Guiot J, Jolly D (2001) The Mediterranean vegetation: what if the atmospheric CO2 increased? Landsc Ecol 16:667–675

Christensen JH, Hewitson B, Busuioc A, Chen A, Gao X, Held I, Jones R, Kolli RK, Kwon W-T, Laprise R, Magaña Rueda V, Mearns L, Menéndez CG, Räisänen J, Rinke A, Sarr A, Whetton P (2007) Regional climate projections. In: Solomon S, Qin D, Manning M, Chen Z, Marquis M, Averyt KB, Tignor M, Miller HL (eds) Climate change 2007: the physical science basis. Contribution of Working Group I to the fourth assessment re-port of the intergovernmental panel on climate change. Cambridge University Press, Cambridge, NY

Cody ML, Mooney HA (1978) Convergence versus non-convergence in Mediterranean-climate ecosystems. Annu Rev Ecol Syst 9:265–321

Correia OA, Martins AC, Catarino FM (1992) Comparative phenology and seasonal foliar nitrogen variation in Mediterranean species of Portugal. Ecol Mediterr 18:7–18

Cowling RM, Rundel PW, Lamont BB, Arroyo MK, Arianoutsou M (1996) Plant diversity in Mediterranean-climate regions. Trends Ecol Evol 11:352–360

Cubasch U, Meehl GA, Boer GJ, Stouffer RJ, Dix M, Noda A, Senior CA, Raper S, Yap KS (2001) Projections on future climate change. In: Houghton JT, Ding Y, Griggs DJ, Noguer M, van der Linden P, Dai X, Maskell K, Johnson CI (eds) Climate change 2001: the scientific basis, contribution of Working Group I to the third assessment report of the intergovernmental panel on climate change. Cambridge University Press, Cambridge

Davis GW, Richardson DM (1995) Mediterranean-type ecosystems. The function of biodiversity. Ecological studies, vol 109. Springer, Berlin

de Lillis M, Fontanella A (1992) Comparative phenology and growth in different species of the Mediterranean maquis of central Italy. Vegetatio 99–100:83–96

De Luis M, Gonzalez-Hidalgo JC, Raventos J (2003) Effects of fire and torrential rainfall on erosion in a Mediterranean gorse community. Land Degrad Dev 14:203–213

di Castri F (1973) Climatographical comparison between Chile and the western coast of North America. In: di Castri F, Mooney HA (eds) Mediterranean type ecosystems, origin and structure. Springer, Berlin/Heidelberg

di Castri F, Goodall DW, Specht RL (eds) (1981) Ecosystems of the world: Mediterranean-type shrublands. Elsevier Scientific Publishing Company, Amsterdam

Dios Miranda JD, Padilla FM, Pugnaire FI (2009) Response of a Mediterranean semiarid community to changing patterns of water supply. Perspect Plant Ecol 11:255–266

Doi H, Takahashi M (2008) Latitudinal patterns in the phenological response of leaf colouring and leaf fall to climate change in Japan. Glob Ecol Biogeogr 17:556–561

Duce P, Spano D, Asunis C, Cesaraccio C, Sirca C, Motroni A (2000) Effect of climate variability on phenology and physiology of Mediterranean vegetation. In: 3rd European conference on Applied Climatology, Pisa

Duce P, Cesaraccio C, Spano D, Snyder RL (2002) Weather variability effect on phenological events in a Mediterranean-type climate. In: 15th conference on biometeorology/aero-biology and 16th international congress of biometeorology, Kansas City

Emberger L (1962) Comment comprendre le territoire phytogéographique méditerranéen francaiş et la position "systématique" de celui-ci. Nationalia Monspeliensia. Série Botanica 14:47–54

Fischlin A, Midgley GF, Price JT, Leemans R, Gopal B, Turley C, Rounsevell MDA, Dube OP, Tarazona J, Velichko AA (2007) In: Parry ML, Canziani OF, Palutikof JP, van der Linden PJ, Hanson CE (eds) Ecosystems, their properties, goods, and services. Climate change 2007: impacts, adaptation and vulnerability. Contribution of Working Group II to the fourth assessment report of the intergovernmental panel on climate change. Cambridge University Press, Cambridge

Fried JS, Torn MS, Mills E (2004) The impact of climate change on wildfire severity: a regional forecast for northern California. Clim Chang 64:169–191

Gao XJ, Giorgi F (2008) Increased aridity in the Mediterranean region under greenhouse gas forcing estimated from high resolution simulations with a regional climate model. Glob Planet Chang 62:195–209

Gao XJ, Pal JS, Giorgi F (2006) Projected changes in mean and extreme precipitation over the Mediterranean region from a high resolution double nested RCM simulation. Geophys Res Lett 33:L03706. doi:10.1029/2005GL024954

García-de-Lomas J, Cózar A, Dana ED, Hernández I, Sánchez-García Í, García CM (2010) Invasiveness of *Galenia pubescens* (Aizoaceae): a new threat to Mediterranean-climate coastal ecosystems. Acta Oecol 36:39–45

García-Mozo H, Galán C, Aira MJ, Belmonte J, Díaz de la Guardia C, Fernández D, Gutierrez AM, Rodriguez FJ, Trigo MM, Dominguez-Vilches E (2002) Modelling start of oak pollen season in different climatic zones in Spain. Agric Forest Meteorol 110:247–257

García-Mozo H, Chuine I, Aira MJ, Belmonte J, Bermejo D, Díaz de la Guardia C, Elvira B, Gutiérrez M, Rodríguez-Rajo J, Ruiz L, Trigo MM, Tormo R, Valencia R, Galán C (2008) Regional phenological models for forecasting the start and peak of the Quercus pollen season in Spain. Agric Forest Meteorol 148:372–380

García-Mozo H, Mestre A, Galán C (2010) Phenological trends in southern Spain: a response to climate change. Agric Forest Meteorol 150(4):575–580

Giorgi F, Lionello P (2008) Climate change projections for the Mediterranean region. Glob Planet Chang 63:90–104

Giorgi F, Bi X, Pal JS (2004) Mean, interannual variability and trends in a regional climate change experiment over Europe. II: climate change scenarios (2071–2100). Clim Dyn 23:839–858

Goldstein AH, Hultman NE, Fracheboud JM, Bauer MR, Panek JA, Xu M, Qi Y, Guenther AB, Baugh W (2000) Effects of climate variability on the carbon dioxide, water and sensible heat fluxes above a ponderosa pine plantation in the Sierra Nevada (CA). Agric Forest Meteorol 101:113–129

Gordo O, Sanz JJ (2005) Phenology and climate change: a long-term study in a Mediterranean locality. Oecologia 146:484–495

Gordo O, Sanz JJ (2009) Long-term temporal changes of plant phenology in the Western Mediterranean. Glob Chang Biol 15:1930–1948

Gordo O, Sanz JJ (2010) Impact of climate change on plant phenology in Mediterranean ecosystems. Glob Chang Biol 16:1082–1106

Graciano C, Guiamet JJ, Goya JF (2005) Impact of nitrogen and phosphorus fertilization on drought responses in Eucalyptus grandis seedlings. Forest Ecol Manag 212:40–49

Gratani L, Varone L (2004) Leaf key traits of *Erica arborea* L., *Erica multiflora* L. and *Rosmarinus officinalis* L. co-occurring in the Mediterranean maquis. Flora 199:58–69

Gritti ES, Smith B, Sykes MT (2006) Vulnerability of Mediterranean Basin ecosystems to climate change and invasion by exotic plant species. J Biogeogr 33:145–157

Gulmon SL (1977) A comparative study of grasslands of California and Chile. Flora 166:261–278

Hanes TL (1981) California chaparral. In: di Castri F, Goodall DW, Specht RL (eds) Eco-systems of the world: Mediterranean-type shrublands. Elsevier Scientific Publishing Company, Amsterdam

Hänninen H (1990) Modeling bud dormancy release in trees from cool and temperate regions. Acta For Fenn 213:1–47

Hayhoe K, Cayan D, Field CB, Frumhoff PC, Maurer EP, Miller NL, Moser SC, Schneider SH, Cahill KN, Cleland EE, Dale L, Drapek R, Hanemann RM, Kalkstein LS, Lenihan J, Lunch CK, Neilson RP, Sheridan SC, Verville JH (2004) Emissions pathways, climate change, and impacts on California. Proc Natl Acad Sci U S A 101:12422–12427

Hollister RD, Webber PJ, Tweedie CE (2005) The response of Alaskan arctic tundra to experimental warming: differences between short- and long-term responses. Glob Chang Biol 11:525–536

Holmgren M, Stapp P, Dickman CR et al (2006) Extreme climatic events shape arid and semiarid ecosystems. Front Ecol Environ 4:87–95

Houghton JT, Ding Y, Griggs DJ, Noguer M, Van der Linden PJ, Xiaosu D (2001) Climate change 2001: the scientific basis. Contribution of Working Group I to the third as-sessment report of the intergovernmental panel on climate change (IPCC). Cambridge University Press, Cambridge

Intergovernmental Panel on Climate Change (2007) Climate Change 2007: Synthesis report, contribution of Working Groups I, II and III to the fourth assessment report of the in-tergovernmental panel on climate change. IPCC, Geneva

Intergovernmental Panel on Climate Change, McCarthy JJ, Canziani OF, Leary NA, Dokken DJ, White KS (eds) (2001) Climate change 2001: impacts, adaptation and vulnerability. Cambridge University Press, Cambridge

International Union for the Conservation of Nature (1999) Biological diversity of dry land, Mediterranean, arid, semiarid, savanna, and grassland ecosystems, World Conservation Union: fourth meeting of subsidiary body on scientific, technical, and technological advice, Montreal

Jato V, Rodríguez-Rajo FJ, Aira MJ (2007) Use of *Quercus ilex* subsp. ballota and pollen-production data for interpreting Quercus pollen curves. Aerobiologia 23:91–105

Jensen KD, Beier C, Michelsen A, Emmett BA (2003) Effects of experimental drought on microbial processes in two temperate heathlands at contrasting water conditions. Appl Soil Ecol 24:165–176

Joffre R, Rambal S (2002) Mediterranean ecosystems. In: Encyclopedia of life sciences. Wiley, Chichester. http://www.els.net. doi: 10.1038/npg.els.0003196

Joffre R, Rambal S, Ratte JP (1999) The dehesa system of southern Spain and Portugal as a natural ecosystem mimic. Agrofor Syst 45:57–79

Körner C, Sarris D, Christodoulakis D (2005) Long-term increase in climatic dryness in the East Mediterranean as evidenced for the island of Samos. Reg Environ Chang 5:27–36

Kozlowski TT, Pallardy SG (1997) Physiology of woody plants. Academic, San Diego

Kramer K (1994) Selecting a model to predict the onset of growth of *Fagus sylvatica*. J Appl Ecol 31:172–181

Kramer K, Leinonen I, Loustau D (2000) The importance of phenology for the evaluation of impact of climate change on growth of boreal, temperate and Mediterranean forest ecosystems: an overview. Int J Biometeorol 44:67–75

Kummerow J (1981) Structure of roots and root systems. In: Di Castri F, Goodal DW, Specht RL (eds) Ecosystems of the world – Mediterranean-type shrublands. Elsevier, Amsterdam

Kuzucuoglu C (1989) Fires in Mediterranean region. Blue Planet Ecol 72:371–412

Lenihan JM, Drapek R, Bachelet D, Neilson RP (2003) Climate change effects on vegetation distribution, carbon, and fire in California. Ecol Appl 13:1667

Llorens L, Penuelas J (2005) Experimental evidence of future drier and warmer conditions affecting flowering of two co-occurring Mediterranean shrubs. Int J Plant Sci 166:235–245

Llorens L, Penuelas J, Estiarte M (2003) Ecophysiological responses of two Mediterranean shrubs, *Erica multiflora* and *Globularia alypum*, to experimentally drier and warmer conditions. Physiol Plant 119:231–243

Llorens L, Penuelas J, Estiarte M, Bruna P (2004) Contrasting growth changes in two dominant species of a Mediterranean shrubland submitted to experimental drought and warming. Ann Bot 94:843–853

Malcolm JR, Markham A, Neilson RP, Garaci M (2002) Estimated migration rates under scenarios of global climate change. J Biogeogr 29:835–849

Manes F, Capogna F, Puppi G, Vitale M (2002) Ecophysiological characterization of Phillirea angustifolia L. and response of resprouts to different fire disturbance intensities. In: Trabaud L, Prodon R (eds) Fire and biological processes. Backhuys Publishers, Leiden

Martinez-Vilalta J, Pinol J (2002) Drought-induced mortality and hydraulic architecture in pine populations of the NE Iberian Peninsula. Forest Ecol Manag 161:247–256

Matias L, Castro J, Zamora R (2011) Soil nutrient availability under a global change scenario in a Mediterranean mountain ecosystem. Glob Chang Biol 17:1646–1657

Mendoza I, Zamora R, Castro J (2009) A seeding experiment for testing tree-community recruitment under variable environments: implications for forest regeneration and conservation in Mediterranean habitats. Biol Conserv 142:1491–1499

Menzel A, Estrella N, Testka A (2005) Temperature response rates from long-term phenological records. Clim Res 30:21–28

Menzel A, Sparks TH, Estrella N et al (2006) European phenological response to climate change matches the warming pattern. Glob Chang Biol 12:1969–1976

Midgley GF, Chapman RA, Hewitson B, Johnston P, De Wit M, Ziervogel G, Mukheibir P, Van Niekerk L, Tadross M, Van Wilgen BW, Kgope B, Morant PD, Theron A, Scholes RJ, Forsyth GG (2005) A status quo, vulnerability and adaptation assessment of the physical and socio-economic effects of climate change in the western Cape. Report to the Western Cape Government, Cape Town, South Africa. CSIR Report No. ENV-S-C 2005–073, CSIR Environmentek, Stellenbosch

Milla R, Castro-Díez P, Montserrat-Martí G (2010) Phenology of Mediterranean woody plants from NE Spain: synchrony, seasonality, and relationships among phenophases. Flora 205:190–199

Minnich RA (1983) Fire mosaics in southern California and northern Baja California. Science 219:1287–1294

Miranda P, Coelho FES, Tomé AR, Valente MA (2002) 20th century Portuguese climate and climate change scenarios. In: Santos FD, Forbes K, Moita R (eds) Climate change in Portugal. Scenarios, impacts and adaptation measures. Gradiva, Lisboa

Moll EJ (1987) Phenology of Mediterranean plants in relation to fire season: with special reference to the Cape Province South Africa. In: Tenhunen JD, Catarino FM, Lange OL, Oechel WC (eds) Plant response to stress: functional analysis in Mediterranean ecosystems, NATO Adv Sci Inst Ser G Ecol Sci. Springer, Berlin/Heidelberg

Moll EJ, Campbell BM, Cowling RM, Bossi L, Karman ML, Boucher C (1984) A description of major vegetation categories in and adjacent to the Fynbos biome. Report No. 83, S Afr Nat Sci Program

Montenegro G (1987) Quantification of Mediterranean plant phenology and growth. In: Tenhunen JD, Catarino FM, Lange OL, Oechel WC (eds) Plant response to stress: functional analysis in Mediterranean ecosystems, NATO Adv Sci Inst Ser G Ecol Sci. Springer, Berlin/Heidelberg

Mooney HA, Conrad CE (1977) Symposium on the environmental consequences of fire and fuel management in Mediterranean ecosystem. USDA Forest Service General, Technical report WO-3, U.S. Government Printing Office

Mooney HA, Kummerow J (1981) Phenological development of plants in Mediterranean climate regions. In: di Castri F, Goodall DW, Specht RL (eds) Ecosystems of the world: Mediterranean-type shrublands. Elsevier Scientific Publishing Company, Amsterdam

Mooney HA, Johnson A, Parson D, Keeley S, Hoffman A, Hays R, Giliberto J, Chu C (1977) The producers-their resources and adaptive response. In: Mooney HA (ed) Convergent evolution in Chile and California Mediterranean climate ecosystems. Dowden Hutchinson & Ross, Stroudsburg

Morin X, Roy J, Sonié L, Chuine I (2010) Changes in leaf phenology of three European oak species in response to experimental climate change. New Phytol 186(4):900–910

Mouillot F, Rambal S, Joffre R (2002) Simulating climate change impacts on fire frequency and vegetation dynamics in a Mediterranean-type ecosystem. Glob Chang Biol 8:423–437

Mouillot F, Ratte JP, Joffre R, Moreno JM, Rambal S (2003) Some determinants of the spatio-temporal fire cycle in a Mediterranean landscape (Corsica, France). Landsc Ecol 18:665–674

Munné-Bosch S, Nogués S, Alegre L (1999) Diurnal variations of photosynthesis and dew absorption by leaves in two evergreen shrubs growing in Mediterranean field conditions. New Phytol 144:109–119

Mutke S, Gordo J, Climent J, Gil L (2003) Shoot growth and phenology modeling of grafted stone pine (Pinus pinea L.) in inner Spain. Ann Forest Sci 60:527–537

Naveh Z (1990) Fire in the Mediterranean: a landscape perspective. In: Goldhammer JG, Jenkins MJ (eds) Fire in ecosystem dynamics. SPB Academic Publ, The Hague

Oechel WC, Moreno MJ (1994) The role of fire in Mediterranean ecosystems. Springer, Berlin/Heidelberg

Ogaya R, Peñuelas J (2004) Phenological patterns of *Quercus ilex*, *Phillyrea latifolia*, and *Arbutus unedo* growing under a field experimental drought. Ecoscience 11:263–270

Ogaya R, Peñuelas J (2007) Tree growth, mortality, and above-ground biomass accumulation in a holm oak forest under a five-year experimental field drought. Plant Ecol 189:291–299

Orshan G (1989) Plant pheno-morphological studies in Mediterranean type ecosystems. Kluwer Acad Pub, Dordrecht

Osborne CP, Chuine I, Viner D, Woodward FI (2000) Olive phenology as a sensitive indicator of future climatic warming in the Mediterranean. Plant Cell Environ 23:701–710

Ovalle C, Aronson J, Del Pozo A, Avendano J (1990) The espinal: agroforestry system of the Mediterranean-type climate region of Chile. Agrofor Syst 10:213–239

Ovalle C, Aronson J, Del Pozo A, Avendano J (1996) Land occupation patterns and vegetation structure of the anthropogenic savannas (espinales) of central Chile. For Ecol Manage 86:129–139

Parry ML (ed) (2000) Assessment of potential effects and adaptations to climate change in Europe: the Europe Acacia Project. Report of concerted action of the environment programme of the Research Directorate General of the Commission of the European Communities, Jackson Environmental Institute, University of East Anglia, Norwich

Pausas JG, Abdel Malak D (2004) Spatial and temporal patterns of fire and climate change in the eastern Iberian Peninsula (Mediterranean Basin). In: Arianoutsou M, Papanastasis VP (eds) Ecology, conservation and management of Mediterranean climate ecosystems of the world. 10th international conference on Mediterranean climate ecosystems, Rhodes, Greece. Millpress, Rotterdam

Pellizzaro G, Cesaraccio C, Duce P, Ventura A, Zara P (2007) Relationships between seasonal patterns of live fuel moisture and meteorological drought indices for Mediterranean shrubland species. Int J Wildland Fire 16:232–241

Peñuelas J (2001) Cambios atmosféricos y climáticos y sus consecuencias sobre el fun-cionamiento y la estructura de los ecosistemas terrestres mediterráneos. In: Zamora R, Pugnaire FI (eds) Ecosistemas mediterráneos Análisis functional. CSIC-AEET Press, Granada

Peñuelas J, Boada M (2003) A global change-induced biome shift in the Montseny mountains (NE Spain). Glob Chang Biol 9:131–140

Peñuelas J, Filella I (2001) Responses to a warming world. Science 294:793–795

Peñuelas J, Filella I, Comas P (2002) Changed plant and animal life cycles from 1952 to 2000 in the Mediterranean region. Glob Chang Biol 8:532–544

Peñuelas J, Filella I, Zhang X, Llorens L, Ogaya R, Lloret F, Comas P, Estiarte M, Terradas J (2004) Complex spatiotemporal phenological shifts as a response to rainfall changes. New Phytol 161:837–846

Peñuelas J, Prieto P, Beier C, Cesaraccio C, De Angelis P, De Dato G, Emmett BA, Estiarte M, Garadnai J, Gorissen A, Kovács-Láng E, Kröel-Dulay G, Llorens L, Pellizzaro G, Riis-Nielsen T, Schmidt IK, Sirca C, Sowerby A, Spano D, Tietema A (2007) Response of plant species richness and primary productivity in shrublands along a north–south gradient in Europe to seven years of experimental warming and drought. Reductions in primary productivity in the heat and drought year of 2003. Glob Chang Biol 13:2563–2581

Peñuelas J, Lloret F, Montoya R (2001) Severe drought effects on mediterranean woody flora in Spain. Forest Sci 47:214–218

Pereira JS, Beyschlag G, Lange OL, Beyschlag W, Tenhunen JD (1987) Comparative phenology of four Mediterranean shrub species growing in Portugal. In: Plant response to stress: functional analysis in Mediterranean ecosystems, NATO Adv Sci Inst Ser G Ecol Sci. Springer, Berlin/Heidelberg

Pinto CA, Henriques MO, Figueiredo JP, David JS, Abreu FG, Pereira JS, Correia I, David TS (2011) Forest phenology and growth dynamics in Mediterranean evergreen oaks: effects of environmental conditions and water relations. Ecol Manag 262(3):500–508

Prieto P (2007) Phenology, biomass and community composition changes in a Mediterranean shrubland submitted to experimental warming and drought. Dissertation, Universitat Autònoma de Barcelona, Barcelona

Prieto P, Peñuelas J, Niinemets Ü, Ogaya R, Schmidt IK, Beier C, Tietema A, Sowerby A, Emmett BA, Kovács Láng E, Kröel-Dulay G, Lhotsky B, Cesaraccio C, Pellizzaro G, de Dato G, Sirca C, Estiarte M (2009) Changes in the onset of spring growth in shrubland species in response to experimental warming along a north–south gradient in Europe. Glob Ecol Biogeogr 18:473–484

Quezel P (1977) Forests of the Mediterranean basin. In: Mediterranean forests and maquis: ecology conservation and management. UNESCO, Paris

Rambal S, Ourcival JM, Joffre R, Mouillot F, Nouvellon Y, Reichstein M, Rocheteau A (2003) Drought controls over conductance and assimilation of a Mediterranean evergreen ecosystem: scaling from leaf to canopy. Glob Chang Biol 9:1813–1824

Ramos MC, Martinez-Casasnovas JA (2004) Nutrient losses from a vineyard soil in North-eastern Spain caused by an extraordinary rainfall event. Catena 55:79–90

Reichstein M, Tenhunen JD, Roupsard O, Ourcival JM, Rambal S, Miglietta F, Peressotti A, Pecchiari M, Tirone G, Valentini R (2002) Severe drought effects on ecosystem CO2 and H2O fluxes at three Mediterranean evergreen sites: revision of current hypotheses? Glob Chang Biol 8:999–1017

Rodrigo FS (2002) Changes in climate variability and seasonal rainfall extremes: a case study from San Fernando (Spain), 1821–2000. Theor Appl Climatol 72:193–207

Rossiter RC, Ozanne PG (1970) South-western temperate forests, woodlands and heaths. In: Moore RM (ed) Australian grassland. Australian National University Press, Canberra

Rundel PW (1981) The matorral zone of central Chile. In: di Castri F, Goodall DW, Specht RL (eds) Ecosystems of the world: Mediterranean-type shrublands. Elsevier Scientific Publishing Company, Amsterdam

Rundel PW (1983) Impact of fire on nutrient cycles in Mediterranean-type ecosystems, with reference to chaparral. In: Kruger FJ, Mitchell DT, Jarvis JUM (eds) Mediterranean-type ecosystems: the role of nutrients. Springer, Berlin/Heidelberg

Rundel PW (1995) Adaptive significance of some morphological and physiological characteristics in Mediterranean plants: facts and fallacies. In: Roy J, Aronson J, di Castri F (eds) Time scales of biological responses to water constraints: the case of Mediterranean biota. SPB Academic Publishers, Amsterdam

Rundel PW (1998) Landscape disturbance in Mediterranean-type ecosystems: an overview. In: Rundel PW, Montenegro G, Jaksic FM (eds) Ecological studies: landscape degradation and biodiversity in Mediterranean-type ecosystems. Springer, Berlin/Heidelberg

Rundel PW (2007) Mediterranean-climate ecosystems. In: Levin S (ed) Encyclopedia of biodiversity. Academic, Elsevier

Rundel PW, Vankat JL (1989) Chaparral communities and ecosystems. In: Keeley S (ed) The California chaparral: paradigms reexamined. Los Angeles County Museum of Natural History, Los Angeles

Sala OE, Chapin IFS, Armesto JJ, Berlow E, Bloomfield J, Dirzo R, Huber Sanwald E, Huenneke LF, Jackson RB, Kinzig A, Leemans R, Lodge DH, Mooney HA, Oesterheld M, Leroy Poff N, Sykes MT, Walker BH, Walker M, Wall DH (2000) Global biodiversity scenarios for the year 2100. Science 287:1770–1774

Sanz-Pérez V, Castro-Díez P, Valladares F (2009) Differential and interactive effects of tem-perature and photoperiod on budburst and carbon reserves in two co-occurring Mediterranean oaks. Plant Biol 11(2):142–151

Sardans J, Penuelas J (2004) Increasing drought decreases phosphorus availability in an evergreen Mediterranean forest. Plant Soil 267:367–377

Sardans J, Penuelas J (2007) Drought changes phosphorus and potassium accumulation patterns in an evergreen Mediterranean forest. Funct Ecol 21:191–201

Sheffield J, Wood EF (2008) Projected changes in drought occurrence under future global warming from multi-model, multi-scenario, IPCC AR4 simulations. Clim Dyn 31:79–105

Simões MP, Madeira M, Gazarini L (2008) The role of phenology, growth and nutrient re-tention during leaf fall in the competitive potential of two species of Mediterranean shrubs in the context of global climate changes. Flora 203:578–589

Somot S, Sevault F, Deque M, Crepon M (2008) 21st century climate change scenario for the Mediterranean using a couple atmosphere ocean regional climate model. Glob Planet Chang 63:112–126

Spano D, Cesaraccio C, Duce P, Snyder RL (1999) Phenological stages of natural species and their use as climate indicators. Int J Biometeorol 42:124–133

Specht RL (1973) Structure and functional response of ecosystems in the Mediterranean climate of Australia. In: di Castri F, Mooney HA (eds) Mediterranean-type ecosystems, origin and structure. Springer, Berlin/Heidelberg

Specht RL (1979) Ecosystems of the world: heathlands and related shrublands. Elsevier, Amsterdam

Specht RL (1981) Mallee ecosystem in southern Australia. In: Castri F, Goodall DW, Specht RL (eds) Mediterranean-type shrublands. Elsevier, Amsterdam

Tenhunen JD, Catarino FM, Lange OL, Oechel WC (1987) Plant response to stress: functional analysis in Mediterranean ecosystems, NATO Adv Sci Inst Ser G Ecol Sci. Springer, Berlin/Heidelberg

Thomas CD, Williams SE, Cameron A, Green RE, Bakkenes M, Beaumont LJ, Collingham YC, Erasmus BFN, de Siqueira MF, Grainger A, Hannah L, Hughes L, Huntley B, van Jaarsveld AS, Midgley GF, Miles L, Ortega-Huerta MA, Peterson AT, Philipps OL (2004) Biodiversity conservation: uncertainty in predictions of extinction risk/effects of changes in climate and land use/climate change and extinction risk (reply). Nature 430:34

Thrower NJW, Bradbury DE (1973) The physiography of the Mediterranean lands with special emphasis on California and Chile. In: di Castri F, Mooney HA (eds) Mediterranean-type ecosystems, origin and structure. Springer, Berlin/Heidelberg

Thuiller W, Lavorel S, Araújo MB, Sykes MT, Prentice IC (2005) Climate change threats to plant diversity in Europe. Proc Natl Acad Sci U S A 102:8245–8250

Trabaud L, Prodon R (1993) Fire in Mediterranean ecosystems. Commission of European Communities, Brussels

Tryjanowski P, Panek M, Sparks TH (2006) Phenological response of plants to temperature varies at the same latitude: case study of dog violet and horse chestnut in England and Poland. Clim Res 32:89–93

Valladares F, Vilagrosa A, Peñuelas J, Ogaya R, Camarero JJ, Corcuera L, Siso S, Gil Pelegrin E (2004) Estres hídrico: ecofisiología y escalas de la sequía. In: Valladares F (ed) Ecologia del bosque mediterráneo en un mundo cambiante. Ministerio de Medio Ambiente,EGRAF, S.A, Madrid

Viegas DX, Viegas MT, Ferreira AD (1992) Moisture content of fine forest fuels and fire occurrence in central Portugal. Int J Wildland Fire 2:69–86

Viegas DX, Piñol J, Viegas MT, Ogaya R (2001) Estimating live fine fuels moisture content using meteorologically-based indexes. Int J Wildland Fire 10:223–240

Zinke PJ (1973) Analogies between the soil and vegetation types in Italy, Greece and California. In: di Castri F, Mooney HA (eds) Mediterranean-type ecosystems, origin and structure. Springer, Berlin/Heidelberg

Chapter 11
Phenologies of North American Grasslands and Grasses

Geoffrey M. Henebry

Abstract Inquiry into the phenologies of grasslands and grasses in North America has progressed substantially in the past decade. Four themes of the recent phenological research are surveyed: (1) the role of exotic and invasive species in affecting grasslands phenology; (2) the role of water and belowground dynamics on phenologies; (3) how experimental manipulations of grasslands have affected constitutive phenologies; and (4) advances in the remote sensing of grasslands. The phyllochron concept used in ontogenetic studies of grass species is discussed in light of grasslands phenology and its link between daylength and thermal time.

11.1 Introduction

In the decade since the first edition of this chapter was written (Henebry 2003), there has been a significant increase in phenological research on grasses and grasslands in North America on multiple fronts. Thus, rather than revise the earlier effort, this new chapter surveys the research since 2002 with an emphasis on four interrelated themes: (1) phenologies of exotic and invasive species, (2) phenology, water and belowground dynamics, (3) experimental manipulations, and (4) remote sensing of grasslands. The chapter concludes with a discussion of the phyllochron concept and its relationship to phenology, revisiting the role of daylength in grasslands phenology.

G.M. Henebry (✉)
Geographic Information Science Center of Excellence, South Dakota State University, Brookings, SD 57007, USA
e-mail: Geoffrey.Henebry@sdstate.edu

M.D. Schwartz (ed.), *Phenology: An Integrative Environmental Science*,
DOI 10.1007/978-94-007-6925-0_11,

11.2 Phenological Shifts in Grasslands

Phenology figures prominently in the first chapter (Rosenzweig et al. 2007) of Working Group II's contribution to the Fourth Assessment Report (AR4) of the the Intergovernmental Panel on Climate Change (IPCC):

> Phenology... is perhaps the simplest process in which to track changes in the ecology of species in response to climate change.

The terms "phenology" or "phenological" are mentioned 91 times in this opening chapter that aims to summarize the state of knowledge on observed changes and responses in natural and managed systems in the face of a variable and changing climate.

Shifts in plant and animal phenologies are sensitive indicators of biological responses to environmental variation (Rosenzweig et al. 2007; Morisette et al. 2008; Hudson and Keatley 2010). Moreover, changes in phenologies can also serve both as forcing and as constraint on ecological, biogeochemical, and meteorological dynamics. Changes in phenologies of both plant and animal species (Schwartz 1994, 1998; Parmesan and Yohe 2003; Root et al. 2003; Menzel et al. 2006; Schwartz et al. 2006) and in the seasonalities of abiotic phenomena (Cayan et al. 2001; Westerling et al. 2006) have already manifested myriad effects of directional climate change.

Grasslands are characterized by high interannual variation in climatic forcings (cf. Fig. 11.1) and by high spatio-temporal variation of aboveground net primary

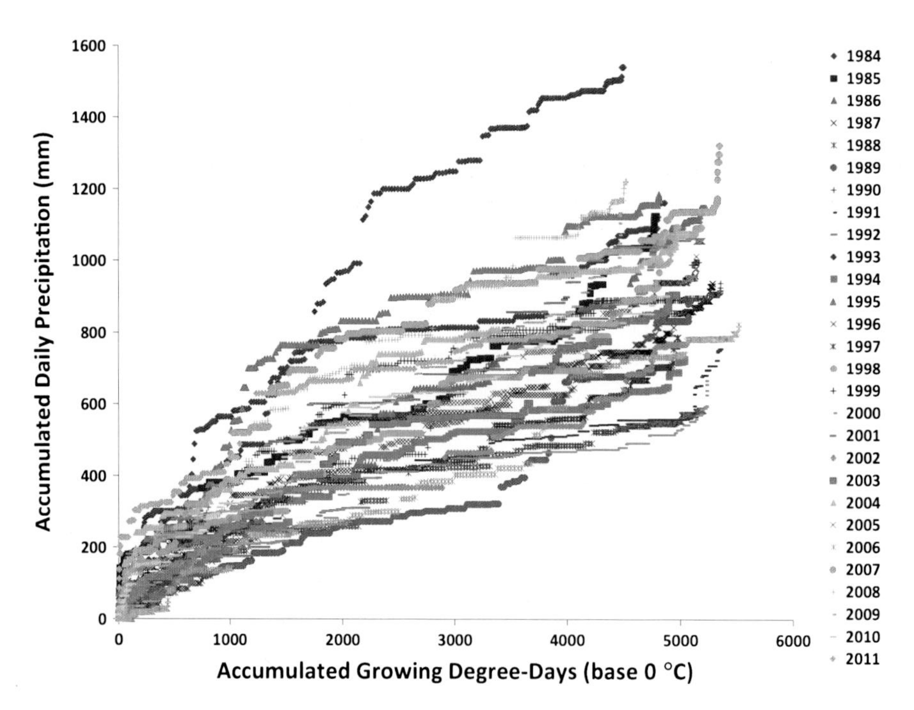

Fig. 11.1 The climatic envelope at Manhattan, Kansas (USA) from 1984 through 2011 characterized by accumulated daily precipitation (mm) and accumulated growing degree-days (base 0 °C). Note the both the high degree of interannual variation and the extreme years of 1989 (dry) and 1993 (wet)

production (Henebry 1993; Knapp and Smith 2001; Craine et al. 2012a). Accordingly, grasslands are likely to reveal shifts in phenology induced by climatic change only over the longer term (Knapp et al. 1998).

11.3 Phenologies of Exotic and Invasive Species

Recognizing time as a key dimension of niche space, Wolkovich and Cleland (2011) articulated a framework of four hypotheses about how species may use phenology to invade and thereby shape ecological communities. The vacant niche hypothesis posits the presence of unused periods of resource availability. The seasonal priority effects hypothesis argues that species taking advantage of available resources earlier than competitors may benefit, if the counterpoising risks of negative effects are not too high. The hypothesis of niche breadth suggests that invasive species may have longer phenophases than native species. The fourth hypothesis is of greater phenotypic plasticity in the face of changing climatic conditions. Evaluating the first two hypotheses against a limit set of phenological data, the authors found support for the seasonal priority effects hypothesis, i.e., the first to arrive captures the resource.

The suite of alternative hypotheses provides a useful framework for evaluating invasion studies. Invasion of perennial grasslands by exotic annual grasses can alter the seasonal pattern of evapotranspiration (Obrist et al. 2003; Prater and DeLucia 2006) and the seasonal availability and pool sizes of soil nutrients (Adair and Burke 2010; Parker and Schimel 2010), lending support to the priority effects hypothesis. In contrast, Enloe et al. (2004) found that a deep-rooted late-maturing invasive forb altered the soil moisture profile leading to drier soils than under the native annual grasslands, suggesting it invades through exploiting a broader spatio-temporal niche. Cleland et al. (2012) explored how phenological overlap could be used in restoration plantings to resist invasion of exotic annual grasses in California grasslands.

Goergen et al. (2011) reported an apparent evolutionary response by native grasses to competition from invasive cheatgrass (*Bromus tectorum*): plants from invaded populations demonstrated consistently earlier phenology in greenhouse experiments than those from uninvaded populations. This response on the part of the invaded rather than the invader lends support to phenotypic plasticity hypothesis.

It is perhaps useful to point out here that remote sensing data alone cannot distinguish between the vacant niche and priority effects hypothesis, despite the efficacy with which patches of invasive species can be detected through phenological asynchronies (Bradley and Mustard 2005, 2006, 2008; Peterson 2005; Huang and Geiger 2008; Huang et al. 2009; Foody and Dash 2010).

11.4 Phenology, Water and Belowground Dynamics

Most competition in grasslands occurs belowground (Weaver 1954; Coffin and Lauenroth 1989). Accessibility has long hindered belowground studies, but in the past few years there have been more papers focusing on the belowground dynamics of phenology.

Nord and Lynch (2009), starting from the premise that acquisition of mobile resources (e.g., N) in the rhizosphere would be approximately proportional to total transpiration, speculated that phenological shifts increasing transpiration would increase acquisition of belowground resources and the converse. For the immobile belowground resources (e.g., P, space), they suggested that acquisition would be approximately proportional to root length duration and, thus, phenological shifts increasing the period of root growth would favor increased acquisition of immobile resources. Finally, they pointed out that, since temporal windows of resource availability and acquisition capability must be aligned for acquisition to occur, shifts in phenologies that induce asynchronies may disrupt resource acquisition.

Root production increases significantly with soil temperature, but shoot and root phenologies are not coincident (Steinaker and Wilson 2008; Steinaker et al. 2010). Spatial niche differentiation in rooting depth and temporal niche separation in phenology can promote coexistence of grass species (Fargione and Tilman 2005), but these traits can also facilitate invasion by non-native species (Adair and Burke 2010) as well as by woody species (Steinaker and Wilson 2008).

The seasonality of water available for evapotranspiration and thus plant growth may influence the distribution of C_3 and C_4 grasses (Winslow et al. 2003; Flanagan 2009). The timing of precipitation was found to be more influential on ecosystem productivity in a California grassland than total rainfall amount (Xu and Baldocchi 2004). Moreover, evapotranspiration in this grassland was closely linked to soil moisture availability during the phenophase transition to senescence (Ryu et al. 2008). Conversely, shifts in phenologies due to experimental warming were shown to alter the ecohydrology in other California grasslands (Zavaleta et al. 2003b). The spatio-temporal heterogeneity of pulsed moisture in dryland ecosystems has been shown to be affected by the phenology of deciduous species (Villegas et al. 2010).

Craine et al. (2010) found flowering biomass of three tallgrass species to be weakly linked to late season soil moisture status, which serves as an indicator of the seasonal moisture regime rather than direct forcing on flowering. Examining the community sequence of flowering for 430 herbaceous species during a single year, Craine et al. (2012b) found soil moisture and landscape position to influence flowering dates but discovered no links to a range of plant functional traits.

11.5 Experimental Manipulations

Although grassland canopies are structurally simple, grassland communities can exhibit complicated dynamics in space and time (Knapp et al. 1998, 2004). The most straightforward manipulation of environmental variables to mimic projected future climates is elevation of surface temperatures. Warming and snow removal experiments in a subalpine meadow in Colorado (USA) eliminated the snow cover well in advance of controls (Dunne et al. 2003). Both warming and snow removal resulted in advanced flowering times and extended the duration of the flowering period.

A more complicated experimental design at a well-studied grassland in California (USA) manipulated temperature, precipitation, ambient CO_2, and N deposition (Zavaleta et al. 2003a, b). The grassland community was dominated by naturalized Eurasian annual grasses and containing both native and exotic perennial grasses and forbs. Initial results of the experiment revealed the counter-intuitive result that warming increased spring soil moisture, likely due to surface drying that induced earlier plant senescence thereby reducing seasonal evapotranspiration (Zavaleta et al. 2003a). Results of the manipulations after a subsequent year of data showed a stronger response from annual grasses and forbs than from perennials. Three years of enhanced N deposition increased grass production and depressed forb production and abundance, thereby suppressing plant community diversity (Zavaleta et al. 2003b). Both warming and increased precipitation increased the forbs but had little impact on the grasses. Elevated CO_2 reduced diversity without large impacts on the grass or forb production or relative abundance (Zavaleta et al. 2003b).

Working with 5 years of data from the same experiment, Dukes et al. (2005) concluded that grassland production and, thereby, the potential for C sequestration under future climates would be minimally impacted by changes in CO_2 concentration, winter precipitation, and modest warming. Increased N deposition, however, would have stronger effects on grassland productivity. A concurrent study on the phenological responses to the same manipulations found, as expected, that warming accelerated canopy development of the grass matrix (Cleland et al. 2006). In contrast, elevated CO_2 delayed both canopy development and time of flowering. Increased N delayed flowering in grasses, but accelerated it in forbs.

Manipulating temperature and precipitation in a tallgrass prairie in central Oklahoma, Sherry et al. (2007) found an apparent warming-induced divergence for the reproductive phenophases of annual and perennial grasses and forbs. For those species initiating flowering prior to peak summer heat, experimental warming advanced the time of flowering, but it delayed flowering in those species flowering after peak heat. This divergence likely arose from differential developmental responses to the warming, rather than differences in soil moisture or size-dependent floral induction.

Fay et al. (2000) demonstrated after the first year of a long term experimental manipulation of tallgrass prairie in Kansas (USA) how changes in the timing of seasonal precipitation could strongly alter ecosystem properties, such as soil CO_2 efflux. Durations of flowering period for the dominant warm-season grasses were significantly affected only by changes in both precipitation timing and quantity, but not with either change alone. A warming manipulation was added to this experimental system in 2003 and Fay et al. (2011) reported the effects on two phenophases: spring green-up and flowering. Warming advanced green-up significantly in the early growing season, but there were no significant effects of altered rainfall patterns until early June, when reduced soil moisture translated into reduced canopy greenness (Fay et al. 2011). No warming effect on senescence was detected. Warming produced no significant effects on production of flowering culms of two dominant tallgrass matrix species: *Andropogon gerardii* and *Sorghastrum nutans*. However, rainfall manipulation had a significant impact on *S. nutans*, but not on *A. gerardii*.

Suttle et al. (2007) manipulated the seasonality and intensity of precipitation in a grassland in California (USA) over 5 years and found that short term effects at the scale of individual species were eventually overshadowed by lagged community responses leading to a more productive grassland with lower consumer abundance.

All of the ecological responses observed in the manipulative experiments could lead, over the longer term, to changes in community structure and ecosystem services. As such these manipulative experiments can offer some insight into ecological process and pattern; however, it may be imprudent to rely solely on these short-term responses to project the myriad ecological consequences of changing environmental conditions. Indeed, a recent study reviewed the results from warming experiments with a comprehensive set of observational studies using a common metric: temperature sensitivity in terms of phenophase change in days per degree Celsius (Wolkovich et al. 2012). The review found significantly greater temperature sensitivity of spring leafing (4X) and flowering (8.5X) in the observational studies than in the warming experiments. These large discrepancies cast doubt on the reliability of manipulative experiments and raise concerns about the use of experimentally derived temperature sensitivities in ecosystem models (Wolkovich et al. 2012).

11.6 Remote Sensing of Grasslands

Grasslands are highly responsive to local environmental conditions and recent weather and grassland canopies are structurally simple; thus, grasslands are of particular interest for the remote sensing of land surface phenologies (cf. Henebry and de Beurs, Chap. 21). Differences in photosynthetic pathways lead to different seasonal niches. This temporal segregation makes detection and classification of grasslands dominated by C_3 versus C_4 species a relatively easier remote sensing challenge (Briggs et al. 1997; Goodin and Henebry 1997; Tieszen et al. 1997; Davidson and Csillag 2001). Yet, there are aspects of the problem that require further refinements, including improved statistical estimation of ground cover for calibration and validation (Davidson and Csillag 2003) and the use of spectral indices other than the long-used normalized difference vegetation index or NDVI (Foody and Dash 2010; Wang et al. 2010).

Remote sensing of invasive species in grasslands is another relatively simpler remote sensing problem. Invasive grasses tend to flourish during seasonal interstices before there is much if any competition from the local flora. Two invasive grasses of dryland ecosystems have received significant attention in the remote sensing literature: *Bromus tectorum* or cheatgrass (Bradley and Mustard 2005, 2006, 2008; Peterson 2005) and *Eragrostis lehmanniana* or Lehmann lovegrass (Huang and Geiger 2008; Huang et al. 2009).

Synoptic imagery from polar-orbiting sensors designed for weather monitoring has demonstrated sensitivity to seasonal patterns in the vegetated land surface. Yet, sophisticated data processing and statistical modeling are needed to extract phenological change from other sources of variability in grasslands (de Beurs and Henebry 2004, 2005, 2010; Bradley et al. 2007; Hermance et al. 2007; Bradley and Mustard 2008). However, others have cast doubt on the utility of the long—but noisy—AVHRR records for phenological studies (Kathuroju et al. 2007; de Beurs and Henebry 2008; Alcaraz-Segura et al. 2010; Beck et al. 2011).

11.7 Phyllochron and Phenology

Phyllochron is "the time interval between appearance of successive leaves on a culm" (Wilhelm and McMaster 1995) or, more generally, the rate of leaf appearance (McMaster 2005). The phyllochron concept has been used extensively to understand the ontogenetic development of economically important grasses, in particular wheat (McMaster and Simka 1988; McMaster 1997, 2005; McMaster and Wilhelm 1998) and rice (Miyamoto et al. 2004; Itoh and Sano 2006; Itoh and Shimizu 2012).

Phyllochron relates to growth rate; phenology relates to development, specifically to the timing of phenophase transitions. Thus, influences on phyllochron may also shed light on controls on phenology in grass canopies. Phyllochrons have been shown to be affected by genotype/variety, daylength, and temperature (McMaster and Smika 1988; Itoh and Sano 2006) and have been modeled with varying levels of success (McMaster and Wilhelm 1998; Fournier et al. 2005). Temperature is a primary environmental determinant of phyllochron in wheat and barley; phyllochron is "quite predictable based on temperature, with other abiotic factors having secondary effects and often only after a threshold value is reached" (McMaster 2005, p. 140). Moreover, "[e]xtrapolations from winter wheat to winter barley and spring barley to spring wheat are easily made as primarily only the thermal time between phenological events changes, mainly due to difference in the phyllochron" (McMaster 2005, p. 142).

The thermal time is readily measured in accumulated growing degree-days with the growing degree-day increment being that portion of the daily average temperature exceeding a specific temperature threshold or base, with only positive increments accumulating (Fig. 11.2). Thus, the passage of days is weighted by the quantity of growing degrees occurring that day, with zero (but not negative) degrees being a permissible weight. Drawing from extensive study on wheat phenology (McMaster and Smika 1988; McMaster and Wilhelm 1998), McMaster (2005) concludes that using a constant base of 0 °C is "very robust and often sufficient for most purposes" (p. 144).

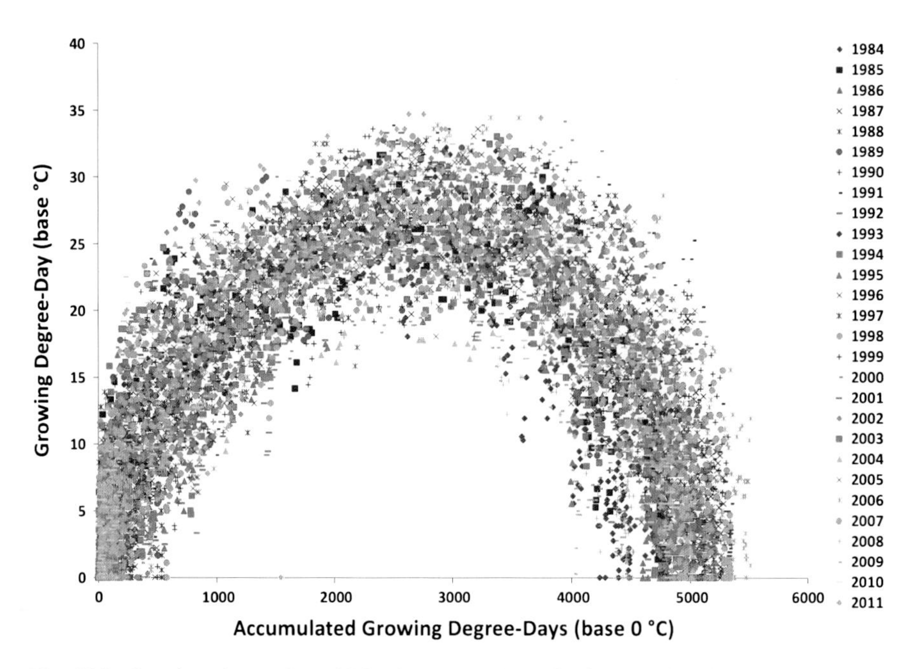

Fig. 11.2 Growing degree-day (GDD) increment (base 0 °C) as a function of accumulated growing degree-day (AGDD) for Manhattan, Kansas (USA) from 1984 through 2011. Notice the high interannual variability around a clear unimodal annual pattern

The convex quadratic model of land surface phenology that is described in Henebry and de Beurs (2013; Chap. 21 of this volume) is grounded in such an understanding of thermal time as a key determinant of herbaceous phenology in temperate environments. Although the growing degree-days are noisy at submonthly time scales in any particular year (Fig. 11.2), there is an underlying seasonal distribution that is parsimoniously approximated by a convex quadratic equation (Fig. 11.3).

A secondary consideration for the phenology of temperate North American grasslands is daylength as it relates to irradiance and the availability of photosynthetically active radiation (PAR) to drive carbon fixation. In a recent review Richardson et al. (2013) call for a re-examination of the role of photoperiod in phenologies. As first summarized in Henebry (2003), a significant body of research in the middle twentieth century focused on the effect of photoperiod or daylength on phenology of rangeland grasses and the variation in that effect in geographic clones (Benedict 1941; Olmsted 1943, 1944, 1945; Rice 1950). McMillan found a strong gradient of more rapid phenological development from south to north and to a lesser extent from east to west (McMillan 1956a, b, 1957, 1959a, b). The longer growing seasons and more mesic environments toward the south and east of the Great Plains allowed grasses to initiate growth earlier and flower later than more northerly or westerly clones. McMillan (1960) argued that these patterns of ecotypic variation within species emerge from natural selection through the continuing

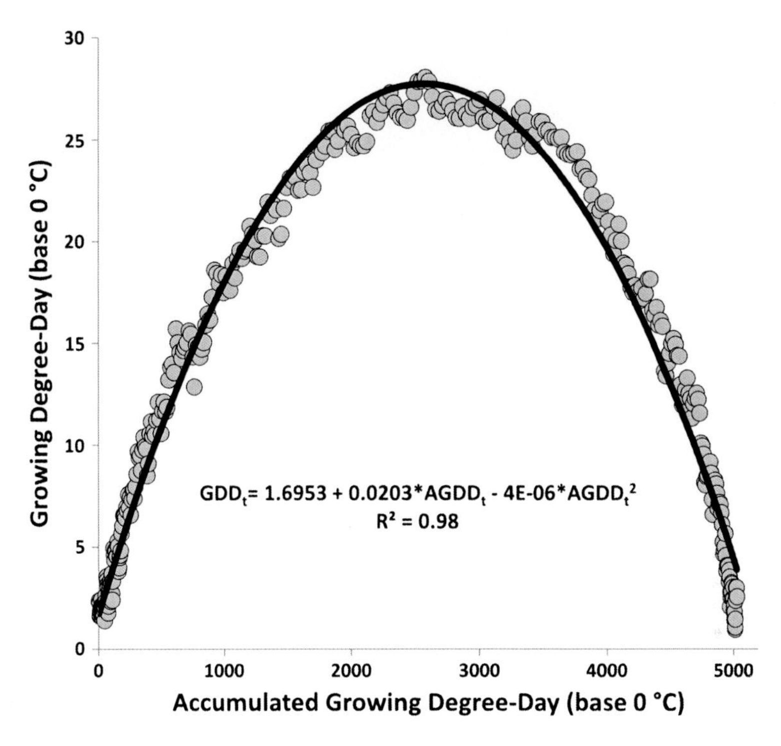

Fig. 11.3 Mean daily growing degree-day (GDD) increments (base 0 °C) from 1984 through 2011 as a function of accumulated growing degree-day for Manhattan, Kansas (USA). Notice how the interannual variation in Fig. 11.2 reduces to a very strong fit. Median daily values (data not shown) also exhibit an excellent fit ($R^2 = 0.975$) and an intercept closer to zero (0.358)

interaction between the "habitat variable", i.e., variation in local environmental forcings and constraints, and the "genetic variable", i.e., the differential responses to the same habitat arising from the differential genetic potential among individual members of a population.

A foundation of convex quadratic model is the expectation that growing degree-days relate to daylength in a quasi-parabolic manner (Fig. 11.4). A fundamental driver of grasslands productivity is PAR availability, but several other forcings and constraints limit productivity at the local scale, such as moisture regime, landscape position, nutrient availability, fire history, and grazing above and belowground (Knapp and Seastedt 1986; Briggs and Knapp 1995; Craine et al. 2012a). Daylength and top-of-atmosphere insolation are constant functions of latitude and thus vary relatively slowly in space. However, the amount of PAR reaching the photosynthetic tissue in the grassland canopy varies with the weather, modulated primarily by cloudiness, and with canopy composition, due to shading from live or senesced biomass (Knapp and Seastedt 1986).

Daylength affects the seasonal distribution of PAR as well as the seasonal thermal regime and its broad predictability points to why a convergence in potential

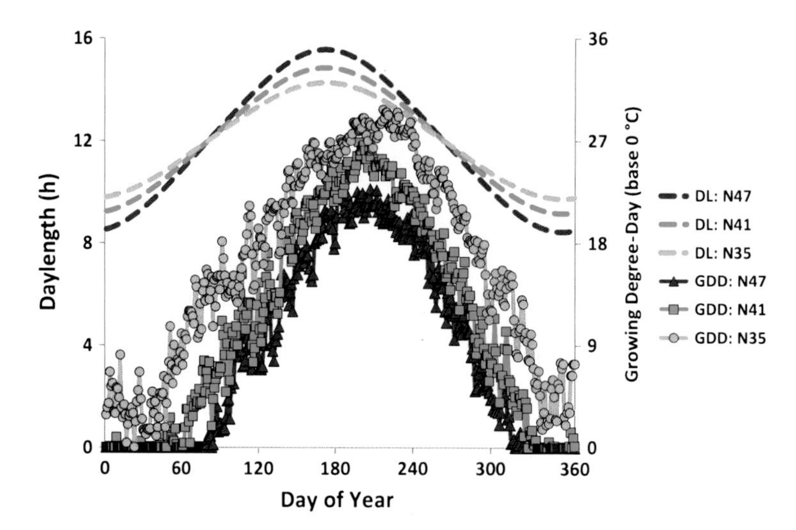

Fig. 11.4 The seasonal relationship between daylength (DL) and mean growing degree-days (GDD) at three locations (47, 41, and 35° North latitude) that comprise a transect in the central Great Plains of North America. Note how peak heat lags peak daylength by 3–4 weeks

net ecosystem production in C_3 grasslands has been observed (Peichl et al. 2013). A comparable study in mesic C_4 grasslands is needed. Peak daylength and peak heat are separated as an approximate geographic function, but if this relationship should change under a warming climate, it will have implications for phenology in grasslands and associated ecosystem services.

Acknowledgments Research was supported in part by NASA grant NNX12AM89G.

References

Adair EC, Burke IC (2010) Plant phenology and life span influence soil pool dynamics: *Bromus tectorum* invasion of perennial C3–C4 grass communities. Plant Soil 335(1–2):255–269

Alcaraz-Segura D, Chuvieco E, Epstein HE, Kasischke ES, Trishchenko A (2010) Debating the greening vs. browning of the North American boreal forest: differences between satellite datasets. Glob Change Biol 16(2):760–770

Beck HE, McVicar TR, van Dijk AIJM, Schellekens J, de Jeu RAM, Bruijzeel LA (2011) Global evaluation of four AVHRR–NDVI data sets: intercomparison and assessment against Landsat imagery. Remote Sens Environ 115:2547–2563

Benedict HM (1941) Growth of some range grasses in reduced light intensities at Cheyenne, Wyoming. Bot Gaz 102:582–589

Bradley BA, Mustard JF (2005) Identifying land cover variability distinct from land cover change: cheatgrass in the Great Basin. Remote Sens Environ 94(2):204–213

Bradley BA, Mustard JF (2006) Characterizing the landscape dynamics of an invasive plant and risk of invasion using remote sensing. Ecol Appl 16:1132–1147

Bradley BA, Mustard JF (2008) Comparison of phenology trends by land cover class: a case study in the Great Basin, U.S.A. Glob Change Biol 14(2):334–346

Bradley BA, Jacob RW, Hermance JF, Mustard JF (2007) A curve-fitting technique to derive interannual phenologies from time series of noisy satellite data. Remote Sens Environ 106:137–145

Briggs JM, Knapp AK (1995) Interannual variability in primary production in tallgrass prairie: climate, soil moisture, topographic position, and fire as determinants of aboveground biomass. Am J Bot 82(8):1024–1030

Briggs JM, Rieck DR, Turner CL, Henebry GM, Goodin DG, Nellis MD (1997) Spatial and temporal patterns of vegetation in the Flint Hills. Trans Kans Acad Sci 100:10–20

Cayan DR, Kammerdiener SA, Dettinger MD, Caprio JM, Peterson DH (2001) Changes in the onset of spring in the western United States. Bull Am Meteorol Soc 82(3):399–415

Cleland EE, Chiariello NR, Loarie SR, Mooney HA, Field CB (2006) Diverse responses of phenology to global changes in a grassland ecosystem. PNAS 103(37):13740–13744

Cleland EE, Larios L, Suding KN (2012) Strengthening invasion filters to reassemble native plant communities: soil resources and phenological overlap. Restor Ecol early online. doi: 10.1111/j.1526-100X.2012.00896.x

Coffin DP, Lauenroth WK (1989) Disturbances and gap dynamics in a semiarid grassland: a landscape-level approach. Landsc Ecol 3(1):19–27

Craine JM, Towne EG, Nippert JB (2010) Climate controls on grass culm production over a quarter century in a tallgrass prairie. Ecology 91:2132–2140

Craine JM, Nippert JB, Elmore AJ, Skibbe AM, Hutchinson SL, Brunsell NA (2012a) Timing of climate variability and grassland productivity. PNAS 109(9):3401–3405

Craine JM, Wolkovish EM, Towne EG, Kembel SW (2012b) Flowering phenology as a functional trait in a tallgrass prairie. New Phytol 193:673–682

Davidson A, Csillag F (2001) The influence of vegetation index and spatial resolution on a two-date remote sensing derived relation to C4 species coverage. Remote Sens Environ 75:138–151

Davidson A, Csillag F (2003) A comparison of three approaches for predicting C_4 species cover of northern mixed grass prairie. Remote Sens Environ 86:70–82

de Beurs KM, Henebry GM (2004) Land surface phenology, climatic variation, and institutional change: analyzing agricultural land cover change in Kazakhstan. Remote Sens Environ 89 (4):497–509. doi:10.1016/j.rse.2003.11.006

de Beurs KM, Henebry GM (2005) A statistical framework for the analysis of long image time series. Int J Remote Sens 26(8):1551–1573

de Beurs KM, Henebry GM (2008) Northern annular mode effects on the land surface phenologies of northern Eurasia. J Clim 21:4257–4279

de Beurs KM, Henebry GM (2010) Spatio-temporal statistical methods for modeling land surface phenology. In: Hudson IL, Keatley MR (eds) Phenological research: methods for environmental and climate change analysis. Springer, New York

Dukes JS, Chiariello NR, Cleland EE, Moore LA, Shaw MR, Thayer S, Tobeck T, Mooney HA, Field CB (2005) Responses of grassland production to single and multiple global environmental changes. PLoS Biol 3(10):e319

Dunne JA, Harte J, Taylor KJ (2003) Subalpine meadow flowering phenology responses to climate change: integrating experimental and gradient methods. Ecol Monogr 73(1):69–86

Enloe SF, DiTomaso JM, Orloff SB, Drake DJ (2004) Soil water dynamics differ among rangeland plant communities dominated by yellow starthistle (*Centaurea solstitialis*), annual grasses, or perennial grasses. Weed Sci 52(6):929–935

Fargione J, Tilman D (2005) Niche differences in phenology and rooting depth promote coexistence with a dominant C_4 bunchgrass. Oecologia 143(4):598–606

Fay PA, Carlisle JD, Knapp AK, Blair JM, Collins SL (2000) Altering rainfall timing and quantity in a mesic grassland ecosystem: design and performance of rainfall manipulation shelters. Ecosystems 3(3):308–319

Fay PA, Blair JM, Smith MD, Nippert JB, Carlisle JD, Knapp AK (2011) Relative effects of precipitation variability and warming on tallgrass prairie ecosystem function. Biogeosciences 8:3053–3068

Flanagan LB (2009) Phenology of plant production in the northwestern Great Plains: relationships with carbon isotope discrimination, net ecosystem productivity and ecosystem respiration. In: Noormets A (ed) Phenology of ecosystem processes. Springer, New York

Foody GM, Dash J (2010) Estimating the relative abundance of C_3 and C_4 grasses in the Great Plains from multi-temporal MTCI data: issues of compositing period and spatial generalizability. Int J Remote Sens 31(2):351–362

Fournier C, Durand JL, Ljutovac S, Schäufele R, Gastal F, Andrieu B (2005) A functiona-structural model of elongation of the grass leaf and its relationships with the phyllochron. New Phytol 166:881–894

Goergen EM, Leger EA, Espeland EK (2011) Native perennial grasses show evolutionary response to *Bromus tectorum* (Cheatgrass) invasiȯn. PLoS One 6(3):e18145

Goodin DG, Henebry GM (1997) Monitoring ecological disturbance in tallgrass prairie using seasonal NDVI trajectories and a discriminant function mixture model. Remote Sens Environ 61:270–278

Henebry GM (1993) Detecting change in grasslands using measures of spatial dependence with Landsat TM data. Remote Sens Environ 46:223–234

Henebry GM (2003) Grasslands of the north American Great Plains. In: Schwartz MD (ed) Phenology: an integrative environmental science. Kluwer, New York

Henebry GM, de Beurs KM (2013) Remote sensing of land surface phenology: A prospectus. In: Schwartz MD (ed) Phenology: an integrative environmental science, 2nd edn. Elsevier, Dordrecht

Hermance JF, Jacob RW, Bradley BA, Mustard JF (2007) Extracting phenological signals from multi-year AVHRR NDVI time series: framework for applying high-order annual splines with roughness damping. IEEE Trans Geosci Remote Sens 45:3264–3276

Huang C-Y, Geiger EL (2008) Climate anomalies provide opportunities for large-scale mapping of non-native plant abundance in desert grasslands. Divers Distrib 14(5):875–884

Huang C, Geiger EL, Van Leeuwen WJD, Marsh SE (2009) Discrimination of invaded and native species sites in a semi-desert grassland using MODIS multi-temporal data. Int J Remote Sens 30(4):897–917

Hudson IL, Keatley MR (2010) Phenological research: methods for environmental and climate change analysis. Springer, New York

Itoh Y, Sano Y (2006) Phyllochron dynamics under controlled environments in rice (*Oryza sativa* L.). Euphytica 150:87–95

Itoh Y, Shimizu H (2012) Phyllochron dynamics during the course of late shoot development might be affected by reproductive development in rice (*Oryza sativa* L.). Dev Genes Evol 222:341–350

Kathuroju N, White MA, Symanzik J, Schwartz MD, Powell JA, Nemani RR (2007) On the use of the advanced very high resolution radiometer for development of prognostic land surface phenology models. Ecol Model 201(1):144–156

Knapp AK, Seastedt TR (1986) Detritus accumulation limits productivity of tallgrass prairie. BioScience 36(10):662–668

Knapp AK, Smith MD (2001) Variation among biomes in temporal dynamics of aboveground primary production. Science 291:481–484

Knapp AK, Briggs JM, Hartnett DC, Collins SL (1998) Grassland dynamics: long-term ecological research in Tallgrass Prairie. Oxford University Press, New York

Knapp AK, Smith MD, Collins SL, Zambatis N, Peel M, Emery S, Wojdak J, Horner-Devine MC, Biggs H, Kruger J, Andelman SJ (2004) Generality in ecology: testing North American grassland rules in South African savannas. Front Ecol Environ 2:483–491

McMaster GS (1997) Phenology, development, and growth of the wheat (*Triticum aestivum* L.) shoot apex: a review. Adv Agron 59:63–118

McMaster GS (2005) Phytomers, phyllochrons, phenology and temperate cereal development. J Agric Sci 143:137–150

McMaster GS, Smika DE (1988) Estimation and evaluation of winter wheat phenology in the central Great Plains. Agric For Meteorol 43(1):1–18

McMaster GS, Wilhelm WW (1998) Is soil temperature better than air temperature for predicting winter wheat phenology? Agron J 90:602–607
McMillan C (1956a) Nature of the plant community. I. Uniform garden and light period studies of five grass taxa in Nebraska. Ecology 37:330–340
McMillan C (1956b) Nature of the plant community. II. Variation in flowering behavior within populations of *Andropogon scoparius*. Ecology 43:429–436
McMillan C (1957) Nature of the plant community. III. Flowering behavior within two grassland communities under reciprocal transplanting. Am J Bot 44:144–153
McMillan C (1959a) Nature of the plant community. V. Variation within the true prairie community-type. Am J Bot 46:418–424
McMillan C (1959b) The role of ecotypic variation in the distribution of the central grassland of North America. Ecol Monogr 29:285–308
McMillan C (1960) Ecotypes and community function. Am Nat 94:245–255
Menzel A, Sparks TH, Estrella N, Koch E, Aasa A, Ahas R, Alm-Kübler K, Bissolli P, Braslavská O, Briede S, Chmielewski FM, Crepinsek Z, Curnel Y, Dahl Å, Defila C, Donnelly A, Filella Y, Jatczak K, Måge F, Mestre A, Nordli Ø, Peñuelas J, Pirinen P, Remišová V, Scheifinger H, Striz M, Susnik A, Van Vliet AJH, Wielgolaski F-E, Zach S, Zust A (2006) European phenological response to climate change matches the warming pattern. Glob Change Biol 12:1969–1976. doi:10.1111/J.1365-2486.2006.01193.X
Miyamoto N, Goto Y, Matsui M, Ukai Y, Morita M, Nemoto K (2004) Quantitative trait loci for phyllochron and tillering in rice. Theor Appl Genet 109:700–706
Morisette JT, Richardson AD, Knapp AK, Fisher JI, Graham E, Abatzoglou J, Wilson BE, Breshears DD, Henebry GM, Hanes JM, Liang L (2008) Unlocking the rhythm of the seasons in the face of global change: challenges and opportunities for phenological research in the 21st century. Front Ecol Environ 5(7):253–260. doi:10.1890/070217
Nord EA, Lynch JP (2009) Plant phenology: a critical controller of soil resource acquisition. J Exp Bot 60(7):1927–1937
Obrist D, Verburg PSJ, Young MH, Coleman JS, Schorran DE, Arnone JA III (2003) Quantifying the effects of phenology on ecosystem evapotranspiration in planted grassland mesocosms using EcoCELL technology. Agric For Meteorol 118(3–4):173–183
Olmsted CE (1943) Growth and development in range grasses. III. Photoperiodic responses in the genus Bouteloua. Bot Gaz 105:165–181
Olmsted CE (1944) Growth and development in range grasses. IV. Photoperiodic responses in twelve geographic strains of side-oats gramma. Bot Gaz 106:46–74
Olmsted CE (1945) Growth and development in range grasses. V. Photoperiodic responses of clonal divisions of three latitudinal strains of side-oats gramma. Bot Gaz 106:382–401
Parker SS, Schimel JP (2010) Invasive grasses increase nitrogen availability in California grassland soils. Invasive Plant Sci Manag 3(1):40–47
Parmesan C, Yohe G (2003) A globally coherent fingerprint of climate change impacts across natural systems. Nature 421:37–42
Peichl M, Sonnentag O, Wohlfahrt G, Flanagan LB, Baldocchi DD, Kiely G, Galvagno M, Gianelle D, Marcolla B, Pio C, Migliavacca M, Jones MB, Saunders M (2013) Convergence of potential net ecosystem production among contrasting C_3 grasslands. Ecol Lett 16(4):502–512. doi:10.1111/ele.12075
Peterson EB (2005) Estimating cover of an invasive grass (*Bromus tectorum*) using tobit regression and phenology derived from two dates of Landsat ETM + data. Int J Remote Sens 26 (12):2491–2507
Prater MR, DeLucia EH (2006) Non-native grasses alter evapotranspiration and energy balance in Great Basin sagebrush communities. Agric For Meteorol 139(1–2):154–163
Rice EL (1950) Growth and floral development of five species of range grass in central Oklahoma. Bot Gaz 111:361–377
Richardson AD, Keenan TF, Migliavacca M, Ryu Y, Sonnentag O, Toomey M (2013) Climate change, phenology, and phenological control of vegetation feedbacks to the climate system. Agric For Meteorol 169:156–173. doi:10.1016/j.agrformet.2012.09.012

Root TL, Price JT, Hall KR, Schneider SH, Rosenzweig C, Pounds JA (2003) Fingerprints of global warming on wild animals and plants. Nature 421(6918):57–60

Rosenzweig C, Casassa G, Karoly DJ, Imeson A, Liu C, Menzel A, Rawlins S, Root TL, Seguin B, Tryjanowski P (2007) Assessment of observed changes and responses in natural and managed systems. In: Parry ML, Canziani OF, Palutikof JP, van der Linden PJ, Hanson CE (eds) Climate change 2007: impacts, adaptation and vulnerability. Contribution of Working Group II to the fourth assessment report of the Intergovernmental Panel on Climate Change. Cambridge University Press, New York

Ryu Y, Baldocchi DD, Ma S, Hehn T (2008) Interannual variability of evapotranspiration and energy exchange over an annual grassland in California. J Geophys Res 113:D09104

Schwartz MD (1994) Monitoring global change with phenology: the case of the spring green wave. Int J Biometeorol 38(1):18–22

Schwartz MD (1998) Green-wave phenology. Nature 394(6696):839–840

Schwartz MD, Ahas R, Aasa A (2006) Onset of spring starting earlier across the northern Hemisphere. Glob Change Biol 12:343–351

Sherry RA, Zhou X, Gu S, Arnone JA III, Schimel DS, Verburg PS, Wallace LL, Luo Y (2007) Divergence of reproductive phenology under climate warming. PNAS 104 (1):198–202

Steinaker DF, Wilson SD (2008) Phenology of fine roots and leaves in forest and grassland. J Ecol 96(6):1222–1229

Steinaker DF, Wilson SD, Peltzer DA (2010) Asynchronicity in root and shoot phenology in grasses and woody plants. Glob Change Biol 16(8):2242–2251

Suttle KB, Thomsen MA, Power ME (2007) Species interactions reverse grassland responses to changing climate. Science 315(5812):640–642

Tieszen LL, Reed BC, Bliss NB, Wylie BK, DeJong DD (1997) NDVI, C_3 and C_4 production and distributions in Great Plains grassland land cover classes. Ecol Appl 7:59–78

Villegas JC, Breshears DD, Zou CB, Royer PD (2010) Seasonally pulsed heterogeneity in microclimate: phenology and cover effects along deciduous grassland-forest. Vadose Zone J 9(3):537–547

Wang C, Jamison BE, Spicci AA (2010) Trajectory-based warm season grassland mapping in Missouri prairies with multi-temporal ASTER imagery. Remote Sens Environ 114(3):531–539

Weaver JE (1954) North American Prairie. Johnsen Publishing, Lincoln

Westerling AL, Hidalgo HG, Cayan DR, Swetnam TW (2006) Warming and earlier spring increases western U.S. forest wildfire activity. Science 313:940–943

Wilhelm WW, McMaster GS (1995) Importance of the phyllochron in studying development and growth in grasses. Crop Sci 35:1–3

Winslow JC, Hunt ER Jr, Piper SC (2003) A phenological model of the global C3 and C4 grass distribution with application to the United States Great Plains under a VEMAP climatic change scenario, Ecological Modelling 163:153–173

Wolkovich EM, Cleland EE (2011) The phenology of plant invasions: a community ecology perspective. Front Ecol Environ 9:287–294

Wolkovich EM, Cook BI, Allen JM, Crimmins TM, Betancourt JL, Travers SE, Pau S, Regetz J, Davies TJ, Kraft NJB, Ault TR, Bolmgren K, Mazer SJ, McCabe GJ, McGill BJ, Parmesan C, Salamin N, Schwartz MD, Cleland EE (2012) Warming experiments underpredict plant phenological responses to climate change. Nature 485:494–497

Xu L, Baldocchi DD (2004) Seasonal variation in carbon dioxide exchange over a Mediterranean annual grassland in California. Agric For Meteorol 123(1–2):79–96

Zavaleta ES, Shaw MR, Chiariello NR, Thomas BD, Cleland EE, Field CB, Mooney HA (2003a) Grassland responses to three years of elevated temperature, CO_2, precipitation, and N deposition. Ecol Monogr 73:585–604

Zavaleta ES, Thomas BD, Chiariello NR, Asner GP, Shaw MR, Field CB (2003b) Plants reverse warming effect on ecosystem water balance. PNAS 100(17):9892–9893

Chapter 12
Mesic Temperate Deciduous Forest Phenology

Jonathan M. Hanes, Andrew D. Richardson, and Stephen Klosterman

Abstract Deciduous forests in temperate climates are characterized by significant seasonal changes in ecological and biogeochemical processes that are directly linked to forest phenology. The timing of spring leaf emergence and autumn leaf senescence is heavily determined by weather and climate, and these phenological events influence the seasonal cycles of water, energy, and carbon fluxes. In addition to its role in ecological interactions and in regulating ecosystem processes, deciduous forest phenology has also been shown to be a robust indicator of the biological impacts of climate change on forest ecosystems. With an emphasis on spring leaf emergence and autumn leaf senescence, this chapter highlights the phenology of canopy trees in mesic temperate deciduous forests by describing the climate of these forests, environmental drivers of phenology, feedback of phenology on lower atmospheric processes, impacts of climate change on phenology, and future research directions.

12.1 Introduction

Deciduous forests in temperate climates are among the most dynamic ecosystems on our planet. They are characterized by significant seasonal changes in ecological and biogeochemical processes that are directly linked to forest phenology. The timing of spring leaf emergence and autumn leaf senescence is heavily determined by weather and climate, and these phenological events influence the seasonal cycles of water, energy, and carbon fluxes (Baldocchi et al. 2005; Morisette et al. 2009;

J.M. Hanes (✉)
Department of Geography, University of Wisconsin-Milwaukee,
Milwaukee, WI 53211, USA
e-mail: jmhanes@uwm.edu

A.D. Richardson • S. Klosterman
Department of Organismic and Evolutionary Biology,
Harvard University, Cambridge, MA 02138, USA

M.D. Schwartz (ed.), *Phenology: An Integrative Environmental Science*,
DOI 10.1007/978-94-007-6925-0_12, © Springer Science+Business Media B.V. 2013

Richardson et al. 2009b; Schwartz and Hanes 2010; Dragoni et al. 2011). In addition to its role in ecological interactions and in regulating ecosystem processes, deciduous forest phenology has also been shown to be a robust indicator of the biological impacts of climate change on forest ecosystems (Myneni et al. 1997; Peñuelas and Filella 2001; Richardson et al. 2006; Morin et al. 2009).

The significance of phenology as an indicator of climate change has motivated the development of methods to understand spatiotemporal trends in deciduous forest phenology. Generally, research on temperate deciduous trees utilizes three methods of phenological observation. In-situ, visual assessments of canopy trees during the spring and autumn are used to document discrete phenophases (Liang and Schwartz 2009; Richardson and O'Keefe 2009). These ground observations are made using predefined protocols that characterize the stages of leaf growth and senescence. Digital "webcam" images are used to continuously monitor the "greenness" of forest canopies within a selected area (Richardson et al. 2007, 2009a; Ahrends et al. 2009). The timing of phenological transitions in the canopy can be identified using multiple images collected throughout the year. Time series of satellite data from two or more spectral bands can also be used to identify seasonal transitions in vegetation activity (Zhang et al. 2003; Fisher et al. 2006; Friedl et al. 2006). In addition to these methods of phenological monitoring, the evident relationships between deciduous forest phenology and temperature have facilitated the creation of models to better understand phenological differences between species and forecast future changes in phenology due to climate change (Hunter and Lechowicz 1992; Richardson et al. 2006; Morin et al. 2009; Vitasse et al. 2011).

12.2 Characterizing Climate in Mesic Temperate Deciduous Forests

Generally, mesic temperate deciduous forests are located within the "temperate broadleaf and mixed forests" biome described in Olson et al. (2001) (Fig. 12.1). The climate of mesic temperate deciduous forests is marked by seasonal fluctuations in temperature and a precipitation regime that is suitable for the continued survival of deciduous tree species. This biome has an average annual temperature of 8.60 °C (±4.76 °C) and an average annual temperature range of 24.52 °C (±8.15 °C). The average summer and winter temperatures across the biome are 19.27 °C (±3.98 °C) and −2.86 °C (±7.76 °C), respectively.[1] This biome is located predominantly in the midlatitudes of the northern hemisphere, which accounts for the temperature seasonality.

[1] Climate statistics were calculated using the high resolution climate data (version 2.1) produced by the Climate Research Unit at the University of East Anglia and supplied by the Data Distribution Centre (www.ipcc-data.org). Further details about the data can be found in Mitchell et al. (2004) and Mitchell and Jones (2005). Average summer temperature was calculated using temperature data for June, July, and August. Average winter temperature was calculated using temperature data for December, January, and February.

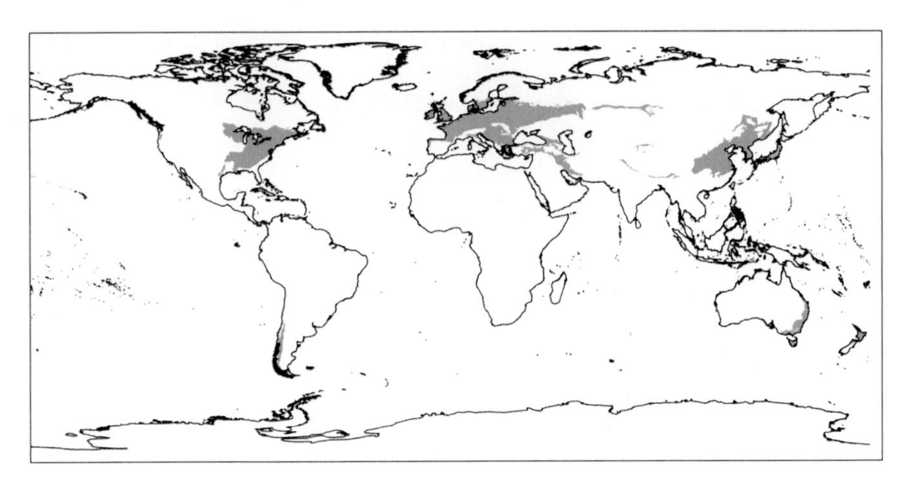

Fig. 12.1 Geographic extent of the Temperate Broadleaf and Mixed Forests biome (From Olson et al. 2001; biome map acquired from the World Wildlife Fund and world basemap acquired from VDS Technologies)

12.3 Characterizing Temperate Deciduous Forest Phenology

12.3.1 Spring Phenology

During the end of summer and autumn, decreasing photoperiods induce an initial phase of dormancy in temperate deciduous tree species (Kramer 1936; Perry 1971; Hänninen and Tanino 2011). Low nighttime temperatures can also induce dormancy in northern ecotypes under longer photoperiods (Tanino et al. 2010). As this initial dormant phase (called endodormancy) continues into the winter months, physiological conditions in the buds prevent leaf development and growth from occurring (as described by Kramer 1994a and Horvath et al. 2003; see Fig. 12.2). To emerge from endodormancy, buds must be exposed to temperatures below a certain threshold temperature for a duration of time. This required exposure to "chilling" temperatures is referred to as the chilling requirement. The chilling requirement, which can vary by species (Morin et al. 2009), minimizes the likelihood of exposure to unfavorable temperatures after the leaves emerge from the buds. The specific threshold below which temperatures contribute to the chilling requirement can also vary by species. However, it is generally thought that temperatures in the range of 0–10 °C contribute to this requirement (Perry 1971). In addition to the chilling requirement, some deciduous tree species, particularly late successional species, require exposure to a critical photoperiod to be released from endodormancy (Körner and Basler 2010; Migliavacca et al. 2012).

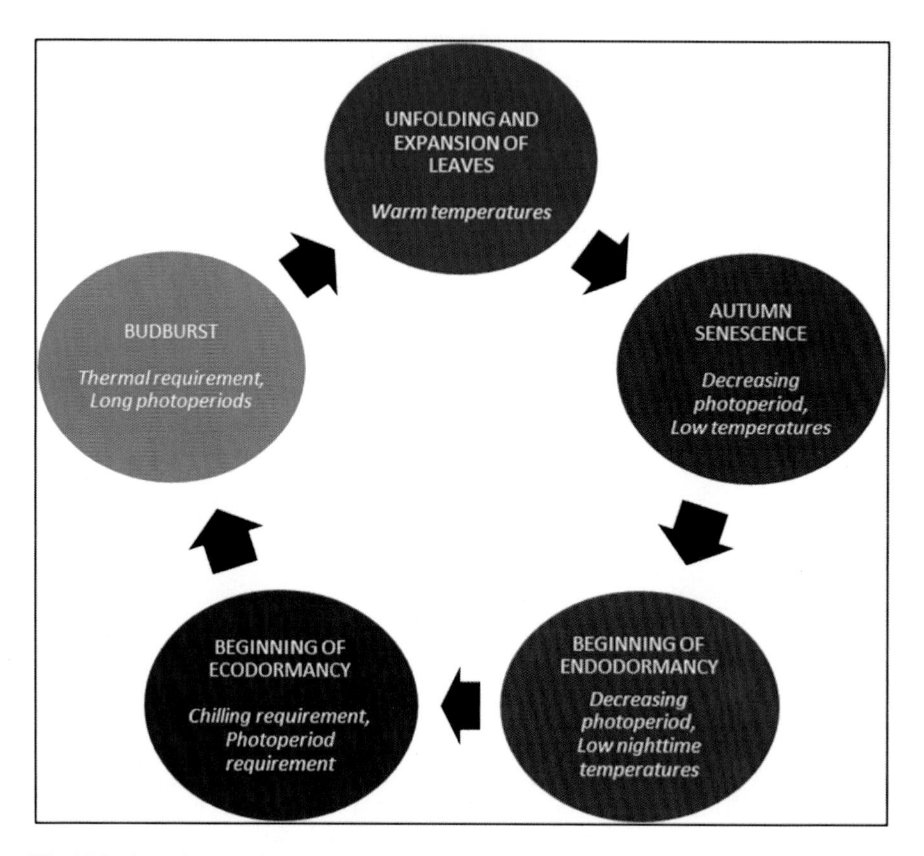

Fig. 12.2 Annual cycle of leaf growth and dormancy (environmental drivers that can contribute to each stage of the cycle are listed in *italics*)

After the buds are released from endodormancy, they enter a second dormant phase called ecodormancy. This phase of dormancy involves the cessation of growth in the buds due to adverse environmental conditions and ends when the buds burst. Buds must be exposed to warmer temperatures and heat accumulation before they burst and the leaves emerge. Thus, in addition to the chilling requirement, thermal requirements exist that govern the spring phenology of temperate deciduous trees. Temperatures above a threshold contribute to the thermal requirement (Cannell and Smith 1986; Murray et al. 1989; Hunter and Lechowicz 1992; Heide 1993; Rötzer et al. 2004; Rousi and Pusenius 2005; Richardson et al. 2006). Like the chilling requirement, both the threshold temperature (Heide 1993) and the thermal requirement for budburst can vary by species (Murray et al. 1989). Past studies have noted an inverse, exponential relationship between the duration of chilling and the thermal accumulation required to initiate leaf emergence (Heide 1993; Ghelardini et al. 2010). This relationship is most pronounced at low amounts of chilling (as noted by Murray et al. 1989; Hunter and Lechowicz 1992). Furthermore, when chilling durations are short, long photoperiods can reduce the thermal requirement necessary for budburst in some species (Heide 1993; Myking and Heide 1995; Myking 1999; Caffarra and Donnelly 2011). After the leaves emerge from the buds, their growth and expansion are driven by temperature (Richardson et al. 2006; Hanes and Schwartz 2011).

Both the chilling duration and thermal requirements of deciduous forests exhibit spatial patterns over large latitudinal gradients. Moving from south to north in the temperate zone of the northern hemisphere, the duration of chilling increases and the thermal requirement for leaf emergence decreases (Zhang et al. 2004, 2007). As a result of the long duration of winter chilling in northerly latitudes, the chilling requirements of deciduous trees are satisfied easily. The lower thermal requirements in northern temperate deciduous forests maximize the length of the growing season and the period of photosynthetic carbon uptake. In more southerly locations in the temperate zone, the shorter duration of chilling can lead to inadequate chilling, which necessitates a larger accumulation of heat to initiate budburst and leaf emergence (Murray et al. 1989; Zhang et al. 2007).

Models predicting budburst have been developed using a variety of approaches (e.g. theoretical, statistical, process-based) that incorporate photoperiod and chilling and warming temperatures. For a detailed description of the categories of phenological models created for the spring season see Hunter and Lechowicz (1992), Chuine (2000), and Morin et al. (2009).

12.3.2 Autumn Phenology

In autumn, leaf abscission marks the end of the growing season in temperate deciduous forests. During this transitional season, the color of the leaves progressively changes from green to either red or yellow as nutrients are withdrawn from the senescing leaves (Lee et al. 2003), which are ultimately shed from the tree. Similar to spring leaf phenology, the timing of leaf senescence in temperate deciduous forests during the autumn is influenced by environmental conditions. However, in contrast to spring phenology, the precise nature of the relationships between autumn phenology and environmental factors is not understood completely (Estrella and Menzel 2006).

Both temperature (Vitasse et al. 2009) and photoperiod (Keskitalo et al. 2005) can regulate the autumn phenology of temperate deciduous trees. For some species, leaf senescence can be driven more by temperature than photoperiod [e.g. *Fagus grandifolia* and *Acer saccharum* (Richardson et al. 2006), *Fagus sylvatica* and *Quercus petraea* (Vitasse et al. 2009)]. For these species and potentially others, a warmer autumn appears to delay leaf senescence. Estrella and Menzel (2006) also noted the potential influence of dryness on leaf senescence for silver birch (*Betula pendula*), with a greater number of rainless days in September coinciding with an advance in the timing of senescence. Despite our understanding that photoperiod and temperature appear to be the primary environmental variables that regulate autumn phenology, our lack of knowledge regarding the exact nature of these influences and how they interact across space and time complicates efforts to create autumn phenology models with strong explanatory power.

12.4 Feedback Effects of Deciduous Forest Phenology

Deciduous forest phenology regulates biological processes affecting the lower atmosphere, including the timing and amount of carbon uptake and water loss via photosynthesis, and the partitioning of the energy budget at the earth's surface. The seasonal signal of carbon uptake by deciduous forests in the northern hemisphere (Keeling et al. 1996) provides context for the importance of phenological controls on global climate. Field and modeling studies have improved understanding of the feedbacks of phenology to climate, however further work is necessary to understand how these feedbacks may react to future climate.

The eddy covariance technique enables field observations of the carbon and water cycles at temporal and spatial scales useful for phenological studies of deciduous forest stands. The high frequency of these measurements makes it possible to observe rapid changes when phenological events, such as leaf emergence and senescence, occur during the growing season. At Harvard Forest in central Massachusetts, relatively short climate anomalies in April and May were found to cause significant interannual variability in gross ecosystem exchange of carbon by affecting the timing of leaf unfolding (Goulden et al. 1996). Due to the clear seasonal signal of CO_2 uptake resulting from the presence of leaves, eddy flux measurements were used to calculate growing season length, which was found to be correlated with annual sums of net ecosystem exchange in a mixed species broadleaf deciduous forest near Oak Ridge, Tennessee (Baldocchi and Wilson 2001).

While these initial studies helped establish the intuitive relation between longer growing season and increased carbon uptake, a more nuanced view of the effect of phenology on the carbon cycle is developing as longer eddy covariance records become available. While a longer presence of leaves on trees theoretically makes carbon assimilation possible for a greater portion of the year, which might result in a negative phenological feedback mediated through increased total annual photosynthesis in plants, drought stress may decrease photosynthetic activity, diminishing or negating this feedback (Angert et al. 2005; Zhao and Running 2010). Understanding of the temporal interplay between phenology and carbon uptake has also advanced. Besides simply extending a given growing season due to early leaf emergence, warmer spring temperatures may also impact carbon uptake later in the year, in the peak of the growing season, possibly due to lagged effects on nutrient availability (Richardson et al. 2009b). A delayed fall was observed to extend the growing season and contribute to increased annual sums of carbon uptake in a deciduous forest in Indiana (Dragoni et al. 2011). However, the effect of fall phenology on late growing season assimilation may be more difficult to deduce due to relatively high uncertainty in determining dates of senescence and abscission from optical phenology observation techniques (Garrity et al. 2011).

In addition to affecting temporal dynamics of the carbon cycle, phenology also has an impact on the surface energy balance through regulation of albedo and surface-atmosphere exchanges of latent and sensible heat. Seasonal changes in albedo have been observed to relate to plant phenological dynamics (Schaaf et al. 2002). Through the process of transpiration, leaf unfolding controls the release of latent heat in the form of water vapor. Consequently, as deciduous forests green up in the spring, atmospheric water vapor concentrations increase and the seasonal trend of increasing temperature may slow down (Hayden 1998; Fitzjarrald et al. 2001). Leaf emergence also affects sensible heat exchange of the land surface, which has been shown to affect the depth of the convective boundary layer (CBL) through changing buoyancy at the surface and turbulent kinetic energy in the CBL (Pielke et al. 1998). The seasonality of the ratio of sensible to latent heat exchange, known as the Bowen ratio, was observed to correlate with the trajectory of leaf unfolding and senescence at Harvard Forest (Moore et al. 1996). Consequently, accurate modeling of deciduous forest phenology is important for calculating temporal changes in the surface energy budget.

Phenology components of biosphere-atmosphere models exhibit a range of methods, from prescriptive approaches for assigning phenological events, to predictive models involving chilling, warming, and photoperiod requirements as discussed above. A comparison of carbon cycle models and eddy covariance data conducted for the North American Carbon Program (NACP) Site Synthesis found that the length of the growing season was consistently overestimated across models with varying phenology representations, through both early beginning and late ending (Richardson et al. 2012). Levis and Bonan (2004) found that a prescribed representation of phenology resulted in an unrealistic decoupling of phenology from photosynthesis and transpiration early in the growing season, when compared to a predictive phenology scheme. However the NACP study also found that models which attempted to predict past phenology performed worse than models where phenology was prescribed by satellite data (Schwalm et al. 2010). Predictive models enable forecasts of how future changes in phenology, driven by environmental conditions, will feed back to affect the environment through the processes discussed in this section. Improved predictive ability of phenological models will therefore result in better forecasts of the feedback effects of deciduous forest phenology.

12.5 Deciduous Forest Phenology and Climate Change

What is the impact of climate change on the phenology of temperate deciduous forests? Here we provide a brief review of some observational studies, and examine forecasts of phenological shifts in response to future climate change.

The key driver of both interannual variability and long-term trends in phenology in these ecosystems is believed to be air temperature, and thus warmer temperatures are expected to increase growing season length by advancing spring onset and delaying autumn senescence. For example, Menzel and Fabian (1999) and Menzel (2000) analyzed phenological data collected over four decades (1959–1996) at the International Phenological Gardens (IPG), a European common garden network. Standardized observations on genetically identical clones of a variety of temperate species indicated that spring phenology had advanced at a mean rate of 2.1 days/decade, while autumn phenology was delayed by 1.5 days/decade. Consequently, an overall increase of 10.8 days in growing season length was observed since the 1960s, and a modeling analysis showed that these trends could be linked to concurrent increases in air temperature. Although these trends were statistically robust, Menzel (2000) noted substantial geographic variability across the IPG network, as well as differences among species. Using a similar IPG dataset, Chmielewski and Rötzer (2001) estimated that a 1 °C increase in mean annual temperature was associated with a 5 day increase in growing season length. Observations of ginkgo phenology at meteorological stations across Japan show a similar pattern (Matsumoto et al. 2003); over the period 1953–2000, spring budding dates were observed to advance by 4 days, whereas autumn leaf-fall dates were delayed by 8 days. In this study, a 1 °C increase in mean annual temperature was associated with a 10 day increase in growing season length. Results of other analyses of temperate deciduous tree phenology confirm these findings (e.g. Miller-Rushing et al. 2007; Menzel et al. 2008; Rutishauser et al. 2008; Ibáñez et al. 2010), although there is variability across species in response to warming, and not all areas are warming at the same rate (Schwartz et al. 2006).

Given the substantial year-to-year variability in observed plant phenology, time series on the order of a decade or two in length are typically not long enough to exhibit statistically significant trends. However, such data can be used to parameterize phenological models, which when run using either historical data or climate projections can be used to infer potential changes in phenology in response to changing climatic drivers. For example, Richardson et al. (2006) conducted a hindcasting analysis for a northern hardwood forest in the northeastern USA. Simple temperature-based models for spring development and autumn senescence were calibrated to 14 years of field observations; the models were then run using almost five decades of daily meteorological data. Consistent and statistically significant trends were predicted across the three study species. For maple, reconstructed dates of spring onset were predicted to have advanced at a rate of 1.8 days/decade, and the growing season was predicted to have increased in length at a rate of 2.1 days/decade.

Similarly, Lebourgeois et al. (2010) combined a decade of ground observations from a network of forest plots across France with an empirical modeling approach. This analysis predicted that for oak and beech trees, spring budburst would advance by 7 days, while autumn leaf coloring would be delayed by 13 days, by 2100. Vitasse et al. (2011) used several years of observational data collected along an elevational gradient in the French Pyrenees, to constrain spring and autumn phenology models for six European tree species. Rates of change in spring phenology over the course of the twenty-first century varied among species, from 0 to 2.4 days/decade. Autumn senescence could only be successfully modeled in two of the study species; in both oak (1.4 days/decade) and beech (2.3 days/decade), delayed senescence was predicted. These and other modeling studies are consistent with the observational studies summarized above.

There may, however, be exceptions to these general patterns, or uncertainties that still need to be resolved. For example, some studies (Morin et al. 2009; Körner and Basler 2010; Migliavacca et al. 2012) have suggested that future shifts in spring phenology may be limited by photoperiod or winter chilling requirements in some species. Additionally, we do not yet have a strong mechanistic understanding of how warm vs. cold temperatures influence autumn phenology (Estrella and Menzel 2006; Archetti et al. 2013), so model forecasts should be interpreted cautiously. Finally, it has also been proposed that earlier budburst and delayed senescence may increase the likelihood of frost damage in temperate species under future climate scenarios (Norby et al. 2003, but see Kramer 1994b), complicating ecological interpretation of increases in growing season length.

Future changes in environmental factors other than temperature are generally considered to be of secondary importance, although impacts on spring and autumn phenological events may be different. For example, precipitation is not considered a major driver of spring phenology in temperate ecosystems (Polgar and Primack 2011), although precipitation has been linked to interannual variability in the timing and intensity of autumn colors (Archetti et al. in review). Similarly, while budburst is not thought to be highly sensitive to ambient CO_2 concentrations (Badeck et al. 2004), there is some evidence that elevated CO_2 delays autumn coloration and senescence (Taylor et al. 2008, but see Norby et al. 2003).

12.6 Future Developments

Our knowledge of the environmental conditions that regulate the timing of dormancy and leaf emergence is extensive. We know that photoperiod and temperature can regulate endo- and ecodormancy, and the timing of leaf emergence, for multiple species. Furthermore, we are aware of the feedback effects of canopy phenology on water, carbon, and energy fluxes and the recent shifts in spring and autumn phenology due to climate change. However, our understanding of the sensitivity of multiple species to environmental conditions in autumn is lacking. Without this knowledge, we are limited in our ability to forecast future changes in autumn phenology and growing season length caused by climate change. The need for future research on autumn phenology is especially pressing because it is requisite for a more complete understanding of how biogeochemical cycles and other ecosystem processes will be impacted by climate change.

Future efforts to understand autumn phenology in temperate deciduous forests will require a focus on the influence of genetics and environmental conditions on physiological processes through the employment of an experimental research design, where environmental variables (e.g. temperature, photoperiod) can be controlled and analyzed separately and collectively.

Acknowledgments A.D.R. acknowledges support from the National Science Foundation, through the Macrosystems Biology program, award EF-1065029; the Northeastern States

Research Cooperative; and the US Geological Survey Status and Trends Program, the US National Park Service Inventory and Monitoring Program, and the USA National Phenology Network through grant number G10AP00129 from the United States Geological Survey. Any opinions, findings, and conclusions or recommendations expressed in this material are those of the authors and do not necessarily reflect the views of the National Science Foundation or USGS.

References

Ahrends HE, Etzold S, Kutsch WL, Stoeckli R, Bruegger R, Jeanneret F, Wanner H, Buchmann N, Eugster W (2009) Tree phenology and carbon dioxide fluxes: use of digital photography for process-based interpretation at the ecosystem scale. Clim Res 39:261–274. doi:10.3354/cr00811

Angert A, Biraud S, Bonfils C, Henning CC, Buermann W, Pinzon J, Tucker CJ, Fung I (2005) Drier summers cancel out the CO_2 uptake enhancement induced by warmer springs. PNAS USA 102:10823–10827. doi:10.1073/pnas.0501647102

Archetti M, Richardson AD, O'Keefe J, Delpierre N (2013) Predicting climate change impacts on the amount and duration of autumn colors in a New England forest. PLoS One 8(3):e57373. doi:10.1371/journal.pone.0057373.

Badeck FW, Bondeau A, Bottcher K, Doktor D, Lucht W, Schaber J, Sitch S (2004) Responses of spring phenology to climate change. New Phytol 162:295–309. doi:10.1111/j.1469-8137.2004.01059.x

Baldocchi DD, Wilson KB (2001) Modeling CO_2 and water vapor exchange of a temperate broadleaved forest across hourly to decadal time scales. Ecol Model 142:155–184. doi:10.1016/S0304-3800(01)00287-3

Baldocchi DD, Black TA, Curtis PS, Falge E, Fuentes JD, Granier A, Gu L, Knohl A, Pilegaard K, Schmid HP, Valentini R, Wilson K, Wofsy S, Xu L, Yamamoto S (2005) Predicting the onset of net carbon uptake by deciduous forests with soil temperature and climate data: a synthesis of FLUXNET data. Int J Biometeorol 49:377–387. doi:10.1007/s00484-005-0256-4

Caffarra A, Donnelly A (2011) The ecological significance of phenology in four different tree species: effects of light and temperature on bud burst. Int J Biometeorol 55:711–721. doi:10.1007/s00484-010-0386-1

Cannell MGR, Smith RI (1986) Climatic warming, spring budburst and frost damage on trees. J Appl Ecol 23:177–191

Chmielewski FM, Rötzer T (2001) Response of tree phenology to climate change across Europe. Agric For Meteorol 108:101–112. doi:10.1016/S0168-1923(01)00233-7

Chuine I (2000) A unified model for budburst of trees. J Theor Biol 207:337–347. doi:10.1006/jtbi.2000.2178

Dragoni D, Schmid HP, Wayson CA, Potter H, Grimmond CSB, Randolph JC (2011) Evidence of increased net ecosystem productivity associated with a longer vegetated season in a deciduous forest in south-central Indiana, USA. Glob Change Biol 17:886–897. doi:10.1111/j.1365-2486.2010.02281.x

Estrella N, Menzel A (2006) Responses of leaf colouring in four deciduous tree species to climate and weather in Germany. Clim Res 32:253–267. doi:10.3354/cr032253

Fisher JI, Mustard JF, Vadeboncoeur MA (2006) Green leaf phenology at Landsat resolution: scaling from the field to the satellite. Remote Sens Environ 100:265–279. doi:10.1016/j.rse.2005.10.022

Fitzjarrald D, Acevedo OC, Moore KE (2001) Climatic consequences of leaf presence in the eastern United States. J Climate 14:598–614. doi:10.1175/1520-0442(2001)014<0598:CCOLPI>2.0.CO;2

Friedl M, Henebry G, Reed B, Huete A, White M, Morisette J, Nemani R, Zhang X, Myneni R (2006) Land surface phenology. http://landportal.gsfc.nasa.gov/Documents/ESDR/Phenology_Friedl_whitepaper.pdf. Accessed 27 Jan 2012

Garrity SR, Bohrer G, Maurer KD, Mueller KL, Vogel CS, Curtis PS (2011) A comparison of multiple phenology data sources for estimating seasonal transitions in deciduous forest carbon exchange. Agric For Meteorol 151:1741–1752. doi:10.1016/j.agrformet.2011.07.008

Ghelardini L, Santini A, Black-Samuelsson S, Myking T, Falusi M (2010) Bud dormancy release in elm (*Ulmus* spp.) clones-a case study of photoperiod and temperature responses. Tree Physiol 30:264–274. doi:10.1093/treephys/tpp110

Goulden ML, Munger JW, Fan SM, Daube BC, Wofsy SC (1996) Exchange of carbon dioxide by a deciduous forest: response to interannual climate variability. Science 271:1576–1578. doi:10.1126/science.271.5255.1576

Hanes JM, Schwartz MD (2011) Modeling land surface phenology in a mixed temperate forest using MODIS measurements of leaf area index and land surface temperature. Theor Appl Climatol 105:37–50. doi:10.1007/s00704-010-0374-8

Hänninen H, Tanino K (2011) Tree seasonality in a warming climate. Trends Plant Sci 18:1380–1385. doi:10.1016/j.tplants.2011.05.001

Hayden BP (1998) Ecosystem feedbacks on climate at the landscape scale. Philos Trans R Soc Lond B 353:5–18. doi:10.1098/rstb.1998.0186

Heide OM (1993) Daylength and thermal time responses of budburst during dormancy release in some northern deciduous trees. Physiol Plant 88:531–540

Horvath DP, Anderson JV, Chao WS, Foley ME (2003) Knowing when to grow: signals regulating bud dormancy. Trends Plant Sci 8:534–540. doi:10.1016/j.tplants.2003.09.013

Hunter AF, Lechowicz MJ (1992) Predicting the timing of budburst in temperate trees. J Appl Ecol 29:597–604

Ibáñez I, Primack RB, Miller-Rushing AJ, Ellwood E, Higuchi H, Lee SD, Kobori H, Silander JA (2010) Forecasting phenology under global warming. Philos Trans R Soc Lond B 365:3247–3260. doi:10.1098/rstb.2010.0120

Keeling CD, Chin JFS, Whorf TP (1996) Increased activity of northern vegetation inferred from atmospheric CO_2 measurements. Nature 382:146–149. doi:10.1038/382146a0

Keskitalo J, Bergquist G, Gardestrom P, Jannson S (2005) A cellular timetable of autumn senescence. Plant Physiol 139:1635–1648. doi:10.1104/pp. 105.066845

Körner C, Basler D (2010) Phenology under global warming. Science 327:1461–1462. doi:10.1126/science.1186473

Kramer PJ (1936) Effect of variation in length of day on growth and dormancy of trees. Plant Physiol 11:127–137

Kramer K (1994a) Selecting a model to predict the onset of growth of *Fagus sylvatica*. J Appl Ecol 31:172–181

Kramer K (1994b) A modeling analysis of the effects of climatic warming on the probability of spring frost damage to tree species in the Netherlands and Germany. Plant Cell Environ 17:367–377. doi:10.1111/j.1365-3040.1994.tb00305.x

Lebourgeois F, Pierrat JC, Perez V, Piedallu C, Cecchini S, Ulrich E (2010) Simulating phenological shifts in French temperate forests under two climatic change scenarios and four driving global circulation models. Int J Biometeorol 54:563–581. doi:10.1007/s00484-010-0305-5

Lee DW, O'Keefe J, Holbrook NM, Feild TS (2003) Pigment dynamics and autumn leaf senescence in a New England deciduous forest, eastern USA. Ecol Res 18:677–694. doi:10.1111/j.1440-1703.2003.00588.x

Levis S, Bonan G (2004) Simulating springtime temperature patterns in the community atmosphere model coupled to the community land model using prognostic leaf area. J Clim 2:4531–4540. doi:10.1175/3218.1

Liang L, Schwartz MD (2009) Landscape phenology: an integrative approach to seasonal vegetation dynamics. Landsc Ecol 24:465–472. doi:10.1007/s10980-009-9328-x

Matsumoto K, Ohta T, Irasawa M, Nakamura T (2003) Climate change and extension of the *Ginkgo biloba* L. growing season in Japan. Glob Change Biol 9:1634–1642. doi:10.1046/j.1365-2486.2003.00688.x

Menzel A (2000) Trends in phenological phases in Europe between 1951 and 1996. Int J Biometeorol 44:76–81. doi:10.1007/s004840000054

Menzel A, Fabian P (1999) Growing season extended in Europe. Nature 397:659. doi:10.1038/17709

Menzel A, Estrella N, Heitland W, Susnik A, Schleip C, Dose V (2008) Bayesian analysis of the species-specific lengthening of the growing season in two European countries and the influence of an insect pest. Int J Biometeorol 52:209–218. doi:10.1007/s00484-007-0113-8

Migliavacca M, Sonnentag O, Keenan TF, Cescatti A, O'Keefe J, Richardson AD (2012) On the uncertainty of phenological responses to climate change and its implication for terrestrial biosphere models. Biogeosci Discuss 9:879–926. doi:10.5194/bgd-9-879-2012

Miller-Rushing AJ, Katsuki T, Primack RB, Ishii Y, Lee SD, Higuchi H (2007) Impact of global warming on a group of related species and their hybrids: cherry tree (Rosaceae) flowering at Mt. Takao, Japan. Am J Bot 94:1470–1478. doi:10.3732/ajb.94.9.1470

Mitchell TD, Jones PD (2005) An improved method of constructing a database of monthly climate observations and associated high-resolution grids. Int J Climatol 25:693–712. doi:10.1002/joc.1181

Mitchell TD, Carter TR, Jones PD, Hulme M, New M (2004) A comprehensive set of high-resolution grids of monthly climate for Europe and the globe: the observed record (1901–2000) and 16 scenarios (2001–2100). Tyndall Centre working paper No. 55. http://www.tyndall.ac.uk/sites/default/files/wp55.pdf. Accessed 12 June 2012

Moore KE, Fitzjarrald DR, Sakai RK, Goulden ML, Munger JW, Wofsy SC (1996) Seasonal variation in radiative and turbulent exchange at a deciduous forest in central Massachusetts. J Appl Meteorol 35:122–134

Morin X, Lechowicz MJ, Augspurger C, O' Keefe J, Viner D, Chuine I (2009) Leaf phenology in 22 North American tree species during the 21st century. Glob Change Biol 15:961–975. doi:10.1111/j.1365-2486.2008.01735.x

Morisette JT, Richardson AD, Knapp AK, Fisher JI, Graham EA, Abatzoglou J, Wilson BE, Breshears DD, Henebry GM, Hanes JM, Liang L (2009) Tracking the rhythm of the seasons in the face of global change: phenological research in the 21st century. Front Ecol Environ 7:253–260. doi:10.1890/070217

Murray MB, Cannell MGR, Smith RI (1989) Date of budburst of fifteen tree species in Britain following climatic warming. J Appl Ecol 26:693–700

Myking T (1999) Winter dormancy release and budburst in *Betula pendula* Roth and *B. pubescens* Ehrh. ecotypes. Phyton 39:139–146

Myking T, Heide OM (1995) Dormancy release and chilling requirement of buds and latitudinal ecotypes of *Betula pendula* and *B. pubescens*. Tree Physiol 15:697–704

Myneni RB, Keeling CD, Tucker CJ, Asrar G, Nemani RR (1997) Increased plant growth in the northern high latitudes from 1981 to 1991. Nature 386:698–702. doi:10.1038/386698a0

Norby RJ, Hartz-Rubin JS, Verbrugge MJ (2003) Phenological responses in maple to experimental atmospheric warming and CO_2 enrichment. Glob Change Biol 9:1792–1801. doi:10.1111/j.1365-2486.2003.00714.x

Olson DM, Dinerstein E, Wikramanayake ED, Burgess ND, Powell GVN, Underwood EC, D'Amico JA, Itoua I, Strand HE, Morrison JC, Loucks CJ, Allnutt TF, Ricketts TH, Kura Y, Lamoreux JF, Wettengel WW, Hedao P, Kassem KR (2001) Terrestrial ecoregions of the world: a new map of life on earth. Bioscience 51:933–938. doi:10.1641/0006-3568(2001)051[0933:TEOTWA]2.0.CO;2

Peñuelas J, Filella I (2001) Responses to a warming world. Science 294:793. doi:10.1126/science.1066860

Perry TO (1971) Dormancy of trees in winter. Science 171:29–36

Pielke R, Avissar R, Raupach M (1998) Interactions between the atmosphere and terrestrial ecosystems: influence on weather and climate. Glob Change Biol 4:461–475. doi:10.1046/j.1365-2486.1998.t01-1-00176.x

Polgar CA, Primack RB (2011) Leaf-out phenology of temperate woody plants: from trees to ecosystems. New Phytol 191:926–941. doi:10.1111/j.1469-8137.2011.03803.x

Richardson AD, O'Keefe J (2009) Phenological differences between understory and overstory: a case study using the long-term Harvard Forest records. In: Noormets A (ed) Phenology of ecosystem processes. Springer, Dordrecht/Heidelberg/London/New York

Richardson AD, Bailey AS, Denny EG, Martin CW, O'Keefe J (2006) Phenology of a northern hardwood forest canopy. Glob Change Biol 12:1174–1188. doi:10.1111/j.1365-2486.2006.01164.x

Richardson AD, Jenkins JP, Braswell BH, Hollinger DY, Ollinger SV, Smith M (2007) Use of digital webcam images to track spring green-up in a deciduous forest. Oecologia 152:323–334. doi:10.1007/s00442-006-0657-z

Richardson AD, Braswell BH, Hollinger DY, Jenkins JP, Ollinger SV (2009a) Near-surface remote sensing of spatial and temporal variation in canopy phenology. Ecol Appl 19:1417–1428. doi:10.1890/08-2022.1

Richardson AD, Hollinger DY, Dail DB, Lee JT, Munger JW, O'Keefe J (2009b) Influence of spring phenology on seasonal and annual carbon balance in two contrasting New England forests. Tree Physiol 29:321–331. doi:10.1093/treephys/tpn040

Richardson AD, Anderson RS, Arain MA, Barr AG, Bohrer G, Chen G, Chen JM, Ciais P, Davis KJ, Desai AR, Dietze MC, Dragoni D, Garrity SR, Gough CM, Grant R, Hollinger DY, Margolis HA, McCaughey H, Migliavacca M, Monson RK, Munger JW, Poulter B, Raczka BM, Ricciuto DM, Sahoo AK, Schaefer K, Tian H, Vargas R, Verbeeck H, Xiao J, Xue Y (2012) Terrestrial biosphere models need better representation of vegetation phenology: results from the North American Carbon Program site synthesis. Glob Change Biol 18:566–584. doi:10.1111/j.1365-2486.2011.02562.x

Rötzer T, Grote R, Pretzsch H (2004) The timing of bud burst and its effect on tree growth. Int J Biometeorol 48:109–118. doi:10.1007/s00484-003-0191-1

Rousi M, Pusenius J (2005) Variations in phenology and growth of European white birch (*Betula pendula*) clones. Tree Physiol 25:201–210. doi:10.1093/treephys/25.2.201

Rutishauser T, Luterbacher J, Defila C, Frank D, Wanner H (2008) Swiss spring plant phenology 2007: extremes, a multi-century perspective, and changes in temperature sensitivity. Geophys Res Lett 35:L05703. doi:10.1029/2007GL032545

Schaaf CB, Gao F, Strahler AH, Lucht W, Li X, Tsang T, Strugnell NC, Zhang X, Jin Y, Muller J, Lewis P, Barnsley M, Hobson P, Disney M, Roberts G, Dunderdale M, Doll C, d'Entremont RP, Hu B, Liang S, Privette JL, Roy D (2002) First operational BRDF, albedo nadir reflectance products from MODIS. Remote Sens Environ 83:135–148. doi:10.1016/S0034-4257(02)00091-3

Schwalm CR, Williams CA, Schaefer K, Anderson R, Arain MA, Baker I, Barr A, Black TA, Chen G, Chen JM, Ciais P, Davis KJ, Desai A, Dietze M, Dragoni D, Fischer ML, Flanagan LB, Grant R, Gu L, Hollinger D, Izaurralde RC, Kucharik C, Lafleur P, Law BE, Li L, Li Z, Liu S, Lokupitiya E, Luo Y, Ma S, Margolis H, Matamala R, McCaughey H, Monson RK, Oechel WC, Peng C, Poulter B, Price DT, Riciutto DM, Riley W, Sahoo AK, Sprintsin M, Sun J, Tian H, Tonitto C, Verbeeck H, Verma SB (2010) A model-data intercomparison of CO_2 exchange across North America: results from the North American Carbon Program site synthesis. J Geophys Res 115:G00H05. doi:10.1029/2009JG001229

Schwartz MD, Hanes JM (2010) Intercomparing multiple measures of the onset of spring in eastern North America. Int J Climatol 30:1614–1626. doi:10.1002/joc.2008

Schwartz MD, Ahas R, Aasa A (2006) Onset of spring starting earlier across the northern Hemisphere. Glob Change Biol 12:343–351. doi:10.1111/j.1365-2486.2005.01097.x

Tanino KK, Kalcsits L, Silim S, Kendall E, Gray G (2010) Temperature-driven plasticity in growth cessation in deciduous woody plants: a working hypothesis suggesting how molecular and cellular function is affected by temperature during dormancy induction. Plant Mol Biol 73:49–65. doi:10.1007/s11103-010-9610-y

Taylor G, Tallis MJ, Giardina CP, Percy KE, Miglietta F, Gupta PS, Gioli B, Calfapietra C, Gielen B, Kubiske ME, Scarascia-Mugnozza GE, Kets K, Long SP, Karnosky DF (2008) Future atmospheric CO_2 leads to delayed autumnal senescence. Glob Change Biol 14:264–275. doi:10.1111/j.1365-2486.2007.01473.x

Vitasse Y, Porte AJ, Kremer A, Michalet R, Delzon S (2009) Responses of canopy duration to temperature changes in four temperate tree species: relative contributions of spring and autumn leaf phenology. Oecologia 161:187–198. doi:10.1007/s00442-009-1363-4

Vitasse Y, Francois C, Delpierre N, Dufrene E, Kremer A, Chuine I, Delzon S (2011) Assessing the effects of climate change on the phenology of European temperate trees. Agric For Meteorol 151:969–980. doi:10.1016/j.agrformet.2011.03.003

Zhang X, Friedl MA, Schaaf CB, Strahler AH, Hodges JCF, Gao F, Reed BC, Huete A (2003) Monitoring vegetation phenology using MODIS. Remote Sens Environ 84:471–475. doi:10.1016/S0034-4257(02)00135-9

Zhang X, Friedl MA, Schaaf CB, Strahler AH (2004) Climate controls on vegetation phenological patterns in northern mid- and high latitudes inferred from MODIS data. Glob Change Biol 10:1133–1145. doi:10.1111/j.1365-2486.2004.00784.x

Zhang X, Tarpley D, Sullivan JT (2007) Diverse responses of vegetation phenology to a warming climate. Geophys Res Lett 34, L19405. doi:10.1029/2007GL031447

Zhao M, Running SW (2010) Drought-induced reduction in global terrestrial net primary production from 2000 through 2009. Science 329:940–943. doi:10.1126/science.1192666

Chapter 13
Phenology at High Latitudes

Frans E. Wielgolaski and David W. Inouye

Abstract Phenology, mainly on plants, in the Northern Hemisphere north of 60°N in the "Old" World and of 50°N in the "New" World is described in the present chapter, both historically and in modern times. Experiments, field work and satellite data are discussed. Phenological observations related to recent climate change at high latitudes are discussed in detail, and they are generally found to be good indicators of such changes during long-term studies. However, various organisms in the food chain, or sometimes even different individuals of the same species, do not react in exactly the same way on climate change. The response may also vary with the continentality of a region and with the time of the year. An increasing mismatch may be seen between production and consumption at various trophic levels both in terrestrial and aquatic ecosystems.

13.1 Introduction

High latitudes are characterized by strong variation in day-length during different seasons of the year. North of the Arctic Circle there is sun 24 h of the day near the Summer Solstice, but no sun at all 6 months earlier or later. The sun angle is always low compared to further south, which means that the aspect of slopes strongly influences light conditions. Temperatures generally decrease towards the poles and the growing seasons are shorter; e.g., at the northernmost coast of Norway there are fewer than 100 days with a daily mean temperature above 5 °C (Aune 1993). In many parts of northern lowland Fennoscandia, the snow-free period is less than

F.E. Wielgolaski (✉)
Department of Bioscience, University of Oslo, Oslo, Norway
e-mail: f.e.wielgolaski@ibv.uio.no

D.W. Inouye
Department of Biology and Rocky Mountain Biological Laboratory, University of Maryland, College Park, MD 20742, USA

M.D. Schwartz (ed.), *Phenology: An Integrative Environmental Science*,
DOI 10.1007/978-94-007-6925-0_13,

120 days (Björbekk 1993). Therefore, organisms living at high latitudes have to be adapted to these conditions. This means that plants have to flower relatively soon after snowmelt (Bliss 1971) in order to ripen seeds successfully. Growth of many plant species may start even before all snow has disappeared, if the upper soil layers have melted, as observed for birch leafing in maritime Norway and in northern Russia (Shutova et al. 2006), and also for herbaceous plants when light penetrates a thin snow layer. Heide (1985) stated that the more severe the environment, the more important survival adaptations seemed to be, while biological competition tended to be less important.

It is difficult to decide where to place the southern limit of "High latitudes." The Arctic Circle (66° 33′) might be one possibility, but in many ways this definition is too narrow. Summer days are long even further south, e.g. about 18 h of sun by summer solstice even at 60°N, and plant photosynthetic activity during the important early summer goes on for many hours every day. In Europe, warm ocean currents keep temperatures higher, particularly near the west coast, compared to North America, and this of course is very important for phenology, especially in spring. Therefore in this paper "High latitudes" are arbitrarily set to 60°N in western Eurasia, but close to 50°N in North America, including alpine areas north of the same limits (see also Chap. 14 in this volume). Generally, this definition covers the boreal and arctic climate zones (e.g. Breckle 2002).

"Modern" phenological observations have been made on plants (bud break, flowering, fruit ripening and leaf coloring in autumn) mainly as inexpensive supplements to meteorological measurements for delineating biological zones, particularly important in sparsely populated high-latitude regions of the world. However, mainly after the First World War, phenological information was also used in agrometeorology for selection of districts for cultivation of certain crops and fruits, especially in Central Europe (Schnelle 1955), but also at higher latitudes.

Many phenological networks were established after the Second World War, e.g. the International Phenological Gardens (IPG) in Europe from the early 1960s (see Chap. 8 in this volume). Some stations in Fennoscandia were active in this network. Particularly during the 1970s and the 1980s, the interest in phenology decreased in large parts of the world, including Fennoscandia. Therefore, IPG observations were often not continued. A change came in the 1990s because of interest in the effect of global change (e.g., Lieth 1997). In this context the old phenological observations, even in remote areas of sparsely populated districts of high latitudes, are very important (e.g., Klaveness and Wielgolaski 1996; Wielgolaski et al. 2011), providing baselines for new phenological information, for example by satellite techniques (Myneni et al. 1997; Delbart et al. 2005; de Beurs and Henebry 2010). Within the International Tundra Experiment (ITEX) several studies have been carried out in various Arctic areas to see whether experimentally-changed microclimate would influence phenology and growth of plants over a few years (e.g., Arft et al. 1999; Walker et al. 2006).

The recent climate change seems to be extremely important at high latitudes, and this warming is expected to continue also in the twenty-first century, particularly over land (IPCC 2007). Basal work on the Fifth Assessment Report (AR5) from the

Intergovernmental Panel on Climate Change (IPCC) is going on in 2012 and 2013. Probably, phenology of both plants, fungi and animals will play a similar important role to show responses in natural and managed systems at high latitudes of the ongoing climate change as reported in AR4 (Rosenzweig et al. 2007).

13.2 History

The first organized phenological observation program, or network, in the world was established at 18 stations in the High-Latitude countries Sweden and Finland (Linne 1751) by the Swedish botanist Carl von Linné (Johansson 1946, 1953). Most of the older data collections at high latitudes did not last for long periods, and often years are missing between the periods of observations. However, in Finland phenological data of both plants and animals from the last part of the 1700s and the first part of the 1800s were recovered by Moberg (1857, 1894).

Phenological observations at high latitudes in Europe (including Russia) have been more common since the mid-1850s. Finnish phenological plant observations for the period 1846–2005 are discussed in relation to climate change by Linkosalo et al. (2009), while Holopainen et al. (2006) have analyzed data from Finland back to 1750. Apparently, phenological data in Finland have been collected more widely than in other countries in Fennoscandia, although in Sweden such data have been presented monthly for many years in reports from the Swedish Meteorological and Hydrological Institute (SMHI). Mean phenological values for various parts of Sweden from 1873 to the 1920s are presented by H.W. Arnell (1923), K. Arnell (1927) and K. Arnell and S. Arnell (1930).

In Norway, the first observations of plant phenology for more scientific purpose took place starting about 1850 (Printz 1865). During the period 1851–1859 phenological data of both plants and migratory birds were collected in north easternmost Norway (about 70°N). Both these data, and data from the capitol of Norway (now Oslo) for the period 1860–1884, were published by Schübeler (1885). For the end of the 1800s and beginning of the 1900s, long-term phenological observations are mainly known from the southern part of the country (Moe 1928; Lie 1931), but some observations were carried out in Troms County in northern Norway in the same period and particularly from 1910 to 1911 (Holmboe 1913). Later, however, in 1928, a phenological network was established that lasted to 1952 with some continuation to 1977 (Lauscher et al. 1955, 1959, 1978; Lauscher 1980; Lauscher and Lauscher 1990), and at a few Norwegian sites in the European Phenological Garden using vegetatively propagated plants (Lauscher 1985). Parts of the Lauscher network material are published in more details by Wielgolaski et al. (2011).

There are no references to older phenological studies in Iceland, but recently Thórhallsdóttir (1998) has studied flowering phenology for 11 years. The best-known phenological observations on Greenland were carried out by Sörensen (1941) in the Northeast, mainly through 3 years in the 1930s. However, Böcher

(1938) has also reported some phenological information in his plant studies from Greenland.

At North American higher latitudes, the only older long-term phenological study on plants and birds was a survey in western Canada by the Royal Society of Canada from the 1890s to 1922, published annually in Proceedings and Transactions of the Royal Society (Beaubien and Freeland 2000). Since then, phenological studies at higher North American latitudes were carried out in eastern Canada with the lilac/honeysuckle surveys of eastern USA (Schwartz and Reiter 2000), and in western Canada (Beaubien and Johnson 1994; Beaubien 1996; Beaubien and Hamann 2011). However, more local phenological observations have been performed (e. g. brief overview in Erskine 1985).

13.3 Recent Phenological Studies

Most of more recent phenological studies and networks at high latitudes are still on plants but valuable information is also available for other organism groups (e.g. invertebrates and vertebrates, as well as fungi). Some studies are mainly descriptive and form traditional biocalendars. These can be used for agricultural planning and for educational purposes as described in more detail in other chapters of the present volume. Others are mainly experimental studies in controlled chambers and open top chambers, or transplant studies of various ecotypes, or modeling studies. Climate change is a key reason behind studies of phenology today, and the old data are of the greatest importance for comparisons with newer ones collected in a traditional way or by satellites.

The time of snowmelt is often considered to be the primary initiator of phenological events in tundra plants (Böcher 1938; Sörensen 1941; Wielgolaski and Kärenlampi 1975; Eriksen et al. 1993; Odland 2011; Chap. 14 in this volume). This seems particularly to be true in the less oceanic alpine and arctic regions, as was shown experimentally by Woodley and Svoboda (1994) by snow removal at sites on Ellesmere Island, Arctic Canada and by Cooper et al. (2011) at the Svalbard archipelago north of Norway. In more oceanic, snow-rich regions, however, such as the outer Troms County in northern Norway, and also at the Russian Kola Peninsula (Shutova et al. 2006), mountain birch may have green leaves before melting of the winter snow in spring. Growth in herbaceous plants may also start before snowmelt when there is enough light through the snow.

Odland (2011) observed that on exposed sites with a sparse snow cover, more than 2 months from snowmelt was needed to reach high enough soil temperature for flowering, while it took less than 6 days in snow-beds that melted out much later. He also found that the differences in the length of growing seasons between exposed sites and snow-beds separated only by a few meters could be more than 2 months. Cooper et al. (2011) found 3–5 weeks from the time of snow-melt to the earliest flowering in arctic areas. Thórhallsdóttir (1998) stated that time of snowmelt was likely to influence flowering only after very cold springs with exceptionally late ablation. She says that flowering in oceanic cold climates is normally not linked to snow-free conditions at all.

In a survey of the flora of sub-arctic Sweden, Molau (1993) found that populations with normal pollination and seed setting flowered early, while apomictic and viviparous species were found among the later-flowering ones. In short and cool summers, he found there was a low proportion of seeds ripening and a reduced seed quality and power of germination. Therefore, vegetative reproduction is very common at high latitudes (e.g. Bliss 1971) and annuals are few, as in High-Altitude regions (see Chap. 14 in this volume). Iversen et al. (2009) studied growth forms in a high-latitude alpine area and found they are also important for the phenology.

13.3.1 Phenological Biocalendars and Their Applications

In the tundra projects of the International Biological Program (IBP) in the early 1970, some phenological information was collected at northern high altitude sites (Wielgolaski 1974a; Bliss 1977). More recently, there have also been phenological studies in the Canadian Arctic (e.g., Woodley and Svoboda 1994) and in forested land in northern Canada (e.g. Colombo 1998). In Russia, phenological spectra or phenograms are given for some plant species on the Kola Peninsula for the period 1994–2000 (Makarova et al. 2001).

Most of the current international phenological networks, such as the European Phenological Gardens (Chmielewski and Rötzer 2001, 2002), have few high latitude sites. One reason for this is that many of the plant species chosen for international use cannot stand the harsh climate at high latitudes. However, regional networks have been established in northern latitudes, e.g. in Europe, like the Norwegian Environmental Education Network launched in 2001, with over 400 participating schools by the end of 2011. In Sweden a phenological network covering many plant species and several volunteering observers has successfully been operating since 2010 from south of 60°N to the far north of the country north of the Arctic Circle. In Finland phenological biocalendars have been presented particularly for tree species at least since 1997 by the Finnish Forest Research Institute (Pudas et al. 2008). In Alberta, Canada, more than 25 plant species have been watched phenologically by volunteers since 1987.

Most often, the newer time series in phenology at higher latitudes have been relatively short (e.g. Woodley and Svoboda 1994; Diekmann 1996; Arft et al. 1999; Wielgolaski 1999, 2001, 2003; Karlsen et al. 2008). It is then important that the observations be combined with other approaches, e.g. experiments, to provide insights into phenology. Another valuable approach is to collect information from stations with great differences in climate, which has to be studied at each site, and in other environmental factors, to facilitate correlations between phenology and environmental variables.

However, in Norway a 50-year phenological time series at several sites and covering many plant species was analyzed recently (Wielgolaski et al. 2011). Other

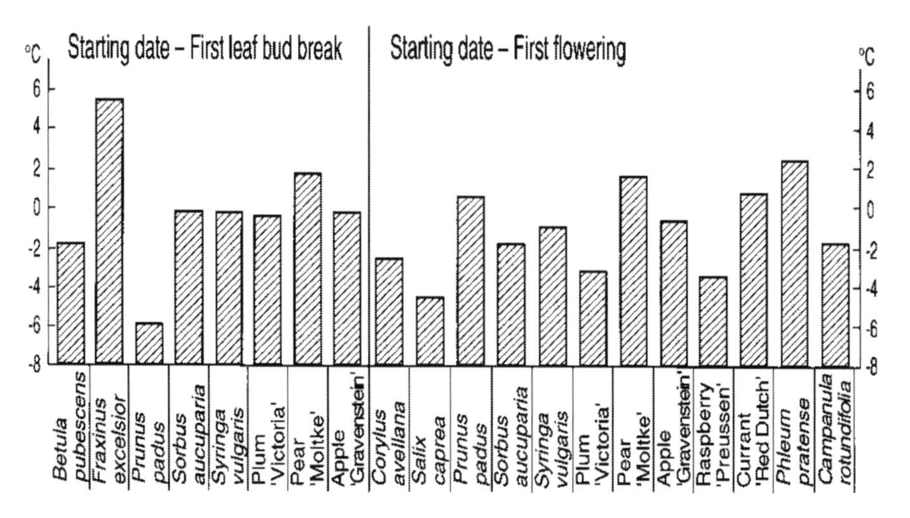

Fig. 13.1 Basic air temperatures found for several plant species in western Norway from the starting date in spring, to leaf bud break and flowering (Based on Wielgolaski 1999)

long time series from the same region at fewer sites and including fewer plant species were published by Shutova et al. (2006) and by Nordli et al. (2008).

As expected, various plants clearly showed different temperature requirements for development, both by plant types and phenophases, but temperature requirements were generally lowest for spring phases of early plants. The first author found the lowest air temperature for development (basic or threshold temperature) based on statistical methods (Wielgolaski 1999) to leaf bud burst in *Prunus padus* and flowering in *Salix caprea* (Fig. 13.1). Leaf bud break of the late-sprouting *Fraxinus excelsior* on the other hand was found to have considerably higher basic air temperature, as did both bud break and flowering of pear ('Moltke'). While *Prunus padus* showed a low basic temperature for leaf bud break, it needed a relatively high basic air temperature for flowering.

Plants growing at low temperatures in high latitudes and altitudes have to be adapted to relatively low basic temperatures to be able to finish their life cycle in a short period or they will be restricted to the warmest places of a district. When the same crop plant variety is cultivated at different latitudes, mean annual temperature sums to reach a specific phenophase normally decreased with increasing latitude (e.g. Strand 1965). In addition to temperature, the photoperiod might have an influence, as observed in specific varieties of grass species (e.g., Skjellvåg 1998). Most likely the photoperiod is important both for spring and autumn phenophases at exposed sites, while plants in snow-beds are more controlled by timing of snowmelt and temperatures.

Although temperature clearly was the most important environmental factor for phenology in all plants studied at high latitudes, edaphic factors (Wielgolaski 2001) also played some role. Strand (1965) pointed out that in Norway heat sums for plant development in agriculture were higher in clay than in sandy soil and that fertilization also influenced the necessary heat sums. In Arctic Canada, Woodley and Svoboda (1994) found that fertilization caused an earlier flowering of *Salix arctica*.

Water conditions are also of some importance to phenology even in high latitudes. For instance a higher number of days with precipitation caused an acceleration of bud break in *Betula pubescens* (Wielgolaski 2003). This is probably due to softening of the bud scale in moist air as was also observed by Junttila et al. (1983). Flowering of the extremely early *Corylus avellana* and the early *Salix caprea* also seemed to be favored by increased precipitation, probably for the same reasons. In later flowering species, increased precipitation most often delayed the blooming, particularly in plants found to have high basic air temperatures (Fig. 13.1). A 12-year study of herb phenology in a Swedish temperate deciduous forest showed that precipitation may be even more important than temperature for the abundance of flowering the year after (Tyler 2001). In most cases high precipitation during the previous autumn was favorable for flowering, but in *Anemone nemorosa* low precipitation resulted in more flowers. In Arctic Canada, Woodley and Svoboda (1994) found that irrigation of a dry riverside during the growing season caused a phenological shift in *Papaver lapponicum*, and an increased length of the budding and flowering periods.

It is clear that various plant species react differently to diverse environmental factors (Köppen 1927), sometimes called "phenological interception", and even in different phenophases within the same species. Therefore, timing of one specific phase in a species can vary between districts of diverse day lengths (i.e., latitude), but also different continentality (e.g., climates with high humidity and moderate temperatures can have different timing for the same phase than climates with high temperatures and lower humidity). This was obvious in a study along a fjord of western Norway (Wielgolaski 2003).

Studies like this are important in phenology of all parts of the world, but more so in high latitude and high altitude districts with short growing seasons and low temperatures, than in districts with less environmental variation between nearby areas. Phenological interception is found both in native plants, e. g. in leaf bud break of *Fraxinus excelsior* in relation to *Quercus robur* (Batta 1969), and in agricultural plants. Knowledge about such variation in response to environmental factors can be a valuable tool to indicate districts with the best climate for certain plants (e.g. Wielgolaski 2003).

In temperate regions, leaf and flower buds of woody plants are normally initiated in the last part of the previous summer (Kramer 1922; Guimond et al. 1998). It might, therefore, be possible to predict the timing of leaf bud burst and flowering in spring based on late phenophases the year before. Wielgolaski (2000) made predictions of leaf bud break and flowering of native and cultivated woody plants in his 3-year study at several sites in western Norway. In Arctic Greenland, Böcher (1938) and Sörensen (1941) found that flower buds were often initiated two and more years before flowering. The same was observed already in the beginning of the 1900s for the alpine/arctic *Ranunculus glacialis* by Resvoll (1918). This indicates that energy is built up in the short Arctic growing seasons through one or more years before it is high enough to start the flower initiation.

13.3.2 Experimental Phenology

In forestry it has been known for many years that transplantation of coniferous tree provenances between southern and northern latitudes has a considerable impact on both phenology and productivity of the trees (Hagem 1931; Kalela 1938; Heikinheimo 1949; Magnesen 1992; Beuker 1994). Planting in the north of southern provenances may cause continuous growth too long in the summer and, therefore, a weak hardening of the shoots by the end of the season. This often has led to frost damage during winter and spring. The well-hardened buds in coniferous trees of the northernmost ecotypes flushed earlier in spring than plants of more southern origin (Beuker 1994) and ended growth earliest in autumn. The photoperiod at the site of origin was found to be a dominant factor in determining the timing of cessation in northern plants (Partanen and Beuker 1999), but Shutova et al. (2006) have reported that also air temperature was important for the time of yellowing of birch leaves at the end of the growing season, as also found by Marchand et al. (2004) for the time of senescence.

Recently, similar transplant studies have been carried out on Nordic mountain birch (*Betula pubescens* ssp. *czerepanovii*) between oceanic and continental districts in northern Fennoscandia, and by transplantation of southern birch provenances to the north. Phenological observations over 10 years have shown that the northernmost mountain birch ecotypes (from 70 to 71°N) also ended growth earliest in the autumn when grown at the same site (Ovaska et al. 2005). Oceanic northern provenances probably were somewhat earlier in ending growth in autumn than the more continental ones from similar latitudes. Both oceanic and relatively continental ecotypes of mountain birch from southern latitudes (60–64°N) showed a longer growing season when planted in more northern districts (e.g. about 68°N) than the northern provenances, being nearly green on September 10 when the northern ones were yellow and red (Ovaska et al. 2005). In transplantation to oceanic districts, survival was better for oceanic provenances, while for transplantation to a continental region the survival rate was lowest in the southernmost and westernmost ecotypes where the height growth was also lower. Also, in an oceanic district, the northernmost plants were tallest after 10 years. As found in coniferous plants the leaf bud break of mountain birch in spring was later in plants of southern and in particular oceanic origins, than in northern provenances transplanted and grown both in oceanic and continental districts at about 68°N (Ovaska et al. 2005).

The bud-burst of *Betula pubescens* provenances of various latitudes and continentality has also been studied in controlled climate (Myking and Heide 1995; Myking 1997). It was found that plants native to mild and unstable winters (as in the south and along the coast of Norway) were released later from dormancy than those from regions with cold and stable winters (Fig. 13.2), probably due to adaptation to avoid frost damage. The early dormancy in northern ecotypes (about 69°N), Myking (1999) stated, could be a decisive adaptation to a short growing season and hardening of the shoots at short days in time to avoid autumn frost. He also found that a period of chilling (below 10°C) was necessary for bud-burst of

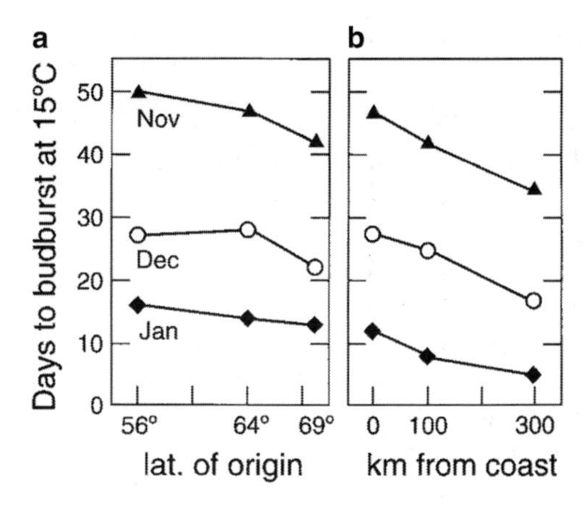

Fig. 13.2 Days to bud-burst in an 8-hour (short-day) photoperiod, after different periods of chilling at 5 °C in *Betula pubescens* provenances along two gradients. (**a**) Latitudinal gradient from 56°N (Denmark), 64°N (Mid – Norway) to 69°N (North – Norway). (**b**) Coastal-inland gradient in Norway below 150 m a.s.l. at about 60°N (Figure 2 from Myking 1999, p. 142, used with permission)

B. pubescens, but the natural chilling period has normally been long enough even in southern and oceanic sites at latitudes above 60°N. Long photoperiods significantly reduced time to leaf bud break in partly dormant buds, but not when dormancy was fully released. However, there may be increasing experimental evidence that light conditions play some role in the timing of spring phenology (Linkosalo et al. 2008). According to Heide (1993) dormancy in *B. pubescens*, as well as in *B. pendula* and *Prunus padus*, was released early in December, while in *Alnus* sp. not until February.

Similarly, Hannerz (1999) suggested that a chilling requirement for *Picea abies* was fulfilled in December even somewhat south of 60°N in Fennoscandia. Leinonen (1996), however, generally observed that chilling requirements increased in *Pinus sylvestris* and *Betula pendula* for coastal populations compared to the more continental ones. Hänninen (1995) concluded that chilling temperature is the major environmental factor regulating rest break, but premature leaf bud break seemed not to be any serious problem for frost damage, according to models of bud-burst phenology of trees from cool and temperate regions.

According to studies in growth chambers, however, chilling probably played a lesser role in bud break at high latitude in nature because the plants there normally met their requirements during winter (Pop et al. 2000). This may change in some districts even at high latitudes because of climate change (Linkosalo et al. 2008), and in some districts also at higher elevations (Vitasse et al. 2011; Chap. 14 in this book).

On the other hand, it has recently been observed in experimental studies at high latitudes that winter warming of plots in the field by air of 2–10 °C for 2–14 days delayed bud break of the deciduous *Vaccinium myrtillus* by more than a week, while the evergreen *V. vitis-idaea* was not affected (Bokhorst et al. 2011). Spring warming of plants advanced the onset of plant growth but it also led to spatial compression of the plant growing season (Post et al. 2008). Field experiments on snowmelt in spring in interior Alaska (Wipf 2010) and on the archipelago of Svalbard north of Norway (Cooper et al. 2011) showed that the sooner after snowmelt a plant species develops, the more its phenology was sensitive to snowmelt.

The ITEX (International Tundra Experiment) project was established at several sites in the Northern Hemisphere to study the influence of temperature and wind shields on vegetative and reproductive growth and development in arctic and alpine plants. Open top chambers with transparent walls were placed in the field for 1–4 years (Arft et al. 1999). It was observed that in the warmer, Low Arctic sites, the strongest response was in vegetative growth, particularly by herbaceous plants, while colder High Arctic sites produced a greater reproductive response. The better opportunities to use energy for investment in flowering and development of seeds afforded by increased temperatures in the High Arctic may provide an opportunity for species to colonize patches of bare ground (Robinson et al. 1998). In a conclusive report after the end of the ITEX project Walker et al. (2006) stated that the passive warming treatment in the study increased the plant-level air temperature by 1–3 °C but also altered the light, moisture and gas exchange somewhat. They concluded that temperature responses were rapid and detected in the plant communities after only two growing seasons. However, the night temperatures in the open top chambers used in the ITEX project were lower than outside, which generally weakened the effects of increased day temperatures (Pieper et al. 2011), although plants may react differently to day and night temperatures in various developmental phenophases (Wielgolaski 1974b). At the semi-desert Svalbard site, there was a strong effect of temperature increment on the flowering of *Dryas octopetala* and also on seed setting (Wookey et al. 1993). While leaf bud break and flowering occurred earlier throughout the whole study period, there was little impact on growth cessation at the end of the season. Some studies, however, may also indicate a possible delayed senescence in the ITEX chambers (e.g., Molau 1997; Stenström et al. 1997).

13.3.3 Climate Change and Biological Responses

Global warming at high latitudes is expected to cause greater increases in temperatures and precipitation during winter than summer (Dickinson 1986; Maxwell 1997). The most pronounced predicted changes are in models of northern latitudes (IPCC 2007). The North Atlantic Oscillation (NAO) seems to be responsible for a large component of the increased temperature in Europe (Post and Stenseth 1999). According to e.g. Chmielewski and Rötzer (2001) the positive phase of NAO has increased clearly in Europe in the period February-April through several years, leading to prevailing westerly winds and thus to higher temperatures, particularly during the last two decades of the 1900s, which biologically generally resulted in earlier phenophases, especially in spring. In most of the higher latitudes increased winter precipitation has caused greater snow accumulation. Despite that, increased temperature has led to earlier snowmelt and longer annual snow-free season in most regions (Maxwell 1992), an earlier and longer growing season (Bliss and Matveyeva 1992; Oechel and Billings 1992) and increased rates of plant population growth (Carlsson and Callaghan 1994).

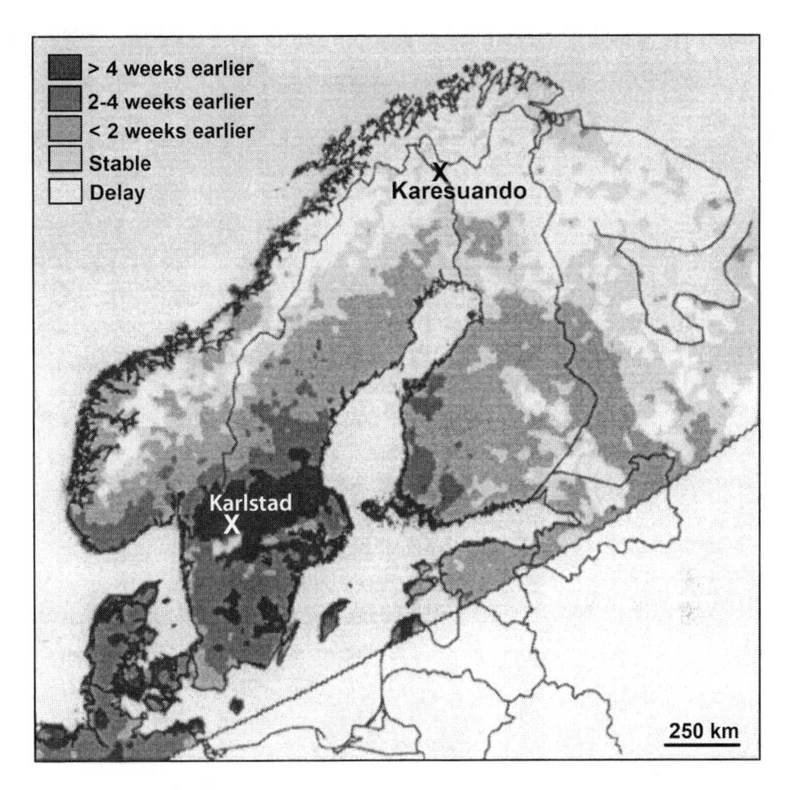

Fig. 13.3 Change in onset of spring in Fennoscandia from 1992 to 1998 on the basis of the GIMMS NDVI dataset (From Høgda et al. 2001, used with permission of the first author, NORUT IT, Tromsö)

However, in areas of Fennoscandia with low winter temperatures, as in high mountain areas of Norway and in some inner parts of northern Fennoscandia, the higher winter precipitation (predicted to increase about 1.5 % per decade at Finnmarksvidda, Hanssen-Bauer 2005; Vikhamar-Schuler et al. 2010) caused a longer snow cover in spring, at least during the last two decades of the 1900s. That was observed by satellite inventory values (Myneni et al. 1997; Delbart et al. 2005; de Beurs and Henebry 2010; Zeng et al. 2011), despite non-significant higher temperatures (predicted as 0.5 °C per decade at Finnmarksvidda, Hanssen-Bauer 2005; Vikhamar-Schuler et al. 2010). Thus, a later onset of spring (Fig. 13.3) was observed in some places during 17 years (1982–1998) of study (Høgda et al. 2001), although there was a somewhat smaller delay during the first decade of the present century (Karlsen et al. 2009a; Høgda personal communication.). The strongest delay (approximately 1 week) occurred in the most continental areas of northern Fennoscandia. It was also demonstrated that starting dates of birch pollen seasons were delayed in the same regions as the delay of spring, while earlier in all other parts of Fennoscandia (Høgda et al. 2002; Karlsen et al. 2009b). A delayed plant development was also found by Kozlov and Berlina (2002) at the Russian Peninsula

Kola for the period 1930–1998 and partly by Shutova et al. (2006) for the period 1964–2003. In most of the high latitudes autumn was delayed in the late 1900s and, therefore, the growing season generally increased in the region, again except for the northern continental section of Europe, where slightly shorter growing seasons sometimes were observed (Høgda et al. 2001; Karlsen et al. 2009a).

Plant responses to climate change in northern latitudes can be predicted by modeling phenological data from experiments studying the effect of temperature changes in growth chambers (e. g., Hänninen 1995; Hannerz 1999; Pop et al. 2000), studies of changes in weather variables as described above, and by models or indices of biosphere response, e. g. based on some average of plant phenology of various species (Schwartz 1997, 1998; Chmielewski and Rötzer 2001; Schwartz et al. 2006). Delbart and Picard (2007) reported that generally leaf appearance in Alaska occurred approximately 10 days earlier from 1975 to the first years of 2000s and similarly in west Siberian tundra since 1965, when based on a model and satellite observations.

At high latitudes a warmer climate may have resulted in a higher-altitude tree line (Skre 2001; Kullman 2010). However, in some places, also a lower grazing pressure by domestic animals may have been very important for that result (Bryn 2008). An increased biodiversity is expected in many districts because of global warming, increasing nutrient availability in the soil by increased decomposition, but also through higher precipitation adding more nutrients, particularly nitrogen.

In the High Arctic, better seed production (Philipp et al. 1990; Arft et al. 1999) is observed, and late-flowering arctic species that previously only rarely ripened their seed may do so more regularly with increasing temperatures (Thórhallsdóttir 1998).

By comparison of plant phenology from old data sets with the same phenophase of more recent observations at a site, it is possible to see changes that may be a result of climate change. In Norway this has been done for the first flowering date of some species in the second half of the 1800s compared with the same species in the second quarter of the 1900s (1928–1952) and the third quarter of the century (1952–1972) e.g. in a south-eastern urban coastal district (Oslo) at about 60°N, and in an elevated rural inland district of southern Norway at about 61°N (Klaveness and Wielgolaski 1996). Most species were flowering earlier in Oslo in the 1900s and particularly in the last period of the century, while there were small changes in flowering dates of the various plant species in the rural inland district between the periods (Fig. 13.4).

Menzel (2000), Chmielewski and Rötzer (2001) and Menzel et al. (2006) have reported from several countries in Europe that leaf bud break in spring of trees advanced on average throughout the last three to four decades of the 1900s by more than 0.2 days/year, while the autumn phases in some cases were delayed by about 0.15 days/year, causing longer growing seasons; in other cases, however, changes were ambiguous. Similarly, Beaubien and Freeland (2000) reported the first bloom of the early-flowering *Populus tremuloides* to be 0.27 days/year earlier in a long term-study (1900–1977) at Edmonton, Canada, and that flowering of the same species advanced by 2 weeks on an average from several sites in Alberta, Canada throughout the period 1936–2006 (Beaubien and Hamann 2011). In Finland

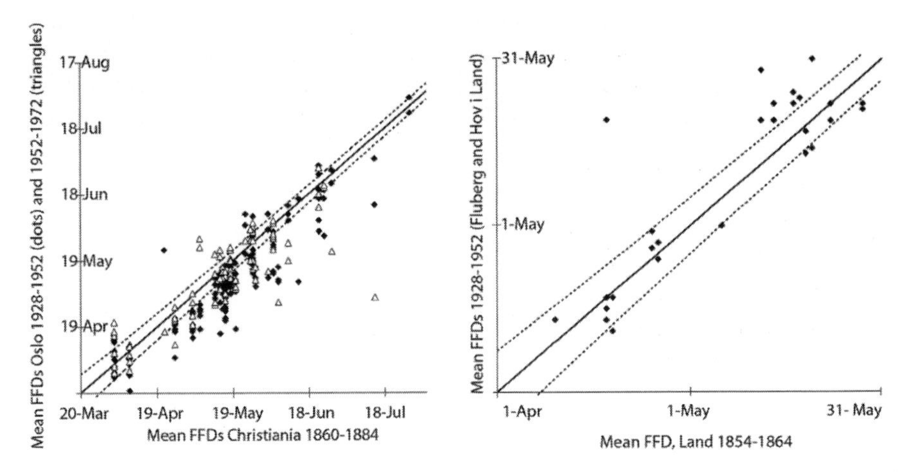

Fig. 13.4 *Left*: Scatter-plot of mean first flowering days (FFDs) in various plant species at ten stations in urban Oslo in the 1900s plotted against mFFDs from Christiania (Oslo) 1860–1884. *Right*: The same from Land, a rural inland district at about 300 m elevation. 95 % C.I. calculated for n = 25 years (From Klaveness and Wielgolaski 1996, used with permission)

phenological data did not permit reliable estimates of the effect of climate change on spring in boreal trees according to some studies reported by Linkosalo (2000), while Pudas et al. (2008) reported an advance of leaf bud break in *Betula pubescens* in the short period 1997–2006 of 0.7 days/year in the south, and 1.4 days/year further north, which is even higher than the average for several countries and species mentioned before (Menzel et al. 2006). Kramer et al. (2000) concluded that phenology of the boreal forests is mainly driven by temperature, affecting the timing of the start of the growing season and thereby its duration, and the level of frost hardiness and thereby the reduction of foliage area and photosynthetic capacity by severe frost events.

Earlier spring resulting from climate change may increase the frost risk of young shoots particularly in maritime regions. However, in more continental high-latitude treeline regions of northern Eurasia it is predicted that the response to frost damage will decrease with further climate change (Bennie et al. 2010). In Greenland earlier snowmelt associated with warmer temperatures has increased risk of frost damage in *Dryas* (Høye et al. 2007a).

Recent long-term Norwegian studies (Nordli et al. 2008; Wielgolaski 2009; Wielgolaski et al. 2011) have shown clear tendencies in decadal variations of phenological timing in several spring phases throughout the country. No clear tendency of influence of climate change, however, was obvious until about 1965 at a southeastern relatively densely populated and polluted site (Ås) and maybe even later, from after 1980, in a western rural district (Njøs), both in flowering of apple ('Gravenstein') (Fig. 13.5) and plum (based on Wielgolaski 2009). This fits well with similar studies in Switzerland throughout more than 100 years, where advances of leaf bud break of *Aesculus hippocastaneum* started

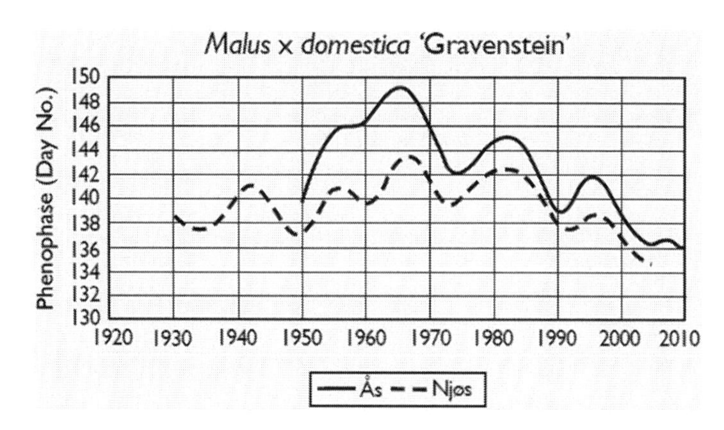

Fig. 13.5 First flowering day (numbers from 1. January) of apple (*Malus × domestica*) 'Gravenstein' during several years at the somewhat polluted site Ås in southeastern Norway and at Njøs in a rural western fjord district of the country. Note the decadal variations calculated in addition to advanced flowering by climate change (Redrawn from Nordli et al. 2008)

already before 1900 in the town of Geneva, while flowering of *Prunus avium* in a rural valley did not advance until about 1985 (Meteo-Swiss 2011). The urban heat-island effect probably explains the urban–rural differences, which are also found by Zhang et al. (2004).

If the climate change with higher temperatures in most parts of the World continues in the coming years as expected by IPCC (2007), at least partly due to human activities, it may cause problems from more hazards at high latitudes, with e.g. more extreme storms followed by flooding. However, longer growing seasons may also cause higher productivity in agriculture further north.

Although phenological studies on plants are mainly reported here, phenological observations have also been made on animals at high latitudes, mainly on migration of birds, but on other animal groups as well (see also other chapters in this book). Forchhammer et al. (2002) reported that in northern Europe, generally migrants arrived earlier in a district after high North Atlantic Oscillation (NAO) winters (mild) than after colder ones, while Antarctic birds were observed to breed later in response to the recent climate change (Barbraud and Weimerskirch 2006). Positive NAO was also reported by Sparks et al. (2005) to be the main reason for some birds having an earlier migration to northern Europe but generally the migration dates also became more elongated and skewed. According to Barrett (2002) there has not been a significant long-term trend in the arrival dates in the north for all birds through the last 30 years of the last century, while Jonzén et al. (2006) observed that long-distance bird migrants advanced their arrival in Scandinavia more than short-distance migrants. More recently (Saino et al. 2011) it is stated that several migrating birds have arrived earlier in the last years at high latitudes, and Both et al. (2004) found they also had earlier egg laying dates.

Earlier arrival is also observed for pink salmon in Alaskan rivers throughout the last 30 years of the previous century (Taylor 2008). Similarly, it is observed that some vertebrates emerged earlier from hibernation in more recent times, while others did not (e.g. Sheriff et al. 2011). Høye et al. (2007b) have reported, however, that earlier snowmelt in high arctic Greenland has advanced spring phenology in plants, arthropods and clutch initiation dates of birds by more than 30 days during the decades around 2000.

There is no reason to expect that all components of food chains will change phenologically at the same rate (Visser et al. 2004). Therefore, mismatch may be seen between early plant growth from climate change at high latitudes and availability of the right food at the best time for herbivores but also for carnivores (so-called trophic interactivity), as stressed already by Harrington et al. (1999). Most data on mismatch between production and consumption are published on birds (e.g. Both et al. 2004), but some information is also available for other vertebrates and invertebrates, maybe particularly for butterflies.

In birds it was earlier expected that short-distance migrants easily could adjust to phenological advancement of plants in spring and therefore be less influenced in their breeding than long-distance migrants. However, Jonzén et al. (2006) found that the opposite was the case in Scandinavia. Today (2013) it is not quite obvious how many migrating bird species are able to keep up with the generally earlier spring at high latitudes. Lehikoinen et al. (2006) stated that milder winters are correlated with earlier migration in many birds in temperate regions. They suggested that earlier ice breaking in the Baltic Sea by climate warming may be expected to increase future breeding success of eiders. On the other hand van der Jeugd et al. (2009) found that Baltic barnacle geese have not shifted their timing of breeding relative to the advancing spring phenology 1984–2007, and there was a clear mismatch with time between food supply and reproduction, despite an advance of egg-laying. Saino et al. (2011) mention that migrants, and particularly those wintering in sub-Saharan Africa and now arriving at the increased temperature at high latitudes, may have an accumulated thermal delay. That will possibly cause an increased mismatch to spring phenology at high latitudes, having a severe impact on their breeding and their populations (see Chap. 24 in this book).

Pettorelli et al. (2005) stated that positive winter NAO in western Norway resulted in spatially more variable phenology offering migrating red deer an extended period with access to high-quality forage leading to increased body weight in the autumn. However, Post et al. (2008) suggested from a short-term experimental study that it is highly relevant to herbivore ecology and in particular to that of reindeer, to consider the manner in which warming by climate change alters the pattern of plant phenology at more immediate spatial scales than that of the regional landscape. Reduced amounts of lichens, important as winter fodder for reindeer in poor heaths at high latitude, are found because of faster nutrient cycling at increased temperature (Vors and Boyce 2009), in contrast to grasses, shrubs and trees (Tømmervik et al. 2005). Another result of increased temperature in the far North is a reduction of sea ice in summer, influencing the availability of arctic bears to be able to catch seals (e.g. Vors and Boyce 2009), as well as influencing changes in the arctic tundra vegetation (Bhatt et al. 2010). Phenological shifts may thus easily result in mismatch in an ecosystem (Donelly et al. 2011), at least as often at high latitudes as closer to the Equator.

Phenologists, therefore, need an empirical understanding of physiological and behavioural adjustments that animals can make in response to seasonal and long-term variation in environmental conditions (Sheriff et al. 2011).

13.4 Conclusions

At high latitudes, there are large differences among species-specific responses to environmental factors. The responses also vary between geographical districts or continentality, and within a species at different times of the year. In many cases phenology may be used in climate change studies, but then there must be a clear description of the sites used in the study: geographically, climatically and edaphically, as well as clear definitions of the phenophases studied and the state of the organism. Long-term studies are also necessary to draw clear conclusions because of decadal variations in both climate and phenological observations.

References

Arft AM, Walker MD, Gurevitch J, Alatalo JM, Bret-Harte MS, Dale M, Diemer M, Gugerl F, Henry GHR, Jones MH, Hollister RD, Jónsdóttir IS, Laine K, Levesque E, Marion GM, Molau U, Mølgaard P, Nordenhäll U, Raszhivin V, Robinson CH, Starr G, Stenström A, Stenström M, Totland Ø, Turner PL, Walker LJ, Webber PJ, Welker JM, Wookey PA (1999) Responses of tundra plants to experimental warming: meta-analyses of the International Tundra Experiment. Ecol Monogr 69:491–511

Arnell HW (1923) Vegetationens aarliga utvecklingsgaang i Svealand. Medd. Statens Meteorologisk – Hydrografiska Anstalt 2:1–80 (German abstr:74–80)

Arnell K (1927) Vegetationens utvecklingsgaang i Norrland. Medd. Statens Meteorologisk – Hydrografiska Anstalt 4:1–28 (German abstr:1–28)

Arnell K, Arnell S (1930) Vegetationens utveckling i Götaland. Medd. Statens Meteorologisk – Hydrografiska Anstalt 6:1–70 (German abstr:69–70)

Aune B (1993) Aarstider og vekstsesong. Kartblad (Map) 3.1.7., scale 1:7 mill. In: Nasjonalatlas for Norge, Det norske meteorologiske institutt – Statens kartverk, Oslo – Hönefoss

Barbraud C, Weimerskirch H (2006) Antarctic birds breed later in response to climate change. PNAS 103:6248–6251

Barrett RT (2002) The phenology of spring bird migration to north Norway. Bird Study 49:270–277

Batta J (1969) Variasjoner i tid for bladsprett hos ask og eik, (Engl summary), Aarsskr. Planteskoledrift Dendrologi 14–15:78–85

Beaubien EG (1996) Plantwatch, a model to initiate phenology in school classes. Phenol Seas 1:33–35

Beaubien EG, Freeland HJ (2000) Spring phenology trends in Alberta, Canada: links to ocean temperature. Int J Biometeorol 44:53–59

Beaubien EG, Hamann A (2011) Spring flowering response to climate change between 1936 and 2006 in Alberta, Canada. Bioscience 61:514–524

Beaubien EG, Johnson D (1994) Flowering plant phenology and weather in Alberta, Canada. Int J Biometeorol 38:23–27

Bennie J, Kubin E, Wiltshire A, Huntley B, Baxter R (2010) Predicting spatial and temporal patterns of bud-burst and spring frost risk in north-west Europe: the implications of local adaptation to climate. Glob Change Biol 16:1503–1514
Beuker E (1994) Adaptation to climatic changes of the timing of bud burst in populations of *Pinus sylvestris* L. and *Picea abies* (L.) Karst. Tree Physiol 14:961–970
Bhatt US, Walker DA, Raynolds MK, Comiso JC, Epstein HE, Jia G, Gens R, Pinzon JE, Tucker CJ, Tweedie CE, Webber PJ (2010) Circulation arctic tundra vegetation change is linked to sea ice decline. Earth Interact 14–008:1–20
Björbekk (1993) Snö. Kartblad (Map) 3.1.4., scale 1.7 mill. In: Nasjonalatlas for Norge, Det norske meteorologiske – Statens kartverk, Oslo – Hönefoss
Bliss LC (1971) Arctic and alpine life cycles. Ann Rev Ecol Syst 2:405–438
Bliss LC (ed) (1977) Truelove Lowland, Devon Island, Canada: a high Arctic ecosystem. University Alberta Press, Edmonton
Bliss LC, Matveyeva NV (1992) Circumpolar arctic vegetation. In: Chapin FS III, Jefferies RL, Reynolds JF, Shaver GR, Svoboda J (eds) Arctic ecosystems in a changing climate. Academic, New York
Böcher TW (1938) Studies on the vegetation of the east coast of Greenland. Medd Grønl 104:1–32
Bokhorst S, Bjerke JW, Street LE, Callaghan TV, Phoenix GK (2011) Impacts of multiple extreme winter warming events on sub-Arctic heathland: phenology, reproduction, growth, and CO_2 flux responses. Glob Change Biol 17:2817–2830
Both C, Artemyev AV, Blaauw B, Cowie RJ, Dekhuijzen AJ, Eeva T, Enemar A, Gustafsson L, Ivankina EV, Järvinen A, Metcalfe NB, Nyholm NEI, Potti J, Ravussin P-A, Sanz JJ, Silverin B, Slater FM, Sokolov LV, Török J, Winkel W, Wright J, Zang H, Visser ME (2004) Large-scale geographical variation confirms that climate change causes birds to lay earlier. Proc R Soc Lond B 271:1657–1662
Breckle SW (2002) Walters vegetation of the earth. Springer, Berlin/Heidelberg
Bryn A (2008) Recent forest limit changes in south-east Norway: effects of climate change or regrowth after abandoned utilization? Nor J Geogr 62:251–270
Carlsson BA, Callaghan TV (1994) Impact of climate change factors on the clonal sedge *Carex bigelowii*: implication for population growth and vegetative spread. Ecography 17:321–330
Chmielewski F-M, Rötzer T (2001) Response of tree phenology to climate change across Europe. Agric For Meteorol 108:101–112
Chmielewski FM, Rötzer T (2002) Annual and spatial variability of the beginning of growing season in Europe in relation to air temperature changes. Clim Res 19:257–264
Colombo SJ (1998) Climatic warming and its effect on bud burst and risk of frost damage to white spruce in Canada. For Chron 74:567–577
Cooper EJ, Dullinger S, Semenchuk P (2011) Late snowmelt delays plant development and results in lower reproductive success in the high Arctic. Plant Sci 180:157–167
de Beurs KM, Henebry GM (2010) A land surface phenology assessment of the northern polar regions using MODIS reflectance time series. Can J Remote Sens 36:S87–S110
Delbart N, Picard G (2007) Modeling the date of leaf appearance in low-arctic tundra. Glob Change Biol 13:2551–2562
Delbart N, Kergoat L, Toan TL, Llermitte J, Picard G (2005) Determination of phenological dates in boreal regions using normalized difference water index. Remote Sens Environ 97:26–38
Dickinson R (1986) The climate system and modelling of future climate. In: Bolin B, Doos B, Jager J, Warrick RA (eds) The greenhouse effect, climate change, and ecosystem. Wiley, Chichester
Diekmann M (1996) Relationship between flowering phenology of perennial herbs and meteorological date in deciduous forests of Sweden. Can J Bot 74:528–537
Donelly A, Caffarra A, O'Neill BF (2011) A review of climate-driven mismatches between interdependent phenophases in terrestrial and aquatic ecosystems. Int J Biometeorol 55:805–817

Eriksen B, Molau U, Svensson M (1993) Reproductive strategies in two arctic *Pedicularis* species (Scrophulariaceae). Ecography 16:154–166
Erskine AJ (1985) Some phenological observations across Canada's boreal regions. Can Field-Nat 99:188–195
Forchhammer MC, Post E, Stenseth NC (2002) North Atlantic oscillation timing of long- and short-distance migration. J Anim Ecol 71:1002–1014
Guimond CM, Andrews PK, Lang GA (1998) Scanning electron microscopy of floral initiation in sweet cherry. J Am Soc Hortic Sci 123:509–512
Hagem O (1931) Forsök med vestamerikanske traeslag, (German summary). Medd Vestl Forstl Forst 12:1–217
Hannerz M (1999) Evaluation of temperature models for predicting bud burst in Norway spruce. Can J For Res 29:9–19
Hänninen H (1995) Effects of climate change on trees from cool and temperate regions: an ecophysiological approach to modelling of bud burst phenology. Can J Bot 73:183–199
Hanssen-Bauer I (2005) Regional temperature and precipitation series for Norway. Comparison from dynamical and empirical downscaling. Met. No. report 15/2005 Climate
Harrington R, Wolwod I, Sparks T (1999) Climate change and trophic interactions. Trends Ecol Evol 14(4):146–150
Heide OM (1985) Physiological aspects of climatic adaptation in plants with special references to high-latitude environments. In: Kaurin A, Junttila O, Nilsen J (eds) Plant production in the north. Norwegian University Press, Tromsö
Heide OM (1993) Daylength and thermal time response of budburst during dormancy release in some northern deciduous trees. Physiol Plant 88:531–540
Heikinheimo O (1949) Results of the experiments on the geographical races of spruce and pine (in Finnish with English summary). Comm Inst For Fenn 37:1–44
Høgda KA, Karlsen SR, Solheim I (2001) Climatic change impact on growing season in Fennoscandia studied by a time series of NOAA AVHRR NDVI data. In: Proceedings of IGARSS 2001, Sydney. ISBN 0-7803-7033-3
Høgda KA, Karlsen SR, Solheim I, Tömmervik H, Ramfjord H (2002) The start dates of birch pollen seasons in Fennoscandia studied by NOAA AVHRR NDVI data. In: Proceedings of IGARSS 2002, Toronto. ISBN 0-7803-7536-X
Holmboe J (1913) Vaarens utvikling i Tromsö amt (in Norwegian). Bergens Mus Aarb 1912:1–248
Holopainen J, Helama S, Timonen M (2006) Plant phenological data and tree-rings as palaeoclimate indicators in south-west Finland since AD 1750. Int J Biometeorol 51:61–72
Høye TT, Ellebjerg SM, Philipp M (2007a) The impact of climate on flowering in the high-Arctic – the case of *Dryas* in a hybrid zone. Arct Antarct Alp Res 39:412–421
Høye TT, Post E, Meltofte H, Schmidt NM, Forchhammer MC (2007b) Rapid advancement of spring in the high Arctic. Curr Biol 17:R449–R451
IPCC (2007) Summary for policymakers. In: Solomon S, Qin D, Manning M, Chen Z, Marquis M, Averyt KB, Tignor M, Miller HL (eds) Climate change 2007: the physical science basis. Contribution of Working Group I to the fourth assessment report of the Intergovernmental Panel on Climate Change. Cambridge University Press, Cambridge/New York
Iversen M, Bråthen KA, Yoccoz NG, Ims RA (2009) Predictors of plant phenology in a diverse high-latitude alpine landscape: growth forms and topography. J Veg Sci 20:903–915
Johansson OV (1946) Det fenologiska observationsmaterialet i Finland och provstudier av detsamma, (in Swedish). Finlands Natur och Folk 88(8):1–118
Johansson OV (1953) Die Phänologie in Finland. Soc Sci Fenn, Commun Biol 11(1):1–55
Jonzén N, Lindén A, Ergon T, Knudsen E, Vik JO, Rubolini D, Piacentini D, Brinch C, Spina F, Karlsson L, Stervander M, Andersson A, Waldenström J, Lehikoinen A, Edvardsen E, Solvang R, Stenseth NC (2006) Rapid advance of spring arrival dates in long-distance migratory birds. Science 312:1959–1961

Junttila O, Stushnoff C, Gusta LV (1983) Dehardening in flower buds of saskatoon-berry, *Amelanchier alnifolia*, in relation to temperature, moisture content, and spring bud development. Can J Bot 61:164–170

Kalela A (1938) Zur Synthese der experimentellen Untersuchungen über Klimarassen der Holzarten. Commun Inst For Fenn 26:1–445

Karlsen SR, Tolvanen A, Kubin E, Poikolainen J, Høgda KA, Johansen B, Danks FS, Aspholm P, Wielgolaski FE, Makarova O (2008) MODIS-NDVI based mapping of the length of the growing season in northern Fennoscandia. Int J Appl Earth Obs Geoinf 10:253–266

Karlsen SR, Høgda KA, Wielgolaski FE, Tolvanen A, Tømmervik H, Poikolainen J, Kubin E (2009a) Growing-season trends in Fennoscandia 1982–2006, determined from satellite and phenology data. Clim Res 39:275–286

Karlsen SR, Ramfjord H, Høgda KA, Johansen B, Danks FS, Brobakk TE (2009b) A satellite-based map of onset of birch (*Betula*) flowering in Norway. Aerobiologia 25:15–25

Klaveness D, Wielgolaski FE (1996) Plant phenology in Norway – a summary of past and present first flowering dates (FFDs) with emphasis on conditions within three different areas. Phenol Seas 1:47–61

Köppen W (1927) Wechsel der phänologischen Zeitenfolge. Meteorol Z 44:175–177

Kozlov MV, Berlina NG (2002) Decline in length of the summer on the Kola Peninsula, Russia. Clim Change 54:387–398

Kramer O (1922) Über die Blütenknospen und der Zeitpunkt der Entstehung von Blütenanlagen bei einigen Obstsorten. Dtsch Obstbauztg 68:306–308

Kramer K, Leinonen I, Loustau D (2000) The importance of phenology for the evaluation of impact of climate change on growth of boreal, temperate and Mediterranean forests ecosystems: an overview. Int J Biometeorol 44:67–75

Kullman L (2010) A richer, greener and smaller alpine world: review and projection of warming-induced plant cover change in Swedish Scandes. Ambio 39:150–169

Lauscher F (1980) Klima, Klimaschwankungen und phänologischer Jahresablauf am europäischen Nordkap. Mitt Österr Geogr Ges 122:193–220

Lauscher F (1985) Zur Phänologie vegetativ vermehrter Pflanzen einheitlicher Herkunft – Beobachtungen in phänologischen Pflanzgärten in Norwegen 1963–1982. Phyton (Horn, Austria) 25:253–272

Lauscher A, Lauscher F (1990) Phänologie Norwegens, Teil IV, Private edition

Lauscher A, Lauscher F, Printz H (1955) Die Phänologie Norwegens, Teil I, Allgemeine Übersicht. Skr Det Norske Videnskaps-Akademi Oslo, 1, Mat-Naturv Kl 1:1–99

Lauscher A, Lauscher F, Printz H (1959) Die Phänologie Norwegens, Teil II, Phänologische Mittelwerte für 260 Orte. Skr Det Norske Videnskaps-Akademi Oslo, 1, Mat-Naturv Kl 1:1–176

Lauscher A, Lauscher F, Printz H (1978) Die Phänologie Norwegens, Teil III, Tabellen-Karten der Mittelwerte. Skr Det Norske Videnskaps-Akademi Oslo, 1, Mat-Naturv Kl 37:1–253

Lehikoinen A, Kilpi M, Öst M (2006) Winter climate affects subsequent breeding success of common eiders. Glob Change Biol 12:1355–1365

Leinonen I (1996) Dependence of dormancy release on temperature in different origins of *Pinus sylvestris* and *Betula pendula* seedlings. Scand J For Res 11:122–128

Lie H (1931) Faenologiske noteringar fraa Telemark (in Norwegian). Tidsskr Norske Landbruk 38:204–206

Lieth H (1997) Aims and methods in phenological monitoring. In: Lieth H, Schwartz MD (eds) Phenology in seasonal climates I. Backhuys Publication, Leiden

Linkosalo T (2000) Analyses of the spring phenology of boreal trees and its response to climate change. Univ Hels Dept For Ecol 22:1–55

Linkosalo T, Lappalainen HK, Hari P (2008) A comparison of phenological models of leaf bud burst and flowering of boreal trees using independent observations. Tree Physiol 28:1873–1882

Linkosalo T, Häkkinen R, Terhivuo J, Tuomenvirta H, Hari P (2009) The time series of flowering and leaf bud burst of boreal trees (1846–2005) support the direct temperature observations of climate warming. Agric For Meteorol 149:453–461

Linne C (1751) Philosophia Botanica (in Latin). Kiesewetter, Stockholm

Magnesen S (1992) Injuries on forest trees related to choice of the species and provenances: a literature survey of a one hundred year epoch in Norwegian forestry. Rep Skogforsk 7:1–46

Makarova OA, Pohilko AA, Kushel JA (2001) Seasonal life of the nature in Kola Peninsula (in Russian, translated in English). Murmansk. ISBN 5-7744-0102-2

Marchand FF, Nijs I, Heuer M, Mertens S, Kockelberg F, Pontailler JY, Impens I, Beyens L (2004) Climate warming postpones senescence in high arctic tundra. Arct Antarct Alp Res 36:390–394

Maxwell B (1992) Arctic climate: potential for change under global warming. In: Chapin FS, Jeffries RL, Reynolds JF, Shaver GR, Svoboda J (eds) Arctic ecosystems in a changing climate. Academic Press, New York

Maxwell B (1997) Recent climate patterns in the Arctic. In: Oechel WC, Callaghan T, Gilmanov T, Holten JI, Maxwell B, Molau U, Sveinbjörnsson B (eds) Global change and arctic terrestrial ecosystems. Springer, Heidelberg

Menzel A (2000) Trends in phenological phases in Europe between 1951 and 1996. Int J Biometeorol 44:76–81

Menzel A, Sparks TH, Estrella N, Koch E, Aasa A, Ahas R, Alm-Kübler K, Bissolli P, Braslavska O, Briede A, Chmielewski FM, Crepinsek Z, Curnel Y, Dahl Å, Defila C, Donnelly A, Filella Y, Jatczak K, Måge F, Mestre A, Nordli Ø, Penuelas J, Pirinen P, Remisová V, Scheifinger H, Striz M, Susnik A, van Vliet AJH, Wielgolaski FE, Zach S, Zust A (2006) European phenological response to climate change matches the warming pattern. Glob Change Biol 12:1969–1976

MeteoSwiss (2011) http://www.meteoschweiz.admin.ch/web/de/klima/klima_schweiz/phaenologie/Phaenobeobachtungen_seit1808.html. Accessed 30 May 2011

Moberg A (1857) Naturalhistoriska daganteckningar gjorda i Finland aaren 1750–1845 (in Swedish). Förh Sällsk Fauna Flora Fenn 3:95–250

Moberg A (1894) Fenologiska iakttagelser i Finland aaren 1750–1845 (in Swedish). Finlands Nat Folk 55:1–165

Moe A (1928) Dates of flowering for native and garden plants at Stavanger 1897–1926. Skr Det Norske Videnskaps-Akademi Oslo, 1, Mat-Naturv Kl 3:1–50

Molau U (1993) Relationship between flowering phenology and life history strategies in tundra plants. Arctic Alp Res 25:391–402

Molau U (1997) Responses to natural climatic variation and experimental warming in two tundra plant species with contrasting life forms: *Cassiope tetragona* and *Ranunculus nivalis*. Glob Change Biol 3(suppl 1):97–107

Myking T (1997) Dormancy, budburst and impacts of climatic warming in coastal-inland and altitudinal *Betula pendula* and *B. pubescens* ecotypes. In: Lieth H, Schwartz MD (eds) Phenology in seasonal climates I. Backhuys Publication, Leiden

Myking T (1999) Winter dormancy release and budburst in *Betula pendula* ROTH and *B. pubescens* EHRH. Ecotypes. Phyton (Horn, Austria) 39(4):139–145

Myking T, Heide OM (1995) Dormancy release and chilling requirement of buds of latitudinal ecotypes of *Betula pendula* and *B. pubescens*. Tree Physiol 15:697–704

Myneni RB, Keeling CD, Tucker CJ, Asrar G, Nemani RR (1997) Increased plant growth in the northern latitudes from 1981 to 1991. Nature 386:698–702

Nordli Ø, Wielgolaski FE, Bakken AK, Hjeltnes SH, Måge F, Sivle A, Skre O (2008) Regional trends for bud burst and flowering of woody plants in Norway as related to climate change. Int J Biometeorol 52:625–639

Odland A (2011) Estimation of the growing season length in alpine areas: effects of snow and temperatures. In: Scmidt JG (ed) Alpine environment: geology, ecology and conservation. Nova Science Publication, New York

Oechel WC, Billings WD (1992) Effects of global change on the carbon balance of arctic plants and ecosystems. In: Chapin FS, Jefferies RL, Reynolds JF, Shaver GR, Svoboda J (eds) Arctic ecosystems in a changing climate. Academic Press, New York

Ovaska JA, Nilsen J, Wielgolaski FE, Kauhanen H, Partanen R, Neuvonen S, Kapari L, Skre O, Laine K (2005) Phenology and performance of mountain birch provenances in transplant gardens: latitudinal, altitudinal and oceanity-continentality gradients. In: Wielgolaski FE (ed) Plant ecology, herbivory and human impact in Nordic mountain birch forests. Springer, Heidelberg

Partanen J, Beuker E (1999) Effects of photoperiod and thermal time on the growth rhythm of *Pinus sylvestris* seedlings. Scand J For Res 14:487–497

Pettorelli N, Mysterud A, Yoccoz NG, Langvatn R, Stenseth NC (2005) Importance of climatological downscaling and plant phenology for red deer in heterogenous landscapes. Proc R Soc B 272:2357–2364

Philipp M, Böcher J, Mattson O, Woodell SRJ (1990) A quantitative approach to the sexual reproductive biology and population structure in some arctic flowering plants: *Dryas integrifolia, Silene acaulis* and *Ranunculus nivalis*. Medd Grønland Biosci 34:1–60

Pieper SJ, Loeven V, Gill M, Johnstone JF (2011) Plant responses to natural and experimental variations in temperature in alpine tundra, southern Yukon, Canada. Arct Antarct Alp Res 43:442–456

Pop EW, Oberbauer SF, Starr G (2000) Predicting vegetative bud break in two arctic deciduous shrub species, *Salix pulchra* and *Betula nana*. Oecologia 124:176–184

Post E, Stenseth NC (1999) Climatic variability, plant phenology, and northern ungulates. Ecology 80:1322–1339

Post E, Pedersen C, Wilmers CC, Forchhammer MC (2008) Warming, plant phenology and the spatial dimension of trophic mismatch for large herbivores. Proc R Soc B 275:2005–2013

Printz HC (1865) Beretning om en i Sommeren 1864 foretagen botanisk Reise i Valders (in Norwegian). Nyt Mag Naturv 14:51–96

Pudas E, Leppälä M, Tolvanen A, Poikolainen J, Venäläinen A, Kubin E (2008) Trends in phenology of *Betula pubescens* across the boreal zone in Finland. Int J Biometeorol 52:251–259

Resvoll TR (1918) Om planter som passer til kort og kold sommer, (in Norwegian). In: Helland A, Sars GO, Torup S (eds) Archiv for Mathematik og Naturvidenskab, M. Johansen Boktryukeri, Kristiania

Robinson CH, Wookey PA, Lee JA, Callaghan TV, Press MC (1998) Plant community responses to simulated environmental change at a high Arctic polar semi-desert. Ecology 79:856–866

Rosenzweig C, Casassa G, Karoly DJ, Imeson A, Liu C, Menzel A, Rawlins S, Root TL, Seguin B, Tryjanowski P (2007) Assessment of observed changes and responses in natural and managed systems. In: Parry ML, Canziani OF, Palutikof JP, van der Linden PJ, Hanson CE (eds) Climate change 2007: impacts, adaptation and vulnerability. Contribution of Working Group II to the fourth assessment report of the Intergovernmental Panel on Climate Change. Cambridge University Press, Cambridge

Saino N, Ambrosini R, Rubolini D, von Hardenberg J, Provenzale A, Hüppop K, Hüppop O, Lehikoinen A, Rainio K, Romano M, Sokolov L (2011) Climate warming, ecological mismatch at arrival and population decline in migratory birds. Proc R Soc B 278:835–842

Schnelle F (1955) Pflanzen-Phänologie. Geest & Portig, Leipzig

Schübeler FC (1885) Viridarium Norvegicum, Norges Vaextrige, Et Bidrag til Nord-Europas Natur- og Kulturhistorie, 1ste bind (in Norwegian). W. C. Fabritius, Christiania

Schwartz MD (1997) Spring index models: an approach to connecting satellite and surface phenology. In: Lieth H, Schwartz MD (eds) Phenology in seasonal climates I. Backhuys Publication, Leiden

Schwartz MD (1998) Green-wave phenology. Nature 394:839–840

Schwartz MD, Reiter BE (2000) Changes in North American spring. Int J Clim 20:929–932

Schwartz MD, Ahas R, Aasa A (2006) Onset of spring starting earlier across the northern Hemisphere. Glob Change Biol 12:343–351

Sheriff MJ, Kenagy GJ, Richter M, Lee T, Tøien Ø, Kohl F, Buck CL, Barnes BM (2011) Phenological variation in annual timing of hibernation and breeding in nearby populations of Arctic ground squirrels. Proc R Soc B 278:2369–2375

Shutova E, Wielgolaski FE, Karlsen SR, Makarova O, Berlina N, Filimonova T, Haraldsson E, Aspholm PE, Flø L, Høgda KA (2006) Growing seasons of Nordic mountain birch in northernmost Europe as indicated by long-term field studies and analyses of satellite images. Int J Biometeorol 51:155–166

Skjellvåg AO (1998) Climatic conditions for crop production in Nordic countries. Agric Food Sci Finl 7:149–160

Skre O (2001) Climate change impact on mountain birch ecosystems. In: Wielgolaski FE (ed) Nordic mountain birch ecosystems. UNESCO/Parthenon Publication Group, Paris/New York/ London

Sörensen T (1941) Temperature relations and phenology of the northeast Greenland flowering plants. Medd Grönl 125:1–307

Sparks TH, Bairlein F, Bojarinova JG, Hüppop O, Lehikoinen EA, Rainio K, Sokolov LV, Walker D (2005) Examining the total arrival distribution of migratory birds. Glob Change Biol 11:22–30

Stenström M, Gugerli F, Henry GHR (1997) Response of *Saxifraga oppositifolia* L. to simulated climate change at three contrasting latitudes. Glob Change Biol 3(suppl 1):44–54

Strand E (1965) Forelesning i plantekultur (in Norwegian.) Norges landbrukshögskole, Aas

Taylor SG (2008) Climate warming causes phenological shift in Pink Salmon, *Oncorhynchus gorbuscha*, behavior at Auke Creek, Alaska. Glob Change Biol 14:229–235

Thórhallsdóttir TE (1998) Flowering phenology in the central highland of Iceland and implications for climatic warming in the Arctic. Oecologia 114:43–49

Tømmervik H, Wielgolaski FE, Neuvonen S, Solberg B, Høgda KA (2005) Biomass and production on a landscape level in the mountain birch forests. In: Wielgolaski FE (ed) Plant ecology, herbivory, and human impact in Nordic mountain birch forests. Springer, Berlin/Hiedelberg

Tyler G (2001) Relationships between climate and flowering of eight herbs in a Swedish deciduous forest. Ann Bot 87:623–630

van der Jeugd H, Eichhorn G, Litvins KE, Stahl J, Larsson K, van der Graaf A, Drent RH (2009) Keeping up with early springs: rapid range expansion in avian herbivore incurs a mismatch between reproductive timing and food supply. Glob Change Biol 15:1057–1071

Vikhamar-Schuler D, Hanssen-Bauer I, Førland E (2010) Long-term climate trends of Finnmarksvidda, Northern-Norway. met. no. report 6/2010:1–41

Visser ME, Both C, Lambrechts MM (2004) Global climate change leads to mistimed avian reproduction. Adv Ecol Res 3:89–110

Vitasse Y, Francois C, Delpierre N, Dufrene E, Kremer A, Chuine I, Delzon S (2011) Assessing the effects of climate change on the phenology of European temperate trees. Agric For Meteorol 151:969–980

Vors LS, Boyce MS (2009) Global declines of caribou and reindeer. Glob Change Biol 15:2626–2633

Walker MD, Wahren CH, Hollister RD, Henry GHR, Ahlquist LE, Alatalo JM, Bret-Harte MS, Calef MP, Callaghan TV, Carroll AB, Epstein HE, Jónsdóttir IS, Klein JA, Magnússon B, Molau U, Oberbauer SF, Rewa SP, Robinson CH, Shaver GR, Suding KN, Thompson CC, Tolvanen A, Totland Ø, Turner PL, Tweedie CE, Webber PJ, Wookey PA (2006) Plant community responses to experimental warming across the tundra biome. PNAS 103:1342–1346

Wielgolaski FE (1974a) Phenological studies in tundra. In: Lieth H (ed) Phenology and seasonality modeling. Springer, New York

Wielgolaski FE (1974b) Phenology in agriculture. In: Lieth H (ed) Phenology and seasonality modeling. Springer, New York

Wielgolaski FE (1999) Starting dates and basic temperatures in phenological observations of plants. Int J Biometeorol 42:158–168

Wielgolaski FE (2000) Predictions in plant phenology. Paper presented at International congress: progress in phenology, Freising

Wielgolaski FE (2001) Phenological modifications in plants by various edaphic factors. Int J Biometeorol 45:196–202

Wielgolaski FE (2003) Climatic factors governing plant phenological phases along a Norwegian fjord. Int J Biometeorol 47:213–220

Wielgolaski FE (2009) Old Norwegian phenodata series in relation to recent ones. Int J Agrometeorol 14:33–38

Wielgolaski FE, Kärenlampi L (1975) Plant phenology of Fennoscandian tundra areas. In: Wielgolaski FE (ed) Fennoscandian tundra ecosystems part1: plants and microorganisms. Springer, Heidelberg

Wielgolaski FE, Nordli Ø, Karlsen SR (2011) Plant phenological variation related to temperature in Norway during the period 1928–1977. Int J Biometeorol 55:819–831

Wipf S (2010) Phenology, growth, and fecundity of eight subarctic tundra species in response to snowmelt manipulations. Plant Ecol 207:53–66

Woodley EJ, Svoboda J (1994) Effects of habitat on variations of phenology and nutrient concentration among four common plant species of the Alexandra Fiord Lowland. In: Svoboda J, Freedman B (eds) Ecology of a polar oasis, Alexandra Fiord, Ellesmere Island, Canada. Captus University Press, Toronto

Wookey PA, Parsons AN, Welker JM, Potter JA, Callaghan TV, Press MC (1993) Comparative responses of phenology and reproductive development to simulated environmental change in sub-arctic and high-arctic plants. Oikos 67:490–502

Zeng H, Jia G, Epstein H (2011) Recent changes in phenology over the northern latitudes detected from multi-satellite data. Environ Res Lett 6:1–11. doi:10.1088/1748-9326/6/4/045508

Zhang X, Friedl MA, Schaaf CB, Strahler AH (2004) Climate controls on vegetation phenological patterns in northern mid- and high latitudes inferred from MODIS data. Glob Change Biol 10:1133–1145

Chapter 14
Phenology at High Altitudes

David W. Inouye and Frans E. Wielgolaski

Abstract High-altitude climates, which strongly influence phenology, are determined by global, regional, and local processes and patterns. Phenology at high altitudes in the temperate zones is characterized by a short growing season that begins with snowmelt. As snowmelt is getting earlier, the phenology of most plants and many animal species is responding, but not uniformly. The fact that frost dates are not changing as quickly as snowmelt is resulting in large-scale frost damage in some years (at low altitudes too). The elevational gradients available in mountain regions offer the potential for observations and experimental studies that will improve our understanding of phenology as an important ecological trait. Relatively little is known about high-altitude phenology in the tropics, or about high-altitude animal phenology, and we are only beginning to understand the consequences of how phenological changes are influencing interactions among species.

14.1 Introduction

Phenology at high altitudes differs from that in most other habitats in four significant ways. First, for much (sometimes the majority) of the calendar year these habitats may be under snow or ice, and there is little photosynthetic activity. Consequently (and second), there is a very short growing season delimited by a combination of temperature and snowpack. Third, this may be one of a few habitats where almost all phenology is tied to a single highly variable event, the timing of snowmelt; few high-altitude plants appear to exhibit photoperiodic responses for phenological events

D.W. Inouye (✉)
Department of Biology and Rocky Mountain Biological Laboratory,
University of Maryland, College Park, MD 20742–4415, USA
e-mail: inouye@umd.edu

F.E. Wielgolaski
Department of Bioscience, University of Oslo, Oslo, Norway

M.D. Schwartz (ed.), *Phenology: An Integrative Environmental Science*,
DOI 10.1007/978-94-007-6925-0_14,

(but see Venn and Morgan 2007). And finally, high altitudes may differ from other habitats in the way that global climate change is affecting phenology.

What is a high altitude? The answer is not as obvious as it might seem. It probably makes more sense to use an ecosystem definition rather than an absolute altitude, as what constitutes a high altitude at high latitudes differs from a high altitude at mid- or low latitudes. For the purposes of this chapter we will consider "high altitude" to refer to alpine or montane ecosystems. Alpine is defined as the area above the natural limit of trees, and it extends over a wide latitudinal and altitudinal range. Another way of defining it, in climatic terms, is that its lower elevational limit corresponds well to the 10 °C isotherm for the warmest summer month (Wardle 1974). The alpine shares many characteristics with high-latitude or arctic ecosystems (Bliss 1971, and see Chap. 13 in this book), but from a phenological perspective one big difference is the much longer day lengths during the arctic growing season. Montane ecosystems are less clearly defined, but would include the ecosystem between grasslands on the lower end and the alpine on the upper end; the term subalpine applies to the upper end of the montane.

There are not as many long-term studies of phenology at high altitudes as there are of low altitudes. In fact, there do not appear to be any studies longer than the one described in this chapter that was initiated in 1973. Thus some of the discussion in this chapter will be colored by the fact that Inouye has worked for most of his research career at a single high-altitude field station, the Rocky Mountain Biological Laboratory (RMBL). Perhaps the information in this chapter will help to stimulate the initiation of studies at other high-altitude sites.

14.2 The High-Altitude Climate

Mountains have been described as "generating their own climate", due to the effect of their mass on circulation patterns, precipitation, and radiation. This creates abundant variation of the climate within mountain regions, but also some general patterns that help to differentiate high-altitude climate, and hence phenology, from that at lower altitudes. Kittel et al. (2002) go into detail about elevation dependence of climate, which can be attributed to factors at high altitudes such as closer contact to the free troposphere, decoupling from convective mixing of the lower troposphere, and snow-albedo feedback. Within the Rocky Mountains, there is a classic orographic precipitation pattern that creates increased precipitation on the west (windward) side of the mountains and a rain shadow on the east, although some winter air masses create up-slope conditions on the east side that generate precipitation there. Rain shadows are common in all parts of the world with high mountains.

Total precipitation increases significantly with elevation, with the highest precipitation occurring near the peak elevations, for instance in the Rocky Mountains, in the Alps, in Great Britain and to some degree also in the Scandinavian mountains (Bleasdale and Chan 1972; Førland 1997; Theurillat and Guisan 2001; Kittel et al. 2002). Maximum and minimum temperatures decrease strongly with increasing altitude throughout mountain areas, although there are some interesting variations. For example, variation of maximum temperatures decreases with elevation in the winter while that of minimum temperatures increases during the summer, and at lower elevations in the central Rockies, temperatures can be colder than those at similar elevations in the north (due to the influence of continental vs. maritime air masses).

Another aspect of high-altitude phenology that is unique to areas with great topographic relief is the potential for phenological inversions. Cold air flow in valley bottoms of high altitude regions around the world often causes delayed phenology of plants (inversions) compared to mountain slopes, and also leads to frost injuries. Lynov (1984) reported statistically significant effects of such phenological inversions for eight species of trees and shrubs, with delays of 2–5 days in times of bud opening and flowering. Areas where this cold air collects are also described as frost hollows or frost pockets.

An analysis of historical data shows that there are significant century-scale positive trends for annual and seasonal precipitation and mean minimum temperature in the U.S. Rocky Mountains (Kittel et al. 2002). Some of these are quite striking; in the northern and central Rockies summer precipitation has increased by 30 and 33 % over the past century. There has also been a trend for increasing annual mean minimum temperature (0.7–0.9 °C for the northern and central mountains). It is probable that these changes, and those forecast for the future, will have consequences for phenology at high altitudes. Giorgi et al. (1997) predicted similar changes for temperature over an altitudinal range in the Swiss Alps, and suggested "... that high elevation temperature change could be used as an early detection tool for global warming."

Both large- and small-scale events can affect high-altitude phenology. For example, the El Niño Southern Oscillation (ENSO) and the North Pacific Oscillation (or Pacific Decadal Oscillation) can affect winter precipitation in the Rocky Mountains. At the other extreme, microclimate can have very large effects. Areas where snow is deposited by wind (on the lee side of ridges, trees, etc.) can melt out much later than nearby sites, and deposition by snow slides and avalanches may create such deep snow depths that certain areas may not melt out at all in a given summer. Wagner and Reichegger (1997) found that a north-facing study site subject to deep snow deposition took about a month longer to melt out than sunlit sites. The cold water from melting snowbanks can have an effect on the phenology of plants it reaches. Holway and Ward (1963) found in an experimental study that meltwater resulted in delays of flowering in 12 of 14 species growing in irrigated plots (typically of about a week, but up to a month).

Given the importance of snowmelt date in determining the beginning of the growing season at high altitudes (Kudo and Hirao 2006), changes in this event are likely to have significant phenological consequences. Snowmelt timing can also influence competitive interactions among snowbed plants, so careful study is required to understand the dynamics of these communities (Hülber et al. 2011).

In recent years two factors have been demonstrated to have a big impact on snowmelt. First, increases of dust-on-snow events have been documented in parts of the Rocky Mountains (Painter et al. 2007; Steltzer et al. 2009). This dust changes significantly the albedo of snow, and can result in snow melting weeks earlier. Second, the ratio of annual precipitation falling as rain vs. snow is changing, with many areas now receiving more rain and less snow (Comis 2011; Knowles et al. 2006). A few studies have already documented the negative effects consequent earlier snowmelt can have on flowering (Inouye 2008; Wipf et al. 2009), and a modeling effort suggests that the observed ongoing trend toward earlier snowmelt is likely to continue (Stewart et al. 2004).

In the southern Colorado Rocky Mountains the ecosystem at 2,900 m is sub-alpine, or montane, as trees are still common at this altitude. The date of first permanent winter snowpack at the RMBL averages about 4 November (range 15 October–24 November, data from 1974 to 2011), and the length of snowcover is 199 days (range 159–233, data from 1975 to 2012). The mean date of first bare ground at a permanent snow measurement station is 20 May, with a range of 22 April–19 June (data from 1975 to 2012). Summer precipitation also appears to play a role in some aspects of phenology at the RMBL. Precipitation for June – August at the NOAA weather station in Crested Butte (2,704 m, 9.5 km from RMBL) has averaged 13.2 cm (range 5.3–22.3 cm, data from 1973 to 2011).

14.3 Literature Review

Perhaps more than any other bioclimatic zone except high latitudes, phenological events at high altitudes are constrained by a short growing season, delimited by cold temperatures and snowpack. Time of snowmelt appears to have an almost universal effect on high-altitude phenology, and variation in phenology can usually be linked to variation in accumulation and then melting of snow, whether this is across time or space. This interaction has been reported by many studies, including Bliss (1956), Holway and Ward (1963, 1965), and Mark (1970). Canaday and Fonda (1974) found that the timing and duration of phenophases of a variety of subalpine plants in the Olympic Mountains (WA) were a function of snowmelt. In general terms Ratcliffe and Turkington (1989) found the same results, although they argue that the identity of dominant species and, to a lesser degree, aspect, are responsible for variations they observed in phenology. Some species tended to flower earlier on south-facing slopes, indicating the potential importance of aspect and microenvironment.

Despite the great influence of snow in determining high-altitude phenology, temperature (Molau et al. 2005; Huelber et al. 2006; Hülber et al. 2010), and to a lesser degree photoperiod, have also been found to have an effect in some studies. Migliavacca et al. (2008) found that for larch trees temperature was the most important factor for spring phenology while photoperiod was more important for autumn senescence. Laboratory studies have also investigated the interaction photoperiod and temperature in alpine species of plants (Keller and Körner 2003). They found that about half of the alpine species tested were sensitive to photoperiod in spring and, therefore, probably not able to utilize fully periods of earlier snowmelt.

Field experiments using the protocol of the International Tundra Experiment (ITEX) have been used to document the importance of temperature for high-altitude phenology (Jarrad et al. 2008). Laboratory studies have also investigated the interaction between photoperiod and temperature in alpine species of plants (Keller and Körner 2003). Not many crops are grown at high altitude, but a study of winter wheat grown at 2,351 m in China found that there were significant increases in wheat yield at that altitude over the period 1981–2005 that they attributed to changes in temperature and precipitation; the 4 % increase in yield exceeded the 3.1 % increase at 1,798 m (Xiao et al. 2008). Most studies of phenology at high altitudes have been short (e.g., Wielgolaski and Kärenlampi 1975); such studies can probably define relatively well the spatial pattern of snowmelt and hence phenology in a particular site, but longer studies are required to gain insights into the effects of climate variables on phenology. In one relatively long study, Walker et al. (1995) followed the phenology of two forbs for 6 years in five different plant communities and found significant differences among years and plant communities. More recently, results were compiled from a 47-year dataset for the Swiss Alps, which found that all but 1 of 19 phenophases were trending toward earlier dates, and that the proportion of significant trends was higher in alpine than lowland sites, even though the shift to earlier occurrences was stronger in the lowlands (Defila and Clot 2005).

Phenophase condensation has been observed in several studies of alpine plants, with full development being accomplished more rapidly where snow persists longer. Examples of this have been reported by Knight et al. (1977, and references therein) and Billings and Bliss (1959). Snowmelt gradients are a common phenomenon at high altitude, as some areas will receive more or less snow and receive more or less insolation and result in earlier or later snowmelt. The consequences of these gradients have been investigated in several studies. Kudo (1992) investigated five herbaceous species along a snowmelt gradient on Mt. Kaun in the Taisetsu Mountains of Hokkaido, where the snowfree period ranged from 55 to 95 days. The later snowmelt occurred, the later flowering and fruiting began, and in the plot that melted out last, no species was able to mature all fruits because of the short growing season. In a study of 56 species over 3 years, Kudo also found that a shorter snow-free period reduced flowering and seeding (Kudo 1991).

Plants found in alpine tundra are often remarkable for the speed with which they can flower and fruit, but this adaptation is required for success given the short growing season (Bliss 1971; Wielgolaski and Kärenlampi 1975). This early flowering is facilitated by the fact that floral initiation often occurs one or more years in advance of flowering; preformation of flower buds is characteristic of many high-altitude plants (e.g. Resvoll 1917; Forbis and Diggle 2001; Meloche and Diggle 2001, and see references in Bliss 1971).

The climate constraints of high altitudes could result in significant selection against late flowering, to allow sufficient time for the development of seeds before the first killing frost in the fall, and this may be why annual and even biennial plants are relatively uncommon at high altitudes (Bliss 1971; Jackson and Bliss 1982). Annuals may be restricted to sites that melt out early or that don't dry out early in the summer (Reynolds 1984). Some studies do show that most species in alpine communities initiate flowering rapidly after snowmelt, with relatively few species flowering late (e.g., Holway and Ward 1965; Billings and Mooney 1968; Totland 1993). *Ranunculus glacialis* reacts contrary to most other alpine plant species to experimental warming; for instance the time of snow melt does not influence phenological variables (Totland and Alatalo 2002).

In a few cases, plants may be able to initiate growth under the snow and get a head start on the growing season. Billings and Bliss (1959) found *Geum turbinatum*, *Carex elynoides*, and *Deschampsia caespitosa* growing under 1–5 cm of snow near the edge of a melting snowbank, and observed that the latter two of these had started growth under 50 cm of snow that did not melt for another 4 days. Young red leaves of *Polygonum bistortoides* were found under 110 cm of old snow (Mooney et al. 1981). Williams and Cronin (1968) found that *Delphinium* species could emerge and develop green cotyledons when snow melted to a depth of 30 cm or less, and Spomer (cited in Richardson and Salisbury 1977) found green plants of *Ranunculus adoneus* under 1 m of snow. Arroyo et al. (1981) recorded the earliest-flowering alpine species in the Chilean Andes as actually blooming precociously under 5–6 cm of snow, and Bliss (1971) reported flowering by species of *Caltha* and *Ranunculus* under 10 cm or more of snow.

Theurillat and Schlüssel (2000) studied seven subalpine-alpine species in the Alps and characterized them by the number of degree-days to bud burst and the end of flowering. Each species differed in its requirements, but only *Vaccinium myrtillus* was closely tied to snowmelt, while the others fit heat sum models that depended on degree-days after a chilling requirement, intensity of chilling from a threshold, or constant degree-days following thaw (Heide 1993; Myking and Heide 1995). It would be interesting to compare models of heat sums and time since snowmelt to see which works best for predicting phenology of a variety of species.

In addition to the physiological constraints of flowering at high altitudes, pollen limitation has been shown to be an important factor affecting early-flowering plants. Kudo and Suzuki (2002) found a positive correlation between flowering time and fruit set for ten species of ericaceous shrubs flowering in alpine Japan that was explained by severe pollen limitation in the early-flowering species. Similarly, Thomson (2010) found evidence of pollen limitation in the earliest flowering individuals of *Erythronium grandiflorum*, and Kameyama and Kudo (2009) found that outcrossing rate increased in later-flowering alpine plants. These studies indicate the important effect of mutualistic interactions in shaping the flowering phenology of alpine plants. Kudo and Suzuki (2002) compared very similar plant communities from alpine Japan and subarctic Sweden and found important differences, so it may not be possible to make conclusions about one of those habitats based on results from the other.

Less work has been done on animal phenology compared to plant phenology, but there are a few studies from high altitudes. Inouye et al. (2000) reported that American robins (*Turdus migratorius*) were arriving at their mountain breeding grounds weeks earlier and yellow bellied marmots were emerging from hibernation weeks earlier than they had 25 years earlier. Those trends have continued since that report (Inouye unpublished), and the earlier emergence from hibernation has resulted in larger body mass before the next hibernation and consequent decline in adult mortality (Ozgul et al. 2010). Pereyra (2011) found that the breeding schedule of Dusky Flycatchers (*Empidonax oberholseri*) was tied to snowmelt, and advancing over the 15-year study, and spring snow conditions affected breeding phenology of an alpine population of American pipits (Hendricks 2003).

Mutualistic interactions such as pollination are important at high elevations. If the multiple partners in such interactions respond at different rates to the same changing environmental cues, or if they respond to different environmental cues, a consequence of climate change could be significant changes in the historical synchrony between partners. A study of the phenology of solitary bees and the flowers that they visit suggests that this may be occurring already. Forrest and Thomson (2011) recorded emergence times of solitary bees from trap nests as well as flowering phenology at the same sites, and found that although their phenologies are both influenced by temperature, plants are more likely than insects to advance phenology in response to springtime warming.

14.3.1 Temporal, Spatial, and Altitudinal Gradients

Plants growing in microsites with shorter growing seasons may be able to compensate somewhat by shortening the time before flowering starts, allowing them to complete seed production before the weather becomes unfavorable. Jackson and Bliss (1984) studied *Polygonum minimum*, a very small (<2 cm tall) subalpine annual plant at 3,000 m in the Sierra Nevada. Its indeterminate growth pattern, and the fact that it can dehisce seeds in as little as 2 months after snowmelt, helps it to survive the phenological constraints of its habitat. The indeterminate growth pattern permits plants to take advantage of longer growing seasons for increased seed production. Another example is provided by *Carex* species in the Austrian Alps (Wagner and Reichegger 1997).

However, if the microsite differences in snowmelt lead to significant differences in flowering time, the end result may be fitness differences among individuals flowering at different times, and eventually, divergence. This kind of differentiation has been seen in mountain environments in Japan (Kudo 1991), Colorado (Galen and Stanton 1991) and North Swedish Lapland (Stenström and Molau 1992). Galen and Stanton (1991) found fitness differences related to emergence phenology; plants in the latest-melting parts of snowbeds produced smaller seeds than earlier-flowering plants, and seed size was correlated with seedling survival rates.

Despite the potential advantages of early flowering, some species do seem to be able to withstand the constraints of late flowering. For example, *Gentianella caucasea* is an annual species found growing up to the subniveal zone in the Central Caucasus that can complete its life cycle within 4–5 weeks (Akhalkatsi and Wagner 1996). Different microclimates appear to have resulted in different phenotypes with significant differences in developmental times. Populations at the lower end of the altitudinal range take at least 24–25 days for floral bud formation, flowering, and fruit ripening, while those at high elevations flower about 3 weeks later than the earliest lower population and take only 18 days to complete reproduction. *Gentianella germanica*, a biennial species found at high altitudes in Europe and the Balkans, is unusual for its late flowering (August to the beginning of November when winter begins) (Wagner and Mitterhofer 1998). Phenological differences in flowering between two morphs of this species may be leading to their isolation. The Norwegian alpine species *Leontodon autumnalis* var. *taraxaci* also flowers late (Totland 1997).

Elevational gradients provide an interesting opportunity for phenological studies at high altitudes. Schuster et al. (1989) studied gene flow in limber pine (*Pinus flexilis*) over its altitudinal range (1,650–3,350 m) in Colorado, and found that pollination phenology is strongly affected by altitude. Sites that differed in elevation by more than 400 m did not usually have overlapping pollination periods, so gene flow by pollination was quite restricted. Mitton et al. (1980) found similar results for ponderosa pine trees (*Pinus ponderosa*). Hoffmann and Walker (1980) looked at phenology of two drought-deciduous shrubs along an altitudinal gradient on the Coastal and Andean Ranges of Chile. Vegetative growth at the upper limit (2,000 m) started 9 weeks later than at the lowest altitude (700 m), and lasted longer at the lower altitude, but flowering and fruiting occurred at about the same time at both ends of the gradient (perhaps indicating a dependence on day-length as a cue). Moser et al. (2010) found that the growing season for European larch changes by 3–4 days per 100 m, with the growing season increasing by about 7 days per degree Celsius.

Blionis and Vokou (2002) studied two *Campanula spatulata* subspecies on Mt. Olympos, over a 400–2,500 m gradient. The upland subspecies flowered later, had a longer duration of flowering, and longer flower life span. At the genus level, considering nine *Campanula* species, time of flowering increased with elevation by 2–3 days for every 100 m, while flower longevity also increased with elevation, by 0.2 day/100 m (Blionis et al. 2001). Common garden experiments over elevational gradients can also lead to good insights into the partitioning of phenological responses between plastic and genetic components (Haggerty and Galloway 2011). The observation by Crimmins et al. (2009) of plants shifting their range to higher elevations along an altitudinal gradient is probably representative of what is happening in other mountain areas, so that future studies of flowering at high altitudes are likely to involve greater species diversity than historical work. In a recent study in the Chech treeline ecotone, Treml et al. (2012) found a close relationship between tree-ring widths and growing season temperatures throughout the twentieth century. Widest tree-rings were observed after the end of 1990s, and there was also a climate-induced upward shift of vegetation and enhanced tree vigor in the same period.

Few studies have looked at animal species' phenology over altitudinal ranges, but Nufio et al. (2010) studied responses of grasshopper communities along an elevational gradient to climate change, using a combination of historical and modern data. Grasshoppers at the two highest sites showed significant phenological advancements, apparently in response to a warmer climate. As animal species move up in altitude or latitude in response to the changing climate, it seems likely that these moves will be accompanied by phenological adjustments in the future.

The influence of climate on the growing season and hence on phenology, can also be seen by transplant experiments. Plants transplanted to lower altitudes will typically develop much sooner than those left in their native high-altitude sites (e.g., Wagner and Reichegger 1997). However, differences in the date of bud burst of mountain birch originating from different elevations but approximately the same latitude indicate that adaption to certain temperature regimes is also of importance for the start of growing season (Ovaska et al. 2005; Wielgolaski and Karlsen 2007).

Environmental gradients may produce some of the same effects as elevational gradients; for example, Stanton et al. (1997) suggested that differences in flowering time of *Ranunculus adoneus* along a snowmelt gradient could reduce opportunities for pollen transfer, and presented evidence for some genetic differentiation among early, middle, and late-melting cohorts. The longer growing season in early-melting sites enhanced vegetative growth at all life-history states and increased fecundity of seedlings (Stanton and Galen 1997).

14.3.2 Community-Level Patterns and Temporal Patterns

Relatively few community-level surveys of plant phenology have been made of alpine plants. Douglas and Bliss (1977) conducted weekly samples in 2 × 2 m plots in the North Cascade Range of Washington and recorded times of vegetative growth, flowering, fruiting, seed dispersal, and dormancy of 32 species in weekly surveys. Most species flowered within 14–24 days following initiation of growth, remained in bloom for 8–20 days, and then had a 10–24 day fruiting stage prior to seed dispersal. This is also reported in various communities of the low alpine region in southern Norway (Wielgolaski and Kärenlampi 1975). Ratcliffe and Turkington (1989), using similar methods, looked at 45 vascular plant species in southern British Columbia. Most species flowered between 15 and 40 days after snowmelt. Bauer (1983) reported data on the seasonal flowering phenology of 24 species of tundra plants visited by bumble bees, from weekly surveys.

Molau (1993) conducted a study of the relationships between flowering phenology and life history strategies of plants in high-altitude Sweden. Some of the immense variation among tundra species in reproductive traits (e.g., seed:ovule and fruit:flower ratios) in this habitat is correlated with flowering phenology (which in turn is linked to snowmelt patterns). Early-flowering species show high outbreeding rates and low seed:ovule ratios, while late-flowering species showed the reverse. Apomixis and vivipary were restricted to late-flowering species, and ploidy levels increased from early- to late-flowering times. The fact that Iversen et al. (2009) found a strong link between growth form of 11 high-latitude alpine species and both flowering and vegetative phenology suggests that there may be more to learn about links between phenology and life history.

Relatively little is known about phenology at high altitudes in the southern hemisphere, or in tropical areas. Inouye and Pyke (1988) did report phenological data for alpine Australia from a 1-year study. Ralph (1978) found that plants of

Azorella compacta (Apiaceae) flowered year-round in populations at 3,960 and 4,500 m. Phenological patterns of the high Andean Cordillera of central Chile were studied by Arroyo et al. (1981), who looked at 97 alpine species. They characterized these species as falling into nine different patterns of phenological behavior. The most common category was perennials that overwintered in a dormant state. At all sites they examined, there was a single prominent maximum in flowering activity, which was earlier on north-facing than south-facing slopes. An interesting observation about flowering longevity was that duration of flowering tended to be longer at higher altitudes (individual flowers lasted longer, and the total flowering period was also longer). Melampy (1987) studied flowering phenology of *Befaria resinosa*, an ericaceous shrub of the eastern Andes of Colombia. This species seems to have two peaks of flowering each year, although one is more significant, and there was some suggestion from the data that the two peaks of rainfall might have been responsible for initiation of flowering. The weather at this 2,200 m site is quite dry for much of the year, and 22 and 31 % of the annual precipitation occurs during the two peaks. Williams-Linera (2003) found differences in the phenology of shrub species in tropical montane cloud forest of Veracruz, Mexico, depending on whether they occurred in the understory or at the forest edge. Kudo and Suzuki (2004) found that different species of dwarf alpine trees in Borneo initiated flowering in response to different environmental cues.

There are often sympatric congeneric wildflower species in high-altitude habitats, and it's not surprising that in such cases they might flower at different times as a way to avoid competition for pollinators. Adams (1983) studied five sympatric species of *Pedicularis* (Scrophulariaceae) on Mount Rainier in the Washington Cascade Mountains, all of which are pollinated by bumble bees. Two bloom early in the growing season, one in mid-season, and two in late-season. At RMBL, pairs of *Mertensia* and *Delphinium* species include species that flower early and at mid-season (Miller-Rushing and Inouye 2009).

Gómez (1993) investigated the effects of flowering synchrony on reproductive success of 80 *Hormathophylla spinosa* (Cruciferae) plants in the Sierra Nevada mountains of Spain. He measured flowering synchrony as the number of days that the flowering of an individual overlaps with the flowering of every other plant in the sample, and found that synchrony ranged from 0.25 to 1.00, with more than 70 % of plants having synchrony levels greater than 0.75. Plants with a lower degree of flowering synchrony were visited by more pollinators and also eaten by fewer herbivores, but the phenological traits did not affect female fertility so Gómez concluded that flowering synchrony in this species is not regulated by selective pressures from pollinators or herbivores.

A recent paper by Aldridge et al. (2011) reported a community-level study of flowering phenology and abundance from the Colorado Rocky Mountains. They found a developing trend toward a bimodal distribution of flower abundance during the summer flowering season, with a mid-season reduction in total flower number, instead of a broad uni-modal flowering peak. If this pattern continues to develop, it bodes poorly for pollinators that require a season-long supply of flowers. Another study at the same site found a significant relationship between snowmelt timing and composition of the assemblage of co-flowering species, suggesting that responses to climate change may lead to altered competitive environments for those species (Forrest et al. 2010). That result couldn't have been predicted by examining temporal trends in the dates of peak flowering by dominant species in the community. Although it may be more common than the literature indicates, mast flowering does occur in at least some alpine or montane ecosystems. *Frasera speciosa* (Gentianaceae) has a flowering pattern that has been described as 'sporadic seasonal synchrony', in which significant flowering events of this long-lived monocarp occurs only once every several years (Beattie et al. 1973; Taylor and Inouye 1985). A similar flowering pattern, but with even longer intervals between significant flowering years, is seen in the lily *Veratrum tenuipetalum* (Iler and Inouye 2013). Mast flowering has also been reported from a few alpine species in New Zealand (Kelly et al. 2000; Rees et al. 2002).

14.4 Flowering Phenology in the Colorado Rocky Mountains

Inouye has conducted since 1973 a study of the timing and abundance of flowering by Rocky Mountain wildflowers, in permanent 2 × 2 m plots near the RMBL. This site is located in the West Elk Mountains of Colorado, at an altitude of about 2,900 m (38º57.5′N, 106º59.3′W; map in Inouye 2008). Every other day for most or all of the growing season all flowers in the plots are counted.

Phenology of flowering and many other events at this site is dependent on the amount of snow that fell during the previous winter and when that snow melts, as the growing season does not begin until the snow is gone. From 1975 to 2012 the annual snowfall (measured as daily snowfall or even more frequently during storms; unpublished data courtesy of Billy Barr) has ranged from 474 to 1,641 cm (mean = 1,100). The date on which the permanent snowpack has begun has ranged from 15 October to 24 November (mean = 4 November), and the snow has disappeared from the measurement site as early as 22 April and as late as 19 June (mean = 21 May). Of course there is much variation in these dates across the landscape, with differences in microclimate, slope aspect, and snow deposition through wind activity resulting in some sites that hold or lose snow earlier or later.

One question that arises is whether the range of dates described here is typical of a longer-term period. Although we do not have snowfall data from RMBL earlier than 1975, a proxy for snowfall, the peak runoff in the East River (which runs through the Lab) is measured at Almont, Colorado, and has been recorded continuously since 1935. The correlation between measurements of winter snowfall at RMBL and the peak runoff measured at Almont is strong ($r^2 = .757$, $p = .0001$). If peak runoff is then used to estimate historical snowfall at RMBL, we find that the period 1975–2010 contains both the maximum and minimum snowfall records for the 76 years of data. This could be an indication of the increased variation in precipitation that has been predicted by some models of global climate change.

For all species that we have examined so far, from the earliest flowering (*Claytonia lanceolata*, *Erythronium grandiflorum* (Lambert et al. 2010) and *Androsace septentrionalis* (Inouye et al. 2003)) to the latest (*Artemisia tridentata*), the timing of flowering is strongly linked to the amount of snow that falls during the previous winter, and hence the timing of snowmelt. For some species the relationship is linear (e.g., *Delphinium nuttallianum*; Fig. 14.1a), while for others it is curvilinear (e.g., *Mertensia ciliata*; Fig. 14.1b) (Miller-Rushing and Inouye 2009). The curvilinear relationship seen for some species may indicate an interaction between temperature and number of days since snowmelt; in years with early snowmelt, it takes more days (i.e., more growing degree days) for some heat sum to be achieved before flowering begins. Such curvilinear relationships may characterize later-flowering species (e.g., Inouye et al. 2002; Miller-Rushing and Inouye 2009), as we have not observed them yet in early-flowering species. This strong reliance on snowmelt as an environmental factor determining flowering time seems to characterize all of the herbaceous species at this site, and possibly the relatively few woody ones as well. There do not seem to be species at this altitude that rely on day length for timing of flowering.

Another factor that is important in the abundance of flowers and is linked to phenology is frost (Inouye 2000). In some years frost can kill almost all of the buds of some species of wildflowers at RMBL, such as *Helianthella quinquenervis*, the aspen sunflower. In 15 of the past 37 years there has been significant frost damage (Inouye 2008). The critical factor for this interaction between snow, frost, and flower abundance appears to be phenology. If there is sufficient snow during the winter, snowmelt, and hence the beginning of the growing season, will be delayed compared to those years with light snowpack. Hard frosts (e.g., temperatures down to −7 °C) can occur as late as the third week in June. If plants begin growth too early in the season, they can have buds at a sensitive stage of development by the third week of June, and are then susceptible to frost damage. If there is sufficient snow to delay snowmelt, and hence the growing season, then development of frost-sensitive buds will be delayed beyond the time when frost is likely to occur. There is some evidence that frost risk at lower altitudes (Shimono 2011) and higher latitudes (Bokhorst et al. 2011) may be increasing as well; an interaction between snowmelt date and frost damage similar to what Inouye has found in Colorado has also been reported from a mountain site in Japan (Shimono and Washitani 2007) and in the Swiss Alps (Wipf et al. 2009), as well as lower altitudes (Eccel et al. 2009).

Given that there is so much evidence supporting the dependence of high-altitude phenology on snowpack, it may be possible to use historical data on snowpacks in other areas to draw conclusions about historical patterns of phenology. For example, spring pulse dates of streamflow exhibit trends toward earlier spring timing, matching patterns of flowering by lilacs and honeysuckles in phenology studies (Cayan et al. 2001).

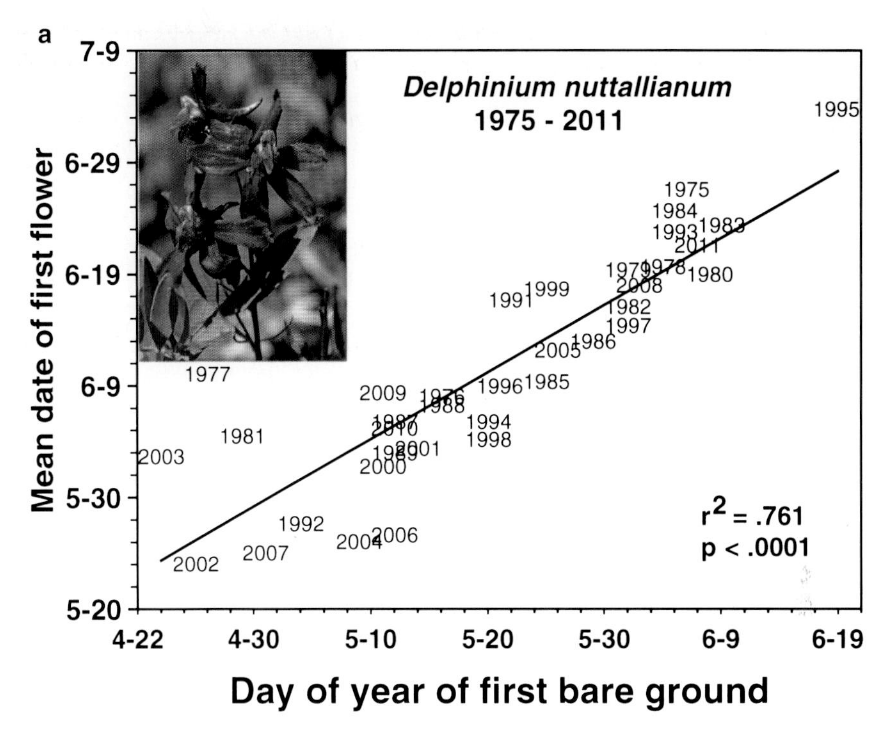

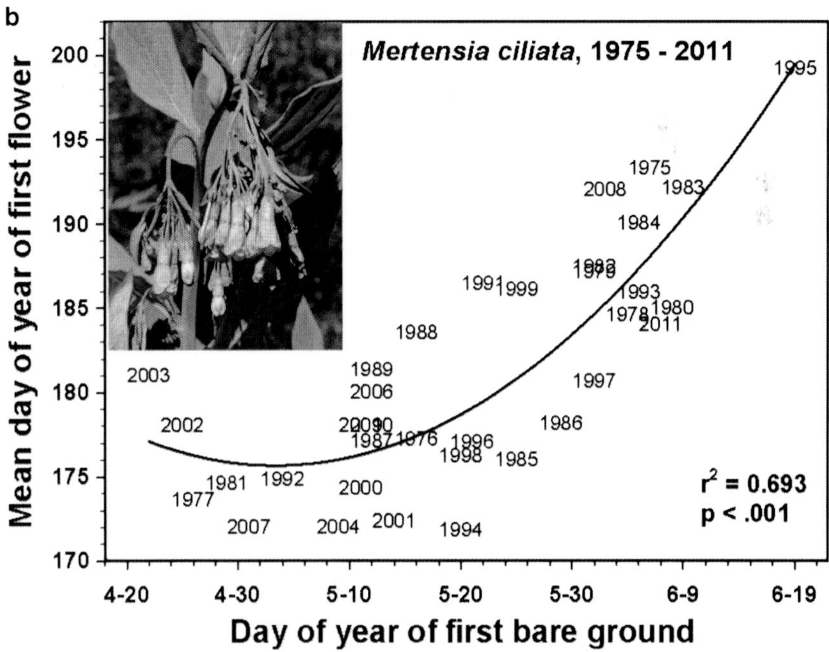

Fig. 14.1 The relationship between the first date of bare ground and the date of first flower for (**a**) *Delphinium nuttallianum* and (**b**) *Mertensia ciliata*

14.5 Phenology and Animal Ecology

Although most of this review has focused on plant phenology, there are indications from the few available studies that animal phenology at high altitudes is also influenced by the same kinds of environmental cues that affect plants. Morton (1994) assessed timing of reproduction for both wild onion (*Allium validum*) and White-crowned Sparrows (*Zonotrichia leucophyrus oriantha*) at the same subalpine meadow in the California Sierra Nevada. Data from 21 years indicated that both flowering date and clutch initiation date were highly correlated with snow conditions (occurring later as snowpack increased). McKinney et al. (2012) reported a growing phenological mismatch since 1975 between the arrival of migrating Broad-tailed Hummingbirds (*Selasphorus platycercus*) and flowering of glacier lilies (*Erythronium grandiflorum*), the earliest-flowering plant they visit at their breeding grounds. If it continues, this trend could result in changes in the migratory phenology of the hummingbirds.

Langvatn et al. (1996) found that plant phenology had important consequences for diet quality, and hence reproduction, for red deer. They found a negative correlation between various fitness measures for female red deer and growing degree-days during the summer. Their interpretation was that when herbage growth is retarded (via cooler weather) the digestibility of plants declines more slowly than in warm summers, resulting in a longer period with good grazing. Although this study was conducted at high latitude rather than high altitude, the same principle may apply at higher altitudes too. Merrill and Boyce (1991) make a similar story for high-elevation summer range for elk in the Yellowstone ecosystem, describing a link between heavy winter snowfall, delayed phenology the next summer, and consequent high-quality forage through late summer. Hebblewhite et al. (2008) found that maximum forage biomass, which is in turn influenced by phenology, was an important factor in foraging and movements of both resident and migratory elk populations at different spatial scales.

14.6 Climate Change and Phenology in the Colorado Rocky Mountains

Inouye et al. (2000) reported that there was no sign of an effect of climate change on the phenology of wildflowers in the Rocky Mountains. This finding contrasts strongly with results from lower altitudes, where many studies have reported that phenological events are happening earlier than they used to (e.g. Bradley et al. 1999; Brown et al. 1999; Roy and Sparks 2000; Peñuelas and Filella 2001; Sagarin and Micheli 2001; Fitter and Fitter 2002; Peñuelas et al. 2002; Walther et al. 2002). The paper by Inouye et al. pointed out that this difference between low and high altitudes in phenological responses to climate change could cause problems for altitudinal migrants, and reported data suggesting that robins (*Turdus migratorius*)

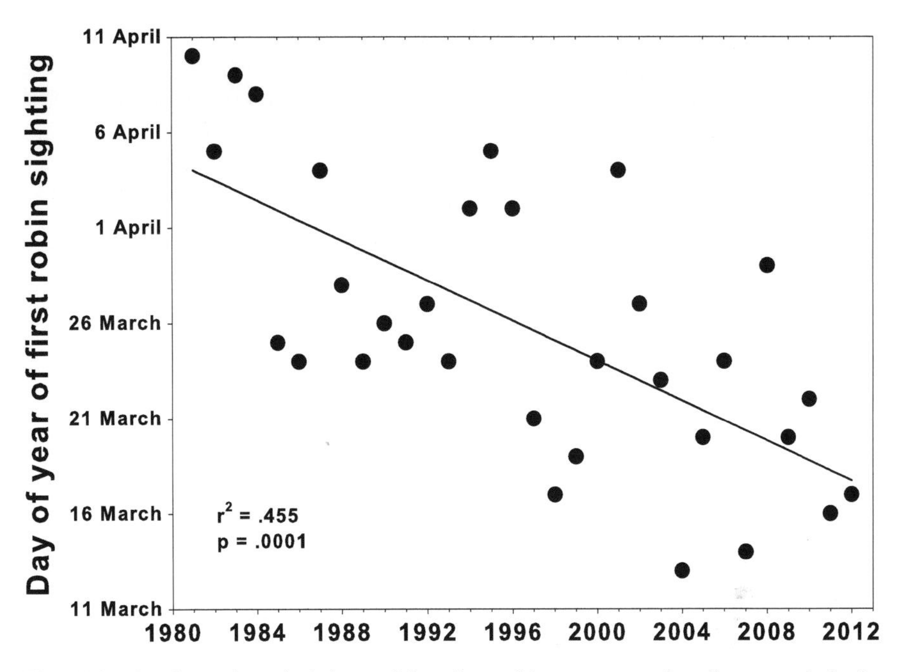

Fig. 14.2 The change in arrival dates of American robins to summer breeding grounds in the Colorado Rocky Mountains (Unpublished data from Billy Barr)

were showing effects of this change in synchrony, and the trend towards earlier arrival has continued to date (Fig. 14.2).

The reason that flowering phenology was not changing at high altitudes is probably related to the trend observed for increased winter precipitation (see Fig. 1 in Inouye et al. 2000, and Fig. 13.4 in this book). Thus, even though air temperatures appeared to be warming, the net result was that there was no trend for earlier snowmelt dates (and hence phenology). Some models of global climate change predict increased precipitation. In contrast to flowering phenology, there is evidence that some animals are responding to warming temperatures with changes in phenology. Inouye et al. (2000) reported that yellow-bellied marmots (*Marmota flaviventris*) were emerging from hibernation significantly earlier than they did a few decades earlier, and therefore emerging when there was much more snow left on the ground than previously. That trend has continued to the present, with subsequent consequences for animal species such as marmots (Ozgul et al. 2010). Emergence dates of ground squirrels and chipmunks from winter hibernation are also changing, but not at the same rate as marmots (Billy Barr unpublished). Apparently they are responding to different cues than the marmots, or they are responding differently to the same cues as marmots. An effect of the warming temperatures has also been reported for a Colorado population of white-tailed ptarmigan *Lagopus leucurus*, in which hatch dates advanced significantly from 1975 to 1999 (Wang et al. 2002).

If the emergence phenology of pollinators and herbivores is changing relative to their host plants and food plants, it seems likely that coevolved relationships may change significantly in the future. This kind of change in synchrony has also been reported from lower altitudes (Visser and Holleman 2001).

High-altitude aquatic habitats may also be showing signs of changing phenology. Caine (2002) reported that spring ice thickness on an alpine lake is declining at a rate of about 2 cm/year, and the duration of ice cover also appears to be declining. Presumably a variety of aspects of plant and animal phenology in such alpine lakes is being influenced by the change in ice cover (which may be a consequence of increasing precipitation).

The changes that are predicted for distribution of alpine plants in the future as a consequence of climate change (e.g., Guisan and Theurillat 2000; Theurillat and Guisan 2001) will be in part a function of changes in phenology. The ultimate result is likely to be an overall trend for reduced availability of suitable habitat, and shifts in the distribution of species richness (either shifting upward or spreading out of patches) (Guisan and Theurillat 2000). There is some evidence that plants in the Swiss Alps may already be showing signs of such a shift upward in distribution (Hofer 1992; Grabherr et al. 1995, and Chap. 13 in this book).

14.7 Conclusions

The disappearance of snow cover appears to be the primary factor influencing phenology at high altitudes in the temperate zone. Not enough is known yet about other high-altitude areas without significant snow cover to confirm what is controlling their phenologies. One consequence of the importance of snow in controlling phenology is that flowering, and other phenological events involving both plants and animals, can be highly variable because of variation across years in snowpack depth and across space because of aspect and microsite differences in snow accumulation and melting. A consequence of this variation may be that no single set of phenological and physiological characteristics is optimally adapted to all of this variability, which would then encourage the evolution and maintenance of a diversity of adaptive strategies in high altitude communities.

Although the timing of events exhibits significant variation among years, their relative timing is much less variable. The sequence of flowering species is typically consistent across years, with the same species flowering each year in early, middle or late season. Variation across space is also consistent, as sites that tend to accumulate snow in one year are likely to do so in other years.

Lest flowering phenology at high altitudes be considered an esoteric topic, consider the economic importance of strawberry production. Strawberries used to be highly seasonal, but it is now possible to produce multiple crops across most temperate climates. This change is the result of developing new multiple-cropping cultivars with cyclic flowering. The genes for this change for all modern day-neutral cultivars come from a single clone of *Fragaria virginiana* ssp. *glauca* from the Wasatch Mountains of Utah (Sakin et al. 1997).

14.8 Avenues for Future Investigation

Experimental studies of phenology are rare, even though snowpack is relatively easily manipulated (e.g., Galen and Stanton 1993; Dunne et al. 2003; Bokhorst et al. 2008), but they offer the potential to elucidate the significance of variation in phenology more quickly than observational studies that rely on natural variation. The International Tundra Experiment has investigated responses of tundra plants to simulated climate warming, but most of its sites are at high latitudes (Henry and Molau 1997; Pieper et al. 2011). Suzuki and Kudo (2000) did carry out an experiment in Japanese alpine tundra using open top chambers and five species of shrubs. Although there was some evidence of earlier leafing and flowering, this was not a consistent response. More recently ITEX chambers have been used in sub-alpine Australia (Hoffmann et al. 2010; Jarrad et al. 2008), and in alpine China, where shrubs had strong phenological responses (Xu et al. 2009). Additional experimental studies such as those using electric heaters (e.g., Dunne et al. 2003) are likely to provide interesting and novel results, although care must be taken in considering how well they mimic actual warming (Wolkovich et al. 2012).

It would be valuable to have additional long-term studies of phenology at high altitudes, not only of flowering and other phenological events for plants, but also for phenology of animals. Not much is known about the phenology of hibernating and migrating species, or insects at high altitudes. There are been some interesting reports of differences in phenological characteristics within species across altitudinal gradients, for example, that flowering may last longer at higher altitudes. Additional studies across such gradients would be useful to confirm this and other aspects of phenology.

The value of herbarium records in preserving phenological data has been demonstrated in a few recent studies, including one of subalpine Australian flora. Gallagher et al. (2009) used this technique to select potential indicator species for future monitoring, and identified locations for such studies, based on changes in historical flowering phenology and how it was changing in response to temperature. Another recent study used data from an eddy flux tower to investigate growing season length and CO^2 uptake in a subalpine forest, reflecting changes in plant activity patterns (Hu et al. 2010). At the largest spatial scale, satellite data were used to document changes in spring phenology on the Tibetan Plateau (Yu et al. 2010). The results of that study were counterintuitive: although spring phenology initially advanced after 1982, it started retreating in the mid-1990s even in spite of continued warming. Warmer springs led to an advance in the growing season, but warm winter conditions caused a delay in spring phenology due to the longer time it took to fulfill chilling requirements.

Studies of the consequences of variation in phenology are uncommon. What are the consequences of flowering early or late, or at the tail vs. the peak of flowering time? What are the consequences of variation in time of emergence from hibernation or in arrival or departure of migrating species? Very little is known about what controls the phenology of high-altitude tropical ecosystems, where snow is not a significant controlling factor. Given that snow may play little or no role in some of those areas, what is the influence of other environmental variables? We suggest that future studies of high-altitude phenology will be particularly interesting in the context of climate change, and hope that this review will help to stimulate additional work on this topic.

References

Adams VD (1983) Temporal patterning of blooming phenology in *Pedicularis* on Mount Rainier. Can J Bot 61:786–6791

Akhalkatsi M, Wagner J (1996) Reproductive phenology and seed development of *Gentianella caucasea* in different habitats in the central Caucasus. Flora 191(2):161–168

Aldridge G, Inouye DW, Forrest JRK, Barr WA, Miller-Rushing AJ (2011) Emergence of a mid-season period of low floral resources in a montane meadow ecosystem associated with climate change. J Ecol 99(4):905–913

Arroyo MTK, Armesto JJ, Villagran C (1981) Plant phenological patterns in the high Andean Cordillera of central Chile. J Ecol 69:205–223

Bauer PJ (1983) Bumblebee pollination relationships on the Beartooth Plateau tundra of southern Montana. Am J Bot 70(1):134–144

Beattie AJ, Breedlove DE, Ehrlich PR (1973) The ecology of the pollinators and predators of *Frasera speciosa*. Ecology 54(1):81–91

Billings WD, Bliss LC (1959) An alpine snowbank environment and its effects on vegetation, plant development, and productivity. Ecology 40:388–397

Billings WD, Mooney HA (1968) The ecology of arctic and alpine plants. Biol Rev Camb Philos Soc 43:481–529

Blionis GJ, Vokou D (2002) Structural and functional divergence of *Campanula spatulata* subspecies on Mt Olympos (Greece). Plant Syst Evol 232(1–2):89–105

Blionis GJ, Halley JM, Vokou D (2001) Flowering phenology of *Campanula* on Mt Olympos, Greece. Ecography 24(6):696–706

Bliss LC (1956) A comparison of plant development in microenvironments of arctic and alpine tundras. Ecol Monogr 26:303–337

Bliss LC (1971) Arctic and alpine plant life cycles. Ann Rev Ecol Syst 2:405–438

Bleasdale A, Chan YK (1972) Orographic influences on the distribution of precipitation. In Distribution of precipitation in mountainous areas, WMO/OMM, vol II. World Meteorological Organization, Geneva

Bokhorst S, Bjerke JW, Bowles FW, Melillo J, Callaghan TV, Phoenix GK (2008) Impacts of extreme winter warming in the sub-Arctic: growing season responses of dwarf shrub heathland. Glob Change Biol 14(11):2603–2612

Bokhorst S, Bjerke JW, Street LE, Callaghan TV, Phoenix GK (2011) Impacts of multiple extreme winter warming events on sub-Arctic heathland: phenology, reproduction, growth, and CO_2 flux responses. Glob Change Biol 17(9):2817–2830

Bradley NL, Leopold AC, Ross J, Huffaker W (1999) Phenological changes reflect climate change in Wisconsin. Proc Natl Acad Sci U S A 96:9701–9704

Brown JL, Li S-H, Bhagabati N (1999) Long-term trend toward earlier breeding in an American bird: a response to global warming? Proc Natl Acad Sci U S A 96(10):5565–5569

Caine N (2002) Declining ice thickness on an alpine lake is generated by increased winter precipitation. Clim Chang 54(4):463–470

Canaday BB, Fonda RW (1974) The influence of subalpine snowbanks on vegetation pattern, production, and phenology. Bull Torrey Bot Club 101(6):340–350

Cayan DR, Kammerdiener SA, Dettinger MD, Caprio JM, Peterson DH (2001) Changes in the onset of spring in the western United States. Bull Am Meteorol Soc 82(3):399–415

Comis D (2011) Global warming in western mountains. Agric Res 59(1):16–17
Crimmins TM, Crimmins MA, David Bertelsen C (2009) Flowering range changes across an elevation gradient in response to warming summer temperatures. Glob Change Biol 15 (5):1141–1152
Defila C, Clot B (2005) Phytophenological trends in the Swiss Alps, 1951–2002. Meteorol Z 14:191–196
Douglas GW, Bliss LC (1977) Alpine and high subalpine plant communities of the North Cascades Range, Washington and British Columbia. Ecol Monogr 47(3):113–150
Dunne JA, Harte J, Taylor KJ (2003) Subalpine meadow flowering phenology responses to climate change: integrating experimental and gradient methods. Ecol Monogr 73(1):69–86
Eccel E, Rea R, Caffarra A, Crisci A (2009) Risk of spring frost to apple production under future climate scenarios: the role of phenological acclimation. Int J Biometeorol 53(3):273–286
Fitter AH, Fitter RSR (2002) Rapid changes in flowering time in British plants. Science 296:1689–1691
Forbis TA, Diggle PK (2001) Subnivean embryo development in the alpine herb *Caltha leptosepala* (Ranunculaceae). Can J Bot Revue Canadienne de Botanique 79(5):635–642
Forrest JRK, Thomson JD (2011) An examination of synchrony between insect emergence and flowering in Rocky Mountain meadows. Ecol Monogr 81(3):469–491
Forrest J, Inouye DW, Thomson JD (2010) Flowering phenology in subalpine meadows: does climate variation influence community co-flowering patterns? Ecology 91(2):431–440
Førland EJ (1997) Precipitation and topography [in Norwegian with English summary]. Klima 2/ 79:23–24
Galen C, Stanton ML (1991) Consequences of emergence phenology for reproductive success in *Ranunculus adoneus* (Ranunculaceae). Am J Bot 78(7):978–988
Galen C, Stanton ML (1993) Short-term responses of alpine buttercups to experimental manipulations of growing season length. Ecology 74(4):1052–1058
Gallagher RV, Hughes L, Leishman MR (2009) Phenological trends among Australian alpine species: using herbarium records to identify climate-change indicators. Aust J Bot 57(1):1–9
Giorgi F, Hurrell W, Marinucci M, Beniston M (1997) Elevation dependency of the surface climate change signal: a model study. J Clim 10:288–296
Gómez JM (1993) Phenotypic selection on flowering synchrony in a high mountain plant, *Hormathophylla spinosa* (Cruciferae). J Ecol 81(4):605–613
Grabherr G, Gottfried M, Gruber A, Pauli H (1995) Patterns and current changes in alpine plant diversity. In: Chapin FS III, Koerner C (eds) Arctic and alpine biodiversity: patterns, causes and ecosystem consequences. Ecological studies. Springer, Berlin
Guisan A, Theurillat J-P (2000) Equilibrium modeling of alpine plant distribution: how far can we go? Phytocoenologia 30(3–4):353–384
Haggerty BP, Galloway LF (2011) Response of individual components of reproductive phenology to growing season length in a monocarpic herb. J Ecol 99(1):242–253
Hebblewhite M, Merrill E, McDermid G (2008) A multi-scale test of the forage maturation hypothesis in a partially migratory ungulate population. Ecol Monogr 78(2):141–166
Heide O (1993) Daylength and thermal time responses of budburst during dormancy release in some northern deciduous trees. Physiol Plant 88(4):531–540
Hendricks P (2003) Spring snow conditions, laying date, and clutch size in an alpine population of American Pipits. J Field Ornithol 74(4):423–429
Henry GHR, Molau U (1997) Tundra plants and climate change: the International Tundra Experiment (ITEX). Glob Change Biol 3(suppl 1):1–9
Hofer HR (1992) Veränderungen in der Vegetation von 14 Gipfeln des Berninagebietes zwischen 1905 und 1985. Berichte des Geobotanischen Instituts der ETH Zürich, Stiftung Rüubel 58:39–54
Hoffmann AJ, Walker MJ (1980) Growth habits and phenology of drought-deciduous species in an altitudinal gradient. Can J Bot 58:1789–1796

Hoffmann AA, Camac JS, Williams RJ, Papst W, Jarrad FC, Wahren CH (2010) Phenological changes in six Australian subalpine plants in response to experimental warming and year-to-year variation. J Ecol 98(4):927–937

Holway JG, Ward RT (1963) Snow and meltwater effects in an area of Colorado alpine. Am Midl Nat 69:189–197

Holway JG, Ward RT (1965) Phenology of alpine plants in northern Colorado. Ecology 46:73–83

Hu JIA, Moore DJP, Burns SP, Monson RK (2010) Longer growing seasons lead to less carbon sequestration by a subalpine forest. Glob Change Biol 16(2):771–783

Huelber K, Gottfried M, Pauli H, Reiter K, Winkler M, Grabherr G (2006) Phenological responses of snowbed species to snow removal dates in the Central Alps: implications for climate warming. Arct Antarct Alp Res 38(1):99–103

Hülber K, Winkler M, Grabherr G (2010) Intraseasonal and habitat-specific variability in phenological control of high alpine plants. Funct Ecol 24(2):245–252

Hülber K, Bardy K, Dullinger S (2011) Effects of snowmelt timing and competition on the performance of alpine snowbed plants. Perspect Plant Ecol 13(1):15–26

Iler AM, Inouye DW (2013) Effects of climate change on mast-flowering cues in a clonal montane herb, *Veratrum tenuipetalum* (Melanthiaceae). Am J Bot 100:1–7

Inouye DW (2000) The ecological and evolutionary significance of frost in the context of climate change. Ecol Lett 3(5):457–463

Inouye DW (2008) Effects of climate change on phenology, frost damage, and floral abundance of montane wildflowers. Ecology 89(2):353–362

Inouye DW, Pyke GH (1988) Pollination biology in the Snowy Mts. of Australia, with comparisons with montane Colorado, U. S. A. Aust J Ecol 13:191–210

Inouye DW, Barr B, Armitage KB, Inouye BD (2000) Climate change is affecting altitudinal migrants and hibernating species. Proc Natl Acad Sci USA 97(4):1630–1633

Inouye DW, Morales MA, Dodge GJ (2002) Variation in timing and abundance of flowering by *Delphinium barbeyi* Huth (Ranunculaceae): the roles of snowpack, frost, and La Niña, in the context of climate change. Oecologia 139:543–550

Inouye DW, Saavedra F, Lee W (2003) Environmental influences on the phenology and abundance of flowering by *Androsace septentrionalis* L. (Primulaceae). Am J Bot 90(6):905–910

Iversen M, Bråthen KA, Yoccoz NG, Ims RA (2009) Predictors of plant phenology in a diverse high-latitude alpine landscape: growth forms and topography. J Veg Sci 20(5):903–915

Jackson LE, Bliss LC (1982) Distribution of ephemeral herbaceous plants near treeline in the Sierra Nevada, California, U.S.A. Arct Alp Res 14:33–42

Jackson LE, Bliss LC (1984) Phenology and water relations of three plant life-forms in a dry treeline meadow. Ecology 65:1302–1314

Jarrad FC, Wahren C-H, Williams RJ, Burgman MA (2008) Impacts of experimental warming and fire on phenology of subalpine open-heath species. Aust J Bot 56(8):617–629

Kameyama Y, Kudo G (2009) Flowering phenology influences seed production and outcrossing rate in populations of an alpine snowbed shrub, *Phyllodoce aleutica*: effects of pollinators and self-incompatibility. Ann Bot 103(9):1385–1394

Keller F, Körner C (2003) The role of photoperiodism in alpine plant development. Arct Antarct Alp Res 35:361–368

Kelly D, Harrison AL, Lee WG, Payton IJ, Wilson PR, Schauber EM (2000) Predator satiation and extreme mast seeding in 11 species of *Chionochloa* (Poaceae). Oikos 90(3):477–488

Kittel TGF, Thornton PE, Royle JA, Chase TN (2002) Climates of the Rocky Mountains: historical and future patterns. In: Baron J, Fagre D, Hauer R (eds) Rocky mountain futures: an ecological perspective. Island Press, Covelo

Knight DH, Rogers BS, Kyte CR (1977) Understory plant growth in relation to snow duration in Wyoming subalpine forest. Bull Torrey Bot Club 104(4):314–319

Knowles N, Dettinger MD, Cayan DR (2006) Trends in snowfall versus rainfall in the western United States. J Clim 19(18):4545–4559

Kudo G (1991) Effects of snow-free period on the phenology of alpine plants inhabiting snow patches. Arct Alp Res 23(4):436–443
Kudo G (1992) Performance and phenology of alpine herbs along a snow-melting gradient. Ecol Res 7(3):297–304
Kudo G, Hirao A (2006) Habitat-specific responses in the flowering phenology and seed set of alpine plants to climate variation: implications for global-change impacts. Popul Ecol 48:49–58
Kudo G, Suzuki S (2002) Relationships between flowering phenology and fruit-set of dwarf shrubs in Alpine Fellfields in northern Japan: a comparison with a subarctic heathland in northern Sweden. Arct Antarct Alp Res 34(2):185–190
Kudo G, Suzuki S (2004) Flowering phenology of tropical-alpine dwarf trees on Mount Kinabalu, Borneo. J Trop Ecol 20:563–571
Lambert AM, Miller-Rushing AJ, Inouye DW (2010) Changes in snowmelt date and summer precipitation affect the flowering phenology of *Erythronium grandiflorum* (glacier lily; Liliaceae). Am J Bot 97(9):1431–1437
Langvatn R, Albon SD, Burkey T, CluttonBrock TH (1996) Climate, plant phenology and variation in age of first reproduction in a temperate herbivore. J Anim Ecol 65(5):653–670
Lynov YS (1984) Phenological inversions in alpine terrain (western Tien Shan). Ékologiya 4:29–33
Mark AF (1970) Floral initiation and development in New Zealand alpine plants. N Z J Bot 8:67–75
McKinney AM, CaraDonna PJ, Inouye DW, Barr B, Bertelsen CD, Waser NM (2012) Asynchronous changes in phenology of migrating Broad-tailed Hummingbirds and their early-season nectar resources. Ecology 93(9):1987–1993
Melampy MN (1987) Flowering phenology, pollen flow and fruit production in the Andean shrub *Befaria resinosa*. Oecologia 73:293–300
Meloche CG, Diggle PK (2001) Preformation, architectural complexity, and developmental flexibility in *Acomastylis rossii* (Rosaceae). Am J Bot 88(6):980–991
Merrill EH, Boyce MS (1991) Summer range and elk population dynamics in Yellowstone National Park. In: Keiter RB, Boyce MS (eds) The greater Yellowstone ecosystem: redfining America's wildlife heritage. Yale University Press, New Haven
Migliavacca M, Cremonese E, Colombo R, Busetto L, Galvagno M, Ganis L, Meroni M, Pari E, Rossini M, Siniscalco C, Morra di Cella U (2008) European larch phenology in the Alps: can we grasp the role of ecological factors by combining field observations and inverse modelling? Int J Biometeorol 52(7):587–605
Miller-Rushing AJ, Inouye DW (2009) Variation in the impact of climate change on flowering phenology and abundance: an examination of two pairs of closely related wildflower species. Am J Bot 96(10):1821–1829
Mitton JB, Sturgeon KB, Davis ML (1980) Genetic differentiation in ponderosa pine along a steep elevational transect. Silvae Genet 29:100–103
Molau U (1993) Relationships between flowering phenology and life history strategies in tundra plants. Arct Alp Res 25(4):391–402
Molau U, Nordenhäll U, Eriksen B (2005) Onset of flowering and climate variability in an alpine landscape: a 10-year study from Swedish Lapland. Am J Bot 92(3):422–431
Mooney HA, Williams KS, Lincoln DE, Ehrlich PR (1981) Temporal and spatial variability in the interaction between the checkerspot butterfly, *Euphydryas calcedona* and its principal food source, the Californian shrub, *Diplacus aurantiacus*. Oecologia 50:195–198
Morton ML (1994) Comparison of reproductive timing to snow conditions in wild onions and White-Crowned Sparrows at high altitude. Gt Basin Nat 54(4):371–375
Moser L, Fonti P, Büntgen U, Esper J, Luterbacher J, Franzen J, Frank D (2010) Timing and duration of European larch growing season along altitudinal gradients in the Swiss Alps. Tree Physiol 30(2):225–233

Myking T, Heide OM (1995) Dormancy release and chilling requirement of buds of latitudinal ecotypes of *Betula pendula* and *B. pubescens*. Tree Physiol 15:697–704

Nufio CR, McGuire CR, Bowers MD, Guralnick RP (2010) Grasshopper community response to climatic change: variation along an elevational gradient. PLoS One 5(9):e12977

Ovaska JA, Nilsen J, Wielgolaski FE, Kauhanen H, Partanen R, Neuvonen S, Kapari L, Skre O, Laine K (2005) Phenology and performance of mountain birch provenances in transplant gardens: latitudinal, altitudinal and oceanity-continentality gradients. In: Wielgolaski FE (ed) Plant ecology, herbivory and human impact in Nordic mountain birch forests. Springer, Heidelberg

Ozgul A, Childs DZ, Oli MK, Armitage KB, Blumstein DT, Olson LE, Tuljapurkar S, Coulson T (2010) Coupled dynamics of body mass and population growth in response to environmental change. Nature 466(7305):482–485

Painter TH, Barrett AP, Landry CC, Neff JC, Cassidy MP, Lawrence CR, McBride KE, Farmer GL (2007) Impact of disturbed desert soils on duration of mountain snow cover. Geophys Res Lett 34(L12502)

Peñuelas J, Filella I (2001) Responses to a warming world. Science 294(5543):793–795

Peñuelas J, Filella I, Comas P (2002) Changed plant and animal life cycles from 1952 to 2000 in the Mediterranean region. Glob Change Biol 8(6):531–544

Pereyra ME (2011) Effects of snow-related environmental variation on breeding schedules and productivity of a high-altitude population of Dusky Flycatchers (*Empidonax oberholseri*). Auk 128(4):746–758. doi:10.1525/auk.2011.10144

Pieper SJ, Loewen V, Gill M, Johnstone JF (2011) Plant responses to natural and experimental variations in temperature in alpine tundra, southern Yukon, Canada. Arct Antarct Alp Res 43 (3):442–456

Ralph CP (1978) Observations on *Azorella compacta* (Umbelliferae), a tropical Andean cushion plant. Biotropica 10(1):62–67

Ratcliffe MJ, Turkington R (1989) Comparative phenology of some alpine vascular plant species on Lakeview Mountain, Southern British Columbia. Can Field Nat 103:348–352

Rees M, Kelly D, Bjornstad ON (2002) Snow tussocks, chaos, and the evolution of mast seeding. Am Nat 160(1):44–59

Resvoll TR (1917) Om planter som passer til kort og kold sommer. Arciv Mathematik Naturvidenskab 35(6):1–224

Reynolds DN (1984) Alpine annual plants: phenology, germination, photosynthesis, and growth in three Rocky Mountain species. Ecology 65:759–766

Richardson SG, Salisbury FB (1977) Plant responses to the light penetrating snow. Ecology 58:1152–1158

Roy DB, Sparks TH (2000) Phenology of British butterflies and climate change. Glob Change Biol 6:407–416

Sagarin R, Micheli F (2001) Climate change in nontraditional data sets. Science 294(543):811

Sakin M, Hancock JF, Luby JJ (1997) Identifying new sources of genes that determine cyclic flowering in Rocky Mountain populations of *Fragaria virginiana* ssp *glauca* Staudt. J Am Soc Hortic Sci 122(2):205–210

Schuster WS, Alles DL, Mitton JB (1989) Gene flow in limber pine: evidence from pollination phenology and genetic differentiation along an elevational transect. Am J Bot 76:1395–1403

Shimono H (2011) Earlier rice phenology as a result of climate change can increase the risk of cold damage during reproductive growth in northern Japan. Agric Ecosyst Environ 144(1):201–207

Shimono A, Washitani I (2007) Factors affecting variation in seed production in the heterostylous herb *Primula modesta*. Plant Species Biol 22(2):65–76

Stanton ML, Galen C (1997) Life on the edge: adaptation versus environmentally mediated gene flow in the snow buttercup, *Ranunculus adoneus*. Am Nat 150(2):143–178

Stanton ML, Galen C, Shore J (1997) Population structure along a steep environmental gradient: consequences of flowering time and habitat variation in the snow buttercup, *Ranunculus adoneus*. Evolution 51(1):79–94

Steltzer H, Landry C, Painter TH, Anderson J, Ayres E (2009) Biological consequences of earlier snowmelt from desert dust deposition in alpine landscapes. Proc Natl Acad Sci USA 106 (28):11629–11634

Stenström M, Molau U (1992) Reproductive ecology of *Saxifraga oppositifolia*: phenology, mating system, and reproductive success. Arct Alp Res 24(4):337–343

Stewart IT, Cayan DR, Dettinger MD (2004) Changes in snowmelt runoff timing in western North America under a 'Business as usual' climate change scenario. Clim Change 62(1):217–232

Suzuki S, Kudo G (2000) Responses of alpine shrubs to simulated environmental change during three years in the mid-latitude mountain, northern Japan. Ecography 25:553–564

Taylor OR Jr, Inouye DW (1985) Synchrony and periodicity of flowering in *Frasera speciosa* (Gentianaceae). Ecology 66(2):521–527

Theurillat J-P, Guisan A (2001) Potential impact of climate change on vegetation in the European Alps: a review. Clim Chang 50(1):77–109. doi:10.1023/a:1010632015572

Theurillat J-P, Schlüssel A (2000) Phenology and distribution strategy of key plant species within the subalpine-alpine ecocline in the Valaisan Alps (Switzerland). Phytocoenologia 30 (3–4):439–456

Thomson JD (2010) Flowering phenology, fruiting success and progressive deterioration of pollination in an early-flowering geophyte. Philos Trans R Soc B 365(1555):3187–3199

Totland Ø (1993) Pollination in alpine Norway: flowering phenology, insect visitors, and visitation rates in two plant communities. Can J Bot – Revue Canadienne de Botanique 71(8):1072–1079

Totland Ø, Alatalo JM (2002) Effects of temperature and date of snowmelt on growth, reproduction, and flowering phenology in the arctic/alpine herb, Ranunculus glacialis. Oecologia 133:168–175

Totland Ø (1997) Effects of flowering time and temperature on growth and reproduction in *Leontodon autumnalis* var. *taraxaci* a late-flowering alpine plant. Arct Alp Res 29(3):285–290

Treml V, Ponocna T, Büntgen U (2012) Growth trends and temperature responses of treeline Norway spruce in the Czech-Polish Sudetes Mountains. Clim Res 55:91–103. doi:10.3354/cr01122

Venn SE, Morgan JW (2007) Phytomass and phenology of three alpine snowpatch species across a natural snowmelt gradient. Aust J Bot 55(4):450–456. doi:10.1071/BT06003

Visser ME, Holleman LJM (2001) Warmer springs disrupt the synchrony of oak and winter moth phenology. Proc R Soc B 268:1–6

Wagner J, Mitterhofer E (1998) Phenology, seed development, and reproductive success of an alpine population of *Gentianella germanica* in climatically varying years. Bot Acta 111 (2):159–166

Wagner J, Reichegger B (1997) Phenology and seed development of the alpine sedges *Carex curvula* and *Carex firma* in response to contrasting topoclimates. Arct Alp Res 29(3):291–299

Walker MD, Ingersoll RC, Webber PJ (1995) Effects of interannual climate variation on phenology and growth of two alpine forbs. Ecology 76(4):1067–1083

Walther G-R, Post E, Convey P, Menzel A, Pamesan C, Beebee TJC, Fromentin J-M, Hoegh-Guldberg O, Bairlein F (2002) Ecological responses to recent climate change. Nature 416:389–395

Wang GM, Hobbs NT, Giesen KM, Galbraith H, Ojima DS, Braun CE (2002) Relationships between climate and population dynamics of white-tailed ptarmigan *Lagopus leucurus* in Rocky Mountain National Park, Colorado, USA. Clim Res 23(1):81–87

Wardle P (1974) Alpine timberlines. In: Ives JD, Barry RG (eds) Arctic and alpine environments. Methuen, London

Wielgolaski FE, Kärenlampi L (1975) Plant phenology of Fennoscandinan tundra areas. In: Wielgolaski FE (ed) Fennoscandian tundra ecosystems, Part 1, vol 16, Ecological studies. Analysis and synthesis. Springer, Berlin

Wielgolaski FE, Karlsen SR (2007) Some views on plants in polar and alpine regions. Rev Environ Sci Biotechnol 6:33–45

Williams MC, Cronin EH (1968) Dormancy, longevity, and germination of seeds of three larkspurs and western false hellebore. Weeds 8:452–461

Williams-Linera G (2003) Temporal and spatial phenological variation of understory shrubs in a tropical montane cloud forest. Biotropica 35(1):28–36

Wipf S, Stoeckli V, Bebi P (2009) Winter climate change in alpine tundra: plant responses to changes in snow depth and snowmelt timing. Clim Chang 94(1–2):105–121

Wolkovich EM, Cook BI, Allen JM, Crimmins TM, Betancourt JL, Travers SE, Pau S, Regetz J, Davies TJ, Kraft NJB, Ault TR, Bolmgren K, Mazer SJ, McCabe GJ, McGill BJ, Parmesan C, Salamin N, Schwartz MD, Cleland EE (2012) Warming experiments underpredict plant phenological responses to climate change. Nature 485(7399):494–497

Xiao G, Zhang Q, Yao Y, Zhao H, Wang R, Bai H, Zhang F (2008) Impact of recent climatic change on the yield of winter wheat at low and high altitudes in semi-arid northwestern China. Agric Ecosyst Environ 127(1–2):37–42

Xu Z-F, Hu T-X, Wang K-Y, Zhang Y-B, Xian J-R (2009) Short-term responses of phenology, shoot growth and leaf traits of four alpine shrubs in a timberline ecotone to simulated global warming, Eastern Tibetan Plateau, China. Plant Species Biol 24(1):27–34

Yu HY, Luedeling E, Xu JC (2010) Winter and spring warming result in delayed spring phenology on the Tibetan Plateau. Proc Natl Acad Sci U S A 107(51):22151–22156

Part III
Phenological Models and Techniques

Chapter 15
Plant Development Models

Isabelle Chuine, Iñaki Garcia de Cortazar-Atauri, Koen Kramer, and Heikki Hänninen

Abstract In this chapter we provide a brief overview of plant phenology modeling, focusing on mechanistic phenological models. After a brief history of plant phenology modeling, we present the different models which have been described in the literature so far and highlight the main differences between them, i.e. their degree of complexity and the different types of response function to temperature they use. We also discuss the different approaches used to build and parameterize such models. Finally, we provide a few examples of applications mechanistic plant phenological models have been successfully used for, such as frost hardiness modeling, tree growth modeling, tree species distribution modeling and temperature reconstruction of the last millennium.

15.1 An Overview of Phenology Modeling During the Last Three Centuries

Phenology modeling has a long history starting in 1735 with a publication by de Reaumur (1735). Reaumur suggested that differences between years and locations in the date of phenological events could be explained by differences in daily temperatures from an arbitrary date to the date of the phenological event considered.

I. Chuine (✉)
Centre d'Ecologie Fonctionnelle et Evolutive, CNRS, Montpellier, France
e-mail: isabelle.chuine@cefe.cnrs.fr

I. Garcia de Cortazar-Atauri
AGROCLIM INRA, Avignon, France

K. Kramer
Alterra - Green World Research, Wageningen University and Research Centre, The Netherlands

H. Hänninen
Department of Biosciences, University of Helsinki, Helsinki, Finland

M.D. Schwartz (ed.), *Phenology: An Integrative Environmental Science*,
DOI 10.1007/978-94-007-6925-0_15, © Springer Science+Business Media B.V. 2013

While this is still the most important assumption in plant phenology modeling, major advances took place in the late twentieth century for two main reasons: (i) the revolution in computer science, and (ii) concerns about global climate change. Global warming is expected to have major impacts on plant functions and fitness, as increasing temperatures change the timing of phenological events (Cleland et al. 2007; Chuine 2010).

Most of plant phenology modeling studies have focused on leaf unfolding and flowering, and much fewer on fruit maturation, growth cessation or leaf senescence. This is in part due to the fact that leaf unfolding and flowering are the most widely observed phenophases, and the timing of these events can be observed accurately. This is much less the case for fruit maturation and leaf coloration. In addition, leaf unfolding is very important for primary production (Piao et al. 2007; Richardson et al. 2010) and flowering largely determines plant reproductive success (see Sect. 15.4).

Most phenology models were developed for tree species, rather than non-woody species (Table 15.1). Many more phenological modeling studies focus on temperate biota than on boreal, tropical or sub-tropical biota (Kupias and Mäkinen 1980; Phillipp et al. 1990; Reich 1994; Thorhallsdottir 1998).

Phenological observations used to develop and test phenology models have two main origins: historical observations in wild populations or phenological gardens and experimental results. The accompanying meteorological data consequently often comes from different sources. In the case of phenological observations in gardens or experiments, accurate meteorological observations are often made on site. In the case of phenological observations on wild populations, meteorological data are usually obtained from weather stations some distance away. Both data types are useful for phenology modeling, but imply different methodologies as we discuss in Sect. 15.3.

15.2 An Overview of Plant Phenology Models

Reaumur (1735) first introduced the concept of degree-day summation, i.e. daily average temperatures accumulated between an arbitrary date of onset and the date of an observed phenological event. Reaumur found that a lower sum of degree-days was accumulated in April, May and June in 1735, when harvesting of crop and grapes took place late, than in 1734 when the harvest took place early. He realized that a plant develops quicker at a higher temperature, thus shortening the interval between sowing and crop harvest, or flowering and vine harvest. He proposed that plant development is proportional to the sum of temperature over time rather than to temperature during the phenological event itself. In many studies since Reaumur, accumulated temperature is recognized as the main factor influencing year-to-year variation in phenology. Three main types of phenology models exist: analytical, statistical, and mechanistic. Analytical models are based on the cost/benefit tradeoff of producing leaves to optimize resource acquisition (Kikuzawa 1991, 1995a, b, 1996; Kikuzawa and Kudo 1995) and

Table 15.1 List of the different phenological models described in the literature (EcoD, ecodormancy; EndoD, endodormancy, DI, dormancy induction)

Model name	Phenophase	Plant types	Reference
Growing degree hours	EcoD	Fruit trees	Anderson et al. (1986)
Chilling hour	EndoD	Fruit trees	Bennett (1949) and Weinberger (1950)
Bidabé	EndoD + EcoD	Apple tree	Bidabé (1967)
Smoothed Utah	EndoD	Fruit trees	Bonhomme et al. (2010)
Dormphot	DI, EndoD, EcoD	Forest trees	Caffarra et al. (2011)
Alternating	EndoD + EcoD	Forest trees	Cannell and Smith (1983) and Kramer (1994b)
Unified	EndoD + EcoD	Trees	Chuine (2000)
Unichill, Uniforc	EndoD + EcoD	Trees	Chuine et al. (1999)
Growing degree-days	EncoD	Crops, grapevine	de Reaumur (1735) in Wang (1960)
Delpierre	Senescence	Forest trees	Delpierre et al. (2009)
Dynamic model	EndoD	Fruit trees	Erez et al. (1990) and Fishman et al. (1987)
Hartkamp	EcoD	Velvet bean	Hartkamp et al. (2002)
Triangle GDH	EcoD	Crops	Hammer et al. (1993)
Sequential	EndoD + EcoD	Forest trees	Richardson et al. (1974) and Hänninen (1987)
Hänninen	EndoD + EcoD	Forest trees	Hänninen (1990) and Hänninen (1995)
Biological days	EcoD + EndoD	Crops	Hunt and Pararajasingham (1995)
Kobayashi and Fuchigami	EcoD	Red osier dogwood	Kobayashi and Fuchigami (1983a)
Deepening rest	EndoD + EcoD	Red osier dogwood	Kobayashi et al. (1982)
Kramer	EndoD + EcoD	Forest trees	Kramer (1994b)
Parallel	EndoD + EcoD	Forest trees	Hänninen (1987) and Landsberg (1974)
Promoting inhibition model	EndoD + EcoD	Fruit trees	Linkosalo et al. (2008)
Positive Utah	EndoD	Fruit trees	Linsley-Noakes et al. (1995)
Logarithmic (Action days)	EcoD	Grapevine	Pouget (1968)
Utah + GDH	EndoD + EcoD	Fruit trees	Richardson et al. (1974)
Asymcur	EndoD	Fruit trees	Richardson et al. (1982)
Sinusoidal-parabolic GDH	EcoD	Grapevine	Riou (1994)
Robertson	EcoD	Crops	Robertson (1968)
Sinclair	EcoD	Crops	Sinclair et al. (1991)
Soltani	EcoD	Chickpea	Soltani et al. (2006)
Four phase	EndoD + EcoD	Trees	Hänninen (1990) and Vegis (1964)
Asymetric unimodal function	Flowering and fruit maturation	Wheat	Wang and Engel (1998)
White	Senescence	Land surface	White et al. (1997)

(continued)

Table 15.1 (continued)

Model name	Phenophase	Plant types	Reference
Asymetric unimodal function	Flowering and fruit maturation	Annual crops	Yan and Hunt (1999)
Asymetric unimodal function	Flowering and fruit maturation	Annual crops	Yin et al. (1995)

are designed to understand the evolution of leaf lifespan strategies in trees, rather than the annual variation in plant phenology.

Statistical phenology models relate the timing of phenological events to climatic factors. Their parameters are estimated from data using various statistical fitting methods. While most of these models do not consider specific biological processes, some are more mechanistic than others. Some are simple correlations with average temperature in different periods of the year (Boyer 1973; Spieksma et al. 1995; Emberlin et al. 1997; Ruml et al. 2012). Others are more complex. For example, the Spring Indices Models (Schwartz and Marotz 1986, 1988; Schwartz 1997; Schwartz et al. 2012) have been successfully used to predict the start of the growing season in North America. It is a multiple regression model of the type:

$$y^{-1} = C + \sum_{k=1}^{n} A_k X_k \tag{15.1}$$

where y is the day-of-the-year date of the phenological event, C and A_k are constants and X_k are the predictor variables, degree-day sums for two threshold temperatures (−0.6 or 5 °C), mean temperature, and number of synoptic weather events (in particular warm air advection).

Mechanistic models formally describe known or hypothetical cause-effect relationships between physiological processes and some driving factors in the plant's environment. New relationships should be introduced in a mechanistic model only if information on their impacts on the process is available. It is important to note that parameters of mechanistic models have physical dimensions that can, in principle (see Sect. 15.3), be measured directly instead of being estimated by fitting. However, this is rarely possible in plant phenology models. As most models described in the literature are of this type, the following paragraphs provide a detailed overview of their hypotheses. The structure of mechanistic models is usually based on systems theory rather than statistical inference (Hänninen and Kramer 2007; Chuine 2010).

From experimental evidence we know at least three things. First, higher temperature accelerates cell growth during ecodormancy, during which dormancy is caused by external factors (Lamb 1948; Sarvas 1972, 1974; Landsberg 1974; Campbell and Sugano 1975; Lang et al. 1985; Caffarra et al. 2011b). Second, endodormancy, during which dormancy is caused by internal factors (Lang et al. 1985) must be broken by a chilling period at cool temperatures before plants enter the phase of ecodormancy (Sarvas 1974; Hänninen 1990; Caffarra et al. 2011a). Third, the evidence for the role of photoperiod in tree phenology is conflicting, depending on phenophase (a plant's phenological state), species and location (Heide 1993a, b; Kramer 1994b; Falusi and Calamassi 1996). However, there is clear evidence that long photoperiod enhances cell growth, compensating for a lack of chilling during the endodormancy phase (Wareing 1953; Heide 1993b; Myking and Heide 1995; Caffarra et al. 2011a).

Most differences between mechanistic phenological models come from the number of different phases they consider and the response function to temperature or photoperiod they consider for each phase. The general structure of mechanistic phenological models is the following:

$$t_n \text{ such that } S_{n,t} = \sum_{\mathrm{t}_{n-1}}^{t_n} R_{n,t}(Z) = S_n{}^* \tag{15.2}$$

where n is a development phase (e.g., encodormancy, ecodormancy, fruit maturation), $S_{n,t}$ is the state of development on day t in phase n; t_n is the end of phase n and t_{n-1} the end of phase $n - 1$. $R_{n,t}$ is the rate of development during phase n on day t which is a function of one or a set of daily or hourly environmental variables Z (e.g., temperature, photoperiod, water potential), and S_n* the critical state required to reach t_n. Virtually any phenological model can fit into this framework. For example, the growing degree-day model also called Thermal Time model or Spring Warming model, the simplest plant phenology model, requires only three parameters and can be written as a one-phase model as follows:

$$R_{1,t}(x_t) = \begin{cases} 0 & \text{if } x_t \leq T_b \\ x_t - T_b & \text{if } x_t > T_b \end{cases} \tag{15.3}$$

where x_t is daily mean temperature, $t_{n-1} = t_0$ is the day on which summation starts, T_b is the summation threshold temperature, and S_I* is the familiar degree-day sum required to complete the phenophase at t_1.

For tree species, more complex models take into account the endodormancy and dormancy inductions phases. Two-phase models for leaf unfolding and flowering typically take endodormancy into account in addition to ecodormancy (e.g., the Sequential, Parallel, Alternating, Deepening Rest models). Three-phase models for leaf unfolding or flowering typically describe the dormancy induction phase in addition to endodormancy and ecodormancy phases (Dormphot model). The Four-phase model for leaf unfolding describes a dormancy induction and an ecodormancy phase, but splits the latter into two phases (true rest and post-rest). In most models phase n follows sequentially phase $n - 1$, but in some models the processes of the different phases can overlap. For instance, a parallel model allows ontogenetic development (which is the typical process of ecodormancy) to take place at high temperatures even before endodormancy has completed.

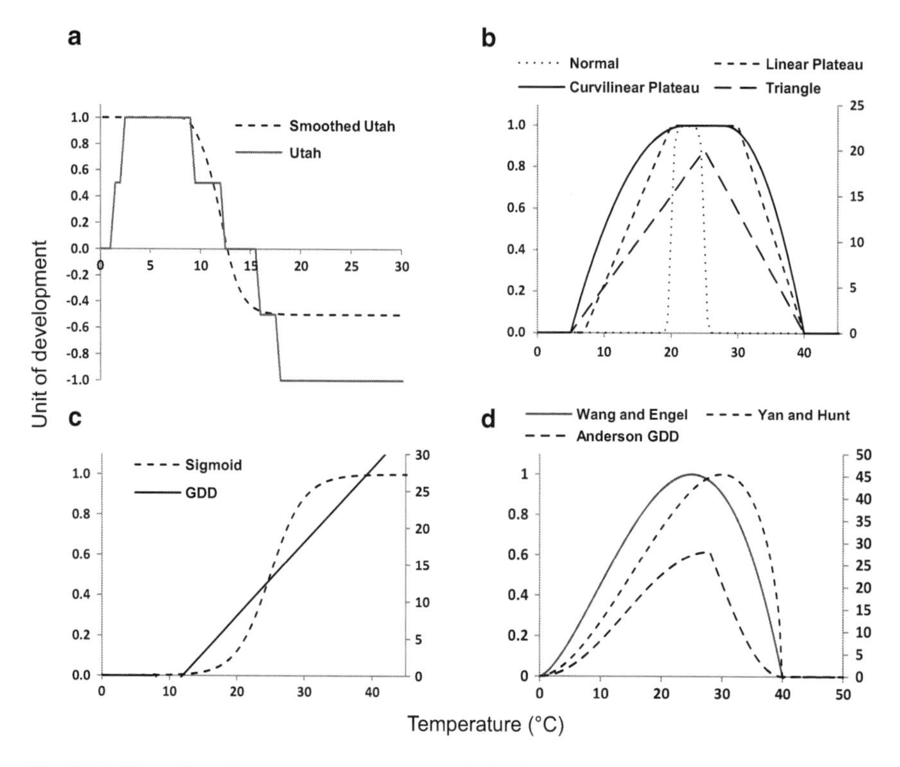

Fig. 15.1 Examples of response to temperature during dormancy induction, endodormancy and ecodormancy. (**a**) Utah and smoothed Utah response functions; (**b**) normal, triangular, linear plateau and curvilinear plateau response functions; (**c**) sigmoid and growing degree-day response functions; (**d**) Wang and Engel, Yan and Hunt and Anderson response functions (see Table 15.1 for the references)

For herbaceous species, especially crops, models can be more or less complex depending on the number of phases they take into account as well, but typically they consider a germination phase that depends on temperature and soil moisture, a flowering phase that depends on temperature, precipitation and photoperiod, a the maturation phase that depends on temperature (Wang and Engel 1998).

Whatever the species, the developmental responses to temperature have been described by various types of functions depending on phase (Fig. 15.1).

The variety of model assumptions and formulations called for a consistent notation and for attempts at unification. This was started by Hänninen (1987) who divided models in two categories ("sequential" and "parallel") based on their ecophysiological distinctions. Hänninen (1990, 1995) introduced a unifying formulation for several model types. In this approach, temperature responses of the sub-models are fixed *a priori*. Kramer (1994a, b) broadened the approach with a fitting procedure to select between different forms of the three sub-models (see Hänninen and Kramer 2007 for a review). Chuine (2000) further generalized model formulation with a Unified model that is based on two general functions to describe the relationships between rates of development and temperature during endodormancy (Fig. 15.1b) and ecodormancy (15.1c). The Unified model is flexible and includes as a special case basically any phenological model where air temperature is the driving factor.

To facilitate the development and the parameterization of mechanistic phenological models, a software package called PMP (Phenology Modeling Platform) has been designed and can be downloaded freely from (http://www.cefe.cnrs.fr/en/logiciels/ressources-documentaires). PMP can be used to parameterize a phenological model using meteorological data and phenological observations, or to run simulations using an available phenological model and meteorological data. The user can parameterize existing models chosen from a library, create a new model from scratch or build from on an existing one. The construction of the model requires determining the number of phases to be considered and for each phase, which meteorological variables have an impact and which response function should be used. The action of a meteorological variable can impact the effect of another variable within a phase, and phases can follow each other either sequentially or in parallel (influencing each other). Model parameter estimates are jointly optimized using the simulated annealing algorithm of Metropolis (Chuine et al. 1998) [see Sect. 15.3.2], and confidence intervals for parameters can be calculated.

15.3 Methodological Considerations

The critical problem with mechanistic phenology models is that the basic biochemistry and biophysics of certain phases is sometimes incompletely understood. This is especially true of the dormancy phase. Therefore, we cannot make direct measurements of model parameter values. Two approaches are used to estimate these values: the experimental approach, that analyzes the temperature response of growth and development under controlled conditions; and the numerical approach, that uses statistical model-fitting techniques.

15.3.1 The Experimental Approach

The experimental approach consists of experiments carried out to analyze the underlying mechanisms of phenological responses, one mechanism at a time. A few studies have followed this method.

Sarvas (1972) determined experimentally the temperature response of development rate during ecodormancy, using observations of meiosis in pollen mother-cells of several forest tree species. He noticed that *developmental time*, i.e. the average time between two meiotic phases declines exponentially with increasing temperature. Thus, the *rate of development* (i.e., the reciprocal of development time) increases with temperature is a sigmoid fashion. He tested this model with the timing of flowering in forest stands. Although the Thermal Time model with a +5 °C threshold temperature approximates quite well the temperature response measured in the laboratory, Sarvas (1972) also found that ontogenetic development could take place between −3 and +5 °C. As the physiological processes of dormancy release were unknown, and still are, Sarvas (1974) determined the progress of dormancy release indirectly using regrowth tests where seedlings were incubated at growth-promoting temperatures following a period of chilling. Both the chilling duration and temperatures were varied systematically. Sarvas found that the duration of chilling required for completion of endodormancy was shortest at +3.5 °C, and concluded that the rate of development (rate of dormancy release) was highest at this temperature. These results led to the triangular temperature response (peaking at 3.5 °C) proposed for the rate of dormancy release (Fig. 15.1b).

Experimental studies also allow the identification of critical temperatures for growth and development of many species and their differences between cultivars (in crops), for example wheat (Porter and Gawith 1999) or grapevines (Pouget 1972), or between provenances (in forest trees).

An experimental approach was also used by Hänninen (1990), who developed the first version of the Unified model to compare various model assumptions concerning the effects of chilling on the response of buds to forcing temperatures. In seedling of *Pinus sylvestris* and *Picea abies*, he found that the effects fell somewhere between the assumptions of the Sequential and the Parallel models.

The Dormphot model of Caffarra et al. (2011a) is also the outcome of experimental results on beech. An interaction between photoperiod and temperature was found during ecodormancy, when longer photoperiod decreases the optimal temperature of development and accelerates bud growth (Caffarra et al. 2011b). These experiments also showed that the accelerating action of photoperiod decreased as the amount of chilling increases.

15.3.2 The Numerical Approach

The statistical approach estimates parameter values with statistical model-fitting techniques. In this approach, field or experimental observations of the timing of phenological events are related to meteorological data gathered at the same location before the event. Two techniques have been used, both estimating parameters using the least squared residuals method.

The easiest method is to fix all but one parameter to a given value, and find the value of the free parameter that minimized the sum of squared residuals. All parameters are varied this way one after the other. This technique has several limits, most importantly (i) a finite number of parameter values can be tested, (ii) parameter values are estimated independently from each other although they are usually not independent, (iii) the least squares function may have several local optima and it is almost impossible to find the global optimum without a more thorough search.

More efficient methods consist in estimating all parameters simultaneously using optimization algorithms. Traditional optimization algorithms such as Downhill

Simplex or Newton methods (Press et al. 1989) rarely converge towards the global optimum because of the strong interdependency of phenological model parameters (Kramer 1994b). The simulated annealing method is more effective in this respect (Chuine et al. 1998, 1999) because it is especially designed for functions with multiple optima. More recently Bayesian approaches have also been used to parameterize phenological models. Bayesian approaches coupled with experimental approaches that provide prior information on the distribution of model parameters can be powerful (Dose and Menzel 2004; Thorsen and Hoglind 2010; Fu et al. 2012).

However, accurate parameter estimation is not sufficient, prediction accuracy is also critical, because phenology models must predict future phenology, whether over the coming year (e.g., for orchard management) or over the next century (e.g., for global warming impact assessment). Cross-validation is an adequate testing method (Chatfield 1988) by which the model is tested by comparing its predictions to observations not used in model fitting. However, this method is data-hungry and it is not always possible to split the dataset into two parts, one to fit the model, the other to test its prediction accuracy. In such case, one can resort to "leave one out" (or jackknife) cross-validation (Stone 1977; Häkkinen 1999).

The above discussion shows that parameter estimates of phenology models can be developed with two quite different approaches. The experimental approach uses detailed ecophysiological laboratory or greenhouse experiments. This is a time-consuming process. Because model parameters may be under genetic control, they often need to be measured for different populations. The statistical approach is much quicker, provided that sufficiently long phenological and temperature records are available, and that adequate statistical methods are used. However, this may be too rough an approach and a combination of the two (experimental and statistical) is probably the best solution to obtain accurate and realistic models.

15.4 Applications of Plant Phenology Models

Plant phenology models are important tools in a wide range of applications such as (1) prediction of the impact of global warming on the phenology of wild and cultivated species (Hänninen et al. 2007; Morin et al. 2009; Hanninen and Tanino 2011), (2) improvement of primary productivity models (Kramer and Mohren 1996; Krinner et al. 2005; Kramer and Hänninen 2009), (3) prediction of the occurrence of pollen in the atmosphere, and thus the occurrence of pollen allergies (Frenguelli and Bricchi 1998; Chuine and Belmonte 2004; Garcia-Mozo et al. 2007, 2008a, b), (4) species distribution modeling (Chuine and Beaubien 2001; Morin et al. 2007, 2008); and (5) climate reconstruction using historical phenological data (Chuine et al. 2004b; Menzel 2005; Meier et al. 2007; García de Cortázar-Atauri et al. 2010; Maurer et al. 2011; Yiou et al. 2012). In the following paragraphs we describe some of these uses.

15.4.1 Frost Hardiness Modeling

Phenology models of bud burst have frequently applied to assess the risk of frost damage to perennial plants (Cannell 1985; Cannell and Smith 1986; Murray et al. 1989; Hänninen 1991; Kramer 1994a; Linkosalo et al. 2000). Bud development and growth is highly correlated to loss of frost hardiness (Sakai and Larcher 1987). Frost hardiness gradually increases while dormancy sets-in and is gradually lost during ecodormancy once endodormancy is broken. The risk of frost damage can be assessed by estimating minimum air temperatures around bud burst (Cannell 1985; Murray et al. 1989; Hänninen 1991). More mechanistic models of cold hardiness have been developed that simulate frost damage over the whole year, and not only around bud burst (Kobayashi and Fuchigami 1983b; Repo et al. 1990; Kellomäki et al. 1995; Leinonen et al. 1995). Leinonen (1996) developed the most complex and probably most accurate frost hardiness model so far. In this model the state of hardiness is regulated by daily air temperature and photoperiod, and the frost hardiness response to these environmental factors depends on the current state of ontogenic development. The minimum temperature that can be withstood without damage varies during the annual cycle. Leinonen introduced an injury that responds to temperature according to the current frost hardiness. It has been known for a long time that frost hardiness is also dependent on water and soluble sugar contents (Siminovitch et al. 1953). Recent studies have tried to model this relationship mechanistically (Poirier et al. 2010).

15.4.2 Forest Growth and Climate Change

An important application of phenology models is their coupling with general models of forest growth to assess climate change impacts. FORGRO uses phenology and frost hardiness models to simulate tree growth and productivity (Kramer 1995; Kramer et al. 1996; Leinonen and Kramer 2002; Kramer and Hänninen 2009). The onset and end of the growing season can be observed either by recording the changes of the canopy such as bud burst, autumn coloration or loss of foliage, or by measuring gas exchanges between the vegetation and the atmosphere. Part of the springtime CO_2 flux is caused by the activity of the understory, which is not described by the FORGRO model (Vesala et al. 1998). The decline of the CO_2 exchange from mid-summer to autumn is mainly the direct effect of decreasing light availability and temperature on photosynthesis (Vesala et al. 1998). A rise in atmospheric CO_2 concentration and temperature influences a multitude of processes in a tree and in a forest stand. FORGRO describes the direct effects of CO_2 and temperature on photosynthesis, and the direct effect of temperature on both plant and soil respiration. The description of these processes can be found in Kramer et al. (1996) and Mohren (1987). Indirect effects of temperature include the duration of the growing season and the level of frost hardiness.

15.4.3 Modeling the Adaptive Response of Phenological Traits to Climate Change, and Species Niche

The seasonal coordination of phenology to local climate conditions has several major impacts on plant survival and reproduction (fitness), as well as on competitive relationships *via* vegetative and reproductive performances (Lechowicz and Koike 1995; Chuine 2010). A recent development in phenological modeling is to assess the adaptive response (in a genetic sense) of traits such as chilling requirement to climate change, and its consequence on the effect of temperature on the timing of budburst. This approach is first described in Kramer et al. (2008) and applied to the northern limits of *Fagus sylvatica* L. (Kramer et al. 2010). The results indicate that adaptation of the timing of budburst in trees is likely to occur even if the rate of climate change occurs in a time span similar to the longevity of individual trees. Moreover, specific forest management may increase the rate at which the timing of bud burst adapts to climate change. The theoretical background in this type of process-based genetic modeling is presented in Kramer and Van der Werf (2010).

Phenological models have also played an important role in species distribution prediction in the last 10 years. Using a species range model (PHENOFIT) Chuine and Beaubien (2001) showed that phenology was a major determinant of species range. PHENOFIT estimates survival and reproductive success based on the match between annual plant development and local seasonal variations of climate. A mismatch between the two may result in frost injury to flowers and leaves, but also in drought injury should the vegetation period occur during the drought season, or in low fecundity should the period between flowering and fall be too short or too cold for fruit to mature (Pigott and Huntley 1981). These mismatches decrease primary productivity, survival and reproductive success.

PHENOFIT has been validated for 22 tree species from North America and Europe (Morin et al. 2007; Gritti et al. 2013) and has been used to predict species distribution changes under different climate scenarios (Morin et al. 2008). Sensitivity analysis showed that the southern boundaries of many species were determined by the inability to fully develop leaves or flowers due to insufficient chilling to break endodormancy, while northern range limits were usually due to the inability to ripen fruit. Western and eastern range limits are usually more sensitive to the ability of the species to resist water stress. Phenological models have played a similar role in insect population dynamics modeling and more recently insect distribution modeling (see Chap. 16).

15.4.4 Climate Reconstruction Using Historical Phenological Data

Phenology observations are a very good proxy for past climate reconstructions (Brazdil et al. 2005). Both correlative (Aono and Omoto 1993; Menzel 2005; Meier et al. 2007; Etien et al. 2008, 2009; Maurer et al. 2009, 2011; Aono and Saito 2010; Možný et al. 2010); and mechanistic phenological models have been

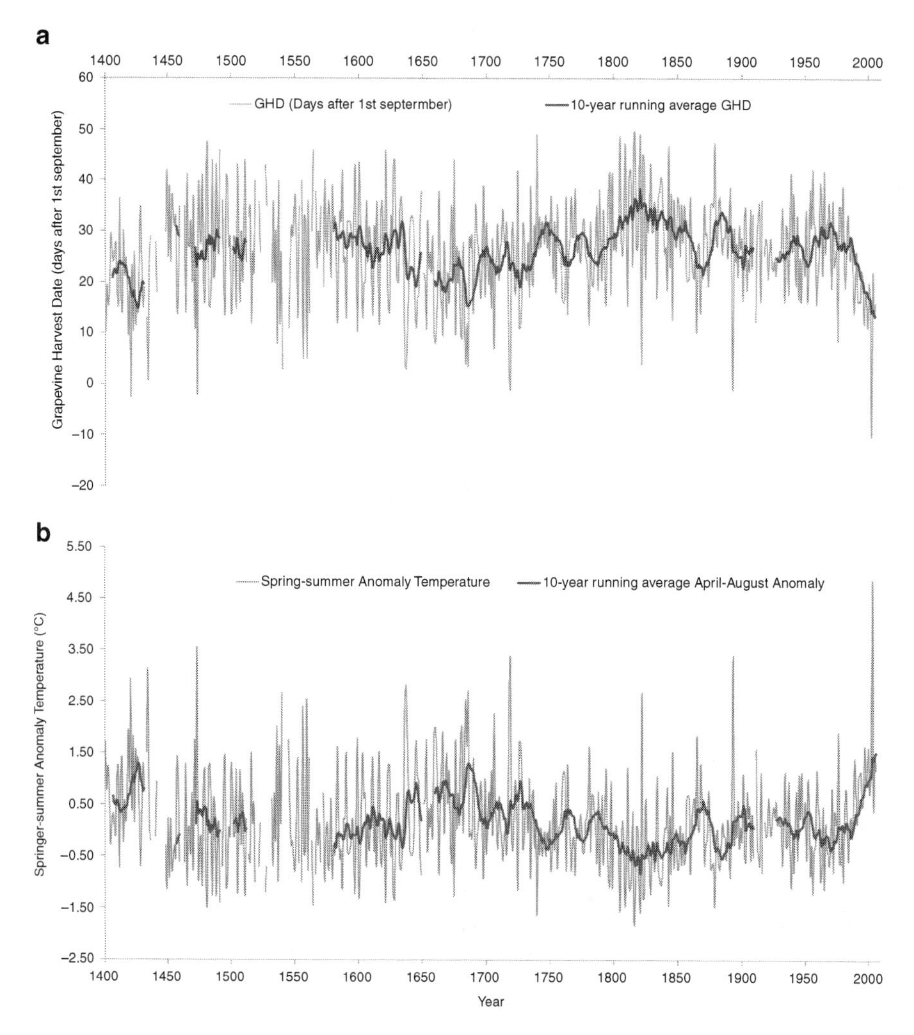

Fig. 15.2 Grapevine harvest date evolution (days after September 1st) in Burgundy (**a**). Reconstructed temperature anomalies of mid-March to August temperature deduced from grapevine harvest dates in Burgundy (**b**). *Bold lines*: 10-year running average

used in this task (Chuine et al. 2004a; García de Cortázar-Atauri et al. 2010; Yiou et al. 2012). The latter approach, in particular, made use of grapevine harvest dates to reconstruct temperature anomalies over the last seven centuries (Fig. 15.2, Chuine et al. 2004), as well as atmospheric pressure anomalies over the last five centuries based on temperature gradients (Yiou et al. 2012). García de Cortázar-Atauri et al. (2010) however warned of the difficulties of such reconstructions. They require robust phenological models parameterized with large data series as well as a good knowledge of the history and denomination of grape varieties, historical events such as wars that can affect harvest dates independently of climate, and of historical changes in agricultural practices that may have impacted on grape harvest dates.

Acknowledgments IC was financially supported by project SCION (ANR-05-BDIV-009) of the French National Research Agency. KK was financially supported by project DynTerra (project no. 5238821) of the Knowledge Base of the Dutch Ministry of Economy, Agriculture and Innovation and the large-scale integrative project MOTIVE (FP7 contract no. 226544). HH was financially supported by the Academy of Finland (project 122194). The authors are most grateful to Jacques Régnière for his thorough review and his corrections which greatly improved the quality of this chapter.

References

Anderson JL, Kesner CD, Richardson EA (1986) Validation of chill unit and flower bud phenology models for Montmorency sour cherry. Acta Hortic 184:71–77

Aono Y, Omoto Y (1993) Variation in the March mean temperature deduced from cherry blossom in Kyoto since the 14th century. J Agric Meteorol 48:635–638

Aono Y, Saito S (2010) Clarifying springtime temperature reconstructions of the medieval period by gap-filling the cherry blossom phenological data series at Kyoto, Japan. Int J Biometeorol 54(2):211–219

Bennett JP (1949) Temperature and bud rest period. Calif Agric 3(11):9–12

Bidabé B (1967) Action de la température sur l'évolution des bourgeons de pommier et comparaison de méthodes de contrôle de l'époque de floraison. Ann Physiol Vég 1:65–86

Bonhomme M, Rageau R, Lacointe A (2010) Optimization of endodormancy release models using series of endodormancy release data collected in France. Acta Hortic 872:51–60

Boyer WD (1973) Air temperature, heat sums, and pollen shedding phenology of longleaf pine. Ecology 54:421–425

Brazdil R, Pfister C, Wanner H, Von Storch H, Luterbacher J (2005) Historical climatology in Europe – the state of the art. Clim Change 70(3):363–430. doi:10.1007/s10584-005-5924-1

Caffarra A, Donnelly A, Chuine I (2011a) Modelling the timing of *Betula pubescens* budburst. II. Integrating complex effects of photoperiod into process-based models. Clim Res 46:159–170. doi:10.3354/cr00983

Caffarra A, Donnelly A, Chuine I, Jones MB (2011b) Modelling the timing of Betula pubescens budburst. I. Temperature and photoperiod: a conceptual model. Clim Res 46:147–157

Campbell RK, Sugano AI (1975) Phenology of bud burst in Douglas-fir related to provenance, photoperiod, chilling and flushing temperature. Bot Gaz 136:290–298

Cannell MGR (1985) Analysis of risks of frost damage to forest trees in Britain. In: Tigerstedt PMA, Puttonen P, Koski V (eds) Crop physiology of forest trees. Helsinki University Press, Helsinki

Cannell MGR, Smith RI (1983) Thermal time, chill days and prediction of budburst in *Picea sitchensis*. J Appl Ecol 20:951–963

Cannell MGR, Smith RI (1986) Climatic warming, spring budburst and frost damage on trees. J Appl Ecol 23:177–191

Chatfield C (1988) Problem solving: a statistician guide. Chapman & Hall, London

Chuine I (2000) A unified model for the budburst of trees. J Theor Biol 207:337–347

Chuine I (2010) Why does phenology drive species distribution? Philos Trans R Soc Lond B 365:3149–3160

Chuine I, Beaubien E (2001) Phenology is a major determinant of temperate tree range. Ecol Lett 4(5):500–510

Chuine I, Belmonte J (2004) Improving prophylaxis for pollen allergies: predicting the time course of the pollen load of the atmosphere of major allergenic plants in France and Spain. Grana 43:1–17

Chuine I, Cour P, Rousseau DD (1998) Fitting models predicting dates of flowering of temperate-zone trees using simulated annealing. Plant Cell Environ 21:455–466

Chuine I, Cour P, Rousseau DD (1999) Selecting models to predict the timing of flowering of temperate trees: implications for tree phenology modelling. Plant Cell Environ 22(1):1–13

Chuine I, Yiou P, Viovy N, Seguin B, Daux V, Ladurie EL (2004) Grape ripening as a past climate indicator. Nature 432:289–290

Chuine I, Yiou P, Viovy N, Seguin B, Daux V, Ladurie EL (2004a) Grape ripening as a past climate indicator. Nature 432(7015):289–290. doi:10.1038/432289a

Chuine I, Yiou P, Viovy N, Seguin B, Daux V, Ladurie ELR (2004b) Grape ripening as an indicator of past climate. Nature 432:289–290

Cleland EE, Chuine I, Menzel A, Mooney HA, Schwartz MD (2007) Changing plant phenology in response to climate change. TREE 22(7):357–365

Delpierre N, Dufrêne E, Soudani K, Ulrich E, Cecchini S, Boé J, François C (2009) Modelling interannual and spatial variability of leaf senescence for three deciduous tree species in France. Agric For Meteorol 149(6):938–948

Dose V, Menzel A (2004) Bayesian analysis of climate change impacts in phenology. Glob Change Biol 10(2):259–272

Emberlin J, Mullins J, Corden J, Millington W, Brooke M, Savage M, Jones S (1997) The trend to earlier Birch pollen season in the U. K.: a biotic response to changes in weather conditions? Grana 36:29–33

Erez A, Fishman S, Linsley-Noakes GC, Allan P (1990) The dynamic model for rest completion in peach buds. Acta Hortic 276:165–174

Etien N, Daux V, Masson-Delmotte V, Stievenard M, Bernard V, Durost S, Guillemin MT, Mestre O, Pierre M (2008) A bi-proxy reconstruction of Fontainebleau (France) growing season temperature from AD 1596 to 2000. Clim Past 4(2):91–106

Etien N, Daux V, Masson-Delmotte V, Mestre O, Stievenard M, Guillemin M, Boettger T, Breda N, Haupt M, Perraud P (2009) Summer maximum temperature in northern France over the past century: instrumental data versus multiple proxies (tree-ring isotopes, grape harvest dates and forest fires). Clim Change 94(3):429–456

Falusi M, Calamassi R (1996) Geographic variation and bud dormancy in beech seedlings (Fagus sylvatica L). Ann Sci Forestieres 53:967–979

Fishman S, Erez A, Couvillon GA (1987) The temperature dependence of dormancy breaking in plants: mathematical analysis of a two-step model involving a cooperative transition. J Theor Biol 124(4):473–483

Frenguelli G, Bricchi E (1998) The use of pheno-climatic model for forecasting the pollination of some arboreal taxa. Aerobiologia 14:39–44

Fu YH, Campioli M, Demaree G, Deckmyn A, Hamdi R, Janssens IA, Deckmyn G (2012) Bayesian calibration of the Unified budburst model in six temperate tree species. Int J Biometeorol 56(1):153–164. doi:10.1007/s00484-011-0408-7

García de Cortázar-Atauri I, Daux V, Garnier E, Yiou P, Viovy N, Seguin B, Boursiquot JM, Parker AK, Van Leeuwen C, Chuine I (2010) Climate reconstructions from grape harvest dates: methodology and uncertainties. Holocene 20(4):599–608

Garcia-Mozo H, Chuine I, Aira M-J, Belmonte J, Bermejo D, Guardia CD, Elvira B, Gutierrez M, Rodriguez-Rajo J, Ruiz L, Trigo MM, Tormo R, Valencia R, Galan C (2007) Regional phenological models for forecasting the start and peak of the Quercus pollen season in Spain. Agric For Manag 148(3):372–380

Garcia-Mozo H, Galán C, Belmonte J, Bermejo D, Candau P, Guardia CD, Elvira B, Gutierrez M, Jato V, Silva I, Trigo MM, Valencia R, Chuine I (2008a) Predicting the start and peak dates of the Poaceae pollen season in Spain using process-based models. Agric For Meteorol 149:256–262

Garcia-Mozo H, Orlandi F, Galan C, Fornaciari M, Romano B, Ruiz L, Guardia CD, Trigo M, Chuine I (2008b) Olive flowering phenology variation between different cultivars in Spain and Italy: modelling analysis. Theor Appl Climatol 95:385–395. doi:10/1007/s00704-008-0016-6
Gritti ES, Duputié A, Massol F, Chuine I (2013) Estimating consensus and associated uncertainty between inherently different species distribution models. Methods in Ecology and Evolution 4:442–452
Häkkinen R (1999) Statistical evaluation of bud development theories: application to bud burst of *Betula pendula* leaves. Tree Physiol 19:613–618
Hammer GL, Carberry PS, Muchow RC (1993) Modelling genotypic and environmental control of leaf area dynamics in grain sorghum. I. Whole plant level. Field Crop Res 33(3):293–310
Hänninen H (1987) Effects of temperature on dormancy release in woody plants: implications of prevailing models. Silva Fenn 21(3):279–299
Hänninen H (1990) Modelling bud dormancy release in trees from cool and temperate regions. Acta For Fenn 213:1–47
Hänninen H (1991) Does climatic warming increase the risk of frost damage in northern trees? Plant Cell Environ 14:449–454
Hänninen H (1995) Effects of climatic change on trees from cool and temperate regions: an ecophysiological approach to modelling of budburst phenology. Can J Bot 73:183–199
Hänninen, H, Kramer K (2007) A framework for modelling the annual cycle of trees in boreal and temperate regions. Silva Fennica 41:167–205
Hanninen H, Tanino K (2011) Tree seasonality in a warming climate. Trends Plant Sci 16 (8):412–416. doi:10.1016/j.tplants.2011.05.001
Hänninen H, Slaney M, Linder S (2007) Dormancy release of Norway spruce under climatic warming: testing ecophysiological models of bud burst with a whole-tree chamber experiment. Tree Physiol 27(2):291–300
Hartkamp AD, Hoogenboom G, White JW (2002) Adaptation of the CROPGRO growth model to velvet bean (*Mucuna pruriens*): I. Model development. Field Crop Res 78(1):9–25
Heide OM (1993a). Daylength and thermal time responses of budburst during dormancy release in some northern deciduous trees. Physiol Plant 88:531–540
Heide OM (1993b) Dormancy release in beech buds (Fagus sylvatica) requires both chilling and long days. Physiol Plant 89:187–191
Hunt LA, Pararajasingham S (1995) CROPSIM – WHEAT: a model describing the growth and development of wheat. Can J Plant Sci 75(3):619–632
Kellomäki S, HÄnninnen H, Kolström M (1995) Computations on frost damage to scots pine under climatic warming in boreal conditions. Ecol Appl 5(1):42–52
Kikuzawa K (1991) A cost-benefit analysis of leaf habit and leaf longevity of trees and their geographical pattern. Am Nat 138:1250–1263
Kikuzawa K (1995a) The basis for variation in leaf longevity of plants. Vegetatio 121:89–100
Kikuzawa K (1995b) Leaf phenology as an optimal strategy for carbon gain in plants. Can J Bot 73:158–163
Kikuzawa K (1996) Geographical distribution of leaf life span and species diversity of trees simulated by a leaf-longevity model. Vegetatio 122:61–67
Kikuzawa K, Kudo G (1995) Effects of the length of the snow-free period on leaf longevity in alpine shrubs: a cost-benefit model. Oikos 73:214–220
Kobayashi KD, Fuchigami LH (1983a) Modeling bud development during the quiescent phase in red-osier dogwood (*Cornus sericea* L.). Agric Meteorol 28:75–84
Kobayashi KD, Fuchigami LH (1983b) Modelling temperature effects in breaking rest in Red-osier Dogwood (*Cornus sericea* L.). Ann Bot 52:205–215
Kobayashi KD, Fuchigami LH, English MJ (1982) Modelling temperature requirements for rest development in *Cornus sericea*. J Am Soc Hortic Sci 107:914–918
Kramer K (1994a) A modelling analysis of the effects of climatic warming on the probability of spring frost damage to tree species in The Netherlands and Germany. Plant Cell Environ 17:367–377

Kramer K (1994b) Selecting a model to predict the onset of growth of *Fagus sylvatica*. J Appl Ecol 31:172–181
Kramer K (1995) Modelling comparison to evaluate the importance of phenology for the effects of climate change in growth of temperate-zone deciduous trees. Clim Res 5:119–130
Kramer K, Hänninen H (2009) The tree's annual cycle of development and the process-based modelling of growth to scale up from the tree to the stand. In: Noormets A (ed) Phenology of ecosystem processes. Applications in global change research. Springer, London
Kramer K, Mohren GMJ (1996) Sensitivity of FORGRO to climatic change scenarios: a case study on *Betula pubescens*, *Fagus sylvatica* and *Quercus robur* in the Netherlands. Clim Change 34:231–237
Kramer K, Van der Werf DC (2010) Equilibrium and non-equilibrium concepts in forest genetic modelling: population- and individually-based approaches. For Syst 19:100–112
Kramer K, Friend A, Leinonen I (1996) Modelling comparison to evaluate the importance of phenology and spring frost damage for the effects of climate change on growth of mixed temperate-zone deciduous forests. Clim Res 7:31–41
Kramer K, Buiteveld J, Forstreuter M, Geburek T, Leonardi S, Menozzi P, Povillon F, Schelhaas M, Cros ET, Vendramin GG, Werf DC (2008) Bridging the gap between ecophysiological and genetic knowledge to assess the adaptive potential of European beech. Ecol Model 216:333–353
Kramer K, Degen B, Buschbom J, Hickler T, Thuiller W, Sykes MT, de Winter W (2010) Modelling exploration of the future of European beech (*Fagus sylvatica* L.) under climate change-range, abundance, genetic diversity and adaptive response. For Ecol Manag 259(11):2213–2222. doi:10.1016/j.foreco.2009.12.023
Krinner G, Viovy N, Noblet-Ducoudrée N, Ogée J, Polcher J, Friedlingstein P, Ciais P, Sitch S, Prentice IC (2005) A dynamic global vegetation model for studies of the coupled atmospheric-biospheric system. Glob Biogeochem Cycles 19:1–33
Kupias R, Mäkinen Y (1980) Correlations of Alder pollen occurrence to climatic variables. In: First international conference on aerobiology, Munich
Lamb RC (1948) Effects of temperature above and below freezing on the breaking of rest in the Latham raspberry. J Am Soc Hortic Sci 51:313–315
Landsberg JJ (1974) Apple fruit bud development and growth; analysis and an empirical model. Ann Bot 38:1013–1023
Lang GA, Early JD, Arroyave NJ, Darnell RL, Martin GC, Stutte GW (1985) Dormancy—toward a reduced, universal terminology. Hortscience 20:809–812
Lechowicz MJ, Koike T (1995) Phenology and seasonality of woody-plants – an unappreciated element in global change research. Can J Bot 73(2):147–148
Leinonen I (1996) A simulation model for the annual frost hardiness and freeze damage of Scots pine. Ann Bot 78(6):687–693
Leinonen I, Kramer K (2002) Applications of phenological models to predict the future carbon sequestration potential of boreal forests. Clim Change 55(1–2):99–113
Leinonen I, Repo T, Hänninen H, Burr K (1995) A second-order dynamics model for the frost hardiness of trees. Ann Bot 76:89–95
Linkosalo T, Carter TR, Hakkinen R, Hari P (2000) Predicting spring phenology and frost damage risk of Betula spp. under climatic warming: a comparison of two models. Tree Physiol 20(17):1175–1182
Linkosalo T, Lappalainen HK, Hari P (2008) A comparison of phenological models of leaf bud burst and flowering of boreal trees using independent observations. Tree Physiol 28(12):1873–1882
Linsley-Noakes GC, Louw M, Allan P (1995) Estimating daily positive Utah chill units from maximum and minimum temperatures. J S Afr Soc Hortic Sci 5:19–24
Maurer C, Koch E, Hammer C, Hammer T, Pokorny E (2009) BACCHUS temperature reconstruction for the period 16th to 18th centuries from Viennese and Klosterneuburg grape harvest dates. J Geophys Res D Atmos 114(22):1–13

Mohren GMJ (1987) Simulation of forest growth, applied to Douglas fir stands in The Netherlands, Landbouw Universiteit Wageningen, Wageningen

Maurer C, Hammerl C, Koch E, Hammerl T, Pokorny E (2011) Extreme grape harvest data of Austria, Switzerland and France from AD 1523 to 2007 compared to corresponding instrumental/reconstructed temperature data and various documentary sources. Theor Appl Climatol 106(1–2):55–68. doi:10.1007/s00704-011-0410-3

Meier N, Rutishauser T, Pfister C, Wanner H, Luterbacher J (2007) Grape harvest dates as a proxy for Swiss April to August temperature reconstructions back to AD 1480. Geophys Res Lett 34:1–6. doi:10.1029/2007GL031381

Menzel A (2005) A 500 year pheno-climatological view on the 2003 heatwave in Europe assessed by grape harvest dates. Meteorol Z 14(1):75–77

Morin X, Augspurger C, Chuine I (2007) Process-based modeling of species' distributions: what limits temperate tree species' range boundaries? Ecology 88(9):2280–2291

Morin X, Viner D, Chuine I (2008) Tree species range shifts at a continental scale: new predictive insights from a process-based model. J Ecol 96:784–794

Morin X, Lechowicz MJ, Augspurger C, Keef JO, Viner D, Chuine I (2009) Leaf phenology in 22 North American tree species during the 21st century. Glob Change Biol 15:961–975

Možný M, Brázdil R, Dobrovolný P, Trnka M (2010) Cereal harvest dates in the Czech Republic between 1501 and 2008 as a proxy for March–June temperature reconstruction. Clim Change 110(3–4):810–821

Murray MB, Cannell MGR, Smith RI (1989) Date of budburst of fifteen tree species in Britain following climatic warming. J Appl Ecol 26:693–700

Myking T, Heide OM (1995) Dormancy release and chilling requirements of buds of latitudinal ecotypes of Betula pendula and B. pubescens. Tree Physiol 15:697–704

Phillipp M, Böcher J, Mattson O, Woodell SLJ (1990) A quantitative approach to the sexual reproductive biology and population structure in some Arctic flowering plants: *Dryas integrifolia, Silene acaulis* and *Ranunculus nivalis*. Medd Grönl Biosci 34:1–60

Piao SL, Ciais P, Friedlingstein P, Peylin P, Reichstein M, Luyssaert S, Margolis H, Fang JY, Barr L, Chen AP, Grelle A, Hollinger D, Laurila T, Lindroth A, Richardson AD, Vesala T (2007) Net carbon dioxide losses of northern ecosystems in response to autumn warming. Nature 451:49–52. doi:10.1038/nature06444

Pigott CD, Huntley JP (1981) Factors controlling the distribution of *Tilia cordata* at the northern limits of its geographical range. III nature and cause of seed sterility. New Phytol 87:817–839

Poirier M, Lacointe A, Améglio T (2010) A semi-physiological model of cold hardening and dehardening in walnut stem. Tree Physiol 30(12):1555–1569

Porter JR, Gawith M (1999) Temperatures and the growth and development of wheat: a review. Eur J Agron 10(1):23–36

Pouget R (1968) Nouvelle conception du seuil de croissance chez la vigne. Vitis 7:201–205

Pouget R (1972) Considérations générales sur le rythme végétatif et la dormance des bourgeons de la vigne. Vitis 11:198–217

Press WH, Flannery BP, Teukolsky SA, Vetterling WT (1989) Numerical recipes in Pascal. Cambridge University Press, Cambridge

Reaumur RAF (1735) Observations du thermomètre, faites à Paris pendant l'année 1735, comparées avec celles qui ont été faites sous la ligne, à l'isle de France, à Alger et quelques unes de nos isles de l'Amérique. Mem Paris Acad Sci 1735:545

Reich PB (1994) Phenology of tropical forests: patterns, causes, and consequences. Can J Bot 73:164–174

Repo T, Mäkelä A, Hänninen H (1990) Modelling frost resistance of trees. Silva Carelica 15:61–74

Richardson EA, Seeley SD, Walker DR (1974) A model for estimating the completion of rest for 'Redhaven' and 'Elberta' peach trees. Hortscience 9:331–332

Richardson EA, Anderson JL, Hatch AH, Seeley SD Asymcur (1982) An asymetric curvilinear fruit tree model. In: 21st international horticultural congress, Hamburg, p 2078

Richardson AD, Black TA, Ciais P, Delbart N, Friedl MA, Gobron N, Hollinger DY, Kutsch WL, Longdoz B, Luyssaert S, Migliavacca M, Montagnani L, Munger JW, Moors E, Piao SL, Rebmann C, Reichstein M, Saigusa N, Tomelleri E, Vargas R, Varlagin A (2010) Influence of spring and autumn phenological transitions on forest ecosystem productivity. Philos Trans R Soc B Biol Sci 365(1555):3227–3246. doi:10.1098/rstb.2010.0102
Riou C (1994) The effect of climate on grape ripening: application to the zoning of sugar content in the European community. Office des Publications Officielles des Communautés Européennes, Luxembourg
Robertson GW (1968) A biometeorological time scale for a cereal crop involving day and night temperatures and photoperiod. Int J Biometeorol 12:191–223
Ruml M, Vuković A, Vujadinović M, Djurdjević V, Ranković-Vasić Z, Atanacković Z, Sivčev B, Marković N, Matijašević S, Petrović N (2012) On the use of regional climate models: implications of climate change for viticulture in Serbia. Agric For Meteorol 158:53–62
Sakai A, Larcher W (1987) Frost survival of plants, vol 62, Ecological studies. Springer, Berlin/Heidelberg
Sarvas R (1972) Investigations on the annual cycle of development on forest trees active period. Commun Inst For Fenn 76(3):110
Sarvas R (1974) Investigations on the annual cycle of development of forest trees. Autumn dormancy and winter dormancy. Commun Inst For Fenn 84:1–101
Schwartz MD (1997) Spring index models: an approach to connecting satellite and surface phenology. In: Lieth H, Schwartz MD (eds) Phenology in seasonal climates. Backhuys Publishers, Leiden, pp 23–38
Schwartz MD, Marotz GA (1986) An approach to examining regional atmosphere-plant interactions with phenological data. J Biogeogr 13:551–560
Schwartz MD, Marotz GA (1988) Synoptic events and spring phenology. Phys Geogr 9:151–161
Schwartz MD, Ault TR, Betancourt JL (2012) Spring onset variations and trends in the continental USA: past and regional assessment using temperature-based indices. Int J Climatol (published online. doi: 10.1002/joc.3625)
Siminovitch D, Wilson CM, Briggs DR (1953) Studies on the chemistry of the living bark of the black locust in relation to its frost hardiness. V. Seasonal transformations and variations in the carbohydrates: starch-sucrose interconversions. Plant Physiol 28:383–400
Sinclair TR, Kitani S, Bruniard J, Horide T (1991) Soybean flowering date: linear and logistic models based on temperature and photoperiod. Crop Sci 31:786–790
Soltani A, Hammer GL, Torabi B, Robertson MJ, Zeinali E (2006) Modeling chickpea growth and development: phenological development. Field Crops Res 99(1):1–13
Spieksma FTH, Emberlin J, Hjelmroos M, Jäger S, Leuschner RM (1995) Atmospheric birch (Betula) pollen in Europe: trends and fluctuations in annual quantities and the starting dates of the seasons. Grana 34:51–57
Stone M (1977) An asymptotic equivalence of choice of model by cross-validation and Akaike's criterion. J R Stat Soc B 38:44–47
Thorhallsdottir TE (1998) Flowering phenology in the central highland of Iceland and implications for climatic warming in the Arctic. Oecologia 114:43–49
Thorsen SM, Hoglind M (2010) Modelling cold hardening and dehardening in timothy. Sensitivity analysis and Bayesian model comparison. Agric For Meteorol 150(12):1529–1542. doi:10.1016/j.agrformet.2010.08.001
Vegis A (1964) Dormancy in higher plants. Annu Rev Plant Physiol 15:185–224
Vesala T, Haataja J, Aalto P, Altimir N, Buzorius G, Garam E, Hämeri K, Ilvesniemi H, Jokinen V, Keronen P, Lahti T, Markkanen T, Mäkelä JM, Nikinmaa E, Palmroth S, Palva L, Pohja T, Pumpanen J, Rannik Ü, Siivola E, Ylitalo H, Hari P, Kulmala M (1998) Long-term field measurements of atmosphere-surface interactions in boreal forest combining forest ecology, micrometeorology, aerosol physics and atmospheric chemistry. Trends Heat Mass Momentum Transf 4:17–35

Wang JY (1960) A critique of the heat unit approach to plant response studies. Ecology 41(4):785–789

Wang E, Engel T (1998) Simulation of phenological development of wheat crops. Agric Syst 58(1):1–24

Wareing PF (1953) Growth studies in woody species. V. Photoperiodism in dormant buds of Fagus sylvatica L. Physiologia Plantarum 6:692–706

Weinberger JH (1950) Chilling requirements of peach varieties. Proc Am Soc Hortic Sci 56:122–128

White MA, Thornton PE, Running SW (1997) A continental phenology model for monitoring vegetation responses to interannual climatic variability. Glob Biogeochem Cycles 11:217–234

Yan W, Hunt LA (1999) An equation for modelling the temperature response of plants using only the cardinal temperatures. Ann Bot 84(5):607–614

Yin X, Kropff MJ, McLaren G, Visperas RM (1995) A nonlinear model for crop development as a function of temperature. Agric For Meteorol 77(1–2):1–16

Yiou P, García de Cortázar-Atauri I, Chuine I, Daux V, Garnier E, Viovy N, Leeuwen C, Parker AK, Boursiquot J-M (2012) Continental atmospheric circulation over Europe during the Little Ice Age inferred from grape harvest dates. Clim Past 8:577–588

Chapter 16
Animal Life Cycle Models (Poikilotherms)

Jacques Régnière and James A. Powell

Abstract This chapter discusses the theoretical basis and application of phenology models for poikilothermic animals, with a particular emphasis on insects. Realistic and accurate models make use of the non-linear, unimodal nature of physiological responses to temperature, using the rate-summation paradigm. In addition, the intrinsic (genetic) variation of developmental rates within populations is described and used to generate simulations where life-cycle events are distributed over time among individuals rather than occurring simultaneously within populations. The usefulness of circle maps to understand the impact of climate on poikilotherm life cycles is illustrated. The application of phenology models at landscape scale, and their use in the study of the impacts of climate and climate change on the distribution of poikilotherms are illustrated with two examples.

16.1 Introduction

Maintaining an appropriate seasonality is a basic ecological requisite for all organisms as critical life-cycle events must be keyed to the appropriate seasonal cycles, whether it be the wet-season/dry-season cycle of the tropics or the summer/winter cycle of temperate zones (Wolda 1988). The more pronounced the seasonal climatic signal, the stronger the requirement for an appropriate timing maintained through phenology. Although the importance of adaptive seasonal timing is no less for warm blooded than for cold blooded animals (e.g. appropriately timed reproduction, hibernation), the primary application of phenology models has been in poikilothermic animals (Nylin and Gotthard 1998) because the timing of events in

J. Régnière (✉)
Natural Resources Canada, Canadian Forest Service, Quebec City, QC, Canada
e-mail: jacques.regniere@rncan-nrcan.gc.ca

J.A. Powell
Department of Mathematics and Statistics, Utah State University, Logan, UT 84322, USA

M.D. Schwartz (ed.), *Phenology: An Integrative Environmental Science*,
DOI 10.1007/978-94-007-6925-0_16,

their life cycles is much more closely related to environmental conditions in these than in warm-blooded animals.

Although the phenology of poikilotherms is affected by many factors, temperature is the strongest determinant and is also the best understood. It deeply affects their metabolic, survival, developmental and reproductive rates. The thermal responses of the various life stages that compose their life cycles differ, and are configured by many selective pressures. Thus, the focus of phenology modeling efforts has been to relate temperature to phenological events and to determine the impact of event timing on the fitness of organisms. There is a large body of scientific literature relating the phenological responses of poikilotherms to temperature. Much of this knowledge pertains to insects or their close relatives. Our focus will therefore be on modeling seasonal life cycles of insects. The underlying physiological mechanisms of response to temperature are universal enough that the methods we describe are applicable to all poikilotherms.

In this chapter, we cover the basic concepts of modeling developmental responses to temperature and their variability (Sect. 16.2). We discuss the analysis of experimental data and the simulation of the effects of daily temperature fluctuations in Sect. 16.3. In Sect. 16.4, we discuss the consequences of annual patterns of temperature fluctuation on life cycle synchronization. In Sect. 16.5, we present applications of animal life cycle models to issues of landscape-wide prediction and climate change.

16.2 Temperature-Dependent Models

Relating temperature to the development of insects requires differentiating between *age* and *stage*. Although both are related to time, age is strictly chronological in nature, while stage is a developmental concept typically defined by distinct morphological characteristics often requiring a molt for transition from one stage to the next. Another time-related concept, *developmental rate*, is the speed of progression through a stage that depends on temperature in a predictable fashion. Assuming that it is constant within a stage, the developmental rate $\rho(T)$ at a constant temperature T is the inverse of the developmental time $\tau(T)$, the time required to complete that life stage at that temperature:

$$\rho(T) = 1/\tau(T) \tag{16.1}$$

Under variable temperatures T_t, where t is time, we define the *physiological age* a_j of an organism in life stage j as the integral:

$$a_j(t) = \int_{t_{j-1}}^{t} \rho_j(T_t)\, dt \tag{16.2}$$

Life stage j begins with $a_j = 0$ at t_{j-1}, which is the time of completion of the previous life stage. Stage j ends when $a_j = 1$, which defines t_j, the time at which stage $j + 1$ starts, and thus $0 \leq a_j \leq 1$. This relationship underlies all models of insect phenology based on rate summation (see Logan and Powell 2001). Once the mathematical relationships between temperature, time, and physiological age are defined, there remains the issue of finding an appropriate functional relationship between temperature and the developmental rate $\rho(T)$.

16.2.1 *Developmental Rate Functions*

The physiological responses of organisms to temperature have received considerable attention in the scientific literature for more than a century. The earliest functional form used to describe the relationship between temperature and rate of development was the *day-degree* model, a concept that dates to the 1700s (Wang 1960), and is still extensively used to model both animal and plant phenology. The degree-day concept is based on the assumption that the relationship between development rate and temperature T is linear once temperature exceeds some threshold T_0:

$$\rho(T) = \begin{cases} \alpha + \beta(T - T_0) & \text{if } T > T_0 \\ 0 & \text{otherwise} \end{cases} \tag{16.3}$$

Typically, a linear regression is fitted between the inverse of mean or median development times and a range of constant temperatures under laboratory conditions (dotted line in Fig. 16.1a). The familiar quantities of degree-day models are the threshold temperature $T_0 = -\alpha/\beta$ below which no development occurs, and the heat sum above this threshold $DD_{>T_0} = 1/\beta$ required for development to complete. Parameters of day-degree models can also be estimated from field observations, by choosing the value of T_0 that reduces variance in the value of $DD_{>T_0}$ at which the phenological event of interest occurs. The degree-day approach is simple both mathematically and in its application. The scientific literature provides degree-day model parameters for a large number of organisms (e.g. Nietschke et al. 2007).

Recently, debate has focused on the Metabolic Theory of Ecology (MTE) in which temperature and body weight primarily determine the rates at which life's central processes occur: metabolism, development, reproduction, population growth, species diversity and even ecosystem processes (Brown et al. 2004). The fundamental response to temperature imbedded in the MTE is the so-called Universal Thermal Dependence (UTD) that is expressed by the Arrhenius equation (Arrhenius 1889):

$$\rho(T) = b_0 \exp(-E/kT) \tag{16.4}$$

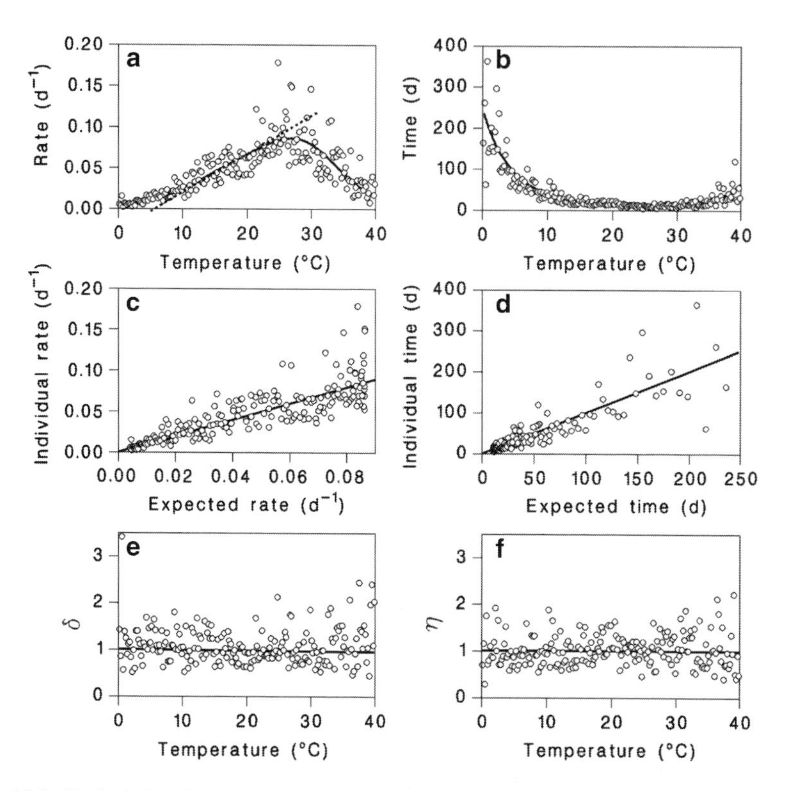

Fig. 16.1 Typical developmental response to temperature, built with the Sharpe-Schoolfield model. (**a**) Development rate as a function of temperature (- - - -: linear degree-day approximation; ——: Eq. (16.5)); ○: individual values generated with normally-distributed parameter variation (Eq. (16.6)). (**b**) Corresponding development times. Relationship between mean (expected) and individual development (**c**) rates (Eq. (16.8)) and (**d**) times (——: equality) (Eq. (16.9)). Relationship between temperature and random deviation from expected development (**e**) rates (Eq. (16.7)) and (**f**) times (——: unity)

where b_0 is a proportionality constant that varies across processes and taxa, E is the activation energy, and k is Boltzmann's constant relating energy to temperature, T in °K.

The Arrhenius equation constitutes a simple null hypothesis for development, as it has only two parameters (one of which, E, is presumed nearly constant), and is easily amenable to theoretical discussions and broad comparisons. However, it fails to represent the unimodal shape of developmental responses (solid line in Fig. 16.1a) over the range of temperatures to which poikilotherms are exposed (Knies and Kingsolver 2010), an observation that was made at least 80 years ago (Janisch 1932). The temperature range, thresholds and optimum temperatures at which this unimodality is expressed, as well as their variability, are critical aspects of thermal responses (Angilletta et al. 2002; de Jong and van der Have 2009; Dixon et al. 2009).

The Sharpe-Schoolfield model, an equation based on enzyme-reaction kinetics, is one truly mechanistic model based on physical principles that is available to explain the fundamental characteristics of physiological rate responses over the entire range of environmentally relevant temperatures (Sharpe and DeMichele 1977). Its numerator is the Eyring equation (Eyring 1935), describing the temperature dependence of chemical reactions, and is the theoretical counterpart of the empirical Arrhenius equation. Its denominator expresses the effect of temperature on the probability of an enzyme being in its active form. These processes generate a unimodal response decreasing to zero at either end of the enzyme's activation temperature range (solid line in Fig. 16.1a):

$$\rho(T) = \frac{T \cdot \exp\left[\frac{1}{R}\left(\varphi - \frac{H_A}{T}\right)\right]}{1 + \exp\left[\frac{H_L}{R}\left(\frac{1}{T_L} - \frac{1}{T}\right)\right] + \exp\left[\frac{H_H}{R}\left(\frac{1}{T_H} - \frac{1}{T}\right)\right]} \tag{16.5}$$

where T is temperature in °K, R is the universal gas constant, φ is the enzyme's entropy of activation, H_A its enthalpy of activation, H_L and H_H are lower and upper energies of enzyme inactivation, and T_L and T_H are the temperatures at which 50 % of this inactivation occurs. Schoolfield et al. (1981) modified the numerator of Eq. (16.5) algebraically to make it more intuitive for the sake of parameter estimation but without changing the model's nature.

A variety of empirical models are available as alternatives to the Sharpe-Schoolfield equation (see Régnière et al. 2012a). Nonlinear rate summation and unimodal response functions are required whenever simulations must cover temperatures over the full range of physiological activity. Life cycle phenomena that involve temperature extremes (diapause for example) also require nonlinear representation. The widely reported acceleration of development under realistic variable temperature is the consequence of this non-linearity (Worner 1992).

16.2.2 Developmental Rate Variability

There are many sources of variation in development within a species. Some are intrinsic (with a genetic basis), some are extrinsic (e.g. microclimate, nutrition) (Yurk and Powell 2010). We briefly discuss microclimatic influences in Sect. 16.5. Intrinsic developmental rate variability between populations or sexes can be handled by applying separate developmental response functions to different populations, or to each sex, when such variation exists. But other intrinsic variation (within populations) can be viewed as individual-level traits that are subject to natural selection under changing environmental conditions. If we assume that the Sharpe-Schoolfield model (Eq. 16.5) represents the essence of mechanisms behind developmental responses to temperature, it seems likely that values of its six parameters vary between individuals (de Jong and van der Have 2009).

Equation (16.5) could be written with each parameter submitted to random variation, such that

$$\varphi = \bar{\varphi} + \varepsilon_{\varphi}, \;\; H_A = \bar{H}_A + \varepsilon_{H_A}, \;\; H_L = \bar{H}_L + \varepsilon_{H_L}, \; \ldots \tag{16.6}$$

Here, each random term ε has its own distribution, with mean 0 and variance ρ_{ε}^2. Let us assume that these distributions are all normal, consistent with quantitative genetic theory. To illustrate this, we assigned a standard deviation of 2 % of the mean values to the four energy parameters φ, H_A, H_L and H_H, and 2 °C to the temperature parameters T_L and T_H. These variations in parameter values generated the variation in development rates and times illustrated in Fig. 16.1.

Because all parameters in Eq. (16.5) appear as arguments in exponential terms, the response to a change ε in a given parameter θ, relative to its mean value $\bar{\theta}$ can be written

$$\frac{\rho(T|\bar{\theta} + \varepsilon)}{\rho(T|\bar{\theta})} \cong a + b\, e^{k\varepsilon} \tag{16.7}$$

where k is a coefficient multiplying the original parameter, θ, including any temperature dependence. Terms a and b depend on which parameter θ is varied. If variation is applied to φ or H_A then $a = 0$ and $b = 1$. If variation is applied to one of the remaining parameters (H_L, T_L, H_H, T_H), then $a = 1 + \gamma$ and $b = -\gamma$, where γ is $e^{k\varepsilon}$ divided by the denominator of Eq. (16.5). Thus, Eq. (16.7) can be rewritten

$$r(T) = \rho(T|\bar{\theta} + \varepsilon) = \delta\rho(T) \tag{16.8}$$

where $\delta = a + b\, e^{k\varepsilon}$ is a lognormally-distributed random term with mean 1 and a variance that has some temperature dependence. If the variation of parameters of the developmental response function (Eq. 16.5) is not large relative to their means this temperature dependence would be difficult to detect in actual data and we can assume that δ is independent of temperature for all practical purposes. A more thorough discussion of these issues is beyond the scope of this chapter, but our example in Fig. 16.1 exhibits lognormal distribution of development rates and times with variance that is apparently constant with temperature (Fig. 16.1e, f). These conclusions hold even if the distribution of developmental parameters themselves is non-normal (e.g. uniform).

Finally, because $\delta = \mathrm{e}^{k\varepsilon}$ and $\eta = 1/\delta = \mathrm{e}^{-k\varepsilon}$ are both lognormally distributed, the distribution of development times is also lognormal, with the same multiplicative variance as development rates ($\sigma_{\eta}^2 = \sigma_{\delta}^2$: Fig. 16.1f). Because of Eq. (16.1), we get a very useful result that can be applied in the development of simulation models:

$$t(T) = \frac{1}{\delta\rho(T)} = \eta\tau(T) \tag{16.9}$$

Incorporating the variability of development rates increases the realism and power of phenology models. It can be applied directly using either individual-based or cohort-based models (see Sect. 16.3).

16.3 Data Analysis and Simulation Modeling

To illustrate this section, we performed experiments on overwintering larvae of the spruce budworm, *Choristoneura fumifarana* (Clem.). Eggs of this nearctic tortricid moth hatch in late summer, and neonate larvae immediately seek suitable locations on host conifers where they spin a silk shelter called a hibernaculum. There, without having fed, they molt to the second instar and enter diapause, a state of arrested development that allows them to resist winter conditions. Once this diapause is completed, by mid-winter (late January), the overwintered larvae lay in a dormant state waiting for warm temperatures. In spring (usually early May), the larvae emerge from the hibernaculum and start feeding on host needles. While it is not known what development processes actually occur during the time between the end of diapause and emergence of these larvae from the hibernaculum, the duration of this life stage is temperature dependent. We obtained several hundred overwintering larvae from the eggs collected in the wild in late summer 2011, and overwintered them under natural conditions in an outdoor insectary. In mid-February 2012, after their diapause had been completed but before they had been exposed to above-zero temperatures, subsamples were placed in controlled-temperature cabinets at one of eight constant temperatures between 8 ° and 28 °C, at 14L:10D photoperiods. Emergence was then monitored daily. Another subset of overwintered larvae remained in the insectary to emerge under natural conditions (Fig. 16.2b). Temperature in the insectary was recorded hourly to provide input for the simulation models (Fig. 16.2c).

Developmental time data from individuals emerged under constant temperatures were fitted to Eq. (16.5) by the maximum-likelihood method of Régnière et al. (2012a). This method provided not only parameter values for mean developmental times (Fig. 16.2a), but also the variance of the lognormal distribution used to describe the intrinsic variation between individuals (the values of δ in Eq. (16.8); see inset, Fig. 16.2a).

16.3.1 Individual-Based Models

Individual-based models simulate development for a collection of individuals, each having its own randomly-assigned traits and going through successive life stages at its own individual pace. At initiation, each individual is assigned random values of δ for each life stage using the lognormal distribution with mean 1 and appropriate variance (with our spruce budworm example, $\sigma_\delta = 0.214$). The model then submits each individual to the same temperature regime (here, the insectary's temperature

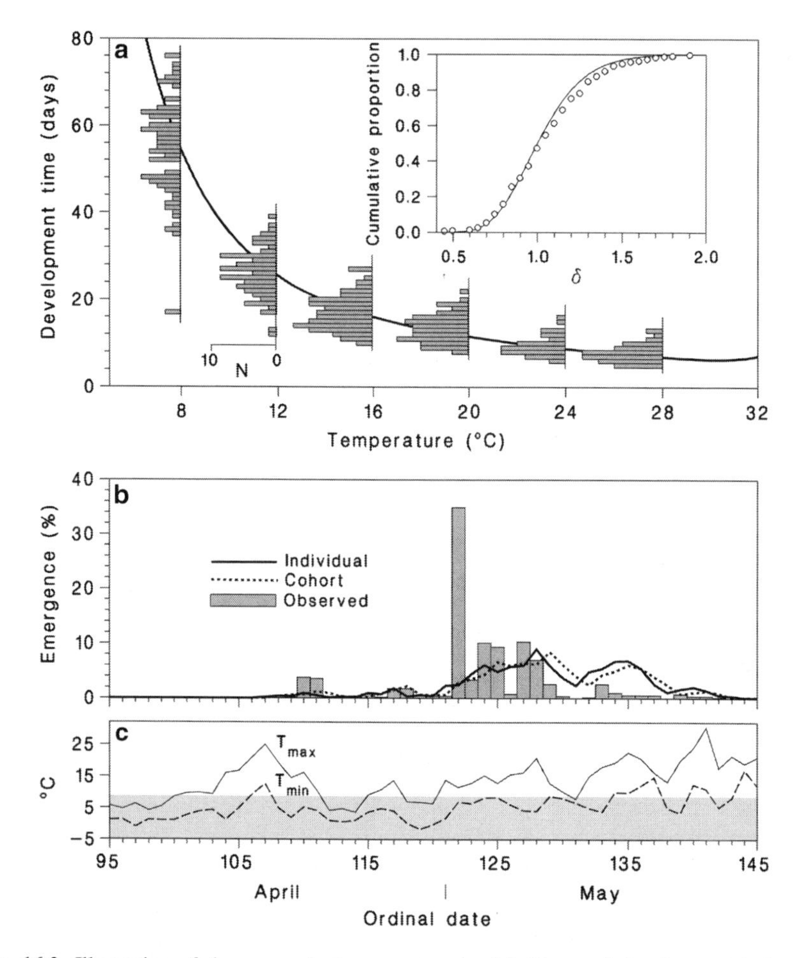

Fig. 16.2 Illustration of the spruce budworm example. (**a**) Observed developmental times of individuals (frequency histograms) at each of eight constant temperatures, and mean developmental time calculated from Eq. (16.5). Inset: observed and expected (lognormal) distribution of individual times relative to the mean, (δ). (**b**) Observed (*bars*) and simulated emergence of larvae from overwintering (*solid*: individual-based model; *dotted*: cohort-based model). (**c**) Daily minimum and maximum temperatures recorded in the insectary

records), evaluating Eq. (16.2) for each and compiling the number in each life stage at each time step (in our example: dormant or emerged larva). Output from this individual-based model is illustrated in Fig. 16.2b (solid line).

One advantage of the individual-based approach is the simplicity with which complex behaviors can be modeled. Among these, the transmission of traits from parent to progeny (e.g., development rates) offers the possibility of explicitly modeling natural selection on developmental response parameters, as inheritance is fundamental in object-oriented programming. Disadvantages of individual-based models include their stochastic nature (in assigning traits to individuals) and high computing demands imposed by the number of individuals simulated and the need for replication.

16.3.2 Cohort-Based Models

Cohort-based models keep track of numbers of individuals entering and exiting the various stages of the species' life history using cohorts, defined as groups of individuals that enter a given life stage at the same time step (most often, a single day). These models calculate the proportion of a cohort completing the life stage they are in during each time step. By specifying initial conditions (number of individuals entering the first life stage over time), the model cascades cohorts through successive life stages, generating realistic time distributions of life-stage frequencies under variable temperature regimes.

Sharpe et al. (1977) pointed out that when the developmental rate distribution within a population is known one can predict the resulting emergence time distribution. At constant temperature, if $f(r|T)$ is the distribution of developmental rates within the population, then the distribution of emergence times is $p(t|T) = f(t^{-1}|T)\ t^{-2}$, using the fact that $r = t^{-1}$ and $dr = -t^{-2}dt$. However, under varying temperature regimes, it is unclear how this applies unless the distribution of development times is stationary across temperatures (Gilbert et al. 2004). Gilbert et al. (2004) and later Yurk and Powell (2010) explicitly write down distributional assumptions on developmental rates and their consequences on developmental times for a range of possibilities. Here, we focus on how our assumption of log-normality may be projected onto a cohort model.

Returning to Eq. (16.9), the relationship between mean and observed development times, an individual's actual development time t in stage j, under variable temperatures $T(t)$ starting at time t_0, satisfies the equation

$$1 = \int_{t_0}^{t} \delta r_j(T(s))ds \text{ or } \int_{t_0}^{t} \rho(T(s))ds = a_j(t) - a_j(t_0) = \eta \tag{16.10}$$

where $a_j(t)$ is the age (or cumulative development since the time stage j started t_0), η is a random variable with lognormal distribution and mean one. The probability distribution function of η can then be written

$$p(\eta) = \frac{1}{\sqrt{2\pi\sigma_\eta^2\eta^2}} \exp\left[-\frac{\left(\ln(\eta) + \frac{1}{2}\sigma_\eta^2\right)^2}{2\sigma_\eta^2}\right] \tag{16.11}$$

Consequently, the probability of emerging at time t, given starting time of t_0, is

$$p(t|t_0) = \frac{1}{\sqrt{2\pi\sigma_\eta^2\left(a_j(t) - a_j(t_0)\right)^2}} \exp\left[-\frac{\left(\ln\left(a_j(t) - a_j(t_0)\right) + \frac{1}{2}\sigma_\eta^2\right)^2}{2\sigma_\eta^2}\right] \tag{16.12}$$

Equation (16.12) is the blueprint for a cohort model for predicting stage completion distributions. If the development data underlying the thermal responses of each life stage were analyzed by the method outlined by Régnière et al. (2012a), the estimates of σ_η^2 are readily available. Given starting numbers of individuals, n_d, in a cohort that ended stage $j-1$ at time d, the total number of individuals, N_t, completing stage j at time t can be written as a sum of individuals from all cohorts of stage j starting at times $d < t$ and completing the stage t:

$$N_t = \sum_{d<t} n_d p(t|d) \tag{16.13}$$

using expression (16.12) to calculate $p(t|d)$ for each cohort of stage j with $d < t$. A good illustration of the cohort approach appears in Fig. 6 of Logan (1988).

In our spruce budworm example, initial conditions are that all individuals start in the same dormant state on 1 February, and there are only two life stages: dormant or emerged. Because all parameters are the same for both individual- and cohort-based models in this example, the output of the cohort-based model (dotted line in Fig. 16.2b) is very similar to that of the individual-based model. The main difference between the two is that a new cohort is created once per day, which introduces a slight delay in simulated emergence frequencies. While the models are very good at predicting the general pattern of emergence (start, mid-point, end), there are differences between observed and simulated emergence. These are undoubtedly caused by not understanding and describing the details of processes that motivate a budworm larva to emerge at a particular time under fluctuating temperature conditions.

16.4 Phenology Models as Circle Maps

Diapause is the ultimate synchronizer of seasonality in species that live in temperate climates. It is often obligatory, as in the spruce budworm, although in many species it is induced by changes in photoperiod. In species that require diapause, only diapausing individuals survive winter to resume development all in the same physiological state in the following spring. There are few species for which a model is available that describes the thermal responses during diapause (e.g. the gypsy moth, discussed below) but most models assume that diapause completes some time after winter solstice.

Interestingly, even in the absence of diapause, the mechanisms of differential development responses (especially lower threshold temperatures) between successive life stages can synchronize the phenology of a species experiencing periodic temperature regimes (Grist and Gurney 1995; Powell et al. 2000; Jenkins et al. 2001). This can be seen by considering the G-function, which maps life-cycle end-times to start-times for a given annual temperature time series.

Each set of yearly temperatures, T_t, generates a unique completion date (t_j) for median individuals in life stage j, given an end date for the previous life stage (t_{j-1}). This time is determined by $a_j = 1$ in Eq. (16.2). Thus, t_j is a function of the completion date of the previous life stage, $t_j = g_j(t_{j-1})$. If the organism has J life stages, the timing of completion of the last pre-reproductive stage is then a function, G, of the median date of entry in the very first life stage. G is determined by nested application of completion functions from successive stages,

$$t_j = G(t_0) = g_j(t_{j-1} = g_{j-1}(t_{j-2} = g_{j-2}(t_{j-3} = g_{j-3}(t_{j-4} = \ldots)))) \tag{16.14}$$

If the temperatures are periodic for the year this defines a circle map from the cycle of starting day in one generation (from 1 to 365) to starting day in the next generation. Changing notation slightly, if t_J^n is the median completion day in generation n, and reproduction (creation of the next generation's first life stage) occurs immediately as that last stage is reached, then

$${t_0}^{n+1} = t_J^n = G({t_0}^n = {t_0}^{n+1}) \mod (365), \tag{16.15}$$

where mod(365) means "remainder after division by 365". Because this function maps start days from one generation to the next around the yearly circle, it is called a circle map (Powell and Logan 2005).

The G function is defined only implicitly through integration. However, in practice it is easy to calculate. Using our definition of $a_j = 1$ in Eq. (16.2), but this time allowing a to accumulate one "age" unit in each life stage without resetting it to zero, then the dates at which the successive life stages occur is obtained by:

$$t_j = \min_t\left[\left(a_j(t) - a_j(t_{j-1})\right) \geq 1\right] \tag{16.16}$$

where $\min_t$ means "find the smallest t at which", providing a simple computational algorithm for calculating t_j in terms of t_{j-1}.

When circle maps are iterated, their outputs (values of t_J^n in successive years or generations) are attracted either to fixed points or to orbits. Asymptotically, some iterate, G^n, of the G function must cross a line $y = t + m \times 365$, where m is a number of years and t is a date. The date, t^*, such that $G^n(t^*) = t^* + m \times 365$, where m and n are minimal, must be one date in an orbit of n dates taking n generations and m years to repeat. The simplest case, $m = 1$ and $n = 1$ is a univoltine fixed point (Fig. 16.3). Under such conditions, no matter when in the year the first generation starts, the life cycle converges from generation to generation to a single, stable initiation date (dashed arrows in the inset of Fig. 16.3). When $m = 2$ and $n = 1$ the cycle is semivoltine, while $m = 1$ and $n = 2$ represent a bivoltine cycle.

The theory of circle maps guarantees that any sequence of start dates across generations will eventually be attracted to such a cycle, and only one type of cycle can exist for a given temperature regime. Moreover, over a range of such regimes,

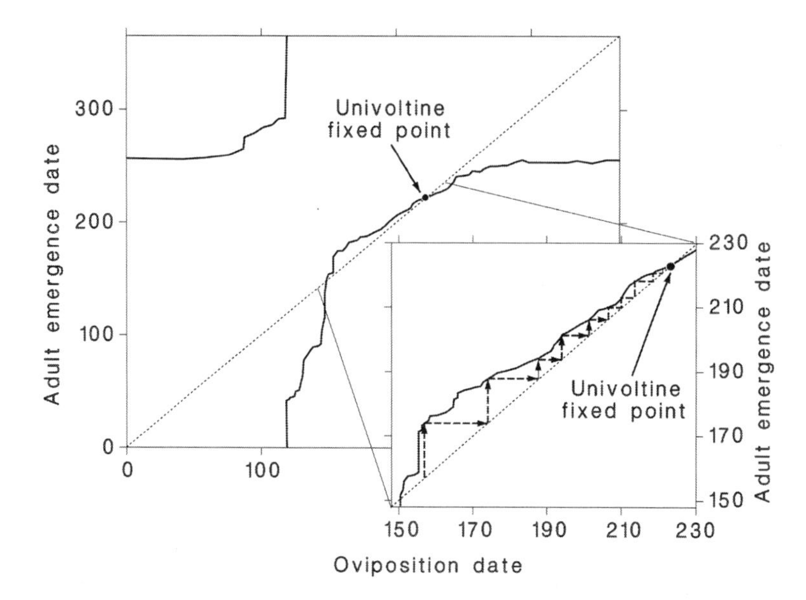

Fig. 16.3 Circle map showing the date of adult emergence of mountain pine beetle as a function of oviposition date under a temperature regime leading to univoltinism. *Dotted diagonal*: line of equality. *Inset*: high-resolution section, with *dashed arrows* showing the trajectory from one generation to the next towards the univoltine fixed point

there is a stability structure that can be stated as follow: the smaller and more relatively prime m and n are, the broader the range of temperatures for which m–n cycles will be attractive. Thus, a univoltine life history will occupy the broadest thermal range, followed by semi- and bi-voltine life histories. Between such bands of stable phenology, start dates approach cycles of asynchronous fractional voltinism (i.e. neither m nor n are 1, so population start dates are divided among different dates in the orbit, out of sequence).

To illustrate these concepts, we developed an example for the mountain pine beetle (*Dendroctonus ponderosae* Hopkins), using development rate functions parameterized in Régnière et al. (2012a). Temperatures were collected in the phloem of infested pine trees in central Idaho, and used as a basic seasonal temperature signal. A continuum of temperature regimes was generated by adding or subtracting a fixed amount to 2002 temperatures. For each, the G-function was calculated, iterating from a starting oviposition date (day 150) for 200 generations. The last 24 generations were graphed with respect to temperature (Fig. 16.4a). The final 24 generations lock onto a unique cycle for each temperature; the G-functions of some of these orbits are depicted in Fig. 16.4b–e. The thermal band of univoltine fixed points occupies a range of 1.5 °C around 2002 temperatures; beyond this the fixed point is lost and complicated, asynchronous cycles predominate. For mountain pine beetle, an appropriate seasonality requires emergence and oviposition between roughly day 145 and day 270. Adaptive, univoltine seasonality is suddenly lost if temperatures increase more than 1 °C above the 2002 mean. As is seen in Fig. 16.4a, the asynchronous

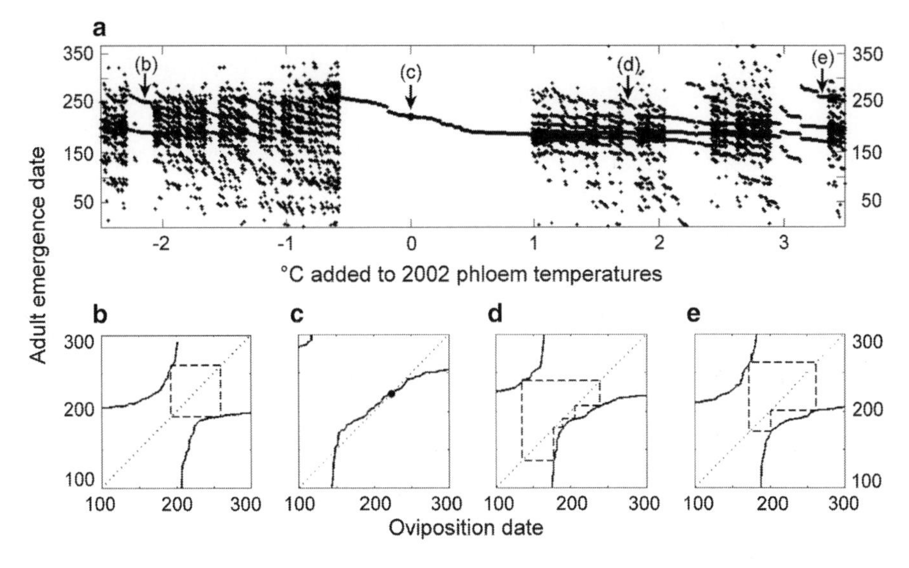

Fig. 16.4 Bifurcation plot and associated circle maps of mountain pine beetle development. (**a**) Adult emergence dates from the last 24 of 200 generations, under different temperature regimes (°C added to observed 2002 phloem temperatures). (**b**) A 2-generation, 3-year cycle appears near −2.2 °C (*solid line*: circle map; *dotted line*: equality; *dashed line*: 2-cycle seasonality). (**c**) A 1-generation, 1-year (univoltine fixed point) life cycle appears at temperatures within [−0.5, 1 °C] of those recorded in 2002. (**d**) A 5 generation, 4-year cycle appears at +1.8 °C. (**e**) A 3-cycle appears at +3.3 °C, in which two generations (emerging at 171 and 267) occur in a single year, followed by a single generation emerging on day 205

cycles outside of the univoltine band would be particularly maladaptive; not only would the population be divided among separate start dates at different positions in the cycle, but almost always some of the attractive dates are well outside an adaptive range of dates.

16.5 Predicting Phenology at Landscape Level

Having a good phenology model is one thing. Providing it with adequate weather inputs is another. This requires consideration of extrinsic sources of phenological variation, especially microclimatic and mesoclimatic. Often, temperature in the microhabitat where animals such as insects live is significantly different from ambient air temperature, due to factors such as exposure to sunlight, as under the bark on the south side of trees, in the surface layers of soil, or in the foliage of plants. Under such circumstances, temperatures in the animal's microhabitat can be many degrees warmer than ambient, especially during the day. When there is no solar input (at night, on the north side of trees or in the shade) temperatures fluctuations may be damped to some extent, so that the developing animal experiences a reduced thermal range. In either case, using incorrect temperatures (e.g. ambient, interpolated ambient extremes, or average daily temperatures) can radically change phenological predictions (Powell and Logan 2005; Powell and Bentz 2009).

One of the areas of central interest in animal ecology is the influence of landscape on ecological processes as determined by environmental conditions, plant communities, and movement (Haila 1995; McGarigal and Cushman 2002). Modeling these influences is key to improving our predictive understanding of these outcomes and improving area-wide management of pests, resources and ecosystems (Ryszkowski 2001; Boutin and Hébert 2002). In poikilotherm ecology, landscapes are key to determining patterns of abundance through their influence on local climate (Chen et al. 1999).

Gas physics, moisture content and solar radiation explain many of the effects of latitude and elevation on landscape air temperatures (Bolstad et al. 1998; Whiteman 2000). Many of these influences are modeled with thermal gradients. A dry (unsaturated) parcel of air cools by 0.98 °C per 100 m elevation (the adiabatic lapse rate). In addition, air temperature near 45°N or S drops by about 0.5 °C per degree of latitude away from the equator. Actual thermal gradients are usually smaller than the adiabatic lapse rate, varying with location, time of year, and even time of day (Régnière and Bolstad 1994). These variations are due in large part to the amount of moisture in the air, its general temperature, air circulation patterns and the proximity of large water bodies.

Several methods have been devised to interpolate climatic variables from a number of point data sources over a surrounding landscape. Local gradients (GIDS; Nalder and Wein 1998) are multiple linear regressions fitted to data from a number of nearby weather stations:

$$Y = a + m_E E + m_N N + m_W W \tag{16.17}$$

where Y is a climate value (e.g., minimum January air temperature or total precipitation), E is elevation, N is latitude and W is longitude of the region's weather stations; a is an intercept constant, and m_E, m_N and m_W are regional gradients for elevation, latitude and longitude. These are applied to differences in elevation (ΔE), latitude (ΔN), and longitude (ΔW) between the unsampled locations and a number of the nearest weather stations. Inverse square distances ($1/d^2$) between the n nearest stations and unsampled locations are used as weights in the estimation of the climate datum (Y_u):

$$Y_u = \frac{\sum_{i=1}^{n} \frac{1}{d_i^2}(Y_i + m_E \Delta E_i + m_N \Delta N_i + m_W \Delta W_i)}{\sum_{i=1}^{n} \frac{1}{d_i^2}} \tag{16.18}$$

This process can be applied equally well to monthly climate statistics (normals) or daily records to obtain air temperature and precipitation information for any number of unsampled points across a landscape.

16.5.1 Disaggregating Monthly Normals

Generation of daily estimates of temperature (and other weather variables such as precipitation) is important to investigation of climatic influences on animal ecology because of the cumulative nature of daily or even hourly conditions. This level of detail is especially important to the ecology of fast-developing poikilotherms, such as insects. As discussed above, thermal responses are strongly non-linear even when described by degree-day approximations, so average outcomes cannot be obtained from average inputs. Thirty days of monthly average temperature does not produce the same phenology as 30 days of variable temperature with the same average. This has been called the Kaufmann effect (Worner 1992).

For general questions concerning the effects of past climate, answers can be obtained by providing models with past weather records. However, such approaches have limitations. First, weather records usually cover a limited period at any given location (especially in North America). Second, it is never clear just how "general" a conclusion actually is about a given ecological process because of the limited amount of historical data available, especially in view of weather variability.

Daily weather generation provides a general approach that can be applied equally well to past, present and future (climate-changed) conditions. Several daily weather generators have been developed (Richardson 1981; Richardson and Wright 1984; Racsko et al. 1991; Hutchinson 1995; Wilks 1999), but many require substantial information inputs and re-parameterization for application in specific geographical areas. Régnière and Bolstad (1994) and Régnière and St-Amant (2007) developed a generally-applicable algorithm (TempGen) for simulation of daily minimum and maximum air temperatures and precipitation using monthly normals (30-year statistics updated on a decadal basis). Because TempGen uses monthly averages as input, it is well suited to accept the climate-change scenarios generated by Global Circulation Models. Output from TempGen, based on climate-changed normals, can therefore be used readily to simulate the impact of global warming on ecological processes modeled from daily climate inputs.

Running simulation models of animal development that use daily weather inputs can be computationally demanding. It may be prohibitively time consuming to produce model output for each unit (pixel, or raster) of a landscape (output map), except with the simplest of degree-day models (e.g. Russo et al. 1993). A solution is to run models for a relatively small number (a few thousand) of randomly-located points and use spatial interpolation to estimate model output at other locations. Régnière and Sharov (1999) used universal kriging with elevation as external drift variable as an interpolation method (see Isaaks and Srivastava 1989 for methodological details). Many other interpolation techniques exist, a subject that is outside the scope of this chapter.

16.5.2 Spruce Budworm Phenology

The spruce budworm defoliates firs and spruces in boreal forests of North America on a cycle of 30–40 years (Royama 1984). Following an obligate winter diapause, larvae emerge over a 2-week period in spring, and the entire development from emergence in the spring to entry of the next generation into diapause can be as short as 60 days. For the remaining time each year, the diapausing insect lives on energy reserves placed in the egg by the female moth.

Régnière et al. (2012b) used an object-oriented, individual-based model of the insect's phenology and overwinter survival to predict potential spruce budworm population growth rates in response to climate change. This model predicts that to the south and at lower elevations the spruce budworm's geographical range is limited by high overwinter mortality. This high mortality occurs when the insect exhausts its energy reserves during a prolonged diapause (i.e. non-diapause development is too rapid). The limiting factor to the north of its range and at higher elevations is the ability of eggs laid in late summer to hatch before being killed by early frost.

Here, we used the same model (Régnière et al. 2012b) to illustrate the geographical variation in the date at which peak egg hatch is expected to occur in Canada east of Ontario, under current climate using normals from the most recent Standard Normal Generating Period (SNGP), 1981–2010, as well as under climate change. Future climate normals (SNGP 2011–2040) were derived from daily output of the Canadian Regional Climate Model (CRCM) version 4.2.0 runs ADJ and ADL (Music and Caya 2007) under the IPCC A2 emissions scenario.

We ran the individual-based spruce budworm seasonal biology model at 1,000 randomly-located simulation points over a landscape ranging from 42 to 52°N and −79.5 to −89.5°E, using the BioSIM simulation control environment. This model is stochastic, and input daily temperature series are also stochastically generated from normals, so each run was replicated ten times and predicted dates of peak egg abundance averaged for each location. This was done using both sets of normals (actual 1981–2010 and predicted 2011–2041). We chose kriging with elevation as external drift variable as interpolation method.

Resulting maps are shown in Fig. 16.5, illustrating the adaptation issues that this insect species faces. To the south, especially in low-elevation valleys of northern New York State, Vermont and New Hampshire, hatch normally occurs prior to the end of June, forcing the species to spend much of the warmest period of the year (July, August) wasting precious energy stores available to diapausing larvae in maintenance. Under climate change, such conditions are expected to spread northward into southern Quebec. To the north and at higher elevations, eggs hatch after the second week of August. In most years this is probably fine, but the insect faces

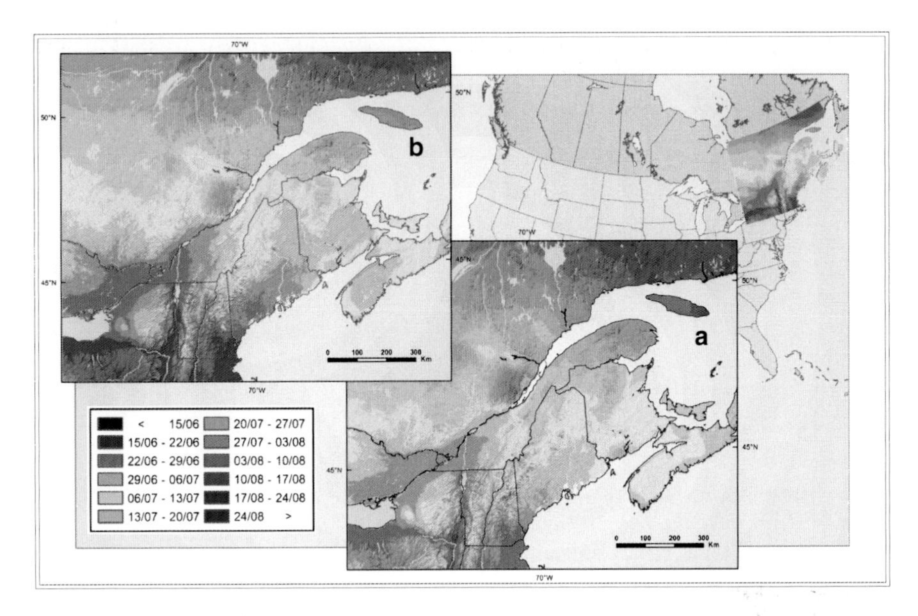

Fig. 16.5 Date of peak spruce budworm egg hatch predicted by the spruce budworm seasonal biology model in eastern Canada. (**a**) With current climate normals (1971–2010). (**b**) With climate changed normals, period 2011–2040

significant risk of exposing this cold-intolerant life stage to early frost. This is believed to be the main factor limiting spruce budworm's distribution in the northern reaches of the boreal forest. Under climate change, we expect this frost-kill risk to drop considerably, especially in the central and eastern mountains and along the Upper North Shore of the St-Lawrence River in Quebec.

16.5.3 Gypsy Moth Phenology

The gypsy moth, *Lymantria dispar* (L.) was accidentally introduced in eastern North America in 1869, and has spread gradually to the north, west and south (Liebhold et al. 1992). It is a pest of deciduous trees, especially oaks (*Quercus*) (Montgomery 1990). This insect overwinters in obligate diapause as an embryonated egg. Hypotheses about the determinants of its rate of spread and eventual range in North America have focused on egg mortality due to low winter temperatures, or on forest susceptibility (Sharov et al. 1999). Limitations that the insect will encounter in establishing to the west and south of its current distribution are less well understood (see Allen et al. 1993).

A detailed cohort-based model of gypsy moth phenology is available that can simulate the entire life cycle of the insect through successive generations in any climate; it was used to determine the areas of Canada most likely to support

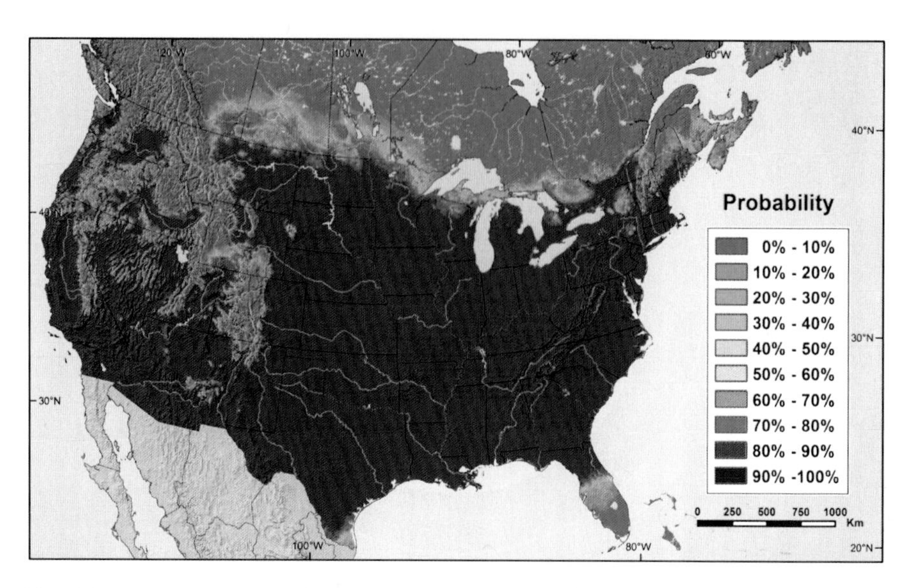

Fig. 16.6 Probability of gypsy moth establishment in North America based on its ability to achieve an adaptive seasonality

establishment of this exotic insect purely on the basis of local climate (Régnière et al. 2009). This analysis was based on whether or not the model predicted a biologically feasible life cycle for the insect in a given location under normal climatic conditions and under climate change. The feasibility of the life cycle was determined from a circle map (*G* function) of peak oviposition dates in 20 successive generations on the same weather input. If a fixed date was reached, no later than the end of October (a time when temperatures are too cold for eggs to undergo embryonic development and enter diapause successfully), the gypsy moth was presumed to have the potential of establishing there.

The same approach was used here to produce a map of the probability of establishment of gypsy moth throughout North America, north of Mexico (Fig. 16.6). For each of several thousand simulation points in North America, the outcome of the model was rated as 0 (seasonality did not remain viable for 20 generations) or 1 (seasonality did remain viable), with 30 replications. The average outcome for each point was used as an estimate of the probability of gypsy moth establishment at that location:

$$P_i = \frac{1}{n+2}\left(1 + \sum_{j=1}^{n} p_{ij}\right) \tag{16.19}$$

where p_{ij} is the simulation outcome for point i and replicate j. Logit-transformed probabilities were interpolated spatially by universal kriging using elevation as a drift variable over a digital elevation model of North America at 30 arc sec ($\approx$1 km) resolution. The resulting map (Fig. 16.6), back-transformed to a probability scale, shows the importance of climate as a limiting factor at the northern edge of the insect's distribution in the south of Canada and in the western mountains of the United States as well as in the southernmost portions of Texas and Florida. We emphasize that these predictions are based solely on feasible seasonality, not considering climatic influences such as overwinter mortality due to cold or drought.

16.6 Conclusions

The life cycles of poikilothermic animals can be modeled with great realism and accuracy both in terms of the timing and variability of key events. For this reason, phenology models form an ideal framework for simulation of seasonal population dynamics based on detailed descriptions of the interactions between the animals and their environment, host resources (or prey) and natural enemies. Phenology models are a cornerstone of integrated pest management, as they enable accurate timing of monitoring operations and application of control strategies for the target life stages. Once costs and benefits of various life history strategies are included, these models can be used to predict the geographical range of species on the basis of their physiological responses to temperature and other key environmental factors that affect their development, survival and reproduction. One of the most exciting and promising avenues for development of individual-based phenology models is their application to evolutionary biology. By their very nature, these models keep track of individuals with specific and inheritable developmental response traits (in the form of parameter values), and constitute an ideal template for simulation of adaptation to changing environmental conditions.

A wide variety of modeling techniques are available, from simple degree-day approaches derived from field observations to the more sophisticated non-linear and stochastic approaches. The latter require experimental data for parameter estimation, as well as field observations for model calibration and validation, but they also are the models that offer the most versatility and reliability.

In this chapter, we have described the theoretical basis for the development of these models, their fundamental structure, and their application to the study of life cycles. Computer technologies allow us to generate landscape-wide phenological projections that are useful in the conduct of area-wide integrated pest management activities. They also provide us with the ability to study and better understand the ecology and distribution of organisms based on comparison of observations with model predictions. We can also use these tools to analyze the probable reactions of indigenous species to changing climate, as was done for the spruce budworm (Régnière et al. 2012b) and the mountain pine beetle (Bentz et al. 2010; Safranyik et al. 2010), as well as to predict the probable distribution and thriving of invasive species, such as gypsy moth (Régnière et al. 2009).

References

Allen JC, Foltz JL, Dixon WN, Liebhold AM, Colbert JJ, Régnière J, Gray DR, Wilder JW, Christie I (1993) Will the gypsy moth become a pest in Florida? Fla Entomol 76:102–113
Angilletta MJ, Niewiarowski PH, Navas CA (2002) The evolution of thermal physiology in ectotherms. J Therm Biol 27:249–268
Arrhenius S (1889) Uber die reaktionsgeschwindigkeit bei der inversion von rohrzucker durcj sauren. Zeitschrift for Physik Chemique 4:226–248
Bentz BJ, Régnière J, Fettig CJ, Hansen EM, Hayes JL, Hicke JA, Kelsey RG, Negrón JF, Seybold SJ (2010) Climate change and bark beetles of the western United States and Canada: direct and indirect effects. Bioscience 60:602–613
Bolstad PV, Swift L, Collons F, Régnière J (1998) Measured and predicted air temperatures at basin to regional scales in the southern Appalachian mountains. Agric For Meteorol 91:161–176
Boutin S, Hébert D (2002) Landscape ecology and forest management: developing an effective partnership. Ecol Appl 12:390–397
Brown JH, Gillooly JF, Allen AP, Savage VM, West GB (2004) Toward a metabolic theory of ecology. Ecology 85:1771–1789
Chen J, Saunders SC, Crow TR, Naiman RJ, Brosofske KD, Mroz GD, Brookshire BL, Franklin JF (1999) Microclimate in forest ecosystem and landscape ecology: variations in local climate can be used to monitor and compare the effects of different management regimes. Bioscience 49:288–297
de Jong G, van der Have TM (2009) Temperature dependence of development rate, growth rate and size: from biophysics to adaptation. In: Whitman DW, Ananthakrishnan TN (eds) Phenotypic plasticity of insects: mechanisms and consequences. Science Publishers, Enfield
Dixon AF, Honek GA, Keil P, Kotela MAA, Sizling AL, Jarosik V (2009) Relationship between the minimum and maximum temperature thresholds for development in insects. Funct Ecol 23:257–264
Eyring H (1935) The activated complex and the absolute rate of chemical reactions. Chem Rev 17:65–77
Gilbert E, Powell JA, Logan JA, Bentz BJ (2004) Comparison of three models predicting developmental milestones given environmental and individual variation. Bull Math Biol 66:1821–1850
Grist EPM, Gurney WSC (1995) Stage-specificity and the synchronization of life cycles to periodic environmental variation. J Math Biol 34:123–147
Haila Y (1995) A conceptual genealogy of fragmentation research: from island biogeography to landscape ecology. Ecol Appl 12:321–334
Hutchinson MF (1995) Stochastic space-time weather models from ground-based data. Agric For Meteorol 73:237–264
Isaaks EH, Srivastava RM (1989) An introduction to geostatistics. Oxford University Press, New York
Janisch E (1932) The influence of temperature on the life-history of insects. Trans R Soc Entomol Lond 80:137–168
Jenkins JL, Powell JA, Logan JA, Bentz BJ (2001) Low seasonal temperatures promote life cycle synchronization. Bull Math Biol 63:573–595
Knies JI, Kingsolver JG (2010) Erroneous Arrhenius: modified Arrhenius model best explains the temperature dependence of ectotherm fitness. Am Nat 176:227–233
Liebhold AM, Halverson JA, Elmes GA (1992) Gypsy moth invasion in North America: a quantitative analysis. J Biogeogr 19:513–520
Logan JA (1988) Toward an expert system for development of pest simulation models. Environ Entomol 17:359–376
Logan JA, Powell JA (2001) Ghost forests, global warming, and the mountain pine beetle. Am Entomol 47:160–173

McGarigal K, Cushman SA (2002) Comparative evaluation of experimental approaches to the study of habitat fragmentation effects. Ecol Appl 12:335–345

Montgomery ME (1990) Variation in the suitability of tree species for the gypsy moth. In: Gottschalk KW, Tivery MJ, Smith SI (eds) Proceedings U.S. Department of Agriculture Interagency Gypsy Moth Research Review. USDA Forest Service General technical report NE 146

Music B, Caya D (2007) Evaluation of the hydrological cycle over the Mississippi River basin as simulated by the Canadian regional climate model (CRCM). J Hydrometeorol 8:969–988

Nalder IA, Wein RW (1998) Spatial interpolation of climatic normals: test of a new method in the Canadian boreal forest. Agric For Meteorol 9:211–225

Nietschke BS, Magarey RD, Bochert DM, Calvin DD, Jones E (2007) A developmental database to support insect phenology models. Crop Prot 26:1444–1448

Nylin S, Gotthard K (1998) Plasticity in life-history traits. Ann Rev Entomol 63:63–84

Powell JA, Bentz BJ (2009) Connecting phenological predictions with population growth rates for mountain pine beetle, an outbreak insect. Landsc Ecol 24:657–672

Powell JA, Logan JA (2005) Insect seasonality: circle map analysis of temperature-driven life cycles. Theor Popul Biol 67:161–179

Powell JA, Jenkins J, Logan JA, Bentz BJ (2000) Seasonal temperature alone can synchronize life cycles. Bull Math Biol 62:977–998

Racsko P, Szeidl L, Semonov M (1991) A serial approach to local stochastic weather models. Ecol Model 57:27–41

Régnière J, Bolstad PV (1994) Statistical simulation of daily air temperature patterns in eastern North America to forecast seasonal events in insect pest management. Environ Entomol 23:1368–1380

Régnière J, Sharov A (1999) Simulating temperature-dependent ecological processes at the sub-continental scale: male gypsy moth flight phenology as an example. Int J Biometeorol 42:146–152

Régnière J, St-Amant R (2007) Stochastic simulation of daily air temperature and precipitation from monthly normals in North America north of Mexico. Int J Biometeorol 51:415–430

Régnière J, Nealis VG, Porter K (2009) Climate suitability and management of the gypsy moth invasion into Canada. Biol Invasions 11:135–148

Régnière J, Powell JA, Bentz BJ, Nealis VG (2012a) Effects of temperature on development, survival and reproduction of insects: experimental design, data analysis and modeling. J Insect Physiol 58:634–647

Régnière J, St-Amant R, Duval P (2012b) Predicting insect distributions under climate change for physiological responses: *Spruce budworm* as an example. Biol Invasions 14:1571–1586

Richardson CW (1981) Stochastic simulation of daily precipitation, temperature and solar radiation. Water Resour Res 17:182–190

Richardson CW, Wright DA (1984) WGEN: a model for generating daily weather variables. US Department of Agriculture, Washington, DC, Agricultural Research Service 8

Royama T (1984) Population dynamics of the *spruce budworm*, *Choristoneura fumiferana* (Clem.). Ecol Monogr 54:429–462

Russo JM, Liebhold AW, Kelley AGW (1993) Mesoscale weather data as input to a gypsy moth (Lepidoptera: Lymantriidae) phenology model. J Econ Entomol 86:838–844

Ryszkowski L (2001) Landscape ecology in agroecosystems management. Advances in agroecology. CRC Press, Boca Raton

Safranyik L, Carroll AL, Régnière J, Langor DW, Riel WG, Shore TL, Peter B, Cooke BJ, Nealis V, Taylor SW (2010) Potential for range expansion of mountain pine beetle into the boreal forest of North America. Can Entomol 142:415–442

Schoolfield RM, Sharpe PJH, Magnuson CE (1981) Non-linear regression of biological temperature-dependent rate models based on absolute reaction-rate theory. J Theor Biol 88:719–731

Sharov AA, Pijanowski BC, Liebhold AM, Gage SH (1999) What affects the rate of gypsy moth (Lepidoptera: Lymantriidae) spread: winter temperature or forest susceptibility? Agric For Entomol 1:37–45
Sharpe PJH, DeMichele DW (1977) Reaction kinetics of poikilotherm development. J Theor Biol 64:649–670
Sharpe PJH, Curry GL, DeMichele DW, Cole CL (1977) Distribution model of organism development times. J Theor Biol 66:21–38
Wang JY (1960) A critique of the heat unit approach to plant response studies. Ecology 41:785–790
Whiteman CD (2000) Mountain meteorology: fundamentals and applications. Oxford University Press, New York
Wilks DS (1999) Simultaneous stochastic simulation of daily precipitation, temperature and solar radiation at multiple sites in complex terrain. Agric For Meteorol 96:85–101
Wolda H (1988) Insect seasonality: why? Annu Rev Ecol Syst 19:1–18
Worner SP (1992) Performance of phenological models under variable temperature regimes: consequences of the Kaufmann or rate summation effect. Environ Entomol 21:689–699
Yurk BP, Powell JA (2010) Modeling the effects of developmental variation on insect phenology. Bull Math Biol 76:1334–1360

Chapter 17
Daily Temperature-Based Temporal and Spatial Modeling of Tree Phenology

Xiaoqiu Chen

Abstract Using *Ulmus pumila* leaf unfolding and leaf fall data at 46 stations during the 1986–2005 period in China's temperate zone, daily temperature-based temporal and spatial phenology models were constructed. The daily temperature-based temporal phenology model provides a more precise and rational tool than the monthly or multi-monthly mean temperature-based phenology model in detecting responses of tree phenology to temperature. For the entire China's temperate zone, a 1 °C increase in spring and autumn daily temperatures during the optimum length periods may induce an advancement of 2.8 days in the beginning date and a delay of 2.1 days in the end date of the *Ulmus pumila* growing season, respectively. Meanwhile, the daily temperature-based spatial phenology model provides a more robust tool than the geo-location based spatial phenology model in simulating and predicting spatial patterns of tree phenology. Regarding 20-year mean growing season modeling, a spatial shift in mean spring and autumn daily temperatures by 1 °C may cause a spatial shift in mean beginning and end dates of the *Ulmus pumila* growing season by −3.1 and 2.6 days, respectively.

17.1 Introduction

Simulating temporal and spatial relationships between occurrence dates of plant phenophases and climatic factors is crucial not only for predicting phenological responses to climate change but also for identifying the carbon-uptake period (Goulden et al. 1996; Black et al. 2000; White and Nemani 2003; Barr et al. 2004; Churkina et al. 2005; Baldocchi 2008) and examining the seasonal exchanges of water and energy between land surface and atmosphere (Wilson and Baldocchi 2000; Kljun et al. 2007). So far, many studies have focused on time series simulation of plant

X. Chen (✉)
College of Urban and Environmental Sciences, Peking University, Beijing, China
e-mail: cxq@pku.edu.cn

M.D. Schwartz (ed.), *Phenology: An Integrative Environmental Science*,
DOI 10.1007/978-94-007-6925-0_17,

phenology using statistical models (Chen 1994; Chmielewski and Rötzer 2001; Schwartz and Chen 2002; Matsumoto et al. 2003; Menzel 2003; Askeyev et al. 2010) at individual sites or in specific areas. The traditional method for detecting temporal phenological responses to temperature was carried out by computing mean temperature during several months, normally including the month in which the mean phenological event occurred as well as preceding months (Chen 1994; Sparks et al. 2000; Chmielewski and Rötzer 2001; Menzel 2003; Gordo and Sanz 2010). Using preceding mean monthly temperature as the independent variable is simple and practical but may not be precise enough because the phenological event is not likely induced by the integral mean monthly temperature exactly, but by mean daily temperature during a certain length period (*LP*) of days. Matsumoto et al. (2003) used mean daily temperature during an optimum *LP* prior to the mean occurrence date of a phenological event to estimate phenological response to temperature. This method is detailed and precise but may not be rational sufficiently because using the mean occurrence date of the phenological event as the end date of *LP* does not consider temperature effects in the period after the mean occurrence date of the phenological event, and excludes temperature effects in the period before the flexible beginning date of *LP* on earlier phenological occurrence dates. As an alternative method, a new approach was proposed to identify phenological response to mean daily temperature.

In contrast with temperature-based temporal phenology modeling, few studies have focused on temperature-based spatial phenology modeling. The traditional approach for simulating spatial series of plant phenology was implemented by establishing a multiple linear regression equation between multiyear mean occurrence dates of a plant phenophase (y) at individual sites and geo-location parameters, such as longitudes (x_1), latitudes (x_2) and altitudes (x_3) of the utilized sites (Nakahara 1948; Park-Ono et al. 1993; Rötzer and Chmielewski 2001; Hense et al. 2002). This approach has two major disadvantages: (1) geo-location parameters are not climatic factors, so that they cannot explain the essential environmental causes nor detect the climatic differences driving spatial variations of plant phenology; and (2) since geo-location parameters are constant at a given site, the multiple linear regression equation cannot represent the interannual variation of plant phenological spatial pattern related to climate change. Because air temperature is the most important factor influencing spatial variations of plant phenology (Chen 1994; Chmielewski and Rötzer 2001; Schwartz and Chen 2002; Menzel 2003; Matsumoto et al. 2003), the spatial series of multiyear mean monthly temperature within a length period (*LP*) of months at individual sites (replacing geo-locations) were used by Chen et al. (2005) for simulating the spatial pattern of growing season beginning date (BGS) and end date (EGS) derived from surface phenology and remote sensing data. However, the correlations between BGS and March–May temperature spatial series and between EGS and August–October temperature spatial series may not reflect the spatial relationship between plant phenology and temperature precisely enough because spatial variations of phenological events are not likely induced by the integral mean monthly temperature exactly, but by mean daily temperature within a certain *LP* of days. In order to simulate the spatial pattern of plant phenology more precisely, mean daily temperature within a *LP* of days should be identified.

17.2 Daily Temperature-Based Temporal Modeling of *Ulmus pumila* Phenology

17.2.1 Study Area and Plant Species

The study area is located in China's temperate zone, including warm, middle and cold temperate zones from south to north. The dominant vegetation types include deciduous conifer forest, deciduous broad-leaved and coniferous mixed forest, deciduous broad-leaved forest, temperate steppe, temperate desert, etc. (Compilation Committee of the Vegetation of China 1980). *Ulmus pumila* (Siberian Elm) is a local deciduous tree and has its natural distribution from the northern subtropical zone to the cold temperate zone which is approximately at 32°–53° north latitude, 74°–134° east longitude, and 3–3,650 m above sea level. In this range, the annual mean temperature is about −4 to +15 °C and annual precipitation is between 16 and 1,200 mm (Ma 1989). Thus, it can serve as an "indicator species" for the phenological study according to the criteria proposed by Newman and Beard (1962). Recently, Ghelardini and Santini (2009) found a good relationship between budburst date of some European elm species and winter-spring temperature. This shows that phenology of the genus *Ulmus* can be used as a sensitive indicator of climate change. Therefore, selecting *Ulmus pumila* as the sample species for studying the phenological growing season and its response to climate change is appropriate and representative.

17.2.2 Phenological and Climate Data

So far, almost all conventional phenological studies in China have been based on discontinuous time series from a few stations in a data set of the Chinese Academy of Sciences (Lu et al. 2006; Zheng et al. 2006). In order to reveal phenological performances of broad geographic coverage and continuous time series in detail, my study group has recently digitized the largest phenological data set in China with permission from the China Meteorological Administration (CMA). Using the new data set, a representative plant species, *Ulmus pumila* was selected to analyze temporal and spatial variability of the growing season and its relation to temperature. The *Ulmus pumila* growing season was defined as the period between the beginning date of leaf unfolding (BGS) and the end date of leaf fall (EGS). The leaf unfolding beginning was identified when a few leaves are fully spreading in spring, whereas the leaf fall end was determined when almost all leaves fall on the ground (China Meteorological Administration 1993). Forty-six stations were chosen for the analysis, which were mostly distributed across all of China's temperate zone, except desert and high mountain areas (Fig. 17.1). The phenological time series are more than 16 years during the period 1986–2005.

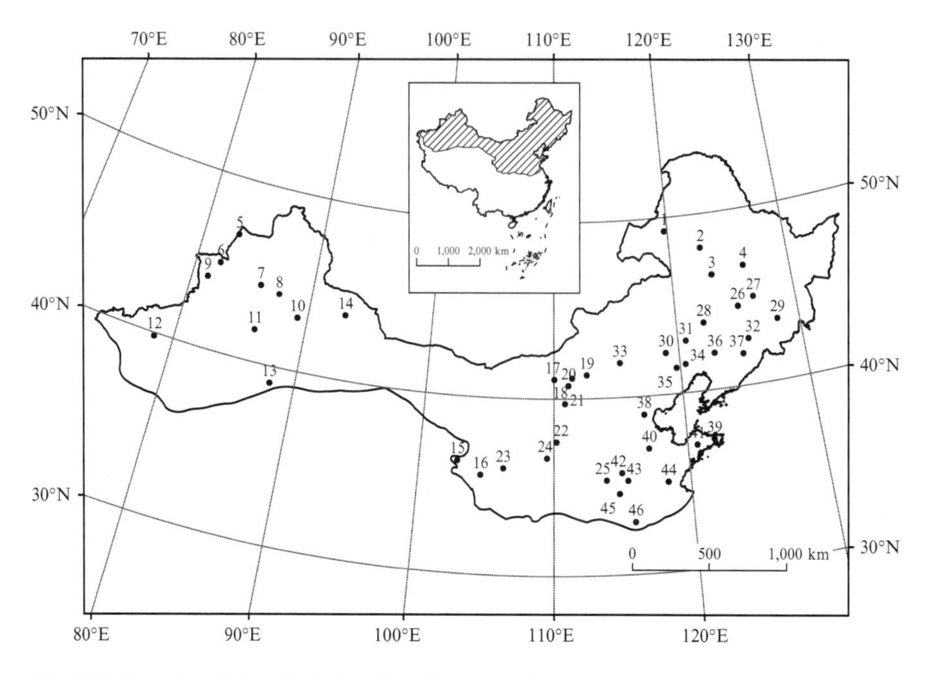

Fig. 17.1 Location of phenological stations for temporal modeling

Air temperature data were acquired from the China Meteorological Data Sharing Service System, including daily mean air temperature at 343 stations in China's temperate zone from 1986 to 2005. Since several phenological stations were not located nearby meteorological stations, ANUSPLIN 4.2 (Hutchinson 2002) and Digital Elevation Model (DEM) data were used to interpolate the daily mean air temperature into 8 km × 8 km grids over China's temperate zone. Thus, gridded daily mean air temperature data were obtained at the phenological stations without meteorological observations.

17.2.3 Temporal Phenology Model (Methodology)

The basic hypothesis of the temporal phenology model was that the occurrence dates (day of year, DOY) of a phenological event are mainly influenced by mean daily temperature within a particular *LP* during and before its occurrence (Chen and Xu 2012a). In order to determine the optimum *LP* during which mean daily temperature affects the *Ulmus pumila* BGS (EGS) most remarkably at a station, first, the interval between the earliest and latest date in BGS (EGS) time series was calculated at the station, respectively, as the *basic LP* (*bLP*). Then, the mean daily temperature time series during the *basic LP* plus a moving *LP* (*mLP*) prior to the earliest date in BGS (EGS) time series were computed by step length of 1 day at the

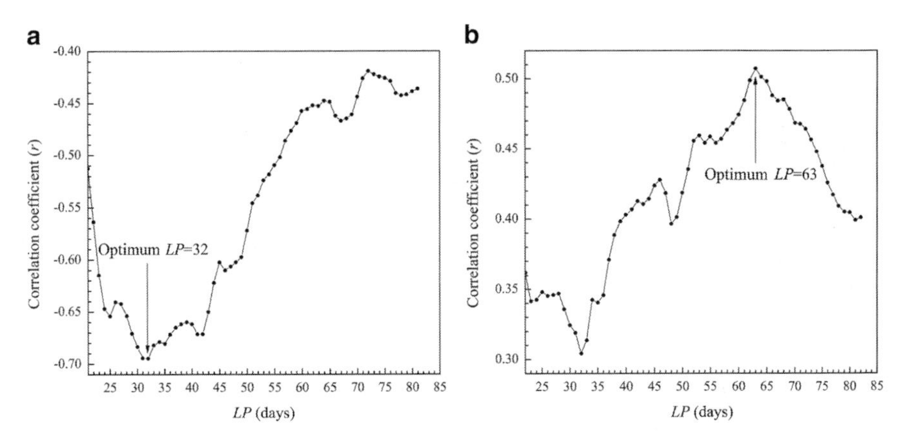

Fig. 17.2 Schematic demonstration for determining the optimum (**a**) spring *LP* and (**b**) autumn *LP* based on correlation coefficients between BGS (EGS) time series and mean daily temperature time series during the different *LPs* at station 28, *bLP* (BGS) = 21 days, *bLP* (EGS) = 22 days

station from 1986 to 2005, namely, during *bLP* + 1 day, *bLP* + 2 days, *bLP* + 3 days, etc. The maximum *mLP* was limited to 60 days. Thus, the *LP* is defined as follows:

$$LP = bLP + mLP \tag{17.1}$$

Moreover, correlation coefficients between BGS (EGS) time series and mean daily temperature time series during the different *LPs* were calculated at the station, respectively. Finally, the optimum *LP* with the largest correlation coefficient between BGS (EGS) and mean daily temperature was obtained at the station, respectively. Figure 17.2 shows an example for determining the optimum spring and autumn *LP* at station 28. The curves illustrate variation of correlation coefficients between BGS (EGS) time series and mean daily temperature time series during the different *LPs*. The optimum *LP* with the largest correlation coefficient is 32 days for BGS (negative correlation) and 63 days for EGS (positive correlation). The above four step procedure for looking for the optimum *LP* at individual stations can also be applied to the entire region. For the latter case, the regional mean BGS (EGS) time series and regional mean daily temperature time series (based on the 46 stations) from 1986 to 2005 were used.

17.2.4 Response of the Growing Season to Interannual Temperature Variation

At individual stations, a significantly negative correlation between BGS and the optimum spring *LP* temperature was apparent at 41 stations (89 %), namely, the higher the spring *LP* temperature in a year, the earlier the BGS. Slopes of significant linear regression lines between BGS and spring *LP* temperature ranged from −1.02 to −7.63 days °C^{-1} (Table 17.1). However, a significantly positive correlation between EGS and the optimum autumn *LP* temperature was found at only 12 stations (26 %), where the higher the autumn *LP* temperature in a year, the later the EGS. Slopes of significant linear regression lines between EGS and autumn *LP* temperature ranged from 1.96 to 7.65 days °C^{-1} (Table 17.1).

For entire China's temperate zone, regional linear regression equations were calculated between the regional mean BGS (EGS) time series and the regional optimum spring (autumn) *LP* (73 days for BGS, 65 days for EGS) temperature time series during 1986 to 2005. The result shows that the regional mean BGS correlates negatively with the regional optimum spring *LP* temperature and the regional mean EGS correlates positively with the regional optimum autumn *LP* temperature, respectively. It is worth noting that the correlation coefficient for BGS is much larger than that for EGS. The regional regression equations indicate that a 1 °C increase in the regional optimum spring *LP* temperature may induce an advancement of 2.8 days in the regional mean BGS, while a 1 °C increase in the regional optimum autumn *LP* temperature may cause a delay of 2.1 days in the regional mean EGS (Fig. 17.3).

17.2.5 Spatial Dependence of the Growing Season Response to Temperature

Table 17.1 shows that the regression slopes (days °C^{-1}) between growing season parameters (BGS, EGS) and temperatures have obvious differences among the stations, which may relate to thermal condition differences among the stations. In order to detect the spatial dependence of the growing season response to temperature, the long-term annual mean temperature from 1986 to 2005 was used as the indicator of thermal condition at individual stations, and a correlation analysis was carried out between growing season-temperature regression slopes and long-term annual mean temperatures at the 46 stations. The result shows that a negative correlation was found between BGS-spring *LP* temperature regression slope and long-term annual mean temperature at individual stations ($P < 0.1$), whereas a positive correlation was detected between EGS-autumn *LP* temperature regression slope and long-term annual mean temperature at individual stations ($P < 0.05$). Generally speaking, either the negative response of BGS to spring *LP* temperature or the positive response of EGS to autumn *LP* temperature (dots above the dashed horizontal line) was stronger at warmer locations than at colder locations (Fig. 17.4).

Table 17.1 Correlation and regression analysis between BGS and spring *LP* temperature and EGS-autumn *LP* temperature at each phenological station

	BGS		EGS	
Station number	*r*	Slope (days °C^{-1})	*r*	Slope (days °C^{-1})
1	−0.76	−1.95***	−0.16	−0.88
2	−0.37	−1.28	−0.36	−0.70
3	−0.86	−2.47***	0.50	1.96*
4	−0.56	−1.65*	0.35	1.83
5	−0.59	−1.46**	0.35	1.83
6	−0.42	−2.54*	0.55	2.08*
7	−0.65	−4.12**	0.46	3.17*
8	−0.82	−2.28***	0.41	3.13
9	−0.74	−4.35***	−0.19	−0.83
10	−0.46	−1.02*	0.60	4.50**
11	−0.85	−4.63***	0.20	1.78
12	−0.56	−2.79*	0.47	7.65*
13	−0.68	−2.62**	0.41	2.66
14	−0.69	−3.18***	−0.43	−2.09
15	−0.51	−4.61*	−0.28	−1.92
16	−0.82	−6.37***	0.39	3.92
17	−0.42	−2.25	0.25	1.21
18	−0.83	−3.99***	0.42	3.18
19	−0.57	−2.48**	0.34	0.56
20	−0.50	−4.38*	0.27	3.26
21	−0.52	−2.27*	0.29	2.86
22	−0.79	−3.36***	0.39	1.72
23	−0.80	−4.00***	−0.12	−1.35
24	−0.75	−4.20***	−0.24	−2.05
25	−0.76	−2.73***	0.24	1.97
26	−0.60	−1.61**	−0.20	−0.71
27	−0.61	−2.33**	0.42	2.77
28	−0.69	−2.51***	0.51	2.67*
29	−0.38	−2.18	0.71	5.20***
30	−0.75	−3.55***	0.30	1.36
31	−0.74	−3.97***	0.53	2.22*
32	−0.76	−4.17***	0.31	1.89
33	−0.79	−3.94***	0.25	0.84
34	−0.81	−2.98***	0.61	4.34**
35	−0.66	−7.63**	0.18	1.52
36	−0.46	−2.64*	0.54	4.76*
37	−0.67	−3.39**	0.19	0.95
38	−0.65	−3.29**	0.31	2.54
39	−0.42	−4.17	0.29	2.75
40	−0.81	−5.10***	0.53	2.30*
41	−0.33	−2.69	0.09	0.58
42	−0.82	−6.58***	0.31	4.35
43	−0.85	−5.88***	0.40	6.29
44	−0.86	−4.46***	0.41	2.12
45	−0.89	−2.94*	0.63	6.42**
46	−0.54	−2.68*	−0.25	−1.74

*$P < 0.05$; **$P < 0.01$; ***$P < 0.001$

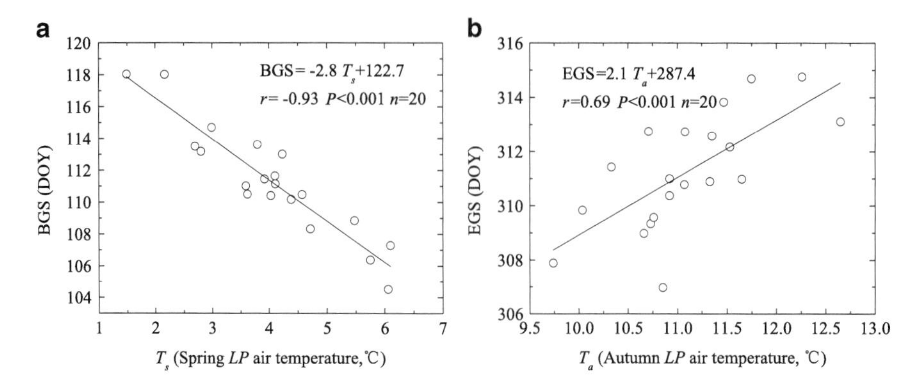

Fig. 17.3 Correlation and regression analysis (**a**) between regional mean BGS and regional spring *LP* temperature (T_s), (**b**) between regional mean EGS and regional autumn *LP* temperature (T_a) from 1986 to 2005

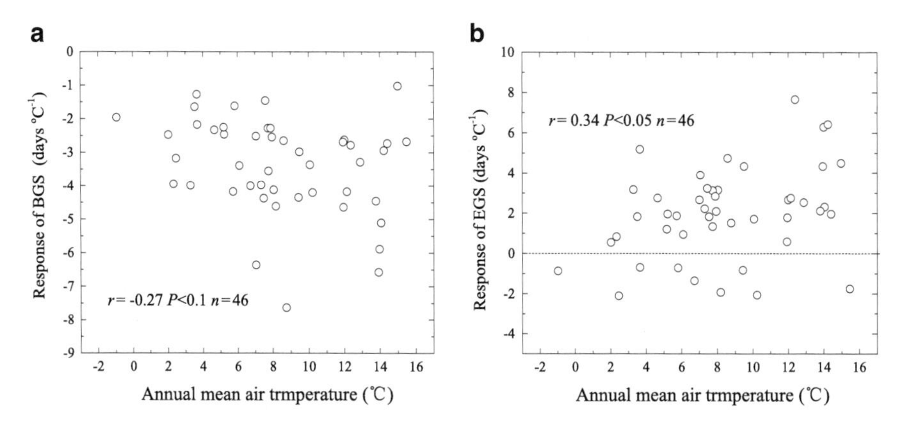

Fig. 17.4 Relationship between growing season response to temperature and long-term annual mean temperature at 46 stations: (**a**) BGS; (**b**) EGS

17.3 Daily Temperature-Based Spatial Modeling of *Ulmus pumila* Phenology

17.3.1 Spatial Phenology Model (Methodology)

Similarly, the basic hypothesis of the spatial phenology model is that the station-to-station variation of a phenological event occurrence date over an area is mainly influenced by station-to-station variation of mean daily temperature within a particular *LP* of days during and before its occurrence over the area (Chen and Xu 2012b). In order to determine the *LP* during which the station-to-station variation of

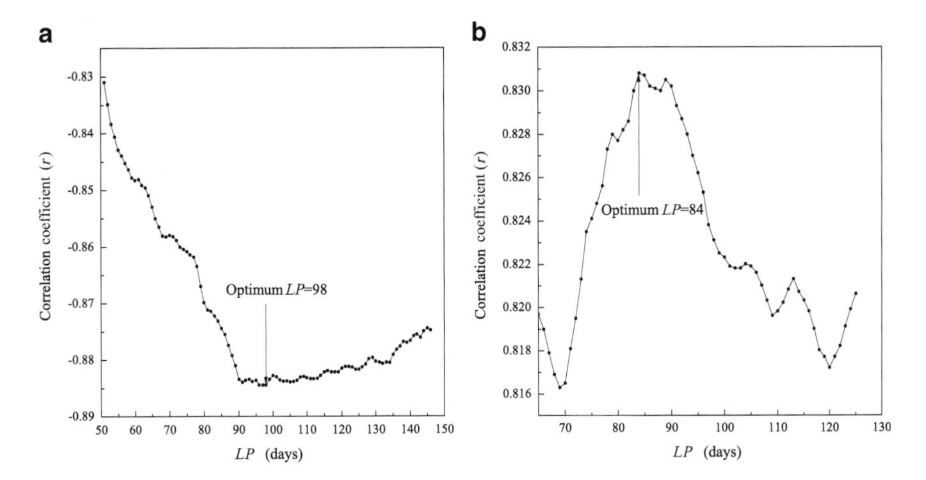

Fig. 17.5 Schematic demonstration of determination of the optimum (**a**) spring length period (*LP*) and (**b**) autumn *LP* based on correlation coefficients between BGS (EGS) spatial series and spring (autumn) daily temperature spatial series within the different *LPs* (*LP* − *bLP* + *mLP*) at the 46 stations in 1986

mean daily temperature affects station-to-station variation of *Ulmus pumila* BGS (EGS) most remarkably in a year, the number of days between the earliest and latest date in BGS (EGS) spatial series was first calculated across the 46 stations in the year, calling this the basic *LP* (*bLP*). Then, the mean daily temperature spatial series of the 46 stations was computed in the year during the *bLP* plus a moving *LP* (*mLP*) prior to the earliest date in BGS (EGS) spatial series by step length of 1 day, namely, during *bLP* + 1 day, *bLP* + 2 days, *bLP* + 3 days, etc. Further, correlation coefficients were calculated between BGS (EGS) spatial series and mean daily temperature spatial series during different spring (autumn) *LPs* (*bLP* + 1 day, *bLP* + 2 days, *bLP* + 3 days, etc.) at the 46 stations in the year. Finally, the optimum spring (autumn) *LP* with the largest correlation coefficient between BGS (EGS) spatial series and spring (autumn) mean daily temperature spatial series was obtained in the year. Figure 17.5 shows an example for determining the optimum spring and autumn *LP* in 1986. The curves illustrate variation of correlation coefficients between BGS (EGS) spatial series and mean daily temperature spatial series within different spring (autumn) *LPs*. The optimum *LP* with the largest correlation coefficient is 98 days for BGS (negative correlation) and 84 days for EGS (positive correlation). The above four step procedure for looking for the optimum *LP* in a year was also applied to the entire study period. In that case, the 20-year mean BGS (EGS) spatial series and 20-year mean spring (autumn) daily temperature spatial series at the 46 stations during the period 1986–2005 were used to calculate the 20-year mean optimum spring (autumn) *LP*.

To evaluate the model performance in the spatial simulation and extrapolation, Root Mean Square Error (RMSE) between predicted and observed BGS or EGS and

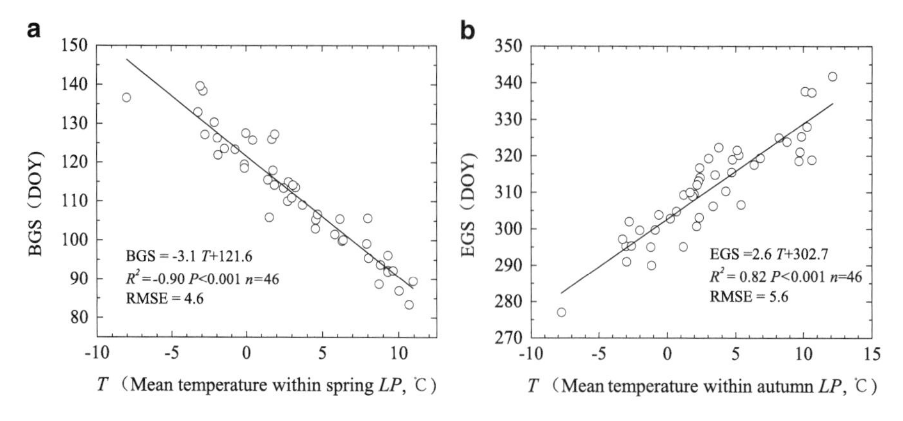

Fig. 17.6 Spatial correlation and regression analyses (**a**) between 20-year mean daily temperature within the optimum spring *LP* and 20-year mean BGS and (**b**) between 20-year mean daily temperature within the optimum autumn *LP* and 20-year mean EGS at the 46 stations

explained variance (R^2) were used. The RMSE was calculated by the following formula:

$$RMSE = \sqrt{\frac{\sum_{i=1}^{n} (Pre_i - Obs_i)^2}{n}} \tag{17.2}$$

where Obs_i denotes the observed BGS or EGS at the station i; Pre_i denotes the predicted BGS or EGS at the station i; n is the number of stations.

17.3.2 Spatial Simulation and Validation of Multiyear Mean BGS and EGS

The mean spring spatial phenology model shows that mean *Ulmus pumila* BGS correlates negatively with the mean daily temperature within the optimum spring *LP* (123 days) at the 46 stations over the period 1986–2005 ($P < 0.001$). That is, the higher the 20-year mean daily temperature within the optimum spring *LP* at a station, the earlier the 20-year mean BGS. Mean spring spatial phenology model explained 90 % of the BGS variance ($P < 0.001$) and the RMSE for differences between observed and simulated BGS is 4.6 days. On average, a spatial shift in the 20-year mean daily temperature within the optimum spring *LP* by 1 °C may induce a spatial shift in the 20-year mean BGS by −3.1 days (Fig. 17.6). Other than BGS, mean *Ulmus pumila* EGS correlates positively with the mean daily temperature within the optimum autumn *LP* (66 days) at the 46 stations over the period 1986–2005 ($P < 0.001$), namely, the higher the 20-year mean daily temperature

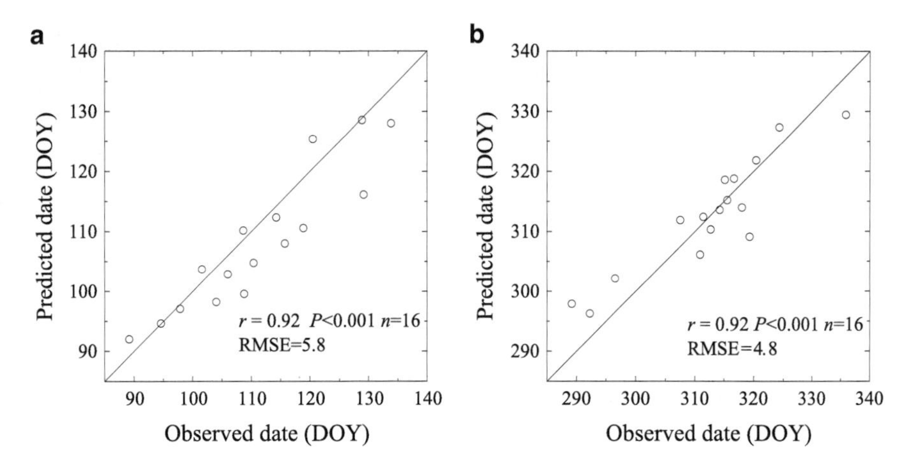

Fig. 17.7 External validation of mean spatial phenology models by the spatial extrapolation of mean (**a**) BGS and (**b**) EGS

within the optimum autumn *LP* at a station, the later the 20-year mean EGS. Mean autumn spatial phenology model explained 82 % of the EGS variance ($P < 0.001$) and the RMSE for differences between observed and simulated EGS is 5.6 days. On average, a spatial shift in the 20-year mean daily temperature within the optimum autumn *LP* by 1 °C may cause a spatial shift in the 20-year mean EGS by 2.6 days (Fig. 17.6). It is worth noting that the simulation accuracy of 20-year mean BGS is higher than that of 20-year mean EGS.

To validate the mean spatial phenology models, multiyear mean daily temperatures within the optimum spring and autumn *LP* at 16 external stations were substituted into the corresponding mean spring and autumn spatial phenology models (Fig. 17.6), respectively, and predicted multiyear mean *Ulmus pumila* BGS and EGS were obtained at the 16 stations. The results show that the RMSE*s* for differences between observed and predicted BGS and between observed and predicted EGS were 5.8 and 4.8 days, respectively (Fig. 17.7). The comparison between model validations (Fig. 17.7) and model simulations (Fig. 17.6) shows that RMSE of the model validation for BGS is larger (by 1.2 days) than that of the model simulation but RMSE of the model validation for EGS is smaller (by 0.8 days) than that of the model simulation. Thus, the mean spatial phenology models indicate a strong spatial extrapolation capability to multiyear mean BGS and EGS.

17.3.3 Spatial Simulation and Validation of Yearly BGS and EGS

In order to explore the spatial relationship between BGS (EGS) and daily temperature during the optimum spring (autumn) *LP* in each year, yearly spatial phenology models were created from 1986 to 2005, resulting in 40 models (20 models for BGS and 20 models for EGS). Similarly, BGS correlates negatively with mean daily temperature within the optimum spring *LP* in each year at the 46 stations ($P < 0.001$), namely, the higher the mean daily temperature within the optimum spring *LP* in a specific year at a station, the earlier the BGS in the year at the station. The explained variances of yearly spring spatial phenology models to BGS are between 73 and 86 % ($P < 0.001$) and the RMSEs of yearly BGS simulations are between 5.4 and 9.8 days. Slopes of significant linear regression equations show that a spatial shift in mean daily temperature within the optimum spring *LP* by 1 °C in a year may induce a spatial shift in BGS between −4.28 and −2.75 days in the year. In contrast, EGS correlates positively with mean daily temperature within the optimum autumn *LP* in each year at the 46 stations ($P < 0.001$). That is, the higher the mean daily temperature within the optimum autumn *LP* in a specific year at a station, the later the EGS in the year at the station. The explained variances of yearly autumn spatial phenology models to EGS are between 49 and 77 % ($P < 0.001$) and the RMSE*s* of yearly EGS simulations are between 7.4 and 11.2 days. Slopes of significant linear regression equations indicate that a spatial shift in mean daily temperature within the optimum autumn *LP* by 1 °C in a year may cause a spatial shift in EGS between 2.17 and 3.16 days in the year. The average simulation accuracy of yearly BGS is higher than that of yearly EGS (Table 17.2).

The model validation was implemented by substituting mean daily temperatures within the optimum spring and autumn *LP* in each year at external stations into the corresponding yearly spring and autumn spatial phenology models, respectively. The results show that the average RMSEs of yearly BGS and EGS predictions were 7.5 and 8.7 days, respectively (Table 17.3), which are close to the average RMSEs of yearly BGS and EGS simulations (Table 17.2). Thus, the yearly spatial phenology models also show a strong spatial extrapolation capability to yearly BGS and EGS.

17.3.4 Climatic Controls of the Spatial Response of BGS and EGS to Temperature

As mentioned above, the spatial response of BGS and EGS to temperature shows obvious interannual variation (Table 17.2). This may be associated with the interannual variation of thermal condition at regional scales. In order to detect the relationship between spatial phenology response and regional temperature regime with regard to the interannual variation, the correlation coefficients were calculated between BGS-temperature spatial regression slope and regional February–April mean temperature and between EGS-temperature spatial regression slope and regional September–November mean temperature from 1986 to 2005, respectively. Regional February–April and September–November mean temperatures were

Table 17.2 Spatial correlation and regression analyses between daily temperature within the optimum spring *LP* and BGS and between daily temperature within the optimum autumn *LP* and EGS in each year

	BGS simulation				EGS simulation			
Year	Number of stations	Slope (days °C^{-1})	R^2	RMSE (days)	Number of stations	Slope (days °C^{-1})	R^2	RMSE (days)
1986	35	−3.07	0.78*	6.1	35	2.83	0.69*	7.7
1987	40	−3.06	0.86*	5.5	40	2.77	0.54*	9.8
1988	41	−3.08	0.81*	6.2	39	2.46	0.61*	8.0
1989	40	−3.36	0.87*	5.4	40	3.00	0.69*	8.2
1990	45	−3.30	0.79*	6.6	45	2.44	0.49*	9.9
1991	46	−2.75	0.73*	7.2	46	2.17	0.58*	8.7
1992	46	−3.01	0.80*	6.6	46	2.17	0.51*	8.9
1993	46	−3.12	0.81*	7.0	46	2.56	0.59*	9.8
1994	46	−2.94	0.79*	6.5	46	3.14	0.68*	9.0
1995	46	−3.25	0.82*	6.9	46	2.70	0.50*	11.2
1996	46	−3.39	0.76*	7.9	46	2.55	0.73*	7.4
1997	45	−3.43	0.85*	6.1	46	2.89	0.66*	9.2
1998	46	−3.71	0.79*	7.0	45	2.58	0.71*	7.9
1999	46	−2.93	0.78*	7.5	46	2.84	0.73*	8.2
2000	46	−3.04	0.79*	7.3	46	2.34	0.73*	7.6
2001	46	−3.43	0.80*	8.1	46	3.09	0.70*	8.4
2002	46	−4.28	0.79*	9.3	46	2.95	0.77*	7.6
2003	46	−3.71	0.72*	9.8	45	2.41	0.67*	8.0
2004	46	−3.33	0.83*	7.6	46	2.98	0.68*	8.5
2005	45	−2.85	0.82*	6.9	45	3.16	0.74*	7.8
Average	–	–	–	7.1	–	–	–	8.6

*$P < 0.001$

selected as the independent variables because they can represent regional thermal status when BGS and EGS occurred over the study area, respectively, and are convenient for comparing regional thermal status among different years. The results show that a significantly negative correlation appeared between BGS-temperature spatial regression slope and regional February–April mean temperature ($P < 0.01$). In general, the negative spatial response of yearly BGS to mean daily temperature within the optimum spring *LP* was stronger in warmer years than in colder years (Fig. 17.8). In contrast, there was no significant correlation between EGS-temperature spatial regression slope and regional September-–November mean temperature.

Further analysis displays that the spatial variability of BGS (in spatial standard deviation of BGS) was also larger in warmer years than in colder years but the spatial variability of mean daily temperature within the optimum spring *LP* (in spatial standard deviation of temperature) was not associated with year-to-year temperature variation. That is, the spatial shift range of mean daily temperature within the optimum spring *LP* remained approximately the same from year to year, whereas the spatial shift range of BGS fluctuated noticeably along with year-to-year

Table 17.3 External validation of yearly spatial phenology models by the spatial extrapolation of yearly BGS and EGS

Year	BGS validation Number of stations	r	RMSE (days)	EGS validation Number of stations	r	RMSE (days)
1986	30	0.86**	6.3	28	0.90**	8.6
1987	32	0.78**	8.9	31	0.70**	9.4
1988	32	0.87**	5.7	32	0.87**	8.5
1989	35	0.78**	7.8	34	0.83**	11.0
1990	39	0.76**	7.9	39	0.78**	11.1
1991	38	0.83**	7.4	37	0.87**	8.7
1992	37	0.90**	6.4	36	0.79**	8.5
1993	33	0.92**	5.8	33	0.83**	8.5
1994	18	0.90**	6.0	19	0.66*	11.9
1995	20	0.92**	5.5	19	0.72**	9.0
1996	17	0.90**	5.2	17	0.88**	7.7
1997	16	0.90**	7.0	16	0.84**	6.9
1998	14	0.91**	7.3	16	0.92**	5.0
1999	15	0.91**	6.3	16	0.86**	7.3
2000	12	0.90**	6.7	15	0.89**	6.0
2001	12	0.89**	9.2	14	0.93**	7.7
2002	12	0.89**	11.4	15	0.92**	6.3
2003	13	0.88**	7.2	16	0.64*	11.9
2004	12	0.80*	11.5	15	0.77**	10.2
2005	13	0.73*	11.0	16	0.81**	10.6
Average	–	–	7.5	–	–	8.7

$*P < 0.01$; $**P < 0.001$

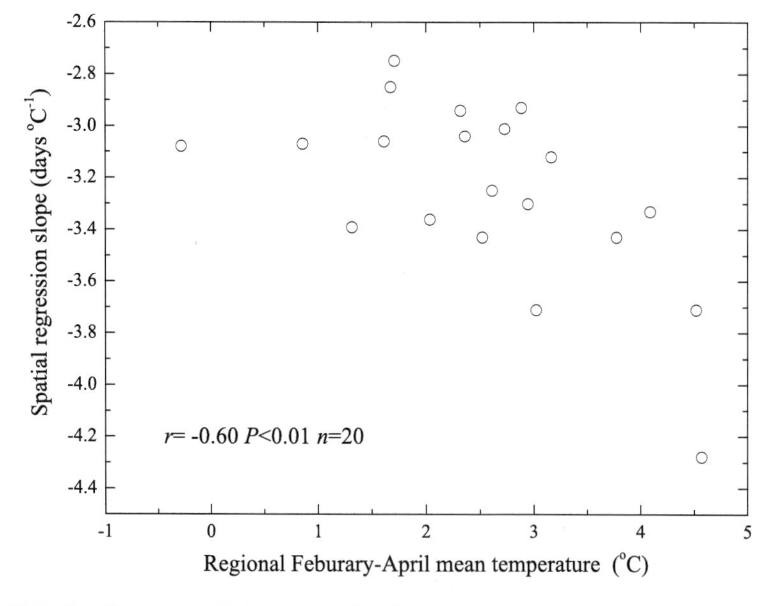

Fig. 17.8 Correlation analysis between regional February–April mean temperature and spatial regression slope (days °C^{-1}) across China's temperate zone from 1986 to 2005

temperature variation. Therefore, the interannual variation of the spatial response of BGS to temperature is associated mainly with spatial variability of BGS from year to year. Namely, climate warming will enhance sensitivity of the spatial response of BGS to temperature through increasing BGS spatial variability. Because climate warming may significantly increase the spatial variability of spring tree phenology and consequently speed up the spatial response of spring tree phenology to temperature, the terrestrial ecosystem under global climate change scenarios will likely become more sensitive and uncertain than at present.

17.4 Conclusions

The daily temperature-based temporal phenology model provides a more precise and rational tool than the monthly or multi-monthly mean temperature-based phenology model in detecting responses of tree phenology to temperature. For entire China's temperate zone, a 1 °C increase in spring temperature during the optimum length period may induce an advancement of 2.8 days in the beginning date of the *Ulmus pumila* growing season, whereas a 1 °C increase in autumn temperature during the optimum length period may cause a delay of 2.1 days in the end date of the growing season. Therefore, the response of the beginning date to temperature is more sensitive than the response of the end date. At individual stations, the sensitivity of the growing season response to temperature depends obviously on the local thermal condition, namely, either the negative response of the beginning date or the positive response of the end date to temperatures was stronger at warmer locations than at colder locations. Thus, future regional climate warming may enhance sensitivity of plant phenological response to temperature, especially in the colder regions with rapid temperature increase.

The daily temperature-based spatial phenology model provides a robust tool for simulating and predicting spatial patterns of tree phenology. The model reveals that spatial patterns of daily temperatures within the optimum spring and autumn length periods control spatial patterns of growing season beginning and end dates of *Ulmus pumila* across China's temperate zone, respectively. Regarding 20-year mean growing season modeling, mean growing season beginning date correlates negatively with mean daily temperature within the optimum spring length period at the 46 stations, whereas mean growing season end date correlates positively with mean daily temperature within the optimum autumn length period. On average, a spatial shift in mean spring and autumn daily temperatures by 1 °C may induce a spatial shift in mean beginning and end dates by −3.1 and 2.6 days, respectively. Similarly, a significant negative and positive correlation was detectable between beginning date and spring daily temperature and between end date and autumn daily temperature at the 46 stations for each year. On average, a spatial shift in spring and autumn daily temperatures by 1 °C in a year may induce a spatial shift in beginning and end dates between −4.28 and −2.75 days and between 2.17 and 3.16 days in the year. Moreover, both mean and yearly spatial phenology models perform satisfactorily in predicting *Ulmus pumila* BGS and EGS at external stations. Further analysis shows that the negative spatial response of yearly beginning date to spring daily temperature was stronger in warmer years than in colder years. This finding suggests that regional climate warming in late winter and spring may enhance sensitivity of the growing season's spatial response to temperature.

References

Askeyev OV, Sparks TH, Askeyev IV, Tishin DV, Tryjanowski P (2010) East versus West: contrasts in phenological patterns? Glob Ecol Biogeogr 19(6):783–793. doi:10.1111/j.1466-8238.2010.00566.x

Baldocchi D (2008) Breathing of the terrestrial biosphere: lessons learned from a global network of carbon dioxide flux measurement systems. Aust J Bot 56(1):1–26

Barr AG, Black TA, Hogg EH, Kljun N, Morgenstern K, Nesic Z (2004) Inter-annual variability in the leaf area index of a boreal aspen-hazelnut forest in relation to net ecosystem production. Agric For Meteorol 126(3–4):237–255. doi:10.1016/j.agrformet.2004.06.011

Black TA, Chen WJ, Barr AG, Arain MA, Chen Z, Nesic Z, Hogg EH, Neumann HH, Yang PC (2000) Increased carbon sequestration by a boreal deciduous forest in years with a warm spring. Geophys Res Lett 27:1271–1274

Chen XQ (1994) Untersuchung zur zeitlich-raeumlichen Aehnlichkeit von phaenologischen und klimatologischen Parametern in Westdeutschland und zum Einfluss geooekologischer Faktoren auf die phaenologische Entwicklung im Gebiet des Taunus. Selbstverlag des Deutschen Wetterdienstes, Offenbach am Main

Chen XQ, Xu L (2012a) Phenological responses of *Ulmus pumila* (Siberian Elm) to climate change in the temperate zone of China. Int J Biometeorol 56(4):695–706. doi:10.1007/s00484-011-0471-0

Chen X, Xu L (2012b) Temperature controls on the spatial pattern of tree phenology in China's temperate zone. Agric For Meteorol 154–155:195–202. doi:10.1016/j.agrformet.2011.11.006

Chen XQ, Hu B, Yu R (2005) Spatial and temporal variation of phenological growing season and climate change impacts in temperate eastern China. Glob Change Biol 11(7):1118–1130. doi:10.1111/j.1365-2486.2005.00974.x

China Meteorological Administration (1993) Observation criterion of agricultural meteorology. China Meteorological Press, Beijing (in Chinese)

Chmielewski FM, Rötzer T (2001) Response of tree phenology to climate change across Europe. Agric For Meteorol 108(2):101–112

Churkina G, Schimel D, Braswell BH, Xiao XM (2005) Spatial analysis of growing season length control over net ecosystem exchange. Glob Change Biol 11(10):1777–1787

Compilation Committee of the Vegetation of China (1980) The vegetation of China. Science Press, Beijing (in Chinese)

Ghelardini L, Santini A (2009) Avoidance by early flushing: a new perspective on Dutch elm disease research. iForest 2(1):143–153

Gordo O, Sanz JJ (2010) Impact of climate change on plant phenology in Mediterranean ecosystems. Glob Change Biol 16(3):1082–1106. doi:10.1111/j.1365-2486.2009.02084.x

Goulden ML, Munger JW, Fan SM, Daube BC, Wofsy SC (1996) Exchange of carbon dioxide by a deciduous forest: response to interannual climate variability. Science 271:1576–1578

Hense A, Glowienka-Hense R, Müller M, Braun P (2002) Spatial modelling of phenological observations to analyse their interannual variations in Germany. Agric For Meteorol 112 (3–4):161–178

Hutchinson MF (2002) Anusplin version 4.2 user guide. Australian National University, Canberra

Kljun N, Black TA, Griffis TJ, Barr AG, Gaumont-Guay D, Morgenstern K, McCaughey JH, Nesic Z (2007) Response of net ecosystem productivity of three boreal forest stands to drought. Ecosystems 10(6):1039–1055. doi:10.1007/s10021-007-9088-x

Lu PL, Yu Q, Liu JD, Lee XH (2006) Advance of tree-flowering dates in response to urban climate change. Agr Forest Meteorol 138(1–4):120–131. doi:10.1016/j.agrformet.2006.04.002

Ma CG (1989) A provenance test of white elm (*Ulmus pumila L.*) in China. Silvae Genet 38:37–44

Matsumoto K, Ohta T, Irasawa M, Nakamura T (2003) Climate change and extension of the *Ginkgo biloba L.* growing season in Japan. Glob Change Biol 9(11):1634–1642. doi:10.1046/j.1529-8817.2003.00688.x

Menzel A (2003) Plant phenological anomalies in Germany and their relation to air temperature and NAO. Clim Change 57(3):243–263

Nakahara M (1948) Phenology . Kawadesyobo Press, Tokyo (in Japanese)

Newman JE, Beard JB (1962) Phenological observations: the dependent variable in bioclimatic and agrometeorological studies. Agron J 54(5):399–403

Park-Ono HS, Kawamura T, Yoshino M (1993) Relationships between flowering date of cherry blossom (*Prumus yedoensis*) and air temperature in East Asia. In: Proceedings of the 13th International Congress of Biometerology, Calgary

Rötzer T, Chmielewski FM (2001) Phenological maps of Europe. Clim Res 18(3):249–257

Schwartz MD, Chen ZQ (2002) Examining the onset of spring in China. Clim Res 21(2):157–164

Sparks TH, Jeffree EP, Jeffree CE (2000) An examination of the relationship between flowering times and temperature at the national scale using long-term phenological records from the UK. Int J Biometeorol 44(2):82–87

White MA, Nemani AR (2003) Canopy duration has little influence on annual carbon storage in the deciduous broad leaf forest. Glob Change Biol 9(7):967–972

Wilson KB, Baldocchi DD (2000) Seasonal and interannual variability of energy fluxes over a broadleaved temperate deciduous forest in North America. Agric For Meteorol 100(1):1–18

Zheng JY, Ge QS, Hao ZX, Wang WC (2006) Spring phenophases in recent decades over eastern China and its possible link to climate changes. Clim Change 77(3–4):449–462. doi:10.1007/s10584-005-9038-6

Chapter 18
Plant Phenological "Fingerprints"

Annette Menzel

Abstract Given the sensitivity of plant phenology to small changes in temperature and the fact that this relationship is universally understood by politicians and the general public it has been widely used in recent years as an indicator of anthropogenically driven climate warming. The timing of plant life-cycle events particularly in spring, in a range of different environments and locations across the world has clearly demonstrated an advance attributable to global warming. Here, a review of trends in direct observations of the response of plant phenology to warming is presented together with a comparison with other methods of landscape level phenological observations such as remote sensing. In addition, the need for meaningful correlations between satellite data and *in situ* phenological observations is highlighted.

18.1 Introduction

Observation of phenological phases is probably the simplest way to track changes in the ecology of species in response to climate change. Other possible responses, such as altered species distribution, population sizes, and community composition, are much harder and more expensive to detect (Walther et al. 2002). Thus, during recent years, phenology has received increasing attention as a bio-indicator for global change. "Snow drops as bearer of bad tidings" was, for example, the title of an article about climate change in a German newspaper in January 2002. In fact, almost each season seems to witness record early onset dates demonstrating clear impacts of climate change. Indeed, phenology constitutes an ideal climate indicator at regional to international levels, because it is easily understood by the general

A. Menzel (✉)
Department of Ecology and Ecosystem Management, Technische Universität München, Freising, Germany
e-mail: amenzel@wzw.tum.de

M.D. Schwartz (ed.), *Phenology: An Integrative Environmental Science*,
DOI 10.1007/978-94-007-6925-0_18,

public, allows the study of changes at a smaller scale, raises awareness of climate change issues, engages the public in the climate change debate, and reconnects people with their natural world (Sparks and Smithers 2002). Consequently, new citizen science based phenological networks have been established (see Chap. 4, Europe). However, the scientific community also welcomes phenology as a tool for global change research. Among others, the length of the growing season and timing of spring events are proposed by the European Environment Agency as global change indicators.

Recent reviews about observed phenological changes, e.g. Menzel and Estrella (2001), Walther et al. (2002), Sparks and Menzel (2002), Root et al. (2003), Parmesan and Yohe (2003), Parmesan (2006), Cleland et al. (2007) summarized various indications of shifts in plant and animal phenology which have been reported for the boreal and temperate zones of the northern hemisphere. Especially in the fourth assessment report of the Intergovernmental Panel on Climate Change (AR4, IPCC 2007), a chapter on the assessment of observed changes and responses in natural and managed systems underlined the predominant role of phenology and its long-term monitoring data for fingerprinting climate change (impacts) (Rosenzweig et al. 2007). This chapter, however, will give an insight to plant phenological changes, since responses of animal life are described in the Sect. 18.6. Special emphasis will be given to the formal attribution of phenological changes to anthropogenic climate warming.

18.2 Responses to a Warming World

The picture of current climate change effects on seasonal plant activity is consistently demonstrated and well documented: An increasing number of studies (see reviews above) report an advance of leaf unfolding and flowering of 2–5 days decade^{-1} and, at times, a delay of leaf colouring and leaf fall of 1–2 days decade^{-1}, and a subsequent lengthening of the growing season in the last five decades. For example, the mean advancement of spring in the International Phenological Gardens network (see Chap. 4) in Europe was 2.0–2.7 days decade^{-1} (Menzel and Fabian 1999; Menzel 2000; Chmielewski and Rötzer 2001). An advance of 1.9 days decade^{-1} (1951–1998) was reported by Defila and Clot (2001) for the Swiss phenological network and 1.6 days decade^{-1} (1951–2000) by Menzel (2003) for the German phenological network. Flowering of lilac and honeysuckle in the western U.S. (1957/1968–1994, Cayan et al. 2001) advanced by 1.5/3.5 days decade^{-1}. A pan European study based on more than 125.000 time series (1971–2000) revealed a mean advance of 2.5 days decade^{-1} for leaf unfolding/flowering, (Menzel et al. 2006). Table 18.1 summarises new key examples of unmistakable evidence from phenological networks as well as remote sensing data and abiotic measures.

These changes are spatially consistent and significantly correlated with concurrent climatic warming, and are simply too numerous and too robust to be

Table 18.1 Key examples of observed change in plant phenology, the CO_2 signal, NDVI (satellite derived greenness), climatological indices, and animal phenology

Geographic range	Time span	Indicator (observed change, days decade^{-1})	References
USA	1936–1998	Mean advance of 55 phenophases at one site in Wisconsin (−1.2)	Bradley et al. (1999)
USA	1970–1999	Earlier flowering dates of 89 plant species in Washington (−2.4)	Abu-Asab et al. (2001)
Canada	1936–1996	Earlier spring flowering index in Alberta (−1.3)	Beaubien and Freeland (2000)
Japan	1953–2005	Advance in winter flowering of apricot at 32 sites (−1.3)	Doi (2007)
UK	1954–2000	General advance in flowering of 385 species (−1.0)	Fitter and Fitter (2002)
UK	1950–2005	315 fungi species show substantial changes in autumn fruiting	Gange et al. (2007)
Germany	1951–2000	Earlier spring phenology (−1.6), last frosts (−2.4), later autumn (2.5)	Menzel et al. (2003)
Germany	1951–2004	Considerable changes in agricultural phases (−1.2)	Estrella et al. (2007)
Germany	1961–2000	Clear advances in fruit tree phenology (−2.3)	Chmielewski et al. (2004)
Switzerland	1951–1998	Advances in pollen season	Clot (2003)
Switzerland	1965–2004	Advances in spring phenology (−1.5), but differ by location, altitude	Studer et al. (2005)
Europ. Alps	1971–2000	Advances in spring phenology and the influence of altitude	Ziello et al. (2009)
Norway, UK	1960–2007	Delay in autumn fruiting	Kauserud et al. (2008)
NW Russia	1930–1998	Local pollution may have advanced autumn (−3.2 earlier)	Kozlov and Berlina (2002)
Estonia	1948–1999	Advances in plant, bird and fish phenology	Ahas and Aasa (2006)
Europe IPG	1959–1996	Earlier spring (−2.1), later autumn (1.5) on cloned species	Menzel and Fabian (1999)
Europe	1951–1995	Effect of urban heat islands on flowering (−4.0)	Rötzer et al. (2000)
Europe	1971–2000	Europe-wide study of spring (−2.5) and autumn phases	Menzel et al. (2006)
Japan	1953–2000	Longer growing season (2.5) of *Gingko biloba*	Matsumoto et al. (2003)
Japan, Korea	1953–2005	Widespread changes determined by local conditions	Primack et al. (2009)

(continued)

Table 18.1 (continued)

Geographic range	Time span	Indicator (observed change, days decade^{-1})	References
N Pacific,	1970–1994	Earlier spring (−3.5) and longer growing season from CO_2 signal	Keeling et al. (1996)
Globe >45 °N	1982–1990	Longer growing season (NDVI) by 12 days over the 1980s	Myneni et al. (1997)
N America	1981–1999	NDVI data suggest a longer active growing season by 12 days	Zhou et al. (2001)
Eurasia	1981–1999	NDVI data suggest a longer active growing season by 18 days	Zhou et al. (2001)
Fennoscand.	1982–2006	GIMMS NDVI data suggest a longer growing season (6.4)	Karlsen et al. (2009)
Northern Hemisphere.	1955–2002	Quicker onset of spring from climatological indices, such as last spring day below 5 °C (−1.4) or last spring freeze date (−1.5)	Schwartz et al. (2006)
Greater Baltic	1951–2000	Climatological growing season lengthened by 7.4 days	Linderholm (2006)
Western USA	1951–2000	Considerable change in agrometeorological indices, e.g. less frost days, longer growing season and increase in growing degree days	Feng and Hu (2004)
Colorado	1975–2009	Earlier end to hibernation of yellow-bellied marmots	Inouye et al. (2000)
Spain	1952–2000	Considerable changes in spring and autumn events	Peñuelas et al. (2002)
N America	1959–1991	Breeding dates of tree swallows advance	Dunn and Winkler (1999)
SE Australia	1970–2000	Advances in bird migration timing	Beaumont et al. (2006)
Antarctica	1950–2004	Later arrival and breeding of seabirds with no change in temperature	Barbraud and Weimerskirch (2006)
Northern high latitudes	1950–2005	Meta-analysis showing general advance in spring migration phenology of birds	Lehikoinen and Sparks (2010)
USA, Poland	1900–1999	Earlier breeding activity in amphibians, e.g. 4 of 6 - species now calling earlier by 10–13 days near New York	Gibbs and Breisch (2001) Tryjanowski et al. (2003)
UK	1976–1998	Earlier flights of butterflies	Roy and Sparks (2000)

(continued)

Table 18.1 (continued)

Geographic range	Time span	Indicator (observed change, days decade^{-1})	References
Spain	1952–2004	Earlier appearance of honeybee, butterfly in warmer springs	Gordo and Sanz (2006)
Oceans	1958–2002	Species–specific changes, but generally advancing	Edwards and Richardson (2004)
UK	1976–2005	Advance in 726 taxa in terrestrial, marine, freshwater ecosystems	Thackeray et al. (2010)

random or due to other factors such as natural climate variability or land-use change. On the continental scale geographical differences are evident with delayed rather than earlier onset of spring phases in southeastern Europe (Menzel and Fabian 1999), advances in spring in Western/Central Europe compared to delays in Eastern Europe (Ahas et al. 2002), and notable differences between states in the USA (Schwartz and Reiter 2000).

Variability within countries, e.g. Germany (Menzel et al. 2001) is not a simple regression on latitude or longitude, and observed differences between regions in Switzerland may also be due to clear altitudinal gradients of changes (Defila and Clot 2001). Nevertheless, a high spatial variability of trends is observed on smaller scales (e.g. Schwartz and Reiter 2000; Menzel et al. 2001) with few stations revealing delayed onset of spring (due to local microclimate conditions, natural variation, genetic differences or other non-climatic factors). Only around one third of stations demonstrated statistically significant advancing trends (see Menzel et al. 2006).

Multi-species studies and experiments reveal a considerable diversity in temperature response, either inter-specific, temporal, by functional traits or life-forms, or due to variations in community composition or functional diversity. Thus, average changes are not valid for all species as shown by many studies analyzing multi-phased phenological data at a particular site (e.g. no or delayed trend in spring: 50 % of species in Wisconsin (Bradley et al. 1999), 11 % of species in Washington D.C. (Abu-Asab et al. 2001), 24 % of species in southern-central England [Fitter and Fitter 2002]). Most recently, those species exhibiting insignificant or delayed trends have been defined as non-responding species (Cook et al. 2012), influenced by other abiotic cues (e. g., photoperiod) or by declines in required fall/winter chilling (vernalization) interfering with acceleration by spring warming.

Considerable natural variability may be related to intra-annual variation with greatest advances in early spring, and only notable advances of succeeding phenophases (e.g. Bradley et al. 1999; Defila and Clot 2001; Menzel et al. 2001; Sherry et al. 2007). Sparks and Smithers (2002) suggest higher temperature changes early in the season as the main reason for this seasonal differentiation. However, Peñuelas et al. (2002) did not find any seasonal differences at their Mediterranean region study site.

Despite this seasonal variation, large inter-specific plant reactions have been reported for example by Menzel and Fabian (1999), Abu-Asab et al. (2001), Defila and Clot (2001), Menzel et al. (2001), Fitter and Fitter (2002), Kozlov and Berlina (2002), Peñuelas et al. (2002), and e.g., Cleland et al. (2006). Fitter and Fitter (2002) note that annuals are more likely to flower earlier that congeneric perennials, and insect-pollinated species more than wind-pollinated species, which is in contrast to Ziello et al. (2012). However, Peñuelas et al. (2002) observed no differences among Raunkiaer life forms, or among plants of different origin.

In contrast to a clear warming response of spring phenophases, the heterogeneous and less pronounced changes in autumn phenology are poorly understood, especially regarding their link to various climate triggers (Rosenzweig et al. 2007). On average the end of season phases have been delayed by 0.3–1.6 days decade^{-1} (Chmielewski and Rötzer 2001; Defila and Clot 2001; Menzel 2000, 2003; Menzel and Fabian 1999; Menzel et al. 2001). An extreme delay was detected by Peñuelas et al. (2002) at one site in northeastern Spain, where leaf fall was 13 days later in 2000 compared to 1952, resulting in a mean change of 2.6 days decade^{-1}. Kozlov and Berlina's study (2002) represents a counter example of a strong advance of birch first leaf fall (by 22 days) compared to the 1930s probably reflecting the severity of environmental pollution on the Kola Peninsula. The pan European study by Menzel et al. (2006) revealed a slight mean delay of autumn phases (0.2 days decade^{-1}).

Fruit ripening in early autumn reveals a clear temperature response, with warmer spring and summer temperature advancing this phase. Several examples of advanced fruit ripening have been reported, such as significantly earlier harvest dates of grapevine in the Bordeaux area (Jones and Davis 2000), 9 days earlier fruiting in 2000 than 1974 in NE Spain (Peñuelas et al. 2002), earlier fruit ripening of *Sambucus nigra* and *Aesculus hippocastanum* in Germany (−2 and −0.5 days decade^{-1}, Menzel 2003) and 2.4 days decade^{-1} for all fruit ripening phases in Europe (Menzel et al. 2006).

In total, the length of the growing season has increased by up to 3.6 days decade^{-1} over the last 50 years in Europe (Menzel and Fabian 1999; Menzel 2000, 2003; Chmielewski and Rötzer 2001; Defila and Clot 2001; Menzel et al. 2001; Walther et al. 2002). Almost twice the typical lengthening was reported by Peñuelas et al. (2002) with an advance of 6.7 days decade^{-1} for the period 1952–2000.

18.3 Cherry Picking and Other Inconsistencies

Phenology observations are extremely suitable to illustrate and communicate climate change impacts. However, when phenological changes are critically questioned, the issue of data accuracy is often raised since no calibrated instruments are used and records may be full of gaps (Menzel 2002). Despite detailed observer manuals, phenological data inevitably depend on human subjectivity and effort. Many discussions have taken place on the scientific value of first events in plant communities which seem to depend on observer numbers and sites/ranges, their efforts as well as on target species density. Thus, it has to be questioned whether more emphasis should be given to mean and peak dates representing the behaviour of entire populations (Miller-Rushing et al. 2008).

In addition, the observed species and events themselves are subject of selection. Sparks and Smithers (2002) proposed that species with a strong temperature response, widespread distribution, and recognition have been preferred for observations.

The theoretical danger that strong trends favouring climate change impacts are more likely to be published in high ranking journals is labelled as publication bias (see Hughes 2000). In order to demonstrate that there is a real systematic shift in phenology associated with warming, the COST 725 consortium analysed more than 125.000 phenological series covering 1971–2000 in Europe. The result (78 % of all spring/summer records advanced, 30 % significantly, only 3 % significantly delayed, average advance of 2.5 days decade^{-1}, overall response to temperature of 2.5 days °C^{-1}) confirmed previous studies (Menzel et al. 2006, see Fig. 4.1). A comprehensive study across trophic levels and environments in the UK further eliminates the suspicion of bias (Thackeray et al. 2010).

Very long-term records suggest that recent decades experienced the greatest ever recorded changes in phenology and accelerated warming in the last two decades caused phenological trends to strengthen, e.g. the advance of spring phases is especially clear for time series' ending in the warm years of the 1990s (1989 and later, Rapp 2002; Scheifinger et al. 2002). However, changes are mainly reported as being linear, e.g. as absolute change (days) during the period analyzed or as the slope of a linear regression line (days per year). Since there is plenty of evidence that the natural world is far from linear in this respect, Dose and Menzel (2004) suggested Bayesian analyses to properly describe phenological time series and assess discontinuities in change over time (change points), often linked to abrupt changes in circulation patterns (Schleip et al. 2008). For Europe, discontinuous change is by far the best option to describe recent phenological time series with a high change point probability occurring in the late 1980s, probably linked to the North Atlantic Oscillation (Studer et al. 2005). Figure 18.1 illustrates this fact by examining the recorded onset of flowering of grapevine at Geisenheim during the last 100 years.

Other contributing factors may be the aging of clones or trees under long-term observation (Menzel and Fabian 1999), a strong genetic control over phenology (e.g. up to 3 week differences can occur between directly neighboring trees) (e.g. Morin et al. 2010), effects of the urban heat island (Menzel and Fabian 1999; Rötzer et al. 2000; Jochner et al. 2011), or the use of different cultivars and varieties in observation of agricultural crops. The majority of agricultural events are driven by farm management and are therefore dubbed 'false' phenological phases. Thus, sowing and subsequent emergence of crops clearly differ from spring/summer events in nature (e.g., mean advance of farmers' activities in Europe of 0.4 days decade^{-1}, Menzel et al. 2006). Different methods have been proposed to disentangle effects of farm management, technological advance and climate change on these changes in phenology (Estrella et al. 2007). In contrast, changes in 'true' phases, such as earlier crop flowering and maturity in recent decades have been attributed to warmer temperatures (Rosenzweig et al. 2007).

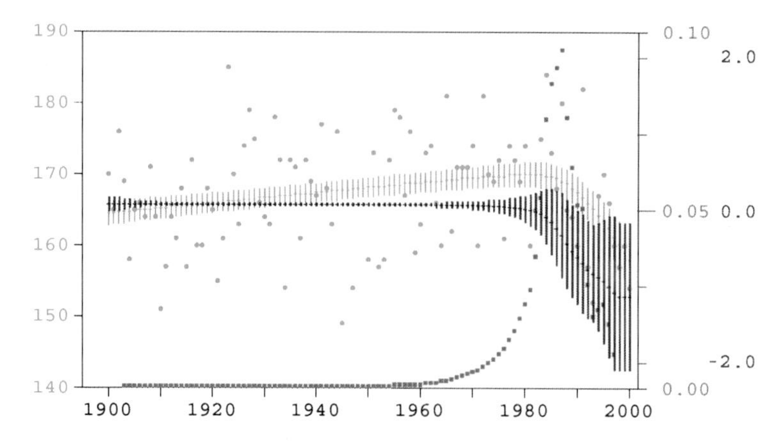

Fig. 18.1 Start of grapevine flowering at Geisenheim, Germany: Observed annual onset dates [DOY] are displayed as *green full dots* and averaged functional behaviour estimated from the one change point model with confidence range as *blue bars* on the *left* y-axis. The probability of the change point for the one change point model is shown as the *blue squares* (*right vertical scale*). The resulting rate of change (trend) is displayed in days per year (*red bars, second right vertical scale*)

18.4 Environmental Links

Although plant phenology is influenced by several factors, evidence for the response or sensitivity of spring events to temperature, often reported as change in days °C^{-1}, is overwhelming (Rosenzweig et al. 2007). In contrast to the climatic factors controlling autumn phenology, the climate signal driving spring phenology is fairly well understood. Nearly all phenophases correlate with spring temperatures in the preceding months (e.g. Sparks and Carey 1995; Sparks et al. 2000; Abu-Asab et al. 2001; Chmielewski and Rötzer 2001, 2002; Menzel 2003). Some spring events, such as the start of the plant growing season in northern and central Europe, also correlate with the North Atlantic Oscillation (NAO) index corresponding to winter climatic conditions (Post and Stenseth 1999; Ottersen et al. 2001; Chmielewski and Rötzer 2001; Menzel 2003). Broadly, warming advances spring and summer phases and delays autumn events such as leaf colouring and leaf fall. Species' responses mostly vary from 1 to 7 days °C^{-1}, but up to 15 days for early spring (Sparks and Carey 1995). National averages in Europe suggest ranges of 2–5 days °C^{-1} for spring and summer phases, and often reverse effects in autumn (Delpierre et al. 2009). Warmer countries tend to exhibit stronger responses apart from a general decreasing sensitivity from spring into summer (Menzel et al. 2006). Available models for spring phases constitute another intelligent possibility to demonstrate the close relationship between spring phases and temperature (see phenological modeling review in Chap. 15). The close relationship between spring phases and air temperature is also the basis of phenological spring indices for flowering and bud burst which are commonly used to describe phenological changes (e.g., Beaubien and Freeland 2000; Schwartz and Reiter 2000; Schwartz and Chen 2002).

Autumn leaf coloring is considerably less directly explained by temperature. Two opposing influencing factors are detected, with warm late summers delaying leaf coloring, and higher temperatures in May and June promoting leaf coloring.

The conclusion of Sparks and Smithers (2002) for the UK—that so far no sizeable effects of rainfall on phenology have been detected—might be true for the whole of Central Europe. However, in their study of Mediterranean vegetation, Peñuelas et al. (2002) found that phenological events of drought sensitive species or non-irrigated agricultural plants correlated with precipitation. In the future (additional) photoperiodic control, CO_2 effects, irrigation, fertilization, and farming practices cannot be disregarded.

18.5 Attribution to Anthropogenic Warming

Not all observed changes are necessarily due to human induced warming. Therefore, the IPCC clearly distinguishes between change detection as the statistical process without considering reasons for this change and attribution of such an observed change to anthropogenic climate change. To formally attribute, more is needed than a significant correlation with local temperature. Several lines of evidence have been pursued when attributing the observed global fingerprint of climate change in nature (Rosenzweig et al. 2007, 2008): firstly, an impressive number of significant observed impacts linked to regional temperature change across the globe of which more than 90 % were in the direction expected from warming (Fig. 18.2). Secondly, the spatial patterns of warming and impacts on nature are consistent across the globe. There are also examples of this spatial matching on finer scales, such as for spring phenological changes (Menzel et al. 2006). Temperature changes on continental scales, in turn, have been successfully attributed to anthropogenic influences. Thirdly, few so-called joint attribution studies could directly link observed changes in nature to modelled climatic variables and the best correlations were found when anthropogenic climate triggers were included (e.g. Root et al. 2003).

18.6 Comparison to Other Fingerprints

Observed start-of-season changes are not restricted to plants. They also include changes in the timing of other spring activities, such as earlier breeding or first singing of birds, earlier arrival of migrant birds, earlier appearance of butterflies and other insects, and earlier choruses and spawning in amphibians. Thus various indications for shifts in plant and animal phenology have already been observed in the boreal and temperate zones of the northern hemisphere (Walther et al. 2002, for discussion of animal phenology see Sect. 18.6). Phenological changes in birds and plants are often similar, as described in some cross-system studies

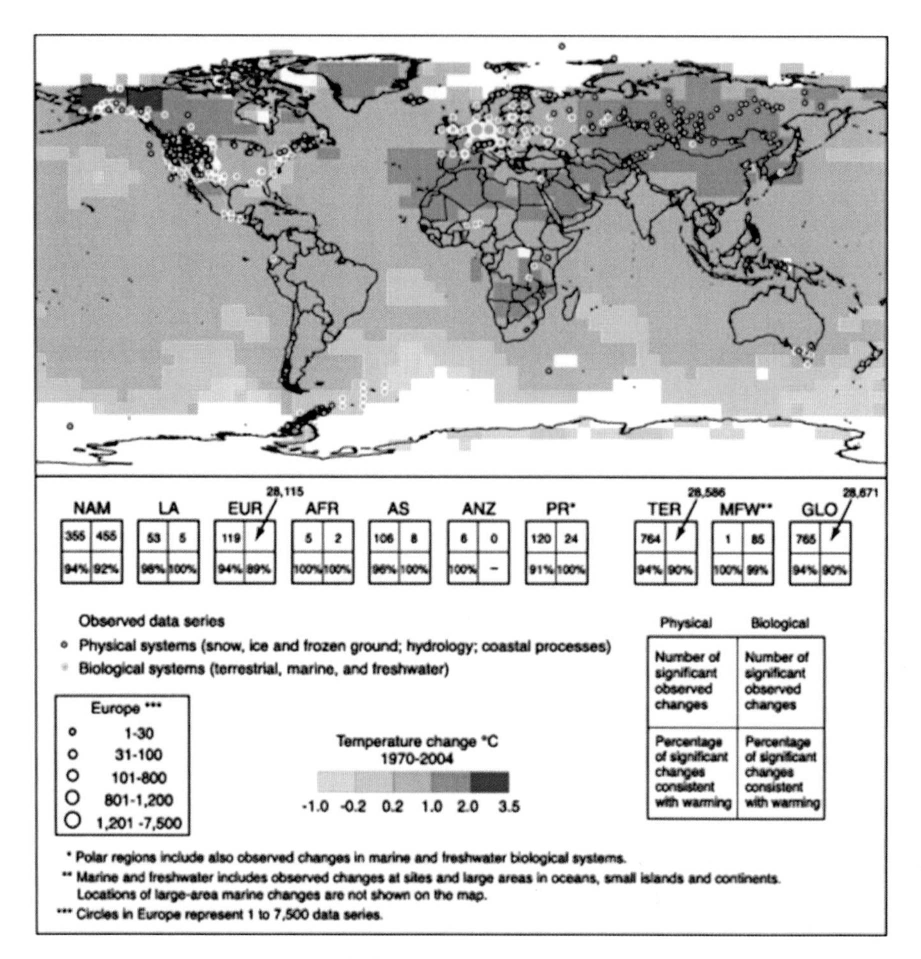

Fig. 18.2 Locations of significant changes in data series of physical systems (snow, ice and frozen ground; hydrology; and coastal processes) and biological systems (terrestrial, marine and freshwater biological systems), are shown together with surface air temperature changes over the period 1970–2004. A subset of about 29,000 data series was selected from about 80,000 data series from 577 studies. These met the following criteria: (i) ending in 1990 or later; (ii) spanning a period of at least 20 years; and (iii) showing a significant change in either direction, as assessed in individual studies. These data series are from about 75 studies (of which about 70 are new since the Third Assessment) and contain about 29,000 data series, of which about 28,000 are from European studies. *White* areas do not contain sufficient observational climate data to estimate a temperature trend. The 2 × 2 boxes show the total number of data series with significant changes (*top row*) and the percentage of those consistent with warming (*bottom row*) for (i) continental regions: North America (*NAM*), Latin America (*LA*), Europe (*EUR*), Africa (*AFR*), Asia (*AS*), Australia and New Zealand (*ANZ*), and Polar Regions (*PR*); and (ii) global scale: Terrestrial (*TER*), Marine and Freshwater (*MFW*), and Global (*GLO*). The numbers of studies from the seven regional boxes (NAM, ..., PR) do not add up to the global (*GLO*) totals because numbers from regions except Polar do not include the numbers related to Marine and Freshwater (*MFR*) systems. Locations of large-area marine changes are not shown on the map. [F1.8, F1.9; Working Group I AR4 F3.9b] (Parry et al. 2007, figure TS.1. on page 30)

(Ahas 1999; Bradley et al. 1999; Peñuelas et al. 2002). An example in Walther et al. (2002) shows that anomalies of leaf unfolding of trees, and arrival/hatching of birds exhibit parallel trends in Germany, with both demonstrating advancing tendencies since the 1990s; air temperature and the spring North-Atlantic Oscillation index (NAO) are negatively correlated with these phenophases.

Warning of potential asynchrony is based on marked differences in the phenological responses of different plant and animal taxa, thus phenological change will have major impacts on ecosystem structure and functioning, and biodiversity. Many interspecific interactions in ecosystems depend on this synchrony of events, such as food web structures (e.g., timing of predators, prey peak abundance, herbivores and host plants), migration and breeding dates, phenology of resources, and pollination. Decoupling of species fine-tuned interactions and relations within ecosystems may thus have profound ecological consequences (e.g., reviews of examples by Walther et al. 2002; Donnelly et al. 2011). More research will be needed to understand why temperature responses of phenology assessed in experimental plots do not match results from long-term monitoring (Wolkovich et al. 2012).

The lengthening of the growing season in recent decades, which is apparent from phenological "ground truth" in mid- and higher latitudes of the Northern Hemisphere, is also found in other data sets, such as satellite images, CO_2 records, and in the temperature records. Most of the numerous existing definitions of the climatological growing season use the dates when air temperature exceeds a threshold in spring and falls below a threshold in autumn (see Linderholm 2006). These meteorological measures also reveal lengthening of the warm season (e.g., Rapp and Schönwiese 1994) or of the ice-free season (Magnuson et al. 2000; Sagarin and Micheli 2001). Several studies have analyzed in detail the lengthening of the frost-free season in several countries of the northern hemisphere (e.g., Robeson 2002; Schwartz and Chen 2002; Menzel et al. 2003; Scheifinger et al. 2003). However, geographical differences in climate change and corresponding plant responses are quite common.

In contrast to phenological ground observations on individual plant species, remote sensing provides a multi-species view at a heterogeneous landscape level, facilitating a continuous assessment and modelling of landscape scale leaf presence. Non-climate effects, such as urbanisation, shifts in agricultural practices, and other disturbances are included and this might explain the sometimes weak correlations between ground and satellite estimates (see White et al. 2009). However, multi-temporal satellite data, such as NOAA AVHRR NDVI time series (see Chap. 5), also reveal a lengthening of the growing season in mid- and higher latitudes of the northern hemisphere (e.g., Myneni et al. 1997; Tucker et al. 2001). Zhou et al. (2001) found a lengthening of the growing season, both due to an earlier start and later end, of 18 $\pm$ 4 days for Eurasia, and 12 $\pm$ 5 days for North America based on normalized difference vegetation index estimates between 1982 and 1999 for 40–70°N. For the temperate vegetation of the Northern hemisphere, Jeong et al. (2011) reported for the period 1982–2008 an earlier start of the growing season of 1.3 days decade^{-1} and a later ending of 2.8 days decade^{-1}. Both reported that extensions to the growing season do not exactly match the results from phenological "ground truthing" However, satellite and ground results are well in accord with an advance of the seasonal cycle by −2 to −2.8 days decade^{-1} for the last two–three decades and an increase in amplitude of the annual CO_2 cycle since the 1960s derived from long-term measurements of CO_2 concentration (Keeling et al. 1996).

A greater degree of confidence in satellite phenological metrics is expected to arise when current challenges in understanding the ecological meaning of the estimates are overcome, and when results can be confidently correlated to *in situ* observations (Reed and Schwartz 2009).

References

Abu-Asab MS, Peterson PM, Shelter SG, Orli SS (2001) Earlier plant flowering in spring as a response to global warming in the Washington, DC, area. Biodivers Conserv 10:597–612

Ahas R (1999) Long-term phyto-, ornitho- and ichthyophenological time-series analyses in Estonia. Int J Biometeorol 42:119–123

Ahas R, Aasa A (2006) The effects of climate change on the phenology of selected Estonian plant, bird and fish populations. Int J Biometeorol 51:17–26

Ahas R, Aasa A, Menzel A, Fedotova VG, Scheifinger H (2002) Changes in European spring phenology. Int J Climatol 22:1727–1738

Barbraud C, Weimerskirch H (2006) Antarctic birds breed later in response to climate change. Proc Natl Acad Sci U S A 103:6248–6251

Beaubien EG, Freeland HJ (2000) Spring phenology trends in Alberta, Canada: links to ocean temperature. Int J Biometeorol 44:53–59

Beaumont LJ, McAllan IAW, Hughes LA (2006) Matter of timing: changes in the first date of arrival and last date of departure of Australian migratory birds. Glob Change Biol 12:1339–1354

Bradley NL, Leopold AC, Ross J, Huffaker W (1999) Phenological changes reflect climate change in Wisconsin. Proc Natl Acad Sci U S A 96:9701–9704

Cayan DR, Kammerdiener SA, Dettinger MD, Caprio JM, Peterson DH (2001) Changes in the onset of spring in the western United States. Bull Am Meteorol Soc 82:399–415

Chmielewski FM, Rötzer T (2001) Response of tree phenology to climate changes across Europe. Agric For Meteorol 108:101–112

Chmielewski FM, Rötzer T (2002) Annual and spatial variability of the beginning of growing season in Europe in relation to air temperature changes. Clim Res 19:257–264

Chmielewski FM, Müller A, Bruns E (2004) Climate changes and trends in phenology of fruit trees and field crops in Germany, 1961–2000. Agric For Meteorol 121:69–78

Cleland EE et al (2006) Diverse responses of phenology to global changes in a grassland ecosystem. Proc Natl Acad Sci U S A 103:13740–13744

Cleland EE, Chuine I, Menzel A, Mooney HA, Schwartz MD (2007) Shifting plant phenology in response to global change. Trends Ecol Evol 22:357–365

Clot B (2003) Trends in airborne pollen: An overview of 21 years of data in Neuchatel (Switzerland). Aerobiologia 19:227–234

Cook BI, Wolkovich EM, Parmesan C (2012) Divergent responses to spring and winter warming drive community level flowering trends. Proc Natl Acad Sci U S A 109(23):9000–9005

Defila C, Clot B (2001) Phytophenological trends in Switzerland. Int J Biometeorol 45:203–207

Delpierre N et al (2009) Modelling interannual and spatial variability of leaf senescence for three deciduous tree species in France. Agric For Meteorol 149:938–948

Doi H (2007) Winter flowering phenology of Japanese apricot *Prunus mume* reflects climate change across Japan. Clim Res 34:99–104

Donnelly A, Caffarra A, O'Neill BF (2011) A review of climate-driven mismatches between interdependent phenophases in terrestrial, aquatic and agricultural ecosystems. Int J Biometeorol 55 (6):805–817

Dose V, Menzel A (2004) Bayesian analysis of climate change impacts in phenology. Glob Change Biol 10:259–272

Dunn PO, Winkler DW (1999) Climate change has affected the breeding date of tree swallows throughout North America. Proc R Soc Lond B 266:2487–2490

Edwards M, Richardson AJ (2004) Impact of climate change on marine pelagic phenology and trophic mismatch. Nature 430:881–884

Estrella N, Sparks TH, Menzel A (2007) Trends and temperature response in the phenology of crops in Germany. Glob Change Biol 13:1737–1747

Feng S, Hu Q (2004) Changes in agro-meteorological indicators in the contiguous United States: 1951–2000. Theor Appl Climatol 78:247–264

Fitter AH, Fitter RSR (2002) Rapid changes in flowering time in British plants. Science 296:1689–1691

Gange AC, Gange EG, Sparks TH, Boddy L (2007) Rapid and recent changes in fungal fruiting patterns. Science 316:71

Gibbs JP, Breisch AR (2001) Climate warming and calling phenology of frogs near Ithaca, New York, 1900–1999. Conserv Biol 15:1175–1178

Gordo O, Sanz JJ (2006) Temporal trends in phenology of the honey bee *Apis mellifera* (L.) and the small white *Pieris rapae* (L.) in the Iberian Peninsula (1952–2004). Ecol Ent 31:261–268

Hughes L (2000) Biological consequences of global warming: is the signal already apparent? Trends Ecol Evol 15:56–61

Inouye DW et al (2000) Climate change is affecting altitudinal migrants and hibernating species. Proc Natl Acad Sci U S A 97:1630–1633

IPCC (2007) Summary for Policymakers. In: Solomon S, Qin D, Manning M, Chen Z, Marquis M, Averyt KB, Tignor M, Miller HL (eds) Climate change 2007: the physical science basis. Contribution of Working Group I to the fourth assessment report of the Intergovernmental Panel on Climate Change. Cambridge University Press, Cambridge/New York

Jeong SJ, Ho GCH, Gim HJ, Brown ME (2011) Phenology shifts at start vs. end of growing season in temperate vegetation over the northern Hemisphere for the period 1982–2008. Glob Change Biol 17(7):2385–2399

Jochner S, Sparks TH, Estrella N, Menzel A (2011) The influence of altitude and urbanisation on trends and mean dates in phenology (1980–2009). Int J Biometeorol 56:387–394

Jones GV, Davis RE (2000) Climate influences on grapevine phenology, grape composition, and wine production and quality for Bordeaux, France. Am J Enol Vitic 51:249–261

Karlsen SR et al (2009) Growing-season trends in Fennoscandia 1982–2006, determined from satellite and phenology data. Clim Res 39:275–286

Kauserud H et al (2008) Mushroom fruiting and climate change. Proc Natl Acad Sci USA 105:3811–3814

Keeling CD, Chin FJS, Whorf TP (1996) Increased activity of northern vegetation inferred from atmospheric CO_2 measurements. Nature 382:146–149

Kozlov M, Berlina N (2002) Decline in the length of the summer season on the Kola peninsula, Russia. Clim Change 54:387–398

Lehikoinen E, Sparks TH (2010) Changes in migration. In: Møller AP, Fiedler W, Berthold P (eds) Effects of climate change on birds. Oxford University Press, Oxford

Linderholm HW (2006) Growing season changes in the last century. Agric For Meteorol 137:1–14

Magnuson JJ, Robertson DM, Benson BJ, Wynne RH, Livingstone DM, Arai T, Assel RA, Barry RG, Card V, Kuusisto E, Granin NG, Prowse TD, Stewart KM, Vuglinski VS (2000) Historical trends in lake and river ice cover in the northern hemisphere. Science 289:1743–1746

Matsumoto K, Ohta T, Irasawa M, Nakamura T (2003) Climate change and extension of the *Ginkgo biloba* L. growing season in Japan. Glob Change Biol 9:1634–1642
Menzel A (2000) Trends in phenological phases in Europe between 1951 and 1996. Int J Biometeorol 44:76–81
Menzel A (2002) Phenology: its importance to the global change community. Clim Change 54:379–385
Menzel A (2003) Plant phenological anomalies in Germany and their relation to air temperature and NAO. Clim Change 57:243–263
Menzel A, Estrella N (2001) Plant phenological changes. In: Walther GR, Burga CA, Edwards PJ (eds) "Fingerprints" of climate change – adapted behaviour and shifting species ranges. Kluwer Academic/Plenum Publishers, New York
Menzel A, Fabian P (1999) Growing season extended in Europe. Nature 397:659
Menzel A, Estrella N, Fabian P (2001) Spatial and temporal variability of the phenological seasons in Germany from 1951–1996. Glob Change Biol 7:657–666
Menzel A, Jakobi G, Ahas R, Scheifinger H, Estrella N (2003) Variations of the climatological growing season (1951–2000) in Germany compared to other countries. Int J Climatol 23:793–812
Menzel A, Sparks T, Estrella N, Koch E et al (2006) European phenological response to climate change matches the warming pattern. Glob Change Biol 12:1969–1976
Miller-Rushing AJ, Inouye DW, Primack RB (2008) How well do first flowering dates measure plant responses to climate change? The effect of population size and sampling frequency. J Ecol 96:1289–1296
Morin X et al (2010) Changes in leaf phenology of three European oak species in response to experimental climate change. New Phytol 186:900–910
Myneni RB, Keeling CD, Tucker CJ, Asrar G, Nemani RR (1997) Increased plant growth in the northern high latitudes from 1981 to 1991. Nature 386:698–702
Ottersen G, Planque B, Belgrano A, Post E, Reid PC, Stenseth NC (2001) Ecological effects of the North Atlantic Oscillation. Oecologia 128:1–14
Parmesan C (2006) Ecological and evolutionary responses to recent climate change. Ann Rev Ecol Evol Syst 37:637–669
Parmesan C, Yohe G (2003) A globally coherent fingerprint of climate change impacts across natural systems. Nature 421:37–42
Parry ML, Canziani OF, Palutikof JP, and Co-authors (2007) Technical summary. Climate change 2007: impacts, adaptation and vulnerability. In: Parry ML, Canziani OF, Palutikof JP, van der Linden PJ, Hanson CE (eds) Contribution of Working Group II to the fourth assessment report of the Intergovernmental Panel on Climate Change. Cambridge University Press, Cambridge
Peñuelas J, Filella I, Comas P (2002) Changed plant and animal life cycles from 1952 to 2000 in the Mediterranean region. Glob Change Biol 8:531–544
Post E, Stenseth NC (1999) Climatic variability, plant phenology, and northern ungulates. Ecology 80:1322–1339
Primack RB et al (2009) Spatial and interspecific variability in phenological responses to warming temperatures. Biol Conserv 142:2569–2577
Rapp J (2002) Konzeption, Problematik and Ergebnisse klimatologischer Trendanalysen für Europa und Deutschland. Berichte des Deutschen Wetterdienstes
Rapp J, Schönwiese CD (1994) "Thermische Jahreszeiten" als anschauliche Charakteristik klimatischer Trends. Meteorol Z 3:91–94
Reed BC, Schwartz MD (2009) Remote sensing phenology: status and the way forward. In: Noortmets A (ed) Phenology of ecosystem processes, applications in global change research. Springer, Dordrecht
Robeson SM (2002) Increasing growing-season length in Illinois during the 20th century. Clim Change 52:219–238
Root TL, Price JT, Hall KR, Schneider SH, Rosenzweig C, Pounds A (2003) Fingerprints of global warming on wild animals and plants. Nature 421:57–60

Rosenzweig C, Casassa G, Karoly DJ, Imeson A, Liu C, Menzel A, Rawlins S, Root TL, Seguin B, Tryjanowski P (2007) Assessment of observed changes and responses in natural and managed systems. In: Parry ML, Canziani OF, Palutikof JP, van der Linden PJ, Hanson CE (eds) Climate change 2007: impacts, adaptation and vulnerability. Contribution of Working Group II to the fourth assessment report of the Intergovernmental Panel on Climate Change. Cambridge University Press, Cambridge

Rosenzweig C, Karoly D, Vicarelli M, Neofotis P, Wu Q, Casassa G, Menzel A, Root TL, Estrella N, Seguin B, Tryjanowski P, Liu C, Rawlins S, Imeson A (2008) Attributing physical and biological impacts to anthropogenic climate change. Nature 453:353–357

Rötzer T, Wittenzeller M, Haeckel H, Nekovar J (2000) Phenology in central Europe – differences and trends of spring phenophases in urban and rural areas. Int J Biometeorol 44:60–66

Roy DB, Sparks TH (2000) Phenology of British butterflies and climate change. Glob Change Biol 6:407–416

Sagarin R, Micheli F (2001) Climate change in nontraditional data sets. Science 294:811

Scheifinger H, Menzel A, Koch E, Peter C, Ahas R (2002) Atmospheric mechanisms governing the spatial and temporal variability of phenological observations in central Europe. Int J Climatol 22:1739–1755

Scheifinger H, Menzel A, Koch E, Peter C (2003) Trends of springtime frost events and phenological dates in central Europe. Theor Appl Climatol 74:41–51

Schleip C et al (2008) Time series modeling and central European temperature impact assessment of phenological records over the last 250 years. J Geophys Res 113:G04026

Schwartz MD, Chen X (2002) Examining the onset of spring in China. Clim Res 21:157–164

Schwartz MD, Reiter BE (2000) Changes in North American spring. Int J Climatol 20(8):929–932

Schwartz MD, Ahas R, Aasa A (2006) Onset of spring starting earlier across the northern hemisphere. Glob Change Biol 12:343–351

Sherry RA et al (2007) Divergence of reproductive phenology under climate warming. Proc Natl Acad Sci U S A 104:198–202

Sparks TH, Carey PD (1995) The responses of species to climate over two centuries: an analysis of the Marsham phenological record. J Ecol 83:321–329

Sparks TH, Menzel A (2002) Observed changes in seasons: an overview. Int J Climatol 22:1715–1725

Sparks TH, Smithers RJ (2002) Is spring getting earlier? Weather 57:157–166

Sparks TH, Jeffree EP, Jeffree CE (2000) An examination of the relationship between flowering times and temperature at the national scale using long-term phenological records from the UK. Int J Biometeorol 44:82–87

Studer S, Appenzeller C, Defila C (2005) Inter-annual variability and decadal trends in alpine spring phenology: a multivariate analysis approach. Clim Change 73:395–414

Thackeray SJ et al (2010) Trophic level asynchrony in rates of phenological change for marine, freshwater and terrestrial environments. Glob Change Biol 16:3304–3313

Tryjanowski P, Rybacki M, Sparks TH (2003) Changes in the first spawning dates of common frogs and common toads in western Poland in 1978–2002. Ann Zool Fenn 40:459–464

Tucker CJ, Slayback DA, Pinzon JE, Los SO, Myneni RB, Taylor MG (2001) Higher northern latitude normalized difference vegetation index and growing season trends from 1982–1999. Int J Biometeorol 45:184–190

Walther GR, Post E, Convey P, Menzel A, Parmesan C, Beebee TJC, Frometin JM, Hoegh-Guldberg O, Bairlein F (2002) Ecological responses to recent climate change. Nature 416:389–395

White MA, de Beurs KM, Didan K et al (2009) Intercomparison, interpretation, and assessment of spring phenology in North America estimated from remote sensing for 1982–2006. Glob Change Biol 15(10):2335–2359

Wolkovich EM, Cook BI, Allen JM et al (2012) Warming experiments underpredict plant phenological responses to climate change. Nature 485:494–497

Zhou L, Tucker CJ, Kaufmann RK, Slayback D, Shabanov NV, Myneni RB (2001) Variations in northern vegetations activity inferred from satellite data of vegetation index during 1981 to 1999. J Geophys Res 106(D17):20,069–20,083

Ziello C et al (2009) Influence of altitude on phenology of selected plant species in the Alpine region (1971–2000). Clim Res 39:227–234

Ziello C, Böck A, Estrella N, Ankerst D, Menzel A (2012) First flowering of wind- pollinated species with greatest phenological advances in Europe. Ecography 35(11):1017–1023. doi:10.1111/j.1600-0587.2012.07607.x

Chapter 19
High-Resolution Phenological Data

Mark D. Schwartz and Liang Liang

Abstract Measurements of visual plant phenology at both high-spatial and high-temporal resolutions have many applications, but are especially useful for bridging the gap between ground-based phenological measurements and moderate-resolution satellite-derived measures of phenology. Results have demonstrated that satellite-derived phenology does present a reasonable representation of spring growth in a northern mixed forest environment (Wisconsin, USA), given the known temporal limitations. Other applications of high-resolution phenological data, including measurements during the autumn season are under development.

19.1 Introduction

Phenological measurements of plant development are valuable for many applications, especially when they can be compared among species and across spatial scales. Satellite sensor-derived phenological metrics offer exceptional opportunities to facilitate such comparisons from regional to continental scales. Yet, in order to fully realize this potential, these sensor-derived measurements must be related to comprehensive observations of individual plant species, so as to unlock the detailed ecological significance contained within them. Unfortunately, making such comparisons is problematic using conventional ground-based phenological observations, since they are generally only obtained for a small number of individuals and species at specific sites, and almost never sampled across space within an area comparable in size to moderate resolution satellite pixels (Schwartz et al. 2013). Further, conventional phenology is usually measured using a

M.D. Schwartz (✉)
Department of Geography, University of Wisconsin-Milwaukee, Milwaukee, WI 53211, USA
e-mail: mds@uwm.edu

L. Liang
Department of Geography, University of Kentucky, Lexington, KY 40506, USA

M.D. Schwartz (ed.), *Phenology: An Integrative Environmental Science*,

DOI 10.1007/978-94-007-6925-0_19,

Table 19.1 Spring phenological protocol for deciduous trees (Schwartz et al. 2013)

Code	Deciduous phenophase	Percentage
0	No buds visible	0
100	Buds visible	<10 %
110	Buds visible	10–50 %
150	Buds visible	50–90 %
190	Buds visible	>90 %
200	Buds swollen	<10 %
210	Buds swollen	10–50 %
250	Buds swollen	50–90 %
290	Buds swollen	>90 %
300	Bud open (leaf visible)	<10 %
310	Bud open (leaf visible)	10–50 %
350	Bud open (leaf visible)	50–90 %
390	Bud open (leaf visible)	>90 %
400	Leaf out (not fully unfolded)	<10 %
410	Leaf out (not fully unfolded)	10–50 %
450	Leaf out (not fully unfolded)	50–90 %
490	Leaf out (not fully unfolded)	>90 %
500	Full leaf unfolded	<10 %
510	Full leaf unfolded	10–50 %
550	Full leaf unfolded	50–90 %
590	Full leaf unfolded	>90 %
600	Leaf expansion	Size < 25 % of full
625	Leaf expansion	Size = 25–50 % of full
650	Leaf expansion	Size = 50–75 % of full
675	Leaf expansion	Size > 75 % of full

discontinuous event-based scale, rather than in a continuous fashion, which would be more comparable to sensor-derived measurements. Indeed, given the inherent difficulties and spatial mismatches, it has even been suggested that direct comparisons of ground-based and satellite-derived measures cannot be accomplished in general, and are only possible under very limited circumstances (White et al. 2005).

Thus, with these potential benefits and concerns in mind, in 2005 we initiated a research program aimed at measuring ground-based phenology both across space in areas comparable to medium-resolution satellite sensor pixels (250–500 m) with a dense spatial sampling scheme (i.e., high spatial resolution) and using a continuous scale, taking measurements every other day, in order to achieve high temporal resolution observations. The spring and autumn protocols for deciduous trees are presented in Tables 19.1 and 19.2 respectively.

The spring protocol has been applied in the vicinity of the WLEF/Park Falls Ameriflux tall tower in northern Wisconsin, USA (45.946°N, 90.272°W). The approach to transform these high-resolution ground-based visual phenological measurements into information which can be effectively compared to moderate-resolution satellite-derived phenology has been extensively tested. Other applications of the methodology are just beginning. The autumn phenological protocol was first developed and tested in Downer Woods, a small woodlot on the

Table 19.2 Autumn phenological protocol for deciduous trees

Event number[a]	Description
800	<10 % of leaves colored (including those on the ground)
810	10–50 % of leaves colored (including those of the ground)
850	50–90 % of leaves colored (including those on the ground)
890	>90 % of leaves colored (including those on the ground)
900	<10 % of leaves fallen
910	10–50 % of leaves fallen
950	50–90 % of leaves fallen
990	>90 % of leaves fallen

[a]Coloring [800-level] and leaf fall [900-level] are recorded simultaneously

UW-Milwaukee campus in the northeast portion of the city of Milwaukee (43.081°N, 87.882°W). Additionally, at WLEF/Park Falls, the approach has been used to show that: (1) positive carbon assimilation trends in spring are caused by the phenology of the dominant trees; (2) average tree species-level phenological variations within a microclimate can be adequately represented with a sample size of 20–30; (3) high-resolution visual phenological measurements can be related to under-canopy light sensor data; and (4) sampling tree phenology every 4 days minimizes data uncertainty and field work expenses in a northern mixed forest environment (Schwartz et al. 2013). Work is currently underway to test similar applications during autumn. The most extensively developed methodology to-date, the approach to develop landscape phenology from high-resolution visual phenological data, and an overview of the associated results will now be presented in the remainder of this chapter. Additional details and theoretical considerations are provided here to illustrate the application of high-resolution phenology based upon results from our earlier publications (e.g., Liang and Schwartz 2009; Liang et al. 2011).

19.2 Landscape Scaling of High Resolution Phenology

19.2.1 The Need for and Practicality of Landscape Scaling

Oftentimes, the interest in plant phenology is not only with individual plants but also the aggregated phenological behaviors such as those on the levels of populations, communities and ecosystems. With detailed information from high resolution phenological measurements, a unique opportunity is open for investigating phenological patterns at these integrated scales. A default application of such aggregated information would be to allow connecting field-based in situ phenological observations with measurements from satellites as well as from flux towers. Such practical concern has made better-integrating of individual plant phenology data a necessity. But we perceive that scaling phenological observations to integrated levels has further significance than merely addressing this technical need. As the processes and patterns of ecosystem functions are dictated by different sets of rules at different spatial and organizational scales, adequate scaling would also facilitate phenological investigations within these respective scale domains.

A further question is whether it is feasible to conduct such scaling. It is clear that with limited phenological data (as collected in a traditionally extensive manner), detailed information about landscapes are largely missing in the spatial and temporal data gaps. Therefore performing scaling becomes less meaningful and impractical because the heterogeneous changes over a study area of interest are not adequately accounted for. However, with high resolution phenology data (both in terms of high spatial sampling density and high temporal observation frequency), the ability to incorporate spatial changes across individuals, species, and communities is greatly expanded. In this case study, a sample unit size of 25 m and an observation frequency of 2–4 days were employed, thus optimizing the precision to capture phenological variations over time and space. As prescribed clearly by Wu et al. (2006), an ideal scaling entails that sampling extent is at least as large as the extent of the phenomenon under study, which implies that the comprehensiveness of sampling (which would incorporate spatial heterogeneity of the entire landscape under study) is critical for generating useful scaled products. Hence, in the case of our study, the level of detail and extent of coverage completely satisfy this requirement. Expanding the same criterion to the temporal dimension, high resolution phenology indeed makes the landscape scaling feasible and the resulting products practically comparable to satellite-derived phenology.

19.2.2 The Rationale of Landscape Scaling

Landscape scaling involves incorporating field observations with landscape characteristics. Observation of individual tree phenology is represented as point features within a sampled area. These point features are attributed with phenological development scores (according to the above-mentioned protocol) along with species information such as diameter at breast height (DBH). In order to scale up these spatially discrete phenological measurements to the landscape scale, additional information on the distribution and abundance of species within landscape patches are required. There is no direct comparability between phenology of individual trees and phenology of a forest unless rough assessment (not considering the scale discrepancy) is all that is needed. Further, from the point of view to bridge spatial scales between field and satellite observations, major challenges include both landscape heterogeneity as well as the differences between observation methods (observer vs. instrument). Sampled phenology on the ground needs to be interpolated and sometimes extrapolated to the entire study area (comparable with satellite pixels) in order to deal with spatial heterogeneity that strongly affects landscape-level signals. This bottom-up approach is rooted in the detailed in situ phenological observations and needs to be realized through substantial understanding of landscape characteristics.

An ecologically coherent and technically efficient approach to landscape scaling may be best carried out in a stepwise manner; and one default option is to follow the preexisting hierarchy of nature. A typical ecological hierarchy predefines the organizational levels of individual organism, population, community, and ecosystem. Because information of a lower level, e.g. population, is fully encompassed by a higher level, e.g. community, the ecological hierarchy is inherently nested. This nested hierarchy is adopted as a fundamental reference to design our phenology scaling procedures. Further, according to the Hierarchical Patch Dynamics (HPD) paradigm (Wu and Loucks 1995; Wu 1999), the corporate levels (population, community, ecosystem, or landscape) can be represented as patches which possess unique characteristics in contrast to their backgrounds or contexts. Given that domains with unique spatial scales often demonstrate unique controls of processes and patterns, the scaling accuracy may be optimized if such coherent relationships are accounted for in the scaling design. Plant phenology, like other physiological attributes of organisms may be best integrated if the existing ecological scales are considered. Besides, the notion of landscape is rather more inclusive than population, community, or ecosystem. Depending on the simplicity/complexity of a study area, a landscape could be comprised of a single patch or multiple patches, which are in turn defined by spatially distinguishable mixes of populations, communities, and ecosystems. An incomplete but more succinct way of understanding landscape may be to equal it with an ecosystem patch in which a varying degree of complexity is further specified. We developed a scaling ladder based on this individual-population-community-landscape (ecosystem patch) structure and characterized the landscape patches within the study areas akin to the HPD paradigm.

19.2.3 Detailed Procedures of Landscape Scaling

The implementation of a stepwise and nested hierarchical scaling ladder involves integrating observations of individual tree phenology first to the population phenology, then from population phenology to community phenology, and finally from community phenology to landscape or ecosystem patch phenology. Each step requires a unique set of landscape properties to be characterized and supplied in order to adequately bring the measurements/indices to a higher level. The complete scaling process with respective steps for our case study is illustrated in Fig. 19.1.

19.2.3.1 From Individual Phenology to Population Phenology

The first step of scaling involves estimating population level phenology from the phenology of plant individuals. We assumed that intraspecific (within species) phenology is autocorrelated genetically and/or environmentally in our study area

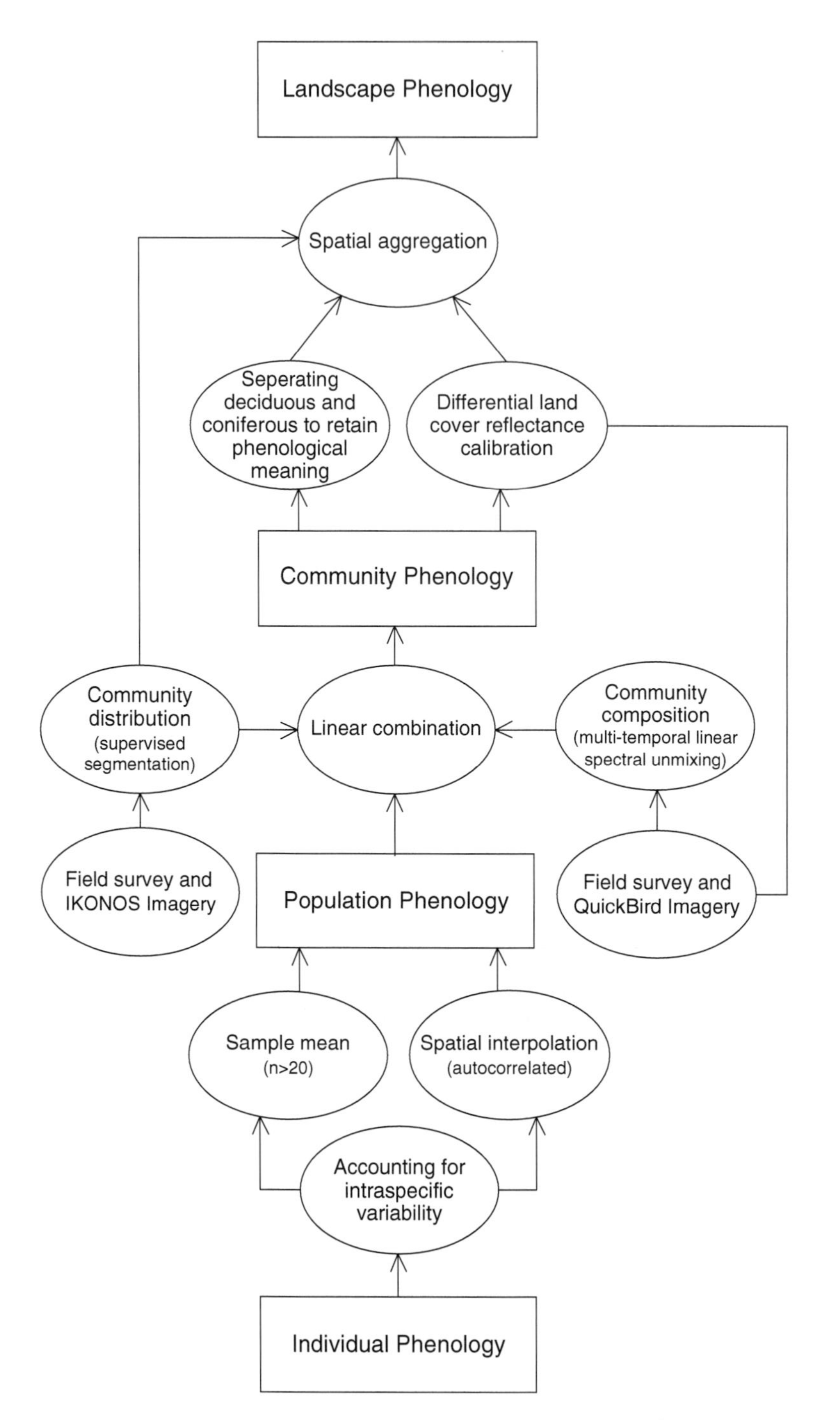

Fig. 19.1 Conceptual diagram and flow chart of hierarchical phenological organization and up-scaling towards landscape phenology; landscape phenology as for an ecosystem patch sits on the same level as ecosystems; the procedure of up-scaling is composed of three primary stages connecting the four levels of phenological representation (Reprinted from Liang et al. (2011), Figure 2, with permission from Elsevier)

because the tree regeneration (e.g., via seed dispersal) and habitat selection/occupation processes were spatially constrained during the history of forest regrowth after logging. Hence, the most accurate way of generating population level estimate would be to conduct a spatial interpolation from all sampled points, on the condition that observations are spatially continuous and dependent. In our case study, data was diagnosed using Moran's I statistics and semivariogram analysis, but no consistent spatial autocorrelations were found among individual phenology observations. On the contrary, individual trees of the same species expressed a mostly discrete and individualistic phenological pattern over the study areas. The reason for this pattern may require further investigation especially through testing multiple sampling schemes. Regardless, this lack of spatial dependency in our data invalidates the use of a spatial interpolation approach, and it calls for an alternative approach for population phenology estimation.

Conventional statistics using an arithmetic mean of samples was thus adopted to calculate population phenology. Given that samples have relatively extensive spatial coverage of respective niches for major species, we can assume their averaged behaviors will capture the behavior of the general population as a whole. In order to further justify this approach in considering sample size effects on the representativeness of sample means, a sample size sensitivity analysis was performed using 2006 trembling aspen phenology in reference to the field protocol. Trembling aspen has the largest sample size (71) in our pilot study area (a smaller area used in 2006 and 2007). We calculated the absolute errors from the population mean (estimated using all samples) of phenological measures for a full range of possible sample size (1–71). For each sample size, ten random samples were taken from the tree phenological readings for every observation day (for 15 days), and the results from the total 150 datasets were averaged. The relationship between sample size and absolute errors from the population mean (approximated using the mean of all samples) is shown in Fig. 19.2. Previous studies suggested minimum sample sizes of between 15 and 20 individual plants were sufficient to accurately represent population estimates of plant phenology (West and Wein 1971; Morellato et al. 2010). Our initial analysis shows that when the sample size surpasses 20, absolute errors are consistently below half of the maximum resolution of the employed phenology protocol (40/2 = 20, equivalent to 20 % of phenological change within a development stage). This criterion allows the population estimate errors to be controlled within the protocol precision. Our follow-up study using data from the larger study areas (since 2008) suggested that a sample size of 30 would allow adequate representations of population level phenology (Schwartz et al. 2013). Therefore, using a minimum sample size (20–30) appears to be a valid empirical approach for generating population phenology estimates.

19.2.3.2 From Population Phenology to Community Phenology

In addition to intraspecific differences, presence and abundance of different species vary significantly across space (interspecific variations). The secondary forest

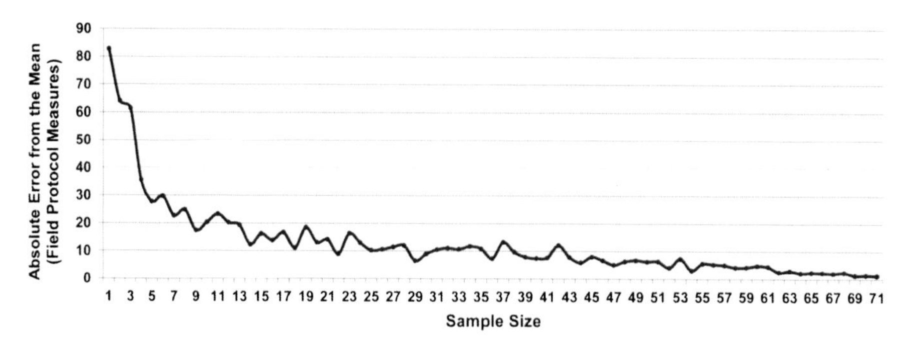

Fig. 19.2 Sample size study using 2006 trembling aspen phenology as an example; each value point was averaged from 10 random samples from 71 observations for all 15-day periods at the corresponding sample size

under study is relatively complex in composition, with multiple plant communities existing in fairly small areas (two complete 625 × 625 m areas). Varying microclimates strongly affect the distribution and abundance of species, and has led to plant communities occupying different microenvironmental niches. Hence, to scale population phenology to the community level, detailed spatial information on the distribution and composition of communities are required. Due to the non-exhaustive survey of the study areas (given the abundance of trees and limited resources), we employed high spatial resolution remote sensing data for precise community characterization. As detailed below, two primary tasks are involved with this effort: (1) to delineate the boundaries of plant communities; and (2) to estimate the abundance of species within each delineated community.

Community boundary delineation—Forest communities within the study areas are delineated using a post-supervised segmentation approach with high resolution satellite imagery. A summertime (2002) IKONOS image set (with both multispectral and panchromatic bands) for the study area was available from NASA. The 4 m multispectral image was first resolution-merged with the 1 m panchromatic image. The resulting image was then processed with an object-based segmentation (feature extraction) classification. The segmentation approach is more advantageous than the traditional pixel-based classification because the plant communities are continuous and self-contained, and they can be better identified with texture, shape, and association in addition to spectral variance. We allowed a segmentation process with the level of differentiation that is more detailed than the intended community boundaries. Therefore, the segmented image generated was composed of a wall-to-wall combination of polygons which are subsections of communities. Then, by using collected tree species/location information, field survey notes, and knowledge about the microenvironmental gradients, segments (polygons) belonging to the same community were merged respectively. This post-supervision classification builds upon the segment boundaries as delineated from the satellite imagery and incorporated necessary ground references. The final results for both study areas (north and south) are shown in Fig. 19.3.

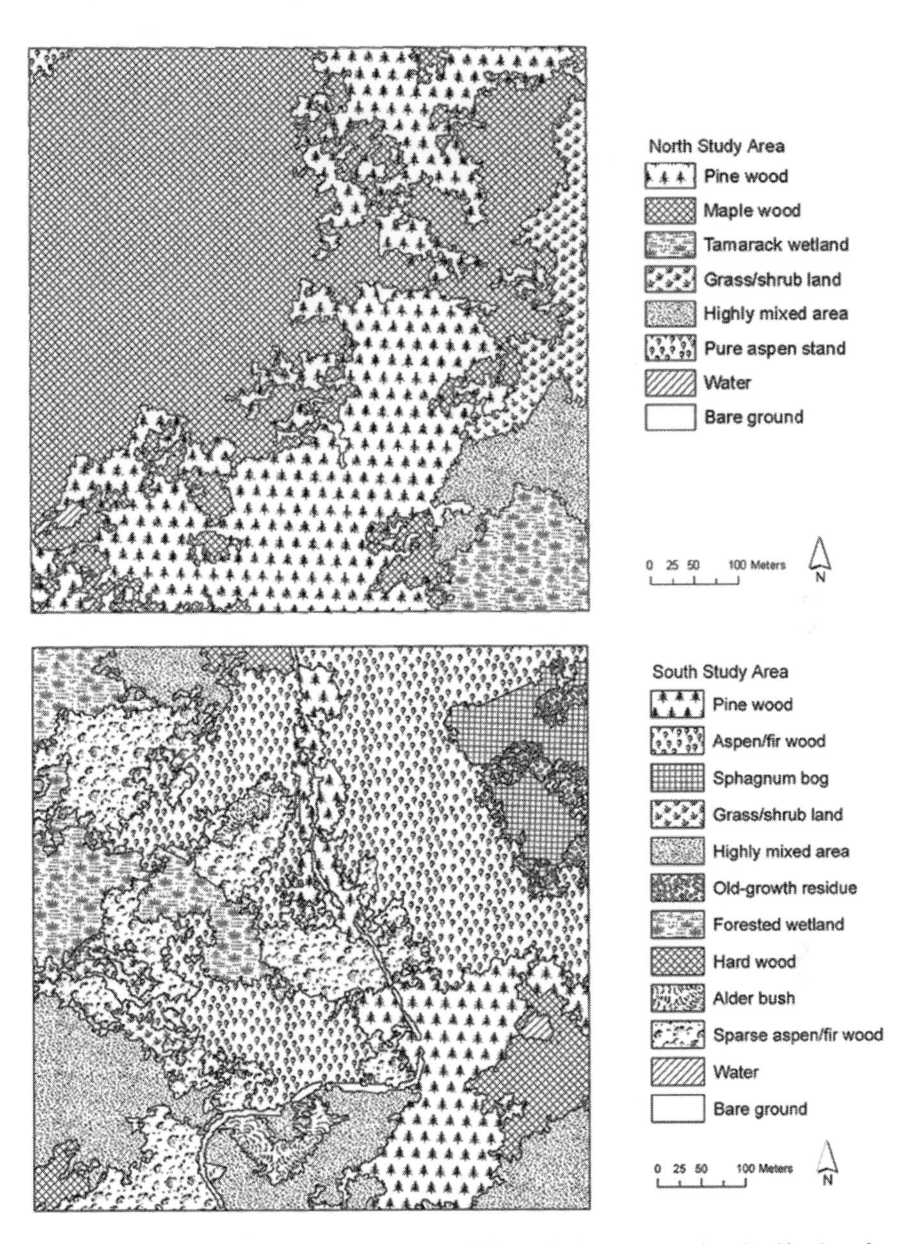

Fig. 19.3 Forest communities in the study areas, delineated using segmentation classification of a summer IKONOS image and manual post-processing based on field survey; the study areas appear tilted due to the UTM projection (Reprinted from Liang et al. (2011), Figure 3, with permission from Elsevier)

Table 19.3 Major species found in forest communities in the northern (N) and southern (S) study areas

Community	Dominant species
Pine wood (N)	Red pine, sugar maple
Maple wood (N)	Sugar maple, red maple, bass wood
Tamarack wetland (N)	Tamarack, black spruce
Grass/shrub land (N)	Grass species, bigtooth aspen, sugar maple, red maple
Highly mixed area (N)	Trembling aspen, speckled alder, balsam fir
Pure aspen stand (N)	Trembling aspen
Pine wood (S)	Red pine, red maple, trembling aspen
Aspen/fir wood (S)	Trembling aspen, balsam fir
Sphagnum bog (S)	Speckled alder, red maple, balsam fir
Grass/shrub land (S)	Grass species, speckled alder, balsam fir, red maple
Highly mixed area (S)	White birch, trembling aspen, speckled alder, red maple, balsam fir
Old-growth residue (S)	Speckled alder, red maple, trembling aspen, balsam fir
Forested wetland (S)	Speckled alder, red maple, balsam fir
Hard wood (S)	Red maple, trembling aspen
Alder bush (S)	Speckled alder
Sparse aspen/fir wood (S)	Trembling aspen, balsam fir, speckled alder, grass species

Community composition estimation—Numerically dominant species of each forest community are known from field sampling (Table 19.3), but only as rough estimates. Thus, the remaining challenge is to accurately quantify the relative abundance of species within each community. Again we use remotely sensed data to complement the limited field survey. Instead of estimating the abundance of each single species (which would be difficult), we adopted an approach estimating the proportions of deciduous and coniferous species within each community as a more achievable and efficient alternative with the available remote sensing data. This approach is feasible due to the differential phenological patterns of the two groups of tree species (i.e., deciduous and coniferous). Specifically, before the onset of land surface greening, deciduous species have no green foliage compared to evergreen coniferous covers. After deciduous canopies develop in spring, reflectance from conifers is still separable from deciduous stands in satellite imagery ("needle-leaf green" vs. "broad-leaf green"). Based on such spatiotemporal differences in the phenological patterns of deciduous and coniferous species and their respective contributions to land surface reflectance, multi-temporal satellite images can be used to separate them, as well as non-vegetated areas (which show minimum variability throughout the season) within a mixed forest environment.

We employed a multi-temporal spectral unmixing approach to high resolution images in order to distinguish between deciduous and coniferous cover types. Three 2.4 m multispectral QuickBird images were provided by NASA. Two of the images were taken simultaneously during the field work in 2007 (May 3rd and May 18th). The remaining image was acquired in 2006 (April 17th) before the field work commenced. The May 3rd image was contaminated with clouds (10.1 %) which affected portions of the study areas (mainly through cloud shadows) even after

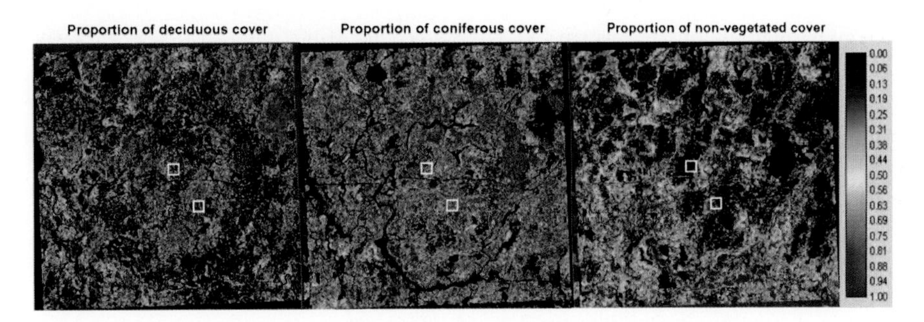

Fig. 19.4 Sub-pixel proportion maps of deciduous, coniferous, and non-vegetated covers generated from multi-temporal linear spectral unmixing; represented in color ramp showing proportions from 0 (*black*) to 1 (*red*); the maps show surrounding areas of the study site with the WLEF flux tower/station visible (a circular feature) in the center of scene; the study areas are located near the WLEF tower, as outlined with white squares (Reprinted from Liang et al. (2011), Figure 4, with permission from Elsevier)

atmospheric correction. Therefore, two cloud-free images were used to represent conditions of deciduous leaf-off (i.e., before leaf bud-break, April 17th, 2006) and leaf-on (i.e., near full leaf expansion, May 18th, 2007) respectively. These two images were co-registered to the IKONOS image (geolocation accuracy visually assessed with GPS measured ground reference points). Iterative experimental processes were undertaken to evaluate ground control points (GCPs) and to find appropriate geometric correction models. The co-registration was performed with 25 GCPs for each image, using 2nd order polynomial models and the nearest neighbor resampling method. The root mean square (RMS) errors were below 0.3 pixel size (2.4 m) or approximately 1 m. This co-registration is a necessary step to ensure the spatial coherency of the multi-temporal analysis.

A spectral unmixing approach estimates the proportions of identified pure land covers (end-members) within each pixel (Roberts et al. 1998; Wu and Murray 2003). Multi-temporal spectral unmixing models have been used to classify vegetation cover types with improved accuracy over traditional methods (Defries et al. 2000; Lu et al. 2004). In order to implement this approach in our study, we first identified homogeneous deciduous and coniferous stands, as well as non-vegetated covers given their distinct patterns in color, texture, and phenology. The signatures of these end-members were then used to estimate their respective contributions across the study area at the sub-pixel level. A linear approach was adopted and the observed reflectance was modeled with a linear combination of the reflectance from end-members. The signatures of these end-members were reviewed with spectral analysis using principal component analysis (PCA) and minimum noise fraction (MNF). The separability of their signatures was validated with feature spaces (scatter plots) of the first three PCA and MNF components (Wu and Murray 2003). Processed images showing proportions of end-members for the entire QuickBird scene (larger than the study areas) are shown in Fig. 19.4.

Community phenology compositing—Detailed information on community distribution and composition, as derived from the above mentioned methods, allows community phenology to be estimated at the pixel level. Community phenology is hence mathematically represented as a linear combination of population phenology and community composition. For each pixel, this relationship is expressed as:

$$\text{Phe\_\_community} = \text{Phe\_deciduous} \times \text{Prop\_deciduous} + \text{Phe\_coniferous} \times \text{Prop\_coniferous} \quad (19.1)$$

where Phe is phenology and Prop represents proportion. Therefore, population phenology estimates of major deciduous species (Table 19.3) for each community were multiplied by their respective proportions to estimate the deciduous component of community phenology. When more than one species is dominant, their weighted averages should be used. The weights were estimated according to field observations. Balsam fir phenology was uniformly applied to compute the coniferous portion of community phenology. Especially, for pine woods where spring pine phenology is hardly noticeable, balsam fir was used as a substitute. Such a compositing can be conducted at the pixel level, and then aggregated for each community, or it can be performed at the community level by calculating the averaged end-member proportions for each community.

Since grass phenology affects the landscape greening before deciduous leaf onset, grass should be incorporated into the deciduous portion. So for all cases with phenological readings below 300 level (initial bud burst), deciduous phenology was estimated with grass phenology only; and for communities such as grass/shrub land, sparse aspen/fir wood and sphagnum bog, grass phenology was used along with tree phenology when trees surpassed phenological stages beyond the 300 level. Grass phenology was estimated using hue-saturation-luminance based percentage greenness index for spring 2006 (Liang et al. 2012). As grass phenology was not measured in subsequent years, speckled alder phenology was used as a substitute given their high correlation with each other (0.935, $\alpha = 0.01$). Speckled alder is also a dominant shrub species with early phenology present in most parts of the study areas. A regression model was built between grass phenology and speckled alder phenology in order to translate the former for use in community phenology estimation.

19.2.3.3 From Community Phenology to Landscape Phenology

Landscape phenology can be generated with simple spatial aggregation of community phenologies, but some further considerations are needed for specific applications. As discussed earlier, in order to compare landscape phenology with satellite-derived phenology, additional calibrations are required. Such calibration involves translating the integrated landscape phenology into an index that is comparable in nature and dimension with land surface reflectance and vegetation indices caused by phenological processes. On the other hand, it is useful to retain the biological inference of phenology measures (as specified in the field protocol) at the aggregated levels. The scaled phenology that is biologically meaningful is especially needed for extracting plant development events, such as full bud burst and full leaf unfolding. Therefore, we derived two sets of landscape phenology indices to address these two events.

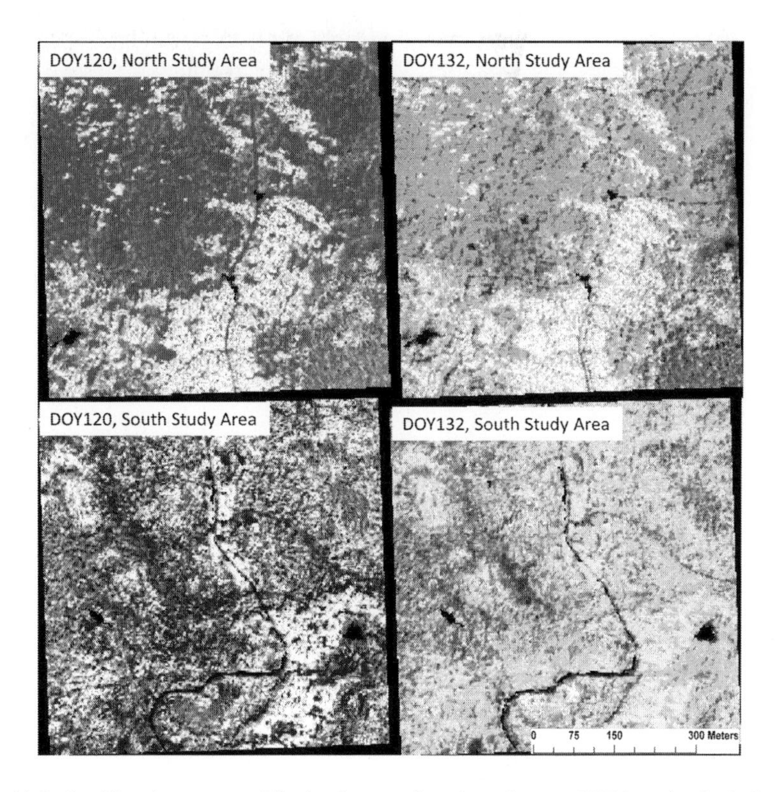

Fig. 19.5 Satellite data compatible landscape phenology layers (UTM projection) for two selected dates in 2008, showing both the northern and southern study areas. The dates are in day of year (DOY). Digital number based grayscale is used to present the calibrated landscape phenology index (in ratio of 0–1, akin to that of vegetation indices)

Landscape phenology indices compatible with satellite signals—Calibration of community level phenology estimates with different reflectance characteristics for deciduous, coniferous, and grass cover, as well as bare ground is employed to generate landscape phenology indices that are compatible with satellite signals. We employed multi-temporal QuickBird images to convert field derived estimates to values compatible to satellite-based vegetation indices. This conversion is achieved through building linear regression models between QuickBird derived normalized difference vegetation index (NDVI) and community level phenology estimates. Averaged NDVI for pure deciduous stands, coniferous stands, and grassy openings were extracted and regressed respectively against averaged deciduous phenology, coniferous (balsam fir) phenology, and grass phenology. Grass phenology was substituted with speckled alder phenology due to reasons detailed earlier. In addition, bare ground (non-vegetated portion) was attributed to an averaged background NDVI value (0.15). The calibration was applied at the sub-pixel level and thus the reflectance variations (from phenological gradients) are represented with pixel level details across the entire study areas (Fig. 19.5).

Landscape phenology indices retaining field protocol meanings—Landscape phenology indices retaining biological significance of phenology as contained in the field protocol is useful for relating to satellite-based biophysical indicators, i.e. phenologically important dates such as start of season or end of season. Given that phenologies of deciduous and coniferous species are different and have different field protocols, landscape phenology indices were derived separately for the two vegetation groups. Rather than employing complex calibration methods for the former index, simple homogeneous landscape scenarios were formed with either deciduous or coniferous covers occupying the entire study areas. For each pixel, deciduous and coniferous components of community phenology directly represent their contributions to landscape phenology respectively. When each portion is spatially aggregated and averaged across a study area or a satellite pixel, a single value for the landscape phenology index for a given date can be generated. Therefore, as this type of landscape phenology index retains the original field protocol meanings, leaf out dates can be determined when the index value (for specified communities or pixels) reaches the 400 level, according to the field protocol.

19.2.4 Summary

The landscape scaling of high resolution phenology as described above using our case study as an example has two levels of applications. One application is related to calibrating coarse-scale ecosystem measures such as satellite-based phenology and biosphere-atmosphere exchange of carbon and water. The above approach was successfully used to validate satellite phenology derived from MODIS vegetation indices (Liang et al. 2011). The second application is to allow phenology to be represented at aggregated levels, and thus to facilitate investigation of phenological patterns and processes at different scales and levels of ecosystem structure. As summarized in the concept of landscape phenology (Liang and Schwartz 2009), such scaling techniques supports implementation of the research perspective in line with deepening and broadening phenology as an integrative environmental science. High-resolution phenology provides an unprecedented opportunity for conducting future detailed multi-scaled phenological analyses within the context of ecological complexity.

References

Defries RS, Hansen MC, Townshend JRG (2000) Global continuous fields of vegetation characteristics: a linear mixture model applied to multi-year 8 km AVHRR data. Int J Remote Sens 21(6–7):1389–1414

Liang L, Schwartz MD (2009) Landscape phenology: an integrative approach to seasonal vegetation dynamics. Landsc Ecol 24(4):465–472

Liang L, Schwartz MD, Fei S (2011) Validating satellite phenology through intensive ground observation and landscape scaling in a mixed seasonal forest. Remote Sens Environ 115:143–157

Liang L, Schwartz MD, Fei S (2012) Photographic assessment of temperate forest understory phenology in relation to springtime meteorological drivers. Int J Biometeorol 56(2):343–355

Lu DS, Batistella M, Moran E (2004) Multitemporal spectral mixture analysis for Amazonian land-cover change detection. Can J Remote Sens 30(1):87–100

Morellato LPC, Camargo MGG, D'Eça Neves FF, Luize BG, Mantovani A, Hudson IL (2010) The influence of sampling method, sample size, and frequency of observations on plant phenological patterns and interpretation in tropical forest trees. In: Hudson IL, Keatley MR (eds) Phenological research. Springer, Dordrecht

Roberts D, Gardner M, Church R, Ustin S, Scheer G, Green R (1998) Mapping chaparral in the Santa Monica Mountains using multiple endmember spectral mixture models. Remote Sens Environ 65(3):267–279

Schwartz MD, Hanes JM, Liang L (2013) Comparing carbon flux and high-resolution spring phenological measurements in a northern mixed forest. Agric For Meteorol 169:136–147

West N, Wein R (1971) A plant phenological index technique. Bioscience 21(3):116–117

White MA, Hoffman F, Hargrove WW, Nemani RR (2005) A global framework for monitoring phenological responses to climate change. Geophys Res Lett 32:L04705

Wu J (1999) Hierarchy and scacling: extrapolating information along a scaling ladder. Can J Remote Sens 25(4):367–380

Wu J, Loucks OL (1995) From balance of nature to hierarchical patch dynamics: a paradigm shift in ecology. Q Rev Biol 70(4):439–466

Wu CS, Murray AT (2003) Estimating impervious surface distribution by spectral mixture analysis. Remote Sens Environ 84(4):493–505

Wu J, Jones KB, Li H, Loucks OL (2006) Scaling and uncertainty analysis in ecology: methods and applications. Springer, Dordrecht

Chapter 20
Weather Station Siting: Effects on Phenological Models

Richard L. Snyder, Donatella Spano, and Pierpaolo Duce

Abstract Accurate temperature data is important for both the development and use of phenological models, and this chapter discusses topics related to temperature measurement for use in phenological models. The chapter presents a short history of temperature measurement, the theory of temperature sensors, radiation shielding, and guidelines on weather sensor placement. Physiological time (degree day) calculation and its application are described. The impact of underlying surface, fetch, and surrounding environment on temperature and phenological time are discussed and some guidelines on measurement are presented.

20.1 Introduction

The collection of accurate temperature data is critically important for phenological model development as well as for predictions when a model is being used. This is true everywhere, but especially in an arid climate where advection and surface wetting (i.e., by rainfall or irrigation) can affect temperature measurements. In fact, the use of bad temperature data can lead to errors as large as differences resulting from climate change. Therefore, accurate temperature measurement is critical for both the application and development of phenological models. Temperature

R.L. Snyder (✉)
Department of Land, Air, and Water Resources, University of California, Davis, CA 95616, USA
e-mail: rlsnyder@ucdavis.edu

D. Spano
Department of Science for Nature and Environmental Resources (DipNet), University of Sassari, Sassari, Italy

Euro-Mediterranean Centre for Climate Change (CMCC), Sassari, Italy

P. Duce
Institute of Biometeorology, National Research Council, Sassari, Italy

M.D. Schwartz (ed.), *Phenology: An Integrative Environmental Science*,
DOI 10.1007/978-94-007-6925-0_20, © Springer Science+Business Media B.V. 2013

readings are affected by the local energy balance, so wetting frequency by irrigation or rainfall can affect readings and therefore standardization and management of the underlying surface is critical to obtain useful models that can be universally applied. In this chapter, the history and proper measurement of temperature is discussed including sensors, shielding, weather station siting, and surface management.

20.2 Background and History of Temperature Measurement

Although everyone intuitively understands the meaning of temperature, it is really not easily defined. One could describe it as a measure of the sensible heat content of the air. However, heat is an even more vague term than temperature. According to Horstmeyer (2001), who reported the following statistics, heat does not really exist. What does exist are unimaginable numbers of air molecules (about 2.69×10^{25} molecules m^{-3} at sea level) that are moving at incredible speeds (increasing from about 1,600 to 1,700 km h^{-1} for dry air as the temperature increases from 0 to 40 °C). The mass of individual air molecules is small and the total volume of the molecules occupies less than 0.1 % of the air volume, but there are about 5.95×10^{14} collisions s^{-1} between molecules. When molecules collide, energy is not created or lost, but kinetic energy is transferred between the molecules. In thermodynamics, temperature is considered a measure of the average kinetic energy of the air molecules expressed as: $T = 0.5M\overline{v^2}/R'$, where $\overline{v^2}$ is the mean of the squared molecular velocity, M is the molar mass, and R' is the molar gas constant. Therefore, when the speed of the molecules increases, the temperature increases proportionally to the mean of the squared velocity.

As air molecules strike our skin, some kinetic energy is absorbed and conducted from molecule to molecule into our bodies. We describe this "sensation" as "heat" and the term "sensible heat" is used to indicate that it is energy that we "sense". If the temperature is higher, the molecules move faster, more hit our skin, more kinetic energy is received, and the hotter we feel. When a thermometer is placed in the air, its surface is also bombarded with air molecules at near sonic speeds. These collisions transfer kinetic energy to the thermometer and some is conducted inside. When the conducted energy reaches and heats the liquid in the thermometer, it expands and moves up the capillary tube indicating a higher temperature.

Historically, temperature measurement was closely linked to the development of thermometers. The earliest records pertaining to the concept of temperature were based on the writings of Aristotle in about 300 B.C. The first evidence of temperature measurement was in the late 1500s in Europe, with temperature scales appearing in the early 1600s. The first meteorological network, called ReteMedicea (i.e. Medici Network) was active during 1654–1670 with temperature readings taken 6–8 times per day in a number of stations (Camuffo et al. 2011). The Grand Duke of Tuscany supported the network and provided thermometers with the same calibration and an operational protocol to obtain comparable readings. The thermometers were known as the "Little Florentine Thermometer" and they were entirely made of glass except for the spirit.

In the mid 1600s, the first scales based on the freezing point of distilled water appeared (Christopher Wren and Robert Hooke). By the 1700s, Fahrenheit's scale, which defined 32° as the ice point and 96° as the human body temperature was developed. Fahrenheit did not use the boiling point in his definition, but the boiling point (212 °F) was used to manufacture thermometers. The Réaumur scale, which set 0° as freezing and 80° as boiling, was proposed by René Antoine Ferchault de Réaumur in 1730, and it was used in Europe for many years. The scale was based on the use of alcohol rather than mercury thermometers and eventually it was replaced by the mercury thermometer using the Celsius scale. Celsius used the ice point and the boiling point to define a temperature scale starting with 0° at the boiling point and 100° as the ice point. However, shortly after Celsius proposed his temperature scale, others began to use degrees centigrade for the scale defined with 100 gradations between the ice point and boiling point (i.e., the reverse of the original Celsius scale). Because there were several temperature scales in usage during the late 1800s, Callendar recommended a single practical temperature using the ice and boiling points of water with standard platinum resistance thermometers for interpolation.

During the twentieth century, several conferences on the international temperature scale were held in 1927, 1948, 1968, and 1990 (i.e., ITS-27, ITS-48, ITS-68, and ITS-90). ITS-27 adopted the concepts of Callender, with the addition of thermocouples being the standard instrument. ITS-48 changed the name of the scale from centigrade to Celsius to credit Celsius for developing the scale and because the word centigrade represents a unit of angle measure in French. ITS-48 also defined the absolute temperature scale as having exactly 100 gradations between ice point and the boiling point of water. ITS-68 fixed the Kelvin temperature scale to absolute zero by setting 1.0 K = 1/273.16 of the triple point (i.e., the equilibrium between the phases of ice, liquid water, and water vapor). The triple point of water was set at exactly 273.16 K and the ice point was given the value of 273.15 K = 0 °C. The boiling point was still set at 373.15 K = 100 °C. They also eliminated the use of the term degrees for the absolute temperature scale (i.e., the unit is Kelvin rather than degrees Kelvin). Later, the ITS-90 reported the boiling point was really 99.975 °C, so the definition of the boiling point was eliminated from the absolute temperature scale. However, at standard temperature and pressure, the Celsius scale still is defined as going from 0 °C at the melting point to 100 °C at the boiling point, and the Fahrenheit scale goes from 32 °F at the melting point to 212 °F at the boiling point. Today, absolute temperature (K) and temperature in Celsius, where (0 °C = 273.15 K), are commonly used in science.

20.3 Measurement Theory

20.3.1 *Sensors*

Primary thermometers measure temperature in the sense that they measure variables, which are directly dependent on temperature, with coefficients that are virtually independent of temperature. For example, a gas thermometer, which measures temperature by the thermodynamic relationship between gaseous pressure, volume and temperature, is a primary thermometer. Secondary thermometers (e.g., thermistors, diodes, transistors, thermocouples, liquid-in-glass, liquid-in-metal, and metal deformation thermometers) measure variables that depend on temperature with coefficients that may be highly dependent on temperature. Secondary thermometers are the main types used for meteorological standards and operational meteorological instruments. Because they are the most commonly used in phenological studies, mercury-in-glass thermometers and thermistors will be discussed in this chapter.

A mercury-in-glass thermometer is a secondary type thermometer because the liquid expansion involves coefficients, which are temperature dependent and not necessarily theoretically predictable. Most meteorological thermometers are full immersion types, so the entire glass thermometer is fully exposed to the air. By making the capillary (the bore in which the liquid rises) small relative to the liquid bulb (reservoir), the sensitivity and resolution of the thermometer increase. However, a limit is reached when the capillary surface tension forces degrade performance.

Maximum and minimum thermometers are two main types of liquid-in-glass thermometers that operate on different principles (Fig. 20.1). The maximum thermometer has a constriction in the capillary bore at a short distance from the fluid reservoir. The liquid expands past the constriction, and expands to a maximum temperature. On cooling, the liquid above the constriction contracts leaving a vacuum barrier, while the liquid below the constriction contracts into the reservoir. This leaves the upper end of the liquid at the maximum temperature. The minimum thermometer has a small rider, which is pushed by the surface forces of the meniscus of the liquid within the capillary bore. The upper end of the rider is left at a minimum temperature position when the liquid rises again.

Thermistors are made of sintered semiconductor materials (e.g. manganese, nickel, copper, iron, cobalt, and uranium oxides) that are pressed into the thermistor form and aged to promote stability. The resistance to electrical current decreases as the temperature rises and simple electronics are used to monitor the resistance. The physics involved in resistance sensor operation is described in Quinn (1985). With a thermistor, a curvilinear relationship between resistance and temperature is needed to calibrate the sensors. Older thermistors were somewhat unstable and, during their lifetime, the resistance would drift requiring frequent calibration. However, now thermistors are aged before calibration to minimize drift. Because the bead thermistors are manufactured with their leads inside the powdered material before

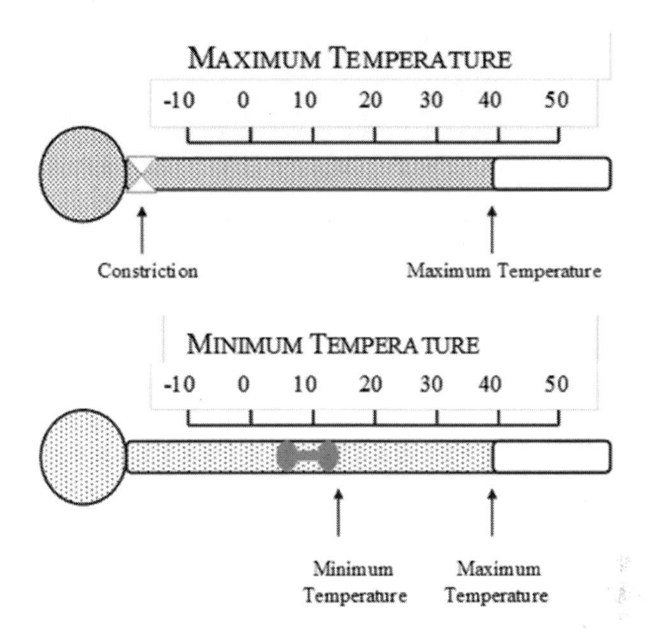

Fig. 20.1 Maximum and minimum liquid-in-glass thermometers

they are pressed, whereas disk thermistors are manufactured prior to the leads being sprayed or printed on the material, thermistors made in the shape of beads are more stable than thermistors pressed into disk shapes. A major advantage of thermistor is that they are small sensors. The wire/metal resistors may be made very thin, but they still must have an appreciable length (several millimeters at the minimum) to have a practical level of resistance (tens of ohms) for temperature measurement. Thermistors can be made as small as 0.03 mm.

20.3.2 Shielding

Although having accurate sensors solves part of the problem with temperature measurement, proper shielding and ventilation are needed to minimize radiation effects on the sensor energy budget. The three main energy forms that affect energy balance on a thermometer are radiant, convective/sensible heat transfer, and thermal conduction. To avoid energy transfer from the mount to the sensor, the contact area should be small and insulated with plastic, ceramic, or cardboard materials that reduce conduction. For electronic sensors, the leads should be small, non-conductive, and maintained in the same shelter environment as the sensor to avoid direct exposure to radiation or other heat sources.

Temperature sensors are frequently shielded from short wave radiation by putting them in a Stevenson Screen (Fig. 20.2), which is made of wood and painted

Fig. 20.2 A standard Stevenson Screen (*left*) and a Gill radiation shield (*right*)

white to reflect away most solar radiation. For electronic sensors, white plastic shields like the Gill shield (Gill 1983) are often used (Fig. 20.2). Long wave radiation is unavoidable, because it is emitted by all surfaces at terrestrial temperatures.

While the goal is to measure air temperature with a thermometer, the actual temperature recorded depends on the air temperature and the housing or shield temperature around the sensor. Assuming that short wave radiation reaching the thermometer is negligible, Monteith and Unsworth (1990) expressed thermometer temperature (T_t) in terms of the shield temperature (T_s), air temperature (T), and resistances to sensible (r_H) and radiation (r_R) heat transfer as:

$$T_t = \frac{r_H T_s + r_R T}{r_R + r_H} \tag{20.1}$$

by assuming the net radiation and convective heat transfer are equal ($R_n = C$) as illustrated in Fig. 20.3. When estimating air temperature, the goal is to have T_t approach T by making r_H much smaller than r_R and/or by making T_s very close to T. To make r_H much smaller than r_R, one can maximize convective heat transfer by ventilation or making the sensor very small. Generally, ventilation of a few meters per second is adequate. When ventilating, the air should be pulled over the sensor to avoid heating from the fan motor. In naturally ventilated Stevenson Screens or Gill shields, errors can result when wind speed is low. Double walls and roofs and painting the shelter white to reflect solar radiation makes T_s closer to T.

20.3.3 Height

When using temperature data in phenological models the height of the temperature measurement can have a big effect on the results. In the USA, the National Weather

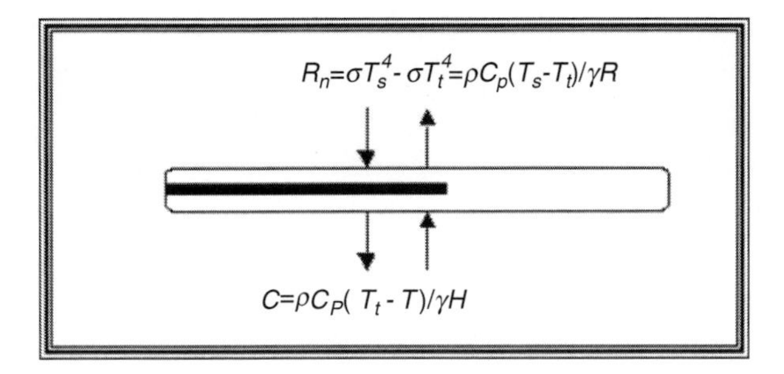

Fig. 20.3 Drawing of the energy balance between a thermometer and its surrounding radiation shield where R_n is the net radiation, C is convective heat transfer, σ is the Stefan-Boltzmann constant, T_s, T_t, and T are the shield, thermometer, and air temperatures (K), ρ is the air density, C_p is the specific heat of air at constant pressure, and r_R and r_H are resistances to radiative and sensible heat transfer

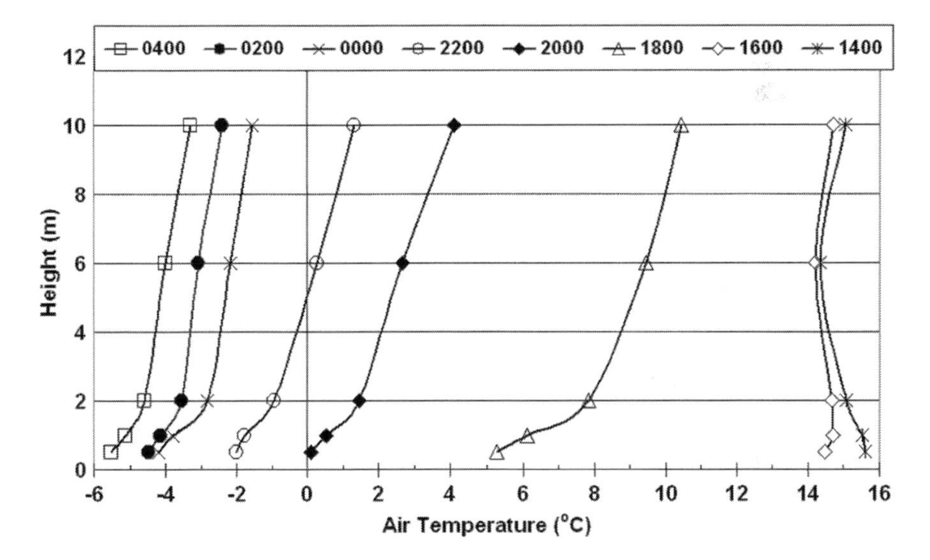

Fig. 20.4 Temperature profile data from a walnut orchard at 0.5, 1.0, 2.0, 6.0, and 10.0 m height prior to and during a radiation freeze night near Ladoga, California, which is in Indian Valley in the Coastal Mountain Range about 200 km north of San Francisco

Service typically reports 10.0 m temperatures, whereas many climate and agricultural weather networks report temperatures for heights varying between 1.5 and 3.0 m. To illustrate the problem, Fig. 20.4 shows the vertical temperature profile changes during the day before and during the night of a spring freeze event in a northern California mountain valley. Clearly, big differences exist between data taken at 0.5 and 10.0 m heights. Therefore, when developing phenological models, temperature data should be used from a station with a temperature measurement height similar to the sensor heights of stations where it is likely to be used.

Similarly, models from the literature should report the temperature measurement height and the model should only be used with data collected at the same height.

20.4 Physiological Time

For many applications (e.g., pest management, crop modeling, and irrigation scheduling) it is useful to predict when a crop or pest will develop to a particular phenological stage. When temperatures are higher, organisms develop in fewer days because they are exposed to the greater heat accumulation than organisms grown under cooler conditions. The accumulation of heat is called "physiological time," which typically is better than "calendar time" for predicting days between phenological stages.

Most organisms show little growth or development below a lower threshold temperature (T_L) and often there is little increase in development rate above an upper threshold (T_U), but the number of hours between T_L and T_U provide a measure of heat units for estimating the days between phenological stages. Each hour of time that the organism is exposed to temperature between the two thresholds is called a "degree-hour." The sum of the degree-hours during a day divided by 24 provides the degree-days on that date, and the degree-days are commonly used in phenological models. With the increased use of electronic sensors and data loggers, calculating the sum of degree-hours and dividing by 24 h per day is now widely practiced. However, hourly data are not available everywhere, and phenological studies using daily maximum and minimum temperature climate data are still commonly used to study past and future climate impacts on phenology.

There are numerous papers on methods to estimate degree-days from daily maximum and minimum temperature data and on how to determine threshold temperatures and the cumulative degree-day requirement for various crops and pests (Baskerville and Emin 1969; Allen 1976; Johnson and Fitzpatrick 1977; Parton and Logan 1981; Kline et al. 1982; Snyder 1985; Wann et al. 1985; Reicosky et al. 1989; de Gaetano and Knapp 1993; Kramer 1994; Yin et al. 1995; Pellizzaro et al. 1996; Roltsch et al. 1999; Snyder et al. 1999, 2001; Cesaraccio et al. 2001). Prior to development of inexpensive data loggers for monitoring hourly temperature, researchers attempted to predict degree-days from daily maximum and minimum temperatures using mathematical models with a range of complexity (Zalom et al. 1983; Snyder et al. 1999).

As illustrated in Fig. 20.5, the cumulative degree-days during a growth stage of interest tend to be nearly identical in different years if the threshold temperature is properly determined. On the other hand, the mean temperature during the growth period can vary dramatically from year-to-year. It is clear that a plot of cumulative degree days during a growth period versus the mean temperature during the same growth period should give a horizontal line with slope equal to zero. Therefore, if sufficient seasonal data are available, one can easily use trial and error to vary the lower and upper threshold temperatures to calculate the regression of the

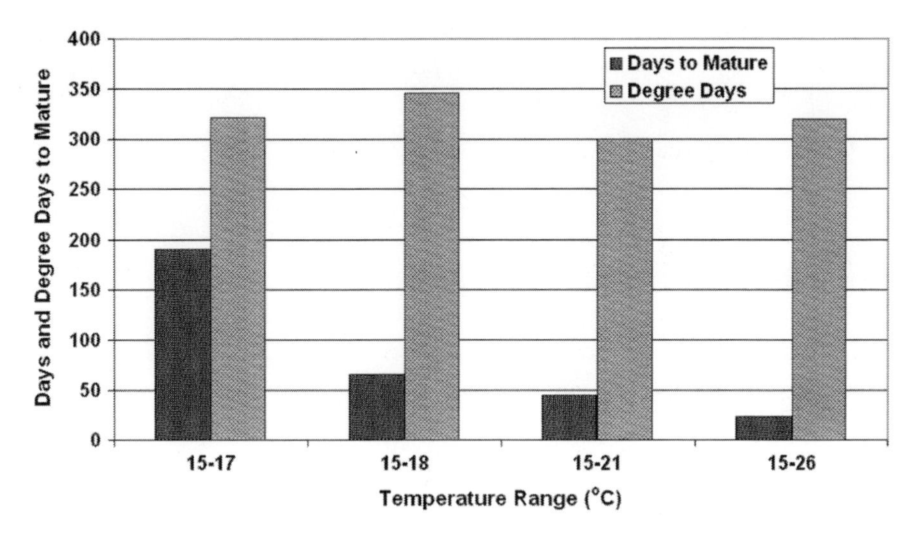

Fig. 20.5 Days and cumulative degree-days for mature cotton bollworm to develop from neonate larvae to adult stage for different temperature ranges (After Wilson and Barnett 1983)

cumulative degree-days versus mean temperature until the thresholds giving the slope is closest to zero. When the slope is zero, the intercept is the cumulative degree-day requirement for predicting the end of the growth period. By definition, the number of cumulative growing degree days is nearly the same each day, so other methods to determine the threshold temperatures are unlikely to provide better estimates.

20.5 Siting Effects

20.5.1 Underlying Surface

The ultimate source of energy for heating the air is the sum of direct and diffuse short-wave (solar) radiation minus that reflected away from the surface. Most of the short-wave radiation is absorbed by plants and soil on the surface and long-wave upward radiation from the surface is the main source of energy directly warming the air. Because the air is heated by interception of long-wave upward radiation, the atmosphere also radiates energy downward to the surface. It is the interception of long-wave upward radiation by green house gases that warms the atmosphere, and increasing greenhouse gas concentrations are slowly increasing the atmospheric temperature.

By convention, downward radiation that adds energy to the surface is positive, and radiation away from the surface is negative. During a clear, midsummer day at noon between about 30° and 50 ° latitude, the downward short-wave radiation is on the order of $R_S = +1000\ W\ m^{-2}$ and, from a grass surface, the albedo is about 25 %

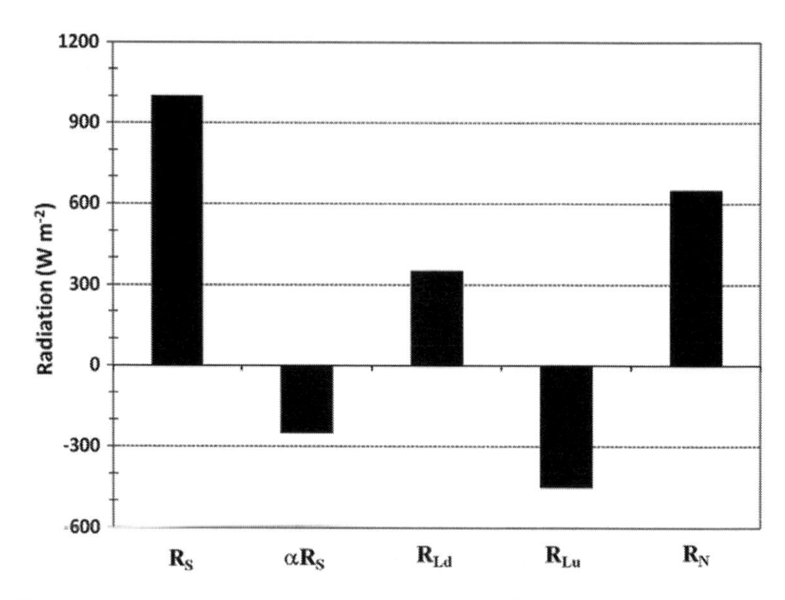

Fig. 20.6 The net radiation (R_N) is the sum of downward (R_S) and upward (aR_S) short waveband radiation and upward (R_{Lu}) and downward (R_{Ld}) long waveband radiation. The albedo (a) is the fraction of Rs that is reflected. This *box* shows the magnitude and sign of midday, clear-sky radiation

of the energy received, so the upward reflected short wave radiation is about $aR_S = -250$ W m^{-2} (Fig. 20.6). Then, the net short-wave balance is about +750 W m^{-2}. Under clear skies at the same latitude, the long-wave upward radiation energy loss is about $R_{Lu} = -450$ W m^{-2} and the downward long-wave radiation is about $R_{Ld} = +350$ W m^{-2}. Thus, the approximate net long-wave radiation balance is $R_{Lu} + R_{Ld} = -100$ W m^{-2}. Combining the short- and long-wave balances, the mid-latitude net radiation near noon on a clear summer day is about $R_n = 650$ W m^{-2} over a grass surface. Under clear skies, the net long-wave radiation changes little with time, but, under overcast skies, because the clouds are warmer than clear sky, it decreases to about $R_{Lu} + R_{Ld} = -10$ W m^{-2}. Short-wave radiation varies considerably over the day with the angle of the sun above the horizon and, of course, it decreases with cloud cover.

Positive short- or long-waveband radiation does one of the following:

1. Heats the soil surface and conducts downward into the soil,
2. Vaporizes water through a phase change that leads to latent heat flux,
3. Heats air near the surface that convects upwards as sensible heat flux,
4. Miscellaneous consumption for heating the plants and photosynthesis.

However, energy flux can reverse direction and do one of the following:

1. Conduct heat upward from the soil to the surface,
2. Condense water vapor as dew or frost and converts the latent energy to sensible heat,

3. Transfer sensible heat to a colder soil surface,
4. Miscellaneous losses due to cooling plants and respiration.

Using the convention that positive Rn at the surface is partitioned into energy that heats the soil (*G*), vaporizes water (λE), heats the air (*H*), or contributes to miscellaneous energy consumption (*M*) used for respiration and photosyn-thesis, the energy balance equation is written as:

$$R_n = G + \lambda E + H + M \qquad (20.2)$$

The miscellaneous term (*M*) is relatively small and is ignored for energy flux calculations. If the soil temperature decreases with depth below the surface, then *G* is positive. Similarly, if the air temperature decreases with height above the surface then *H* is positive. If the temperature profiles are reversed then *G* or *H* are negative and heat is transferred to the surface. If more water vapor is vaporized than condensed on the surface, then the water vapor flux is upwards and λE is positive. If more water condenses than vaporizes, then λE is negative.

Temperature is a measure of the sensible heat content of air at the height of the sensor. Because the temperature gradient with height is typically much greater than with horizontal distance, vertical fluxes of sensible heat are generally much more important than horizontal fluxes. Therefore, energy balance at the surface, which largely determines the sensible heat content at the sensor level, is extremely important. If the surface is warmer than the air above, a positive *H* will likely increase the sensible heat content and hence the temperature of air at the sensor level. If the surface is colder than the air above, a negative *H* will tend to lower the sensible heat content and temperature at the sensor height.

Assuming no horizontal advection and the same incoming long-wave and solar irradiance, the outgoing long-wave and, hence, net radiation depend mainly on the surface albedo and temperature, which is affected by G, H, and λE. Albedo is affected by surface properties and the angle of incidence of the solar radiation. The relative partitioning of absorbed incoming radiation to *G*, *H*, and λE determines the surface temperature. Soil heating is affected by thermal conductivity and soil heat capacity. Latent heat flux is mainly affected by the presence of water to vaporize and secondarily by the water vapor content of the air and turbulence, which transfers sensible and latent heat to and from the surface. Sensible heat flux is the residual energy after the R_n contributions to heating the soil (*G*) and evaporating water (λE) are removed. Sensible heat content of air near the surface determines the air temperature. For a given incoming flux density of radiation, the presence or absence of water is the main factor affecting changes in sensible heat content of air near the surface and, hence, temperature. For example, recording temperature above a transpiring grass surface will generally result in lower temperature than measurements above a dry, bare ground surface. In addition, the difference between temperature recordings over the two surfaces is greater as the climate becomes more arid.

Generally, most climate and weather forecast stations in the USA are located over plots of land with the natural vegetation of the region. Most agricultural weather stations are located over grass or irrigated grass surfaces to standardize the site for comparisons and to allow for model development that can be transferred to other locations. In arid climates, variations in the energy balance due to intermittent rainfall, can greatly affect temperature readings. For example, Snyder et al. (2001) compared annual degree-day calculations, with a 10 °C lower threshold, from stations located over irrigated grass and over bare soil in four regions of California and found that the cumulative degree-days were between 3.2 and 10.7 % higher for the non-irrigated, bare soil. The difference is likely due to differences in energy balance over bare soil that is similar to the grass during the rainy season but can be quite different when the soil dries and the grass continues to transpire. It is probable that differences in vegetation around the weather stations contributed to differences in the degree-day accumulation.

The type of underlying surface around a weather station should be selected depending on the purpose of the data. If the purpose is to characterize phenological development of natural vegetation, the stations should be sited in a natural setting that represents the conditions of the vegetation being studied. However, if the data are used to model phenology of irrigated crops, it is best to site the stations over an irrigated grass surface that removes variability due to intermittent rainfall.

20.5.2 Fetch

Fetch is the upwind distance of the same vegetation underlying a weather station. Inadequate fetch is a problem for some micrometeorological measurements and it can lead to errors in temperature measurements if the purpose is to collect data over a standard surface (e.g., irrigated grass). Under extremely arid conditions, the authors have observed systematic over-estimation of temperature by about 4 % from a station downwind from a desert when compared to a station over irrigated grass 177 m from the edge of the desert (Fig. 20.7). While this amounts to only about 1.6 °C higher temperature at 40 °C, the readings were systematic and they led to considerable differences in degree-day accumulations (Snyder et al. 2001).

20.5.3 Surrounding Environment

Even when temperatures are recorded over the same standard surface, sometimes, big differences are recorded due to the surrounding environment. For example, Figs. 20.7 and 20.8 show the corresponding temperatures recorded over irrigated grass surfaces at Torrey Pines and Miramar Naval Air Station near San Diego, California. Torrey Pines is on the coast and Miramar is located about 30 km inland. Clearly, although they are not far apart, there were big

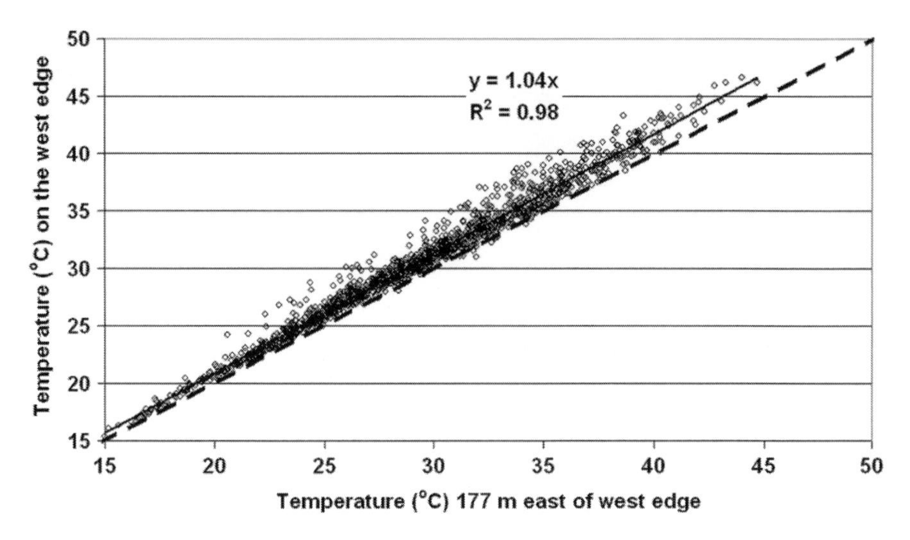

Fig. 20.7 Corresponding hourly temperature data (May–June 2000) measured on the west edge of a grass field and 177 m east of the west edge of a large grass field surrounded by desert near Indio, California. The prevailing wind was from the west. The *solid line* is the least squares linear regression line and the *dashed line* is the 1:1 relationship line

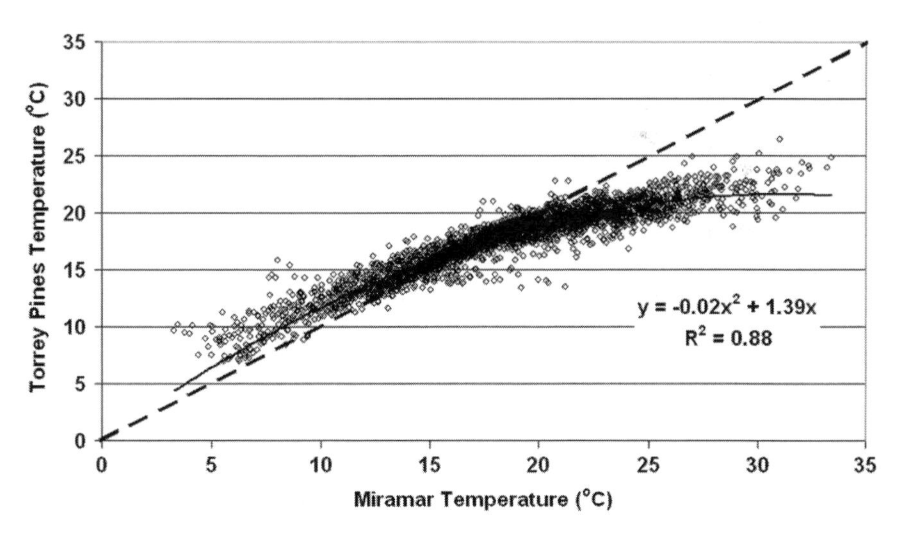

Fig. 20.8 Corresponding hourly temperature data (June–November 2000) from Torrey Pines on the coast and Miramar Naval Air Station, which is 30 km inland. The *solid line* is the least squares regression of the best fit quadratic equation for the data and the *dashed line* shows the 1:1 relationship line

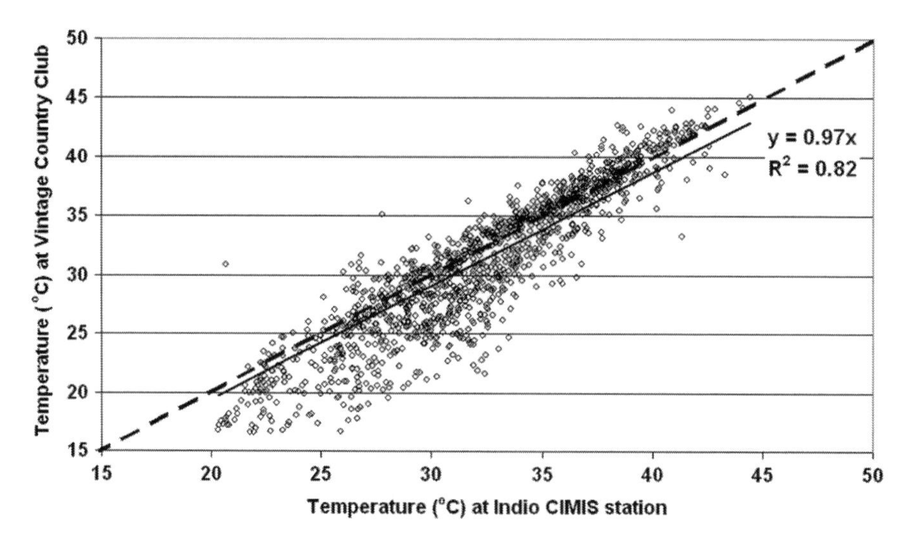

Fig. 20.9 Corresponding hourly temperature data (June – November 2000) measured at the Vintage Country Club near Indio California and the Indio CIMIS station. The wind speed at the Vintage Country Club averaged about 8 % of the Indio CIMIS site. The *solid line* is the best fit linear regression for the data and the *dashed line* represents the 1:1 relationship line

differences in the temperature data with the coastal site having warmer temperatures when they are low and the inland site having warmer temperatures when they are high. This phenomenon is most likely related to the sea surface temperature having a strong influence on air temperature near the coast. If phenological models were to be applied in this region, many stations would be needed to account the temperature effects of the ocean and how they change with distance from the coast.

In areas that are far from the ocean, there are often big differences in temperature over similar underlying surfaces. For example, Fig. 20.9 shows the corresponding temperatures measured at the Indio CIMIS station and the Vintage Country Club, which are only about 30 km apart in the below sea level desert in southern California. In this example, the regression statistics indicate that the temperature at the Vintage Country Club is about 3 % less than at the Indio CIMIS station, but that is because the regression was forced through the origin and the regression line is strongly influenced by the high temperatures during midday, which are similar. However, at lower temperatures, during the morning and afternoon, the temperatures at the Vintage Country Club were considerably more variable and lower. The temperature differences are mainly due to lower wind speeds, resulting from blockage by buildings and vegetation, and mountains that shade the station in the afternoon at the Vintage Country Club. Because the diurnal temperature curves are different, the same temperature curve model cannot be used to estimate degree-days at both sites. Therefore, the use of temperature driven phenological models is problematic in an area with multiple microclimate zones.

20.6 Conclusions

Temperature is the driving factor in most phenological models, and proper measurement is critical for both development and use of the models. In addition to selecting accurate sensors, they should be mounted at an appropriate height and properly shielded from short-wave radiation (double shielding is best). Choosing small sensors that respond rapidly, protecting electronic leads, and ventilation (in areas with little wind) can improve accuracy of the temperature measurements. Data should be collected at a height that is typical of other weather stations in the area where the model will be used. Generally, agricultural weather stations collect temperature data at 1.5–2.0 m height and weather services tend to measure at 10.0 m height. For phenological models of natural vegetation, it is best to site the weather station in a similar environment without irrigation. However, when the models are used for irrigated crops, the stations should be sited over an irrigated grass surface to avoid temperature fluctuations due to intermittent rainfall at the measurement site. Strong temperature gradients can occur near large water bodies (e.g., the ocean or large lakes) and in hilly or mountainous regions where sunlight is blocked during part of the day. In such regions, more weather stations are needed to better characterize microclimate differences. However, even when the temperature data are accurately determined, inaccuracies in model predictions can occur because it is plant temperature rather than air temperature that truly drives the phenological development.

References

Allen JC (1976) A modified sine wave method for calculating degree days. Environ Entomol 5:388–396

Baskerville GL, Emin P (1969) Rapid estimation of heat accumulation from maximum and minimum temperatures. Ecology 50:514–517

Camuffo D, Bertolin C, della Valle A, Cocheo C, Diodato N, Enzi S, Sghedoni M, Barriendos M, Rodriguez R, Dominguez-Castro F, Garnier E, Alcoforado MJ, Nunes MF (2011) Climate change in the Mediterranean over the last five hundred years. In: Carayannis EG (ed) Planet earth 2011 – global warming challenges and opportunities for policy and practice. InTech, Rijeka

Cesaraccio C, Spano D, Duce P, Snyder RL (2001) An improved model for degree-days from temperature data. Int J Biometeorol 45:161–169

de Gaetano A, Knapp WW (1993) Standardization of weekly growing degree day accumulations based on differences in temperature observation and method. Agric For Meteorol 66:1–19

Gill GC (1983) Comparison testing of selected naturally ventilated solar radiation shields, Final report contract # NA-82-OA-A-266, NOAA, St. Louis

Horstmeyer S (2001) Building blocks – what goes on in a cubic meter of air? Weatherwise 54 (5):20–27

Johnson ME, Fitzpatrick EA (1977) A comparison of methods of estimating a mean diurnal temperature curve during the daylight hours. Arch Meteorol Geophys Bioklimatol Ser B 25:251–263

Kline DE, Reid JF, Woeste FE (1982) Computer simulation of hourly dry-bulb temperatures. Virginia Agric Exp Sta, No 82–5. Virginia Politechnical Institute and State University, Blacksburg

Kramer K (1994) Selecting a model to predict the onset of growth of *Fagus sylvatica*. J Appl Ecol 31:172–181

Monteith JL, Unsworth MH (1990) Principles of environmental physics, 2nd edn. Edward Arnold, London

Parton WJ, Logan JA (1981) A model for diurnal variation in soil and air temperature. Agric For Meteorol 23:205–216

Pellizzaro G, Spano D, Canu A, Cesaraccio C (1996) Calcolo dei gradi–giorno per la previsione delle fasi fenologiche nell'actinidia. Italus Hortus 5:24–30

Quinn TJ (1985) Temperature. Academic, New York

Reicosky LJ, Winkelman JM, Baker JM, Baker DG (1989) Accuracy of hourly air temperatures calculated from daily minima and maxima. Agric For Meteorol 46:193–209

Roltsch JW, Zalom FG, Strawn AJ, Strand JF, Pitcairn MJ (1999) Evaluation of several degree day estimation methods in California climates. Int J Biometeorol 42:169–176

Snyder RL (1985) Hand calculating degree-days. Agric For Meteorol 35:353–358

Snyder RL, Spano D, Cesaraccio C, Duce P (1999) Determining degree-day threshold from field observations. Int J Biometeorol 42:177–182

Snyder RL, Spano D, Duce P, Cesaraccio C (2001) Temperature data for phenological models. Int J Biometeorol 45:178–183

Wann M, Yen D, Gold HJ (1985) Evaluation and calibration of three models for daily cycle of air temperature. Agric For Meteorol 34:121–128

Wilson LT, Barnett WW (1983) Degree-days: an aid in crop and pest management. Calif Agric 37:4–7

Yin X, Kropff MJ, McLaren G, Visperas RM (1995) A nonlinear model for crop development as a function of temperature. Agric For Meteorol 77:1–16

Zalom FG, Goodell PB, Wilson WW, Bentley WJ (1983) Degree-days: the calculation and the use of heat units in pest management, Leaflet 21373. Division of Agriculture and Natural Resources, University of California, Davis

Part IV
Sensor-Derived Phenology

Chapter 21
Remote Sensing of Land Surface Phenology: A Prospectus

Geoffrey M. Henebry and Kirsten M. de Beurs

Abstract The process of observing land surface phenology (or LSP) using remote sensing satellites is fundamentally different from ground level observation of phenophase transition of specific organisms. The scale disparity between the spatial extent of the organisms and the spatial resolution of the sensor leads to an ill-defined mixture of target and background or signal and noise. Much progress has been made in the monitoring and modeling of land surface phenologies over the past decade. The chapter first provides a brief overview of land surface phenology, starting with the Landsat 1 in 1972, and then proceeds to a survey of current LSP products. The problem of indistinct phenometrics in remote sensing data is considered and the alternative phenometrics derived from the convex quadratic model are presented with an application in the North American Great Plains using MODIS data from 2001 to 2012. The chapter concludes with a view forward to outstanding challenges for LSP research in the coming decade.

21.1 Introduction

In the first edition of this book, the chapter on "remote sensing phenology" provided a synopsis of that epoch's state-of-the-practice (Reed et al. 2003). Over the past decade there has been explosive growth in basic and applied research on land surface phenology. As there have been recent synopses of the current state-of-the-practice (Reed et al. 2009; de Beurs and Henebry 2010b), we turn here instead

G.M. Henebry (✉)
Geographic Information Science Center of Excellence, South Dakota State University, Brookings, SD 57007, USA
e-mail: Geoffrey.Henebry@sdstate.edu

K.M. de Beurs
Department of Geography and Environmental Sustainability, The University of Oklahoma, Norman, OK 73019, USA

M.D. Schwartz (ed.), *Phenology: An Integrative Environmental Science*,
DOI 10.1007/978-94-007-6925-0_21,

to diagnose current challenges and to point toward potential paths forward. But first it is instructive to consider from where we have come.

21.2 A Brief Historical Overview

Efforts to observe from space the dynamics of the vegetated land surface commenced at the opening of Landsat era. A project in support of the first Earth Resource Technology Satellite (ERTS-1; later named Landsat 1) entitled "Phenology Satellite Experiment" sought to capture and study both the "green wave" of foliage onset and development and the "brown wave" of senescent foliage coloration, drying and/or abscission (Dethier et al. 1973; Knapp and Dethier 1976). Since ERTS-1 was launched July 23, 1972, the research presented in March 1973 in the first open forum for ERTS data users described the brown wave of senescence using ERTS-1 Multi-Spectral Scanner (MSS) data (Dethier et al. 1973). Given the intense focus on the vernal green wave in subsequent decades, it is ironic that what may well be the first presentation on remote sensing of land surface phenology focused on the autumnal brown wave. The authors concluded: "Satellite data, such as that received from ERTS-1, will make world-wide phenological monitoring possible" (Dethier et al. 1973, p. 164).

Much of the effort in the following years focused on the use of crop phenology to advance agricultural monitoring and yield estimation (Kanemasu 1974; Kanemasu et al. 1974; Rea and Ashley 1976; Heilman et al. 1977; Bauer et al. 1979; Tucker et al. 1979; Hlavka et al. 1980; Badhwar 1984). Despite the relatively high spatial resolution of these multispectral data (<60 m), a key limitation was the long return interval relative to vegetation dynamics.

The synoptic imagery available from the Advanced Very High Resolution Radiometer (AVHRR) onboard Polar Orbiting Environmental Satellites (POES) offered, in contrast, much finer temporal resolution (sub-daily) at the cost of much coarser (1–4 km) spatial resolution (Yates et al. 1986). This space-for-time tradeoff made it possible to capture seasonal vegetation dynamics at the continental scale (Goward et al. 1985; Justice et al. 1985; Tucker et al. 1985). Thus, the AVHRRs became the workhorse platform for the study of land surface phenologies (Justice et al. 1986; Lloyd 1990; Ehrlich et al. 1994; Reed et al. 1994; Loveland et al. 1995; Moulin et al. 1997; White et al. 1997), until the advent of the MODIS era at the turn of the century, which brought multiple data products generated from sensors with higher spatial and spectral resolution (Justice et al. 1998, 2002; Zhang et al. 2003).

Land surface phenology is a neologism that first occurs, as far as we can discover, in an obscure conference paper from the mid-1990s:

> Scaling up from watershed to regional observations also requires a conceptual shift. At a spatial resolution of 30 m, the canopy is still meaningful as a resolvable scene object; thus, the phenology of the canopy can be observed. At a spatial resolution of 1 km, however, landscape details typically are too blurred to enable canopies is to be well-resolved; thus, we introduce the concept of phenology of the land surface. Implicit to this concept is the

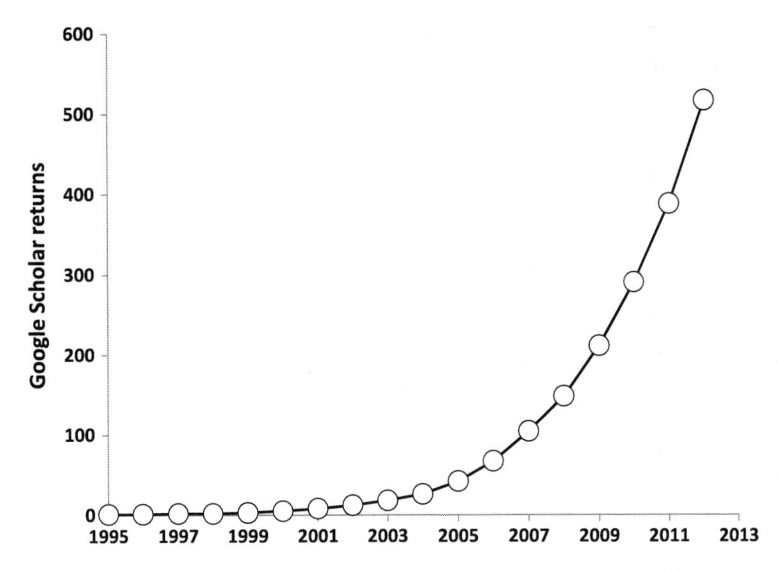

Fig. 21.1 Temporal profile in Google Scholar cumulative returns for the phrase "land surface phenology" from 1995 to 2012 (first return is dated 1995)

> assumption of a significant degree of spectral mixing from different land covers. This is not a traditional phenology associated with specific events in a plant's life history; rather, land surface phenology describes the seasonality of reflectance characteristics that are associated with stages of vegetation development. (Henebry and Su 1995, p 143)

Its use was re-introduced in the grasslands chapter of the first edition of this book (Henebry 2003) and more widely through an article bearing the phrase in its title (de Beurs and Henebry 2004). The fitness of the term to differentiate between ground-level observations of plant phenology and space-borne observations of the vegetated land surface has been confirmed by the community (Fig. 21.1; Friedl et al. 2006; Morisette et al. 2008; Reed et al. 2009). To quote from a community white paper requested by NASA on the topic of an earth system data record for phenology (Friedl et al. 2006):

> Land surface phenology is defined as the seasonal pattern of variation in vegetated land surfaces observed from remote sensing. While the observed patterns are related to biological phenomena, land surface phenology is distinct from traditional definitions of vegetation phenology, which refer to specific life cycle events such as budbreak, flowering, or leaf senescence using in-situ observations of individual plants or species.

Both the observation of land surface phenologies (hereafter LSPs) achieved with remote sensors and the ground-level observers of life cycle events of particular species share the fundamental act of ordering events in time; indeed, phenology is principally about the timing of events that can have significance for organisms, populations, and ecological communities as well as for biogeochemical and hydrological cycles and ecosystem services. However, as alluded to in the quotations above, the process of observing LSPs is burdened with some challenges that set it apart from the traditional phenological observing that is the subject of most of the chapters in this book.

21.3 LSP Products

Most studies of land surface phenologies have used optical image time series, exploiting the spectral contrast between the red region of the electromagnetic spectrum, which green vegetation strongly absorbs, and the near infrared region, which green vegetation strongly reflects. It is this spectral contrast that forms the basis for many vegetation indices (VIs), e.g., Normalized Difference Vegetation Index (NDVI: Tucker 1979), Enhanced Vegetation Index (EVI: Huete et al. 2002), Wide Dynamic Range Vegetation Index (WDRVI: Gitelson 2004), and retrieved canopy variables, e.g., Leaf Area Index (LAI: Myneni et al. 2002), Fraction of Absorbed Photosynthetically Active Radiation (FPAR: Myneni et al. 2002); and Fractional Vegetation Cover (FVeg: Carlson and Ripley 1997).

Despite the significance of phenology for earth system monitoring and modeling (Morisette et al. 2008; Richardson et al. 2012), there are few datasets that explicitly derive phenological metrics (or phenometrics). By our count just five exist, one is a standard global product and four have been developed for projects specific to North America or the conterminous US (CONUS). The global product relies on the time series from the MODIS (Moderate Resolution Imaging Spectroradiometer) on the Terra and Aqua satellites. Three of the project-specific products also rely on MODIS data, but from different sources for different purposes. The fifth phenologically explicit product uses AVHRR data.

21.3.1 MODIS Land Change Dynamics Product

The sole global product (MCD12Q2; https://lpdaac.usgs.gov/products/modis_products_table/mcd12q2) is generated by NASA from time series of the MODIS (Moderate Resolution Imaging Spectroradiometer) on the Terra and Aqua satellites (Ganguly et al. 2010). The product is built up from a time series of the Enhanced Vegetation Index (EVI; Huete et al. 2002) calculated from the MODIS NBAR (Nadir Bidirectional Reflectance Distribution Function (BRDF) Adjusted Reflectance) dataset (MCD43C1). Several phenometrics are derived through derivatives of double logistic equations (Zhang et al. 2003, 2006) fitted to the EVI time series for each pixel at 500 m resolution every 8 days within a 16-day moving window. These phenometrics include: the day of "onset greenness increase", the day of "onset greenness maximum", the day of "onset greenness decrease", the day of "onset greenness minimum", the EVI value at the "onset greenness minimum", the EVI value at the "onset greenness maximum", and the growing season area under the curve calculated as the summation of EVI values from "onset greenness increase" to "onset greenness minimum". The product can handle two growing season peaks as might occur in a double-cropping system and the northern and southern hemispheres are reported separately with 6 month phase difference (Ganguly et al. 2010).

Most applications of the MCD12Q2 product have been exploratory in nature, seeking to discover what kinds of spatio-temporal patterns emerge from the data (Peñuelas et al. 2004; Zhang et al. 2004, 2005; McManus et al. 2010), how these might relate to differences in land cover, land use, and vegetation community (Zhang et al. 2004; Stohlgren et al. 2010), and how these data relate to ground level observations (Fisher and Mustard 2007; Soudani et al. 2008; Zhang et al. 2009).

21.3.2 MODIS for NACP Products

The MODIS for NACP (North American Carbon Program) project (http://accweb.nascom.nasa.gov/index.html) used the TIMESAT software package (Jonsson and Eklundh 2004) to fit asymmetric Gaussian functions (Jonsson and Eklundh 2002) to pixel-level time series from two different MODIS products: MODIS surface reflectance data (MOD09), and the MODIS leaf area index (LAI) data (MCD15) (Gao et al. 2008). From the MOD09 data two pairs of vegetation index time series were calculated at 250 and 500 m: the EVI and the NDVI. The NDVI has a longer heritage, but tends to lose sensitivity over denser canopies (Gitelson 2004). The EVI tends to have a larger dynamic range than the NDVI and is more resistant to atmospheric and soil background effects (Huete et al. 2002). For the LAI time series, the MCD15A2 8-day product was used.

Eleven phenometrics were generated for each fitted time series: (1) time of start of season; (2) time of end of season; (3) length of season; (4) base level (mean of off-season minima); (5) time of mid-season; (6) largest data value for the fitted function during the season; (7) seasonal amplitude; (8) rate of increase at start of season; (9) rate of decrease at end of season; (10) integral of amplitude from zero across the season length; and (11) integral restricted to amplitude between base across the season length. The products captured up to two seasonal cycles for these 11 phenometrics. Three additional variables were included to capture the annual signal amplitude: (12) annual maximum; (13) annual minimum; and (14) annual mean. The products also report the root mean squared error (RMSE) of the fits and associated quality flags (Tan et al. 2011). Phenometrics exist for 2001 to 2010; these data are accessible through http://accweb.nascom.nasa.gov/data/search.html.

The phenometrics from the MODIS for NACP product have been used to study phenological variation within aspen populations (Li et al. 2010; Gray et al. 2011), and the autumnal phenologies of deciduous species found in northern hardwood forests (Isaacson et al. 2012).

21.3.3 USFS ForWarn Products

In an effort to develop a near-real time system to detect and track significant changes in forests across CONUS (Hargrove et al. 2009), the ForWarn project of the US Forest Service Eastern Forest Environmental Threat Assessment Center (http://forwarn.forestthreats.org/) has produced a series of phenometrics from MODIS data that has been processed—not with the standard NASA system—but instead using the eMODIS system at the USGS Center for Earth Resource Observations and Science (EROS). The eMODIS system arose out of a need for provisioning of MODIS data after the Direct Broadcast System installation at EROS was discontinued in 2006 (Jenkerson and Schmidt 2008; Jenkerson et al. 2010). ForWarn processed the eMODIS feed further to generate LSP datasets for 2003–2009 that includes a large set basic and derived phenometrics, including the rates of vernal green-up and autumnal brown-down, the duration of spring, fall, and growing season, and interannual descriptive statistics (Christie 2012). The phenometrics have been used to develop and refine the change detection and tracking algorithms within the ForWarn system, which has shown its utility to resource managers in multiple instances (http://forwarn.forestthreats.org/highlights; Worley-Firley 2012), including detection of gypsy moth outbreaks (Spruce et al. 2011). These data are available at http://forwarn.forestthreats.org/data/data-access.

21.3.4 USGS Remote Sensing Phenology Products

The USGS Center for EROS Remote Sensing Phenology website (http://phenology.cr.usgs.gov) distributes LSP products for CONUS from two different sensors (AVHRR and MODIS). The method USGS uses to derive phenometrics from NDVI time series is the delayed moving average (DMA; Reed et al. 1994), which compares observed to smoothed data (Swets et al. 1999) to detect departures from local trends. This suite of phenometrics includes timing of start and end of season and their corresponding NDVI values, the timing and NDVI value of the annual maximum, duration of the growing season, the amplitude of the growing season (difference between the NDVI at maximum and the NDVI at start of season), and time-integrated NDVI. These AVHRR phenometrics are the longest record available at 1 km, 23 years, from 1989 through 2011. The DMA method has also been applied to eMODIS data at 250 m resolution for 2001 through 2011. These data are available at http://phenology.cr.usgs.gov/get_data_main.php.

The USGS phenometrics have been applied in a variety of contexts, including phenometric intercomparisons (Schwartz et al. 2006; White et al. 2009), trend analysis (Reed 2006; Dragoni and Rahman 2012), agricultural land use mapping (Pervez and Brown 2010; Howard et al. 2012), and monitoring of vegetation productivity (Gu et al. 2013) and vegetation stress (Brown et al. 2008).

21.3.5 Implicit LSPs

This paucity of phenologically explicit products stems in part from the fact that any time series of observations with sufficient duration and resolution may be phenologically implicit, i.e., it may contain sufficient information to enable the characterization, estimation, or modeling of LSPs. There are a large number of LSP studies that use image time series from one or more of several possible sensors to craft a "local solution" to LSP characterization, e.g., minimizing snow effects in Siberia (Delbart et al. 2005), detecting phenological responses to climatic change (White et al. 2005), distinguishing grass from tree green-up dynamics in African savanna (Archibald and Scholes 2007), attenuating snow effects in Scandinavia (Beck et al. 2007), monitoring cropping in West Africa (Brown and de Beurs 2008), mapping crop types in North America (Wardlow and Egbert 2008), assessing conflict impacts on agriculture in Afghanistan (de Beurs and Henebry 2008b), or mapping winter wheat in China (Pan et al. 2012). In fact, these studies exceed the number of those that rely on the phenologically explicit LSP products. Why should this be the case? There are multiple possible reasons.

First, as noted before, there is only one global product and it offers a single solution that may not perform optimally in regional to local applications; thus, there is an impetus to provide refined solutions. For example, the temporal resolution available from MCD12Q2 may be not sufficiently fine to capture rapid phenophase transitions. Second, extracting phenometrics from a time series is not difficult, particularly with the advent of software packages such as TIMESAT (Jonsson and Eklundh 2004). Third, there are other MODIS product time series that offer different looks at the land surface: MOD09 (surface reflectance), MOD13 (vegetation indices), MOD15 (LAI & FPAR). Fourth, LSPs can be extracted from time series of other optical sensors, including AVHRR GIMMS (Tucker et al. 2005), AVHRR LTDR (Pedelty et al. 2007), SPOT VEGETATION (Weiss et al. 2007), WELD (Roy et al. 2010), among others. A fifth and perhaps the fundamental reason why most researchers do not use any of the extant LSP products is the fact that land surface phenologies rarely have distinct phenophase transitions.

While the concept of a "start-of-season" (or SOS) has strong intellectual appeal (likewise for "end-of-season" or EOS), defining how it appears is surprisingly slippery in practice. Each of the phenologically explicit LSP products mentioned above uses a different metric for SOS: in MCD12Q2, rate of change in curvature of fitted logistic; in MODIS for NACP, the third derivative of a fitted Gaussian; in USFS ForWarn, a threshold of 20% of seasonal amplitude; and for USGS Remote Sensing Phenology, an adaptive threshold based on a delayed moving average. Note that each of these SOS phenometrics rely on relationships that are functions of time of observation, with units of days packaged within compositing periods of varying durations. An intercomparison effort sought to identify, using the same AVHRR time series, which among ten techniques corresponded more closely and consistently to a diverse set of ground level observations of SOS (White et al. 2009). The divergence in correspondence was remarkable:

> Compared with an ensemble of the 10 SOS methods, we found that individual methods differed in average day-of-year estimates by ±60 days and in standard deviation by ±20 days. The ability of the satellite methods to retrieve SOS estimates was highest in northern latitudes and lowest in arid, tropical, and Mediterranean ecoregions. (White et al. 2009, p 2335)

The key lesson we learned from the intercomparison experiment is that SOS is not a feature of LSPs that can be retrieved reliably. This conclusion should not be too surprising. The need for distinctiveness in phenological items has long been recognized.

In a seminal work on phenological observations, the noted ecologist Aldo Leopold outlined several characteristics that distinguish "good" phenological items (Leopold and Jones 1947):

- low labor cost/simple to observe
- sharp/distinct to minimize error among observers
- common/abundant
- high degree of accessibility (visibility or audibility)
- reliability of recurrence
- continuity
- evidence of newness
- locally-determined dynamics
- sufficient prior knowledge exists to identify the unusual

This list was derived from field experience of ecologists making observations on the ground at a scale where individuals and species are clearly recognizable. Yet, these characteristics are useful when considering the fitness of metrics for LSPs. Of particular importance is the requirement for the item to be sharp or distinct to minimize error among observers. In field observations of phenologies of individual species, this requirement can be met. But in remote sensing observations, where the scales of observation blend together vegetation and background, SOS (likewise EOS) is not distinct; thus, these are various competing definitions of what constitutes SOS phenomenologically. It is also notable that each LSP product applies smoothing operations prior to fitting model curves, which attenuate noise and blur signal.

21.4 Alternative Phenometrics

There are alternative metrics for LSPs. Several flexible parametric models have been developed to describe the time course of the VI as a function of a simple temporal index (e.g., day of year, compositing period). Badhwar (1984) developed a method to parameterize NDVI temporal profiles that was later used by Tucker et al. (2001). The temporal profile is divided at the peak VI into two time series.

Four parameters' coefficients need to be estimated for each part of the curve, or eight parameter coefficients in total. Jönsson and Eklundh (2002) developed a model fit consisting of a number of local model functions that are merged into a global function. Merging of multiple local functions increases fitting flexibility and enables a fitted function to track complex temporal behavior that is not possible using a simple Gaussian model or low-order Fourier estimates (Jönsson and Eklundh 2002). The fit of the local function alone requires the estimation of seven parameter coefficients. Zhang et al. (2003) fitted a double logistic model of vegetation growth to the EVI and, like Tucker et al. (2001), divided the annual VI curve in two parts—growth and senescence—to fit the models separately. For both the growth and senescence phases to be fitted, a total of eight parameter coefficients need to be estimated. It can be argued that these models are unnecessarily complicated. Although increasing model complexity can increase the value of the coefficient of determination (r^2), more parameters to fit translates into fewer degrees of freedom and lower statistical power. It is unclear how the parameter estimation is influenced by the temporal resolution of the data (Ahl et al. 2006; Zhang et al. 2009). Finally, high model complexity can inhibit straightforward ecological or biogeophysical interpretations of the parameters or the resulting phenometrics.

Another class of parametric models take their cue from traditional phenological models. Plant phenology models relate thermal regimes of the growing season with events in plant development (Schwartz 2003). The thermal regime of the growing season can be measured as accumulated growing degree-days (AGDDs) by summing between defined periods the average daily temperature above some base temperature value (Eq. 21.1):

$$\mathrm{AGDD_t} = \mathrm{AGDD_{t-1}} + \max[\mathrm{AvgTemp_t} - \mathrm{BaseTemp}, 0] \qquad (21.1)$$

where $\mathrm{AvgTemp_t}$ is the average daily air temperature at day t, and BaseTemp is a temperature baseline related to vegetation type (e.g., 0 °C for perennials, 4 °C for cool-season annuals, 10 °C for warm-season annuals; de Beurs and Henebry 2010b). The max function restricts the AGDD from going negative, but allows it not to advance if the average temperature does not exceed the baseline. Thermal-based regression models using AGDDs as the explanatory variable have long been used in crop phenology studies to describe and predict the green-up, flowering, fruiting, and senescence stages of crops and to compare among crop varieties (e.g., Nuttonson 1955; Hodges 1990). More recently, AGDDs have been used to track temperate grassland phenologies (Goodin and Henebry 1997; Mitchell et al. 2001; Smart et al. 2001; Davidson and Csillag 2003; Henebry 2003). It is a small conceptual step to tracking LSPs using AGDD.

Studies have shown that the VI time course of herbaceous vegetation in temperate and boreal ecosystems can be well approximated as a quadratic function of AGDD (Eq. 21.2):

$$\mathrm{VI_t} = \alpha + \beta \mathrm{AGDD_t} + \gamma \mathrm{AGDD_t}^2 \qquad (21.2)$$

Interpretation of these three parameters is straightforward. The intercept α indicates the background VI value at the beginning of the observation period. The linear parameter β affects the slope and the quadratic parameter γ the curvature. When the fitted model is convex in shape, the sign of the β is positive and the sign of the γ is negative, which produces a piece of a parabola that first rises and then falls as thermal time advances. Since the convex quadratic provides the appropriate shape for LSP, we restrict our attention to this model form.

The parameter coefficients of the fitted convex quadratic model yield two useful phenometrics: (1) the peak height (PH), which is the maximum VI value in the fitted model or the vertex of the parabola (Eq. 21.3); and (2) the thermal time to peak (TTP), which is the amount of accumulated heat units (as degree-days) required to reach the peak height (Eq. 21.4):

$$PH = \alpha - (\beta^2/4Y) \tag{21.3}$$

$$TTP = -\beta/2y \tag{21.4}$$

Note that these phenometrics are derived from the parametric model fitted to the data rather than directly from the data. This approach to LSP modeling has been applied to image time series from the AVHRR (de Beurs and Henebry 2004, 2005a, b, 2008a), MODIS (de Beurs et al. 2009; de Beurs and Henebry 2010a), and SPOT VEGETATION sensors (Brown and de Beurs 2008; Brown et al. 2010).

21.5 Application of the Convex Quadratic LSP Model

To illustrate the utility of the convex quadratic LSP model, we fitted it to each pixel time series for every year from 2001 to 2012. The underlying image time series data are the NDVI calculated from the MODIS NBAR product (MCD43C4) at 0.05° (~5.6 km) spatial resolution and 16 day temporal resolution and the MODIS land surface temperature (LST) product (MOD11C2) at 0.05° spatial resolution and 8 day temporal resolution. The four panels of Fig. 21.2 display, clockwise from upper left, averages across the 12 years of data for (1) the peak height or largest seasonal NDVI value, (2) the coefficients of determination (r^2) that indicate the goodness of the model fit, (3) the day of year at which the peak height occurs, and (4) the thermal time to peak height in accumulated growing degree-days. Note in the average peak height map the expected longitudinal gradient of decreasing peak NDVI values from east to west. Notice also that larger urban areas appear as paler islands amid a matrix of high NDVI. The average coefficient of determination map shows that the vast majority of pixels have r^2 values above 0.85, but the spatial heterogeneity of higher values suggests some influence of vegetation type on goodness of fit. The average day of year (DOY) map reveals areas that reach peak NDVI much earlier in the year than the summer croplands. These dark

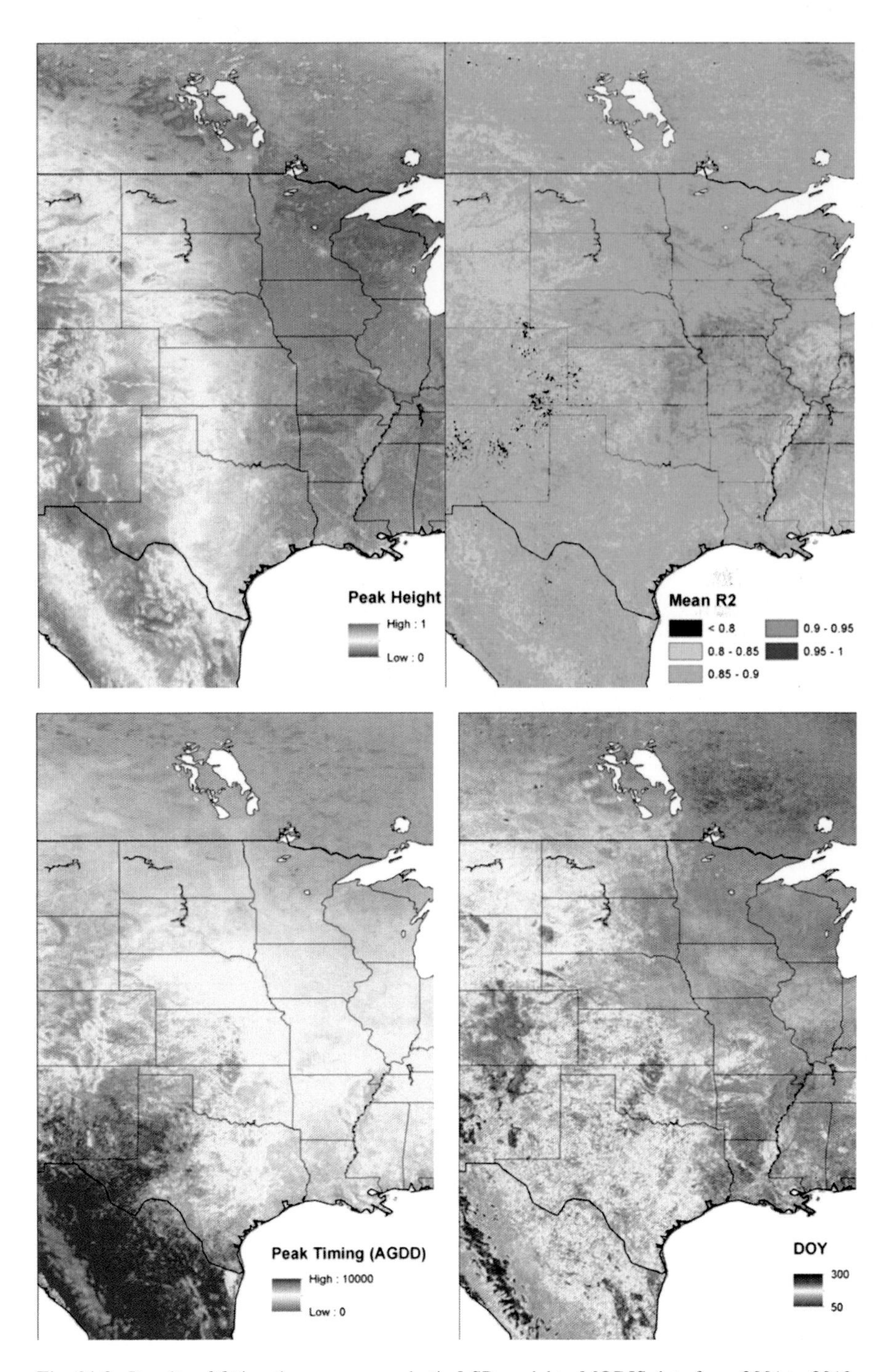

Fig. 21.2 Results of fitting the convex quadratic LSP model to MODIS data from 2001 to 2012: average peak heights (*upper left*); average coefficients of determination (*upper right*); average thermal times to peak in accumulated growing degree-days (*lower left*); average day of year at peak height (*lower right*)

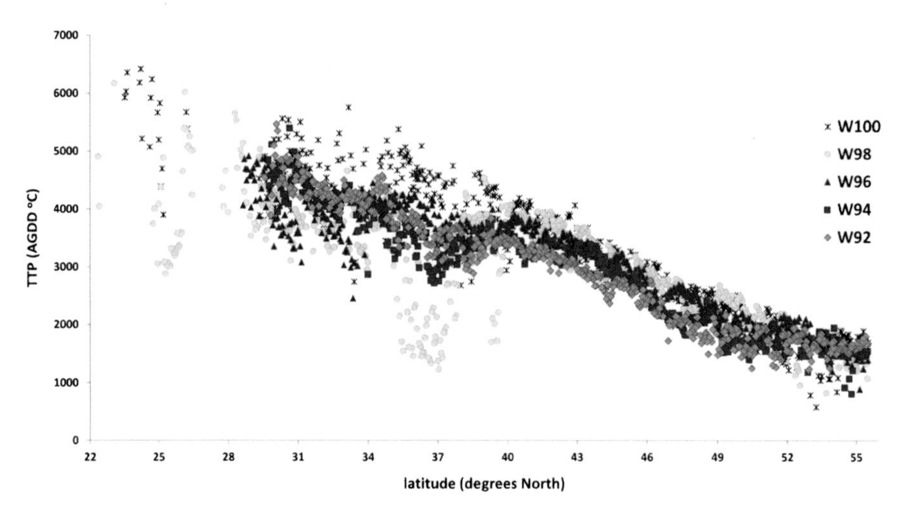

Fig. 21.3 Average values of thermal time to peak along latitudinal transects across a longitudinal gradient from W92° to W100°

brown areas include winter wheat belts in central and western Kansas and Oklahoma, the Nebraska panhandle, and eastern Colorado, a large area of fall/winter sorghum south of Harlingen, Texas in the state of Tamaulipas, Mexico, as well as other agricultural lands on either side of the national border. The final panel shows the average amount of accumulated growing degree-days required to each peak NDVI. While this map also reveals the winter wheat belts, it captures the expected latitudinal decrease in thermal time to peak (TTP) values.

A closer look at latitudinal transects of the TTP values reveals significant geographic patterns in the phenometric. There is a remarkably strong latitudinal trend of increasing TTP with decreasing latitude (Fig. 21.3); however, there are no values south of about N29° due to the absence of land at longitudes W92° to W96°. The variability in this latitudinal trend increases substantially south of N40°, and there is also a weaker longitudinal trend showing increasing TTP toward the west. These trends result from a combination of climatic factors and land use. The transect at W98° passes through the Oklahoma winter wheat belt and thus shows lower TTP values from N35° to N37°. This transect also passes into a patch of fall/winter sorghum in Tamaulipas, Mexico from N25° to N26°, where TTP values are again lower than in adjacent areas.

The western corn and soybean belts are the cropland regions with the strongest latitudinal trends in TTP (Fig. 21.4): a simple linear regression relating TTP to latitude explains 87–94% of the variance between N40° and N49° at W92° to W100°. The southern border of Nebraska is N40° and the northern border of North Dakota is N49°. Large fractions of US grain harvest and bioethanol production occurs within this bounding box.

It is also possible to interrogate the parameter coefficients directly to highlight areas of change. Figure 21.5 shows the difference in parameter coefficients for two

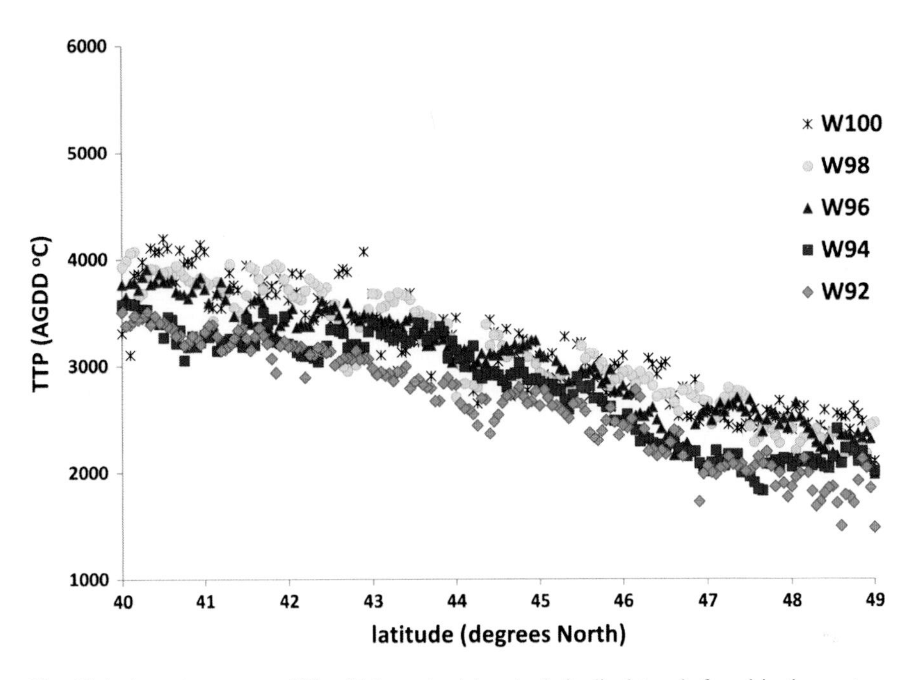

Fig. 21.4 An enlargement of Fig. 21.3 emphasizing the latitudinal trends found in the western corn and soy belts

periods within the data record: 2001–2005 and 2007–2011. Both 2006 and 2012 were omitted from the analysis as years with widespread drought. Note the concomitant decreases in β and increases in γ coefficients and the resulting decreases in TTPs in the region bounded by N46°W97° and N43°W96° (eastern Nebraska and South Dakota). These changes are due partly to climatic variability and partly to changes in land use. Government mandates to increase ethanol production for blending with gasoline led in the late 2000s to large increases in corn harvested for biofuels (US Census Bureau 2012), which has led within this region to changes in cropping patterns and land cover (Wright and Wimberly 2013) and to the loss of wetlands (Johnston 2013).

Although the spatial resolution of our analysis is rather too coarse for an effective land cover land use change study, it does demonstrate how the convex quadratic LSP model can be used to characterize the dynamics of the vegetated land surface and quantify spatio-temporal change. The scale of analysis is appropriate, however, for linkages between the land surface and boundary layer processes. The convex quadratic LSP model is both computationally simple and can be used predictively, but it does not work everywhere. Application of this model to an image time series can be considered a form of functional data analysis (Ramsay and Silverman 2005). Where it fits well, it provides a reasonable approximation to the proximate cause of the LSP. Where it fails, it indicates the need for another kind of forcing function. Although most of the applications to date have used AGDD as that

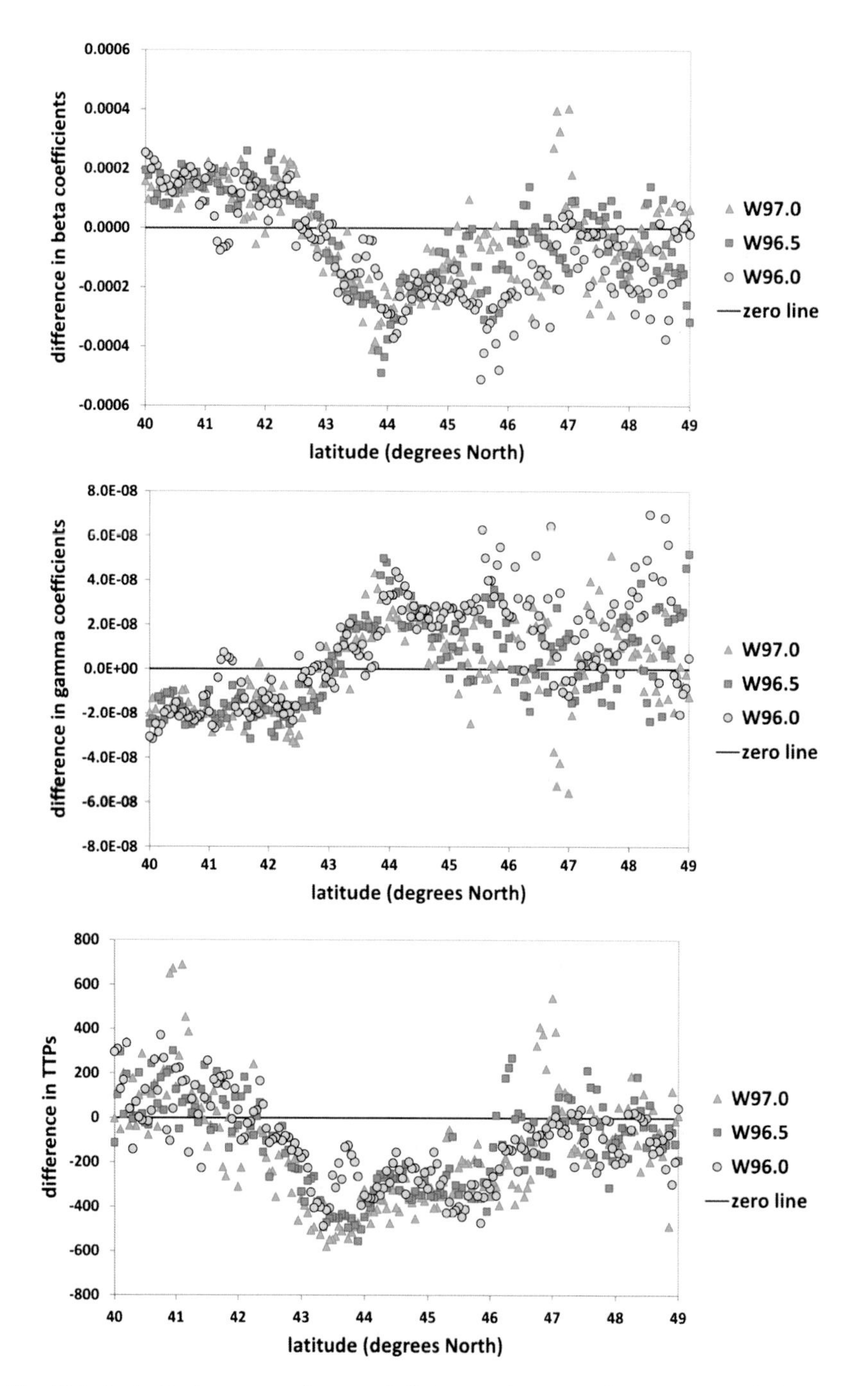

Fig. 21.5 Latitudinal transects of temporal differences in average β parameter coefficients (*top panel*), average γ parameter coefficients (*middle panel*), and average TTPs (*bottom panel*). Differences in each panel were calculated as the average for 2007–2011 minus the average for 2001–2005

forcing (calculated from reanalysis data, station data, or land surface temperature data), it is appropriate to emphasize that cumulative moisture functions have also proved useful for tropical agriculture (Brown and de Beurs 2008; Brown et al. 2010, 2012).

21.6 Looking Forward

The rapid pace of LSP monitoring and modeling has positioned the field to make significant advances in the coming decade. To close this chapter, we will review some lessons learned, recommend new directions, and point to some encouraging new developments.

The primary lesson to be learned from the past decade of LSP research is that there is no single approach to LSP that fits all. We urge the community to explore approaches to LSP monitoring and modeling that embrace suites of sensors and algorithms toward developing biome-tuned LSP models. Perhaps it is appropriate here to introduce the concept of land surface seasonality (LSS) that could be used in tandem with LSPs to tackle biome-specific monitoring and modeling. For example, the seasonality of soil freeze/thaw is a key transition in ecosystem processes and one that can be monitored effectively using microwaves (Frokling et al. 1999; McDonald et al. 2004; Kim et al. 2011). Although synoptic observations from passive microwave radiometers have been studied for as long as optical sensors (e.g., Wang 1985; Choudhury et al. 1987; Neale et al. 1990; Prigent et al. 1997), early comparative studies indicated that microwaves were less responsive to LSP than observations in the optical region (e.g., Justice et al. 1989; Townshend et al. 1989). A welcome development over the past decade is the application of active (Frolking et al. 2006; Bartsch et al. 2007) and passive microwaves (Jones et al. 2011, 2012) to study LSPs and LSSs.

A corollary to the primary lesson is that LSP research needs to focus on timing of status changes and sharp phenophase transitions rather than ill-defined events, such as the start (or end) of season. Alternative phenometrics need to be investigated. The TTP metric associated with the convex quadratic model is but one example.

Most of the LSP literature has focused on vernal dynamics, but it is necessary to move beyond spring green-up. Recent work in autumnal dynamics has shown some areas of tractability (Richardson et al. 2006, 2009; Dragoni and Rahman 2012; Zhang et al. 2012), but there is much more to explore in monitoring and modeling the processes of canopy coloring, drying, and foliage abscission.

Most of the LSP literature has focused on optical imagery and on a handful of vegetation indices derived therefrom. It is time to move beyond the visible to near infrared (VNIR) region and incorporate multiple remote sensing modalities, including the thermal infrared (Anderson et al. 1997, 2011). It is also time to move beyond sole reliance on NDVI and EVI and discover how other VNIR indices can capture canopy and landscape scale dynamics (Viña et al. 2004; Delbart et al. 2005, 2006; de Beurs et al. 2009; Tuanmu et al. 2010; Viña and Gitelson 2011).

Most of the LSP literature has focused on natural landscapes and ecosystems. It should be possible to leverage our understanding of human-managed systems to advance LSP monitoring and modeling. Although some work has been done with LSPs within agricultural settings, it is often to enable crop type mapping or crop yield estimation (Jakubauskas et al. 2002; Sakamoto et al. 2005, 2010; Wardlow and Egbert 2008); agricultural land uses present the opportunity to enhance our knowledge about phenophase transition timing (Kovalskyy and Henebry 2012a,b). Likewise, there has been limited work on LSPs in urban settings (White et al. 2002; Zhang et al. 2004; Fisher et al. 2006; Gazal et al. 2008), but climatological contrasts between the urban core and periphery offer another approach to gain insights on modulation of LSP by land use and land cover.

Very little has been explored to date on the influence of LSPs and LSSs on the spatial structure of surface characteristics. Field studies have shown how spatial structure of reflectance changes during the growing season (Goodin and Henebry 1998; Davidson and Csillag 2003; Goodin et al. 2004). Spatial analyses of image time series have revealed characteristic seasonal patterns in reflectance (Henebry and Su 1993, 1995; Viña and Henebry 2005) and backscattering (Henebry and Kux 1995, 1997) that enable the detection and evaluation of change. With the increased accessibility of the Landsat archive, this avenue of LSP research may be very fruitful area in the coming decade.

Cross-calibration of LSP metrics with other indicators of phenology has been studied since the Phenology Satellite Experiment with ERTS (Dethier et al. 1973). More recent efforts to cross-calibrate estimates of phenophases have found a tendency for LSP timings to be early relative to a suite of bioclimatic indicators (Schwartz and Reed 1999; Schwartz et al. 2002). What constitutes an appropriate reference set for ground level phenological observations remains an open question. However, it is clear at this junction that the community needs coordinated observations at multiple scales to link landscape heterogeneity to pixel variability (Liang and Schwartz 2009; Liang et al. 2011). The use of flux tower observations (Noormets 2009; Gonsamo et al. 2012a, b; Kovalskyy et al. 2012) and "phenocams" (Richardson et al. 2007; Ahrends et al. 2008; Chap. 22, this volume) for cross-calibration are critical, but there is another source of finer spatial resolution remote sensing data that promises a rich source for cross-calibration efforts, viz., the Landsat data record that is newly accessible through the WELD project (Roy et al. 2010; http://landsat.usgs.gov/WELD.php). Currently, WELD covers only the conterminous US and Alaska using Landsat 7 data after May 2003, when a failure in the scan-line corrector (SLC) generated gaps in the scene. Despite these missing data, the "SLC-off" imagery contains a wealth of information (Loveland et al. 2008). Moreover, the opening of the entire USGS Landsat data archive to free access (Wulder et al. 2012) and the effort consolidate disparate national holdings into a consolidated global archive (http://landsat.usgs.gov/Landsat_Global_Archive_Consolidation.php) set the stage for the production of a global Landsat data record that spans multiple sensors and epochs. As of late 2012 NASA has committed to fund a Global Long-Term Multi-Sensor WELD project; thus, we anticipate a boom in 30-m LSP studies in the next decade, particularly for conservation ecology and invasive species research (Bradley et al. 2012; Walker et al. 2012).

Validation of land surface products is the proverbial "elephant in the room". Note that we say land surface products and not land surface phenology products. The challenge facing the remote sensing community is larger than validation of just LSPs. The Land Product Validation Subgroup (LPVS) of the Committee on Earth Observation Systems (CEOS) Working Group on Calibration and Validation (WGCV) has been active in a number of areas (http://lpvs.gsfc.nasa.gov), including land surface phenology (http://lpvs.gsfc.nasa.gov/pheno_background.html). Despite an effort to self-organize (Morisette et al. 2010), progress in bringing the LSP community together to engage in validation exercises has been slow, compared to what has been accomplished for leaf area index (LAI) retrievals (Morisette et al. 2006; Garrigues et al. 2008). This situation is due, in large part, to a lack of funding for a validation campaign, but it is also attributable to (1) the relative scale-invariance of intensive variables like vegetation indices, (2) the sensitivity of vegetation indices to sensor band centers and bandwidths, (3) the lack of sharply defined phenometrics previously discussed, and (4) the wide variety of different ways that phenometrics can be extruded from image time series (White et al. 2009).

A necessary companion subject to validation is uncertainty. Advances in georeferencing, geographic information systems, and online mapping services make the geolocation of phenological observations relatively straightforward. We have already spoken about the need for well-defined phenometrics in LSP research. An outstanding challenge is the treatment of temporal uncertainties in phenological observations, particularly with remote sensing data that are frequently composited to minimize the obscuring effects of cloud cover. How can we pierce the temporal veil of composited imagery to increase temporal precision of observations?

A method—aoristic analysis—was developed in the late 1990s in the field of quantitative criminology to leverage temporally imprecise geospatial data of crime reports (Ratcliffe and McCullagh 1998). Aoristic analysis enables a better treatment of data for which temporal coordinates are poorly specified or only partially bounded. The term "aorist" refers to the simple past tense in Ancient Greek and indicates an action that was completed in the past. The fundamental method of aoristic analysis is to distribute the temporal uncertainty of an event—in the form of temporal weights—evenly across temporal units from one definite observation to another and then sum across all events within a designated area (Ratcliffe 2000, 2002). The shape of the resulting histogram of temporal weights can increase the temporal precision of when events occurred. Aoristic analysis has recently been extended to archaeology where temporally imprecise but spatially located data are the norm (Crema et al. 2010).

So what is the relevance of aoristic analysis to phenological monitoring and modeling? Phenological observations are frequently records of discrete events rather than continuing reports of phenophase status. Temporal imprecision in phenophase transitions arise from observations occurring less frequently than daily. Multiple temporal observations are possible within a specified area, whether this be on the ground or within a scene of a sensor. By spreading out the temporal uncertainty between subsequent observations of a pixel it is possible to recover information below interval of the compositing period. This approach has the possibility of enabling inference of phenophase onset, cessation, and duration at higher temporal precision, improving LSP model fit and predictive skill, and improving cross-scale calibration and validation procedures. However, for aoristic analysis to work, it is necessary that the composited product retain tags indicating the temporal coordinates for each pixel included in the composite. As the NBAR data underlying the global MODIS LSP product represents a modeled value contingent on an array of observations, aoristic analysis cannot be implemented. A similar situation occurs with smoothed and gap-filled data products since the smoothing already spreads the temporal uncertainty, but in an unrecoverable manner. Aoristic analysis has yet to gain widespread application in the LSP community, but we foresee its increasing use with finer spatial resolution data such as Landsat.

Finally, there remains the challenge of LSP forecasting. There are three types of prediction for phenophase transitions. First, there is the statistical prediction embedded within a climatological expectation: "Based on many years of observation at this location, the average date of occurrence of phenophase Z is $Y \pm X$ days". This approach includes both direct statistical summaries of remote sensing data and phenometrics derived from statistical modeling of the remote sensing data. While this type of forecasting is relatively easy, it is hobbled by the relatively short duration of satellite records as well as the assumption of a single underlying climatic regime to which the local LSPs respond. Nevertheless, it is effective as a first approximation. Second, there is a rule-based approach motivated by limiting factors (e.g., Jolly et al. 2005). This approach requires some level of ecological understanding which is lacking for many ecosystems (Richardson et al. 2013), but it may provide effective seasonal bounds. However, it may not be sensitive to spatio-temporal modulation of limiting factors due to land cover and land use. Third, there is prognostic modeling of phenophase transitions (e.g., Kathuroju et al. 2007), that may also incorporate data assimilation (e.g., Stöckli et al. 2008, 2011; Kovalskyy and Henebry 2012b). There is a clear need to move toward prognostic modeling to improve the representation of phenophase transitions within the land surface models that link to regional and global climate models (Pitman et al. 2009; Richardson et al. 2012).

In conclusion, the monitoring and modeling of land surface phenologies has advanced considerably during the past decade as more sensors, more data, and more investigators have probed how the vegetated land surface responds to seasonality and to anthropogenic influences. We expect that the coming decade of research involving even more sensors, data, and investigators will yield some elegant solutions to the outstanding challenges.

Acknowledgments Research was supported in part by NASA grants NNX11AB77G to KMdB and NNX12AM89G to GMH.

References

Ahl DE, Gower ST, Burrows SN, Shabanov NV, Myneni RB, Knyazikhin Y (2006) Monitoring spring canopy phenology of a deciduous broadleaf forest using MODIS. Remote Sens Environ 104(1):88–95

Ahrends HE, Brügger R, Stöckli R, Schenk J, Michna P, Jeanneret F, Wanner H, Eugster W (2008) Quantitative phenological observations of a mixed beech forest in northern Switzerland with digital photography. J Geophys Res 113, G04004

Anderson MC, Norman JM, Diak GR, Kustas WP, Mecikalski JR (1997) A two-source time-integrated model for estimating surface fluxes using thermal infrared remote sensing. Remote Sens Environ 60:195–216

Anderson MC, Kustas WP, Norman JM, Hain CR, Mecikalski JR, Schultz L, Gonzalez-Dugo MP, Cammalleri C, d'Urso G, Pimstein A, Gao F (2011) Mapping daily evapotranspiration at field to continental scales using geostationary and polar orbiting satellite imagery. Hydrol Earth Syst Sci 15:223–239

Archibald S, Scholes RJ (2007) Leaf green-up in a semi-arid African savanna -separating tree and grass responses to environmental cues. J Veg Sci 18:583–594

Badhwar GD (1984) Automatic corn-soybean classification using Landsat MSS data, I, near-harvest crop proportion estimation. Remote Sens Environ 14:15–29

Bartsch A, Kidd RA, Wagner W, Bartalis Z (2007) Temporal and spatial variability of the beginning and end of daily spring freeze/thaw cycles derived from scatterometer data. Remote Sens Environ 106(3):360–374

Bauer ME, Cipra JE, Anuta PE, Etheridge JB (1979) Identification and area estimation of agricultural crops by computer classification of Landsat MSS data. Remote Sens Environ 8:77–92

Beck PSA, Atzberger C, Høgda KA, Johansen B, Skidmore AK (2007) Improved monitoring of vegetation dynamics at very high latitudes: a new method using MODIS NDVI. Remote Sens Environ 100:321–334

Bradley BA, Olsson AD, Wang O, Dickson BG, Pelech L, Sesnie SE, Zachmann LJ (2012) Species detection vs. Habitat suitability: Are we biasing habitat suitability models with remotely sensed data? Ecol Model 244:57–64

Brown ME, de Beurs KM (2008) Evaluation of multi-sensor semi-arid crop season parameters based on NDVI and rainfall. Remote Sens Environ 112(5):2261–2271

Brown ME, de Beurs KM, Marshall M (2012) Global phenological response to climate change in crop areas using satellite remote sensing of vegetation, humidity and temperature over 26 years. Remote Sensing of Environment 126:174–183

Brown JF, Wardlow BD, Tadesse T, Hayes MJ, Reed BC (2008) The vegetation drought response index (VegDRI): a new integrated approach for monitoring drought stress in vegetation. GISci Remote Sens 45(1):16–46

Brown ME, de Beurs KM, Vrieling A (2010) The response of African land surface phenology to large scale climate oscillations. Remote Sens Environ 114:2286–2296

Carlson TN, Ripley DA (1997) On the relation between NDVI, fractional vegetation cover, and leaf area index. Remote Sens Environ 62(3):241–252

Choudhury BJ, Tucker CJ, Golus RE, Newcomb WW (1987) Monitoring vegetation using Nimbus-7 scanning multichannel microwave radiometer's data. Int J Remote Sens 8 (3):533–538

Christie B (2012) ForWarn's phenology datasets. http://forwarn.forestthreats.org/sites/default/files/ForWarn%20Phenology%20Data.pdf

Crema ER, Bevan A, Lake MW (2010) A probabilistic framework for assessing spatio-temporal point patterns in the archaeological record. J Archaeol Sci 37(5):1118–1130

Davidson A, Csillag F (2003) A comparison of three approaches for predicting C_4 species cover of northern mixed grass prairie. Remote Sens Environ 86:70–82 de Beurs KM, Henebry GM (2004) Land surface phenology, climatic variation, and institutional change: analyzing agricultural land cover change in Kazakhstan. Remote Sens Environ 89 (4):497–509. doi:10.1016/j.rse.2003.11.006

de Beurs KM, Henebry GM (2005a) A statistical framework for the analysis of long image time series. Int J Remote Sens 26(8):1551–1573

de Beurs KM, Henebry GM (2005b) Land surface phenology and temperature variation in the IGBP high-latitude transects. Glob Chang Biol 11(5):779–790

de Beurs KM, Henebry GM (2008a) Northern annular mode effects on the land surface phenologies of northern Eurasia. J Clim 21:4257–4279

de Beurs KM, Henebry GM (2008b) War, drought, and phenology: changes in the land surface phenology of Afghanistan since 1982. J Land Use Sci 3(2–3):95–111

de Beurs KM, Henebry GM (2010a) A land surface phenology assessment of the northern polar regions using MODIS reflectance time series. Can J Remote Sens 36(suppl 1): S87–S110

de Beurs KM, Henebry GM (2010b) Spatio-temporal statistical methods for modeling land surface phenology. In: Hudson IL, Keatley MR (eds) Phenological research: methods for environmental and climate change analysis. Springer, Dordrecht

de Beurs KM, Wright CK, Henebry GM (2009) Dual scale trend analysis distinguishes climatic from anthropogenic effects on the vegetated land surface. Environ Res Lett 4:045012

Delbart N, Kergoat L, Le Toan T, L'Hermitte J, Picard G (2005) Determination of phenological dates in boreal regions using normalized difference water index. Remote Sens Environ 97:26–38

Delbart N, Le Toan T, Kergoat L, Fedotova V (2006) Remote sensing of spring phenology in boreal regions: a free of snow-effect method using NOAA-AVHRR and SPOT-VGT data (1982–2004). Remote Sens Environ 100:52–62

Dethier B, Ashley MD, Blair B, Hopp RJ (1973) Phenology satellite experiment. In: Freden SC, EP Mercanti, MA Becker (eds) Symposium on significant results obtained from the Earth Resources Technology Satellite—1, vol I. Technical presentations, section A. NASA: Washington, DC, GPO NAS 1.21:327

Dragoni D, Rahman AF (2012) Trends in fall phenology across the deciduous forests of the eastern USA. Agric For Meteorol 157:96–105

Ehrlich D, Estes JE, Singh A (1994) Applications of NOAA-AVHRR 1 km data for environmental monitoring. Int J Remote Sens 15(1):145–161

Fisher JI, Mustard JF (2007) Cross-scalar satellite phenology from ground, Landsat, and MODIS data. Remote Sens Environ 109(3):261–273

Fisher JI, Mustard JF, Vadeboncoeur MA (2006) Green leaf phenology at Landsat resolution: scaling from the field to the satellite. Remote Sens Environ 100(2):265–279

Friedl M, Henebry G, Reed B, Huete A, White M, Morisette J, Nemani R, Zhang X, Myneni R (2006) Land surface phenology. A Community White Paper requested by NASA. April 10. http://cce.nasa.gov/mtg2008_ab_presentations/Phenology_Friedl_whitepaper.pdf

Frolking S, McDonald KC, Kimball JS, Way JB, Zimmermann R, Running SW (1999) Using the space-borne NASA scatterometer (NSCAT) to determine the frozen and thawed seasons. J Geophys Res 104(D22):27895–27907

Frolking S, Milliman T, McDonald K, Kimball J, Zhao M, Fahnestock M (2006) Evaluation of the SeaWinds scatterometer for regional monitoring of vegetation phenology. J Geophys Res 11: D17302

Ganguly S, Friedl MA, Tan B, Zhang X, Verma M (2010) Land surface phenology from MODIS: characterization of the collection 5 global land cover dynamics product. Remote Sens Environ 114(8):1805–1816

Gao F, Morisette JT, Wolfe RE, Ederer G, Pedelty J, Masuoka E, Myneni R, Tan B, Nightingale J (2008) An algorithm to produce temporally and spatially continuous MODIS-LAI time series. IEEE Geosci Remote Sens Lett 5(1):60–64

Garrigues S, Lacaze R, Baret F, Morisette JT, Weiss M, Nickeson JE, Fernandes R, Plummer S, Shabanov NV, Myneni RB, Knyazikhin Y, Yang W (2008) Validation and intercomparison of global Leaf Area Index products derived from remote sensing data. J Geophys Res 113:G02028
Gazal R, White MA, Gillies R, Rodemaker E, Sparrow E, Gordon L (2008) GLOBE students, teachers, and scientists demonstrate variable differences between urban and rural leaf phenology. Glob Chang Biol 14(7):1568–1580
Gitelson AA (2004) Wide dynamic range vegetation index for remote quantification of biophysical characteristics of vegetation. J Plant Physiol 161:165–173
Gonsamo A, Chen JM, Wu C, Dragoni D (2012a) Predicting deciduous forest carbon uptake phenology by upscaling FLUXNET measurements using remote sensing data. Agric For Meteorol 165:127–135
Gonsamo A, Chen JM, Price DT, Kurz WA, Wu C (2012b) Land surface phenology from optical satellite measurement and CO_2 eddy covariance technique. J Geophys Res 117, G03032
Goodin DG, Henebry GM (1997) Monitoring ecological disturbance in tallgrass prairie using seasonal NDVI trajectories and a discriminant function mixture model. Remote Sens Environ 61:270–278
Goodin DG, Henebry GM (1998) Seasonality of finely-resolved spatial structure of NDVI and its component reflectances in tallgrass prairie. Int J Remote Sens 19:3213–3220
Goodin DG, Gao J, Henebry GM (2004) The effect of solar zenith angle and sensor view angle on observed patterns of spatial structure in tallgrass prairie. IEEE Trans Geosci Remote Sens 42 (1):154–165
Goward SN, Tucker CJ, Dye DG (1985) North American vegetation patterns observed with the NOAA-7 advanced very high resolution radiometer. Plant Ecol 64(1):3–14
Gray LK, Gylander T, Mbogga MS, Chen P-Y, Hamann A (2011) Assisted migration to address climate change: recommendations for aspen reforestation in Western Canada. Ecol Appl 21 (5):1591–1603
Gu Y, Wylie BK, Bliss NB (2013) Mapping grassland productivity with 250-m eMODIS NDVI and SSURGO database over the Greater Platte River Basin, USA. Ecol Indic 24:31–36
Hargrove WW, Spruce JP, Gasser GE, Hoffman FM (2009) Toward a national early warning system for forest disturbances using remotely sensed phenology. Photogramm Eng Remote Sens 75(10):1150–1156
Heilman JL, Kanemasu ET, Bagley JO, Rasmussen VP (1977) Evaluating soil moisture and yield of winter wheat in the Great Plains using Landsat data. Remote Sens Environ 6:315–326
Henebry GM (2003) Grasslands of the North American great plains. In: Schwartz MD (ed) Phenology: an integrative environmental science. Kluwer, Dordrecht/Boston
Henebry GM, Kux HJH (1995) Lacunarity as a texture measure for SAR imagery. Int J Remote Sens 16:565–571
Henebry GM, Kux HJH (1997) Spatio-temporal analysis of SAR image series from the Brazilian Pantanal. In: Proceedings of the 3rd ERS symposium on space at the service of our environment, SP-414, ESA, Noordwijk. http://earth.esa.int/workshops/ers97/papers/henebry1/index.html
Henebry GM, Su H (1993) Using landscape trajectories to assess the effects of radiometric rectification. Int J Remote Sens 14:2417–2423
Henebry GM, Su H (1995) Observing spatial structure in the Flint Hills using AVHRR maximum biweekly NDVI composites. In: Proceedings of the 14th North American Prairie Conference. Kansas State University Press, Manhattan. http://images.library.wisc.edu/EcoNatRes/EFacs/NAPC/NAPC14/reference/econatres.napc14.ghenebry.pdf
Hlavka CA, Haralick RM, Carlyle SM, Yokoyama R (1980) The discrimination of winter wheat using a growth-state signature. Remote Sens Environ 9:277–294
Hodges T (1990) Predicting crop phenology. CRC Press, Boca Raton
Howard DM, Wylie BK, Tieszen LL (2012) Crop classification modelling using remote sensing and environmental data in the Greater Platte River Basin, USA. Int J Remote Sens 33 (19):6094–6108

Huete A, Didan K, Miura T, Rodriguez EP, Gao X, Ferreira LG (2002) Overview of the radiometric and biophysical performance of the MODIS vegetation indices. Remote Sens Environ 83(1–2):195–213

Isaacson BN, Serbin SP, Townsend PA (2012) Detections of relative differences in phenology of forest species using Landsat and MODIS. Landsc Ecol 27:529–543

Jakubauskas ME, Legates DR, Kastens JH (2002) Crop identification using harmonic analysis of time-series AVHRR NDVI data. Comput Electron Agric 37(1–3):127–139

Jenkerson CB, Schmidt GL (2008) eMODIS product access for large scale monitoring. In: Proceedings of the 17th Pecora symposium, paper 19. http://www.asprs.org/a/publications/proceedings/pecora17/0019.pdf

Jenkerson C, Maiersperger T, Schmidt G (2010) eMODIS: A user-friendly data source. USGS open-file report 2010–1055. http://pubs.usgs.gov/of/2010/1055/pdf/OF2010-1055.pdf

Johnston CA (2013) Wetland losses due to row crop expansion in the Dakota Prairie Pothole region. Wetlands 33(1):175–182

Jolly WM, Nemani R, Running SW (2005) A generalized, bioclimatic index to predict foliar phenology in response to climate. Glob Chang Biol 11:619–632

Jones MO, Jones LA, Kimball JS, McDonald KS (2011) Satellite passive microwave remote sensing for monitoring global land surface phenology. Remote Sens Environ 115:1102–1114

Jones MO, Kimball JS, Jones LA, McDonald KC (2012) Satellite passive microwave detection of North America start of season. Remote Sens Environ 123:324–333

Jonsson P, Eklundh L (2002) Seasonality extraction by function fitting to time-series of satellite sensor data. IEEE Trans Geosci Remote Sens 40:1824–1832

Jonsson P, Eklundh L (2004) TIMESAT—a program for analyzing time-series of satellite sensor data. Comput Geosci 30(8):833–845

Justice CO, Townshend JRG, Holben BN, Tucker CJ (1985) Analysis of the phenology of global vegetation using meteorological satellite data. Int J Remote Sens 6(8):1271–1318

Justice CO, Holben BN, Gwynne MD (1986) Monitoring east African vegetation using AVHRR data. Int J Remote Sens 7(11):1453–1474

Justice CO, Townshend JRG, Choudhury BJ (1989) Comparison of AVHRR and SMMR data for monitoring vegetation phenology on a continental scale. Int J Remote Sens 10(10):1607–1632

Justice CO, Vermote E, Townshend JRG, DeFries R, Roy DP, Hall DK, Salomonson VV, Privette JL, Riggs G, Strahler A, Lucht W, Myneni RB, Knyazikhin Y, Running SW, Nemani RR, ZhengMing W, Huete AR, van Leeuwen W, Wolfe RE, Giglio L, Muller J, Lewis P, Barnsley MJ (1998) The Moderate Resolution Imaging Spectroradiometer (MODIS): land remote sensing for global change research. IEEE Trans Geosci Remote Sens 36(4):1228–1249

Justice CO, Townshend JRG, Vermote EF, Masuoka E, Wolfe RE, Saleous N, Roy DP, Morisette JT (2002) An overview of MODIS land data processing and product status. Remote Sens Environ 83(1–2):3–15

Kanemasu ET (1974) Seasonal canopy reflectance patterns of wheat, sorghum, and soybean. Remote Sens Environ 3:43–47

Kanemasu ET, Niblett CL, Manges H, Lenhert D, Newman MA (1974) Wheat: its growth and disease severity as deduced from ERTS-1. Remote Sens Environ 3:255–260

Kathuroju N, White MA, Symanzik J, Schwartz MD, Powell JA, Nemani RR (2007) On the use of the Advanced Very High Resolution Radiometer for development of prognostic land surface phenology models. Ecol Model 201(1):144–156

Kim Y, Kimball JS, McDonald KC, Glassy J (2011) Developing a global data record of daily landscape freeze/thaw status using satellite passive microwave remote sensing. IEEE Trans Geosci Remote Sens 49(3):949–960

Knapp WW, Dethier BE (1976) Satellite monitoring of phenological events. Int J Biometeorol 20 (3):230–239

Kovalskyy V, Henebry GM (2012a) A new concept for simulation of vegetated land surface dynamics: the event driven phenology model part I. Bio Geosci 9:141–159

Kovalskyy V, Henebry GM (2012b) Alternative methods to predict actual evapotranspiration illustrate the importance of accounting for phenology: the event driven phenology model part II. Bio Geosci 9:161–177. doi:10.5194/bg-9-161-2012

Kovalskyy V, Roy DP, Zhang X, Ju J (2012) The suitability of multi-temporal web-enabled Landsat data NDVI for phenological monitoring—a comparison with flux tower and MODIS NDVI. Remote Sens Lett 3(4):325–334

Leopold A, Jones SE (1947) A phenological record for Sauk and Dane counties, Wisconsin, 1935–1945. Ecol Monogr 17(1):81–122

Li H, Wang X, Hamann A (2010) Genetic adaptation of aspen (*Populus tremuloides*) populations to spring risk environments: a novel remote sensing approach. Can J For Res 40 (11):2082–2090. doi:10.1139/X10-153

Liang L, Schwartz MD (2009) Landscape phenology: an integrative approach to seasonal vegetation dynamics. Landsc Ecol 24:465–472

Liang L, Schwartz MD, Fei S (2011) Validating satellite phenology through intensive ground observation and landscape scaling in a mixed seasonal forest. Remote Sens Environ 115 (1):143–157

Lloyd D (1990) A phenological classification of terrestrial vegetation cover using shortwave vegetation index imagery. Int J Remote Sens 11(12):2269–2279

Loveland TR, Merchant JW, Brown JF, Ohlen DO, Reed BC, Olson P, Hutchinson J (1995) Seasonal land-cover regions of the United States. Ann Assoc Am Geogr 85(2):339–355

Loveland TR, Cochrane MA, Henebry GM (2008) Landsat still contributing to environmental research. Trends Ecol Evolut 23(4):182–183

McDonald KC, Kimball JS, Njoku E, Zimmermann R, Zhao M (2004) Variability in springtime thaw in the terrestrial high latitudes: monitoring a major control on the biospheric assimilation of atmospheric CO2 with spaceborne microwave remote sensing. Earth Interact 8:1–23

McManus KM, Morton DC, Masek JG, Wang D, Sexton JO, Nagol JR, Ropars P, Boudreau S (2010) Satellite-based evidence for shrub and graminoid tundra expansion in northern Quebec from 1986 to 2010. Glob Chang Biol 18(7):2313–2323

Mitchell R, Fritz J, Moore K, Moser L, Vogel K, Redfearn D, Wester D (2001) Predicting forage quality in switchgrass and big bluestem. Agron J 93:118–124

Morisette JT, Baret F, Privette JL, Myneni RB, Nickeson JE, Garrigues S, Shabanov NV, Weiss M, Fernandes RA, Leblanc DG, Kalacska M, Sanchez-Azofeifa GA, Chubey M, Rivard B, Stenberg P, Rautiainen M, Voipio P, Manninen T, Pilant AN, Lewis TE, Iiames JS, Colombo R, Meroni M, Busetto L, Cohen WB, Turner DP, Warner ED, Petersen GW, Seufert G, Cook R (2006) Validation of global moderate-resolution LAI products: a framework proposed within the CEOS land product validation subgroup. IEEE T Geosci Remote 44 (7):1804–1817

Morisette JT, Richardson AD, Knapp AK, Fisher JI, Graham E, Abatzoglou J, Wilson BE, Breshears DD, Henebry GM, Hanes JM, Liang L (2008) Unlocking the rhythm of the seasons in the face of global change: challenges and opportunities for phenological research in the 21st century. Front Ecol Environ 5(7):253–260. doi:10.1890/070217

Morisette JT, Nightingale J, Nickeson J (2010) Assessing the accuracy of landscape-scale phenology products: an international workshop on the validation of satellite-based phenology products; Dublin, Ireland, 18 June 2010. Eos 91(44):407

Moulin S, Kergoat L, Viovy N, Dedieu G (1997) Global-scale assessment of vegetation phenology using NOAA/AVHRR satellite measurements. J Clim 10:1154–1170

Myneni RB, Hoffman S, Knyazikhin Y, Privette JL, Glassy J, Tian Y, Wang Y, Song X, Zhang Y, Smith GR, Lotsch A, Friedl M, Morisette JT, Votava P, Nemani RR, Running SW (2002) Global products of vegetation leaf area and fraction absorbed PAR from year one of MODIS data. Remote Sens Environ 83(1–2):214–223

Neale CMU, McFarland MJ, Chang K (1990) Land-surface-type classification using microwave brightness temperatures from the Special Sensor Microwave/Imager. IEEE Trans Geosci Remote Sens 28(5):829–838

Noormets A (2009) Phenology of ecosystems processes. Springer, New York
Nuttonson MV (1955) Wheat-climate relationships and the use of phenology in ascertaining the thermal and photo-thermal requirements of wheat. American Institute of Crop Ecology, Washington, DC
Pan Y, Li L, Zhang J, Liang S, Zhu X, Sulla-Menashe D (2012) Winter wheat area estimation from MODIS-EVI time series data using the crop proportion phenology index. Remote Sens Environ 19:232–242
Pedelty J, Devadiga S, Masuoka E, Brown M, Pinzon J, Tucker C, Vermote E, Prince S, Nagol J, Justice C, Roy D, Ju J, Schaaf C, Liu J, Privette J, Pinheiro A (2007) Generating a long-term land data record from the AVHRR and MODIS instruments. In: Proceeding of the IEEE international geoscience and remote sensing symposium 2007 (IGARSS 2007), pp 1021–1025. Available at: http://ltdr.nascom.nasa.gov/ltdr/docs/LTDR_IGARSS2007_paper.pdf
Peñuelas J, Filella I, Zhang X, Llorens L, Ogaya R, Lloret F, Comas P, Estiarte M, Terradas J (2004) Complex spatiotemporal phenological shifts as a response to rainfall changes. New Phytol 161(3):837–846
Pervez MS, Brown JF (2010) Mapping irrigated lands at 250-m scale by merging MODIS data and national agricultural statistics. Remote Sens 2(10):2388–2412
Pitman AJ, Noblet-Ducoudré N, Cruz FT, Davin EL, Bonan GB, Brovkin V, Claussen M, Delire C, Ganzeveld L, Gayler V, van den Hurk BJJM, Lawrence PJ, van der Molen MK, Müller C, Reick CH, Seneviratne SI, Strengers BJ, Voldoire A (2009) Uncertainties in climate responses to past land cover change: first results from the LUCID intercomparison study. Geophys Res Lett 36, L14814
Prigent C, Rossow W, Matthews E (1997) Microwave land surface emissivities estimated from SSM/I observations. J Geophys Res 102(D18):21867–21890
Ramsay JO, Silverman BW (2005) Functional data analysis, 2e. Springer, New York
Ratcliffe JH (2000) Aoristic analysis: the spatial interpretation of unspecific temporal events. Int J Geograp Inf Sci 14(7):669–679
Ratcliffe JH (2002) Aoristic signatures and the spatio-temporal analysis of high volume crime patterns. J Quant Criminol 18(1):23–43
Ratcliffe JH, McCullagh MJ (1998) Aoristic crime analysis. Int J Geog Inf Sci 12(7):751–764
Rea J, Ashley M (1976) Phenological evaluations using landsat-1 sensors. Int J Biometeorol 20 (3):240–248
Reed BC (2006) Trend analysis of time-series phenology of North America derived from satellite data. GISci Remote Sens 43(1):24–38
Reed BC, Brown JF, VanderZee D, Loveland TR, Merchant JW, Olhen DO (1994) Measuring phenological variability from satellite imagery. J Veg Sci 5(5):703–714
Reed BC, White MA, Brown JF (2003) Remote sensing phenology. In: Schwartz MD (ed) Phenology: an integrative environmental science. Kluwer, Dordrecht/Boston
Reed BC, Schwartz MD, Xiao X (2009) Remote sensing phenology. In: Noormets A (ed) Phenology of ecosystem processes. Springer, New York
Richardson AD, Bailey AS, Denny EG, Martin CW, O'Keefe J (2006) Phenology of a northern hardwood forest canopy. Glob Chang Biol 12(7):1174–1188
Richardson AD, Jenkins JP, Braswell BH, Hollinger DY, Ollinger SV, Smith ML (2007) Use of digital webcam images to track spring green-up in a deciduous broadleaf forest. Oecologia 152 (2):323–334
Richardson AD, Braswell BH, Hollinger DY, Jenkins JP, Ollinger SV (2009) Near-surface remote sensing of spatial and temporal variation in canopy phenology. Ecol App 19:1417–1428
Richardson AD, Anderson RC, Arain MA, Barr AG, Bohrer G, Chen G, Chen JM, Ciais P, Davis KJ, Desai AR, Dietze MC, Dragoni D, Garrity SR, Gough CM, Grant R, Hollinger DY, Margolis HA, McCaughey H, Migliavacca M, Monson RK, Munger JW, Poulte B, Raczka BM, Ricciuto DM, Sahoo AK, Schaefer K, Tian H, Vargas R, Verbeeck H, Xiao J, Xue J (2012) Terrestrial biosphere models need better representation of vegetation

phenology: results from the North American Carbon Program Site Synthesis. Glob Chang Biol 18:566–584

Richardson AD, Keenan TF, Migliavacca M, Ryu Y, Sonnentag O, Toomey M (2013) Climate change, phenology, and phenological control of vegetation feedbacks to the climate system. Agric For Meteorol 169:156–173

Roy DP, Ju J, Kline K, Scaramuzza PL, Kovalskyy V, Hansen M, Loveland TR, Vermote E, Zhang C (2010) Web-enabled Landsat Data (WELD): Landsat ETM + composited mosaics of the conterminous United States. Remote Sens Environ 114(1):35–49

Sakamoto T, Yokozawa M, Toritani H, Shibayama M, Ishitsuka N, Ohno H (2005) A crop phenology detection method using time-series MODIS data. Remote Sens Environ 96 (3–4):366–374

Sakamoto T, Wardlow BD, Gitelson AA, Verma SB, Suyker AE, Arkebauer TJ (2010) A two-step filtering approach for detecting maize and soybean phenology with time-series MODIS data. Remote Sens Environ 114(10):2146–2159

Schwartz MD (2003) Phenology: an integrative environmental science. Kluwer, Dordrecht/Boston

Schwartz MD, Reed BC (1999) Surface phenology and satellite sensor-derived onset of greenness: an initial comparison. Int J Remote Sens 20(7):3451–3457

Schwartz MD, Reed BC, White MA (2002) Assessing satellite-derived start of season measures in the conterminous USA. Int J Climatol 22(14):1793–1805

Schwartz MD, Ahas R, Aasa A (2006) Onset of spring starting earlier across the northern hemisphere. Glob Chang Biol 12(2):343–351

Smart AJ, Schacht WH, Moser LE (2001) Predicting leaf/stem ratio and nutritive value in grazed and nongrazed big bluestem. Agron J 93:1243–1249

Soudani K, le Maire G, Dufrêne E, François C, Delpierre N, Ulrich E, Cecchini S (2008) Evaluation of the onset of green-up in temperate deciduous broadleaf forests derived from Moderate Resolution Imaging Spectroradiometer (MODIS) data. Remote Sens Environ 112 (5):2643–2655

Spruce JP, Sader S, Ryan RE, Smoot J, Kuper P, Ross K, Prados D, Russell J, Gasser G, McKellip R, Hargrove WW (2011) Assessment of MODIS NDVI time series data products for detecting forest defoliation from gypsy moth outbreaks. Remote Sens Environ 115:427–437

Stöckli R, Rutishauser T, Dragoni D, O'Keefe J, Thornton PE, Jolly M, Lu L, Denning AS (2008) Remote sensing data assimilation for a prognostic phenology model. J Geophys Res 113, G04021

Stöckli R, Rutishauser T, Baker I, Liniger MA, Denning AS (2011) A global reanalysis of vegetation phenology. J Geophys Res 116, G03020

Stohlgren TJ, Ma P, Kumar S, Rocca M, Morisette JT, Jarnevich CS, Benson N (2010) Ensemble habitat mapping of invasive plant species. Risk Anal 30(2):224–235

Swets DL, Reed BC, Rowland JR, Marko SE (1999) A weighted least-squares approach to temporal smoothing of NDVI. In: Proceedings of the 1999 ASPRS annual conference. http://phenology.cr.usgs.gov/pubs/ASPRS%20Swets%20et%20al%20Smoothing.pdf

Tan B, Morisette J, Wolfe R, Gao F, Nightingale JM, Pedelty J, Ederer G (2011) User guide for MOD09PHN and MOD15PHN. Version 3.0. http://accweb.nascom.nasa.gov/project/docs/User_guide_C5_PHN.pdf. Accessed 3 Feb 2011

Townshend JRG, Justice CO, Choudhury BJ, Tucker CJ, Kalb VT, Goff TE (1989) A comparison of AVHRR and SMMR data for continental land cover characterization. Int J Remote Sens 10 (10):1633–1642

Tuanmu M-N, Viña A, Bearer S, Xu W, Ouyang Z, Zhang H, Liu J (2010) Mapping understory vegetation using phenological characteristics derived from remotely sensed data. Remote Sens Environ 114:1833–1844

Tucker CJ (1979) Red and photographic infrared linear combinations for monitoring vegetation. Remote Sens Environ 8:127–150

Tucker CJ, Elgin JH Jr, McMurtrey JE III, Fan CJ (1979) Monitoring corn and soybean crop development with hand-held radiometer spectral data. Remote Sens Environ 8:237–248

Tucker CJ, Townshend JRG, Goff TE (1985) African land-cover classification using satellite data. Science 227:369–375

Tucker CJ, Slayback DA, Pinzon JE, Los SO, Myneni RB, Taylor MG (2001) Higher northern latitude normalized difference vegetation index and growing season trends from 1982 to 1999. Int J Biometeorol 45:184–190

Tucker CJ, Pinzon JE, Brown ME, Slayback MA, Pak EW, Mahoney R, Vermote EF, El Saleous N (2005) An extended AVHRR 8-km NDVI dataset compatible with MODIS and SPOT vegetation NDVI data. Int J Remote Sens 26(20):4485–4498

US Census Bureau (2012) Table 858. Crops—supply and use: 2000 to 2010. Statistical abstract of the United States. http://www.census.gov/compendia/statab/2012/tables/12s0858.pdf

Vina A, Gitelson AA (2011) Sensitivity to foliar Anthocyanin content of vegetation indices using green reflectance. IEEE Geosci Remote Sens Lett 8(3):464–468

Viña A, Henebry GM (2005) Spatio-temporal change analysis to identify anomalous variation in the vegetated land surface: ENSO effects in tropical South America. Geophys Res Lett 32, L21402. doi:10.1029/2005GL023407

Viña A, Henebry GM, Gitelson AA (2004) Satellite monitoring of vegetation dynamics: sensitivity enhancement by the Wide Dynamic Range Vegetation Index. Geophys Res Lett 31, L04503. doi:10.1029/2003GL019034

Walker J, de Beurs KM, Wynne RH, Gao F (2012) An evaluation of data fusion products for the analysis of dryland forest phenology. Remote Sens Environ 117:381–393

Wang JR (1985) Effect of vegetation on soil moisture sensing observed from orbiting microwave radiometers. Remote Sens Environ 17:141–151

Wardlow BD, Egbert SL (2008) Large-area crop mapping using time-series MODIS 250 m NDVI data: an assessment for the U.S. Central Great Plains. Remote Sens Environ 112 (3):1096–1116

Weiss M, Baret F, Garrigues S, Lacaze R (2007) LAI and fAPAR CYCLOPES global products derived from VEGETATION. Part 2: validation and comparison with MODIS collection 4 products. Remote Sens Environ 110(3):317–331

White MA, Thornton PE, Running SW (1997) A continental phenology model for monitoring vegetation responses to interannual climatic variability. Global Biogeochem Cycle 11 (2):217–234

White MA, Nemani RR, Thornton PE, Running SW (2002) Satellite evidence of phenological differences between urbanized and rural areas of the eastern United States deciduous broadleaf forest. Ecosystems 5:260–277

White MA, Hoffman F, Hargrove WW, Nemani RR (2005) A global framework for monitoring phenological responses to climate change. Geophys Res Lett 32, L04705

White MA, de Beurs KM, Didan K, Inouye DW, Richardson AD, Jensen OP, O'Keefe J, Zhang G, Nemani RR, van Leeuwen WJD, Brown JF, de Wit A, Schaepman M, Lin X, Dettinger M, Bailey AS, Kimball J, Schwartz MD, Baldocchi DD, Lee JT, Lauenroth WK (2009) Intercomparison, interpretation, and assessment of spring phenology in North America estimated from remote sensing for 1982 to 2006. Glob Chang Biol 15(10): 2335–2359

Worley-Firley S (2012) USFS eastern threat center develops forest technology and tools. Natl Woodl 2012(Fall):12–15

Wright CK, Wimberly MC (2013) Recent land cover change in the western corn belt threatens grasslands and wetlands. PNAS (in review following revision) Published online before print February 19, 2013, doi:10.1073/pnas.1215404110

Wulder MA, Masek JG, Cohen WB, Loveland TR, Woodcock CE (2012) Opening the archive: How free data has enabled the science and monitoring promise of Landsat. Remote Sens Environ 122:2–10

Yates H, Strong A, McGinnis D Jr, Tarpley D (1986) Terrestrial observations from NOAA operational satellites. Science 231:463–470

Zhang X, Friedl MA, Schaaf CB, Strahler AH, Hodges JCF, Gao F, Reed BC, Huete A (2003) Monitoring vegetation phenology using MODIS. Remote Sens Environ 84(3):471–475

Zhang X, Friedl MA, Schaaf CB, Strahler AH (2004) Climate controls on vegetation phenological patterns in northern mid- and high latitudes inferred from MODIS data. Glob Chang Biol 10 (7):1133–1145

Zhang X, Friedl MA, Schaaf CB, Strahler AH, Liu Z (2005) Monitoring the response of vegetation phenology to precipitation in Africa by coupling MODIS and TRMM instruments. Journal of Geophysical Research: Atmospheres 110(D12, 27) doi:10.1029/2004JD005263

Zhang X, Friedl MA, Schaaf CB (2006) Global vegetation phenology from Moderate Resolution Imaging Spectroradiometer (MODIS): evaluation of global patterns and comparison with in situ measurements. J Geophys Res 111, G04017

Zhang X, Friedl MA, Schaaf CB (2009) Sensitivity of vegetation phenology detection to the temporal resolution of satellite data. Int J Remote Sens 30(8):2061–2074

Zhang X, Goldberg MD, Yunyue Y (2012) Prototype for monitoring and forecasting fall foliage coloration in real time from satellite data. Agric For Meteorol 158–159:21–29

Chapter 22
Near-Surface Sensor-Derived Phenology

Andrew D. Richardson, Stephen Klosterman, and Michael Toomey

Abstract "Near-surface" remote sensing provides a novel approach to phenological monitoring. Optical sensors mounted in relatively close proximity (typically 50 m or less) to the land surface can be used to quantify, at high temporal frequency, changes in the spectral properties of the surface associated with vegetation development and senescence. The scale of these measurements—intermediate between individual organisms and satellite pixels—is unique and advantageous for a variety of applications. In this chapter, we review and discuss a variety of approaches to near-surface remote sensing of phenology, including methods based on broad- and narrow-band radiometric sensors, and using commercially available digital cameras as inexpensive imaging sensors.

22.1 Introduction

Traditionally, plant phenology data have been recorded by a human observer, based on direct visual inspection of individual organisms in the field (e.g., Sparks and Menzel 2002). This approach is suited to the identification of dates at which specific phenophases (e.g., budburst or flowering) occur. The advent of satellite remote sensing in the 1970s opened up new opportunities for global-scale monitoring of seasonal changes in the spectral properties of vegetation. Vegetation indices, such as the normalized difference vegetation index (NDVI) and enhanced vegetation index (EVI), have been used to quantify vegetation "greenness", and various algorithms have been developed to estimate the timing of phenological transitions (e.g., start- and end-of-season) from time series of these indices (e.g., Zhang et al. 2006; White et al. 2009).

A.D. Richardson (✉) • S. Klosterman • M. Toomey
Department of Organismic and Evolutionary Biology, Harvard University,
Cambridge, MA 02138, USA
e-mail: arichardson@oeb.harvard.edu

M.D. Schwartz (ed.), *Phenology: An Integrative Environmental Science*,
DOI 10.1007/978-94-007-6925-0_22,

Table 22.1 Pros and cons of different approaches to monitoring phenology

Approach	Pros	Cons
Direct field observations	Ability to characterize specific phenophases (e.g., budburst, flowering) Can observe individual plants Can focus on species of interest	Spatial coverage typically limited Temporal resolution of observations often inadequate (≈ weekly) Time consuming if a population or ecosystem is to be adequately sampled Potential for observer bias, subjectivity
Satellite remote sensing	Yields seasonal trajectory of "greenness" Provides spatial integration across pixel (10–1,000 m) Offers global coverage	Tradeoffs between spatial and temporal resolution Cloud cover may obscure land surface Atmospheric corrections are required and may be uncertain
Near-surface remote sensing	Yields quantitative (seasonal trajectory of "greenness") or categorical (visual assessment of specific phenophases) data on phenology Opportunity for spatial integration across instrument footprint or to focus on individual organisms (spatial scale ≈ 1–100 m) Continuous in time (data collected at time interval of minutes to hours) Potential to separate structure (leaf area) and function (photosynthetic capacity) Relatively inexpensive (many options in $100–$5,000 price range)	Instrument footprint may not represent larger areas (ecosystem to region) Some infrastructure required (e.g., structure for mounting, data logging equipment or Internet connectivity) Instruments may fail (lightning strikes, extreme heat/cold, rodents like to chew cables, etc.); large data gaps are possible if instruments are not monitored regularly

An alternative (Table 22.1) to these approaches exists in the form of "near-surface" remote sensing, in which radiometric or imaging sensors, typically fixed to permanent structures (e.g., towers, masts, or buildings), are used to observe and quantify changes in the land surface in a manner analogous to airborne remote sensing, but at a spatial scale similar to ground observations by a human.

One advantage of near-surface remote sensing is its ability to serve as a bridge between direct observations and satellite data, and thus to facilitate scaling from organisms to landscapes. With near-surface remote sensing, spatial integration across the canopy is possible, thereby facilitating comparison with eddy covariance measurements of CO_2 and H_2O fluxes, and simplifying analysis of relationships between phenology and ecosystem processes.

However, there are other reasons why near-surface remote sensing has great potential for routine phenological monitoring (see Table 22.1). For example, compared to either direct or airborne observations, for which data are typically available only at intervals of days-to-weeks, automated sensors can provide data that are essentially continuous in time.

Additionally, near-surface remote sensing data provide quantitative information about the whole seasonal trajectory of vegetation development and senescence for a well-defined group of organisms within the sensor footprint. By comparison, data collected by a human observer are potentially subjective, and inherently difficult to put on an ordinal scale. Satellite pixels, on the other hand, may integrate across heterogeneous species mixtures or different land cover types, which is valuable for some applications but complicates biological interpretation of the observed seasonal patterns or trends.

Finally, by combining different types of near-surface sensor data, it may be possible to separate seasonal changes in canopy structure from changes in canopy function, particularly as related to photosynthetic capacity or efficiency.

In this chapter, we provide an overview of some of the different sensor-based approaches to monitoring phenology. We discuss the instrumentation requirements and data processing, review some of the recent applications of different technologies, evaluate uncertainties and shortcomings, and identify potential opportunities that may be opened up with the development of new technologies.

22.2 Instruments for Sensor-Based Phenology

Near-surface remote sensing technologies can be divided into two broad categories: radiometric sensors and imaging sensors.

Radiometric sensors vary in their field of view, from wide (up to 180°) to narrow (<45°), and their spectral sensitivity, from broadband to narrowband (Balzarolo et al. 2011). Broadband sensors include the Li-Cor LI-190 quantum sensor to measure photosynthetic photon flux density (PPFD, 400–700 nm) and the Kipp and Zonen CMP-6 pyranometer to measure total shortwave solar radiation (285–2,800 nm). Narrowband sensors target specific regions or bands of the electromagnetic spectrum. These range in sophistication from Skye's SKR1800 two-channel sensor with red (625–680 nm) and near infrared (835–890 nm) bands, to multichannel spectrometers (e.g., ASD, Ocean Optics, or PP Systems) that can simultaneously measure hundreds of bands across an entire spectral range. Each of these sensors outputs a single number that averages across the measurement footprint. Thus there is typically no information about spatial variability (but see Hilker et al. 2011), and it is not possible to distinguish among objects or individual organisms within the instrument's field of view.

Imaging sensors, on the other hand, produce digital pictures of a scene with which spatial variability can be analyzed and, if the pictures are of sufficient resolution, different organisms (or even organs) can be identified. Examples include digital cameras (including networked cameras, or "webcams") that record conventional RGB (red, green, blue) imagery and multichannel cameras (e.g., Tetracam MCA) that yield information for six or more specific wavebands. Highly sophisticated (and expensive) hyperspectral imaging spectrometers also fall in this category, but these have been used for mapping of biochemical attributes of vegetation, and classification or mapping of vegetation types, rather than continuous phenological monitoring (Ustin et al. 2004).

A key distinction between radiometric sensors and most imaging sensors is that the former can be calibrated to yield reference-traceable measurements of radiance (flux density, or power per unit area), whereas the latter usually (imaging spectroradiometers are a noted exception) are not.

22.3 Broadband Radiometric Sensors

The foliage of terrestrial vegetation absorbs most (>80 %) incident solar radiation in the visible region (400–700 nm) of the electromagnetic spectrum, because these are the wavelengths that are used to drive photosynthesis. Longer wavelengths in the near infrared (700–1,400 nm), which are of little use for photosynthesis, are only weakly absorbed. Thus both transmittance and reflectance of infrared wavelengths tend to be high. The distinction between low reflectance of visible wavelengths, and high reflectance of near infrared wavelengths, is a spectral signature that can be used to distinguish healthy green vegetation from soil, buildings and roads, senescent foliage, and leaf litter. This is the basis of the commonly used normalized difference vegetation index, NDVI, where r denotes reflectance:

$$\mathrm{NDVI} = \frac{r_{\mathrm{NIR}} - r_{\mathrm{red}}}{r_{\mathrm{NIR}} + r_{\mathrm{red}}}$$

Huemmrich et al. (1999) noted that readily available broadband sensors could be used to measure quantities that would correspond approximately to r_{red} and r_{NIR}. Specifically, by measuring incident ($Q\downarrow$) and reflected ($Q\uparrow$) PPFD with a quantum sensor, r_{VIS} ($=Q\uparrow/Q\downarrow$) could be used as an estimate of r_{red}, while shortwave albedo (measured with upward and downward-pointing pyranometers, $\alpha = R\uparrow/R\downarrow$) provides a measure of $r_{\mathrm{VIS+NIR}}$. Because PPFD is measured in μmol $\mathrm{m}^{-2}\ \mathrm{s}^{-1}$, whereas shortwave radiation is measured in W m^{-2}, estimation of r_{NIR} itself requires some assumptions to standardize the units. These assumptions are discussed and evaluated more fully by Huemmrich et al. (1999) and Jenkins et al. (2007) (and an alternative, the two-band enhanced vegetation index EVI2, is proposed by Rocha and Shaver 2009). One solution is to estimate r_{NIR} as:

$$r_{\mathrm{NIR}} = \frac{R\uparrow - 0.25 \times Q\uparrow}{R\downarrow - 0.25 \times Q\downarrow}$$

Calculated using these estimates of r_{red} and r_{NIR}, "broadband NDVI" has been used to monitor the phenology of both temperate and boreal forest (Fig. 22.1, Huemmrich et al. 1999; Wang et al. 2004; Jenkins et al. 2007), temperate grasslands and crops (Nagy et al. 2007; Wilson and Meyers 2007; Tittebrand et al. 2009), and tropical forest (Doughty and Goulden 2008) ecosystems. In many of these studies,

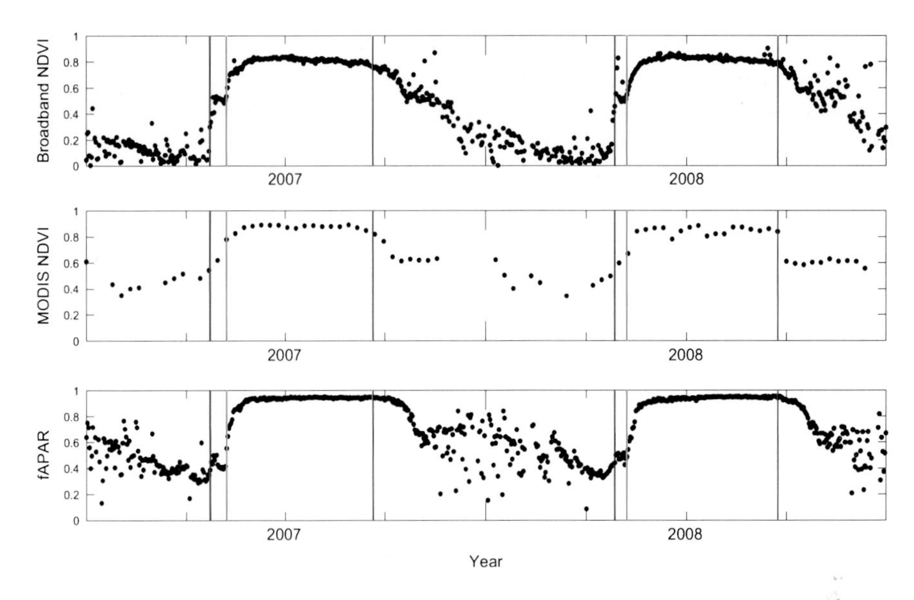

Fig. 22.1 Time series of broadband NDVI, MODIS NDVI, and f_{APAR} at Bartlett Experimental Forest (2007–2008). Data represent averages across mid-day values. Dominant vegetation is comprised of deciduous northern hardwood (maple-beech-birch) species. Snowmelt date (*blue line*), budburst date (*green line*) and onset of leaf coloration (*red line*), based on visual assessment of camera data from the same tower, are indicated

broadband NDVI compared favorably with NDVI calculated from satellite data, with the higher time resolution of broadband NDVI an obvious advantage.

In some studies, the broadband NDVI signal has been shown to be noisy during winter months, probably because of periodic snow on one or more of the upward-looking radiation sensors. Furthermore, there is a two-stage rise in NDVI that occurs with snowmelt (stage 1), and green-up (stage 2); failure to identify these as separate events will bias estimates of phenological transition dates. However, from spring through autumn, the quality of the data is generally sufficient to clearly distinguish leaf-on and leaf-off dates, and to track the rates of canopy development and senescence (Fig. 22.1).

Canopy phenology can also be monitored with continuous measurements of photosynthetically active radiation above ($Q\downarrow$, $Q\uparrow$) and below (transmitted, Q_T) the canopy, thereby permitting the absorbed fraction (f_{APAR}) to be calculated (e.g., Fig. 22.1, Turner et al. 2003; Jenkins et al. 2007; Doughty and Goulden 2008). Ignoring the radiation that is reflected from the forest floor:

$$\mathrm{APAR} = Q\downarrow - Q\uparrow - Q_T;\ f_{\mathrm{APAR}} = \frac{\mathrm{APAR}}{Q\downarrow}$$

APAR (absorbed photosynthetically active radiation) may be calculated in a number of different ways, including (1) using mid-day data that approximately correspond to the timing of satellite overpass (Jenkins et al. 2007), and (2) using daily integrals, effectively averaging across different illumination geometries (Turner et al. 2003). However, both (1) and (2) are subject to the effects of seasonal variation in solar elevation (Doughty and Goulden 2008), which Richardson et al. (2012) proposed could be addressed by (3) using measurements when the solar zenith angle is closest to 57°—at this angle, all leaf inclination distribution functions converge. This latter method is preferred if continuous estimates of leaf area index are to be obtained from f_{APAR} using gap fraction theory (Richardson et al. 2012).

As with broadband NDVI, snow on upward-pointing sensors can be problematic during winter months. Airborne dust may also in some cases necessitate frequent cleaning of sensors. Spatial heterogeneity in the below-canopy light environment (e.g. sunflecks) can be substantial, and adequate sampling is necessary to reduce uncertainties (Fuentes et al. 2006; Garrity et al. 2011). Doughty and Goulden (2008) discuss other potential sources of uncertainty.

We advise against the use of inexpensive light sensors to measure Q_T. Research-grade quantum sensors have a spectral response that is nearly constant across the visible spectrum. Sensors with an unknown spectral response, or that are sensitive to both visible and near-infrared radiation, are not recommended, because of the impact of canopy development and senescence on the spectral distribution of radiation that penetrates to the understory.

To summarize, both broadband NDVI and f_{APAR} are easily measured using sensors that are commonly in use at many research sites. Highly specialized instruments are not needed. The resulting measurements are analogous (but not identical), to similar quantities measured from satellites. For this reason they may be more useful for evaluation of satellite phenology products than ground observations by a human observer. Both quantities also provide canopy-level information that is useful for modeling of CO_2 fluxes (e.g., Jenkins et al. 2007).

22.4 Narrow-Band Radiometric Sensors

Narrow-band radiometric sensors can also be used to measure vegetation indices such as NDVI (Baghzouz et al. 2010; Eklundh et al. 2011; Balzarolo et al. 2011). An advantage to working with narrow-band instruments is that the very same wavebands that are used in satellite remote sensing can be targeted, facilitating comparison between near-surface and satellite data. Alternatively, specific wavelengths may be chosen because of their physiological relevance (Inoue et al. 2008; Ryu et al. 2010). The photochemical reflectance index (PRI), which tracks xanthophyll cycle pigments through narrowband reflectance measurements at 531 and 570 nm, is one such example (Gamon et al. 1992, 1997). In this manner, it should be possible to separate the phenology of vegetation structure (development and senescence of leaves) from the phenology of vegetation function (seasonality of photosynthetic capacity or efficiency). Balzarolo et al. (2011) survey various types of narrow-band radiometric sensors that have been deployed across Europe.

Unfortunately, the cost of narrow-band spectral instruments, which are targeted at a smaller end-user market, tends to be higher than that of broadband instruments. Inexpensive alternatives have been proposed and prototyped, but these are not yet commercially available (Garrity et al. 2010; Ryu et al. 2010).

Various tower-mounted and tram-based spectrometers have been deployed, but these tend to be specialized, custom-built instruments, and research objectives usually extend beyond simple monitoring of phenology (Balzarolo et al. 2011). Gamon et al. (2006), Leuning et al. (2006), and Hilker et al. (2007) each describe the design of a specific system in greater detail.

22.5 Monitoring Phenology with Imaging Sensors

Standard digital cameras have been used for a range of environmental monitoring applications, including agriculture (e.g., Hague et al. 2006; Slaughter et al. 2008) and ecology (e.g., Luscier et al. 2006; Booth and Cox 2008). Graham et al. (2006) deployed a networked digital camera, dubbed the "MossCam", to record images of a moss-covered rock every 15 min. Quantitative image analysis demonstrated that changes in the physiological state of the moss, due to drying and re-wetting, could be detected with this approach. With a specific emphasis on phenology, Richardson et al. (2007) showed how a similar method could be used to monitor spring green-up in a temperate deciduous forest. The seasonal cycle of canopy "greenness" derived from camera imagery clearly reveals information about autumn coloration and senescence as well (e.g., Fig. 22.2). Numerous studies, in a wide variety of ecosystems (including deciduous and evergreen forests, grasslands, wetlands and semi-arid shrublands), have

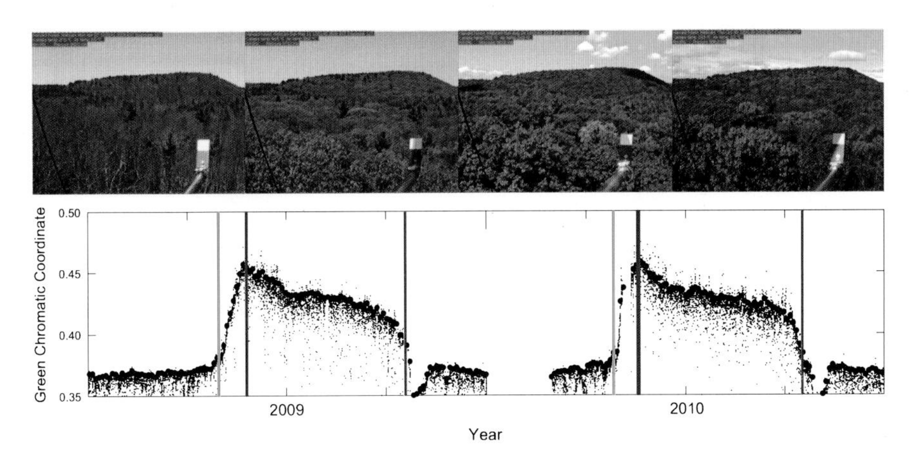

Fig. 22.2 View of the Harvard Forest canopy, from a camera mounted on the EMS AmeriFlux tower (*top*), through winter, early spring, summer, and autumn (L to R). The time series of canopy "greenness" (green chromatic coordinate, filtered (*large black circles*) to a 3-day product using the 90th quantile approach) derived from camera imagery, is compared with ground observations of 50 % budburst (*green line*), 50 % of leaves at 75% of final length (*blue*), and 50 % leaf color (*red*) dates (*bottom*)

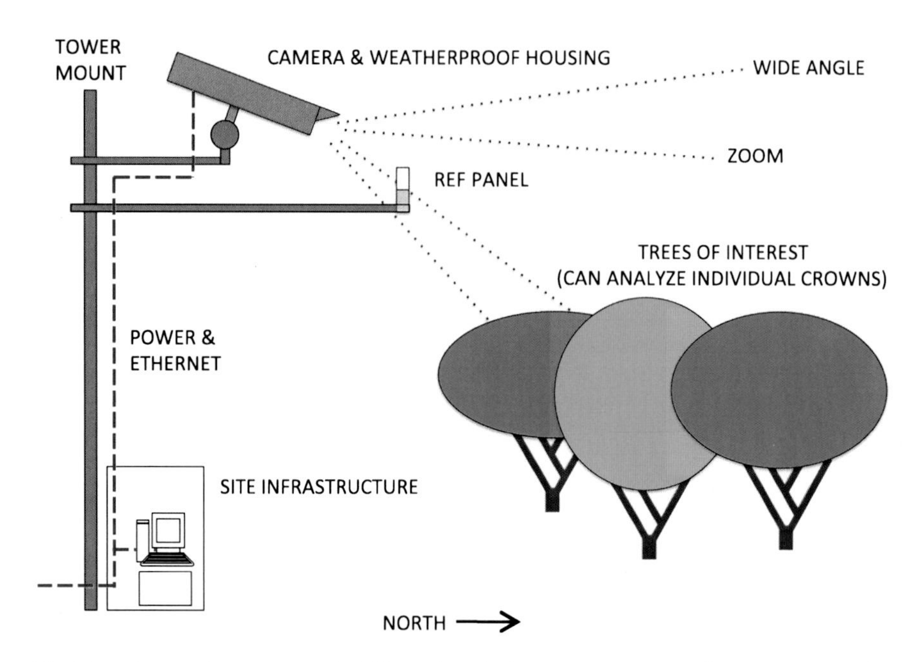

Fig. 22.3 Schematic showing typical PhenoCam deployment strategy at a forested research site. The camera is mounted on a tower, or other structure that is taller than the vegetation, so as to have a view out across the landscape. The camera is pointed north (in the northern hemisphere) to minimize lens flare, shadows, and forward scattering off the canopy. A reference panel, while optional, can be used to monitor long-term stability of the imaging sensor, and to provide information about day-to-day variation in lighting conditions (weather, aerosols, solar elevation and azimuth). Images can be stored locally on a computer, or sent to a remote server via the Internet

followed in recent years (e.g., Ahrends et al. 2008, 2009; Graham et al. 2009; Richardson et al. 2009; Ide and Oguma 2010; Kurc and Benton 2010; Bater et al. 2011a,b; Migliavacca et al. 2011; Nagai et al. 2011; Sonnentag et al. 2011, 2012; Elmore et al. 2012; Hufkens et al. 2012). Additionally, a "PhenoCam" network has been developed (http://phenocam.sr.unh.edu/).

The basic idea of the PhenoCam approach, which has also become known as "digital repeat photography" (Crimmins and Crimmins 2008; Sonnentag et al. 2012) is that with sequential images, taken from a fixed location and with a set field of view and viewing geometry (e.g., Fig. 22.3), changes in the characteristics of the vegetation can be easily detected either by subjective visual assessment (Ahrends et al. 2009; Ide and Oguma 2010), or by quantitative analysis (as shown in Figs. 22.2 and 22.4; Richardson et al. 2009). The archived image constitutes a permanent record of the state of the system at a specific point in time; this image can be inspected or re-analyzed as new questions arise, or as new analytical techniques are developed.

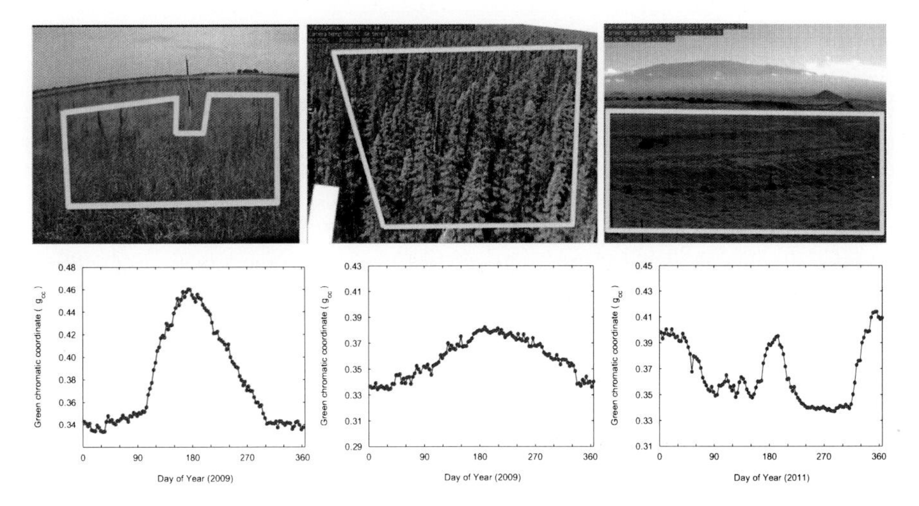

Fig. 22.4 Time series of canopy greenness derived from camera imagery of diverse ecosystems. *Left* to *right*: An agricultural field at the University of Illinois; an evergreen conifer forest near Chibougamau, Quebec; and a C3/C4 grassland near Kamuela, Hawaii. The *yellow polygons* indicate the "region of interest" analyzed within each image. Of particular note are the repeated cycles of greening and browning seen in the Hawaii time series; these are driven by periodic rain events, which are especially pronounced in June and November/December

22.5.1 Camera Selection

Most applications of digital repeat photography to phenology have used relatively inexpensive, commercially available, digital cameras (Sonnentag et al. 2012). Stand-alone networked webcams are widely used (e.g. Richardson et al. 2009) because image capture and off-site archiving can be accomplished without reliance on a local computer or data logger. For example, many webcams have a user interface with which the frequency of image capture can be controlled, and images can be automatically uploaded on a schedule to a remote server over the Internet, via FTP. Non-networked cameras, which typically store images on removable flash memory cards, can also be used. Options include, for example, inexpensive and weatherproof time-lapse cameras targeted at hobbyists, and third-party firmware upgrades that can be used to add time-lapse capabilities to some consumer-grade point-and-shoot cameras (Sonnentag et al. 2012). Hybrid solutions, whereby a high-quality digital camera is connected to and controlled by a local computer, may also be applicable in some cases (Ahrends et al. 2008; Crimmins and Crimmins 2008). While Internet connectivity is not essential, it does allow images to be viewed, or made available on a web page, in near-real time. This has obvious advantages when it comes to monitoring the functionality of the system, and ensuring continuous, gap-free data. Indeed, reliability is of paramount importance, but is non-trivial, as lightning, rodents, and weather extremes can be constant challenges.

There are many ways in which various brands and models of cameras are different. These include, for example, the size and resolution of the imaging sensor, and the optical quality of the lens. Ultimately, however, although images of greater sharpness and detail may be visually appealing, for basic monitoring purposes Sonnentag et al. (2012) concluded, "camera choice might be of secondary importance." And while RAW-format files, which correspond most directly to the radiometric properties of the scene being imaged, are ideally preferred over processed (for white balance, color saturation, contrast, and sharpness) and compressed image formats such as JPEG (Verhoeven 2010), Sonnentag et al. (2012) found little phenological information was lost even when moderately high JPEG compression was used. For some applications, however, there may be instances where ultra-high resolution RAW imagery is desirable.

22.5.2 Field Installation

Careful deployment of the camera (Fig. 22.3) is essential in order to obtain good-quality data. For example, while cameras should obviously be aimed at the vegetation that is of interest, consideration should also be given to minimizing lens flare, forward scattering off the vegetation, and deep shadows. With the exception of automatic exposure, it is recommended that all automatic camera settings (particularly color balance or white balance) be turned off. Some sky in the image may be desirable as it can be used to identify what the weather was like on a particular day. However, if too much sky is included, the vegetation signal may become very noisy as the camera attempts to compensate for dynamic variation in sky brightness caused by changes in weather and clouds (note that with some cameras it is possible to specify the region of the image that will be used to set the exposure). Finally, while images need not be archived every minute, there do seem to be substantial advantages to having relatively frequent imagery (every 30 min), rather than just one image per day (Sonnentag et al. 2012). Unless RAW format files are being archived, bandwidth and storage requirements are relatively minor given current networking and storage capabilities.

22.5.3 Image Processing

Consumer-grade digital cameras record color information in three separate layers: red, green, and blue. According to the standard additive color model, representation of any given color is achieved by varying the intensity of these three primary colors. Each pixel in the image has associated with it a digital number ("DN") triplet, with each element in the triplet corresponding to the intensity of one of the color layers. In 24 bit images, there are 8 bits per channel; black is thus represented as (0,0,0) and white as (255,255,255). Quantitative image analysis consists of extracting the DN triplets for individual pixels, and then averaging the DN triplets across multiple pixels within a user-defined region of interest (corresponding to, for example, an individual tree crown, or all trees in the foreground of the image). This analysis can be conducted in a variety of software packages including R and MATLAB, and a freeware toolkit has also been developed (http://phenocam.sr.unh.edu/webcam/tools/).

Time series of RGB DN triplets are noisy and are of little use for phenological analyses, because both external factors affecting scene illumination (clouds, aerosols, solar elevation/azimuth) and internal processing (including exposure control and color balance adjustment) confound the underlying seasonal signal associated with foliage development and senescence. This variability can be largely suppressed by converting DN triplets (R_{DN}, G_{DN}, B_{DN}) to their respective chromatic coordinates (r_{CC}, g_{CC}, b_{CC}) (Richardson et al. 2007; Sonnentag et al. 2012). The seasonal trajectories of g_{CC} for a number of different vegetation types are shown in Fig. 22.4. While many studies have used g_{CC} to study canopy development and senescence, Richardson et al. (2009), Nagai et al. (2011), and Sonnentag et al. (2012) have noted that the timing, intensity and duration of autumn colors (in deciduous species whose leaves turn yellow, orange and red prior to abscission) can also be quantified using r_{CC}.

$$r_{CC} = \frac{R_{DN}}{R_{DN} + G_{DN} + B_{DN}}; g_{CC} = \frac{G_{DN}}{R_{DN} + G_{DN} + B_{DN}}; b_{CC} = \frac{B_{DN}}{R_{DN} + G_{DN} + B_{DN}}$$

The "excess green" index (G_{EX}), originally developed for agricultural applications by Woebbecke et al. (1995), has also been widely used to minimize effects of changing illumination in phenological studies, and in some cases may accomplish this more effectively than g_{CC}:

$$G_{EX} = 2G_{DN} - (R_{DN} + B_{DN})$$

Other indices can be constructed by various nonlinear transformations of RGB triplets. Sakamoto et al. (2012) used a two-band red-green difference index called VARI (Visible Atmospherically Resistant Index), while Sonnentag et al. (2011) used two-band ratio indices, grR (green-red ratio) and rbR (red-blue ratio). Note that it does not matter whether these ratio indices are calculated with digital numbers or chromatic coordinates, as the normalization factor in the denominator of the chromatic coordinates will cancel out.

$$\mathrm{VARI} = \frac{G_{DN} - R_{DN}}{R_{DN} + G_{DN}}; \mathrm{grR} = \frac{G_{DN}}{R_{DN}}; \mathrm{rbR} = \frac{R_{DN}}{B_{DN}}$$

Sakamoto et al. (2012) suggested working with exposure-adjusted DN values, which take aperture (f-stop), sensitivity (ISO), and exposure time (shutter speed) into account. This is only possible if such data are recorded in the image metadata. However, results suggest that this approach resulted in good agreement with indices calculated from surface reflectance measured using radiometric instruments. It is also possible to convert RGB triplets to other color spaces, and this approach has been used successfully in some phenological analyses (e.g., HSV [hue, saturation and value] and HSL [hue, saturation and lightness]: Crimmins and Crimmins 2008; Graham et al. 2009; Mizunuma et al. 2011).

While a full discussion of color theory is beyond the scope of this chapter, it is important to keep in mind that the RGB triplets extracted from an image represent a simplified description of the actual spectral characteristics of the surface. And, while the indices and transformations described above may help facilitate interpretation of the data, some information is inadvertently lost when two or more channels are reduced to a single index value. Finally, redundancy is unavoidable if too many indices are calculated from the underlying RGB triplets.

Even once RGB DN triplets have been converted to the above indices, substantial variability may remain, both over the course of individual days and due to day-to-day differences in weather (see Fig. 22.2). Various methods have been proposed to extract the best quality phenological signal from these data. These have included absolute DN thresholds (Richardson et al. 2009), weather filtering indices (Ide and Oguma 2010), and recursive outlier removal algorithms (Migliavacca et al. 2011). Most recently, Sonnentag et al. (2012) noted that sub-optimal conditions (clouds, precipitation, low light) tended to reduce the observed vegetation greenness (calculated by either g_{CC} or G_{EX}). A moving-window quantile approach was proposed, whereby index values from all images collected over a 3-day period were binned together, and the 90th quantile greenness value across these was assigned to the middle day. Results from this method were superior to other approaches, including previously recommended methods such as mid-day averages.

22.5.4 Calibration and Long-Term Stability

Calibration and long-term stability of the imaging sensor are issues that have been acknowledged but not yet fully addressed or resolved (Richardson et al. 2009; Ide and Oguma et al. 2010; Migliavacca et al. 2011; Sonnentag et al. 2012). Ide and Oguma (2010) used changes in the apparent color of snow and sky to evaluate sensor degradation over 6 years of deployment in the field; not only was the sensitivity of different channels shown to change over time, but also the drift appeared to be camera-specific. A reference panel has been proposed in some studies (Richardson et al. 2009), but because the color of the panel itself may fade over time, this appears to be more useful for evaluating the day-to-day variability in illumination conditions, rather than long-term stability of the imaging sensor. Finally, as noted by Sonnentag et al. (2012), it is difficult to make absolute comparisons of index values from one camera with those from another camera. Development of an inexpensive reference standard that could easily be deployed across sites would be of enormous value. Such a standard would facilitate multi-site comparisons, and would greatly enhance the value of long-term studies from individual sites. Calibration and standard development are areas where further research is obviously needed.

22.5.5 Other Possibilities with Imaging Sensors

Imaging sensors have been used in a number of other configurations for phenological monitoring. For example, Nagai et al. (2010, 2011) used a conventional digital camera with a fisheye (180°) lens to record nadir views of a deciduous forest canopy. This is the standard arrangement used within the Japanese Phenological Eyes Network (PEN; http://pen.agbi.tsukuba.ac.jp/). Proof-of-concept work has also been conducted using gigapixel camera systems, in which a camera mounted on a pan-tilt head is used to capture hundreds of individual images which are then stitched together into a single panorama (e.g., Brown et al. 2012).

Both CCD (charge-coupled device) and CMOS (complementary metal–oxide–semiconductor) sensors used in conventional cameras are sensitive to infrared radiation, specifically 700–1,000 nm. An "IR cut" filter over the sensor normally blocks these wavelengths, but this filter can be removed altogether, resulting in composite RGB + IR images. With some cameras that are targeted at the security or surveillance market (e.g., the StarDot Netcam SC IR used in the PhenoCam network), the IR cut filter can be triggered by the camera's software, enabling back-to-back RGB and RGB + IR images to be recorded (Richardson et al. *unpublished*). It should be possible to leverage this functionality to compute NDVI-style indices, which could prove useful for monitoring of vegetation stress over the course of the growing season. Alternatively, the IR cut filter can be removed and replaced with a bandpass filter that targets specific wavelengths of interest. Shibayama et al. (2009, 2011) used a two-camera system (one fitted with a red, and the other a near-infrared, bandpass filter) to monitor leaf area development of a rice paddy. Using a slightly different approach, but still making use of a camera where the IR cut filter was replaced with a near-infrared bandpass filter, Sakamoto et al. (2012) obtained infrared images of maize and soybean crops. During the day, the sun provided natural illumination, whereas at night the canopy was illuminated using the camera's built-in flash. A very clear phenological signal, correlated with total aboveground biomass, was apparent in the nighttime relative brightness index ($NBRI_{NIR}$) that was derived from these data.

Finally, there are commercially available multichannel cameras, such as Tetracam's ADC (Agricultural Digital Camera) and MCA (Multi-Camera Array) systems. The ADC records information in three wavebands: red, green and near-infrared. Steltzer and Welker (2006) measured plot level-NDVI in the Arctic at the peak of the growing season with ADC imagery, while Higgins et al. (2011) monitored seasonal changes in NDVI of African savanna with ADC images taken from a helicopter. The MCA is a more sophisticated device, consisting of six separate monochromatic imaging sensors, each of which is fitted with its own bandpass filter. However, there do not appear to be published studies where either of these instruments has been used for automated and continuous phenological monitoring.

22.6 Future Prospects

We can easily envision new technological developments that would benefit a variety of near-surface remote sensing applications, including phenological monitoring. Multichannel imaging sensors (targeting 4, 16, 64 or more narrow wavebands, rather than just the red, green and blue channels of the standard digital camera) would permit the use of more sophisticated spectral indices for tracking phenology, and could also yield real-time information about physiological processes such as photosynthesis. Higher resolution imaging sensors (operating at 1, 10, or even 100 gigapixels), with which it might be possible to visually inspect individual buds or leaves from a distance of tens of meters, would permit more direct linking of observer-based phenology with camera-based approaches. Deployment in remote locations would be facilitated by the development of improved long-distance, high-bandwidth wireless networking, and stand-alone low-power sensors that can operate for months or years without maintenance or recharging.

In reality, much of the technology on this "wish list" is already available—but not in a package that is commercially available, or at a price that is affordable. For the time being, the high cost of cutting-edge technology remains a barrier to the adoption of more sophisticated near-surface remote sensing approaches by the phenological community.

These limitations aside, even relatively simple and inexpensive sensors, as discussed here, can provide high frequency, quantitative data on vegetation phenology at a unique spatial scale that bridges the individual organism scale of traditional ground observations, and the landscape scale of satellite remote sensing.

22.7 Conclusions

Sensor-based approaches to phenological monitoring have been widely adopted over the last decade, and long-term (5–10 years) data sets from many monitoring sites are now available. Both radiometric sensors and imaging sensors can yield high quality data at a temporal frequency that is adequate to resolve phenological transitions with good precision. These data will undoubtedly become increasingly valuable as individual time series become longer in length. While we caution that the lack of a common protocol across sites may hinder synthesis activities, a number of studies have already shown there is great potential for regional-to-continental scale phenological monitoring using data from the thousands of cameras that post imagery to publicly-available Internet web pages in near-real time (Jacobs et al. 2009; Graham et al. 2010).

Acknowledgments We thank Oliver Sonnentag and Youngryel Ryu for assistance with processing the data used in Fig. 22.1, and Koen Hufkens for providing the code used to generate the time series shown in Figs. 22.2 and 22.4. A.D.R. acknowledges support from the National Science Foundation, through the Macrosystems Biology program, award EF-1065029;

the Northeastern States Research Cooperative; and the US Geological Survey Status and Trends Program, the US National Park Service Inventory and Monitoring Program, and the USA National Phenology Network through grant number G10AP00129 from the United States Geological Survey. Any opinions, findings, and conclusions or recommendations expressed in this material are those of the authors and do not necessarily reflect the views of the National Science Foundation or USGS.

References

Ahrends HE, Brugger R, Stockli R, Schenk J, Michna P, Jeanneret F, Wanner H, Eugster W (2008) Quantitative phenological observations of a mixed beech forest in northern Switzerland with digital photography. J Geophys Res-Biogeosci 113:G04004

Ahrends HE, Etzold S, Kutsch WL, Stoeckli R, Bruegger R, Jeanneret F, Wanner H, Buchmann N, Eugster W (2009) Tree phenology and carbon dioxide fluxes: use of digital photography at for process-based interpretation the ecosystem scale. Clim Res 39:261–274

Baghzouz M, Devitt DA, Fenstermaker LF, Young MH (2010) Monitoring vegetation phenological cycles in two different semi-arid environmental settings using a ground-based NDVI system: a potential approach to improve satellite data interpretation. Remote Sens 2:990–1013

Balzarolo M, Anderson K, Nichol C, Rossini M, Vescovo L, Arriga N, Wohlfahrt G, Calvet JC, Carrara A, Cerasoli S, Cogliati S, Daumard F, Eklundh L, Elbers JA, Evrendilek F, Handcock RN, Kaduk J, Klumpp K, Longdoz B, Matteucci G, Meroni M, Montagnani L, Ourcival JM, Sanchez-Canete EP, Pontailler JY, Juszczak R, Scholes B, Martin MP (2011) Ground-based optical measurements at European flux sites: a review of methods, instruments and current controversies. Sensors 11:7954–7981

Bater CW, Coops NC, Wulder MA, Hilker T, Nielsen SE, McDermid G, Stenhouse GB (2011a) Using digital time-lapse cameras to monitor species-specific understorey and overstorey phenology in support of wildlife habitat assessment. Environ Monit Assess 180:1–13

Bater CW, Coops NC, Wulder MA, Nielsen SE, McDermid G, Stenhouse GB (2011b) Design and installation of a camera network across an elevation gradient for habitat assessment. Instrum Sci Technol 39:231–247

Booth DT, Cox SE (2008) Image-based monitoring to measure ecological change in rangeland. Front Ecol Environ 6:185–190

Brown TB, Zimmermann C, Panneton W, Noah N, Borevitz J (2012) High-resolution, time-lapse imaging for ecosystem-scale phenotyping in the field. In: Normanly J (ed) Methods in molecular biology. Springer, New York, pp 71–96

Crimmins MA, Crimmins TM (2008) Monitoring plant phenology using digital repeat photography. Environ Manag 41:949–958

Doughty CE, Goulden ML (2008) Seasonal patterns of tropical forest leaf area index and CO2 exchange. J Geophys Res-Biogeosci 113:G00B06

Eklundh L, Jin HX, Schubert P, Guzinski R, Heliasz M (2011) An optical sensor network for vegetation phenology monitoring and satellite data calibration. Sensors 11:7678–7709

Elmore AJ, Guinn SM, Minsley BJ, Richardson AD (2012) Landscape controls on the timing of spring, autumn, and growing season length in mid-Atlantic forests. Glob Change Biol 18:656–674

Fuentes DA, Gamon JA, Cheng YF, Claudio HC, Qiu HL, Mao ZY, Sims DA, Rahman AF, Oechel W, Luo HY (2006) Mapping carbon and water vapor fluxes in a chaparral ecosystem using vegetation indices derived from AVIRIS. Remote Sens Environ 103:312–323

Gamon JA, Penuelas J, Field CB (1992) A narrow-waveband spectral index that tracks diurnal changes in photosynthetic efficiency. Remote Sens Environ 41:35–44

Gamon JA, Serrano L, Surfus JS (1997) The photochemical reflectance index: an optical indicator of photosynthetic radiation use efficiency across species, functional types, and nutrient levels. Oecologia 112:492–501

Gamon JA, Cheng YF, Claudio H, MacKinney L, Sims DA (2006) A mobile tram system for systematic sampling of ecosystem optical properties. Remote Sens Environ 103:246–254

Garrity SR, Vierling LA, Bickford K (2010) A simple filtered photodiode instrument for continuous measurement of narrowband NDVI and PRI over vegetated canopies. Agr For Meteorol 150:489–496

Garrity SR, Bohrer G, Maurer KD, Mueller KL, Vogel CS, Curtis PS (2011) A comparison of multiple phenology data sources for estimating seasonal transitions in deciduous forest carbon exchange. Agr For Meteorol 151:1741–1752

Graham EA, Hamilton MP, Mishler BD, Rundel PW, Hansen MH (2006) Use of a networked digital camera to estimate net CO2 uptake of a desiccation-tolerant moss. Int J Plant Sci 167:751–758

Graham EA, Yuen EM, Robertson GF, Kaiser WJ, Hamilton MP, Rundel PW (2009) Budburst and leaf area expansion measured with a novel mobile camera system and simple color thresholding. Environ Exp Bot 65:238–244

Graham EA, Riordan EC, Yuen EM, Estrin D, Rundel PW (2010) Public Internet-connected cameras used as a cross-continental ground-based plant phenology monitoring system. Glob Change Biol 16:3014–3023

Hague T, Tillett ND, Wheeler H (2006) Automated crop and weed monitoring in widely spaced cereals. Precis Agric 7:21–32

Higgins SI, Delgado-Cartay MD, February EC, Combrink HJ (2011) Is there a temporal niche separation in the leaf phenology of savanna trees and grasses? J Biogeogr 38: 2165–2175

Hilker T, Coops NC, Nesic Z, Wulder MA, Black AT (2007) Instrumentation and approach for unattended year round tower based measurements of spectral reflectance. Comput Electron Agric 56:72–84

Hilker T, Gitelson A, Coops NC, Hall FG, Black TA (2011) Tracking plant physiological properties from multi-angular tower-based remote sensing. Oecologia 165:865–876

Huemmrich KF, Black TA, Jarvis PG, McCaughey JH, Hall FG (1999) High temporal resolution NDVI phenology from micrometeorological radiation sensors. J Geophys Res-Atmos 104:27935–27944

Hufkens K, Friedl M, Sonnentag O, Braswell BH, Milliman T, Richardson AD (2012) Linking near-surface and satellite remote sensing measurements of deciduous broadleaf forest phenology. Remote Sens Environ 117:307–321

Ide R, Oguma H (2010) Use of digital cameras for phenological observations. Ecol Inform 5:339–347

Inoue Y, Penuelas J, Miyata A, Mano M (2008) Normalized difference spectral indices for estimating photosynthetic efficiency and capacity at a canopy scale derived from hyperspectral and CO2 flux measurements in rice. Rem Sens Environ 112:156–172

Jacobs N, Burgin W, Fridrich N, Abrams A, Miskell K, Braswell BH, Richardson AD, Pless R (2009) The global network of outdoor webcams: properties and applications. In: Proceedings ACM GIS '09, November 4–6, 2009 Seattle, WA, pp 111–120

Jenkins JP, Richardson AD, Braswell BH, Ollinger SV, Hollinger DY, Smith ML (2007) Refining light-use efficiency calculations for a deciduous forest canopy using simultaneous tower-based carbon flux and radiometric measurements. Agr For Meteorol 143:64–79

Kurc SA, Benton LM (2010) Digital image-derived greenness links deep soil moisture to carbon uptake in a creosotebush-dominated shrubland. J Arid Environ 74:585–594

Leuning R, Hughes D, Daniel P, Coops NC, Newnham G (2006) A multi-angle spectrometer for automatic measurement of plant canopy reflectance spectra. Remote Sens Environ 103:236–245

Luscier JD, Thompson WL, Wilson JM, Gorham BE, Dragut LD (2006) Using digital photographs and object-based image analysis to estimate percent ground cover in vegetation plots. Front Ecol Environ 4:408–413

Migliavacca M, Galvagno M, Cremonese E, Rossini M, Meroni M, Sonnentag O, Cogliati S, Manca G, Diotri F, Busetto L, Cescatti A, Colombo R, Fava F, di Celia UM, Pari E, Siniscalco C, Richardson AD (2011) Using digital repeat photography and eddy covariance data to model grassland phenology and photosynthetic CO2 uptake. Agr For Meteorol 151:1325–1337

Mizunuma T, Koyanagi T, Mencuccini M, Nasahara KN, Wingate L, Grace J (2011) The comparison of several colour indices for the photographic recording of canopy phenology of Fagus crenata Blume in eastern Japan. Plant Ecol Divers 4:67–77

Nagai S, Nasahara KN, Muraoka H, Akiyama T, Tsuchida S (2010) Field experiments to test the use of the normalized-difference vegetation index for phenology detection. Agr For Meteorol 150:152–160

Nagai S, Maeda T, Gamo M, Muraoka H, Suzuki R, Nasahara KN (2011) Using digital camera images to detect canopy condition of deciduous broad-leaved trees. Plant Ecol Divers 4:79–89

Nagy Z, Pinter K, Czobel S, Balogh J, Horvath L, Foti S, Barcza Z, Weidinger T, Csintalan Z, Dinh NQ, Grosz B, Tuba Z (2007) The carbon budget of semi-arid grassland in a wet and a dry year in Hungary. Agr Ecosyst Environ 121:21–29

Richardson AD, Jenkins JP, Braswell BH, Hollinger DY, Ollinger SV, Smith ML (2007) Use of digital webcam images to track spring green-up in a deciduous broadleaf forest. Oecologia 152:323–334

Richardson AD, Braswell BH, Hollinger DY, Jenkins JP, Ollinger SV (2009) Near-surface remote sensing of spatial and temporal variation in canopy phenology. Ecol Appl 19:1417–1428

Richardson AD, Anderson RS, Arain MA, Barr AG, Bohrer G, Chen G, Chen JM, Ciais P, Davis KJ, Desai AR, Dietze MC, Dragoni D, Garrity SR, Gough CM, Grant R, Hollinger DY, Margolis HA, McCaughey H, Migliavacca M, Monson RK, Munger JW, Poulter B, Raczka BM, Ricciuto DM, Sahoo AK, Schaefer K, Tian H, Vargas R, Verbeeck H, Xiao J, Xue Y (2012) Terrestrial biosphere models need better representation of vegetation phenology: results from the North American Carbon Program site synthesis. Glob Change Biol 18:566–584

Rocha AV, Shaver GR (2009) Advantages of a two band EVI calculated from solar and photosynthetically active radiation fluxes. Agr For Meteorol 149:1560–1563

Ryu Y, Baldocchi DD, Verfaillie J, Ma S, Falk M, Ruiz-Mercado I, Hehn T, Sonnentag O (2010) Testing the performance of a novel spectral reflectance sensor, built with light emitting diodes (LEDs), to monitor ecosystem metabolism, structure and function. Agr For Meteorol 150:1597–1606

Sakamoto T, Gitelson AA, Nguy-Robertson AL, Arkebauer TJ, Wardlow BD, Suyker AE, Verma SB, Shibayama M (2012) An alternative method using digital cameras for continuous monitoring of crop status. Agr For Meteorol 154–155:113–126

Shibayama M, Sakamoto T, Takada E, Inoue A, Morita K, Takahashi W, Kimura A (2009) Continuous monitoring of visible and near-infrared band reflectance from a rice paddy for determining nitrogen uptake using digital cameras. Plant Prod Sci 12:293–306

Shibayama M, Sakamoto T, Takada E, Inoue A, Morita K, Takahashi W, Kimura A (2011) Estimating paddy rice leaf area index with fixed point continuous observation of near infrared reflectance using a calibrated digital camera. Plant Prod Sci 14:30–46

Slaughter DC, Giles DK, Downey D (2008) Autonomous robotic weed control systems: a review. Comput Electron Agric 61:63–78

Sonnentag O, Detto M, Vargas R, Ryu Y, Runkle BRK, Kelly M, Baldocchi DD (2011) Tracking the structural and functional development of a perennial pepperweed (*Lepidium latifolium* L.) infestation using a multi-year archive of webcam imagery and eddy covariance measurements. Agr For Meteorol 151:916–926

Sonnentag O, Hufkens K, Teshera-Sterne C, Young AM, Friedl M, Braswell BH, Milliman T, O'Keefe J, Richardson AD (2012) Digital repeat photography for phenological research in forest ecosystems. Agr For Meteorol 152:159–177

Sparks TH, Menzel A (2002) Observed changes in seasons: an overview. Int J Climatol 22:1715–1725
Steltzer H, Welker JM (2006) Modeling the effect of photosynthetic vegetation properties on the NDVI-LAI relationship. Ecology 87:2765–2772
Tittebrand A, Spank U, Bernhofer C (2009) Comparison of satellite- and ground-based NDVI above different land-use types. Theor Appl Climatol 98:171–186
Turner DP, Urbanski S, Bremer D, Wofsy SC, Meyers T, Gower ST, Gregory M (2003) A cross-biome comparison of daily light use efficiency for gross primary production. Glob Change Biol 9:383–395
Ustin SL, Roberts DA, Gamon JA, Asner GP, Green RO (2004) Using imaging spectroscopy to study ecosystem processes and properties. Biosci 54:523–534
Verhoeven GJJ (2010) It's all about the format – unleashing the power of RAW aerial photography. Int J Remote Sens 31:2009–2042
Wang Q, Tenhunen J, Dinh NQ, Reichstein M, Vesala T, Keronen P (2004) Similarities in ground- and satellite-based NDVI time series and their relationship to physiological activity of a Scots pine forest in Finland. Rem Sens Environ 93:225–237
White MA, de Beurs KM, Didan K, Inouye DW, Richardson AD, Jensen OP, O'Keefe J, Zhang G, Nemani RR, van Leeuwen WJD, Brown JF, de Wit A, Schaepman M, Lin XM, Dettinger M, Bailey AS, Kimball J, Schwartz MD, Baldocchi DD, Lee JT, Lauenroth WK (2009) Intercomparison, interpretation, and assessment of spring phenology in North America estimated from remote sensing for 1982–2006. Glob Change Biol 15:2335–2359
Wilson TB, Meyers TP (2007) Determining vegetation indices from solar and photosynthetically active radiation fluxes. Agr For Meteorol 144:160–179
Woebbecke DM, Meyer GE, Vonbargen K, Mortensen DA (1995) Color indexes for weed identification under various soil, residue, and lighting conditions. Trans ASAE 38:259–269
Zhang XY, Friedl MA, Schaaf CB (2006) Global vegetation phenology from moderate resolution imaging spectroradiometer (MODIS): evaluation of global patterns and comparison with in situ measurements. J Geophys Res-Biogeosci 111:G04017

Part V
Phenologies of Selected Lifeforms

Chapter 23
Aquatic Plants and Animals

Wulf Greve

Abstract The topic of this chapter is concerned with the greatest volume of the earth's ecosystems. The separate sub-ecosystems and their organisms are introduced and exemplified. The fresh water systems are distinguished from the salt water systems and compared with respect to the affiliated organisms. The subsystems rivers, lakes, estuaries and oceans are housing a multitude of annually repetitive organic processes based on geophysics such as the annual temperature change, the freezing and thawing of lakes and water flow from catchment areas determining the onset of the annual succession, the layering of deeper waters temporarily separating biota linked by the daily vertical migration of zooplankton, the regional migration within the sea and between rivers and shelf seas and the timing of the reproductive season within the biota, in which the short lived plankton displays a multitude of populations succeeding each other all year. Among marine zooplankton the phenology of the start of season (SOS), middle of season (MOS), the end of season (EOS) and the resulting length of season (LOS) permits observations in phenology and seasonality with respect to seasonal preceding temperatures and long term shifts resulting from global warming. Phenology is a useful means for functional definition, determination and prediction of annual and long term seasonality.

23.1 The Hydrosphere

Oceans cover 70.8 % of the earth to a mean depth of 3,729 m and thereby provide a volume of 1,350 million km^3 of inhabitable biosphere (Gerlach 1994). Childress (1983) calculated that this volume forms 99.5 % of the earth's biosphere (Fig. 23.1).

W. Greve (✉)
German Center for Marine Biodiversity Research, Senckenberg Research Institute, Hamburg, Germany
e-mail: wgreve@senckenberg.de

M.D. Schwartz (ed.), *Phenology: An Integrative Environmental Science*,
DOI 10.1007/978-94-007-6925-0_23, © Springer Science+Business Media B.V. 2013

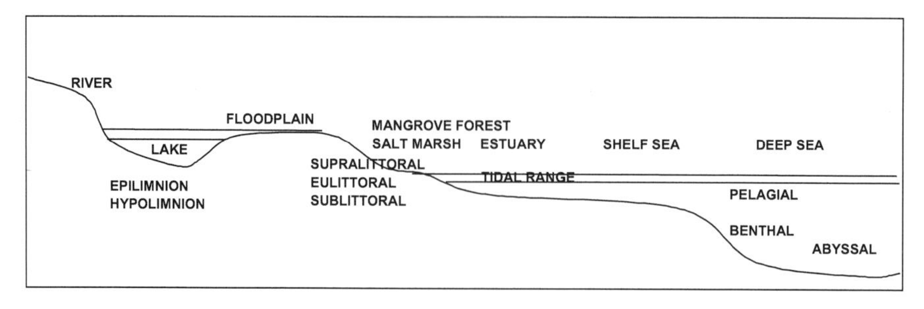

Fig. 23.1 Schematic of the hydrosphere, indicating areas addressed in this chapter

This large volume of the biosphere contributes only about 40 % of global primary production, however, as productivity is limited by light and by nutrients. Light penetrates just the upper 100 m of the water, and the majority of ocean zones lack nutrients, except in regions with upwelling, currents from less productive areas, or run-off from the coasts. These processes depend on wind speed, wind direction, and the rainfall pattern in the catchment areas of rivers (which in turn vary seasonally). Light, wind, temperature, and production conditions are more constant in the depths of lakes and oceans.

The deep-water basins are often linked to surface biology by decaying organic sediments (Lampitt 1985). These nutrients reach the deep-sea benthos in days to weeks, transmitting a seasonal production signal to the abyssal organisms. In some lake and sea basin sediments, such materials contain records of millennia of upper-layer biotic processes, as documented in micropaleontology (Schmiedl et al. 2003) and sedimentology (Alheit and Hagen 1997) literature.

23.2 Organisms of the Hydrosphere

Life first developed in the sea. Organisms of the hydrosphere live permanently within water, or temporarily depend on it for nutrition or propagation. Almost all phyla of the plant and animal kingdom live in the world oceans, while several taxonomic units are not represented in fresh-water or terrestrial ecosystems (Table 23.1).

The marine taxonomic variation spans from diatoms and dinoflagellates, supporting the bulk of primary production in the world oceans, to kelp, sea-grass, and mangroves, adapted to saline water. Many other organisms depending on, but living partially outside the sea (like sea birds and the inhabitants of the salt marshes and mangrove forests), are generally treated as marine organisms. Limnetic systems like the freshwater tidal zone and other areas that occasionally or seasonally are flooded, also become part-time members of the hydrosphere. The marine littoral zone is influenced by the lunar driven tides with ranges up to 15 m, exposing organisms to constant changes in saline water, sunshine, rain, and temperature.

The following are some examples of biota where an intensive interaction of hydrological and terrestrial organisms takes place. The upper surface of oceans and lakes is inhabited by *Pleuston* exposed to the air, next comes the *Neuston*,

Table 23.1 Distribution of plant and animal phyla in the aquatic realm (marine and fresh water, or limnetic) with indication of frequent (■) or seldom (+) occurrence

Taxonomic units	Marine	Fresh	Taxonomic units	Marine	Fresh
Algae			Lamellibranchia	■	■
Cyanophyceae	■	■	Cephalopoda	■	
Rhodophyceae	■	+	Articulata		
Chrysophyceae	+	■	Annelida	■	■
Xanthophyceae	+	■	Pentastomida	■	■
Bacillariophyceae	■	■	Tardigrada	■	■
Phaeophyceae	■	+	Arthropoda		
Coccolithophorida	■	+	Chelicerata		
Dinophyceae	■	■	Merostomata	■	
Euglenophyceae	+	■	Arachnida	+	■
Chlorophyceae	■	■	Pantopoda	■	
Prasinophyceae	■	+	Mandibulata		
Charophyceae		■	Crustacea	■	■
Angiosperma	■	■	Antennata		
Gymnosperma	+	■	Insecta	+	■
Protozoa	■	■	Hemichordata	■	
Porifera	■	+	Echinodermata	■	
Cnidaria	■	+	Pogonophora	■	
Ctenophora	■		Chordata		
Tentaculata	■		Tunicata	■	
Scolecida			Appendicularia	■	
Plathelmintes	■	■	Acrania	■	
Nemertini	■	+	Vertebrata		
Nemathelmintes	■	■	Pisces	■	■
Kamptozoa	■	+	Amphibia		■
Mollusca			Reptilia	■	■
Gastropoda	■	■	Aves	■	■
			Mammalia	■	■

organisms settling or temporarily living below the water surface. The *Pelagos* are organisms of the water, consisting of *Phytoplankton* and *Zooplankton*, drifting organisms, and *Nekton*, actively swimming organisms. The bottom of lakes and seas is inhabited by *Benthos*, the organisms living on or within the sediment (*Phytobenthos*, plants, and *Zoobenthos*, animals, are distinct sub-groups). Additional details are available in the literature (e.g., Parsons et al. 1984).

23.3 Physical Forcing of the Hydrosphere

23.3.1 Light

Solar radiation is the prime light source for any organism. Solar radiation varies in energy intensity, wavelength, and irradiation, at hourly, daily, and seasonal time-scales, and also spatially, depending on earth-sun geometry, cloud cover, turbidity and water depth. These light conditions directly or in-directly cause physiological responses. Lunar light may also render a physiologically relevant signal for reproductive synchronization (Franke 1990).

23.3.2 *Temperature*

Water is a transparent and fluid substance, with a great solubility for ions. It can be transported through the atmosphere in a gaseous state from which it is released as rain or snow under low temperatures. Snow may aggregate to layers of ice in glaciers. Since water at 4 °C has the greatest density/weight, abyssal temperatures are always close to this value. Thermal stratification of water bodies extends from warmer surface waters to colder bottom waters, forming a thermocline from the mixed surface waters to the deeper layers. As water below 4 °C gets lighter again, freezing begins at the water surface and surface temperatures around 4 °C favor the mixing of the water column. The temperature range of the world oceans is narrower than that of terrestrial systems (−2 to 43 °C versus −68 to 65 °C) (Kinne 1963). The annual as well as the daily variation is greatest in the temperate zone, decreasing equatorward and poleward. In limnetic biota, this general order is also affected by altitude as environmental temperature decreases with altitude. Temperature changes with weather, climate, advection and turbulent diffusion. The heat content of a water body is of decisive importance for the organisms living within it (Pohlmann 1996). Changes in seasonal temperatures may be especially useful for understanding the physiological responses in any organism (Orton 1920) and its phenology.

23.3.3 *Moon*

According to regional topography, the depth of ocean water is altered by the gravity of the moon and, to a lesser extent, the sun. If both forces supplement each other at "full moon" and at "new moon" twice monthly "spring-tides" with higher high tides and lower low tides occur. The tidal cycle includes one high tide and one low tide and has a length of ~12.25 h. The tidal range varies from a few centimeters in enclosed seas or the open ocean, to as much as ten or more meters in some shelf seas. Besides being the driving force of the tides, the moon also provides especially bright and especially dark nights at "full moon" and at "new moon." These events also provide cues to animal behaviors that utilize the diversity of spring tides (Caspers 1951; Neumann 1967).

23.3.4 Currents

The motion of water bodies due to: (1) changes in altitude in rivers; (2) varying precipitation or thawing of ice and snow; (3) wind-driven or tidal currents; or (4) the global and regional ocean transport system, dislocates planktonic organisms within the water body and supplies sessile benthos and nekton with possible food sources. Besides environmental forcing within the water body, the ATTZ (aquatic terrestrial transition zone) is greatly modified by current dependent sea-level changes. In the marine environment, tides are the primary forces that form the littoral zone (through standard and spring tides). Extreme, wind driven high tides can cover the supra-littoral zone with salt marshes. Similar longer-lasting current-based events, are the flood pulses forming temporary river flood plains found in the tropics (Junk 1989). Such temporary aquatic habitats are members of both aquatic and terrestrial systems (Lake 1995).

23.4 Physiological Responses of Marine and Limnetic Organisms

Any population of organisms is exposed to a complex pattern of environmental forcing. Which factors determine phenological responses depends on the species-specific physiological reception mode of the population. According to the ecological niche concept (Elton 1927), any environmental parameter may be important to the physiological responses of the population. If the actual value lies beyond the range of the physiological tolerance of the species, it excludes it from the system. If it lies within the tolerance range, the species-specific physiological response profile leads to a phenological response (Fig. 23.2). Many investigations are concerned with the determination of the tolerance zone of marine and limnetic organisms (Sewell 1999).

Single environmental forcing may be observed as well as a combination of environmental cues that influence population responses. Temperature, salinity and oxygen (Alderdice and Forrester 1968) simultaneously determine the decisive ecological niche of fish. Equally, light and temperature determine the reproduction of brown algae (Henry 1988). Threshold transitions (Werner 1962), physiological forcing by parameter aggregations (Lange and Greve 1997), or combinations of forcing (such as temperature and day-length on the reproductive physiology of the viviparous sea-perch, *Cymatogaster aggregata*) are decisive cues (Harrington 1959). Temperature controls physiological processes in poikilotherms (nearly all marine organisms according to the Q10 rule, see (Belehradek 1935). This general functional relationship underlies the species- and process-specific thermal profile, including the limits of temperatures tolerated, and the physiological preference in which differences become especially evident in populations of trophologically sympatric species (Heyen et al. 1998). This element of functional biodiversity deserves increased analysis of its functional and genetic basis (Uhlig 1995).

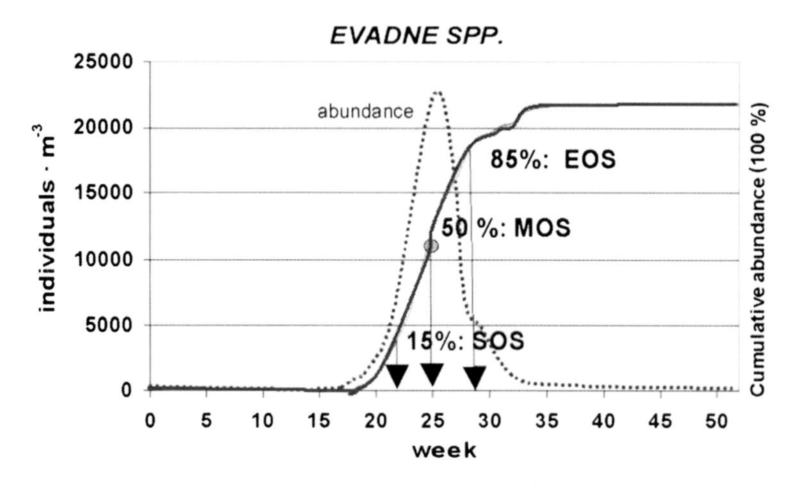

Fig. 23.2 The determination of the phenophases start of season (*SOS*), middle of season (*MOS*) and end of season (*EOS*) for the cladoceran *Evadne* spp. by transforming the population curve into the cumulative population curve and determining the days of the threshold passage of 15, 50 and 85 % cumulative abundance

23.5 Seasonality in the Hydrosphere

Seasonality varies mainly with latitude, altitude or depth, and large scale climatic forcing such as ENSO (El Nino Southern Oscillation) or NAO (North Atlantic Oscillation). Limnetic systems are closely related to the surrounding terrestrial systems, that may permit the correlation of terrestrial (e.g. time of cherry blossom) and hydrological signals (e.g. water temperature) with pike-perch year-class success (estimated from abundance of copepod nauplii, Tesch 1962). The scientific treatment of hydrological biota varies more with respect to the pelagial or benthal than to the marine or limnetic community. In plankton research, the food chain paradigm, and a resulting concentration on the succession of trophic levels or guilds, has suppressed investigation of the specific response profiles to environmental forcing of species populations (such as is seen in benthic and fisheries research).

The seasonal signal resulting from solar radiation and temperature is integrated into the production cycle of the polar and temperate zone. In general, the description of annual succession is based on trophic relationships (Lenz 2000). However, the succession of populations of a comparable ecological guild can also be described by forcing parameters other than trophology (Greve and Reiners 1988). In contrast, benthology, (especially phytobenthos research) has investigated the physiological responses of single species, mainly in the laboratory (Lüning 1989) and in studies of geographic distribution (Molenaar 1997). Sea-grass and mangroves (as other gymnosperms and angiosperms) can live either continuously or temporarily within the hydrosphere. These organisms have a much longer generation time than planktonic populations and can thus follow the traditional seasonal pattern of budding, blossoming, fruiting, and leaf-fall. Benthic and nektonic animals respond to both the seasonal timing of the benthic production and population succession in open water, where many species live during their larval periods, making use of the pulsed system of suitable food production.

The ontogeny of many species includes migration between marine biota and fresh water, sediment and water, or the surrounding terrestrial system and water. Floods or droughts increase variation of the hydrosphere, and organisms have evolved strategies to utilize the new biota to survive unfavorable dry periods. Marine bentho-pelagic coupling includes diurnal migrations from the sediment into the pelagial zone and back, ontogenetic seasonal transfer of sperm (see Greve 1974), eggs and larvae into the pelagial, and the return of recruits to the benthic or nektonic populations.

Further, resting stages such as dormant eggs (Dahms 1995) or phytoplankton cysts, add to this bentho-pelagic exchange, especially in limnetic systems and shelf seas. The alternation of generations between polyp and medusiod stages in hydrozoa is another example of bentho-pelagic coupling. All these processes are seasonal and periodic, related either to the length of the day, lunar driven tides, or seasons of the year. They thus offer options to phenological research, though so far the hierarchy of external or internal functional relationships determining the phenology of the populations is not generally known. It will be a major effort to understand the primary functional relationships determining the phenology of populations. Two research strategies need to be developed separately: traditional plankton research that discriminates between functional groups (Lenz 2000), and population research that discriminates between species-populations with various physiological, trophodynamic and temporal preferences (Giese 1959; Lindley 2002).

23.6 Methods of Aquatic Phenology

The classical phenological criteria of plants can e applied to organisms with an annual ontogenetic cycle or inter-annually lasting biomass, such as those in the aquatic terrestrial transition zone (ATTZ), which includes the benthic environment as well as birds, seals, and tortoises that feed on marine production. The majority of the hydrosphere is inhabited by plankton with generation times of days, weeks, or months, which results in extreme population dynamics (changes of several orders of magnitude within weeks in the phytoplankton and zooplankton). Some of these changes are due to bentho-pelagic coupling through the cysts and dormant eggs of pelagic species (Dahms 1995) or through larvae of benthic organisms or fish (meroplankton). With such high dynamics, the abundance of the population under investigation itself can be a phenological criterion. The passage of absolute or relative threshold values permits the temporal definition of phenological events (Greve et al. 2001) although it requires collection of a complete year's data before starting the analysis. Sufficiently frequent measurements over multiple years are required for investigation of the variance in phenology and its functional relationship to physical forcing (Colebrook 1960; Heyen et al. 1998).

Measurements on the population dynamics of limnetic systems often have been made on a weekly (Monday–Friday) basis. Measurements in marine systems generally require a more intensive effort from sea-going platforms. Light vessels, weather-ships (Oestvedt 1955) and offshore islands (Greve and Reiners 1988) provide alternatives. There is neither a marine biometeorological observation system (as in terrestrial agro-meteorological observation), nor an equivalent to terrestrial phenological gardens (Menzel 2000). The continuous plankton recorder (CPR) program is the most extensive observation series for marine plankton (Colebrook 1978). Measurements cover the North Atlantic and the North Sea, with monthly sections repeated for a periods of 40 or more years. Besides these measurements, local time-series from the British Channel (near Plymouth or at the station Helgoland Roads and other locations, ICES report 2002) have been used for the analysis of long-term population and benthos dynamics (Kröncke et al. 1998; Reid et al. 1998; Greve et al. 2001, 2005; Valdés et al. 2006) and for analysis of climatic forcing within marine populations, including phenological findings (Southward et al. 1975; Edwards and Richardson 2004).

23.7 The State of Phenology in Marine and Limnetic Systems

Organisms of the hydrosphere respond similarly to seasonal and climatic forcing as those of terrestrial systems (Walther et al. 2002). The degree to which these responses are documented varies substantially. This includes a broad collection of population data on the annual timing of life histories with special attention to reproductive synchronization. So far, these studies are only partially related to phenology.

23.7.1 *Thallus Algae*

Often in marine botany, physiological responses of thallus algae (studied in detail from laboratory and field observations) are used to confirm the validity of derived functional relationships (Lüning 1989). With this approach, global latitudinal comparisons of phenological responses could be studied in seaweeds (Wiencke et al. 1994; Molenaar 1997). Examples of botanical phenological observations exist for Chlorophyta (Clifton and Clifton 1999), Rhodophyta (Arasaki 1981; Chamberlain 1985; Breeman et al. 1988), and Phaeophyta (Henry 1988; Deshmukhe and Tatewaki 2001). In these studies reproductive behavior was studied in the field, where tetrasporophytes and gametophytes appeared to be strictly regulated by temperature/day-length responses (Breeman et al. 1988).

23.7.2 Marine Higher Plants

Growth, reproduction, and leaf fall were documented and timed for sea grass (Reyes et al. 1995; Oliveira et al. 1997), and mangroves (Gwada et al. 2000).

23.7.3 Flood Plain Forests

The phenology of tropical flood plain forests has been extensively studied, along with the accompanying rich fauna (Junk et al. 1989). The phase transition of a terrestrial ecosystem being converted into an aquatic habitat for months with fish immigrating into the canopy of trees and back again into a forest system is pulsed by seasonal floods. Phenology is applied here to the terrestrial as to the hydrological system.

23.7.4 Marine Benthic Communities

In zoology, the timing of reproduction was the subject of phenological studies on echinodermata (Stanwell-Smith 1998; Garrido 2001), polychaeta spawning synchrony (Hardege 1997), gonad maturity and spawning in bivalve mollusks. In the sepiolid squid *Rossia pacifica,* the hatching of eggs appeared to be keyed to the new moon, and extended over more than 2 months. Low seawater temperatures may affect this process (Summers 1985).

23.7.5 Marine Vertebrates

In several species of gulls, molt has a species-specific phenology (Howell et al. 1999). The gonadal maturation of fish has been extensively studied. It depends on the ambient sea water temperature which can be influenced by migration. In the mixed water of German Bight the spawning is greatly determined by the preceding (weeks 1–10) SST (Fig. 23.3).

23.7.6 Limnetic Animals

Fresh water systems are rich in populations living part-time in the water such as insects, amphibians, and some crustaceans. Besides hatching information, phenological data are used for life history strategy analysis of Chironomidae (Mihuc and Toetz 1996). In 29 species of aquatic dance flies (Empididae, Clinocerinae and Hemerodromiinae) it was possible to discriminate the variance of temporal orientation for closely related species in the same area (Wagner and Gathmann 1996). Even the autumnal phenology of odonata reveals a species-specific phenology

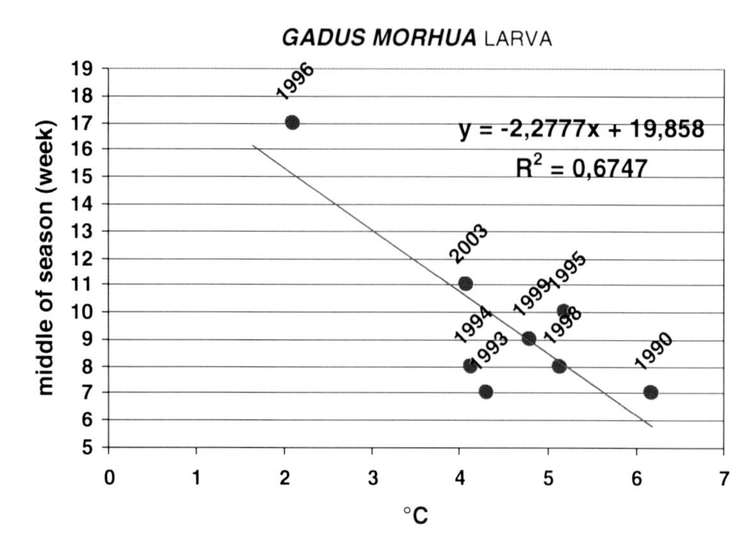

Fig. 23.3 The phenology of the phenophase middle of season (*MOS*) in cod (*Gadus morhua*) larvae at Helgoland Roads dependent on the preceding winter temperatures

(Joedicke 1998). In the amphibian *Rana yavapaiensis*, the timing of reproductive periods corresponds to ecological advantages (Sartorius and Rosen 2000).

23.7.7 Limnetic Plankton

In the epilimnion of lakes, the succession of populations is well, though rarely documented with phenology (Adrian 1997; Winder and Schindler 2004; Adrian et al. 2006). Investigation of the ice-duration period provides information on the phenology of a forcing parameter. The clear water phase (standing for the greatest filtration efficiency of zooplankton) also indicates a phenological timing that responds to spring temperature and dominant forcing of the North Atlantic Oscillation (NAO, Adrian et al. 1999). Consequences of climatic changes were also studied in lakes (Straile 2000; Gerten 2002). Experimental global warming has been the subject of investigations in rivers (Hogg et al. 1995).

23.7.8 Marine Dinoflagellates

The phenology of phytoplankton depends on light, nutrients, temperature and grazing (Wiltshire et al. 2008) and is less dependendent on temperature than zooplankton. The NAO is also correlated with population dynamics of the marine dinoflagellate *Noctiluca scintillans* (Heyen et al. 1998). The seasonality of *N. scintillans* also shows a correlation with seasonal ambient temperatures (Greve et al. 2001).

23.7.9 Marine Zooplankton

In the life history of the highly dynamic populations, triggers or indicators of improved nutritional conditions could be very helpful. In the copepod *Calanus finmarchicus*, Miller et al. (1991) suspected that photoperiod ended the resting period in February-March and induced maturation. An internal long-range timer is also suspected of ending the resting period. However, as the availability of food and the number of predators can hardly be tested in advance, and the production of juveniles is energy- and time-costly, reproduction is an optimization strategy. The match-mismatch hypothesis (Cushing 1990) describes this dilemma, especially for meroplanktonic populations that generally have no parental care.

Animal migration provides a further example of marine phenological optimization, such as in the timing of the first appearance of immigrating *Crangon crangon* and *Pleuronectes platessa* into the spring wadden sea (Van der Veer and Bergmann 1987), the first fish-feeding in competing populations (Juanes et al. 1994), and the mutual predation of *Pleurobrachia pileus* and *Calanus helgolandicus* representing a dichotomy in the system equilibria (Greve 1995). Variability in the phenology of marine populations could help explain and predict the influences of global warming and other forces on ecological equilibrium changes, which in turn decide the fate of the complete biocoenosis (Greve et al. 2001).

This scenario can be observed in North Sea plankton, where high frequency zooplankton time-series are available. Since 1975, every Monday, Wednesday and Friday two plankton samples (150 and 500 μ) are collected at the Helgoland Roads station (54° 1′ 18″ N, 7° 54′ E). The analysis of the population phenology resulted in calculation of an annual 15 % cumulative abundance value, which corresponds with the mean SST (surface salinity temperature) over the April to June period for juvenile ctenophores (Figs. 23.4 and 23.5).

The heat content of the North Sea changed at the end of the 1980s (Pohlmann 1996). SST in the winter months showed this change in correspondence with the NAO. The response of the *P. pileus* population was highly non-linear. Until 1987, the ctenophore showed a distinct spring bloom in May/June, with abundance increases of more than four orders of magnitude. Since then (for the period analyzed), winter abundances increased and summer abundances decreased. Abundance changes also decreased in speed and dimension. Since *P. pileus* is a key species of the German Bight ecosystem (it controls the copepod populations during the times of maximum abundance, see (Greve and Reiners 1988, 1995) the change in phenology has important consequences for the ecosystem equilibrium.

This change is one of the few documented examples for non-linear climate-related ecosystem responses. Similar changes have been analyzed in competing populations, such as clupeid fishes (Alheit 1997). Global warming consequences, such as the lateral displacement of populations (Southward et al. 1975) have been documented, but are not yet a topic in global zooplankton research (Marine Zooplankton Colloquium 2001), even though complete biocoenoses in the temperate zone change their stability regimes. This alteration corresponds with and

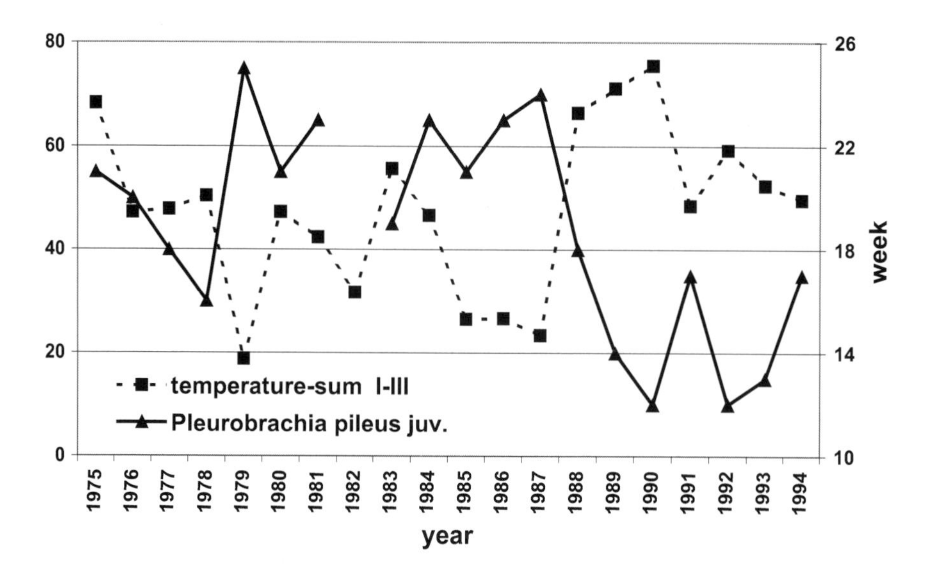

Fig. 23.4 Winter temperature (°C) and start of season of *P. pileus* (week) at Helgoland Roads

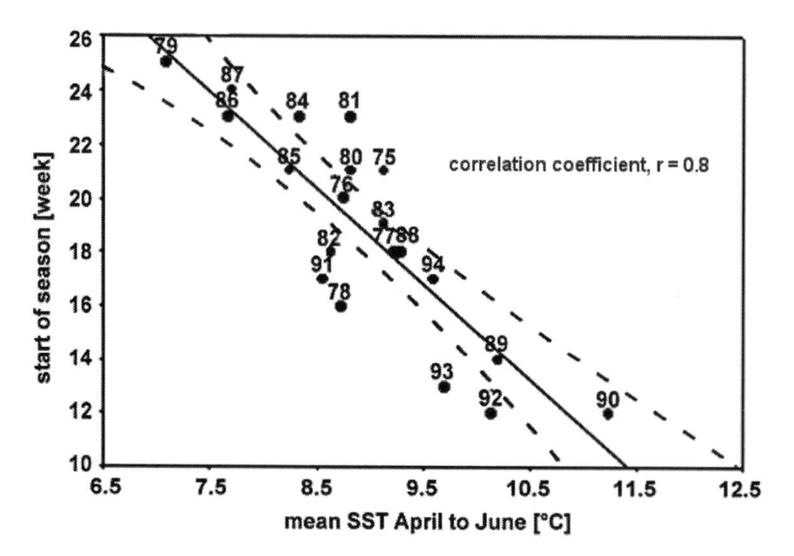

Fig. 23.5 Correlation of the phenophase start of season (*SOS*) of juvenile *Pleurobrachia pileus* with ambient mean water temperature

supports the species-specific phenological results observed and calculated for Helgoland Roads zooplankton (Fig. 23.6, Table 23.2; Greve et al. 2001). The use of sequential phenophases for the determination of seasonal events enables the linkage of phenology with seasonality and ecology via the match mismatch scenario and its effect on population successes.

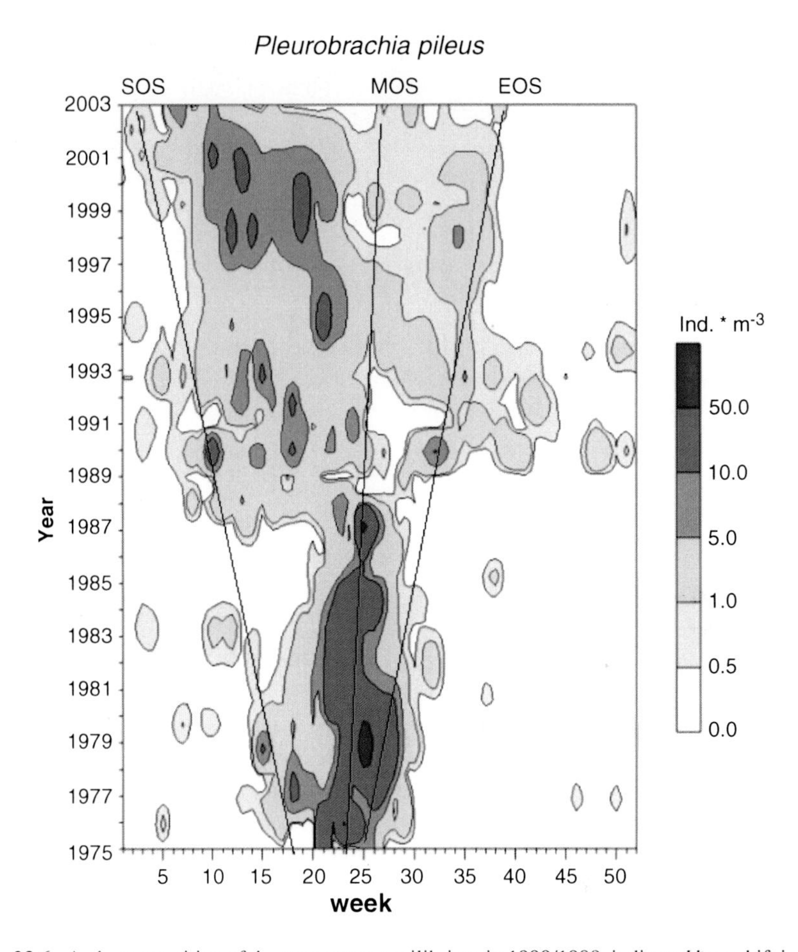

Fig. 23.6 A phase transition of the ecosystem equilibrium in 1988/1989, indicated by a shift in the annual abundance cycle from periodic blooms, to extended mean abundances of the local key–species *P. pileus* in the German Bight. The phenological linear regressions indicating this shift in the drastic change of the length of season (*LOS*) the difference of end of season (*EOS*) – start of season (*SOS*)

Aspects of the phenology of marine zooplankton have been investigated in detail by Mackas et al. (2011) in a benchmark paper covering the seas of the globe and the approaches towards an integrated view of marine zooplankton phenology.

23.8 Phenological Options in Applied Ecology

The phenological analysis of marine and limnetic populations has demonstrated that the timing of life history events also responds to changes in physical forcing (Walther et al. 2002). The expected continuation of global temperature increases

Table 23.2 Decadal mean annual phenophase shifts (weeks) of major groups of zooplankton at Helgoland Roads, ranked according the length of season (LOS) which results from the shifts of start of season (SOS) and end of season (EOS)

	SOS	MOS	EOS	LOS
Appendicularia	−0.187	−0.049	−0.027	0.160
Lamellibranchia	−0.013	0.005	0.126	0.139
Fischlarven	0.086	0.183	0.226	0.139
Hydrozoa	−0.060	−0.024	0.051	0.111
Ctenophora	−0.080	0.017	−0.014	0.066
Cladocera	−0.097	−0.095	−0.035	0.062
Noctiluca scintillans	−0.012	0.033	0.048	0.060
Copepoda	−0.088	0.012	−0.051	0.037
Polychaeta	−0.127	−0.222	−0.092	0.036
Chaetognatha	−0.011	−0.013	−0.020	−0.009
Decapoda	−0.128	−0.155	−0.144	−0.016
Cyclopoida	−0.125	−0.151	−0.157	−0.032
Calanoida	−0.117	−0.066	−0.162	−0.045
Echinodermata	−0.105	−0.177	−0.163	−0.058
Gastrpoda	0.002	−0.086	−0.120	−0.122
Harpacticoida	0.168	0.152	0.045	−0.123

will alter life in the aquatic biocoenoses, for which biometeorological measurements provide a valuable source of information. These observations will have to follow a variety of research strategies, as benthic, planktonic, nektonic, and submerse terrestrial populations will continue to require diverse measurements, in-situ observations of sedentary and plankton organisms, and physiological laboratory studies. Ecosystem management will have to include phenologically determined functional relationships, secondary match/mismatch consequences, lateral displacement of populations, and community changes in the administrative framework (Aksornkoae 1993; Avila et al. 1996).

Observation strategies in the hydrosphere are based on diverse population-specific techniques, which could be improved by inclusion and documentation of phenological criteria. The resulting scientific analyses and standardization of techniques may lead to a more universal approach to hydrological phenology, and systems for collecting those parameters which are of special administrative importance and observational feasibility. Such monitoring systems can be the basis of early warning, operative administrative modeling, and sustainable management of organic resources in a changing world, and could be imbedded into Global Observing Systems (e.g., GOOS).

In the marine zooplankton at least phenology has reached a state of predictive models based on the experience of local time series. These enable the phenological prediction of the timing of populations at Helgoland Roads under www.senckenberg.de/dzmb/plankton based on the current thermal information.

Acknowledgments Rita Adrian and Inka Bartsch, who provided information on their fields of research, supported this study. The Helgoland Roads time-series analysis was undertaken with the support of grants DFG 282/3-1,2 and BMBF 03F181A.

References

Adrian R (1997) Calanoid-cyclopoid interactions: evidence from an 11-year field study in a eutrophic lake. Freshw Biol 38:315–325
Adrian R, Walz N, Hintze T, Hoeg S, Rusche R (1999) Effects of ice duration on plankton succession during spring in a shallow polymictic lake. Freshw Biol 41:621–632
Adrin R, Wilhelm S, Gerten D (2006) Life-history traits of lake plankton species may govern their phenological response to climate. Global Change Biol 12(4):652–661
Aksornkoae S (1993) Ecology and management of mangroves. IUCN Wetlands and Water Resources Programme, Gland, Switzerland
Alderdice DF, Forrester CR (1968) Some effects of salinity and temperature on early development and survival of English sole (*Parophyrus vetulus*). J Fish Res Bd Can 25:495–521
Alheit J, Hagen E (1997) Long-term climate forcing of European herring and sardine populations. Fish Oceanogr 6:130–139
Arasaki S (1981) A comparison of the phenology of intertidal Porphyra on the coasts of Japan and western North America. Proc Int Seaweed Symp 8:273–277
Avila M, Otaiza R, Norambuena R, Nunez M (1996) Biological basis for the management of 'luga negra' (*Sarcothalia crispata* Gigartinales, Rhodophyta) in southern Chile. Hydrobiologia 326–327(1):245–252
Belehradek J (1935) Temperature and living matter. In: Protoplasma monograph. Borntraeger, Berlin
Breeman AM, Meulenhoff EGS, Guiry MD (1988) Life history regulation and phenology of the red alga *Bonnemaisonia hamifera*. Helgoländer Meeresuntersuchungen 42(3–4):535–551
Caspers H (1951) Rhythmische Erscheinungen in der Fortpflanzung von *Clunio marinus* (dipt. Chiron.) und das Problem der lunaren Periodizität bei Organismen. Arch Hydrobiol 18 (suppl Bd):415–594
Chamberlain YM (1985) Trichocyte occurrence and phenology in four species of *Pneophyllum* (Rhodophyta, Corallinaceae) from the British Isles. Br Phycol J 20:375–379
Childress J (1983) Oceanic biology: lost in space? In: Brewer PG (ed) Oceanography, the present and the future. Springer, New York
Clifton KE, Clifton LM (1999) The phenology of sexual reproduction by green algae (Bryopsidales) on Caribbean coral reefs. J Phycol 35:24–34
Colebrook JM (1960) Continuous plankton records: methods of analysis, 1950–1959. Bull Mar Ecol 5:51–64
Colebrook JM (1978) Continuous plankton records: zooplankton and environment, north-east Atlantic and North Sea, 1948–1975. Oceanol Acta 1(1):9–23
Cushing DH (1990) Recent studies on long term changes in the sea. Freshw Biol 23:71–84
Dahms HU (1995) Dormancy in the Copepoda – an overview. Hydrobiologia 306:199–211
Deshmukhe GV, Tatewaki M (2001) Phenology of brown alga *Coilodesme japonica* (Phaeophyta, Dictyosiphonales) with respect to the host-specificity along Muroran coast, North Pacific Ocean, Japan. IJMS 30:161–165
Edwards M, Richardson AJ (2004) Impact of climate change on marine pelagic phenology and trophic mismatch. Nature 430:881–884
Elton C (1927) Animal ecology. University of Chicago Press, Chicago
Franke HD (1990) Photopollution: coastal artificial light affects reproductive synchronisation in a litoral polychaete. Verhandlungen der Deutschen Zoologischen Gesellschaft 83:481
Garrido CL, Barber BJ (2001) Effects of temperature and food ration on gonads and oogenesis of the green sea urchin, *Strongylocentrotus droebachiensis*. Mar Biol 138:447–456
Gerlach SA (1994) Marine systeme. Springer, Berlin
Gerten D, Adrian R (2002) Effects of climate warming, north Atlantic Oscillation, and El Nino/ Southern oscillation thermal conditions and plankton dynamics in European and North American lakes. TSWJ 2:586–606

Giese AC (1959) Comparative physiology: annual reproductive cycles of marine invertebrates. A Rev Physiol 21:547–576

Greve W (1974) Planktonic Spermatophores found in a culture device with spionid Polychaetes. Helgol Wiss Meeresunters 26:370–374

Greve W (1995) Mutual predation causes bifurcations in pelagic ecosystems: the simulation model PLITCH (PLanktonic swITCH), experimental tests, and theory. ICES J Mar Sci 52:505–510

Greve W, Reiners F (1988) Plankton time – space dynamics in German Bight – a systems approach. Oecologia 77:487–496

Greve W, Reiners F (1995) Biocoenotic process patterns in German Bight. In: Eleftheriou A, Ansell A, Smith CJ (eds) Biology and ecology of shallow coastal waters. Olsen & Olsen, Fredensborg, Denmark

Greve W, Lange U, Reiners F, Nast J (2001) Predicting the seasonality of north sea Zooplankton. In: Kröncke I, Türkay M, Sündermann J (eds) Burning issues of north sea ecology, Proceedings of the 14th international Senckenberg Conference North Sea 2000. Senckenbergiana marit, Frankfurt am Main

Greve W, Prinage S, Zidowitz H, Nast J, Reiners F (2005) On the phenology of North Sea ichthyoplankton. ICES J Mar Sci 62:1216–1223

Gwada P, Makoto T, Uezu Y (2000) Leaf phenological traits in the mangrove *Kandelia candel* (L.) Druce. Aquat Bot 68:1–14

Hardege JD, Bentley MG (1997) Spawning synchrony in *Arenicola marina*: evidence for sex pheromonal control. Proc R Soc Lond B Biol 264:1041–1047

Harrington RW (1959) Effects of four combinations of temperature and daylength on the ovogenetic cycle of a low-latitude fish, *Fundulus confluentus* GOODE and BEAN. Zoologica 44:149–168

Henry EC (1988) Regulation of reproduction in brown algae by light and temperature. Bot Mar 31:353–357

Heyen H, Fock H, Greve W (1998) Detecting relationships between the interannual variability in ecological time series and climate using a multivariate statistical approach – a case study on Helgoland Roads zooplankton. Clim Res 10:179–191

Hogg ID, Williams DD, Eadie JM, Butt SA (1995) The consequences of global warming for stream invertebrates: a field simulation. J Therm Biol 20:199–206

Howell SNG, King JR, Corben C (1999) First prebasic molt in herring-, Thayer's-, and glaucous-winged gulls. J F Ornithol 70:543–554

Joedicke R (1998) Autumnal phenology of central European Odonata. 2. Observations in the Lower Rhine Region, Germany. Opusc Zool Flumin 159:1–20

Juanes F, Buckel JA, Conover DO (1994) Accelerating the onset of piscivory: intersection of predator and prey phenologies. J Fish Biol 45:41–54

Junk WJ, Bayley PB, Sparks RE (1989) The flood pulse concept in river-floodplain systems. In: Dodge DP (ed) Proceedings of the international large river symposium. Fisheries and Oceans, Canada (Governmental Agency)

Kinne O (1963) The effects of temperature and salinity on marine and brackish water animals I temperature. Oceanogr Mar Biol A Rev 1:301–340

Kröncke I, Dippner JW, Heyen H, Zeiss B (1998) Long-term changes in macrofaunal communities off Norderney (East Frisia, Germany) in relation to climate variability. Mar Ecol Prog Ser 167:25–36

Lake PS (1995) Of floods and droughts. River and stream ecosystems of Australia. In: Cushing CE, Cummins KW, Minshall GW (eds) River and stream ecosystems. Elsevier, Amsterdam

Lampitt RS (1985) Evidence for the seasonal deposition of detritus to the deep-sea floor and its subsequent resuspension. Deep-Sea Res 32:885–897

Lange U, Greve W (1997) Does temperature influence the spawning time, recruitment and distribution of flatfish via its influence on the rate of gonadal maturation? Deutsche Hydrographische Zeitschrift 49(3):251–263

Lenz J (2000) Introduction. In: Harris R, Weibe P, Lenz J, Skjoldal HR, Huntley M (eds) ICES zooplankton methodology manual. Academic Press, San Diego

Lindley JA, Batten SD (2002) Long-term variability in the diversity of North Sea zooplankton. J Mar Biol Ass UK 82(3):1–40

Lüning K, tom Dieck I (1989) Environmental triggers in algal seasonality. Bot Mar 32(5):389–397

Mackas DI, Greve W, Edwards M, Chiba S, Tadokoro K, Eloire D, Mazzocchi MG, Batten S, Richardson AJ, Joihnson C, Head E, Conversi A, Pelusa T (2011) Changing zooplankton seasonality in a changing ocean: comparing time series of zooplankton phenology. Prog Oceanogr 97:31–62

Marine Zooplankton Colloquium (2001) Future marine zooplankton research- a perspective. Mar Ecol Prog Ser 222:297–308

Menzel A (2000) Trends in phenological phases in Europe between 1951 and 1996. Int J Biometeorol 44:76–81

Mihuc TB, Toetz DW (1996) Phenology of aquatic macroinvertebrates in an alpine wetland. Hydrobiologia 330:131–136

Miller CB, Cowles TJ, Wiebe PH, Copley NJ, Grigg H (1991) Phenology in *Calanus finmarchicus*; hypotheses about control mechanisms. Mar Ecol Prog Ser 72:79–91

Molenaar FJ, Breeman AM (1997) Latitudinal trends in the growth and reproductive seasonality of *Delesseria sanguinea, Membranoptera alata, and Phycodrys rubens* (Rhodophyta). J Phycol 33(3):330–343

Neumann D (1967) Genetic adaption in emergence time of *Clunio* populations to different tidal conditions. Helgoländer wiss Meeresunters 15:163–171

Oestvedt OJ (1955) Zooplankton investigation from weather ship M in the Norwegian Sea, 1948–1949. Hvalradets Skrifter Sci Results Mar Biol Res 40:1–93

Oliveira EC, Corbisier TN, De Eston VR, Ambrosio O (1997) Phenology of a seagrass (*Halodule wrightii*) bed on the southeast coast of Brazil. Aquat Bot 56:25–33

Orton JH (1920) Sea-temperature, breeding and distribution of marine animals. J Mar Biol Ass UK 12:330–366

Parsons TR, Takahashi M, Hargrave B (1984) Biological oceanographic processes, 3rd edn. Pergamon Press, Oxford/New York

Pohlmann T (1996) Simulating the heat storage in the North Sea with a three-dimensional circulation model. Cont Shelf Res 16:195–213

Reid PC, Planque B, Edwards M (1998) Is observed variability in the long-term results of the continuous plankton recorder survey a response to climate change? Fish Oceanogr 7(3/4):282–288

Reyes J, Sanson M, Afonso-Carrillo J (1995) Distribution and reproductive phenology of the seagrass Cymodocea nodosa (Ucria) Ascherson in the Canary Islands. Aquat Bot 50:171–180

Sartorius SS, Rosen PC (2000) Breeding phenology of the lowland leopard frog (*Rana yavapaiensis*): implications for conservation and ecology. Southwest Nat 45:267–273

Schmiedl G, Mitschele A, Beck S, Emeis K, Helleben C, Schulz H, Sperling M (2003) Benthic foraminiferal record of ecosystem variability in the eastern Mediterranean sea during times of saprobel S5 and S6 deposition. Palaeogeogr Palaeocl 190:139–164

Sewell MA, Young CM (1999) Temperature limits to fertilization and early development in the tropical sea urchin *Exhinometra lucunter*. J Exp Mar Biol Ecol 47:291–305

Southward AJ, Butler EI, Pennycuick P (1975) Recent cyclic changes in climate and in abundance of marine life. Nature 253:714–717

Stanwell-Smith D, Peck LS (1998) Temperature and embryonic development in relation to spawning and field occurrence of larvae of three Antarctic echinoderms. Biol Bull Mar Biol Lab Woods Hole 194:44–52

Straile D, Adrian R (2000) The north Atlantic oscillation and plankton dynamics in two European lakes – two variations on a general theme. Global Change Biol 6:663–670

Summers WC (1985) Ecological implications of life stage timing determined from the cultivation of *Rossia pacifica* (Mollusca: Cephalopoda). Vie et Mileu 35(3/4):249–254

Tesch FW (1962) Witterungsabhängigkeit der Brutentwicklung und Nachwuchsförderung bei *Lucioperca lucioperca* L. Kurze Mitteilungen aus dem Institut für Fischereibiologie der Universität Hamburg 12:37–44

Uhlig G, Sahling G (1995) Noctiluca scintillans: zeitliche Verteilung bei Helgoland und räumliche Verbreitung in der Deutschen Bucht (Langzeitreihen 1970–1993). Ber Biol Anst Helgoland 9:1–127

Valdés L, O'Brien T, López-Urrutia A (2006) Zooplankton monitoring results in the ICES area, Summary Status Report 2004/2005. ICES Cooperative Research Report 281

Van der Veer HW, Bergmann MJN (1987) Predation by crustaceans on a newly settled O-group plaice *Pleuronectes platessa* in the Western wadden sea. Mar Ecol Prog Ser 35:203–215

Wagner R, Gathmann O (1996) Long-term studies on aquatic dance flies (Diptera, Empididae) 1883–1993: distribution and size patterns along the stream, abundance changes between years and the influence of environmental factors of the community. Archiv Fuer Hydrobiologie 137:385–410

Walther GR, Post E, Convey P, Menzel A, Parmesan C, Beebee TJC, Fromentin JM, Hoeg-Guldberg O, Bairlein F (2002) Ecological responses to recent climate change. Nature 416:389–395

Werner B (1962) Verbreitung und jahreszeitliches Auftreten *Rathkea octopunctata* (M. Sars) und *Bougainvillia superciliaris* (L. Agassiz), (Athecata-Anthomedusae). Ein Beitrag zur kausalen marinen Tiergeographie. Kieler Meeresforsch 18:55–66

Wiencke C, Bartsch I, Bischoff B, Peters AF, Breeman AM (1994) Temperature requirements and biogeography of Antarctic, Arctic and Amphiequatorial seaweeds. Bot Mar 37(3):247–259

Wiltshire KH, Malzahn AM, Wirtz K, Greve W, Janisch S, Mangelsdorf P, Manly BFJ, Boersma M (2008) Resilience of North Sea phytoplankton spring bloom dynamics: an analysis of long-term data at Helgoland Roads. Limnol Oceanogr 53:1294–1302

Winder M, Schindler DE (2004) Climatic effects on the phenology of lake processes. Global Change Biol 10(11):1844–1856

Chapter 24
Birds

Tim H. Sparks, Humphrey Q.P. Crick, Peter O. Dunn, and Leonid V. Sokolov

Abstract There is a wealth of data on bird phenology, particularly on the timing of spring migration and the timing of breeding. Over the last decade or so there has emerged a large and growing literature examining changes in bird phenology and the likely causes of those changes. Here we give examples of changes in migration and breeding, with data originating from both amateur citizen science schemes and rigorously controlled schemes run by scientists. The overwhelming evidence is that spring migration and breeding have both got earlier. The likely cause of these changes is an increase in temperature, but other factors are also possible. Notwithstanding this general pattern it is also clear that responses are species-specific and may also vary from location to location.

T.H. Sparks (✉)
Institute of Zoology, Poznań University of Life Sciences, Poznań, Poland

Fachgebiet für Ökoklimatologie and Institute for Advanced Study, Technische Universität München, Munich, Germany

Sigma, Coventry University, Coventry, UK
e-mail: thsparks@btopenworld.com

H.Q.P. Crick
Natural England, Peterborough, UK

P.O. Dunn
Department of Biological Sciences, University of Wisconsin-Milwaukee, Milwaukee, WI 53211, USA

L.V. Sokolov
Biological Station Rybachy, Zoological Institute, Russian Academy of Sciences, St. Petersburg, Russia

M.D. Schwartz (ed.), *Phenology: An Integrative Environmental Science*,

DOI 10.1007/978-94-007-6925-0_24,

24.1 Introduction

The Roman mosaic floor at Lullingstone, Kent, UK depicts the four seasons as human characters; that representing spring has a barn swallow *Hirundo rustica* on her shoulder to uniquely identify that season. Thus, for almost 2,000 years people have associated spring with the arrival of migratory birds. In the UK the (currently) oldest phenological record dates to 1703 and refers to the call of the cuckoo *Cuculus canorus*. It is hardly surprising then that the timing of bird activity is a productive area of phenological research.

Birds are probably the most popular group of plants or animals and are keenly watched. In temperate zones their behavior is very seasonal and therefore ideal for phenological study. Indeed, the volume of data on the timing of bird activity is considered to outweigh that of any other form of phenological data. Unfortunately, large amounts of these data are collected at different sampling intensities and without any large-scale coordination. Interpretation of such data must be undertaken with care and should combine empirical findings with the known ecology of species. It must be remembered that birds are highly mobile and often secretive, and, thus, phenological observation relies on the intensity of recording and on the density and visibility of the bird species. The most reliable observational measurements derive from those species that are large and obvious, such as the white stork *Ciconia ciconia*, or occur in large numbers and are associated with human habitation, such as the barn swallow or house martin *Delichion urbica*. There are many types of data that range from the casual observation of birds in an individual's garden, to the intensive daily netting of birds at bird observatories, to the national coordinated schemes on the nest timing of birds. We include examples of several of these types of data in this chapter.

24.2 Changes in the Timing of Migration

24.2.1 Spring Arrival of Summer Visitors

The bulk of avian phenological data relates to the passage in spring of birds that overwinter in warmer environments. These include birds that migrate across continents, as well as species that travel shorter distances. The latter group includes species such as skylark *Alauda arvensis* in central and northern Europe and American robin *Turdus migratorius* in North America (Inouye et al. 2000). A third group of birds are partially migratory, in that a proportion of their population migrates. This last group includes species such as starling *Sturnus vulgaris*. Increasingly we learn of normally migratory birds overwintering, such as chiffchaff *Phylloscopus collybita* in the UK (Geen 2002), white stork in Germany, skylark in Poland and Canada goose *Branta canadensis* in parts of North America.

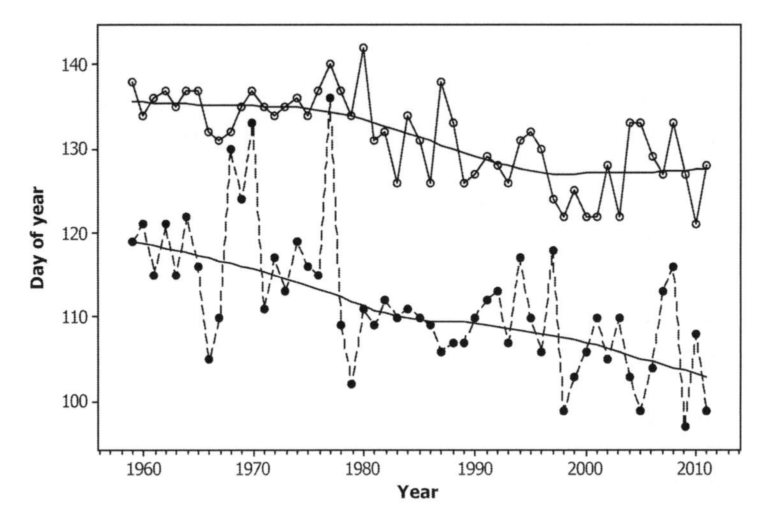

Fig. 24.1 The change in mean arrival date of a short distance migrant, chaffinch (*solid symbols*), and a long distance migrant, willow warbler (*open symbol*), at Rybachy, Russia 1959–2011. Smoothed (LOWESS) lines have been superimposed (Data source: Biological Station Rybachy)

So, different bird species are migratory in different regions and migrate over varying distances. It is not surprising then that different cultures associate spring with different species. In the UK the reporting of the first cuckoo has long been of high media profile, whereas elsewhere in Europe white stork or skylark may be considered better indicators of spring. In North America, the arrival of robins is traditionally associated with the arrival of spring, although one of the most famous local examples is the arrival of swallows at San Juan Capistrano in southern California. Thus, there may be a different emphasis on which species are recorded returning to their breeding grounds. This is particularly so of observational studies undertaken by individuals. Schemes that operate from bird observatories based on observation or netting are less subject to bias in species choice.

In general, short distance migrants return earlier to breeding areas than do long distance migrants, for example in Poland (Tryjanowski et al. 2002), where short distance migrants have shown a greater trend to earlier arrival in recent years than long distance migrants. Short distance migrants have the flexibility of a short passage time to respond quickly to changing environmental conditions. In Fig. 24.1 changes in mean migration timing at Biological Station Rybachy, on the Baltic coast of Russia are shown for chaffinch *Fringilla coelebs*, a short distance migrant, and willow warbler *Phylloscopus trochilus*, a long distance migrant. During the 53 years from 1959 to 2011, chaffinches arrived an average (±SE) of 0.31 ± 0.06 days earlier per year ($p < 0.001$), while willow warblers arrived 0.23 ± 0.03 days per year earlier ($p < 0.001$). This is equivalent, respectively, to 17 and 12 days earlier over the recording period. It is worth emphasizing that these are not first migrating birds but rather mean migration dates of birds caught in stationary traps. Thus they derive from a methodology not subject to observational bias and indicate a

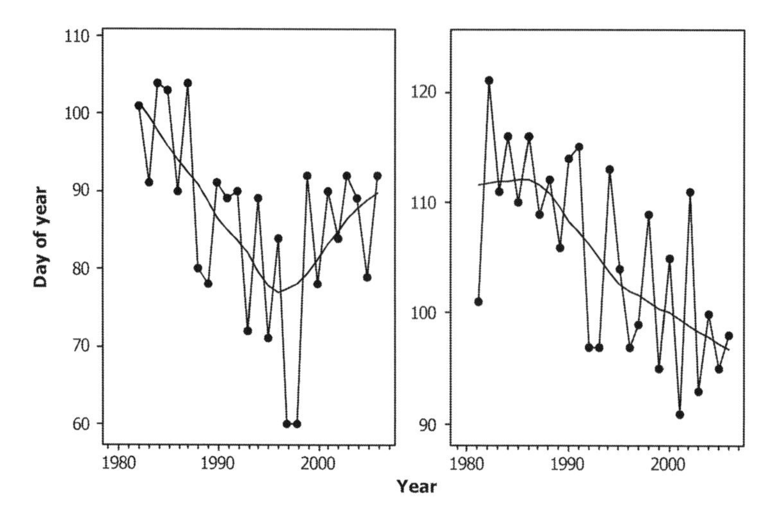

Fig. 24.2 The change in first arrival date of whimbrel (*left*), and hobby (*right*) from Essex, UK 1981–2006. Smoothed (LOWESS) lines have been superimposed (Data source: Essex Bird Reports)

shift in the migration distribution pattern of the species. The greater variability in arrival dates of earlier species is apparent in this graph (see also Mason 1995) and is a characteristic of migration phenology data sets in general.

In addition to the data on the whole migration distribution available from bird observatories there are many sources of data supplying first arrival date only. An example from the Essex (UK) Bird Reports is given in Fig. 24.2. This shows recent trends in first arrival of whimbrel *Numenius phaeopus* and hobby *Falco subutteo*. Both show a marked change in first arrival; hobby has advanced by 0.71 ± 0.17 ($p < 0.001$) days per year equivalent to 18 days earlier over the recording period and whimbrel advanced by a similar amount but at the end of the series got slightly later. These first dates need to be treated with caution as they may reflect changes in population size and thus visibility of the species (Sparks et al. 2001; Tryjanowski et al. 2005). Indeed the hobby population is increasing in the UK and some of the trend towards earliness may be an artifact of a larger population. However these first dates do also change in highly visible species with static populations, so they can reflect a change in at least one aspect of the migration distribution, even if they do not necessarily tell us about changes to the whole arrival distribution (Sparks et al. 2005). In many regions including Wisconsin, USA (Bradley et al. 1999), Russia (Sokolov et al. 1998; Sokolov 2006) and the UK (Sparks et al. 2007) the majority of the bird events recorded have tended to become earlier.

Not all data show a trend towards earliness, for example barn swallow in Slovakia (Sparks and Braslavská 2001), migrants in NE Scotland (Jenkins and Sparks 2010) and in the Southern Urals of Russia (Sokolov and Gordienko 2008). Inevitably the overall picture can be distorted if only significant results get

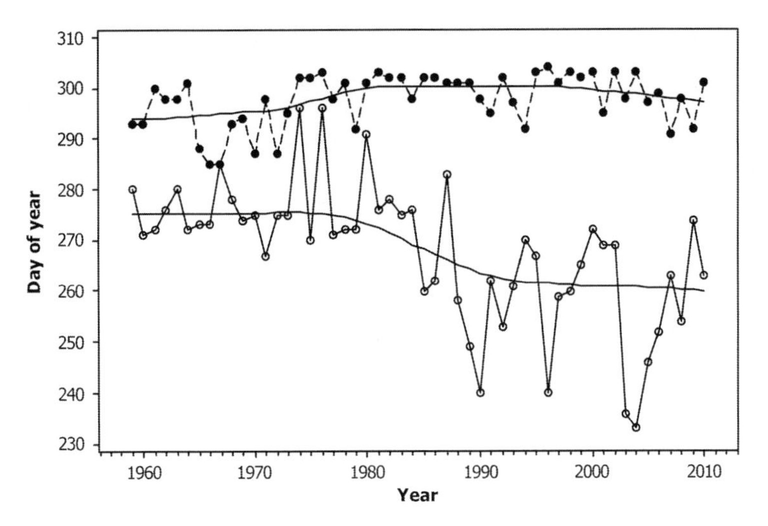

Fig. 24.3 The change in the last departure of tree pipit (*open symbol*) and chiffchaff (*closed symbol*) at Rybachy, Russia from 1959 to 2010. Smoothed (LOWESS) lines have been superimposed (Data source: Biological Station Rybachy)

published. However, a meta-analysis of available data on first and mean/median arrival dates confirmed the balance of evidence towards recent earlier arrivals (Lehikoinen and Sparks 2010). Recently it has been suggested that populations of species not demonstrating phenological flexibility may be in decline as a consequence (Møller et al. 2008).

24.2.2 Autumn Departure of Summer Visitors

Determining the date of autumn departures of breeding birds is more problematic for observational recorders. Where records are present, they usually concern the last observation date of a species, which is generally more difficult to pin down than the first bird in spring. More reliable data derive from netting schemes where birds are trapped on southwards passage. Figure 24.3 shows the last recorded capture date for two species, tree pipit *Anthus trivialis* and chiffchaff, from the Biological Station Rybachy. These demonstrate a trend towards earlier departure in the former species of 0.53 ± 0.10 days per year ($p < 0.001$) and a trend towards later departure of the latter of 0.12 ± 0.05 days per year ($p = 0.012$). These trends equate to 28 days earlier and 6 days later, respectively, over the 52-year record. Changes in the autumn departure of birds at Rybachy give a mixed picture and a similar situation was found within the Russian Arctic Circle with some species departing significantly earlier and some significantly later in a 60+ year record of bird migration (Gilyazov and Sparks 2002). In Rybachy most birds do not show a significant trend one way or the other (Sokolov et al. 1999). However there is plenty of evidence that in years of earlier

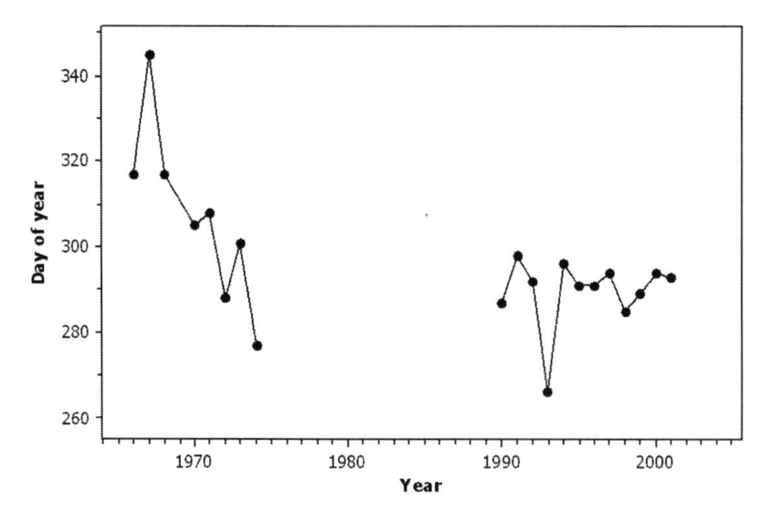

Fig. 24.4 The change in first autumn arrival of Bewick's swan in Essex, UK from 1966 to 2001 (with gaps) (Data source: Essex Bird Reports)

breeding autumn migration gets underway earlier (e.g., Ellegren 1990; Mitchell et al. 2012). In an examination of autumn departures it must be remembered that species are likely to have different optimal strategies and must balance any benefits of earlier departure from the breeding grounds with costs of missed additional breeding opportunities (in multibrooded species) and mistimed arrival at migratory staging areas or wintering grounds (Jenni and Kéri 2003; Tøttrup et al. 2006; Thorup et al. 2007; Filippi-Codacciuoni et al. 2010; Lehikoinen and Sparks 2010; Lehikoinen and Jaatinen 2012).

24.2.3 Winter Visitors

Species from colder environments can overwinter in milder environments to the south. In the Northern Hemisphere the numbers of such species tend to be fewer than summer visitors and the recording effort is lower. In Britain, emphasis has been placed on two species of thrush, redwing *Turdus iliacus* and fieldfare *Turdus pilaris*, that move in from Scandinavia. Fewer records exist for overwintering waders. Sparks and Mason (2004) suggest that greater changes have taken place in the timing of short distance migrants than for long distance migrants, but this conclusion is heavily influenced by substantial timing changes in a small number of raptors. Scarcity of data does not make it easy to make generalizations. An example of changed arrival in Bewick's swan *Cygnus bewickii* is shown in Fig. 24.4. While departure of summer visitors from wintering grounds is considered to be driven by photoperiod (e.g., Kok et al. 1991) the cues for migration of winter visitors have received less attention.

24.3 Influence of Climate on Migration Timing

There is substantial evidence that arrival dates are related to local temperatures. This may be an indirect effect through the supply of insect prey, or it may reflect southerly, and hence warm, tail winds aiding migration. There can be little doubt that en-route temperatures must play their part in migration. Figure 24.5 displays the relationship between the arrival date of barn swallow as recorded at four British observatories and spring temperature. The relationship suggests earlier arrival by 2.8 ± 0.5 days for every 1 °C increase ($p < 0.001$).

A response of migration timing to temperature has been detected in a wide range of studies including those in Poland (Tryjanowski et al. 2002), Slovakia (Sparks and Braslavská 2001), Russia (Sokolov and Payevsky 1998; Sokolov et al. 1999; Gilyazov and Sparks 2002; Sokolov and Kosarev 2003), and France (Sueur and Triplet 2001).

Sokolov and Kosarev (2003) suggested that the onset of spring migration from Africa may be influenced by precipitation. In years with low February precipitation long-distance migrants arrive to the Courish Spit earlier than when precipitation is more abundant. A strong negative relationship was found between arrival dates of long-distance migrants to the Oxford, UK area and mean winter temperature anomalies in Africa at 20° N (Cotton 2003). Balbontín et al. (2009) have found that barn swallows of different ages responded differently to environmental conditions in the winter quarters, as already shown by Saino et al. (2004). Specifically, barn swallows from south-western Spain, especially old birds, delayed arrival

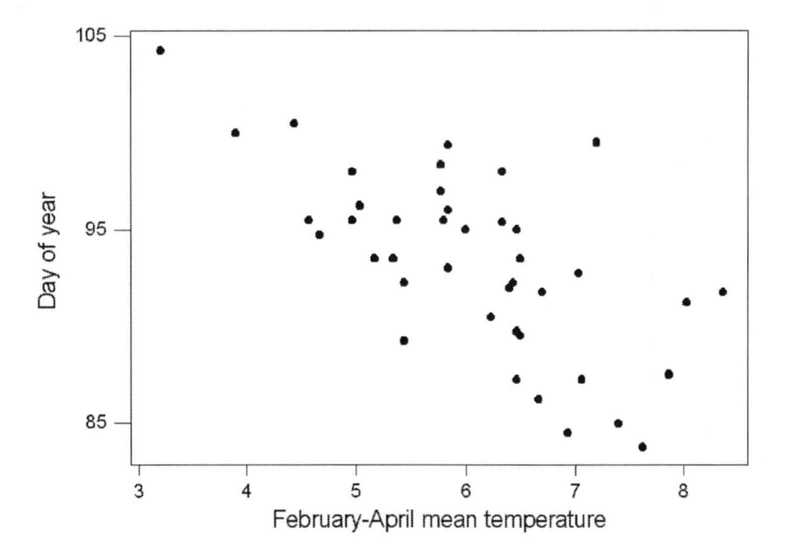

Fig. 24.5 The relationship between mean first arrival date of barn swallow from four British bird observatories 1959–1999 and mean February–April Central England temperature (°C) (Data source: Dungeness, Portland, Bardsey, and Calf of Man observatories)

at the breeding grounds in years of abundant rainfall and high primary production in the winter quarters, which was the opposite to that found in an Italian population. Differences in responses to winter conditions in arrival schedules among populations imply that individuals respond to similar conditions in different ways, perhaps because of differences in winter habitats as suggested above (Ambrosini et al. 2009). Marra et al. (1998) also showed a direct relationship between the onset of spring migration of American redstarts *Setophaga ruticilla* and both foraging opportunities in winter quarters in mangrove swamps and body condition of the birds. The earlier that redstarts gained fuel necessary for migration, the earlier they left wintering areas. Studds and Marra (2011) used longitudinal data on spring departure dates of American redstarts to show that annual variation in tropical rainfall and food resources are associated with marked differences in the timing of spring departure of the same individuals between years. This finding challenges the idea that photoperiod alone regulates the onset of migration, providing evidence that intensifying drought in the tropical winter could hinder adaptive responses to climatic warming in the temperate zone.

It cannot be ruled out that, in recent decades, the rate of spring migration through Africa has accelerated due to changing environmental conditions, rather than changed departure time from winter quarters. Migrants may not traverse arid areas, primarily in northern Africa, more quickly than hitherto. This view is supported by studies showing that passage migrants, in particular whitethroats *Sylvia communis* were mist-netted in Sahel (Lake Chad, Nigeria) in the 2000s earlier than in the late 1960s (Ottosson et al. 2002).

24.4 Changes in the Nest Timing of Birds

Some species of bird form very obvious nests. This is particularly true of species that are large, or build nests associated with human habitation. The white stork builds nests on the tops of trees, chimneys or other vertical structures. The rook *Corvus frugilegus* and grey heron *Ardea cineria* both build nests in treetop colonies that are very obvious. In the UK the rook is such an obvious species that Robert Marsham included it in his eighteenth century *Indications of Spring* (Sparks and Carey 1995) by noting the dates of nest building and the dates on which young could be detected. Hole nesting species can be encouraged to nest in particular locations by providing nest boxes and this makes recording their breeding activity much simpler (e.g., pied flycatcher *Ficedula hypoleuca* (Both et al. 2004)).

For other species it is necessary to search for nests and it is usual practice to follow the eggs during laying and hatching and to record the subsequent growth of nestlings. From these dates it is possible to back-calculate first egg date. In the UK the British Trust for Ornithology (BTO) conducts a nest record scheme that receives approximately 30,000 records each year. From these data, statistics on the timing of nesting can be calculated. Crick et al. (1997) showed that a large number of species from a wide range of guilds (types) are breeding progressively earlier. An example

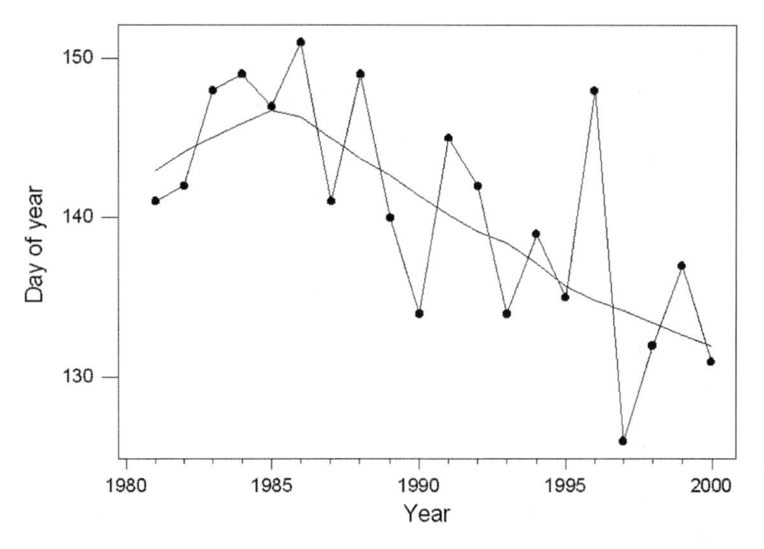

Fig. 24.6 Change in the commencement of nesting (5th percentile) of reed warbler in the UK 1981–2000. Smoothed (LOWESS) line has been superimposed (Data source: British Trust for Ornithology)

from the BTO scheme of reed warbler *Acrocephalus scirpaceus* is given in Fig. 24.6. Over the last 20 years there has been a marked trend towards earlier breeding in this species. Regression against time suggests that the commencement of egg laying has advanced by 0.77 ± 0.21 days per year ($p = 0.002$), that is by 15 days in 20 years.

This advance in nest timing has been reported from a wide range of species and locations including red-necked starlings *Sturnia phillipenisi* in Japan (Koike and Higuchi 2002), pied flycatcher in Russia (Sokolov 2000), tits *Parus* spp. in Germany (Winkel and Hudde 1997), goldeneye *Bucephala clangula* in Germany (Ludwichowski 1997) and tree swallow *Tachycineta bicolor* in North America (Dunn and Winkler 1999). At Rybachy in Russia there has been a strong advance in the breeding time of many species. Figure 24.7 displays the situation for pied flycatcher that indicates a trend towards earlier breeding of 0.22 ± 0.06 days per year ($p < 0.001$) equivalent to an 8-day advance over the period of recording.

24.5 Climate Influence on Timing of Breeding

Bird species time their reproduction to maximize the number of offspring produced within a season. Some species are capable of producing multiple broods per year and will need to balance the success of the first brood with the options for subsequent brood(s) (Crick et al. 1993). Other species that are single brooded aim to hatch their young at a time of optimal food supply, usually in the form of

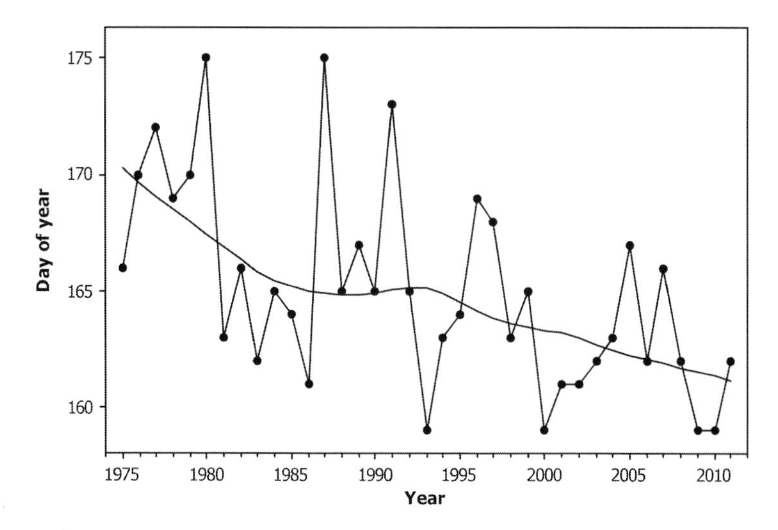

Fig. 24.7 Change in the mean nest timing of pied flycatcher at Rybachy, Russia 1975–2011. Smoothed (LOWESS) line has been superimposed (Data source: Biological Station Rybachy)

invertebrates. It is known that the development times of invertebrates under elevated temperatures can halve, for example from 56 to 23 days in the case of the development of winter moth *Operophtera brumata* (Buse et al. 1999) but birds cannot accelerate the incubation period of individual eggs.

Changes in breeding phenology are among the most studied aspects of the effects of climate change on birds. Most studies (79 % i.e. 44/56) have found that birds lay earlier when it is warmer in the spring (Dunn 2004), and by extension, long-term changes in laying date are often attributed to climate warming. For example, in reed warblers, birds nest an average of 5.0 ± 0.7 days earlier for every degree warmer in spring ($p < 0.001$, Fig. 24.8). Visually, Fig. 24.8 suggests that a straight-line relationship may not be valid over the entire range of temperatures; excluding the colder years would increase the response to temperature detailed above. On the other hand, some bird species do not appear to be changing their breeding phenology in response to changes in temperature, and species responses can vary between locations.

This could be due to differences between species in how temperatures at different times of the breeding season affect food abundance. For example, during 1973–1995 there has been an increase in late spring temperatures in the Netherlands, but there has been neither change in early spring temperature, nor any change in the laying date of great tits *Parus major*, and it is the temperature during early spring that is most closely correlated with laying date (Visser et al. 1998). Thus, birds may only respond to temperatures during particular time periods that affect the abundance of their breeding resources. Interestingly, great tits and pied flycatchers on the same study area in the Netherlands respond differently to climate change (Both and Visser 2001). The tits, which are residents, breed about 2 weeks earlier than the flycatchers and show no significant change in laying date, because their critical temperature

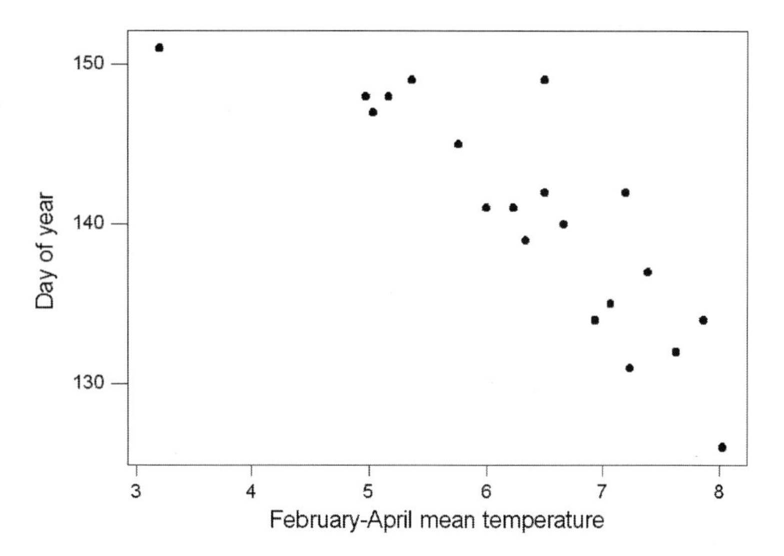

Fig. 24.8 The relationship between the commencement of nesting (5th percentile) in reed warbler in the UK 1981–2000 and mean February–April Central England temperature (°C) (Data source: British Trust for Ornithology)

period has not changed over time. In contrast, the flycatchers, which migrate from wintering grounds in Africa, have advanced their laying date, because their critical temperature period, which is later in spring, has become warmer over time (Both and Visser 2001). Despite the advancement in laying date of flycatchers, selection over the past 20 years has become even stronger for an earlier laying date (i.e., more young are recruited from earlier nests). It appears that flycatchers have not responded to this selection pressure because their timing of arrival is determined mostly by photoperiod. Thus, the cue for initiating migration (photoperiod) has become maladaptive in some respects because it no longer provides a cue for the best time to arrive on the breeding grounds (Both and Visser 2001).

As a consequence, some bird species are not shifting their laying dates at the same rate as shifts in their food supply that are also associated with climate change. This can lead to temporal mismatches between the optimal time of breeding and the food supply that birds need for raising young (Visser and Both 2005). Recent studies have begun to reveal that these mismatches are often due to the different ways that birds and their food supply respond to temperature. In songbirds, it is often thought that earlier laying is caused primarily by birds tracking shifts in the supply of their food, particularly caterpillars. However, recent studies have revealed that temperature can also affect laying date in other ways through effects on thermoregulation, gonadal development and the phenology of vegetation eaten by caterpillars.

In Corsica, for example, the laying date of blue tits *Cyanistes caeruleus* is affected differently by temperature and the phenology of vegetation in deciduous and evergreen habitats (Thomas et al. 2010). Tits feed on caterpillars that eat oak *Quercus* spp. leaves, and in spring, females generally lay earlier in deciduous oak *Q. humilus* forest. In this habitat it is the phenology of the oaks (or insects that feed on oak leaves) that primarily determines the date of laying, rather than temperature (based on path analysis). The effect of temperature in this case is indirect through its effect on tree (and insect) phenology. However, in evergreen oak *Q. ilex* forest, laying generally starts about a month later, and, here, the phenology of oaks (or insect food) is less important and there is a stronger direct effect of temperature on laying date. These differences appear to be related to the reliability and importance of different cues. In the deciduous oak forest where breeding is early, temperatures can be variable and low overnight, which can increase thermoregulatory costs (Thomas et al. 2010). In this situation, the phenology of oaks (and the associated insect food) may give a more reliable cue about the progress of the season that is less affected by short-term changes in temperature. Thus, in deciduous forest, the effect of temperature on timing of laying is primarily through its effect on oak (and insect) phenology, whereas in evergreen habitat there is also a significant direct effect of temperature on date of laying. In both habitats, there are significant correlations between timing of laying and both temperature and insect phenology (Thomas et al. 2010); however, it is important to note that the relative importance of these relationships changes in different habitats and over the course of the breeding season.

In both evergreen and deciduous oak habitats female tits appear to use the phenology of trees to time their laying such that their nestlings develop during the peak of food (caterpillar) abundance (Thomas et al. 2010). Whether the timing of breeding (laying) is related to the timing of peak food abundance in other species is an open question, because only a handful of studies have examined phenology of the food supply and timing of breeding in detail. In contrast to tits in Europe, which feed on caterpillars with a distinct and short seasonal peak in abundance, studies of other species with a more general diet suggest that timing of breeding may be related to food abundance when the eggs are laid, rather than when the nestlings are developing. For example, the laying date of tree swallows in North America is related to the abundance of aerial insects during egg formation, rather than during the period of nestling development (Dunn et al. 2011). In most swallow populations, aerial insect abundance is generally unpredictable at the beginning of the season, and there is no seasonal peak in food abundance as occurs in tits. Instead, there is a general increase in food abundance over the breeding season with no apparent decline, except perhaps later in the summer after the birds have finished breeding (Dunn et al. 2011). Thus, female swallows cannot reliably time their laying based on conditions during the early breeding season as in tits.

Studies of birds and other animals reveal mixed evidence for the mismatch hypothesis, which may be related to: (1) seasonality of the food supply, (2) the strength of selection on synchronized breeding, and (3) the predictability of peaks in food supply later in the season (reviewed by Dunn et al. 2011). For example, the mismatch hypothesis may not apply in species with food supplies that are abundant throughout most of the breeding season, such that there is little benefit to synchronize breeding with shifts in a narrow food peak (see also Halupka et al. 2008).

24.6 Critical Appraisal of the State of Knowledge and What More Needs to Be Done

There can be little doubt that bird phenology has changed in recent decades. This has a most marked effect on the timing of spring migration of summer visitors and on the timing of breeding. The effects on autumn departure of summer visitors and the migration timing of winter visitors are less consistent, possibly reflecting the greater difficulty in collecting these data and possibly reflecting that individual species have their own strategies. We have little doubt that it is easier to collect phenological information on plants than on birds, which are highly mobile and sometimes secretive. In a short chapter like this it has not been possible to discuss other aspects of bird phenology, such as the timing of molt in autumn and the detection of bird song in resident species in spring.

There has been a general trend towards earlier arrival of birds in spring and this has been reported from a wide range of geographic locations. Arrival times in general appear to be related to temperatures with earlier arrival in warmer years. However a comparison of arrival date with temperature at destination has tended to ignore the potential influence of temperatures en-route. The response to destination temperatures is typically 2 days/°C, not enough to explain the sometimes dramatic shifts in the timing of migration. We do need to take into account migration route temperatures, land use change and, possibly, adaptation strategies to fully explain the magnitude of recent changes and to be able to predict the consequences of global warming.

Similarly we have detected some very marked trends towards earlier breeding activity, often associated with warmer weather. Breeding earlier often results in larger clutch sizes or greater survival of offspring (Lack 1968), so we might expect warmer springs to benefit bird populations. Indeed, a 2.5 °C increase is predicted to increase the carrying capacity of a Norwegian dipper *Cinclus cinclus* population by 58 % (Sæther et al. 2000). Similarly, black-throated blue warblers *Dendroica caerulescens* in New Hampshire had higher annual fecundity in La Niña years (Sillett et al. 2000), when May temperatures tend to be warmer in New England. In some species, the particular pattern of climate change is likely to be important. For example, in capercaillie *Tetrao urogallus* the weather most favorable for successful breeding consisted of a quickly rising temperature in April when eggs were laid and a warm, dry period in early June when chicks hatched (Moss et al. 2001). It is also important to keep in mind that even though warmer temperatures may increase reproductive success, this increase could easily be eliminated by increases in mortality on the wintering grounds or at later stages in life. Thus, there may be a number of constraints that limit the responses of bird populations to warmer weather (Winkler et al. 2002). Considering how much we know about bird biology in general, it is somewhat surprising that we know so little about such basic issues as the effects of phenology on reproductive success and population demography. More focused studies of the effects of phenology are needed.

Phenological data vary enormously in their quality. Some data are better than no data at all, but we do need to be cautious in interpreting change. The most reliable data come from schemes that follow a standardized protocol, either through observation or capture, but even these methods can be criticized. Capture relies on landfall of migrants, which increases in poorer weather conditions. Observation requires good visibility and may not take into account nocturnal migrants unless radar is being used. Population size may influence first events, but are not expected to influence the mean/median or standard deviation of the data.

Despite some reservations over data interpretation there can be little doubt that a response to a changing climate is happening in the phenology of bird populations. This evidence is particularly compelling for the timing of breeding and migration arrival where intensive scientific research has taken place. Rich sources of data provide ample opportunity to examine bird phenology and the complex relationships both between and within species and will be a productive area of research in the coming years.

References

Ambrosini R, Møller AP, Saino N (2009) A quantitative measure of migratory connectivity. J Theor Biol 257:203–211

Balbontín J, Møller AP, Hermosell IG, Marzal A, Reviriego M, de Lope F (2009) Individual responses in spring arrival date to ecological conditions during winter and migration in a migratory bird. J Anim Ecol 78:981–989

Both C, Visser ME (2001) Adjustment to climate change is constrained by arrival date in a long-distance migrant bird. Nature 411:296–298

Both C, Artemyev AV, Blaauw B et al (2004) Large-scale geographical variation confirms that climate change causes birds to lay earlier. Proc R Soc Lond B 271:1657–1662

Bradley NL, Leopold AC, Ross J, Huffaker W (1999) Phenological changes reflect climate change in Wisconsin. Proc Natl Acad Sci USA 96:9701–9704

Buse A, Dury SJ, Woodburn RJW, Perrins CM, Good JEG (1999) Effects of elevated temperature on multi-species interactions: the case of Pedunculate Oak, Winter Moth and Tits. Func Ecol 13(suppl 1):74–82

Cotton PA (2003) Avian migration phenology and global climate change. Proc Natl Acad Sci USA 100:12219–12222

Crick HQP, Gibbons DW, Magrath RD (1993) Seasonal variation in clutch size in British birds. J Anim Ecol 62:263–273

Crick HQP, Dudley C, Glue DE, Thomson DL (1997) UK birds are laying eggs earlier. Nature 388:526

Dunn PO (2004) Breeding dates and reproductive performance. In: Møller AP, Fiedler W, Berthold P (eds) Birds and climate change. Elsevier, San Diego

Dunn PO, Winkler DW (1999) Climate change has affected the breeding date of tree swallows throughout North America. Proc R Soc Lond B 266:2487–2490

Dunn P, Winkler D, Whittingham L, Hannon S, Robertson R (2011) A test of the mismatch hypothesis: how is timing of reproduction related to food abundance in an aerial insectivore? Ecology 92:450–461

Ellegren H (1990) Timing of autumn migration in Bluethroats *Luscinia s.svecica* depends on timing of breeding. Ornis Fennica 67:13–17

Filippi-Codacciuoni O, Moussus JP, Urcun JP, Jiguet F (2010) Advanced departure dates in long-distance migratory raptors. J Ornithol 151:687–694

Geen G (2002) Common Chiffchaff (Chiffchaff) *Phylloscopus collybita*. In: Wernham CV, Toms MP, Marchant JH, Clark JA, Siriwardena GM, Baillie SR (eds) The migration atlas: movements of the birds of Britain and Ireland. T and AD Poyser, London

Gilyazov A, Sparks T (2002) Change in the timing of migration of common birds at the Lapland nature reserve (Kola Peninsula, Russia) during 1931–1999. Avian Ecol Behav 8:35–47

Halupka L, Dyrcz A, Borowiec M (2008) Climate change affects breeding of reed warblers *Acrocephalus scirpaceus*. J Avian Biol 39:95–100

Inouye DW, Barr B, Armitage KB, Inouye BD (2000) Climate change is affecting altitudinal migrants and hibernating species. Proc Natl Acad Sci USA 97:1630–1633

Jenkins D, Sparks TH (2010) The changing bird phenology of Mid Deeside, Scotland 1974–2010. Bird Stud 57:407–414

Jenni L, Kéri M (2003) Timing of autumn bird migration under climate change: advances in long distance migrants, delays in short distance migrants. Proc R Soc Lond B 270:1467–1472

Koike S, Higuchi H (2002) Long-term trends in the egg-laying date and clutch size of Red-cheeked Starlings Sturnia philippensis. Ibis 144:150–152

Kok OB, Van Ee CA, Nel DG (1991) Daylength determines departure date of the spotted flycatcher *Muscicapa striata* from its winter quarters. Ardea 79:63–66

Lack D (1968) Ecological adaptations for breeding in birds. Methuen, London

Lehikoinen A, Jaatinen K (2012) Delayed autumn migration in northern European waterfowl. J Ornithol 153:563–570

Lehikoinen E, Sparks TH (2010) Changes in migration. In: Møller AP, Fiedler W, Berthold P (eds) Effects of climate change on birds. Oxford University Press, Oxford

Ludwichowski I (1997) Long-term changes of wing-length, body mass and breeding parameters in first-time breeding females of goldeneyes (*Bucephala clangula clangula*) in Northern Germany. Vogelwarte 39:103–116

Marra PP, Hobson KA, Holmes RT (1998) Linking winter and summer events in a migratory bird by using stable carbon isotopes. Science 282:1884–1886

Mason CF (1995) Long-term trends in the arrival dates of spring migrants. Bird Stud 42:182–189

Mitchell GW, Newman AEM, Wikelski M, Norris DR (2012) Timing of breeding carries over to influence migratory departure in a songbird: an automated radiotracking study. J Anim Ecol. doi:10.1111/j.1365-2656.2012.01978.x

Møller AP, Rubolini D, Lehikoinen E (2008) Populations of migratory bird species that did not show a phenological response to climate change are declining. Proc Natl Acad Sci USA 105:16195–16200

Moss R, Oswald J, Baines D (2001) Climate change and breeding success: decline of the capercaillie in Scotland. J Anim Ecol 70:47–61

Ottosson U, Bairlein F, Hjort C (2002) Migration patterns of Palaearctic Acrocephalus and Sylvia warblers in north-eastern Nigeria. Vogelwarte 41:249–262

Sæther BE, Tufto J, Engen S, Jerstad K, Røstad OW, Skåtan JE (2000) Population dynamical consequences of climate change for a small temperate songbird. Science 287:854–856

Saino N, Szép T, Romano M, Rubolini D, Spina F, Møller AP (2004) Ecological conditions during winter predict arrival date at the breeding quarters in a trans-Saharan migratory bird. Ecol Lett 7:21–25

Sillett TS, Holmes RT, Sherry TW (2000) Impacts of a global climate cycle on population dynamics of a migratory songbird. Science 288:2040–2042

Sokolov LV (2000) Spring ambient temperature as an important factor controlling timing of arrival, breeding, post-fledging dispersal and breeding success of Pied Flycatchers *Ficedula hypoleuca* in Eastern Baltic. Avian Ecol Behav 5:79–104

Sokolov LV (2006) Influence of the global warming on the timing of migration and breeding of passerines in the 20th century. Entomol Rev 86:59–81

Sokolov LV, Gordienko NS (2008) Has recent climate warming affected the dates of bird arrival to the Il'men Reserve in the Southern Urals? Russ J Ecol 39:56–62

Sokolov LV, Kosarev VV (2003) Relationship between timing of arrival of passerines to the Courish Spit and North Atlantic Oscillation index (NAOI) and precipitation in Africa. Proc Zool Inst Russ Acad Sci 299:141–154

Sokolov LV, Payevsky VA (1998) Spring temperatures influence year-to-year variations in the breeding phenology of passerines on the Courish Spit, eastern Baltic. Avian Ecol Behav 1:22–36

Sokolov LV, Markovets MY, Shapoval AP, Morozov YG (1998) Long-term trends in the timing of spring migration of passerines on the Courish spit of the Baltic sea. Avian Ecol Behav 1:1–21

Sokolov LV, Markovets MY, Morozov YG (1999) Long-term dynamics of the mean date of autumn migration in passerines on the Courish spit of the Baltic sea. Avian Ecol Behav 2:1–18

Sparks TH, Braslavská O (2001) The effects of temperature, altitude and latitude on the arrival and departure dates of the swallow *Hirundo rustica* in the Slovak Republic. Int J Biomet 45:212–216

Sparks TH, Carey PD (1995) The responses of species to climate over two centuries: an analysis of the Marsham phenological record, 1736–1947. J Ecol 83:321–329

Sparks TH, Mason CF (2004) Can we detect change in the phenology of winter migrant birds in the UK? Ibis 146(1):57–60

Sparks TH, Roberts DR, Crick HQP (2001) What is the value of first arrival dates of spring migrants in phenology? Avian Ecol Behav 7:75–85

Sparks TH, Bairlein F, Bojarinova JG, Hüppop O, Lehikoinen EA, Rainio K, Sokolov LV, Walker D (2005) Examining the total arrival distribution of migratory birds. Glob Change Biol 11:22–30

Sparks TH, Huber K, Bland RL, Crick HQP, Croxton PJ, Flood J, Loxton RG, Mason CF, Newnham JA, Tryjanowski P (2007) How consistent are trends in arrival (and departure) dates of migrant birds in the UK? J Ornithol 148:503–511

Studds CE, Marra PP (2011) Rainfall-induced changes in food availability modify the spring departure programme of a migratory bird. Proc R Soc Lond B 278:3437–3443

Sueur F, Triplet P (2001) Réchauffement climatique: les passereaux arrivent-ils plus tôt au printemps? Avifaune Picardie 1:111–120

Thomas DW, Bourgault P, Shipley B, Perret P, Blondel J (2010) Context-dependent changes in the weighting of environmental cues that initiate breeding in a temperate passerine, the Corsican blue tit (*Cyanistes caeruleus*). Auk 127:129–139

Thorup K, Tøttrup AP, Rahbek C (2007) Patterns of phenological changes in migratory birds. Oecologia 151:697–703

Tøttrup AP, Thorup K, Rahbek C (2006) Changes in timing of autumn migration in north European songbird populations. Ardea 94:527–536

Tryjanowski P, Kuźniak S, Sparks T (2002) Earlier arrival of some farmland migrants in western Poland. Ibis 144:62–68

Tryjanowski P, Kuźniak S, Sparks TH (2005) What affects the magnitude of change in first arrival dates of migrant birds? J Ornithol 146:200–205

Visser ME, Both C (2005) Shifts in phenology due to global climate change: the need for a yardstick. Proc R Soc Lond B 272:2561–2569

Visser ME, van Noordwijk AJ, Tinbergen JM, Lessells CM (1998) Warmer springs lead to mistimed reproduction in great tits (*Parus major*). Proc R Soc Lond B 265:1867–1870

Winkel W, Hudde H (1997) Long-term trends in reproductive traits of tits (*Parus major, P. caeruleus*) and Pied Flycatchers *Ficedula hypoleuca*. J Avian Biol 28:187–190

Winkler DW, Dunn PO, McCulloch CE (2002) Predicting the effects of climate change on avian life-history traits. Proc Natl Acad Sci USA 99:13595–13599

Chapter 25
Reproductive Phenology of Large Mammals

Jeffrey Kerby and Eric Post

Abstract Many large herbivores, specifically ungulates, display a distinct seasonality in their reproductive phenology. Focusing on empirical studies of caribou/reindeer, moose, and red deer, we illustrate the influence of abiotic (i.e. climatic) and biotic (i.e., density dependent) factors on the timing of calving–an important life-history trait affecting population dynamics. Furthermore, we clarify the distinction between the concepts of timing and synchrony of births, as well as the difference between long-term (i.e., evolutionary) and proximal influences on these population level traits. These distinctions are essential when interpreting the consequences of variation in the timing of parturition, particularly in the context of changing abiotic seasonality caused by climate change.

25.1 Introduction

This discussion of the reproductive phenology of mammals will focus on large herbivores, sometimes referred to as ruminants or ungulates. Large herbivores have been a major focus of theoretical and empirical studies of the influences of biotic and abiotic factors on reproductive phenology since the pioneering study by Estes (1976) on breeding synchrony in wildebeest (*Connochaetes taurinus*). The examples used in this chapter derive mainly from multi-annual studies of caribou/reindeer (both *Rangifer tarandus*) and moose (*Alces alces*) in arctic and sub-arctic environments, and red deer (*Cervus elaphus*) on the north-temperate Isle of Rhum, Scotland. All three species illustrate the influences of abiotic (i.e., climatic) factors on reproductive phenology, while the latter illustrates with striking clarity the influence of population density on timing of calving.

J. Kerby • E. Post (✉)
Department of Biology, The Pennsylvania State University, University Park, PA 16802, USA
e-mail: esp10@psu.edu

M.D. Schwartz (ed.), *Phenology: An Integrative Environmental Science*,
DOI 10.1007/978-94-007-6925-0_25, © Springer Science+Business Media B.V. 2013

The distinction will be made in this chapter between the timing and synchrony of births, as well as between long-term (i.e., evolutionary) and proximal influences on timing and synchrony of births. The distinction between timing and synchrony is important because different forces may act upon the two; moreover, biological and environmental factors may act upon one to influence the other. The distinction between long-term and proximal influences is aimed at clarifying the difference between general patterns of reproductive phenology and interannual variation about those patterns.

25.2 Long-Term Influences and Evolutionary Considerations

In many species of large herbivores, there is an obvious season of births. This is especially evident in seasonal environments, though the length of the birth season is highly variable among species. The season of births is referred to as the timing of parturition, while the length of this season (the number of days over which births occur within a population) is referred to as synchrony of partucxxrition.

Many competing, though not necessarily exclusive, hypotheses have been forwarded to explain why births occur seasonally and with a high degree of synchrony in populations of several species of large herbivores, including wildebeest, bighorn sheep (*Ovis canadensis*, Festa-Bianchet 1988), white-tailed deer (*Odocoileus virginianus*, McGinnis and Downing 1977), Dall's sheep (*Ovis dalli*, Rachlow and Bowyer 1991), Mongolian gazelles (*Procapra gutturosa*, Olson et al. 2005), and mule deer (*Odocoileus hemionus*, Bowyer 1991). Chief among these are the "predation hypothesis" and the "seasonality hypothesis."

The predation hypothesis, forwarded originally to explain synchronous breeding in colonially nesting birds (Darling 1938), predicts that synchronous reproduction should result from selection by predation against early- or late-born offspring. Wildebeest in Ngorongoro Crater, Tanzania, for example, experience predation by spotted hyenas (*Crocuta crocuta*) during the calving season. Wildebeest young born into large groups at the peak of the highly synchronized calving season are much less likely to be killed by hyenas than those born at the beginning or end of the calving season (Estes 1976; Estes and Estes 1979). In part, the advantage of being born during the peak of calving derives from predator swamping—the reduction of risk to the individual of being killed because of the greater numbers of potential prey per predator during the peak—while additional benefit may be derived from being born into groups of vigilant mothers congregating during the peak of calving (Rutberg 1987; Bøving and Post 1997).

Rutberg (1987) suggested, however, that synchronous parturition is unlikely to have evolved from asynchronous parturition solely in response to selection acting through predation on newborns. Instead, predation on early- and late-born neonates might act to increase synchrony of parturition in populations that already display seasonal birth peaks timed to coincide with seasonal peaks in resource availability (Rutberg 1987).

The seasonality hypothesis predicts that selection acting through intra-annual variation in weather and resource availability results in optimization of the timing of parturition by individuals to coincide with the seasonal peak of food availability (Sekulic 1978; Sadleir 1969). In understanding how seasonality might result in the evolution of synchronous parturition, it is important to consider, as Rutberg (1987) pointed out, that natural selection probably does not act on synchrony of parturition because synchrony is a population-level characteristic. Rather, synchrony may result from selection on the timing of parturition by individuals, and the consequent variation in their own reproductive success and that of their offspring as influenced by seasonal variation in resource availability (Rutberg 1987; Ims 1990a).

Hence, seasonality may be an important selective force in the evolution of the annual timing of reproduction, with consequences for synchrony, while predation may reinforce or strengthen synchrony without exerting a clearly discernable influence on the annual timing of parturition (Ims 1990a, b). Indeed, evidence from multi-annual studies of depredated populations of Dall's sheep (Rachlow and Bowyer 1991) and moose (Bowyer et al. 1998) in Alaska, USA, and of mule deer (Bowyer 1991) in California, USA, indicate that, despite predation on newborns, timing and synchrony of parturition in the focal populations appear to relate to long-term climatic patterns and their influence on offspring survival.

Translocation experiments help isolate the influence of evolutionary history on the timing of parturition in reindeer. In a particularly striking example, reindeer introduced from Russia to Canada's Mackenzie Delta have continued to calve 4 – 6 weeks earlier than native caribou living on an adjacent range, even after decades of coexistence (Zhigunov 1968; Leader-Williams 1988), a trait mirrored by other non-native reindeer populations in Alaska and Iceland (Leader-Williams 1988; Kielland personal communication, 2012). Differing migratory strategies and foraging behavior may explain their ability to persist on these seemingly mistimed schedules. Other ungulate populations, including Himalayan tahr, red deer (Caughley 1971), and reindeer from Norway (Leader-Williams 1988), have, at least initially, retained seasonal reproductive traits similar to their source populations even after relocation from northern to southern hemisphere, albeit with a 6-month reversal in dates (Leader-Williams 1988), further evidence that seasonal selection is an underlying driver of the general timing of parturition.

While seasonal reproduction by large mammals may reflect evolutionary strategies by individuals to time parturition to coincide with seasonal peaks in resource availability, multi-annual studies typically reveal variation among years in the onset and median date of parturition. We may, for instance, refer to the birth season of large herbivores in sub-arctic environments as occurring generally in mid to late May, but it would be ill-informed to state that Alaskan moose give birth precisely on May 18 each year. The median date of parturition by female moose in Denali National Park, Alaska, for example, varied by 7 days over a 5-year period (Bowyer et al. 1998), while that of female caribou in the same area varied by 8 days over 9 years (Adams and Dale 1998, Fig. 25.1).

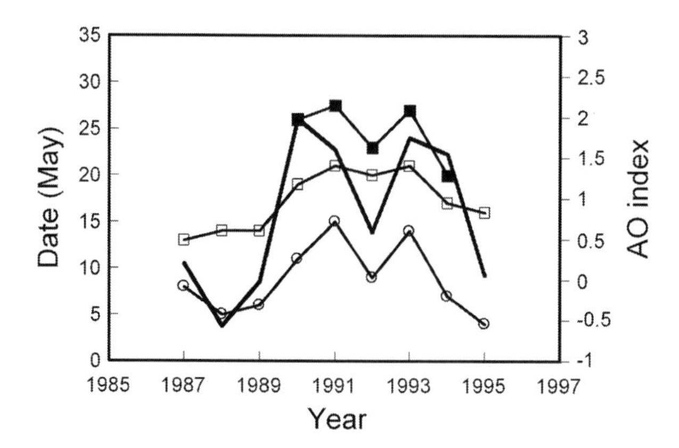

Fig. 25.1 Timing of calving by caribou and moose in Denali National Park, Alaska, USA, in relation to the winter index of the Arctic Oscillation (AO). Shown are annual first (○) and median (□) dates of calving by caribou (Adapted from Adams and Dale 1998), and median dates of calving by moose (■) (Adapted from Bowyer et al. 1998), overlain by the AO index (Thompson and Wallace 1998; *heavy black line*) of the previous winter

25.3 Proximal Influences on Timing of Parturition

25.3.1 Direct Climatic Influences

The distinction between the general timing of the season of births and the exact dates over which parturition occurs in a period of years is one of long-term (i.e., evolutionary) vs. proximal influences on the timing of parturition. In highly seasonal or extreme environments, interannual variation in the timing of parturition may relate directly to variation in weather as it influences condition of reproducing females before conception or during pregnancy. Ostensibly, weather may reduce the physical condition of females to the point where oestrus and, consequently, parturition are delayed. Indeed, the interannual variation in onset and median dates of parturition by Alaskan caribou depicted in Fig. 25.1 was highly correlated with late winter snowfall during the previous year (Adams and Dale 1998).

Alternatively, or additionally, weather may influence the condition of reproductive females during pregnancy, with consequences for the timing of parturition. Keech et al. (2000), for example, noted that pregnant female moose with good body condition (thick rump fat) gave birth earlier than pregnant females with poor body condition, and concluded that this difference reflected environmental influences during pregnancy.

Bowyer et al. (1998) found no relation between the interannual variability in timing of parturition by moose depicted in Fig. 25.1 and local winter weather, but suggested this might have reflected low sample size (5 years of data). Nonetheless, the high degree of correlation between the median date of calving by moose and the median date of calving by caribou ($r = 0.85$, $P = 0.07$), and the median date of calving by moose and the first date of calving by caribou ($r = 0.95$, $P = 0.01$) in Denali park (Fig. 25.1), suggests a common climatic influence on interannual variation in the timing of parturition by both species. The winter index of the Arctic Oscillation (Thompson and Wallace 1998) correlates well with both annual data on cumulative snow depth (Mech et al. 1998) for the period in Denali National Park ($r = 0.62$, $P = 0.04$) and annual dates of first calving ($r = 0.78$, $P = 0.01$) and median calving ($r = 0.78$, $P = 0.01$) by caribou (Fig. 25.1). Although this correlation was not significant for moose ($r = 0.42$, $P = 0.49$), the directions of these relationships suggest later parturition following snowy winters in both populations.

25.3.2 Indirect Climatic Influences: Plant Phenology and Productivity

In addition to acting directly on condition of reproductive females both prior to and during pregnancy, weather can also influence the timing of parturition through its influence on the timing of plant growth. Desert bighorn sheep in California, USA, for example, display a peak in lambing just after winter rains when forage plant productivity peaks (Rubin et al. 2000). Similarly, timing of parturition in southern mule deer in California relates to temperatures and precipitation in the last third of gestation, when forage productivity is highest (Bowyer 1991).

Caribou inhabiting sub-arctic and arctic environments, where the season of plant growth is short and forage plants display a distinct peak in nutrient quality (Klein 1990), also appear to time parturition to coincide with patterns of plant phenology. On the Southern Alaska Peninsula, USA, the progression of the caribou calving season tracks closely the progression of plant phenology on calving ranges of two herds (Fig. 25.2a, Post and Klein 1999). In this example, the proportion of calves observed on each calving range increased rapidly with the number of forage species emerging, despite the fact that the dates of onset of calving differed between these populations, whose nutritional regimes differed (Post and Klein 1999). Similarly, a comparison among arctic and sub-arctic herds of caribou and wild reindeer, some of which were depredated and others not, revealed a close association between the onset of calving and the onset of the season of plant growth (Fig. 25.2b, Skogland 1989). The extent to which timing of parturition tracks plant phenology in *Rangifer* varies within populations from year to year with, as will be explained later in this chapter, consequences for offspring production and survival.

In other species of ungulate, however, such as red deer, timing of parturition appears more flexible and is sensitive to inter-annual variation in climatic conditions (Post 2003a). The long-term, individual-based study of red deer on the Isle of Rhum, Scotland, reveals, for example, that during a density stable period from 1980 to 2007, females advanced their dates of oestrous and parturition by an average of 7.3 and 11.8 days respectively (Moyes et al. 2011). Growing degree-days (GDD), an indirect measure of plant development, explained significant variation in

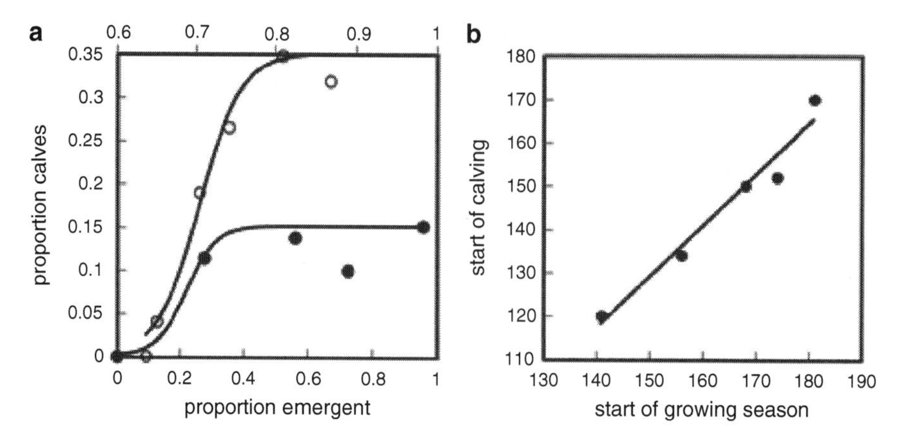

Fig. 25.2 Timing of calving by caribou in relation to plant phenology. In panel (**a**), the proportion of calves in congregations of female caribou of the Southern Alaska Peninsula Herd, USA, increases with the proportion of forage plants emergent on separate calving ranges, Caribou River (○) and Black Hill (●) (Adapted from Post and Klein 1999). In panel (**b**), a cross-population comparison reveals that the start of the calving season is highly correlated with the start of the plant-growing season on ranges in North America and Norway (Adapted from Skogland 1989)

both of these advances, even after accounting for an annual trend toward earlier calving presumably related to direct climate effects (Moyes et al. 2011).

Occasional out-of-season births have been documented in populations of red deer (Guinness et al. 1978), reindeer (Olstad 1930) and elk (Smith 1994) among other species. In Spain, manipulation of red deer fertilization dates via artificial insemination has revealed that, despite equivocal observational evidence of gestation length plasticity (Guinness et al. 1978), females impregnated unusually early in the season are capable of prolonging gestation duration, in some instances by up to 10 days (Garcia et al. 2006). Recent studies suggest this ability may be common to several cervid species (Asher 2011). Garcia et al. (2006) interpret this as evidence of an adaptive plasticity in gestation length designed to lower the costs of parturition mistimed with seasonal peaks in resource availability.

25.3.3 Density-Dependent Influences

Prior to the 1970s, the density of red deer on the Isle of Rhum was carefully managed, but in 1971 culling ceased and the female segment of the population was allowed to increase naturally from 57 to 166 (a three-fold increase) by 1983. Over this period, the median date of calving progressed from June 1st to June 11th (Fig. 25.3, Clutton-Brock et al. 1987). As density increased towards carrying capacity on Rhum, competition for resources increased, reducing the condition of females prior to breeding; consequently, the average date of conception progressed

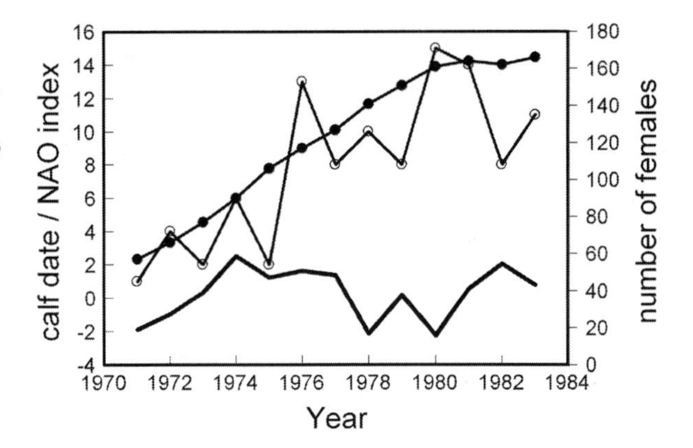

Fig. 25.3 Median dates of calving (○) by female red deer in the population on the Isle of Rhum, Scotland, in relation to female density (●) and the winter NAO index (*heavy line*) of the previous year (Data on calving dates and female density are from Clutton-Brock et al. 1985)

from September 25th to October 9th over the period, resulting in later birth dates (Clutton-Brock and Albon 1989). These patterns have also been observed in a red deer population in Norway, where the effects of high population density have been related to delayed and synchronized ovulation, resulting in similarly delayed dates of parturition (Langvatn et al. 2004).

As Clutton-Brock and Albon (1989) pointed out, the influence of population density on timing of calving on Rhum was so strong that, after accounting for density, the median date of parturition varied by only 4 days between 1971 and 1983. In the latter part of this period, from 1977 onward, there is an apparent negative association between the timing of calving and the wintertime state of the North Atlantic Oscillation (NAO) (Fig. 25.3). Because the NAO correlates positively with late winter temperatures in the north Atlantic region, this would seem to suggest, at higher densities, a tendency toward earlier calving in years with warmer winters and, presumably, earlier springs.

To investigate these relationships, we tested for combined influences of density and climate (the NAO) in a non-linear generalized additive model (GAM) (Hastie and Tibshirani 1999). The results of the GAM support clearly the reported influence of density on timing of calving, and indicate that this relationship tends to level off at the highest densities (Fig. 25.4a). This seems to suggest that although increasing density results in progressively later dates of calving, there is a constraint on this relationship imposed, perhaps, by physiological limits on the timing of oestrus and/or the length of the period of gestation. Moreover, the GAM indicates a tendency toward earlier calving 1 year after warm winters (Fig. 25.4b). Although it is not possible to determine with these data whether this reflects a direct or indirect influence of weather, the timing of plant growth occurs earlier following warm (positive NAO) winters in many regions (Post and Stenseth 1999; Post 2003b).

The influence of density on timing of calving by red deer on Rhum suggests density dependence may influence timing of reproduction in other populations as well. Hence, we used the same approach, together with data on annual estimates of population density for caribou in the Denali Herd (Mech et al. 1998) to test for combined influences of density dependence and the Arctic Oscillation on timing of

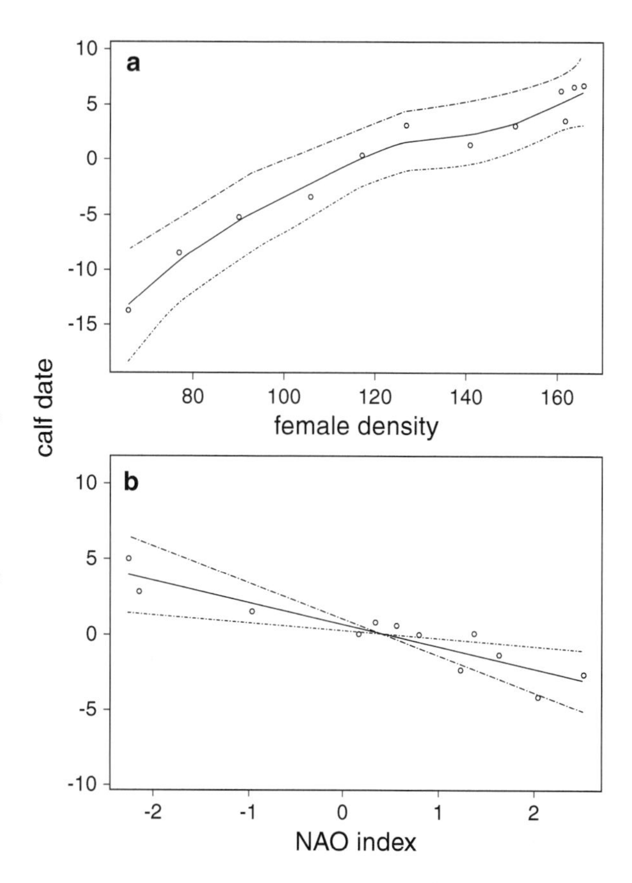

Fig. 25.4 Results of a generalized additive model (GAM) of timing of calving by red deer on the Isle of Rhum in relation to density (**a**) and the winter NAO index of the previous year (**b**). Following the general model of the timing of life history events in relation to climate, density, and resources developed in Post et al. (2001), the GAM also included an autoregressive spline function specifying the influence of the timing of calving in the previous year (not shown). Y-axis values are standardized to the null deviance of the fitted model (*solid lines*), and dashed lines are 95 % confidence bands

calving in that population. This analysis reveals an effect of density on timing of calving by caribou in the Denali herd (Fig. 25.5a) that mirrors the effect of density on timing of calving by red deer on Rhum. Moreover, the influence of the Arctic Oscillation on timing of calving is still evident after incorporating the influence of density (Fig. 25.5b).

25.4 Consequences of Variation in Timing of Parturition

The timing of birth in relation to resources and maternal condition influences the early development of individuals in many species (Lindström 1999). What population-level consequences might arise, though, from environmentally- and biologically-induced variation in timing of parturition?

For red deer on Rhum, survival of calves through their first year is strongly dependent upon their birth weight and birth date (Clutton-Brock et al. 1987). Calf survival through their first summer increases with birth weight, but declines with

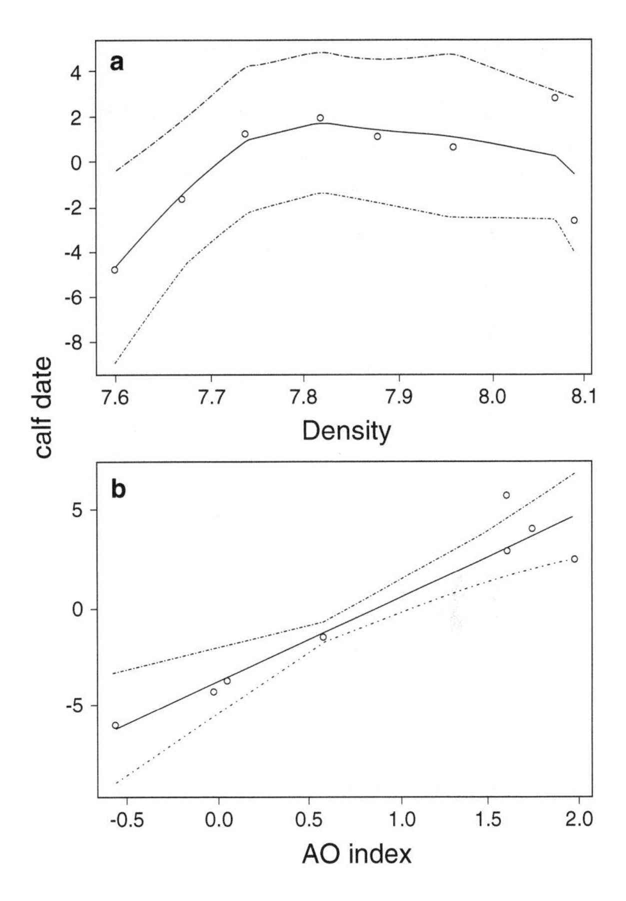

Fig. 25.5 Influences of population density (**a**) and the winter AO index (**b**) of the previous year on timing of calving by caribou in the Denali Herd, as for red deer in Fig. 25.4

later birth dates (Clutton-Brock et al. 1987). Calves born late are also less likely to survive their first winter (Clutton-Brock et al. 1987). The consequences of variation in birth dates for population dynamics are evident in the fact that calf winter mortality is the key factor regulating population size of red deer on Rhum (Clutton-Brock et al. 1985).

Climatic conditions during pregnancy can also result in phenotypic variation among cohorts of individuals that relates to conditions in their year of birth (Post et al. 1997). In Soay sheep (*O. aries*) in the Outer Hebrides of Scotland, the North Atlantic Oscillation influences birth dates and weights of lambs, with consequences for long-term variation among cohorts (Forchhammer et al. 2001, 2002) that contribute to population dynamics (Coulson et al. 2001).

In the Tanana Flats of interior Alaska, USA, birth date influences juvenile survival of moose: later-born individuals die more quickly in their first year (Keech et al. 2000). How birth date contributes to the dynamics of this population is not, however, clear. In Denali National Park, timing of parturition has no apparent influence on survivorship of young moose (Bowyer et al. 1998).

Timing of parturition influences juvenile survival of caribou in the Denali herd indirectly through vulnerability to predation. Caribou calves born during the peak of parturition (5–8 days after the beginning of the calving season) are significantly less likely to be killed by predators than those born before or after the peak (Adams et al. 1995). The contribution of variation in birth dates of caribou calves to their population dynamics is evident in the observations that 98 % of calf mortality in the Denali Herd is due to predation, and that predation during the calving season is the main limiting factor for the Denali Herd (Adams et al. 1995).

25.5 Conclusion: Implications of Climate Change

Given that local weather and large-scale climate exert direct and indirect influences on the timing of parturition in large herbivores, we might expect directional climate change to elicit shifts in the timing of parturition by many species. Such climate-mediated shifts in timing of reproduction may, however, be counteracted by changes in density (Forchhammer et al. 1998). As well, seasonal reproduction in northern ungulates may be cued to photoperiod (Leader-Williams 1988; Lu et al. 2010), which may also constrain long-term shifts in parturition, as is apparently the case with egg laying in some birds (Both and Visser 2001). Moreover, although the GAMs used to analyze timing of parturition among red deer on Rhum and caribou in Denali Park suggest linear relations between climate and birth dates (Figs. 25.4 and 25.5), we might expect constraints on gestation length or timing of onset of oestrus to limit the extent to which timing of parturition can advance in response to climatic warming.

Considering the influence that timing of parturition exerts on juvenile survival, and the link between juvenile survival and population dynamics in many species of large herbivores (Gaillard et al. 1998), it is reasonable to speculate that climate change may influence population dynamics of many species through timing of reproduction. If so, climatic influences on timing of parturition may contribute to spatial synchrony among populations, as in the high degree of correlation between timing of parturition by moose and that by caribou in Denali Park (Fig. 25.1). The dynamics of caribou and muskoxen (*Ovibos moschatus*) on opposite coasts of Greenland, for instance, are highly correlated (Post and Forchhammer 2002), and the dynamics of populations of both species in Greenland relate to the North Atlantic Oscillation at time lags that correspond to time to first reproduction in both species (Forchhammer et al. 2002).

Finally, the broader trophic context of phenological response to warming, or the lack thereof, has bearing on the population level consequences of reproductive phenology (Visser and Both 2005). A long-term study of caribou and plant phenology in a predator free system in western Greenland has documented earlier plant emergence in warmer springs whereas the timing of caribou parturition has remained relatively constant (Thing 1984; Post and Forchhammer 2008). An index quantifying the magnitude of phenological asynchrony between plant emergence and caribou calving over several years of observations is positively correlated with both early calf mortality and low calf recruitment, indicating potential population level consequences of this mismatch (Post and Forchhammer 2008). Unlike red deer on the Isle of Rhum (Moyes et al. 2011), caribou in west Greenland appear unable to track the directional phenological response of their forage plants to changing abiotic conditions (Post and Forchhammer 2008), presumably because, unlike nearly all other mammals, *Rangifer* apparently lack a circadian clock (van Oort et al. 2007; Lu et al. 2010). This perspective, one inclusive of phenological dynamics across multiple trophic levels, must be considered when forecasting the consequences of reproductive phenology under climate change for any large herbivore species. Further research into the implications of climate change for timing of reproduction in large herbivores, with consequences for population dynamics and synchrony, is certainly warranted.

References

Adams LG, Dale BW (1998) Timing and synchrony of parturition in Alaskan caribou. J Mamm 79:287–294
Adams LG, Singer FJ, Dale BW (1995) Caribou calf mortality in Denali National Park Alaska. J Wildl Manag 59:584–594
Asher GW (2011) Reproductive cycles of deer. Anim Repr Sci 124:170–175
Both C, Visser ME (2001) Adjustment to climate change is constrained by arrival date in a long-distance migrant bird. Nature 411:296–298
Bøving PS, Post E (1997) Vigilance and foraging behaviour of female caribou in relation to predation risk. Rangifer 17:55–63
Bowyer RT (1991) Timing of parturition and lactation in southern mule deer. J Mamm 72:138–145
Bowyer RT, Van Ballenberghe V, Kie JG (1998) Timing and synchrony of parturition in Alaskan moose: long-term versus proximal effects of climate. J Mamm 79:1332–1344
Caughley G (1971) Investigation of hybridization between free-ranging wapiti and red deer in New Zealand. N Z J Sci 14:993–1008
Clutton-Brock TH, Albon SD (1989) Red deer in the highlands. Oxford University Press, Oxford
Clutton-Brock TH, Major M, Guinness FE (1985) Population regulation in male and female red deer. J Anim Ecol 54:831–846
Clutton-Brock TH, Major M, Albon SD, Guinness FE (1987) Early development and population dynamics in red deer. I. Demographic consequences of density-dependent changes in birth weight and date. J Anim Ecol 56:53–67
Coulson T, Catchpole EA, Albon SD, Morgan BJT, Pemberton JM, Clutton-Brock TH, Crawley MJ, Grenfell BT (2001) Age, sex, density, winter weather, and population crashes in Soay sheep. Science 292:1528–1531
Darling FF (1938) Bird flocks and breeding cycle. Cambridge University Press, Cambridge
Estes RD (1976) The significance of breeding synchrony in the wildebeest. E African Wildl J 14:135–152
Estes RD, Estes RK (1979) The birth and survival of wildebeest calves. Z Tierpsych 50:45–95
Festa-Bianchet M (1988) Birthdate and survival in bighorn lambs (Ovis canadensis). J Zool 214:653–661
Forchhammer MC, Post E, Stenseth NC (1998) Breeding phenology and climate. Nature 391:29–30

Forchhammer MC, Clutton-Brock TH, Lindström J, Albon SD (2001) Climate and population density induce long-term cohort variation in a northern ungulate. J Anim Ecol 70:721–729
Forchhammer MC, Post E, Stenseth NC, Boertmann D (2002) Long-term responses in arctic ungulate dynamics to variation in climate and trophic processes. Popul Ecol 44:113–120
Gaillard JM, Festa-Bianchet M, Yoccoz NG (1998) Population dynamics of large herbivores: variable recruitment with constant adult survival. Trends Ecol Evol 13:58–63
Garcia AJ, Landete-Castillejos T, Carrion D, Gaspar-Lopez E, Gallego L (2006) Compensatory extension of gestation length with advance of conception in red deer (Cervus elaphus). J Exp Zool 305a:55–61
Guinness FE, Gibson RM, Clutton-Brock TH (1978) Calving times of Red deer (cervus elaphus) on rhum. J Zool Lond 185:105–114
Hastie TJ, Tibshirani RJ (1999) Generalized additive models. Chapman & Hall, New York
Ims RA (1990a) On the adaptive value of reproductive synchrony as a predator-swamping strategy. Am Nat 136:485–498
Ims RA (1990b) The ecology and evolution of reproductive synchrony. Trends Ecol Evol 5:135–140
Keech MA, Bowyer RT, Ver Hoef JM, Boertje RD, Dale BW, Stephenson TR (2000) Life-history consequences of maternal condition in Alaskan moose. J Wildl Manag 64:450–462
Klein DR (1990) Variation in quality of caribou and reindeer forage plants associated with season, plant part, and phenology. Special Issue – Rangifer 3:123–130
Langvatn R, Mysterud A, Stenseth NC, Yoccoz NG (2004) Timing and synchrony of ovulation in red deer constrained by short northern summers. Am Nat 163(5):763–772
Leader-Williams N (1988) Reindeer on South Georgia. Cambridge University Press, Cambridge
Lindström J (1999) Early development and fitness in birds and mammals. Trends Ecol Evol 14:343–348
Lu W, Meng Q-J, Tyler NJC, Stokkan K-A, Loudon ASI (2010) A circadian clock is not required in an arctic mammal. Curr Biol 20:1–5
McGinnis BS, Downing RL (1977) Factors affecting the peak of white-tailed deer fawning in Virginia. J Wildl Manag 41:715–719
Mech LD, Adams LG, Meier TJ, Burch JW, Dale BW (1998) The wolves of Denali. University of Minnesota Press, Minneapolis
Moyes K, Nussey DH, Clements MN, Guinness FE, Morris A, Morris S, Pemberton JM, Kruuk LEB, Clutton-Brock TH (2011) Advancing breeding phenology in response to environmental change in a wild red deer population. Glob Ch Biol 17:2455–2469
Olson KA, Fuller TK, Schaller GB, Lhagvasuren B, Odonkhuu D (2005) Reproduction, neonatal weights, and first-year survival of Mongolian gazelles (Procapra gutturosa). J Zool Lond 265:227–233
Olstad O (1930) Rats and reindeer in the Antarctic. Sci Res Norwegian Antarctic Exped 4:1–20
Post E (2003a) Timing of reproduction in large mammals: climatic and density-dependent influences. In: Schwartz MD (ed) Phenology, an integrative environmental science, 1st edn. Kluwer Academic Publishers, Dordrecht
Post E (2003b) Large-scale climate synchronizes the timing of flowering by multiple species. Ecology 84:277–281
Post E, Forchhammer MC (2002) Synchronization of animal population dynamics by large-scale climate. Nature 420:168–171
Post E, Forchhammer MC (2008) Climate change reduces reproductive success of an Arctic herbivore through trophic mismatch. Philos Trans R Soc B 363:2367–2373
Post E, Klein DR (1999) Caribou calf production and seasonal range quality during a population decline. J Wildl Manag 63:335–345
Post E, Stenseth NC (1999) Climatic variability, plant phenology, and northern ungulates. Ecology 80:1322–1330
Post E, Stenseth NC, Langvatn R, Fromentin J-M (1997) Global climate change and phenotypic variation among red deer cohorts. Proc R Soc Lond B 264:1317–1324

Post E, Forchhammer MC, Stenseth NC, Callaghan TV (2001) The timing of life history events in a changing climate. Proc R Soc Lond B 268:15–23
Rachlow JL, Bowyer RT (1991) Interannual variation in timing and synchrony of parturition in Dall's sheep. J Mamm 72:487–492
Rubin ES, Boyce WM, Bleich VC (2000) Reproductive strategies of desert bighorn sheep. J Mamm 81:769–786
Rutberg AT (1987) Adaptive hypotheses of birth synchrony in ruminants: an interspecific test. Am Nat 130:692–710
Sadleir RMFS (1969) The ecology of reproduction in wild and domesticated mammals. Methuen, London
Sekulic R (1978) Seasonality of reproduction in the sable antelope. E African Wildl J 16:177–182
Skogland T (1989) Comparative social organization of wild reindeer in relation to food, mates, and predator avoidance. Paul Parey Publishers, Berlin
Smith BL (1994) Out-of-season births of elk calves in Wyoming. Prairie Natr 26:131–136
Thing H (1984) Feeding ecology of the West Greenland Caribou (Rangifer tarandus groenlandicus) in the Sisimiut-Kangerlussuaq region. Vildtbiologisk St 12(3):1–53
Thompson DW, Wallace JM (1998) The Arctic oscillation signature in the wintertime geopotential height and temperature fields. Geophys Res Lett 25:1297–1300
van Oort BEH, Tyler NJC, Gerkema MP, Folkow L, Stokkan K-A (2007) Where clocks are redundant: weak circadian mechanisms in reindeer living under polar photic conditions. Naturwissenschaften 94:183–194
Visser ME, Both C (2005) Shifts in phenology due to global climate change: the need for a yardstick. Proc R Soc B 272:2561–2569
Zhigunov PS (1968) Reindeer Husbandry. Israel Program for Scientific Translations, Jerusalem

Part VI
Applications of Phenology

Chapter 26
Vegetation Phenology in Global Change Studies

Kirsten M. de Beurs and Geoffrey M. Henebry

Abstract Changes in the character of the vegetated land surface are frequently expressed in terms of temporal trends in the Normalized Difference Vegetation Index (NDVI) retrieved from spaceborne sensors. In the past these change studies were typically based upon AVHRR data. By the end of 2011, we acquired 11 full years of NASA MODIS data which is a greatly improved dataset compared to extant AVHRR datasets. In this chapter, we present a change analysis based on a global NASA MODIS product (MCD43C4) at a 0.05° (~5.6 km) spatial resolution and a 16-day temporal resolution from 2001 through 2011. This new change map based on 11 years of data presents statistically significant positive and negative changes resulting from both direct and indirect impacts of climatic variability and change, disturbances, and human activity. We found significant negative changes in 8.7 % of the global land area (or 11.8×10^6 km^2), with hotspots in Canada, southeastern USA, Kazakhstan, and Argentina. Significant positive changes appeared in 6.0 % of the global land area (8.0×10^6 km^2) with hotspots in Turkey, China and Western Africa. Attribution is the key challenge in any change analysis. We provide several examples attributable to major modes of change, focusing both on natural disturbances arising from climatic variability and change, and also on changes arising directly from human actions.

K.M. de Beurs (✉)
Department of Geography and Environmental Sustainability, The University of Oklahoma, Norman, OK 73019, USA
e-mail: kdebeurs@ou.edu

G.M. Henebry
Geographic Information Science Center of Excellence, South Dakota State University, Brookings, SD 57007, USA

M.D. Schwartz (ed.), *Phenology: An Integrative Environmental Science*,
DOI 10.1007/978-94-007-6925-0_26,

26.1 Introduction

Widespread increases in plant growth across northern latitudes were first reported in 1997 (Myneni et al. 1997). Currently, there are a vast number of vegetation change analyses based on Advanced Very High Resolution Radiometer (AVHRR) Normalized Difference Vegetation Index (NDVI) data at global scales (Tucker et al. 2001; Zhou et al. 2001; Slayback et al. 2003; Goetz et al. 2005, 2007; Xiao and Moody 2005; Julien et al. 2006; Reed 2006; Huemann et al. 2007; Donohue et al. 2009; Julien and Sobrino 2009b). Many studies confirm changes in spatial and temporal patterns of terrestrial vegetation, linking these changes to warmer winters and springs (Nemani et al. 2003). Most of these large scale vegetation studies are based on the NDVI data derived from reflectance observations acquired by a series of AVHRR orbiting sensors (Myneni et al. 1997). The NDVI exploits a spectral contrast between red and near infrared reflectance to indicate the presence of green vegetation (Tucker 1979). Changes in the character of the vegetated land surface are then expressed in terms of temporal trends in the NDVI retrieved from spaceborne sensors.

The earlier AVHRR NDVI vegetation studies were typically based on just 9 years of satellite data (Myneni et al. 1997, 1998). Later studies (e.g. Julien and Sobrino 2009a, b) used longer AVHRR time series (1981–2003). By the end of 2011, we acquired 11 full years of improved data (2001–2011 in the Northern Hemisphere and July 2000 to June 2011in the Southern Hemisphere) from NASA's Moderate Resolution Imaging Spectroradiometer (MODIS) sensor, which was first launched at the end of 1999 (on the Terra satellite) and again in a different orbit in 2002 (on the Aqua satellite). As a successor to the AVHRR, MODIS provides improved spatial and spectral resolution (Gallo et al. 2005). There are a number of studies that have investigated continuity potential for AVHRR and MODIS (Gallo et al. 2005; Tucker et al. 2005; Brown et al. 2006). Some researchers indicate that the long-term AVHRR data record could be improved to address cross-sensor NDVI continuity (Gallo et al. 2005; Brown et al. 2006). Some recommend a linear transformation between MODIS and AVHRR datasets to improve continuity (Ji et al. 2008). However, cross-sensor NDVI continuity is not straightforward because of differences in processing strategies and sensor-specific spectral band characteristics (van Leeuwen et al. 2006). NASA's land long term data record (LTDR) is a first attempt to produce a consistent long term dataset. However, the data is currently only available until 1999. To date, there are only a few large scale vegetation change analyses based on MODIS data only, especially on a global scale. One recent study evaluated the correlation between fraction of photosynthetically active radiation (FPAR) and the El Nino Southern Oscillation (Potter et al. 2008) for the period 2001–2005. Another evaluated FPAR and leaf area index (LAI) data until 2005 for the tundra and boreal forests in the Northern Hemisphere only (Zhang et al. 2008). Yet another studied droughts in the Great Plains between 2001 and 2005 based on both NDVI and Normalized Difference Water Index (NDWI) data (Gu et al. 2007). MODIS net primary productivity (NPP) data

(2000–2009) have also been used to investigate the drought induced reduction in global terrestrial net primary productivity (Zhao and Running 2010).

A synoptic assessment of vegetation changes can reveal hotspots of change; these hotspots merit closer attention because they indicate significant shifts in local water and carbon fluxes and in the surface energy balance (Kuemmerle et al. 2008; Vuichard et al. 2008; Henebry 2009). There is an urgent need to improve change attribution in studies of land surface dynamics. In particular, it is critical to consider direct human impacts on land surface dynamics originating from land use decisions.

26.2 Data Processing

26.2.1 MODIS Reflectance Data: 0.05° GCM Grid

We selected a global NASA MODIS product (Terra + Aqua Nadir BRDF-Adjusted Reflectance data MCD43C4) at a 0.05° (~5.6 km) spatial resolution and a 16-day temporal resolution from the beginning of 2001 through the last composite in 2011. For the southern hemisphere we used data from the first composite of July 2000 until the last composite of June 2011. The reflectance data in these products have been adjusted using models of bidirectional reflectance distribution functions to simulate reflectance from a nadir view. The MODIS dataset is delivered as 8-day rolling composites, based on a moving window of 16 days.

For each composite in the time series we calculated the Normalized Difference Vegetation Index (NDVI) as follows:

$$NDVI = (NIR - Red)/(NIR + Red) \tag{26.1}$$

where NIR is MODIS band 2, Red is MODIS band 1.

After calculating the NDVI for each composite, we resampled the 8-day product to 16-days by averaging consecutive 8-day composites. The final product consists of 23 NDVI composites for each year (2001 through 2011). We omitted a pixel time series from the analysis when either (1) it lacked more than 40 % of the data or (2) it exhibited low NDVI seasonality, specifically, an average NDVI <0.10 and a seasonal coefficient of variation of NDVI <5 %. As a result of these constraints, we filtered out deserts, inland water bodies, and persistently cloudy and/or hazy areas.

26.2.2 MODIS Land Surface Temperature Data

To calculate accumulated growing degree-days we used the daytime and nighttime land surface temperature (LST) data at 0.05° spatial resolution (MOD11C2). We reconfigured the land surface temperature data from Kelvin to degrees Celsius. Then, we calculated 8-day growing degree-days (AGDD) as follows:

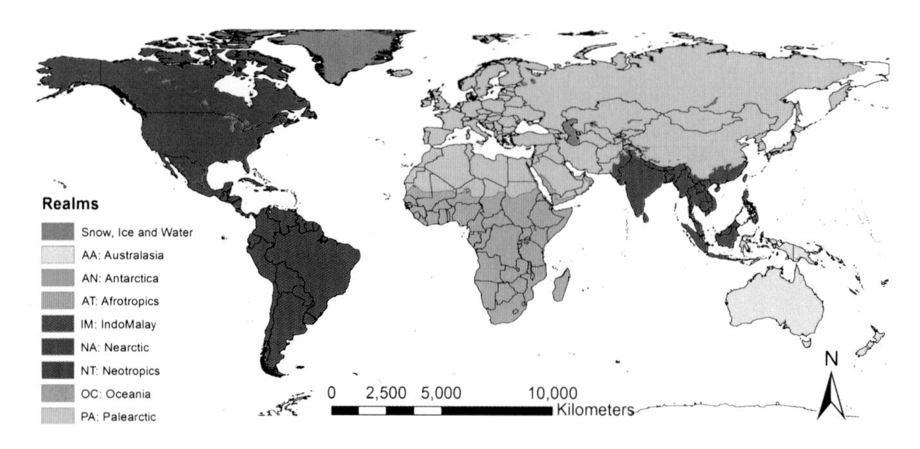

Fig. 26.1 Realms from the WWF Ecoregion scheme used in this study

$$GDD_t = \left(\frac{nighttime\,LST + daytime\,LST}{2} \right) \tag{26.2}$$

We accumulated the GDD by simple summation commencing each 1 January when GDD exceeded the base temperature of 0 °C:

$$AGDD_t(^{\circ}C) = AGDD(^{\circ}C)_{t-1} + \max(GDD_t, 0) \tag{26.3}$$

We only accumulate growing degree-days that are larger than 0. We summarized the data into 16-day composites by taking the maximum of two consecutive composites. We chose a base of 0 °C for the AGDD calculations since this threshold is often used for high-latitude annual crops, such as spring wheat, and for perennial grasslands. We have successfully applied this method several times before (de Beurs and Henebry 2004a, 2005a, b, 2008, 2010b).

26.2.3 WWF Ecoregions

To summarize the results of the change analysis, we used the ecoregional scheme (Olson et al. 2001) of the World Wildlife Fund (WWF). *Ecoregions* as defined by the WWF reflect the *potential* vegetation (Olson et al. 2001). Vegetation, however, can change within a region under anthropogenic influence; indeed, cropland is the dominant land cover in many ecoregions. The terrestrial vegetated land surface is divided into 687 ecoregions. There are two coarser levels of organization within the WWF ecoregions database. The top level is *realm*, roughly comparable to continents (Fig. 26.1). The next finer level is *biome*, which distinguishes major physiognomic types. The WWF scheme defines a total of 8 realms and 14 terrestrial biomes. We use both realms and biomes to summarize the trends.

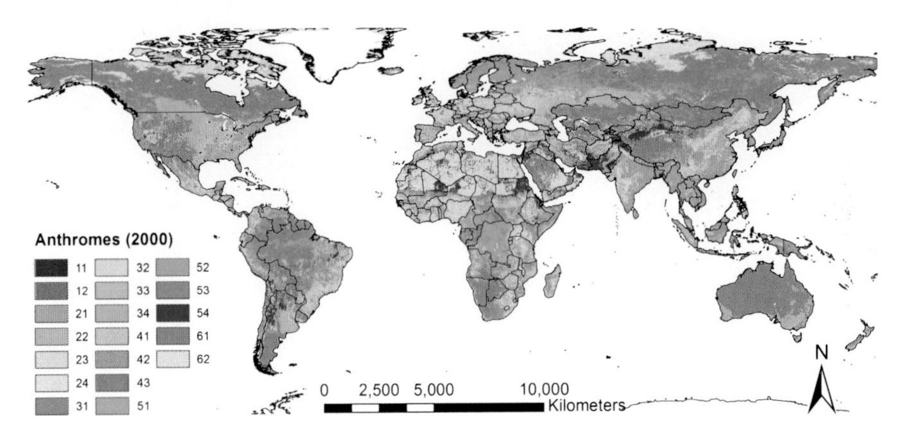

Fig. 26.2 Anthromes in the year 2000. Please see Table 26.3 for a definition of the anthromes

26.2.4 Anthromes

Humans are directly interacting with ecosystems surrounding them, making a profound impact on how the ecosystems are altered. In this study we use the anthromes dataset as described in Ellis and Ramankutty (2008) to identify the changes by human impacted regions. Anthromes are defined as human biomes and describe the terrestrial biosphere in its contemporary form. Estimated population density plays an important role in the delineation of the anthromes. We downloaded the Anthromes v2.0 data from http://ecotope.org/anthromes/v2/data/ and used the data for the year 2000 which was the latest year that the Anthrome data was available (Ellis and Ramankutty 2008). Figure 26.2 gives the global anthromes. The spatial resolution of the data is 0.083° lat/lon. We analyze the vegetation change within each anthrome to determine the effect of people on the amount of vegetation change.

26.3 Methods

26.3.1 Change Analysis

Using simple linear regression to estimate a trend from a time series is a widespread practice in the remote sensing literature. We have stressed previously that NDVI time series typically violate several basic assumptions that validate regression analysis (de Beurs and Henebry 2004b, 2005b; de Beurs et al. 2009). NDVI time series typically violate the assumption that all ordinate values (i.e., mapped on the y-axis) should be mutually independent as there is usually high positive autocorrelation between consecutive observations. We have previously discussed the Seasonal Kendall (aka Seasonal Mann-Kendall) trend test corrected for autocorrelation as a good alternative to change analysis by simple linear regression (de Beurs and Henebry 2004b, 2005b).

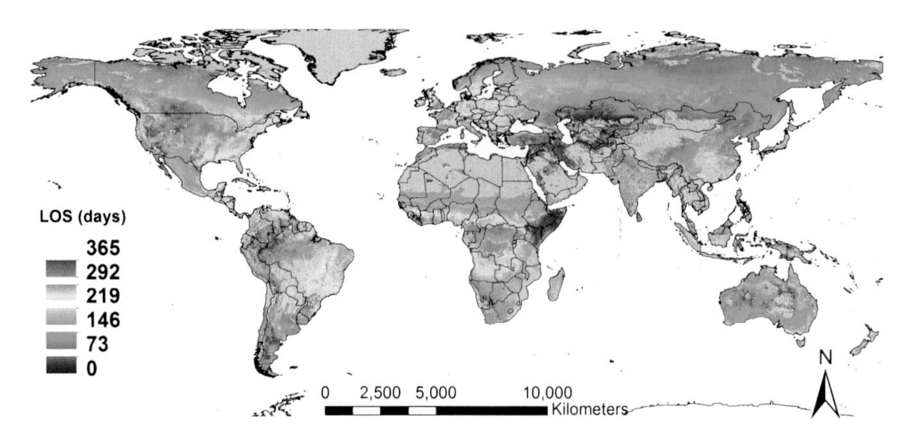

Fig. 26.3 Length of the growing season based on midpoint NDVI method for the averaged years of 2001–2011. *Gray* areas were not calculated

The original Mann-Kendall (MK) trend test is nonparametric and is calculated by summing the number of times a particular observation has a higher value than any of the previous observations (Hirsch et al. 1982). If the value of a particular composite is higher than a previous composite, one is added to the test statistic; if the values are equal, nothing is added; and if the value is lower than a previous composite, one is subtracted. The Seasonal Kendall (SK) trend test for image time series first calculates the MK statistic for each composite separately. The SK statistic for the complete time series consists of the sum of the MK statistics for all composites. The autocorrelation correction is applied to the calculation of the covariance for every combination of seasonal periods (Hirsch and Slack 1984; Hess et al. 2001; de Beurs and Henebry 2004b; de Beurs et al. 2009).

We calculate the SK test for every pixel for all composites during the growing season. The growing season is determined as the average start of season and the average end of season (as determined by the NDVI midpoint method) between 2001 and 2011 for the Northern and Southern Hemisphere (Fig. 26.3). Since the number of composites incorporated in the change analysis for each pixel is allowed to vary in the Northern and Southern Hemisphere, locations that are in the far northern (southern) latitudes generally incorporate fewer composites during the growing season as a result of their longer winters (de Beurs et al. 2009). If the growing season would change drastically between 2001 and 2011, this would be revealed in the trend analysis despite the fact that we maintain an average beginning and ending of the growing season for the trend detection. A changing growing season would result in the observations within the growing season to be higher in the beginning (e.g. earlier growing season) or end (e.g. later growing season). After applying the SK algorithm we accept every pixel with $p < 0.01$ as highly significant.

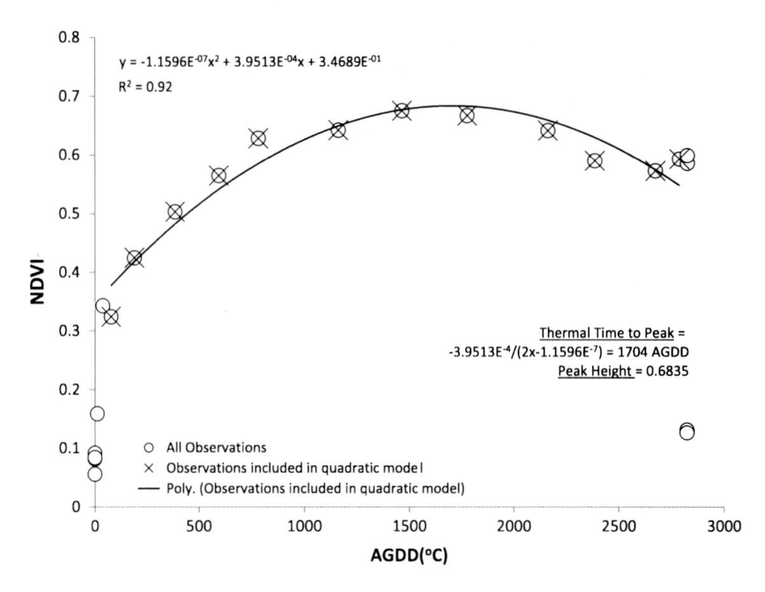

Fig. 26.4 Example of the quadratic growing season model for an area in Canada. The thermal time to peak height can be calculated based on the parameter estimates of the quadratic model

26.3.2 Phenological Analysis

We use quadratic regression models to link MODIS derived AGDD with NDVI and calculate the number of AGDD to reach the peak of the growing season which we have named thermal time to peak and peak NDVI (Fig. 26.4, de Beurs and Henebry 2005b, 2008, 2010c). We only fit the quadratic regression models to the composites that are part of the growing season as determined by the NDVI midpoint method which is the same as we use in Sect. 3.1 (Fig. 26.3, White et al. 1997). We present the percentage change from 2001 to 2011 in the thermal time to peak and the peak NDVI for three regions that reveal a change according to the SK analysis.

26.4 Results

26.4.1 Global Vegetation Changes at Realm Level

Figure 26.3 displays the average length of the growing season as determined using the NDVI midpoint 16-day averaged composites from 2001 until 2011. The gray areas in Fig. 26.3 are masked out as described in Sect. 2.1. In total, 26.1 % of the global land area was masked out. In the Northern Hemisphere, the length of the growing season clearly reveals a north–south gradient with much shorter growing seasons to the north with a minimum of about 60 days. The southern hemisphere reveals a comparable, though less pronounced gradient.

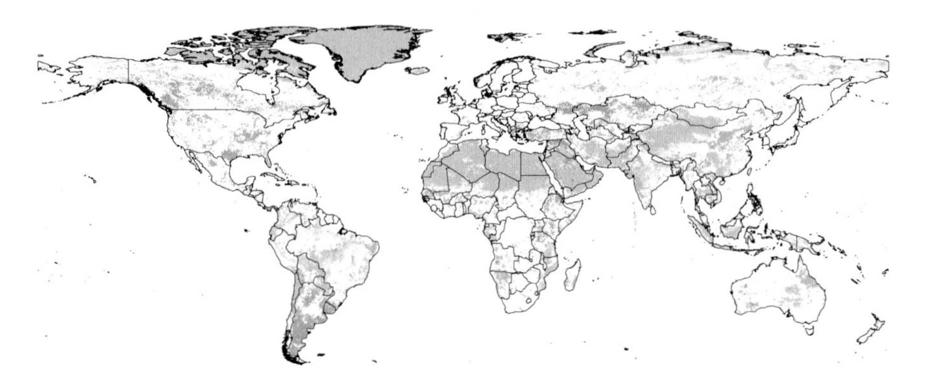

Fig. 26.5 Global vegetation changes in 2001–2011 revealed through the Seasonal Kendall test applied to NASA MODIS time series. Areas outlined in *orange* and *green* indicate highly significant ($p \leq 0.01$) negative and positive changes, respectively. Areas in *gray* were excluded from analysis. Areas in *white* did not exhibit highly significant changes

We report here only highly significant changes ($p < 0.01$) in the vegetated land surface since 2000 (Fig. 26.5). The change results by realm are summarized in Table 26.1. We found significant negative changes in 8.7 % of the global land area (11.8×10^6 km^2), with hotspots in Canada, southeastern USA, Kazakhstan, Argentina, East Africa, the Middle East, and Western Australia. Negative trends especially dominated the Western Hemisphere at the level of WWF realms: with almost 14 % of negative change in the Nearctic (2.8×10^6 km^2) and more than 14 % in the Neotropics (2.8×10^6 km^2, Table 26.1). The Palearctic realm (PA, Eurasia) and the Nearctic realm (NA, North America) reveal 6.8 and 13.7 % negative change, or 3.5×10^6 km^2 and 2.8×10^6 km^2, respectively. Significant positive changes appeared in 6.0 % of the global land area (8.0×10^6 km^2), with the largest percentage of positive change in the IndoMalay (IM; 10.3 %, 0.8 $\times 10^6$ km^2), followed by the Afrotropics (AT; 6.8 %, 1.5×10^6). The tropical realms, especially IM, have a large number of omitted pixels due to cloud cover and/or atmospheric haze (IM, 36.8 %; AT, 14.2 %).

26.4.2 Global Vegetation Changes at Biome Level

We also divided the vegetated land surface by WWF biome type (Table 26.2). We found the largest percentages of negative changes in the ‘Temperate Grasslands, Savannas & Shrublands’ (23.7 %, 2.4×10^6 km^2, biome 8), ‘Temperate Conifer Forests’ (20.0 %, 0.8×10^6 km^2, biome 5) and the ‘Flooded Grasslands & Savannas’ (15.5 %, 0.2×10^6 km^2, biome 12). Other biomes with extensive changes include the ‘Temperate Broadleaf & Mixed Forest’ (biome 4, 8.7 %, 1.1×10^6 km^2), ‘Boreal Forests/Taiga’ (biome 6, 7.6 %, 1.1×10^6 km^2), ‘Tropical and Subtropical Grasslands, Savannas & Shrublands’ (biome 7, 8.1 %, 1.6×10^6 km^2) and ‘Deserts and Xeric Shrublands’ (biome 13, 6.7 %,

Table 26.1 Global trends in vegetation 2001–2011 by Realm

	Positive		Negative		Not significant		Omitted	
Realm	km^2	%	km^2	%	km^2	%	km^2	%
Snow & ice	2,019	<0.1	9,661	0.3	466,534	16.3	2,378,489	83.3
AA	536,599	5.8	387,637	4.2	6,788,949	73.3	1,554,663	16.8
AN	0	0.0	1,146	9.6	4,776	40.0	6,032	50.5
AT	1,472,553	6.8	1,773,194	8.1	15,457,951	70.9	3,083,498	14.2
IM	877,696	10.3	456,998	5.3	4,060,872	47.5	3,147,576	36.8
NA	1,031,126	5.0	2,811,494	13.7	13,687,691	66.9	2,939,100	14.3
NT	879,314	4.5	2,786,218	14.4	10,694,703	55.2	5,022,020	25.9
OC	382	0.9	4,476	10.3	30,326	69.5	8,462	19.4
PA	3,258,275	6.3	3,524,078	6.8	28,299,906	54.3	16,995,150	32.6
Total global	**8,057,967**	**6.0**	**11,754,905**	**8.7**	**79,491,709**	**59.1**	**35,134,992**	**26.1**

Realms: *AA* Australasia, *AN* Antarctica, *AT* Afrotropics, *IM* IndoMalay; *NA* Nearctic; *NT* Neotropics; *OC* Oceania; *PA* Palearctic
Positive trend ($p < 0.01$); Negative trend ($p < 0.01$); The global total land area does not include Antarctica

Table 26.2 Global trends in vegetation 2001–2011 by Biome

	Positive		Negative		Not significant		Omitted	
Biome	km^2	%	km^2	%	km^2	%	km^2	%
1	699,803	3.5	1,308,175	6.6	11,842,949	59.6	6,016,860	30.3
2	274,116	7.5	340,583	9.3	1,789,396	49.0	1,246,349	34.1
3	73,347	10.4	65,649	9.3	547,086	77.4	21,045	3.0
4	1,037,335	8.1	1,118,295	8.7	9,923,565	77.4	741,021	5.8
5	271,709	6.6	817,948	20.0	2,708,819	66.3	288,363	7.1
6	1,427,238	9.5	1,142,315	7.6	12,155,537	80.6	351,788	2.3
7	1,727,611	8.8	1,594,561	8.1	13,586,813	68.9	2,810,799	14.3
8	308,995	3.1	2,397,981	23.7	7,124,421	70.4	282,356	2.8
9	67,368	5.9	177,593	15.5	625,811	54.6	276,054	24.1
10	275,341	5.4	488,639	9.5	2,554,098	49.7	1,823,487	35.4
11	111,257	1.5	92,454	1.2	3,779,187	50.1	3,553,290	47.1
12	445,893	13.8	295,432	9.2	2,171,912	67.3	314,433	9.7
13	1,311,848	4.7	1,871,982	6.7	9,988,047	35.7	14,812,429	52.9
14	22,056	6.3	25,577	7.4	201,725	58.0	98,269	28.3
98	3,958	0.4	17,114	1.6	477,070	45.9	540,669	52.0
99	109	<0.1	628	<0.1	15,233	0.8	1,957,788	99.2
Total global	**8,057,967**	**6.0**	**11,754,905**	**8.7**	**79,491,709**	**59.1**	**35,134,992**	**26.1**

Biomes: 1 Tropical & Subtropical Moist Broadleaf Forests, 2 Tropical & Subtropical Dry Broadleaf Forests, 3 Tropical & Subtropical Coniferous Forests, 4 Temperate Broadleaf & Mixed Forests, 5 Temperate Conifer Forests, 6 Boreal Forests/Taiga, 7 Tropical & Subtropical Grasslands, Savannas & Shrublands, 8 Temperate Grasslands, Savannas & Shrublands, 9 Flooded Grasslands & Savannas, 10 Montane Grasslands & Shrublands, 11 Tundra, 12 Mediterranean Forests, Woodlands & Scrub, 13 Deserts & Xeric Shrublands, 14 Mangroves; 98: Lake; 99: Rock and Ice

Table 26.3 Global trends in vegetation 2001–2011 by Anthrome

Anthrome	% Pos	% Neg	% Not sig	% Omitted
11: Urban	2.8	18.3	68.8	10.1
12: Mixed settlements	6.2	14.1	68.9	10.8
21: Rice villages	5.6	1.8	42.0	50.6
22: Irrigated villages	18.1	6.2	61.1	14.6
23: Rainfed villages	14.3	5.9	64.4	15.4
24: Pastoral villages	12.5	9.2	63.6	14.7
31: Residential irrigated croplands	13.8	10.5	67.4	8.3
32: Residential rainfed croplands	8.8	9.1	74.8	7.3
33: Populated croplands	5.6	15.2	75.1	4.1
34: Remote croplands	5.1	9.9	80.3	4.7
41: Residential rangelands	10.6	9.3	66.1	14.0
42: Populated rangelands	6.4	11.9	64.6	17.1
43: Remote rangelands	4.5	12.0	58.9	24.6
51: Residential woodlands	5.2	9.0	73.6	12.2
52: Populated woodlands	5.6	7.0	75.7	11.7
53: Remote woodlands	5.6	7.2	73.6	13.6
54: Inhabitated treeless and barren	3.6	7.6	28.3	60.5
61: Wild woodlands	6.9	8.4	77.7	7.0
62: Wild treeless and barren	0.8	1.8	26.1	71.3

1.9×10^6 km^2). The largest percentage of positive vegetation change can be found in the ‘Mediterranean Forests, Woods & Scrub’ biome (13.8 %); however, the areal extent of change is low (0.5×10^6 km^2). The second largest percentage of positive change is found in the ‘Tropical & Subtropical Coniferous Forests’ (10.4 %, 0.07×10^6 km^2), which also revealed 9.3 % of negative change. The largest *area* of positive change is found in the ‘Tropical & Subtropical Grasslands, Savannas & Shrublands’ (1.7×10^6 km^2, 8.8 %) this biome also revealed about the same area of negative change (8.1 %). A similar amount of positive change can be found in the ‘Boreal Forest/Taiga’ biome (1.4×10^6 km^2, 9.5 %).

26.4.3 Global Vegetation Changes by Anthrome

Table 26.3 reveals that the anthromes with the largest percentages of positive vegetation change are the villages (except for the rice villages). Irrigated villages experienced the most positive change (18.1 % of the villages are changing positively). Other anthromes with large percentage of positive vegetation change are residential irrigated croplands (13.8 %) and residential rangelands (10.6 %). The most areas with negative change can be found in urban anthromes (18.3 %) and mixed settlements (14.1 %). Populated croplands also revealed large percentages of negative change (15.2 %).

26.5 Discussion

There are few studies that investigate NDVI trends at the global scale based on MODIS data for the same time period. Most trend analyses are based on linear regression instead of on the superior nonparametric SK test as we apply here. That said, our general results correspond well with the results of others who have looked at MODIS NDVI for similar time periods (Fensholt and Proud 2012). Attribution is the key challenge in any change analysis. Previous large-extent change maps have shown the vegetation anomalies attributed to disturbance events (Potter et al. 2003), temperature and precipitation changes (Zhou et al. 2003), and correlations with large-scale climate processes (Nemani et al. 2003; Potter et al. 2008; Julien and Sobrino 2009a). Global climate change has predominantly been linked to increased vegetation productivity (Lashof and Ahuja 1990; Myneni et al. 1997, 1998; Nemani et al. 2003). Some argue that negative trends could be the result of sensor degradation in Terra and Aqua (Wang et al. 2012). This new change map based on 11 years of data presents a mixture of positive and negative changes resulting from both direct and indirect impacts of climatic variability and change. Human land use decisions also drive many of the observed vegetation changes. We provide several examples attributable to these major modes of change, focusing first on natural disturbances arising from climatic variability and change, and then on changes arising directly from human actions.

26.5.1 Climatic Variability and Warming

A warming climate has been invoked previously as the primary cause of terrestrial vegetation activity increases in the Northern Hemisphere (Myneni et al. 1997, 1998; Tucker et al. 2001; Slayback et al. 2003; Jia et al. 2003; Nemani et al. 2003), yet, few regions reveal positive vegetation trends over the span from 2000 through 2011. Several areas in far northeastern Russia, northern Alaska and Canada exhibit significant positive changes, which are likely related to increases in season length as well as decreases in snow cover driven by the Northern Annular Mode (de Beurs and Henebry 2008, 2010a). Most regions with increasing vegetation trends are located in the cooler tundra and taiga areas (Angert et al. 2005; Soja et al. 2007; Goetz et al. 2007), while other areas in North America with noted growing season warming even reveal significant vegetation declines. We found significant vegetation increases in just 0.11×10^6 km^2 of tundra (1.5 %), but 1.4×10^6 km^2 (9.5 %) of taiga/boreal forests biomes, located mainly in the Northeast Siberian Taiga ecoregion, the East Siberian Taiga ecoregion, the Ural Montane Forests and Tundra, and Scandinavian and Russian Taiga ecoregion in Northern Russia. Figure 26.6 provides an overview of far Northeastern Russia that shows a significant amount of positive vegetation change. The percentage change maps for the peak height and the thermal time to peak reveal that, while generally there is an increase in NDVI between 2001 and 2011, this increase is not uniformly expressed

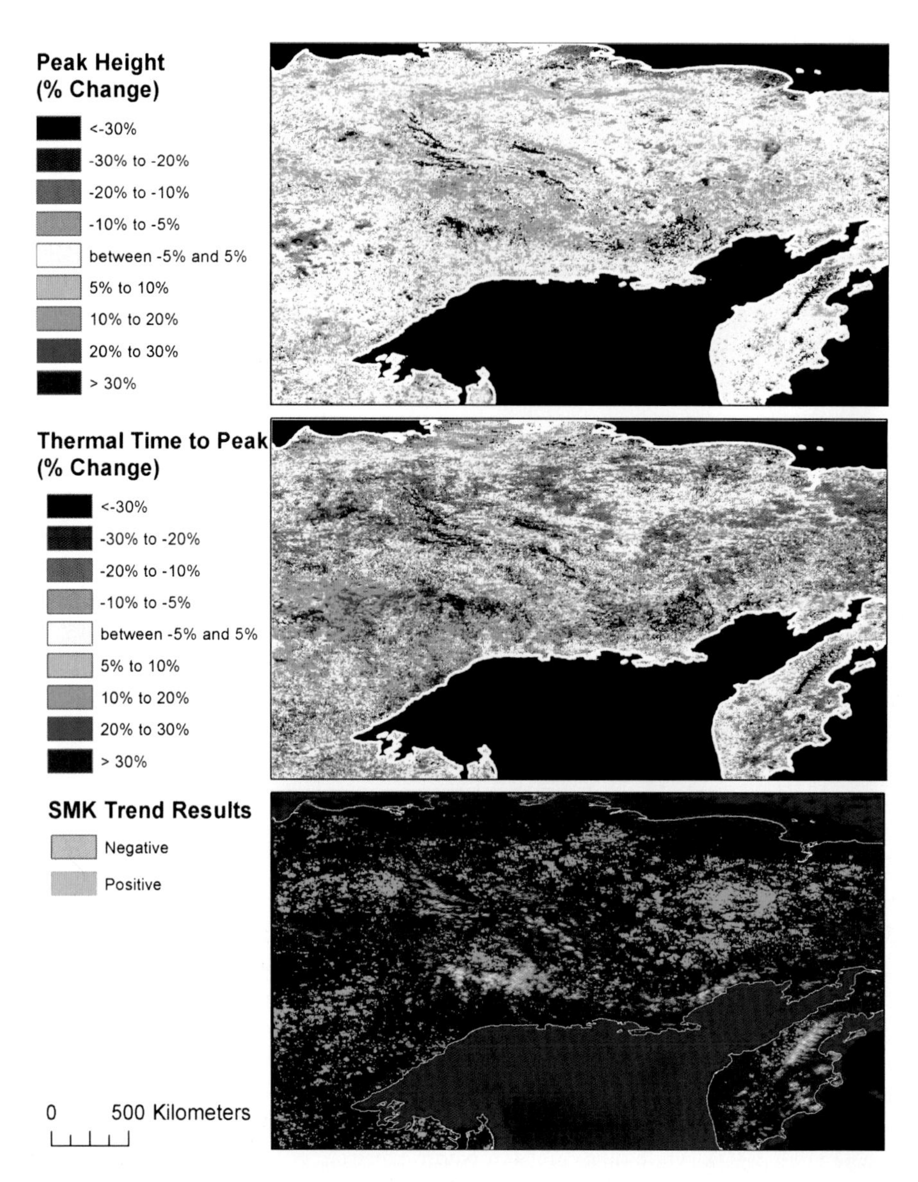

Fig. 26.6 Changes in peak height and thermal time to peak based on the quadratic regression models for northeastern Russia. *Bottom figure* gives the observed trends in this region

in a change in peak height or timing. Most areas with increasing NDVI according to the SK test reveal a slight increase in the peak of the NDVI during the growing season. In addition, most areas reveal a decline in the thermal time to peak, meaning that the peak of the growing season is reached for fewer accumulated growing degree-days. However, some areas reveal an increase in the thermal time to peak (more growing degree-days are necessary to reach the peak NDVI).

In North America the positive changes revealed a patchier pattern predominantly located in the Northwest Territories Taiga, the Midwestern Canadian Shield Forests where they are intermixed with strongly declining areas, and the western part of the Central Canadian Shield Forests. A scattering of positive vegetation changes are found in the Eastern Canadian forests ecoregion as well. These results confirm that the vegetation response to a changing climate is complex and vegetation is not increasing uniformly (Wilmking et al. 2004; Soja et al. 2007).

The El Niño – Southern Oscillation (ENSO) is most likely driving some other observed increasing NDVI trends, such as the large area with significant change in the border area between Mexico and Texas. High positive correlation between the El Niño-3.4 index and the fraction of absorbed photosynthetically active radiation (fAPAR) as measured with MODIS data has been previously observed between 2000 and 2005 (Potter et al. 2008). This increase in vegetation is almost completely contained within the Chihuahua desert ecoregion. This area of increasing vegetation experienced severe drought between 1994 and 2003/2004 (Stahle et al. 2009). These authors indicate that the drought has most likely been continuing into 2009; however, their analysis does not incorporate data past 2004. Our vegetation trend analysis, which spans from 2001 to 2011, indicates strong recovery from drought in this region after 2004. Areas just to the north in the USA reveal systematic declines in vegetation indices emphasized by the drought in 2011.

26.5.2 Climatic Extremes—Drought

Drought conditions during 2004 in the spring wheat regions and arid grasslands of Kazakhstan produced large patches of negative change (Lindeman 2005). In another study we performed a detailed trend analysis for this region based on 500 m MODIS data and found that 15 % of the land surface experienced vegetation declines in Central Asia (Kazakhstan, Kyrgyzstan, Tajikistan, Turkmenistan, and Uzbekistan) over the past few years while only close to 0.5 % experienced vegetation increases (de Beurs et al. 2009). The regions with negative trends were very widespread and spanned several land cover classes (Fig. 26.7). When investigating the phenological metrics we found that the changes were mainly visible in the peak height (Fig. 26.7). The peak timing revealed far fewer changes (data not shown). FAO production statistics for Kazakhstan revealed a decline in wheat area cultivated between 1992 and 2000, illustrating the socio-economic impact of institutional changes following the collapse of the Soviet Union. In contrast, the yield and production variability revealed the effects of weather. Yield and production were down sharply in 2003, 2004, and 2005, as a result of drought conditions. In addition, the years 2009 and 2010 were droughty as well, with an exceptional heat wave occurring in 2010 (Dole et al. 2011). Negative precipitation trends were also confirmed by a station analysis, as well as by analyses of recent station and

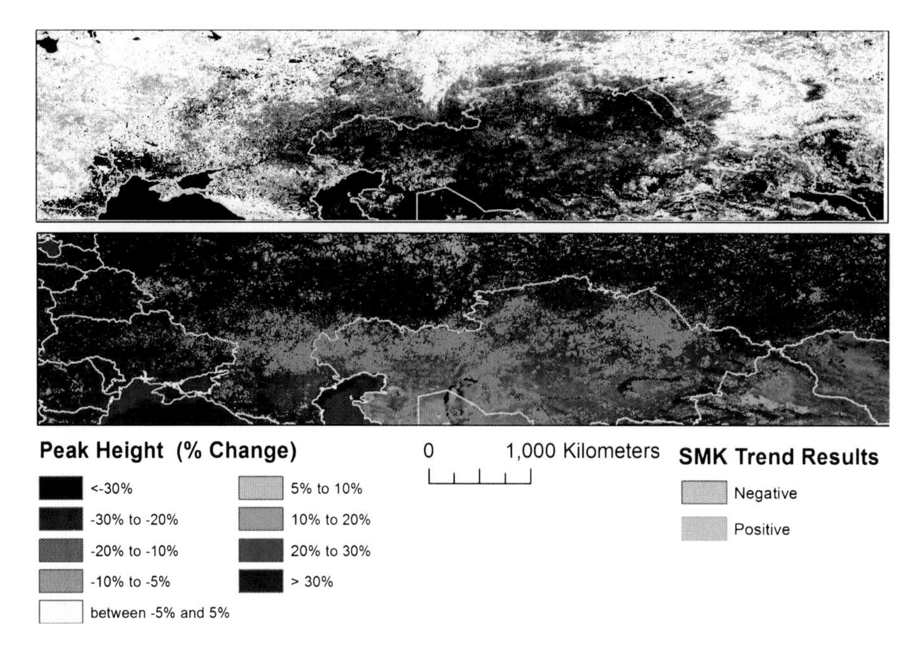

Fig. 26.7 Changes in peak height based on the quadratic regression models for northern Kazakhstan. *Bottom figure* gives the observed trends in the region

gridded precipitation data for Kazakhstan (Akhmadiyeva and Groisman 2008; Wright et al. 2009).

26.5.3 Insect Damage

More favorable climatic conditions have allowed a significant expansion of the suitable habitat for mountain pine beetle and also has affected the intensity of pine beetle infestations in Western Canada and the USA (Goetz et al. 2007; Kurz et al. 2008, Fig. 26.8). The negative vegetation changes in Western Canada are especially dominating in the Fraser Plateau and Basin complex ecoregion and Cascade Mountains leeward forests. The cumulative outbreak area in British Columbia was 130,000 km^2 by the end of 2006 (Kurz et al. 2008). Pine beetle infestations also caused significant disturbances in Idaho, Washington, and Oregon. The area attacked by mountain pine beetles in Oregon increased ten fold from about 200 km^2 in 2001 to well over 2,000 km^2 by 2007 (Nelson et al. 2008). While the mountain pine beetle continues to be the most frequently encountered bark beetle in Idaho, the intensity of the damage has been declining in the last few years due to the lack of suitable hosts following extensive die-offs from previous attacks. We found that the changes were most visible in the peak height of the growing season (Fig. 26.8). Very few changes were evident in the thermal time to peak (data not shown).

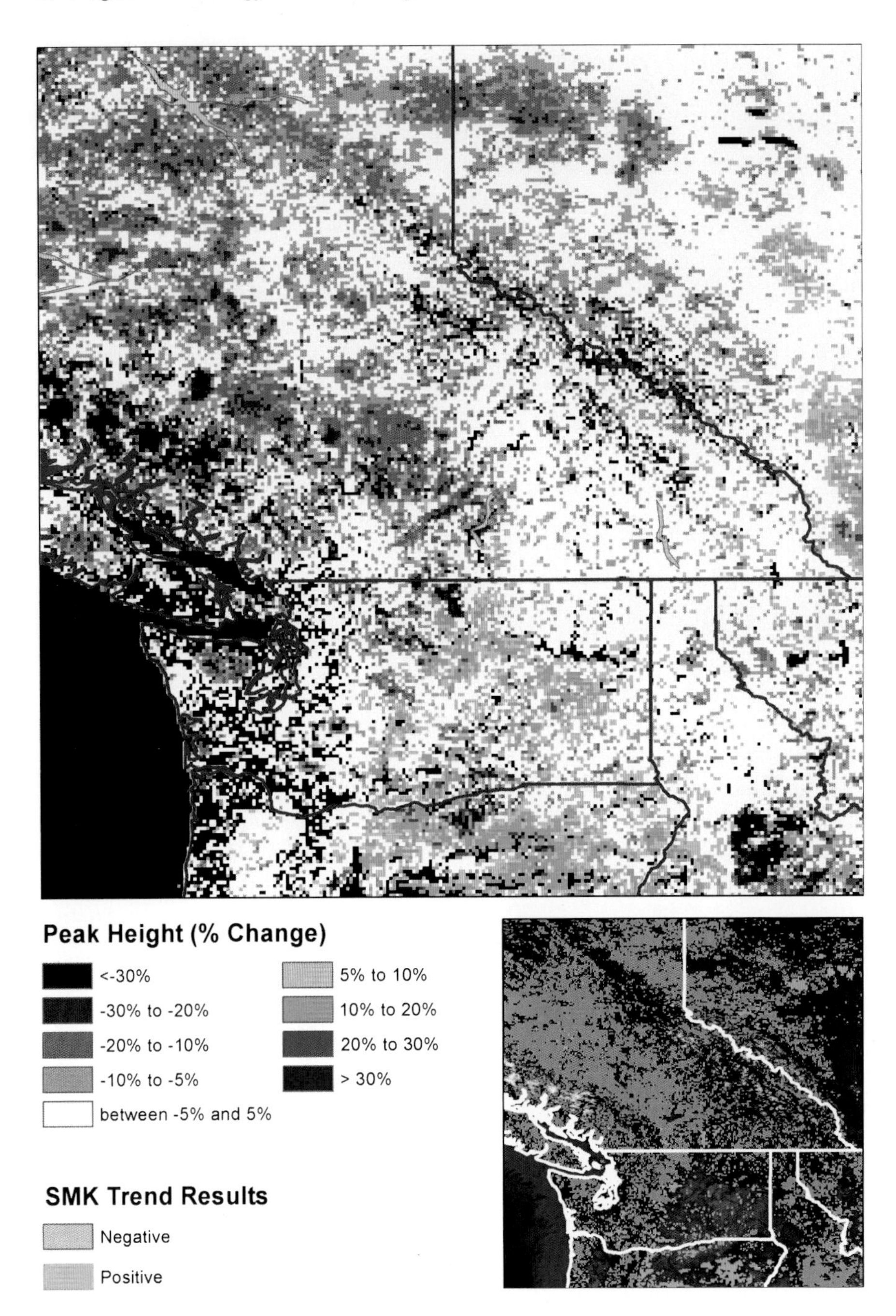

Fig. 26.8 Changes in peak height based on the quadratic regression models for western Canada. *Bottom figure* gives the observed trends in this region

26.5.4 *Urban Dynamics*

Negative changes observed in the eastern USA are associated predominantly with larger metropolitan areas, *e.g.*, Atlanta, Jacksonville, Charlotte, and the Boston-Washington corridor. These changes are especially pronounced in the southeastern USA, most likely related to a combination of rapid (sub)urban expansion and regional drought that led to municipal water use restrictions (Carbone and Dow 2005). Most changes occurred in the urban land cover classes in the southeastern USA. Negative trends due primarily to urban expansion/intensification can be found outside the USA, *e.g.*, St. Petersburg, Russia, and Cairo, Egypt. The amount of negative trends in urban areas is more than three times as large as the amount of positive trends. Most of the positive trends can be found in India and China.

26.5.5 *Institutional Change*

Within China is one of the few regions with widespread increases in vegetation activity (Fig. 26.3). We suspect that these changes are most strongly related to institutional changes that have occurred in China over the past decade (de Beurs et al. 2012). Since 1998, the Chinese government installed two wide reaching programs to conserve the natural environment: the Natural Forest Conservation Program (NFCP) in 1998 (Zhang et al. 2000; Liu et al. 2008), followed by the Grain to Green Program (GTGP) in 1999 (Liu et al. 2008). NFCP's goals for the period 2001–2010 were to restore and protect natural forests by closing off mountainous areas, and to increase timber production in forest plantations (Zhang et al. 2000). As a result, the combined area of mountain closure and plantation grew to almost 110,000 km^2 by 2005 (Liu et al. 2008). GTGP's goals were to increase vegetative cover and convert cropland on steep slopes back to forest and grassland. This program resulted in another 90,000 km^2 of cropland converted into forest or grasslands and 117,000 km^2 of barren land afforested (Liu et al. 2008). These increases in the vegetated land surface are apparent in MODIS imagery from central China.

26.6 Conclusions

The vegetation change analysis we present here is primarily an exploratory tool that can highlight areas of interest, *viz.*, those exhibiting statistically significant change. Attribution of the changes remains a critical but challenging exercise. Here we use multiple lines of evidence to infer a range of causes for observed global positive and negative vegetation changes. The use of phenological metrics helps to identify potential causes of the changes. We conclude that climatic variability and change has led to vegetation disturbance patterns ranging from insect mortality to droughts, while the previously widespread increases of vegetation productivity are no longer evident. Instead, climatic forcings appear to have caused mainly negative changes in the vegetated land surface since 2000. This conclusion is in agreement with another study investigating the terrestrial net primary productivity between 2000 and 2009 (Zhao and Running 2010). Human impacts are revealed as mixed signals. Increasing trends in vegetation were linked to either conservation initiatives (e.g., China) or drought recovery (e.g., Turkey, Australia); decreasing trends in vegetation were linked with continuing deforestation or urban activity. This complex picture of recent vegetation dynamics at the scale of 0.05° (~5.6 km) points to the continuing need for improved representations of land surface dynamics in modeling of weather, climate, and carbon dynamics (Betts et al. 2007).

Acknowledgments This research was supported in part by the NEESPI and NASA LCLUC projects entitled *Evaluating the effects of institutional changes on regional hydrometeorology: Assessing the vulnerability of the Eurasian semi-arid grain belt* (NNG06GC22G) to GMH and *Land abandonment in Russia: Understanding recent trends and assessing future vulnerability and adaptation to changing climate and population dynamics* (NNX09AI29G) to KMdB. We would like to thank P. de Beurs for the application development that allowed us to estimate the trend statistics efficiently.

References

Akhmadiyeva ZK, Groisman PY (2008) General assessment of climate change in Kazakhstan since 1990. Hydrometeorol Environ 2:45–53, in Russian

Angert A, Biraud S, Bonfils C, Henning CC, Buermann W, Pinzon J, Tucker CJ, Fung I (2005) Drier summers cancel out the CO2 uptake enhancement induced by warmer springs. Proc Nat Acad Sci USA 102(31):10823–10827. doi:10.1073/pnas.0501647102

Betts RA, Falloon PD, Goldewijk KK, Ramankutty N (2007) Biogeophysical effects of land use on climate: model simulations of radiative forcings and large-scale temperature change. Agr For Meteorol 142(2–4):216–223

Brown ME, Pinzon JE, Didan K, Morisette JT, Tucker CJ (2006) Evaluation of the consistency of long-term NDVI time series derived from AVHRR, SPOT-vegetation, SeaWiFS, MODIS, and Landsat ETM + sensors. IEEE Trans Geosci Remote Sens 44(7):1787–1793

Carbone GJ, Dow K (2005) Water resource management and drought forecasts in South Carolina. JAWRA. J Am Water Resour As 41(1):145–155. doi:10.1111/j.1752-1688.2005.tb03724.x

de Beurs KM, Henebry GM (2004a) Land surface phenology, climatic variation, and institutional change: analyzing agricultural land cover change in Kazakhstan. Remote Sens Environ 89 (4):497–509. doi:410.1016/j.rse.2003.1011.1006

de Beurs KM, Henebry GM (2004b) Trend analysis of the pathfinder AVHRR land (PAL) NDVI data for the deserts of Central Asia. Geosci Remote Sens Lett 1(4):282–286

de Beurs KM, Henebry GM (2005a) Land surface phenology and temperature variation in the IGBP high-latitude transects. Glob Ch Biol 11:779–790

de Beurs KM, Henebry GM (2005b) A statistical framework for the analysis of long image time series. Int J Remote Sens 26(8):151–1573

de Beurs KM, Henebry GM (2008) Northern annular mode effects on the land surface phenologies of Northern Eurasia. J Clim 21:4257–4279

de Beurs KM, Henebry GM (2010a) A land surface phenology assessment of the northern polar regions using MODIS reflectance time series. Can J Remote Sens 36:S87–S110 de Beurs KM, Henebry GM (2010b) Spatio-temporal statistical methods for modeling land surface phenology. In: Hudson IL, Keatley MR (eds) Phenological research: methods for environmental and climate change analysis. Springer, New York

de Beurs KM, Henebry GM (2010c) Spatio-temporal statistical methods for modelling land surface phenology. Phenological research: methods for environmental and climate change analysis 177–208. doi:10.1007/978-90-481-3335-2_9

de Beurs KM, Wright CK, Henebry GM (2009) Dual scale trend analysis distinguishes climatic from anthropogenic effects on the vegetated land surface. Environ Res Lett 4(4):045012

de Beurs KM, Yan D, Karnieli A (2012) The effect of large scale conservation programs on the vegetative development of China's Loess Plateau. In: Chen J, Wan S, Henebry GM et al (eds) Dryland East Asia (DEA): land dynamics amid social and climate change. HEP - De Gruyer

Dole R, Hoerling M, Perlwitz J, Eischeid J, Pegion P, Zhang T, Quan X-W, Xu T, Murray D (2011) Was there a basis for anticipating the 2010 Russian heat wave? Geophys Res Lett 38(6): L06702. doi:10.1029/2010gl046582

Donohue RJ, McVvicar TR, Roderick ML (2009) Climate-related trends in Australian vegetation cover as inferred from satellite observations, 1981–2006. Glob Ch Biol 15(4):1025–1039. doi:10.1111/j.1365-2486.2008.01746.x

Ellis EC, Ramankutty N (2008) Putting people in the map: anthropogenic biomes of the world. Fr Ecol Environ 6:439–447. doi:10.1890/070062

Fensholt R, Proud SR (2012) Evaluation of earth observation based global long term vegetation trends – comparing GIMMS and MODIS global NDVI time series. Remote Sens Environ 119:131–147. doi:10.1016/j.rse.2011.12.015

Gallo K, Ji L, Reed B, Eidenshink J, Dwyer J (2005) Multi-platform comparisons of MODIS and AVHRR normalized difference vegetation index data. Remote Sens Environ 99(3):221–231. doi:10.1016/j.rse.2005.08.014

Goetz SJ, Bunn AG, Fiske GJ, Houghton RA (2005) Satellite-observed photosynthetic trends across boreal North America associated with climate and fire disturbance. Proc Nat Acad Sci USA 102(38):13521–13525. doi:10.1073/pnas.0506179102

Goetz SJ, Mack MC, Gurney KR, Randerson JT, Houghton RA (2007) Ecosystem responses to recent climate change and fire disturbance at northern high latitudes: observations and model results contrasting northern Eurasia and North America. Environ Res Lett 2(4):1–9. doi:04503110.1088/1748-9326/2/4/045031

Gu Y, Brown JF, Verdin JP, Wardlow B (2007) A five-year analysis of MODIS NDVI and NDWI for grassland drought assessment over the central Great Plains of the United States. Geophys Res Lett 34(6):L06407. doi:10.1029/2006GL029127

Henebry GM (2009) Carbon in idle croplands. Nature 457:1089–1090

Hess A, Iyer H, Malm W (2001) Linear trend analysis: a comparison of methods. Atmos Environ 35:5211–5222

Hirsch RM, Slack JR (1984) A nonparametric trend test for seasonal data with serial dependence. Water Resour Res 20(6):727–732

Hirsch RM, Slack JR, Smith RA (1982) Techniques of trend analysis for monthly water quality data. Water Resour Res 18(1):107–121

Huemann BW, Seaquist JW, Eklundh L, Jönsson P (2007) AVHRR derived phenological change in the Sahel and Soudan, Africa, 1982–2005. Remote Sens Environ 108:385–392

Ji L, Gallo K, Eidenshink JC, Dwyer J (2008) Agreement evaluation of AVHRR and MODIS 16-day composite NDVI data sets. Int J Remote Sens 29(16):4839–4861. doi:10.1080/01431160801927194

Jia GJ, Epstein HE, Walker DA (2003) Greening of arctic Alaska, 1981–2001. Geophys Res Lett 30(20):2067. doi:10.1029/2003GL018268

Julien Y, Sobrino JA (2009a) Global land surface phenology trends from GIMMS database. Int J Remote Sens 30(13):3495–3513. doi:10.1080/01431160802562255

Julien Y, Sobrino JA (2009b) The Yearly Land Cover Dynamics (YLCD) method: an analysis of global vegetation from NDVI and LST parameters. Remote Sens Environ 113(2):329–334. doi:10.1016/j.rse.2008.09.016

Julien Y, Sobrino JA, Verhoef W (2006) Changes in land surface temperatures and NDVI values over Europe between 1982 and 1999. Remote Sens Environ 103:43–55

Kuemmerle T, Hostert P, Radeloff VC, Perzanowski K, Kruglov I (2008) Post-socialist farmland abandonment in the Carpatians. Ecosystems 11:614–628

Kurz WA, Dymond CC, Stinson G, Rampley GJ, Neilson ET, Carroll AL, Ebata T, Safranyik L (2008) Mountain pine beetle and forest carbon feedback to climate change. Nature 452 (7190):987–990. doi:10.1038/nature06777

Lashof DA, Ahuja DR (1990) Relative contributions of greenhouse gas emissions to global warming. Nature 344(6266):529–531

Lindeman M (2005) Kazakhstan: drought reduces 2004 wheat yield. Production Estimates and Crop Assessment Division Foreign Agricultural Service, FAS USDA

Liu JG, Li SX, Ouyang ZY, Tam C, Chen XD (2008) Ecological and socioeconomic effects of China's policies for ecosystem services. Proc Nat Acad Sci USA 105(28):9477–9482. doi:10.1073/pnas.0706436105

Myneni RB, Keeling CD, Tucker CJ, Asrar G, Nemani RR (1997) Increased plant growth in the northern high latitudes from 1981 to 1991. Nature 386(6626):698–702

Myneni RB, Tucker CJ, Asrar G, Keeling CD (1998) Interannual variations in satellite-sensed vegetation index data from 1981 to 1991. J Geophys Res 103(D6):6145–6160

Nelson A, Sprengel K, Flowers R, Kanaskie A, McWilliams M (2008) Forest health highlights in Oregon – 2007. Forest Health Highlights, Portlant

Nemani RR, Keeling CD, Hashimoto H, Jolly WM, Piper SC, Tucker CJ, Myneni RB, Running SW (2003) Climate-driven increases in global terrestrial net primary production from 1982 to 1999. Science 300:1560–1563

Olson DM, Dinerstein E, Wikramanayake ED, Burgess ND, Powel GVN, Underwood EC, D'Amico JA, Itoua I, Strand HE, Morrison JC, Loucks CJ, Allnutt TF, Ricketss TH, Kura YLJF, Wettengel WW, Hedao P, Kassem KR (2001) Terrestrial ecoregions of the world: a new map of life on earth. BioScience 51(11):933–938

Potter C, Tan P, Steinbach M, Klooster S, Kumar V, Myneni R, Genovese V (2003) Major disturbance events in terrestrial ecosystems detected using global satellite data sets. Glob Ch Biol 9:1005–1021

Potter C, Boriah S, Steinbach M, Kumar V, Klooster S (2008) Terrestrial vegetation dynamics and global climate controls. Clim Dyn 31(1):67–78. doi:10.1007/s00382-007-0339-5

Reed BC (2006) Trend analysis of time-series phenology of North America derived from satellite data. GISci Remote Sens 43(1):24–38

Slayback DA, Pinzon JE, Los SO, Tucker CJ (2003) Northern hemisphere photosynthetic trends 1982–1999. Glob Ch Biol 9(1):1–15

Soja AJ, Tchebakova NM, French NHF, Flannigan MD, Shugart HH, Stocks BJ, Sukhinin AI, Parfenova EI, Chapin Iii FS, Stackhouse PW Jr (2007) Climate-induced boreal forest change: predictions versus current observations. Glob Planet Ch 56:274–296

Stahle DW, Cook ER, Villaneuva Diaz J, Fye FK, Burnette DJ, Griffin RD, Acuna Soto R, Seager R, Heim RR Jr (2009) Early 21st-century drought in mexico. EOS, Trans 90(11):89–90

Tucker CJ (1979) Red and photographic infrared linear combinations for monitoring vegetation. Remote Sens Environ 8:127–150

Tucker CJ, Slayback DA, Pinzon JE, Los SO, Myneni RB, Taylor MG (2001) Higher northern latitude normalized difference vegetation index and growing season trends from 1982 to 1999. Int J Biometeorol 45(4):184–190

Tucker CJ, Pinzon JE, Brown ME, Slayback D, Pak EW, Mahoney R, Vermote E, El Saleous N (2005) An extended AVHRR 8-km NDVI dataset compatible with MODIS and SPOT vegetation NDVI data. Int J Remote Sens 26:4485–4498 van Leeuwen WJD, Orr BJ, Marsh SE, Herrmann SM (2006) Multi-sensor NDVI data continuity: uncertainties and implications for vegetation monitoring applications. Remote Sens Environ 100(1):67–81. doi:10.1016/j.rse.2005.10.002

Vuichard N, Ciais P, Belelli L, Smith P, Valentini R (2008) Carbon sequestration due to the abandonment of agriculture in the former USSR since 1990. Glob Biogeochem Cy 22, GB4018

Wang D, Morton D, Masek J, Wu A, Nagol J, Xiong X, Levy R, Vermote E, Wolfe R (2012) Impact of sensor degradation on the MODIS NDVI time series. Remote Sens Environ 119:55–61. doi:10.1016/j.rse.2011.12.001

White MA, Thornton PE, Running SW (1997) A continental phenology model for monitoring vegetation responses to interannual climatic variability. Glob Biogeochem Cy 11(2):217–234

Wilmking M, Juday GP, Barber VA, Zald HSJ (2004) Recent climate warming forces contrasting growth responses of white spruce at treeline in Alaska through temperature thresholds. Glob Change Biol 10(10):1724–1736. doi:10.1111/j.1365-2486.2004.00826.x

Wright CK, de Beurs KM, Akhmadiyeva ZK, Groisman PY, Henebry GM (2009) Recent temperature and precipitation trends in Kazakhstan reveal significant changes in growing season weather. Environ Res Lett 4:045020

Xiao J, Moody A (2005) Geographical distribution of global greening trends and their climatic correlates: 1982–1998. Int J Remote Sens 26:2371–2390

Zhang PC, Shao GF, Zhao G, Le Master DC, Parker GR, Dunning JB, Li QL (2000) Ecology – China's forest policy for the 21st century. Science 288(5474):2135–2136

Zhang K, Kimball JS, Hogg EH, Zhao MS, Oechel WC, Cassano JJ, Running SW (2008) Satellite-based model detection of recent climate-driven changes in northern high-latitude vegetation productivity. J Geophys Res-Biogeosci 113(G3):13. doi:G0303310.1029/2007jg000621

Zhao M, Running S (2010) Drought-induced reduction in global terrestrial net primary productivity from 2000 through 2009. Science 329(5994):940–943

Zhou L, Tucker CJ, Kaufmann RK, Slayback D, Shabanov NV, Myneni RB (2001) Variation in northern vegetation activity inferred from satellite data of vegetation index during 1981 to 1999. J Geophys Res 106(20):69–83

Zhou L, Kaufmann RK, Tian Y, Myneni RB, Tucker CJ (2003) Relation between interannual variations in satellite measures of northern forest greenness and climate between 1982 and 1999. J Geophys Res 108(D1):ACL 3-1–ACL 3-16. doi:10.1029/2002JD002510

Chapter 27
Temperature Sensitivity of Canopy Photosynthesis Phenology in Northern Ecosystems

Shuli Niu, Yuling Fu, Lianhong Gu, and Yiqi Luo

Abstract Northern Hemisphere terrestrial ecosystems have been recognized as areas with large carbon uptake capacity and sinks and are sensitive to temperature change. However, the temperature sensitivity of ecosystem carbon uptake phenology in different biomes of northern ecosystems has not been well explored. In this study, based on our previous effort in characterizing canopy photosynthesis phenology indices, we analyzed how these phenology indices responded to temperature changes by using spatial temperature variability in the temperate and boreal ecosystems in the north hemisphere. Eddy covariance flux measurements of canopy photosynthesis were used to examine the temperature sensitivity of canopy photosynthesis phenology in different biomes and seasons (spring and autumn). Over all the 68 sites, the upturning day, peak recovery day, peak recession day, and senescence day of canopy photosynthesis were all sensitive to mean annual air temperature. Sites with higher mean annual air temperature had earlier carbon uptake and peak recovery day, but later ending of carbon uptake and peak recession

Equal contribution

S. Niu (✉)
Institute of Geographical Sciences and Natural Resources Research,
Chinese Academy of Sciences, Beijing, China

Department of Microbiology and Plant Biology, University of Oklahoma,
Norman, OK 73019, USA
e-mail: sniu@ou.edu

Y. Fu
Institute of Geographical Sciences and Natural Resources Research,
Chinese Academy of Sciences, Beijing, China

L. Gu
Environmental Sciences Division, Oak Ridge National Laboratory,
Oak Ridge, TN 37831, USA

Y. Luo
Department of Microbiology and Plant Biology, University of Oklahoma,
Norman, OK 73019, USA

M.D. Schwartz (ed.), *Phenology: An Integrative Environmental Science*,
DOI 10.1007/978-94-007-6925-0_27,

day. As a consequence, effective growing season length was linearly increased with temperature for all the biomes. Spring phenology indices were more sensitive to temperature change than fall phenology. Besides phenology, peak canopy photosynthesis capacity was also linearly increased with temperature, and contributed even more to annual carbon assimilation changes than growing season length. These findings suggest a predominant temperature controls on annual carbon assimilation in northern ecosystems by changing both canopy photosynthesis phenology and physiology. The temperature sensitivity of canopy photosynthesis phenology and physiology indices revealed in this study are helpful to develop better models to predict impacts of global climate change on vegetation activities.

27.1 Introduction

Temperature strongly influences terrestrial ecosystem carbon cycle by directly changing physiological activities and indirectly mediating phenology. In temperate and boreal ecosystems, phenology play even more important role in controlling the seasonal onset and ending of the carbon uptake, with a consequent impact on net ecosystem production (Goulden et al. 1996; Piao et al. 2007; Baldocchi 2008; Barr et al. 2009). So, it is expected that the unprecedented climate warming will significantly alter growing season length by changing community phenology, with a consequence of driving annual carbon uptake in northern terrestrial ecosystems (Menzel and Fabian 1999; Peñuelas and Filella 2001; Sherry et al. 2007).

Previous studies have documented a correlation between earlier spring phenology and rising temperature in recent years (Cleland et al. 2007). It is reported that the first leaf dates and last frost dates were 1.2 and 1.5 days earlier per decade, respectively, for Northern Hemisphere temperate land areas from 1955 to 2002 (Schwartz et al. 2006). However, very recently, it is reported that spring warming results in delayed spring phenology on the Tibetan Plateau due to later fulfillment of chilling requirements (Yu et al. 2011). Comparing with the spring phenology in response to temperature change, autumn phenology has even more inconsistent response to autumn warming, with early (Piao et al. 2008) or late (Piao et al. 2007; Dragoni et al. 2011) ending of carbon uptake. Most of these previous studies on the relationship between climate change and phenology are based on remote sensing data, model simulation, or observational network of species-specific plant phenophases, while the direct evidence of changes in canopy carbon uptake phenology and its relationship with temperature change are far from clear.

In this chapter, we continued our previous effort (Gu et al. 2003, 2009) and used a systematic methodology to identify the beginning and ending dates of canopy carbon uptake as well as the length of growing season as indicated by canopy carbon uptake period in the northern ecosystems and explored their temperature sensitivities. We derived a series of phenological indices that can be used to characterize canopy photosynthetic phenology. The advance of the eddy covariance technique (Baldocchi and Wilson 2001; Baldocchi 2003) provides a tool amenable for studying the dynamics of plant community photosynthesis (Falge et al. 2002; Gu et al. 2003, 2009). We have used our analytical framework to successfully analyzed plant community photosynthesis (Gu et al. 2003, 2009). In this chapter, we have two objectives : (1) to explore the temperature responses of the critical phenology stages of canopy photosynthesis in different biomes and seasons; (2) to examine the controls of effective growing season length and peak photosynthetic capacity in regulating annual carbon accumulation in different biomes. An extended analysis allows us to develop a comprehensive picture on how the canopy photosynthetic phenology indices of plant community respond to temperature change in different vegetation types.

27.2 Sites and Data Used in the Present Study

We used data from the FLUXNET 'La Thuile' dataset (http://www.fluxdata.org), in which the 30 min and daily eddy covariance measurements (net ecosystem exchange, NEE, of CO_2) have been standardized, gap-filled and partitioned to the component fluxes of ER and GPP using a set of common algorithms (Papale et al. 2006; Moffat et al. 2007). In this dataset, for each site, the meteorological data (air and soil temperature, precipitation, humidity, vapor pressure deficit, global radiation and etc.) as well as the carbon fluxes data (NEE, ecosystem respiration (ER), and gross primary productivity (GPP)) were provided. From the 253 sites with available data, we identified those temperate or boreal ecosystems because in these summer active ecosystems, temperature rather than water availability is considered the main drivers for seasonal variation in phenology. We studied 68 sites with 263 site-years data of three major vegetation types in northern ecosystems, including 22 deciduous broad leaf foresee (DBF), 29 evergreen needle leaf forest (ENF), and 17 grasslands (Table 27.1). These sites range from 30.85°N to 69.14°N (Table 27.1). The sites without active photosynthesis throughout a year or without a full formant season, or with large data gaps during the growing season were not included in this study. Our analysis was based on canopy photosynthetic rates which were derived from NEE in the same way as described in detail in Gu et al. (2003).

27.3 Quantifying Plant Community Photosynthetic Phenology

We used the same method with those in Gu et al. (2009) to quantify the canopy photosynthetic phenology. We first fit the instantaneous canopy photosynthetic rate which was derived from the NEE measurement to the following equation. The canopy photosynthetic capacity (CPC) is defined as the maximal gross photosynthetic rate at the canopy level when the environmental conditions (e.g. light, moisture, and temperature) are non-limiting for the time of a year under consideration (Fig. 27.1).

$$A(t) = y_0 + \frac{a_1}{\left[1 + \exp\left(-\frac{t-t_{01}}{b_1}\right)\right]^{c_1}} - \frac{a_2}{\left[1 + \exp\left(-\frac{t-t_{02}}{b_2}\right)\right]^{c_2}} \tag{27.1}$$

where $A(t)$ is the CPC in day t; y_0, a_1, a_2, b_1, b_2, c_1, c_1, t_{01}, and t_{02} are empirical parameters to be estimated.

Table 27.1 Main site characteristics, climatic indexes, and studied periods of flux sites used in this analysis

Site	Lat.	Long.	Years	Vegetation	Climate
AT-Neu	47.12	11.32	2002–2006	GRA	Temperate
Be-Vie	50.31	6.00	1997–2006	ENF	Temperate
CA-Ca2	49.87	−125.29	2002–2005	ENF	Temperate
CA-Ca3	49.53	−124.90	2002–2004	ENF	Temperate
CA-Let	49.71	−112.94	1999–2005	GRA	Temperate
CA-Man	55.88	−98.48	1995–2003	ENF	Boreal
CA-Mer	45.41	−75.52	1999–2005	GRA	Temperate
CA-NS1	55.88	−98.48	2003–2004	ENF	Boreal
CA-NS2	55.91	−98.52	2002–2004	ENF	Boreal
CA-NS3	55.91	−98.38	2002–2005	ENF	Boreal
CA-NS4	55.91	−98.38	2003–2004	ENF	Boreal
CA-NS5	55.86	−98.49	2002–2005	ENF	Boreal
CA-NS6	55.92	−98.96	2002–2005	DBF	Boreal
CA-NS7	56.64	−99.95	2003–2005	GRA	Boreal
CA-Oas	53.63	−106.20	1997–2005	DBF	Boreal
CA-Obs	53.99	−105.12	2000–2005	ENF	Boreal
CA-Ojp	53.92	−104.69	2000–2005	ENF	Boreal
CA-Qcu	49.27	−74.04	2002–2006	ENF	Boreal
CA-Qfo	49.69	−74.34	2004–2006	ENF	Boreal
CA-TP4	42.71	−80.36	2004–2005	ENF	Temperate
CN-Do1	31.58	121.96	2005	GRA	Subtropical
CN-Du1	42.05	116.67	2004–2005	GRA	Temperate
CN-Xfs	44.13	116.33	2004–2006	GRA	Temperate
DE-Bay	50.14	11.87	1997–1998	ENF	Temperate
DE-Hai	51.08	10.45	2000–2006	DBF	Temperate
DE-Meh	51.28	10.66	2004–2006	GRA	Temperate
DE-Tha	50.96	13.57	1997–2005	ENF	Temperate
DE-Wet	50.45	11.46	2002–2005	ENF	Temperate
DK-Sor	55.49	11.65	1997–2006	DBF	Temperate
FI-Hyy	61.85	24.29	1997–2006	ENF	Boreal
FI-Kaa	69.14	27.30	2000–2006	GRA	Boreal
FI-Sod	67.36	26.64	2000–2006	ENF	Boreal
HU-Bug	46.69	19.60	2003–2006	GRA	Temperate
IT-Lav	45.96	11.28	2001–2006	ENF	Temperate
IT-Mbo	46.02	11.05	2003–2006	GRA	Temperate
IT-Non	44.69	11.09	2001–2006	DBF	Temperate
IT-Ren	46.59	11.43	1999–2006	ENF	Temperate
IT-Ro1	42.41	11.93	2001–2004	DBF	Subtropical
IT-Ro2	42.39	11.92	2002–2006	DBF	Subtropical
JP-Tak	36.15	137.42	1999–2004	DBF	Temperate
JP-Tom	42.74	141.52	2001–2003	DBF	Temperate
NL-Cal	51.97	4.93	2003–2006	GRA	Temperate
RU-Che	68.61	161.34	2003–2005	ENF	Boreal
RU-Fyo	56.46	32.92	1999–2006	ENF	Temperate
RU-Zot	60.80	89.35	2002–2004	ENF	Boreal

(continued)

Table 27.1 (continued)

Site	Lat.	Long.	Years	Vegetation	Climate
SE-Fla	64.11	19.46	1997–2002	ENF	Boreal
SE-Nor	60.09	17.48	1996–2005	ENF	Temperate
UK-Ham	51.12	−0.86	2004–2005	DBF	Temperate
US_Goo	34.25	−89.87	2004–2005	GRA	Subtropical
US-ARB	35.55	−98.04	2005–2006	GRA	Temperate
US-Bar	44.06	−71.29	2004–2005	DBF	Temperate
US-Bn1	63.92	−145.38	2003	ENF	Boreal
US-Dk2	35.97	−79.10	2003–2005	DBF	Subtropical
US-Ha1	43.54	−72.17	1992–2006	DBF	Temperate
US-Ho1	45.20	−68.74	1996–2004	ENF	Temperate
US-Ho2	45.21	−68.75	1999–2004	ENF	Temperate
US-IB2	41.84	−88.24	2006–2007	GRA	Temperate
US-LPH	42.54	−72.19	2003–2004	DBF	Temperate
US-MMS	39.32	−86.41	1999–2005	DBF	Subtropical
US-Moz	38.74	92.2	2005–2006	DBF	Subtropical
US-NC1	35.81	−76.71	2005–2006	GRA	Subtropical
US-NR1	40.03	−105.55	1999–2003	ENF	Boreal
US-Oho	41.55	−83.84	2004–2005	DBF	Temperate
US-Pfa	45.95	−90.27	1997–2003	DBF	Temperate
US-Syv	46.24	−89.35	2002–2005	DBF	Temperate
US-UMB	45.56	−84.71	1999–2003	DBF	Temperate
US-WBW	35.96	−84.29	1995–1999	DBF	Subtropical
US-WCr	45.81	−90.08	1999–2006	DBF	Temperate

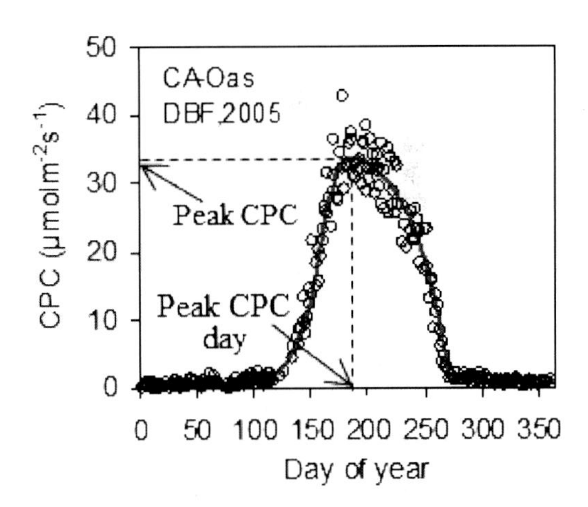

Fig. 27.1 Illustration of the canopy photosynthetic capacity (*CPC*)

In practice, the following iterative procedures were used to estimate the parameters in Eq. (27.1).

a. Compute hourly or half-hourly (depending on observational time steps) canopy photosynthetic rates from NEE measurements.

b. Select the largest value from each day to form a time series of the daily maximal canopy photosynthetic rate. The time series shall cover the complete seasonal cycle.

c. Fit Eq. (27.1) to the obtained time series.

d. For each point in the time series, compute the ratio of the daily maximal canopy photosynthetic rate to the value predicted by Eq. (27.1) for the corresponding day with the fitted parameters.

e. Conduct the Grubb's test (NIST/SEMATECH 2006) to detect if there is an outlier in the obtained ratios.

f. If an outlier is detected, remove this outlier and go to Step *c*.

g. If no outlier is found, remove the data points whose ratios are at least one standard deviation (1σ) less than the mean ratio. The remaining dataset is considered to consist of the canopy photosynthetic capacity at various times of the growing season.

h. Fit Eq. (27.1) to the time series of the CPC. Equation (27.1) with the obtained parameters depict the seasonal cycle of plant community photosynthesis and is then used for further analyses (see the next section).

27.3.1 Characterizing the Dynamics in CPC

The growth rate (k) of the CPC is the derivative of the canopy photosynthetic capacity with respect to the day (t) of year:

$$\begin{aligned} k(t) &= \frac{dA(t)}{dt} \\ &= \frac{a_1 c_1}{b_1} \frac{\exp\left(-\frac{t-t_{01}}{b_{01}}\right)}{\left[1+\exp\left(-\frac{t-t_{01}}{b_{01}}\right)\right]^{1+c_1}} - \frac{a_2 c_2}{b_2} \frac{\exp\left(-\frac{t-t_{02}}{b_{02}}\right)}{\left[1+\exp\left(-\frac{t-t_{02}}{b_{02}}\right)\right]^{1+c_2}} \end{aligned} \quad (27.2)$$

The maximal growth rate of canopy photosynthetic capacity is termed 'Peak Recovery Rate' and denoted by k_{PRR}; the day on which this rate occurs is termed 'Peak Recovery Day' and denoted by t_{PRD} (Fig. 27.2):

$$k_{PRR} = k(t_{PRD}) \quad (27.3)$$

We further define 'Recovery Line' (RL) as the line that passes through the maximum with a slope of k_{PRR}. Its equation can be written as follows:

$$A_{\mathrm{RL}}(t) = k_{PRR} t + A(t_{PRD}) - k_{PRR} t_{PRD} \quad (27.4)$$

where A_{RL} is the canopy photosynthetic capacity predicted by the Recovery Line. Similarly, we term the most negative growth rate of canopy photosynthetic capacity

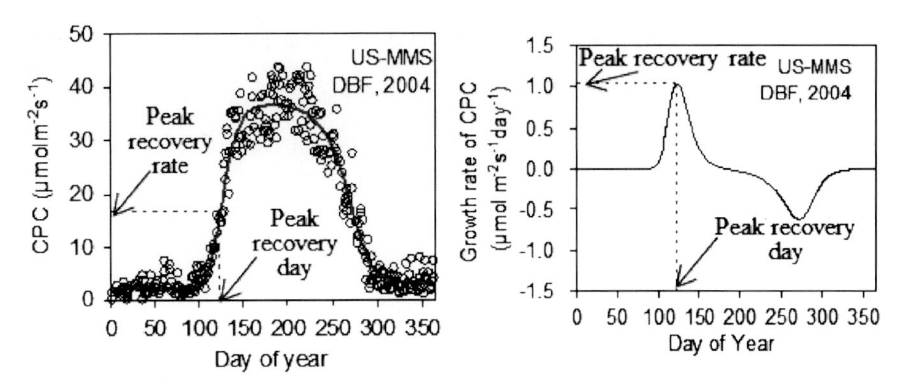

Fig. 27.2 Illustration of the peak recovery rate (k_{PRR}) and peak recovery day (t_{PRD})

'Peak Senescence Rate' and denote it by k_{PSR} and the day on which k_{PSR} occurs 'Peak Senescence Day' and denote it by t_{PSD}:

$$k_{PSR} = k(t_{PSD}) \tag{27.5}$$

Accordingly, we define 'Senescence Line' (SL) as the line that passes through the minimum (the most negative) with a slope of k_{PSR} and describe it by the following equation:

$$A_{SL}(t) = k_{PSR}t + A(t_{PSD}) - k_{PSR}t_{PSD} \tag{27.6}$$

where A_{SL} is the canopy photosynthetic capacity predicted by the Senescence Line.

It is very difficult to determine t_{PRD} and t_{PSD} analytically from the Eq. (27.1). However, they can be approximated by:

$$t_{PRD} \approx t_{01} + b_1 \ln(c_1) \tag{27.7}$$

and

$$t_{PSD} \approx t_{02} + b_2 \ln(c_2) \tag{27.8}$$

Equation (27.7) is obtained by setting the derivative of the first term in Eq. (27.2) with respect to t to zero and solve for t where the first term is at maximum; Eq. (27.8) is obtained by setting the derivative of the second term in Eq. (27.2) with respect to t to zero and solve for t where the second term is at maximum. Equations (27.7) and (27.8) hold because when t is small, the second term in Eq. (27.2) is close to zero and when t is large, the first term is close to zero. Alternatively, one could simply compute the value of k for each day of the year and pick up the maximum and the minimum as we did in this study.

27.3.2 Characterizing Canopy Photosynthetic Potential

We calculated the area under the curve of $A(t)$, which is an indicator of how much carbon dioxide can be potentially assimilated by a plant community over a complete cycle of photosynthesis in a year. As in Gu et al. (2009), we term this area as 'Carbon Assimilation Potential' (u):

$$u = \int_{t_{start}}^{t_{end}} A(t)dt \tag{27.9}$$

For the purpose of calculating the carbon assimilation potential u, it is not necessary to determine t_{start} and t_{end} exactly as long as one whole seasonal cycle of photosynthesis is included between t_{start} and t_{end}. This is because the two tails of A contribute little to u. Therefore we conveniently set $t_{\text{start}} = 1$ and $t_{\text{end}} = 365$ for warm-season vegetation sites Here we don't intend to use t_{start} (t_{end}) to denote the start (end) of the growing season.

The peak canopy photosynthetic capacity over a complete seasonal cycle of plant community photosynthesis and the day on which this peak occurs should contain useful information about the function of the vegetation and its interaction with the climate. We use A_P to denote the peak canopy photosynthetic capacity:

$$A_P = \max\{[A(t), \quad t_{start} < t < t_{end}]\} \tag{27.10}$$

We use t_P to denote the day on which the peak canopy photosynthetic capacity occurs. t_P is called 'Peak Canopy Photosynthetic Capacity Day' or simply 'Peak Capacity Day'.

27.3.3 Transitions Between Phases

We name the transitions between the consecutive phases identified above 'Upturn Day' (t_U), 'Stabilization Day' (t_S), 'Downturn Day' (t_D), and 'Recession Day' (t_R), respectively. We set the upturn day at the intersection between the recovery line and the x-axis and the recession day at the intersection between the senescence line and the x-axis (Fig. 27.3). The upturn day and recession day are calculated from Eqs. (27.4) and (27.6), respectively, as follows:

$$t_U = t_{PRD} - A(t_{PRD})/k_{PRR} \tag{27.11}$$

$$t_R = t_{PSD} - A(t_{PSD})/k_{PSR} \tag{27.12}$$

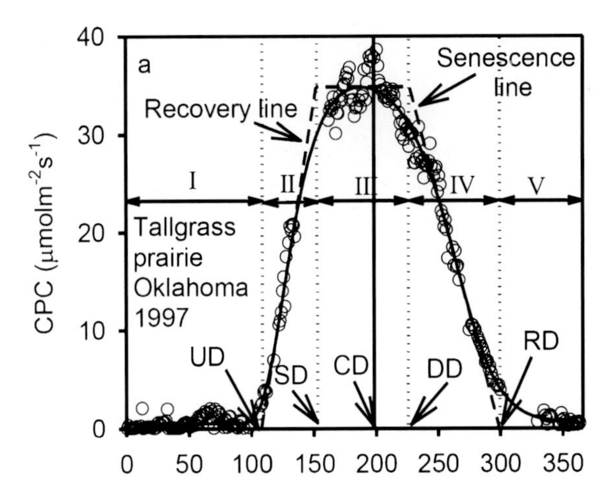

Fig. 27.3 Illustration of the five critical phenology phases for canopy photosynthesis

The stabilization day and downturn day are set at the days on which the peak canopy photosynthetic capacity A_P is predicted to occur based on the RL equation (Eq. 27.4) and the SL equation (Eq. 27.6), respectively. These two dates are given by:

$$t_S = t_{PRD} + [A_P - A(t_{PRD})]/k_{PRR} \tag{27.13}$$

$$t_D = t_{PSD} + [A_P - A(t_{PSD})]/k_{PSR} \tag{27.14}$$

We can also use the standard deviation of the "growing days" to measure the length of the growing season. To do so, we first define the mean or Center Day (t_C) of the growing season as follows:

$$t_C = \frac{\int_{t_{start}}^{t_{end}} tA(t)dt}{u} \tag{27.15}$$

The standard deviation σ of the 'growing days' from the center day of the growing season is:

$$\sigma = \left(\frac{\int_{t_{start}}^{t_{end}} (t - t_C)^2 A(t)dt}{u} \right)^{0.5} \tag{27.16}$$

The length of the growing season can then be measured by the scaled standard deviation:

$$L_E = 2\sqrt{3}\sigma \tag{27.17}$$

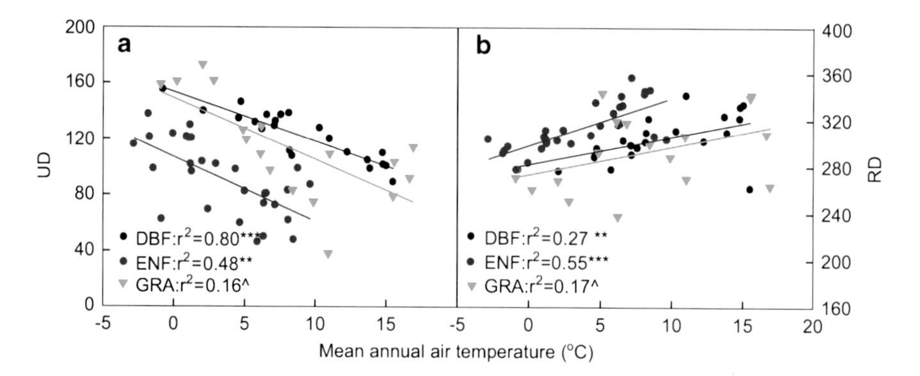

Fig. 27.4 The relationship between mean annual air temperature and upturning day (*UD*) and recession day (*RD*) for the three biomes. *DBF* deciduous broadleaf forest, *ENF* evergreen needle leaf forest, *GRA* grassland. ^, **, and *** represents the relationship was significant at $P < 0.1$, 0.01, and 0.001 levels, respectively

We name the scaled standard deviation the 'Effective Growing Season Length' and denote it by L_E. The scaling factor $2\sqrt{3}$ is introduced so that L_E is exactly the width if the temporal pattern of $A(t)$ is a rectangle (Gu et al. 2003). Gu et al. (2003) defined the center day as the 'center of gravity' of the curve $A(t)$. In the present paper, the center day is defined as a statistical mean and is thus more straightforward.

27.4 Critical Phenology Indices in Response to Temperature Change

Upturning day (UD) of canopy photosynthesis changed a lot across the sites. Air temperature was the dominant factor that controls the spring recovery of canopy photosynthesis in northern ecosystems. For all the three vegetation types, DBF, ENF, and GRS, the UD was negatively correlated with the mean annual air temperature (Fig. 27.4a). Sites with higher mean annual air temperature got to the UD much earlier than those with lower temperature. The slopes of the relationships were not significantly different among the biomes, suggesting a similar sensitivity of the UD in response to temperature change. In contrast, the recession day (RD) of canopy photosynthesis showed positive relationship with mean annual temperature across the sites (Fig. 27.4b). Sites with higher air temperature got to the RD much late than those with lower temperature.

Peak recovery day (PRD) of canopy photosynthesis varied among the three biomes, with the latest for evergreen needle leaf forest, and similarly earlier for deciduous broadleaf forest and grassland (Fig. 27.5a). Similar to the temperature response of UD, PRD was negatively correlated with mean annual air temperature.

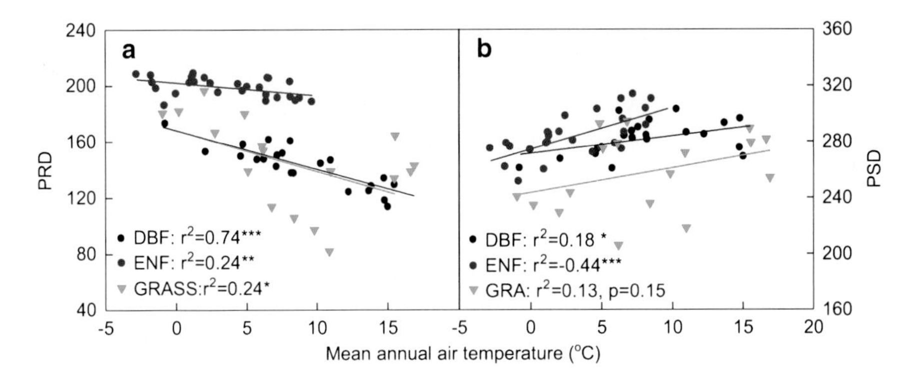

Fig. 27.5 The relationship between mean annual air temperature and peak recovery day (*PRD*) and peak senescence day (*PSD*) for the three biomes (see Fig. 27.4 for abbreviations)

Sites with higher air temperature got to the PRD much earlier than those with lower temperature. In contrast with the temperature response of PRD, peak senescence day (PSD) increased with mean annual air temperature. Sites with higher air temperature got to the PSD much later than those with lower temperature (Fig. 27.5b).

27.5 Effective Growing Season Length and Photosynthetic Capacity in Response to Temperature Change and Controlling Annual GPP

Due to the temperature response of beginning and ending days of canopy photosynthesis, the effective growing season length (Le) in the northern systems was also sensitive to temperature change. It increased with mean annual temperature for all the three biomes (Fig. 27.6a). Sites with higher mean annual air temperature had longer Le than those with lower temperature. The slope was much lower in grassland than that in ENF and DBF (Fig. 27.6a), suggesting that grassland was less sensitive to temperature in growing season length than ENF and DBF. Besides growing season length, peak canopy photosynthesis capacity (CPC) also showed positive linear relationship with mean annual air temperature (Fig. 27.6b), suggesting that photosynthetic capacity was also sensitive to temperature change in northern ecosystems.

In northern systems, both growing season length and photosynthetic capacity determine annual GPP (Fig. 27.7). Comparing with growing season length, peak CPC explained more of annual GPP changes across the sites (Fig. 27.7) for all the three biomes.

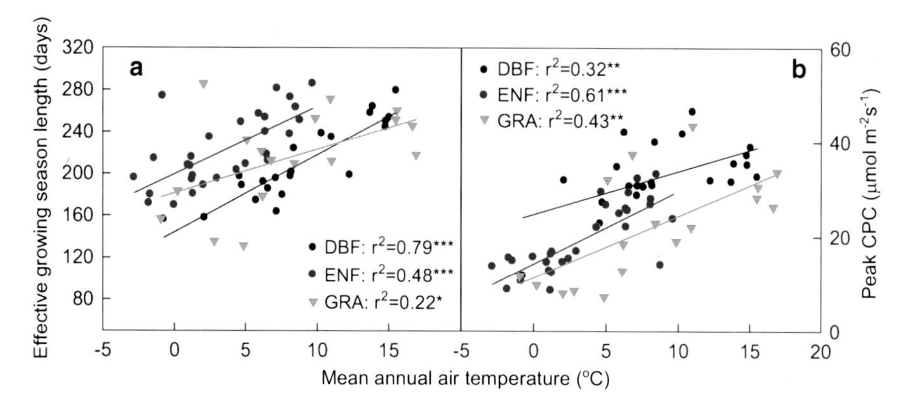

Fig. 27.6 The relationship between mean annual air temperature and effective growing season length (Le) and peak canopy photosynthesis capacity (*CPC*) for the three biomes (see Fig. 27.4 for abbreviations)

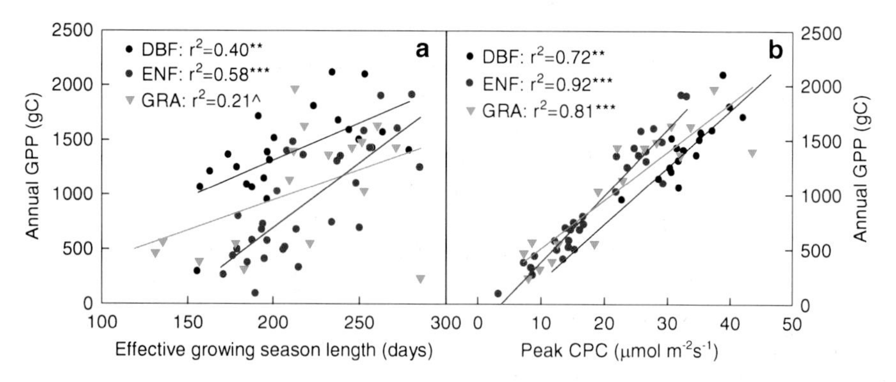

Fig. 27.7 The relationship of annual gross primary productivity (*GPP*) with effective growing season length (Le) and peak canopy photosynthesis capacity (*CPC*) for the three biomes (see Fig. 27.4 for abbreviations)

27.6 Discussion and Conclusions

27.6.1 Temperature Sensitivity of Canopy Photosynthesis Phenology

Phenology change is one of the most important factors affecting future vegetation productivity in response to rising global temperature. In this chapter we continued the effort initiated in Gu et al. (2003a, 2009) and explored the critical canopy photosynthesis phenology in response to temperature change. The spring recovery and fall recession phases derived from the seasonal dynamics of canopy photosynthesis showed the high sensitivity to temperature change. These phases reflect unique functioning of plant communities at different stages of the growing season.

We found that the upturning day, peak recovery day, recession day, and peak senescence day of canopy photosynthesis all showed linear relationship with annual mean air temperature, suggesting the sensitivity of canopy photosynthesis phenology in response to temperature change in northern ecosystems.

Our results suggest that air temperature drives the changes in phenology of carbon uptake in the northern hemisphere. However, the recovery and senescence phenology showed different sensitivity to temperature changes. Spring phenology was more sensitive to temperature than fall phenology, which was reflected by the DOY changes vs temperature changes between the recovery phase and senescence phase. UD vs temperature had higher slopes than RD vs temperature, and PRD vs temperature had higher slopes than PRD vs temperature, for all the three biomes (Fig. 27.4). This suggests that early beginning of spring upturning day is more sensitive to temperature change than fall photosynthesis phenology. The higher temperature sensitivity of spring phenology than autumn was primarily due to that irradiances and temperatures are higher, and water is generally less limiting in spring than in autumn (Niu et al. 2011). Some previous studies also confirmed that the lengthening of the growing season by a certain number of days in spring stimulates ecosystem C uptake more than a lengthening by the same number of days in the fall (Kramer et al. 2000; Piao et al. 2007). This has been attributed also to greater radiative inputs and longer days, as well as better moisture availability as the result of snow melt and relative lower evaporative demand in spring than in fall (Black et al. 2000; Barr et al. 2004). So, the more sensitive spring phenology may favor carbon uptake in a warmer environment. Given the greater increase in the late winter and early spring temperatures than late spring and early summer temperature (Groisman et al. 1994), the advance of spring phenology should be even greater than the delay of fall. The advance of spring carbon uptake phenology and the delay of autumn carbon uptake phenology with increasing temperature indicate a longer growing season. Therefore, we expect a much greater carbon uptake in a warmer environment in the northern ecosystems due to the changes in canopy photosynthesis phenology.

Our results also showed that different vegetation types have different temperature sensitivity in phenology. Deciduous broadleaf forecast (DBF) and grassland was more sensitive to temperature changes when compared with evergreen needle leaf forest (ENF) in terms of peak recovery day. This is probably due to the life history strategy between the deciduous leaf and evergreen leaf. For DBF and grassland, the production of new foliage is a prerequisite for photosynthesis. The early recovery could be advantageous in terms of C uptake. By comparison, ENF strategy is more conservative. The recovery of photosynthesis in ENF is reversible, proceeds slowly and involves multiple steps (Monson et al. 2005). Different temperature sensitivities of biomes have important potential ecological consequences, especially in light of recent effects of climate change on species composition or vegetation changes. It is well known that species differ in their climatic determinants of phenology. Models also predict that because of this, species will respond in different ways to future climate change (Morin et al. 2008). The changes in vegetation or biomes under climate change or land use change will further modify ecosystem carbon cycle in response to temperature change.

27.6.2 Co-determination of Le and Peak CPC on Annual GPP

It is generally assumed that warming will increase the length of the growing season (Penuelas and Filella 2001; Linderholm 2006) and thus stimulate primary productivity in boreal and temperate ecosystems (Baldocchi et al. 2001; Churkina et al. 2005; Baldocchi 2008; Dragoni et al. 2011). The observed links between Le and annual GPP in the northern hemisphere are consistent with the results in previous studies. We found that GPP will increase 5.5–11.8 $gCm^{-2}day^{-1}$ increase in growing season length (Fig. 27.7), which is in accordance with the model simulation (4.9–9.8 $gCm^{-2}day^{-1}$ in (Piao et al. 2007)) and (8.5 $gCm^{-2}day^{-1}$ in (Euskirchen et al. 2006)). Our finding indicates that extending the growing season in a warmer environment partly contributes to the increase of terrestrial biomass storage in Northern Hemisphere. However, early spring and longer growing season does not always result in more carbon uptake. In some ecosystems, especially the water limited ecosystems, earlier onset of growing season may result from a shallow snowpack, leaving less moisture in the soil in summer and limiting plant growth later in the growing season (Hu et al. 2010).

While the growing season length affected how much CO_2 could be potentially assimilated by a plant community over the course of a growing season, physiological processes such as the maximum photosynthetic capacity may also be important. We continued and extended our previous studies on regulation of peak CPC on canopy carbon assimilation potential (Gu et al. 2009). This study shows that both the effective growing season length of carbon uptake and the maximum carbon uptake capacity significantly determined the annual GPP in northern ecosystem, and peak CPC could even explain more changes in annual GPP than Le (Fig. 27.7). This indicates that the climate warming-induced changes in the growing season length largely but not totally contribute to the enhancement of terrestrial carbon uptake. Peak canopy photosynthetic capacity is also a good predictor for the canopy carbon assimilation potential. Peak CPC across the site was more related to temperature ($r^2 = 0.32, 0.78$) than precipitation ($r^2 = 0.14, 0.33$, respectively, for DBF and ENF) in DBF and ENF, and similarly related to temperature ($r^2 = 0.43$) and precipitation ($r^2 = 0.45$) in grassland, suggesting that temperature is also the dominant controlling factor for peak CPC in the northern ecosystems. To the best of our knowledge, this is the first evidence that reveals general controls of temperature on GPP in the northern hemisphere by influencing both phenological and physiological process of canopy photosynthesis.

The expanded analysis in this study allows us to develop a comprehensive understanding on how the photosynthetic phenology and physiology indices of plant community in response to temperature change and how the relationships change with vegetation types and seasons. The emergent patterns we observed of the co-determination of growing season length and canopy photosynthesis potential for carbon assimilation potentials in the northern hemisphere was well confirmed or established across the extended sites. These efforts are the starting point for developing ecosystem emergent properties and better understanding temperature sensitivity of ecosystem carbon uptake. It is worth noting that GPP is not observed and is rather estimated from the observed NEE, which means patterns reported here could be partly influenced by ER. However, it is unlikely that the temperature response of GPP phenology is caused by ER because either over- or under-estimation of ER will not change phenological stages but possibly change peak CPC.

Given the rich datasets contained in the FLUXNET database, our study could easily be expanded to a wider range of sites. Because net ecosystem productivity follow a similar seasonal dynamics with GPP, the algorithm developed by Gu et al. (2003, 2009) could be applicable to examine NEP phenology, which directly reflects the ecosystem net carbon sequestration.

Acknowledgments This work was financially supported by the Terrestrial Carbon Program at the Office of Science, US Department of Energy, Grants DE-FG02-006ER64317 and DE-FG02-01ER63198 and US National Science Foundation (NSF) grant DEB 0444518, DEB 0743778, DEB 0840964, DBI 0850290, and EPS 0919466 to YL. Oak Ridge National Laboratory (ORNL) is managed by UT-Battelle, LLC, for the U.S. Department of Energy under contract DE-AC05-00OR22725.

References

Baldocchi D (2003) Assessing the eddy covariance technique for evaluating carbon dioxide exchange rates of ecosystems: past, present and future. Global Change Biol 9(4):479–492

Baldocchi D (2008) Breathing of the terrestrial biosphere: lessons learned from a global network of carbon dioxide flux measurement systems. Aust J Bot 56:1–26

Baldocchi DD, Wilson KB (2001) Modeling CO2 and water vapor exchange of a temperate broadleaved forest across hourly to decadal time scales. Ecol Model 142(1–2):155–184

Baldocchi D, Falge E, Gu LH, Olson R, Hollinger D, Running S, Anthoni P, Bernhofer C, Davis K, Evans R, Fuentes J, Goldstein A, Katul G, Law B, Lee XH, Malhi Y, Meyers T, Munger W, Oechel W, KTP U, Pilegaard K, Schmid HP, Valentini R, Verma S, Vesala T, Wilson K, Wofsy S (2001) FLUXNET: a new tool to study the temporal and spatial variability of ecosystem-scale carbon dioxide, water vapor, and energy flux densities. Bull Am Meteorol Soc 82(11):2415–2434

Barr AG, Black TA, Hogg EH, Kljun N, Morgenstern K, Nesic Z (2004) Inter-annual variability in the leaf area index of a boreal aspen-hazelnut forest in relation to net ecosystem production. Agric Forest Meteorol 126(3–4):237–255. doi:10.1016/j.agrformet.2004.06.011

Barr A, Black A, McCaughey H (2009) Climatic and phenological controls of the carbon and energy balances of three contrasting boreal forest ecosystems in western Canada. In: Noormets A (ed) Phenology of ecosystem processes. Springer Science, New York

Black TA, Chen WJ, Barr AG, Arain MA, Chen Z, Nesic Z, Hogg EH, Neumann HH, Yang PC (2000) Increased carbon sequestration by a boreal deciduous forest in years with a warm spring. Geophys Res Lett 27(9):1271–1274

Churkina G, Schimel D, Braswell BH, Xiao XM (2005) Spatial analysis of growing season length control over net ecosystem exchange. Global Change Biol 11(10):1777–1787

Cleland EE, Chuine I, Menzel A, Mooney HA, Schwartz MD (2007) Shifting plant phenology in response to global change. Trends Ecol Evol 22(7):357–365. doi:10.1016/j.tree.2007.04.003

Dragoni D, Schmid HP, Wayson CA, Potter H, Grimmond CSB, Randolph JC (2011) Evidence of increased net ecosystem productivity associated with a longer vegetated season in a deciduous forest in south-central Indiana, USA. Global Change Biol 17(2):886–897. doi:10.1111/j.1365-2486.2010.02281.x

Euskirchen ES, McGuire AD, Kicklighter DW, Zhuang Q, Clein JS, Dargaville RJ, Dye DG, Kimball JS, McDonald KC, Melillo JM, Romanovsky VE, Smith NV (2006) Importance of recent shifts in soil thermal dynamics on growing season length, productivity, and carbon sequestration in terrestrial high-latitude ecosystems. Global Change Biol 12(4):731–750. doi:10.1111/j.1365-2486.2006.01113.x

Falge E, Baldocchi D, Tenhunen J, Aubinet M, Bakwin P, Berbigier P, Bernhofer C, Burba G, Clement R, Davis KJ, Elbers JA, Goldstein AH, Grelle A, Granier A, Guomundsson J, Hollinger D, Kowalski AS, Katul G, Law BE, Malhi Y, Meyers T, Monson RK, Munger JW, Oechel W, Paw KT, Pilegaard K, Rannik U, Rebmann C, Suyker A, Valentini R, Wilson K, Wofsy S (2002) Seasonality of ecosystem respiration and gross primary production as derived from FLUXNET measurements. Agr Forest Meteorol 113(1–4):53–74. doi:Pii S0168-1923(02)00102-8

Goulden ML, Munger JW, Fan SM, Daube BC, Wofsy SC (1996) Exchange of carbon dioxide by a deciduous forest: response to interannual climate variability. Science 271(5255):1576–1578

Groisman PY, Karl TR, Knight RW (1994) Observed impact of snow cover on the heat-balance and the rise of continental spring temperatures. Science 263(5144):198–200

Gu L, Post WE, Baldocchi D, Black TA, Verma SB, Vesala T, Wofsy SC (2003) Phenology of vegetation photosynthesis. Phenology: an integrative environmental science, vol 39. Kluwer Academic Publishers, Dordrecht/Boston

Gu L, Post WM, Baldocchi DD, Black TA, Suyker AE, Verma SB, Vesala T, Wofsy SC (2009) Characterizing the seasonal dynamics of plant community photosynthesis across a range of vegetation types. Phenology of ecosystem processes. Springer, Dordrecht/New York

Hu J, Moore DJP, Burns SP, Monson RK (2010) Longer growing seasons lead to less carbon sequestration by a subalpine forest. Global Change Biol 16(2):771–783

Kramer K, Leinonen I, Loustau D (2000) The importance of phenology for the evaluation of impact of climate change on growth of boreal, temperate and Mediterranean forests ecosystems: an overview. Int J Biometeorol 44(2):67–75

Linderholm HW (2006) Growing season changes in the last century. Agr Forest Meteorol 137 (1–2):1–14. doi:10.1016/j.agrformet.2006.03.006

Menzel A, Fabian P (1999) Growing season extended in Europe. Nature 397(6721):659

Moffat AM, Papale D, Reichstein M, Hollinger DY, Richardson AD, Barr AG, Beckstein C, Braswell BH, Churkina G, Desai AR, Falge E, Gove JH, Heimann M, Hui DF, Jarvis AJ, Kattge J, Noormets A, Stauch VJ (2007) Comprehensive comparison of gap-filling techniques for eddy covariance net carbon fluxes. Agr Forest Meteorol 147(3–4):209–232. doi:10.1016/j.agrformet.2007.08.011

Monson RK, Sparks JP, Rosenstiel TN, Scott-Denton LE, Huxman TE, Harley PC, Turnipseed AA, Burns SP, Backlund B, Hu J (2005) Climatic influences on net ecosystem CO2 exchange during the transition from wintertime carbon source to springtime carbon sink in a high-elevation, subalpine forest. Oecologia 146(1):130–147. doi:10.1007/s00442-005-0169-2

Morin X, Viner D, Chuine I (2008) Tree species range shifts at a continental scale: new predictive insights from a process-based model. J Ecol 96(4):784–794. doi:10.1111/j.1365-2745.2008.01369.x

Niu SL, Luo YQ, Fei SF, Montagnani L, Bohrer G, Janssens IA, Gielen B, Rambal S, Moors E, Matteucci G (2011) Seasonal hysteresis of net ecosystem exchange in response to temperature change: patterns and causes. Global Change Biol 17(10):3102–3114. doi:10.1111/j.1365-2486.2011.02459.x

Papale D, Reichstein M, Aubinet M, Canfora E, Bernhofer C, Kutsch W, Longdoz B, Rambal S, Valentini R, Vesala T, Yakir D (2006) Towards a standardized processing of net ecosystem exchange measured with eddy covariance technique: algorithms and uncertainty estimation. Biogeosciences 3(4):571–583

Penuelas J, Filella I (2001) Phenology – responses to a warming world. Science 294 (5543):793–795

Piao SL, Friedlingstein P, Ciais P, Viovy N, Demarty J (2007) Growing season extension and its impact on terrestrial carbon cycle in the Northern Hemisphere over the past 2 decades. Global Biogeochem Cycles 21(3):1–11. doi:10.1029/2006gb002888

Piao SL, Ciais P, Friedlingstein P, Peylin P, Reichstein M, Luyssaert S, Margolis H, Fang JY, Barr A, Chen AP, Grelle A, Hollinger DY, Laurila T, Lindroth A, Richardson AD, Vesala T (2008) Net carbon dioxide losses of northern ecosystems in response to autumn warming. Nature 451(7174):49–U43. doi:10.1038/Nature06444

Schwartz MD, Ahas R, Aasa A (2006) Onset of spring starting earlier across the Northern Hemisphere. Global Change Biol 12(2):343–351. doi:10.1111/j.1365-2486.2005.01097.x

Sherry RA, Zhou X, Gu S, Arnone JA, Schimel DS, Verburg PS, Wallace LL, Luo Y (2007) Divergence of reproductive phenology under climate warming. Proc Natl Acad Sci USA 104 (1):198–202

Yu H, Luedeling E, Xu J (2011) Winter and spring warming result in delayed spring phenology on the Tibetan Plateau. Proc Natl Acad Sci USA 45:1–6. doi:10.1073/pnas.1012490107

Chapter 28
Phenology and Evapotranspiration

Richard L. Snyder and Donatella Spano

Abstract Many phenological models use ambient air temperature to estimate phenological stages during current and projected future climate conditions. However, the difference between ambient air temperature and plant-canopy temperature biases such estimates. Therefore, consideration of how energy balance factors affect evapotranspiration leads to more realistic plant-canopy temperatures, and better prediction of plant phenological development. Evapotranspiration (ET) has a big impact on the relationship between plant-canopy and air temperature, so awareness of ET facilitates understanding of temperature based phenological models, their limitations, and possible changes in response to climate change. This chapter presents information on the estimation of reference ET (ET_o), applying crop coefficient (K_c) values to determine well-watered crop ET (ET_c), and assessing water stress effects on crop ET. It also discusses how to account for water stress effects to determine actual crop ET (ET_a) and it presents some of the problems associated with estimating the ET of natural ecosystems. The difference between plant-canopy and air temperature is shown to depend on energy balance factors that affect ET. The temperature relationships vary spatially and with time, and therefore using air temperature based degree day models can lead to errors when predicting plant phenological development, which depends on plant-canopy temperature. It is shown that the ET effect on temperature will likely decrease the accuracy of plant phenology models as climate changes.

R.L. Snyder (✉)
Department of Land, Air, and Water Resources, University of California, Davis, CA 95616, USA
e-mail: rlsnyder@ucdavis.edu

D. Spano
Department of Science for Nature and Environmental Resources (DipNet), University of Sassari, Sassari, Italy

Euro-Mediterranean Centre for Climate Change (CMCC), Sassari, Italy

M.D. Schwartz (ed.), *Phenology: An Integrative Environmental Science*,
DOI 10.1007/978-94-007-6925-0_28, © Springer Science+Business Media B.V. 2013

28.1 Introduction

Most phenological models use air temperature as the main input variable; however, the plant temperature, which actually controls the development rate, is commonly different from the air temperature. Therefore, degree day calculations from air temperature will commonly differ from those calculated with plant canopy temperature. Since the difference between plant-canopy and air temperature varies spatially and temporally, this makes it difficult to accurately develop air temperature based phenological models that are universally applicable. The relationship between canopy and air temperature could change in a different climate, using air temperature based phenological models to accurately project the impact of climate change on phenology is also somewhat problematic.

Weather variables affect energy balance through their impact on available energy (R_n-G) and partitioning into sensible heat flux (H) and latent heat flux (LE) through the energy balance equation:

$$R_n - G = H + LE \tag{28.1}$$

where R_n is the net radiation, G is the ground heat flux, and LE is the evapotranspiration (ET) expressed in energy units (MJ $m^{-2}h^{-1}$ or MJ $m^{-2}day^{-1}$). Interaction between the weather variables and the plants determines the difference between plant-canopy and air temperature which in turn affects the utility of air-temperature based phenology models. For example, given two large, well-watered grass fields in locations with identical weather except for the wind speed, the grass site with greater wind speed will have higher LE and lower canopy temperature (T_o) because more of the R_n-G contributes to LE and less is partitioned into heating the surface. Similarly, if the weather was identical except for the dew point temperature, the location with higher humidity would have lower ET and a higher canopy temperature.

The partitioning of available energy to ET and heating has a big impact on the relationship between plant-canopy and air temperature, so awareness of ET facilitates understanding of temperature-based phenological models, their limitations, and possible impacts of climate change. This chapter discusses the estimation of ET, how spatial and temporal climate variation affects the canopy resistance and plant temperature relative to air temperature, and how this impacts on phenological models.

28.2 Reference Evapotranspiration

Reference evapotranspiration (ET_{ref}) is a measure of the evaporative demand of the air. While the idea of ET_{ref} was conceived decades ago and was widely distributed in the UN-FAO Irrigation and Drainage Paper 24 (Doorenbos and Pruitt 1977), there was considerable disagreement on the best equation to estimate ET_{ref} until the UN-FAO Irrigation and Drainage Paper 56 (Allen et al. 1998), which was later refined to a “standardized” ET_{ref} equation by the American Society of Civil

Engineers-Environmental Water Resources Institute (Allen et al. 2005). Allen et al. (2005) defined ET_{ref} for short canopies (ET_o) and ET_{ref} for tall canopies (ET_r), but, since ET_r is only used in a few states within the western USA and ET_o is used worldwide, only ET_o is discussed in this chapter.

Technically, ET_{ref} is a virtual ET because it is based on an equation using weather data as an input and assumptions about the canopy and aerodynamic resistance. In reality, the most widely used ET_{ref} equation provides an estimate of the ET of a large field of well-watered cool-season grass that is 0.12 m tall. The ET_{ref} approach is widely used by agronomists, horticulturalists, and engineers to estimate the ET of a well-watered crop (ET_c) using a crop coefficient (K_c) curve versus date and the equation:

$$ET_c = ET_o \times K_c \tag{28.2}$$

where the K_c value accounts for differences between the virtual reference crop and a real crop. A complete discussion of the K_c method for estimating ET_c is beyond the scope of this chapter, but thorough discussions are provided in Allen et al. (2005, 2011). A brief review is included in the next section.

Reference evapotranspiration for short canopies (ET_o), is estimated from daily weather data using a modified version of the Penman-Monteith equation (Allen et al. 1998, 2005).

$$ET_o = \frac{0.408\Delta(R_n - G) + \gamma \frac{900}{T+273} u_2(e_s - e_a)}{\Delta + \gamma(1 + 0.34u_2)} \tag{28.3}$$

where Δ (kPa °C^{-1}) is the slope of the saturation vapor pressure curve at mean air temperature, R_n and G are the net radiation and soil heat flux density in MJ m^{-2}day^{-1}, γ (kPa °C^{-1}) is the psychrometric constant, T (°C) is the daily mean temperature, u_2 (m s^{-1}) is the mean wind speed, e_s (kPa) is the saturation vapor pressure calculated from T, and e_a (kPa) is the actual vapor pressure calculated from T_d (°C), which is the mean daily dew point temperature. For a complete explanation of Eq. 28.3 and the estimation of R_n, e_s, e_a, Δ, and γ from daily or monthly solar radiation, temperature, humidity and wind speed data, see Allen et al. (1998, 2005, 2011). For daily calculations, the ground heat flux is $G = 0$, however, for monthly daily mean calculations the equation:

$$G = 0.07\left(T_{m,i+1} - T_{m,i-1}\right) \tag{28.4}$$

is used, where $T_{m,i-1}$ and $T_{m,i+1}$ are the mean daily temperatures for the previous and next month, respectively.

The aerodynamic resistance (r_a) for ET_o is estimated as:

$$r_a = \frac{208}{u_2} \tag{28.5}$$

for r_a (s m^{-1}) and u_2 (m s^{-1}). Note that the fraction $0.34u_2 \approx \frac{70}{208/u_2} = \frac{r_c}{r_a}$ and Eq. 28.3 assumes that $r_c = 70$ s m^{-1} with the current atmospheric CO_2 concentration.

Based on the FACE studies (Long et al. 2004), the stomatal conductance of C3 species plants is expected to decrease by 20 %, from 10 to 8 mm s^{-1} if the CO_2 concentration increases from 372 to 550 ppm. A linear change in stomatal conductance is likely in response to increasing CO_2 concentration, and a linear regression of the stomatal conductance (g_s) versus ppm of CO_2 gave the equation $g_s = 14.18–0.0112$ CO_2 for g_s in mm s^{-1}. The stomatal resistance equals the inverse of the conductance, so the stomatal resistance (r_s) in s m^{-1} is given by:

$$r_s = \frac{1000}{14.18 - 0.0112CO_2} \quad (28.6)$$

The canopy resistance is estimated as the stomatal resistance divided by half of the leaf area index, where LAI = 2.88 for 0.12 m tall grass. Therefore, the canopy resistance (r_c) for current CO_2 concentration was estimated as:

$$r_c = \frac{r_s}{LAI/2} = \frac{100}{1.44} = 69.4 \approx 70\,s\,m^{-1} \quad (28.7)$$

Thus, r_s is estimated as a function of the ppm of CO_2 using Eq. 28.6 and r_c is calculated using r_s in Eq. 28.7.

28.3 Well-Watered Crop Evapotranspiration

While ET_o is a measure of the "evaporative demand" of the atmosphere, crop coefficients account for the difference between the crop evapotranspiration (ET_c) and ET_o. The main factors affecting the difference are (1) light absorption by the canopy, (2) canopy roughness, which affects turbulence, (3) crop physiology, (4) leaf age, and (5) surface wetness. Because evapotranspiration is the sum of evaporation (*E*) from soil and plant surfaces and transpiration (*T*), which is vaporization that occurs inside of the plant leaves, it is often best to consider the two components separately.

When not limited by water availability, both transpiration and evaporation are limited by the availability of energy to vaporize water. During early growth of crops, ET_c is dominated by soil evaporation and the rate depends on dryness of the soil surface layer. As a canopy develops, interception of radiation by the foliage increases and transpiration rather than soil evaporation dominates ET_c. Field and row crop K_c values generally increase until the canopy ground cover reaches about 75 % and the peak K_c is reached when the canopy of tree and vine crops has reached about 70 % ground cover. The ground cover percentage associated with the peak K_c is slightly lower for tree and vine crops because the taller plants intercept more solar radiation at the same ground cover.

During the off-season and initial crop growth, soil evaporation (E) is the main component of ET_c, so a good estimate of the K_c (or K_e) coefficient for bare soil evaporation is useful to estimate off-season and initial crop growth evaporation. A two-stage method for estimating soil evaporation presented by Stroosnijder (1987) and refined by Snyder et al. (2000) and Ventura et al. (2006) is useful to estimate bare-soil evaporation. Stage-1 evaporation occurs when the E is limited by only energy availability. The equation: $E_1 = ET_o(1.22 - 0.04ET_o)$ provides an estimate of the evaporation rate (E_1) during stage-1 (Ventura et al. 2006). The cumulative evaporation during stage-2 is a linear function of the square root of the cumulative stage-1 evaporation $\left(\sqrt{CE_1}\right)$. Stage-2 evaporation starts when the $\sqrt{CE_1}$ equals the slope (β) of the linear regression through the origin. The β value is found by plotting measured cumulative evaporation (CE) versus $\sqrt{CE_1}$ and increasing the minimum X-axis value until the minimum X value and the slope of the regression line through the origin are approximately equal. Once the β is known, soil evaporation coefficients are calculated as: $K_e = \frac{CE}{\sqrt{CET_o}}$ for a range of wetting frequencies and mean cumulative ET_o (CET_o) rates. A $\beta = 2.6$ gives soil K_e values as a function of wetting frequency and mean ET_o rates that are similar to the widely-used bare soil coefficients published in Doorenbos and Pruitt (1977).

The main sources of K_c information are published in Doorenbos and Pruitt (1977), Allen et al. (1998, 2011). In those publications, crop growth is described in terms of the growth periods (1) initial, (2) rapid, (3) midseason, and (4) late season. The publications provide examples of seasonal K_c curves based on the number of days in each growth period. The general shape of K_c curves, based on a modification of the curves from Doorenbos and Pruitt (1977), is shown for field and row crops (Fig. 28.1) and deciduous orchard and vine crops (Fig. 28.2). While this method does work for many locations, the growth date information is site specific and it depends somewhat on cultural practices and climate. Crop growth and K_c information relative to Figs. 28.1 and 28.2 are provided for a few major California crops (Table 28.1). The K_c and growth information in Table 28.1 is based on local measurements and data from Doorenbos and Pruitt (1977), Allen et al. (2005, 2011).

28.4 Evapotranspiration with Water Deficits

28.4.1 Crops

The actual crop evapotranspiration (ET_a) is always less than or equal to ET_c because one or more stress factors has reduced the ET from its maximum (energy limited) value. The most common stress factors are water and salinity stress. Our current knowledge of salinity stress makes it difficult to estimate the impact of soil and water salinity on plant transpiration, but the UN-FAO approach to account for water stress effects ET are reasonably accurate (Allen et al. 1998). In that method, the

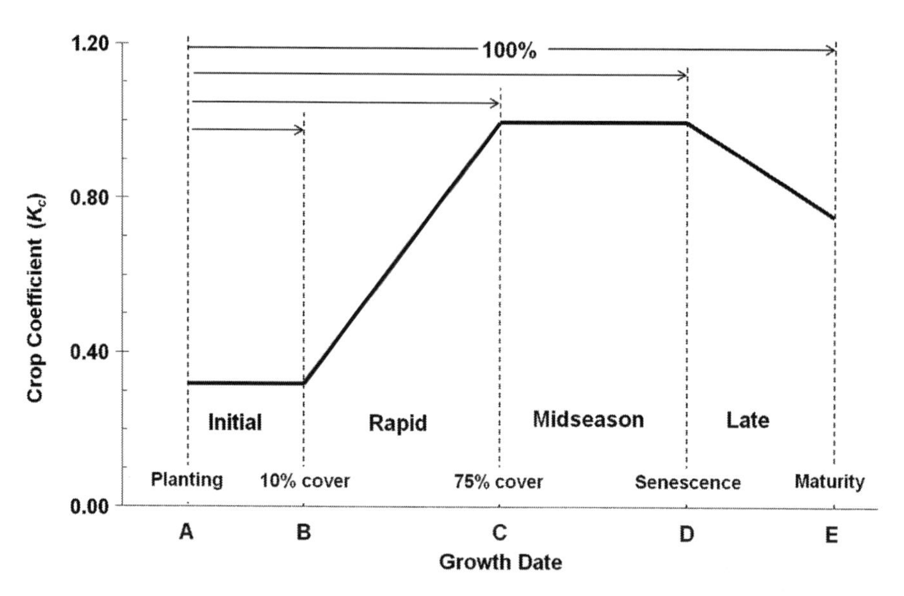

Fig. 28.1 Hypothetical crop coefficient (K_c) curve for typical field and row crops showing the growth stages and percentages of the season from planting to critical growth dates. Inflection points in the K_c curve occur at 10 and 75 % ground cover and at the onset of late season (date D). The season ends when transpiration (T) from the crop ceases ($T \approx 0$)

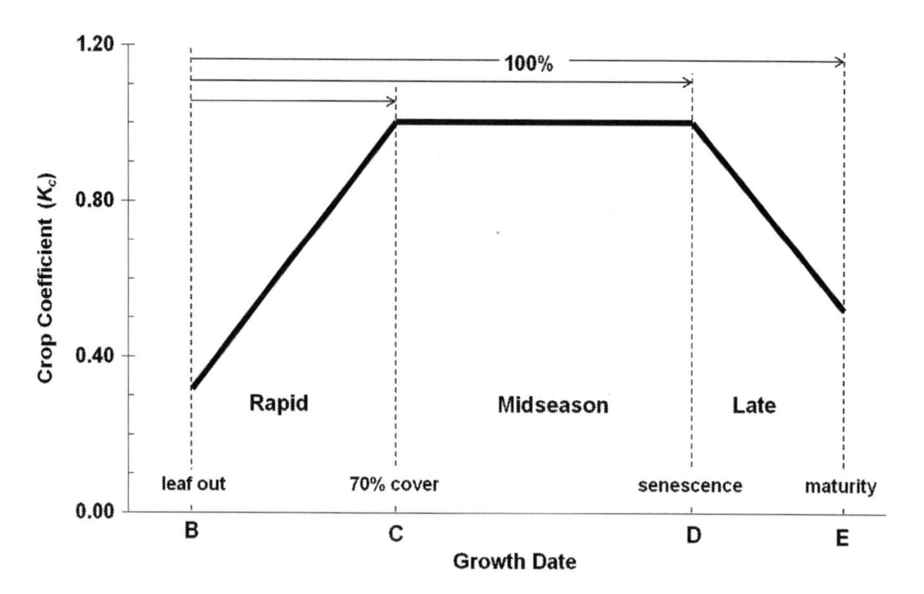

Fig. 28.2 Hypothetical crop coefficient (K_c) curve for typical deciduous orchard and vine crops showing the growth stages and percentages of the season from leaf out to critical growth dates. Inflection points occur at 70 % ground cover and at the onset of late season (date D). The season ends at leaf fall or when transpiration (T) from the crop ceases ($T \approx 0$)

Table 28.1 General crop development and Kc information for a few major crops

	Begin date	End date	Crop coefficients	% of season							
Crop	Mon	Day	Mon	Day	K_cB	K_cC	K_cD	K_cE	A-B	A-C	A-D
Grains (small)	11	1	5	31	0.33	1.15	1.15	0.15	20	45	75
Rice (paddy)	5	15	9	17	1.20	1.05	1.05	0.80	16	52	80
Cotton	4	25	9	18	0.30	1.05	1.05	0.90	15	41	72
Sugar beets	3	7	8	24	0.25	1.15	1.15	0.94	17	42	75
Maize (grain)	5	17	9	17	0.20	1.05	1.05	0.60	17	45	78
Sorghum (grain)	5	15	9	17	0.20	1.05	1.05	0.60	16	44	76
Sunflower	5	1	11	15	0.35	1.10	1.10	0.60	16	44	79
Turfgrass (C3)	1	1	12	31	0.80	0.80	0.80	0.80	25	50	75
Turfgrass (C4)	1	1	12	31	0.60	0.60	0.60	0.60	11	22	92
Tomato (process)	4	30	9	30	0.20	1.20	1.20	0.60	25	50	80
Melons	4	1	7	30	0.50	1.05	1.05	0.75	21	50	83
Onions	6	1	9	9	0.55	1.20	1.20	0.55	10	27	73
Garlic	3	1	5	31	0.60	1.20	1.20	0.60	10	27	73
Potatoes	4	25	10	10	0.40	1.15	1.15	0.75	21	45	79
Decid. orchards	3	1	11	26	0.70	1.20	1.20	0.50	0	33	78
Citrus & subtrop	1	1	12	31	1.00	1.00	1.00	1.00	0	38	86
Grapes (table)	4	1	11	1	0.40	1.10	1.10	0.40	0	31	69
Grapes (wine)	4	1	11	1	0.40	0.80	0.80	0.40	0	25	75

available water content of the soil, which depends on soil characteristics and crop rooting depth and distribution, is defined in terms of "readily available water" (RAW) and "total available water" (TAW) as shown in Fig. 28.3. TAW is the water held within the crop rooting depth that falls between "field capacity" and the "permanent wilting point". RAW is the fraction of TAW that is readily available for the plants to extract from the soil, so it is assumed that, when the water content is within RAW, there is no reduction in plant transpiration when the soil water content resides between field capacity (θ_{fc}) and the threshold water content (θ_t). When soil water is extracted below (θ_t), the transpiration rate decreases linearly as a function of the ratio of $\theta_t - \theta$ to $\theta_t - \theta_{wp}$, where θ is the actual soil water content. A stress coefficient (K_s), which is a function of the soil water content, is used to reduce the ET_a relative to ET_c. The $K_s = 1.00$ when the soil water content is greater than the threshold available water content (θ_t), and the K_s is calculated as:

$$K_s = 1 - \left(\frac{\theta_t - \theta}{\theta_t - \theta_{wp}}\right) \tag{28.8}$$

when the soil water content is less than θ_t. Once the K_s is known on a given date, the ET_a is estimated as:

$$ET_a = ET_o \times K_c \times K_s \tag{28.9}$$

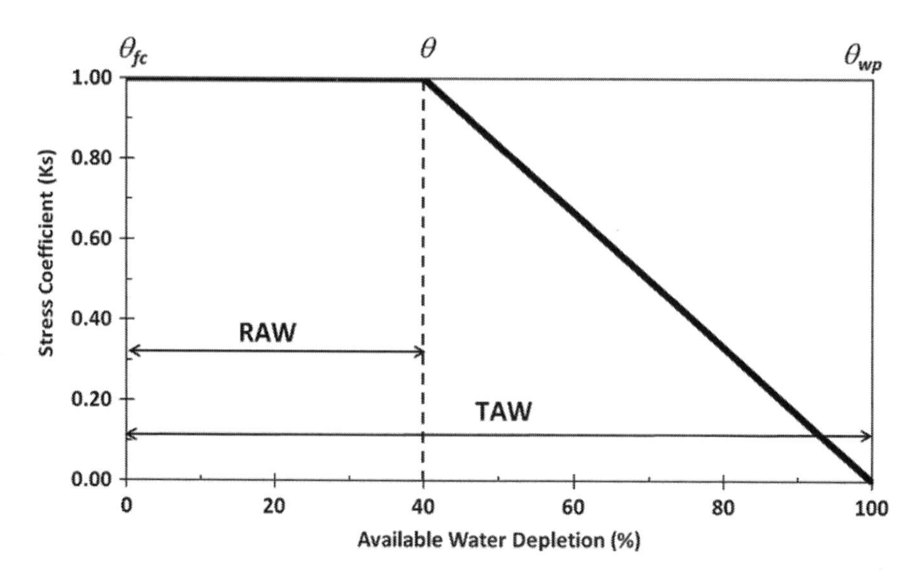

Fig. 28.3 Stress coefficient (K_s) curve as a function of available water depletion (%), where the soil water contents are θ_{fc} at field capacity, θ_t at the threshold water content, and θ_{wp} at the wilting point

28.4.2 Natural Ecosystems

Difficulties in estimating the ET of well-watered vegetation for a natural ecosystem result from (1) local advection and edge effects, (2) non-uniform plant canopies of mixed vegetation, (3) variations in radiation and sensible and ground heat flux due to undulating terrain, (4) heterogeneous soils, and (5) wind blockage or funnelling. Because of stomatal closure and reduced transpiration, estimating natural ecosystem ET and its effects on plant and air temperature relationships under water and/or salinity stress is even more complicated than for well watered crops. It is possible to estimate the ET from a natural ecosystem using the same ET_o and K_c method as is used for crops, but calibration is needed for to adjust for water stress, topography, and water contributions from fog, dew, and water tables. The methodology (ECOWAT model) was presented in Spano et al. (2009). Since estimating ET of natural ecosystems is not easy, it is also difficult to discuss its effect on plant-canopy and air temperature relationships. It is certain, however, that the plant-canopy and air temperature relationships are more complicated for natural ecosystems than for well-watered crops.

28.5 Temperature, Phenology, and Evapotranspiration

Based on the introduction, it is clear that degree days based on plant rather than air temperature are probably a better indicator of plant growth, and plant temperature is affected by transpiration changes diurnally, seasonally, and in the future with climate change. In this section, the calculation of degree days and the relation between air and canopy temperature are discussed and examples are provided to demonstrate how relationships between degree days from air and canopy temperatures might differ today and in the future.

28.5.1 *Analysis Methods*

28.5.1.1 Degree Day Calculations

A plant can develop from one stage to another by experiencing a long period of time at a low temperature or a short period time at a high temperature. The number of days to develop is longer for plants grown at low temperatures than at high temperatures, so those plants have a slower development rate. It is possible to estimate plant development simply as a function of the mean temperature during the development period, but the number of days in the period would vary annually and there would be no way to predict the end of the period. To overcome this problem, researchers have shown that combining time and temperature together as degree days (°D) improves the estimation of phenological development. Using °D as a measure of phenological time is practical because the cumulative °D during a phenological period is approximately the same regardless of the mean temperature during the period.

Degree days are computed as the number of degree hours (°*H*) divided by 24 h per day. The °*H* are computed as $T_H - T_L$ for hourly mean temperature (T_H) and a lower temperature threshold (T_L). If the $T_H < T_L$, then negligible development is expected and °$H = 0$. For some plants, there is an upper threshold temperature (T_U) above which the development rate does not increase. For plants with a T_U, the °*H* are calculated using $T_U - T_L$ whenever $T_H > T_U$. Because hourly temperature is often unavailable for climate studies, several methods to estimate °D from daily maximum and minimum temperatures were developed. All of the methods are based on approximating the daily (24 h) temperature trend from the threshold temperatures and the daily mean maximum and minimum temperatures, calculating °*H* from the hourly temperature approximation, and estimating the °D = °$H/24$. Several of these methods are described in Zalom et al. (1983) and Cesaraccio et al. (2001). In this section, the single triangle method (Zalom et al. 1983) is used to determine °D, the monthly standardized Penman-Monteith equation (Allen et al. 2005) is used to estimate monthly mean daily ET_o, and the Penman-Monteith equation (Monteith and Unsworth 1990) is used to estimate the mean daily maximum plant-canopy temperature by month. Then, using the reference ET (ET_o) for short canopies as a representative plant-canopy, the impact of ET_o on the plant-canopy and air temperature relationships is discussed.

The single triangle method for estimating degree days uses three equations to estimate the °D above a lower threshold (T_L):

$$\begin{aligned} &If\ T_x \leq T_L,\ then\ ^oD = 0 \\ &If\ T_n \geq T_L,\ then\ ^oD = \frac{T_x + T_n}{2} - T_L \\ &Otherwise,\ ^oD = \left(\frac{T_x - T_L}{2}\right)\left(\frac{T_x - T_L}{T_x - T_n}\right) \end{aligned} \quad (28.10)$$

In the analysis presented here, it was assumed that there is no upper threshold and the lower threshold is: $T_L = 10$ °C when calculating degree day examples. A modified Penman-Monteith equation was used to estimate the effect of evaporation on the mean daily maximum plant-canopy temperature (see the next section). The daily minimum plant-canopy temperature was set equal to the daily minimum air temperature.

28.5.1.2 Plant-Canopy and Air Temperature Calculations

Calculation of the maximum plant-canopy air temperature was accomplished using the following modification of the Penman-Monteith equation (Monteith and Unsworth 1990)

$$T_{xc} - T_x = \left(\frac{\gamma\left(1 + \frac{r_c}{r_a}\right)}{\Delta + \gamma\left(1 + \frac{r_c}{r_a}\right)}\right)\left(\frac{r_a}{\rho C_p}\right)(R_n - G) - \left(\frac{e_s(T) - e}{\Delta + \gamma\left(1 + \frac{r_c}{r_a}\right)}\right) \quad (28.11)$$

where T_{xc} is the maximum plant-canopy temperature and T_x is the maximum air temperature. It was assumed that $\frac{R_s}{R_o} = \frac{R_{sd}}{R_{od}}$, where R_s (W m^{-2}) is the monthly mean daily maximum instantaneous solar radiation, $R_o = 1{,}000$ W m^{-2} is the monthly maximum possible instantaneous solar radiation at the surface calculated from extraterrestrial radiation (Allen et al., 2005), R_{sd} (MJ m^{-2}day^{-1}) is the monthly mean daily solar radiation, and R_{od} (MJ m^{-2}day^{-1}) is the maximum possible monthly mean solar radiation. Rearranging terms gives: $R_s = 1000\left(\frac{R_{sd}}{R_{od}}\right)$. The procedures from Allen et al. (2005) were used to compute the maximum R_n (W m^{-2}), the vapor pressure (e) in kPa from the monthly mean daily dew point temperature (T_d) in °C, and ground heat flux (G) in W m^{-2} as 10 % of R_s in W m^{-2}. The saturation vapor pressure [$e_s(T_x)$], the slope of the saturation vapor pressure curve at temperature T_x (Δ) in kPa K^{-1}, and the psychrometric constant (kPa K^{-1}) were also computed following procedures in Allen et al. (2005).

It is well-known that the canopy resistance (r_c) is likely to increase due to partial stomatal closure in a higher CO_2 environment, so a method to estimate the effect was derived. The r_c value used in Eq. 28.3 is an average over a 24 h day, while the r_c

Table 28.2 Site information for CIMIS stations used in this comparison

Station	Latitude	Longitude	Elev. Msl (m)	Record (years)
Castroville	36°46′05″N	121°46′25″W	2.7	29
Davis	38°32′09″N	121°46′32″W	18.3	29
Indio	33°44′46″N	116°15′28″W	12.2	7
Hopland	39°00′25″N	123°04′45″W	353.6	21

in Eq. 28.11 is an instantaneous value during daylight hours. From Allen et al. (2005), the $r_c = 70\ \mathrm{s\ m^{-1}}$ is used for daily calculations, as in Eq. 28.3, and the $r_c = 50\ \mathrm{s\ m^{-1}}$ is employed for hourly calculations during daylight hours as in Eq. 28.11. In the analysis presented here, it was assumed that, at a higher CO_2 concentration, the daytime canopy resistance, which was used to estimate the maximum canopy and air temperatures, will remain $20\ \mathrm{s\ m^{-1}}$ less than the daily mean r_c, which is used to calculate ET_o using Eq. 28.3.

Monthly means of daily climate data from four California Irrigation Management Information System, i.e., CIMIS, stations (Snyder and Pruitt 1992) were used to calculate monthly means of daily ET_o using Eq. 28.3 and the monthly mean daily maximum temperatures were used to estimate the daily maximum canopy temperatures by month using Eq. 28.11. The calculations were done once using the current mean climate data assuming a CO_2 concentration of 375 ppm and again using a CO_2 concentration of 550 ppm and increases of 2 °C for the maximum, 4 °C for the minimum, and 4 °C for dew point temperature. This scenario falls within the range of global climate change expected by the end of this century (IPCC 2007). The monthly means of daily wind speed were used in Eq. 28.11 for the plant-canopy temperature calculations. Location information for the four CIMIS stations is provided in Table 28.2.

28.5.2 *Results*

28.5.2.1 Canopy and Air Temperature Under Current Climate Conditions

Climate data and the results from the ET_o and maximum plant-canopy temperature calculations are shown in Tables 28.3, 28.4, 28.5 and 28.6. Comparing the current conditions air (T_x) and plant-canopy (T_{xc}) temperatures illustrates a fundamental problem with air temperature based °D models. It is generally assumed that °D models based on air temperature will provide good estimates of phenological development. The assumption is that the air and canopy temperatures are similar and consistent over time. The data presented here, however, show that the differences between T_x and T_{xc} vary temporally due to seasonal climate change. For example, in Castroville (Table 28.3), the T_{xc} is less than T_x during December and January, but it exceeds T_x by as much as 7.2 °C in the summer.

Table 28.3 Current climate data with current and projected temperature and °D calculations for the Castroville CIMIS station #19

Mon	R_s (MJ m^{-2})	Tn (°C)	$U2$ (m s^{-1})	Td (°C)	Tx (°C)	Txc (°C)	T'_x (°C)	T'_{xc} (°C)	Dd (°D)	Ddc (°D)	D'_d (°D)	D'_{dc} (°D)	ETo mm	ET'_o mm
1	8.7	4.2	2.3	7.0	15.6	15.3	17.6	18.6	176	171	207	223	1.1	0.9
2	11.3	5.3	2.4	8.0	16.4	17.8	18.4	21.3	155	174	183	223	1.6	1.3
3	15.8	6.3	2.5	9.0	16.4	21.2	18.4	25.0	156	230	187	289	2.0	1.7
4	20.8	6.2	2.8	9.4	20.0	25.0	22.0	29.1	276	375	316	457	3.3	2.9
5	23.2	8.0	3.0	10.6	19.8	26.0	21.8	30.3	183	280	214	346	3.5	3.1
6	23.6	9.4	3.4	11.5	18.4	25.5	20.4	29.8	134	241	164	305	3.3	2.8
7	21.0	11.6	2.8	12.9	18.8	26.0	20.8	30.2	161	272	192	338	3.1	2.7
8	18.1	11.6	2.5	13.5	18.5	25.5	20.5	29.6	156	264	187	328	2.5	2.2
9	16.9	11.2	2.3	13.1	18.8	25.1	20.8	29.1	149	244	179	305	2.4	2.1
10	13.7	9.0	2.1	11.5	19.1	22.6	21.1	26.5	158	212	189	271	1.9	1.7
11	10.5	6.4	2.0	9.2	18.5	19.1	20.5	22.6	181	190	211	243	1.5	1.3
12	8.4	4.1	2.3	6.8	16.2	15.1	18.2	18.4	188	170	219	222	1.2	1.0

Variables with a prime (′) are for the projected climate. ET_o and ET'_o are in mm day^{-1}
Current variables are solar radiation (R_s), minimum temperature (T_n), wind speed (u_2), dew point (T_d), maximum temperature (T_x), maximum canopy temperature (T_{cx}), degree days based on T_x (D_d), degree days based on T_{xc} (D_{dc}), reference ET (ET_o), and cumulative reference ET (CET_o). Projected variables are the maximum temperature (T'_{xc}), the maximum canopy temperature (T'_{xc}), degree days based on $T'_x(D'_d)$, degree days based on T'_{xc} (D'_{dc}), and daily reference ET (ET'_o). Total °D were D_d =2,074, D_{dc} = 2,824, D'_d = 2,449, and D'_{dc} = 3,550

Table 28.4 Current climate data with current and projected temperature and °D calculations for the Davis CIMIS station #6

Mon	R_s (MJ m^{-2})	T_n (°C)	u_2 (m s^{-1})	T_d (°C)	T_x (°C)	T_{xc} (°C)	T'_x (°C)	T'_{xc} (°C)	D_d (°D)	D_{dc} (°D)	D'_d (°D)	D'_{dc} (°D)	ET_o mm	ET'_o mm
1	6.5	3.6	2.6	5.4	12.7	11.3	14.7	14.5	141	120	172	169	0.9	0.6
2	10.4	5.0	2.7	6.6	16.0	16.1	18.0	19.5	154	155	182	203	1.6	1.3
3	15.9	6.0	2.7	7.2	19.0	21.0	21.0	24.7	202	232	233	290	2.8	2.4
4	21.5	7.8	3.0	6.9	22.8	24.7	24.8	28.7	300	338	340	419	4.5	4.1
5	25.5	10.4	3.0	9.2	26.3	28.3	28.3	32.6	259	290	290	356	5.7	5.2
6	28.8	12.7	3.0	10.8	30.1	31.2	32.1	35.6	342	358	372	424	7.0	6.5
7	29.0	13.7	2.7	12.7	32.9	33.1	34.9	37.5	412	416	443	483	7.4	6.9
8	26.0	13.2	2.5	11.5	32.5	31.9	34.5	36.0	398	389	429	453	6.7	6.4
9	20.9	12.1	2.4	10.2	30.8	29.0	32.8	32.9	344	316	374	375	5.5	5.2
10	14.8	9.6	2.4	7.6	26.3	23.6	28.3	27.2	259	217	290	273	3.9	3.7
11	9.4	5.4	2.4	5.3	18.4	16.3	20.4	19.6	195	163	225	213	2.1	1.8
12	6.5	2.7	2.7	4.3	12.8	11.0	14.8	14.1	157	128	188	177	1.0	0.8

Variables with a prime (′) are for the projected climate. ET_o and ET'_o are in mm day^{-1}
Current variables are solar radiation (R_s), minimum temperature (T_n), wind speed (u_2), dew point (T_d), maximum temperature (T_x), maximum canopy temperature (T_{cx}), degree days based on T_x (D_d), degree days based on T_{xc} (D_{dc}), reference ET (ET_o), and cumulative reference ET (CET_o). Projected variables are the maximum temperature (T'_x), the maximum canopy temperature (T'_{xc}), degree days based on T'_x (D'_d), degree days based on $T'_{xc}(D'_{dc})$, and daily reference ET (ET'_o). Total °D were D_d =3,162, D_{dc} = 3,122, D'_d = 3,537, and D'_{dc} = 3,835

Table 28.5 Current climate data with current and projected temperature and °D calculations for the Indio CIMIS station #162

Mon	R_s (MJ m^{-2})	T_n (°C)	u_2 (m s^{-1})	T_d (°C)	T_x (°C)	T_{xc} (°C)	T'_x (°C)	T'_{xc} (°C)	D_d (°D)	D_{dc} (°D)	D'_d (°D)	D'_{dc} (°D)	ET_o mm	ET'_o mm
1	11.6	6.7	2.2	1.0	21.4	18.1	23.4	21.3	228	176	259	227	3.0	2.8
2	14.3	8.0	2.5	1.8	21.3	19.7	23.3	23.1	186	164	214	212	3.6	3.3
3	19.9	11.7	3.1	3.1	25.6	23.9	27.6	27.7	268	241	299	300	5.7	5.2
4	24.8	14.8	3.8	3.2	28.3	26.3	30.3	30.4	463	423	503	505	7.7	7.1
5	27.7	19.4	4.1	5.9	33.3	29.5	35.3	33.7	507	449	538	514	9.8	9.1
6	28.7	22.9	4.1	8.6	37.2	32.0	39.2	36.2	601	523	631	586	11.0	10.3
7	27.1	26.0	3.6	10.0	40.5	33.7	42.5	37.7	721	615	752	677	11.0	10.4
8	25.0	26.0	3.1	13.0	39.8	33.8	41.8	37.7	710	616	741	677	9.6	9.2
9	22.4	24.0	3.2	9.8	38.2	31.5	40.2	35.4	634	534	664	591	8.9	8.5
10	18.3	19.0	2.9	7.2	33.5	27.7	35.5	31.4	504	413	535	470	6.8	6.5
11	13.6	12.7	2.6	4.4	26.1	21.9	28.1	25.3	282	219	312	271	4.4	4.1
12	11.2	6.6	2.2	0.5	21.5	17.8	23.5	21.0	231	173	262	223	3.0	2.8

Variables with a prime (′) are for the projected climate. ET_o and ET'_o are in mm day^{-1} Current variables are solar radiation (R_s), minimum temperature (T_n), wind speed (u_2), dew point (T_d), maximum temperature (T_x), maximum canopy temperature (T_{cx}), degree days based on T_x (D_d), degree days based on T_{xc} (D_{dc}), reference ET (ET_o), and cumulative reference ET (CET_o). Projected variables are the maximum temperature (T'_x), the maximum canopy temperature (T'_{xc}), degree days based on T'_x (D'_d), degree days based on T'_{xc} (D'_{dc}), and daily reference ET (ET'_o). Total °D were $D_d = 5{,}334$, $D_{dc} = 4{,}546$, $D'_d = 5{,}709$, and $D'_{dc} = 5{,}254$

Table 28.6 Current climate data with current and projected temperature and °D calculations for the Hopland CIMIS station #85

Mon	R_s (MJ m^{-2})	T_n (°C)	u_2 (m s^{-1})	T_d (°C)	T_x (°C)	T_{xc} (°C)	T'_x (°C)	T'_{xc} (°C)	D_d (°D)	D_{dc} (°D)	D'_d (°D)	D'_{dc} (°D)	ET_o mm	ET'_o mm
1	7.1	3.5	1.5	3.6	13.2	11.8	15.2	14.9	150	129	181	178	1.1	0.9
2	10.8	4.1	1.7	3.9	14.3	16.1	16.3	19.4	143	167	171	215	1.6	1.5
3	14.6	4.8	1.8	4.6	16.8	20.4	18.8	24.0	186	242	217	298	2.4	2.3
4	19.3	5.3	1.8	4.6	18.4	24.5	20.4	28.3	262	383	302	460	3.3	3.2
5	23.9	7.8	1.8	6.6	23.0	29.2	25.0	33.3	235	332	266	394	4.6	4.4
6	26.7	10.1	1.9	9.0	27.2	31.7	29.2	35.9	259	326	289	389	5.6	5.4
7	28.6	12.9	1.8	10.5	31.6	34.3	33.6	38.5	380	422	411	486	6.5	6.4
8	26.5	13.1	1.7	9.8	32.4	33.5	34.4	37.4	395	411	426	473	6.2	6.1
9	22.3	12.1	1.6	8.4	30.5	30.7	32.5	34.5	339	342	369	399	5.1	5.0
10	16.6	10.1	1.6	6.4	26.8	25.9	28.8	29.4	261	248	292	302	3.6	3.5
11	10.5	6.7	1.4	5.6	19.3	18.8	21.3	22.1	190	182	220	232	1.8	1.8
12	6.9	3.6	1.5	3.2	13.7	11.8	15.7	15.0	157	127	188	176	1.1	1.0

Variables with a prime (′) are for the projected climate. ET_o and ET'_o are in mm day^{-1} Current variables are solar radiation (R_s), minimum temperature (T_n), wind speed (u_2), dew point (T_d), maximum temperature (T_x), maximum canopy temperature (T_{cx}), degree days based on T_x (D_d), degree days based on T_{xc} (D_{dc}), reference ET (ET_o), and cumulative reference ET (CET_o). Projected variables are the maximum temperature (T'_x), the maximum canopy temperature (T'_{xc}), degree days based on T'_{xc} (D'_d), degree days based on T'_{xc} (D'_{dc}), and daily reference ET (ET'_o). Total °D were $D_d = 2{,}957$, $D_{dc} = 3{,}311$, $D'_d = 3{,}332$, and $D'_{dc} = 3{,}999$

In addition to temporal differences, there are big spatial differences due to micro-climate. At Hopland (Table 28.6), the T_{xc} was as much as 8.3 °C warmer than the T_x in the spring months, but the T_{xc} and T_x temperatures were similar in the fall. In Castroville, the T_{xc} was well above T_x during the period from March through September presumably because of the cold ocean breeze blowing over the solar heated surface. In Davis (Table 28.4), the T_{xc} was higher than T_x from March through July, but the T_x was higher than T_{xc} during the remainder of the year. In Indio (Table 28.5), the T_{xc} was always lower than the T_x with big differences in the late summer and fall. Based on observations at the four sites, it is unlikely that a phenological model based on T_x that was developed at one location would apply to the other sites. There is little or no literature on T_{xc} based degree day models because of the difficulty to continuously measure plant-canopy temperature. A model based on plant-canopy temperature, however, could account for the effects of the energy balance and improve phenological development estimates.

28.5.2.2 Projected Climate Change Effect on ET_o

The Hopland station (Table 28.6) showed a small annual drop (53 mm) in ET_o due to the increased air temperature, humidity, and CO_2 concentration, whereas the annual decreases were 110, 133, and 164 mm for Castroville, Davis, and Indio. At the Hopland station, the projected monthly mean daily ET_o was consistently about 0.1–0.2 mm day^{-1} less than for the current climate. The other three stations (Tables 28.3, 28.4 and 28.5) all showed a projected ET_o drop of about 0.2 mm day^{-1} during the winter months. Decreases of 0.4–0.5 mm day^{-1} were observed during summer in Castroville and Davis. The ET_o dropped by as much as 0.7 mm day^{-1} during the summer in Indio. While one often relates higher temperature to elevated ET_o, this is clearly not true when partial stomatal closure due to higher CO_2 concentration and higher humidity counteract the increased temperature effect. There is no evidence for elevated ET_o rates due to higher temperature in the analyzed data.

28.5.2.3 Canopy and Air Temperature with Projected Climate Change

Applying the projected climate change scenario to data from the four locations led to even bigger differences between canopy and air temperatures. For Castroville (Table 28.3), while the T_x values were increased by 2 °C each month, the T_{xc} temperatures rose by 3.3–4.2 °C each month with bigger increases in the summer. For Davis (Table 28.4), the projected increases in monthly T_{xc} were about 0.2 °C bigger than for Castroville. The difference was most likely related to a larger reduction in ET_o in Davis than in Castroville (Tables 28.3 and 28.4). For the hot, dry desert climate in Indio (Table 28.5), the T_{xc} temperatures in the projected

Table 28.7 Annual degree day (°D) accumulations and differences between plant-canopy and air temperature based °D calculations for the current and projected climate

	$°D_x$	$°D_{xc}$	$°D_{xc}$-$°D_x$	$°D_{xc}$-$°D_x$	$°D'_x$	$°D'_{xc}$	$°D'_{xc}$-$°D'_x$	$°D'_{xc}$-$°D'_x$
Location	°D	°D	°D	%	°D	°D	°D	%
Castroville	2,074	2,824	750	36	2,449	3,550	1,101	45
Davis	3,162	3,122	−40	−1	3,537	3,835	298	8
Indio	5,334	4,546	−788	−15	5,709	5,254	−455	−8
Hopland	2,957	3,311	355	12	3,332	3,999	667	20

The percentage differences are also shown. Variables with a prime (′) are for the projected climate
Note: The annual degree day (i.e., $°D_x$, $°D_{xc}$, $°D'_x$, $°D'_{xc}$ and) values were computed using the threshold temperature $T_L = 10\,°C$, monthly mean minimum temperature (T_n) for each site, and the monthly mean maximum air temperature for the current (T_x) and projected (T'_x), and the estimated monthly mean maximum canopy temperatures for the current (T_{xc}) and projected (T'_{xc}) climates

climate rose to where the T_{xc} and T_x temperatures were nearly the same for February through April, but the T_{xc} temperature was still lower than the air temperature from May through December. The increase in T_{xc} relative to T_x was mostly related to the lower ET_o in the projected climate. Summertime increases in T_{xc} in the projected climate were on the order of 4.2 °C. For the Hopland station (Table 28.6), the T_{xc} rise was considerably more than the T_x increase in the projected climateduring the late spring. The T_{xc} rose by about 4.2 °C relative to a 2 °C rise in T_x during April through June. In the fall, there was little difference between the T_{xc} and T_x temperatures. Because the difference between T_{xc} and T_x increased in all locations for the projected climate, it seems that T_x based degree day models will become less accurate with time.

28.5.2.4 Degree Day Comparisons in the Current and Projected Climate

Spatial and temporal changes in the difference between T_{xc} and T_x can lead to inaccuracy in temperature based degree day phenology models. Tables 28.3, 28.4, 28.5 and 28.6 show the monthly °D calculations using the monthly mean daily maximum and minimum temperatures and a threshold temperature $T_L = 10$ °C for both current and projected air and canopy temperatures. The tables also show the cumulative °D data using the current and projected air and canopy temperatures. There are large differences in the annual °D totals calculated with the current T_{xc} and T_x temperatures, and the differences change dramatically in the projected climate.

The annual °D differences between using T_{xc} and T_x in the current (D_{dc}-D_d) and projected (D'_{dc}-D'_d) climate in Table 28.7 are variable; indicating that a °D model based on air temperature is problematic due to the effects of ET on the plant-canopy versus air temperature. For example, Davis had similar annual °D accumulations and the others sites had dissimilar annual °D accumulations using the plant-canopy and the air temperatures. Although the Davis site had similar annual °D accumulation for the canopy and air temperatures, the °D distributions over the year were not alike. Currently, in Davis (Table 28.4), the monthly mean temperatures and the monthly °D figures are mostly higher for the plant-canopy than air temperature during the first half of the year. The plant-canopy based °D values are mainly less than the air temperature values during the last half of the year. For Davis, looking at only the annual total °D, the seasonal differences are lost. In the projected climate for Davis (Table 28.4), the projected T_{xc} values moved upwards relative to T_x, so the canopy temperature is considerably higher than air temperature during the first half of the year, and it is about the same as air temperature during the last half of the year.

In Castroville (Table 28.3), the projected T_{xc} values increased more than T_x temperatures, so using an air temperature based degree day model developed in the current climate will under-estimate phenological development in the projected climate. In the desert at Indio (Table 28.5), the T_{xc} values are less than the T_x temperature during all months, so the °D values are also lower for the plant-canopy based °D model. In the projected climate, the °D from the T_{xc} will increase to near those of the T_x during January through April, but the lower T_{xc} values will lead to lower °D accumulations for May through December. On an annual basis, the difference between the °D calculations from plant-canopy and air temperature will decrease in the projected climate. For the Hopland station, the T_{xc} are higher during February through September, and the difference is likely to increase in the projected climate. Therefore, the air temperature based °D accumulations are likely to under-estimate the canopy °D accumulations during much of the growing season, and the difference will increase in the projected climate.

28.5.3 Stress Effects on Degree Day Phenology Models

Since many crops and native plants are not irrigated, soil water deficits and plant water stress are common. Many factors affect the level of stress, so it is difficult to quantify the effects of water stress on canopy temperature and phenological development. It is well-known that moderate water stress first affects the growth and differentiation of plant cells and severe stress can lead to partial or full stomatal closure and reduced transpiration. The authors have observed water stress to dramatically increase phenological development in maize and soybeans, where the season was shortened by as much as 2–3 weeks in response to severe water stress. It is likely that the water stress induced transpiration reduction led to higher plant temperature and faster development. Although the canopy temperature increased in response to the water stress, it is unlikely that air temperature measured over a well watered weather station site would respond to the water stress experienced by the crop. Any time that transpiration decreases due to (1) water stress, (2) CO_2 increases, (3) higher humidity, or (4) lower wind speed, the canopy temperature is likely to rise relative to air temperature. If this occurs, then air temperature based degree day models will exhibit more error in predicting phenological development. In some cases, the errors can be large.

28.5.4 Improving Degree Day Based Phenology Models

The biggest difficulty with use of degree day based models is that they commonly use air temperature to drive the models when they should be driven by crop canopy temperature. In some climates, canopy and air temperatures are similar, and the °D models will work well using air temperature. In areas with low ET and high radiation, however, the canopy temperature can greatly exceed the air temperature. The difference is even bigger at low wind speed and high humidity. Because micro-climates vary spatially and temporally, either direct measurement of canopy temperature or better estimates of the canopy temperature could improve degree day phenology models. Direct measurement of canopy temperature is possible with infrared thermometers, but there are problems to obtain accurate measurements for sparse canopies (i.e., for growing field crops) and it is difficult to collect long-term continuous readings without errors. If the input parameters are known, then Eq. 28.11 in this chapter can provide a method to estimate the canopy temperature for use in a degree day model.

28.6 Conclusions

This chapter discussed the estimation of reference evapotranspiration (ET_o), crop coefficient (K_c) values to estimate well-watered crop evapotranspiration, water stress coefficient (K_s) factors to determine actual crop evapotranspiration, and the effect of ET on temperature based degree day models for phenological development in both the current climate and in a projected climate having higher temperature, CO_2 concentration, and humidity. The estimation of crop evapotranspiration is difficult and there is uncertainty in the use of K_c and K_s values for estimating crop evapotranspiration from reference evapotranspiration. The effect of water stress on crop development is another confounding factor because of the reduced transpiration effect on plant-canopy temperature. While site specific empirical degree day models can provide good estimates of phenological development, one should use caution when applying empirical models to other micro-climates.

The difference between canopy and air temperature is a function of the surface energy balance, and in this chapter, canopy and air temperature differences were compared for a coastal climate (Castroville), an inland valley climate (Hopland), a hot, dry inland semi-arid climate (Davis), and a desert climate (Indio) using the existing climate and a projected climate. In general, decreasing ET in any manner increases the canopy temperature relative to air temperature. The difference between canopy and air temperature depends on several climate factors that vary spatially and with time, so using air temperature based degree day models can lead to errors when predicting phenology that actually depends on plant-canopy temperature. Perhaps the best way to improve the universality of phenological models is to use measured plant-canopy temperature data or canopy temperature estimates derived from energy balance measurements.

References

Allen RG, Pereira LS, Raes D, Smith M (1998) Crop evapotranspiration: guidelines for computing crop water requirements. FAO Irrigation and Drainage Paper 56. FAO, Rome

Allen RG, Walter IA, Elliott RL, Howell TA, Itenfisu D, Jensen ME, Snyder RL (2005) The ASCE standardized reference evapotranspiration equation. American Society of Civil Engineers, Reston

Allen RG, Howell TA, Snyder RL (2011) Irrigation water requirements. In: Stetson LE, Mecham BQ (eds) Irrigation, 6th edn. Irrigation Association, Falls Church

Cesaraccio C, Spano D, Duce P, Snyder RL (2001) An improved model for degree-days values from temperature data. Int J Biometeorol 45:161–169. doi:10.1007/s004840100104

Doorenbos J, Pruitt WO (1977) Guidelines for predicting crop water requirements. FAO Irrigation and Drainage Paper 24. UN-FAO, Rome

IPCC (2007) Contribution of working group I to the fourth assessment report of the intergovernmental panel on climate change. Cambridge University Press, Cambridge, United Kingdom and New York, NY, USA

Long SP, Ainsworth EA, Rogers A, Ort DR (2004) Rising atmospheric carbon dioxide: plants FACE the future. Ann Rev Plant Biol 55:591–628

Monteith JL, Unsworth MH (1990) Principles of environmental physics, 2nd edn. Edward Arnold, London

Snyder RL, Pruitt WO (1992) Evapotranspiration data management in California. In: Proceedings of the Water Forum '92 – Irrigation & Drainage Session, ASCE, New York

Snyder RL, Bali K, Ventura F, Gomez-MacPherson H (2000) Estimating evaporation from bare or nearly bare soil. J Irrig Drain Eng 126(6):399–403

Spano D, Snyder RL, Sirca C, Duce P (2009) ECOWAT – a model for ecosystem evapotranspiration estimation. Agric For Meteorol 149:1584–1596

Stroosnijder L (1987) Soil evaporation: test of a practical approach under semi-arid conditions. Neth J Agric Sci 35:417–426

Ventura F, Snyder RL, Bali KM (2006) Estimating evapotranspiration from bare soil using soil moisture data. J Irrig Drain Eng 132(2):153–158

Zalom FG, Goodell PB, Wilson WW, Bentley WJ (1983) Degree-days: the calculation and the use of heat units in pest management. Leaflet 21373, Div of Agric and Nat Res, Univ of Calif, Davis

Chapter 29
Phenology in Agriculture and Horticulture

Frank-M. Chmielewski

Abstract In agriculture and horticulture, phenological observations have a long tradition since many management decisions and the timing of field works (planting, fertilizing, irrigating, crop protection, harvesting, etc.) are based on plant development. This chapter deals with both the historical and modern aspects of phenology in agriculture, including the impacts of climate change on plant development. The individual paragraphs give some examples how important are phenological observations to detect changes in the duration of phenological phases, to define the length of growing season – which sets the environmental limits for crop production – to select suitable growing areas for perennial and field crops, and how these data can be used to develop phenological models in agriculture and horticulture. The chapter ends with some applications of phenological models to calculate possible shifts in the timing of ripening and blossoming stages in relation to climate change.

29.1 Introduction

This chapter deals with both traditional aspects of phenology in agriculture (growing season length and different applications of phenological data and models in agriculture) as well as modern aspects, which focus on impacts of climate change on phenophases of field and perennial crops.

Generally, phenology has a long tradition in agriculture and horticulture. The knowledge of the annual timing of phenophases and their variability can help to improve the crop management which leads finally to higher and more stable crop yields and to an improved food quality. Phenological observations are essential for

F.-M. Chmielewski (✉)
Agricultural Climatology, Department of Crop and Animal Sciences, Faculty of Agriculture and Horticulture, Humboldt-University of Berlin, Berlin, Germany
e-mail: chmielew@agrar.hu-berlin.de

M.D. Schwartz (ed.), *Phenology: An Integrative Environmental Science*,
DOI 10.1007/978-94-007-6925-0_29,

many aspects in practical agriculture. The data can be used to define the growing season length in a region. On the basis of the available time in the year, cropping schedules can be developed which include suitable crops and varieties, the organization of crop rotation, and catch cropping. Phenological observations also play an important role in processes that are relevant in practical agriculture, such as the timing of irrigation, fertilization, and crop protection. The data are also necessary to evaluate the risk of frost damage and to make forecasts of plant development and harvest dates. In agrometeorological studies, phenological data are used to analyse crop-weather relationships and to describe or model the phytoclimate.

The individual sections of this chapter give some examples of the uses of phenological data in agriculture and horticulture. The chapter shows that the relatively small changes in air temperature have had already distinct impacts on plant development of fruit trees and field crops. It ends with some modelling aspects for field crops and perennials, which are basic for any projections in the timing of phenological stages due to climate change.

29.2 Phenological Observations in Agriculture and Horticulture

Phenological observations in agriculture and horticulture are common and have great value (see Sect. 29.3). They include such principal growing stages or field work as seeding/planting, germination/bud development, leaf development, formation of side shoots/tillering, stem elongation/rosette growth/shoot development, booting/development of harvestable parts of vegetative plants, inflorescence emergence/heading, flowering/blossoming, development of fruit, ripening and maturity of fruit and seed, senescence, beginning of dormancy, and harvest.

In Germany, the German Weather Service (DWD) is running a network of about 1,500 phenological observers (Bruns 2001). The phenological observations cover natural vegetation species, field crops, fruits, and vines. Observed crops are sugar beets, permanent grassland, oats, maize, sunflowers, and winter cereals such as barley, rape, rye, and wheat. Observations of fruits include apple, pear, red currant, sour/sweet cherry, and gooseberry species. For vines the varieties Mueller-Turgau and Riesling are observed (see also Chap. 30). Additionally, there are some important long-term field experiments in Europe and in Germany (Körschens 1997; MLUV 2009) which also include phenological observations. The Faculty of Agriculture and Horticulture at the Humboldt-University is running in Berlin-Dahlem a long-term *Agrometeorological Field Experiment* since 1953 which is one of the oldest experiments of this kind in Europe (Chmielewski and Köhn 1999a, b).

In order to have comparable phenological observations it is absolutely necessary to define the phenological phases, exactly. For phenological observations, different scales were developed in the past. In agriculture the *Feekes-scale* was introduced (Feekes 1941). This scale was based on 23 phenological phases for winter wheat

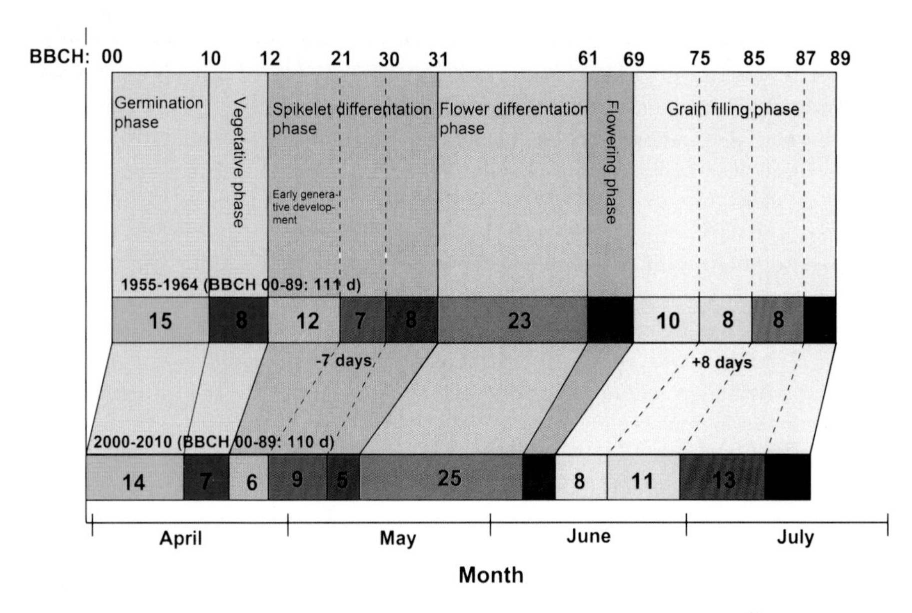

Fig. 29.1 Average timing of phenological stages and duration of phenological phases for spring barley between sowing and full ripeness at the Experimental Station Berlin-Dahlem for two different periods 1955–1964 and 2000–2010 (BBCH 00: sowing, 10: emergency, 12: two-leaf-stage, 21: beginning of tillering, 30: beginning of stem elongation, 31: first-node-stage, 61: beginning of flowering, 69: end of flowering, 75: milky-, 85 dough-, 87: yellow-, 89: full ripeness) (Data calculated by Cevik 2011)

between germination and ripeness. By illustrations of Large (1954) the scale became known worldwide (see also Clive-James 1971). In subsequent years the Feekes-scale was modified by many authors for phenological observations of cereals (Keller and Baggiolini 1954; Petr 1966; Broekhuizen and Zadoks 1967). Later, Zadoks et al. (1974) developed a phenological decimal-coded system for cereals, including rice. This scale was published by the European Association for Plant Breeding (EUCARPIA) and is still known as the *Eucarpia (EC)-scale*. In Germany the EC-scale was adopted by the German Federal Biological Research Centre for Agriculture and Forestry (BBA). In 1981, the BBA also published a decimal code for maize, which was not considered in the EC-scale before. Today the extended *BBCH-scale* (Strauß et al. 1994; Meier 1997) is recommended for phenological observations, worldwide. This decimal code, which is divided into principal and secondary growth stages, is based on the well-known cereal code developed by Zadoks et al. (1974).

Figure 29.1 shows the average timing of phenological stages and the duration of phenological phases of spring barley at the Experimental Station in Berlin-Dahlem for two 10-years periods, a very early (1955–1964) and the most recent one (2000–2010). For both periods the time between sowing (BBCH 00) and full ripeness (BBCH 89) hardly changed (111 vs. 110 days). The sowing date has slightly advanced by 3 days from 2 April (1955–1964) to 30 March (2000–2010).

However, the shift of all following stages is more pronounced, which has advanced up to 14 days for the medium milky- ripeness (BBCH 75). It is also obvious that two fundamental phases, the *spikelet differentiation phase* (BBCH 12–31) and the *grain filling phase* (BBCH 69–89) have changed contrarily by −7 and +8 days, respectively. These observed changes in the duration of developmental stages could have effects on yield formation and thus on the final crop yield, if they would continue. This means that the impact of climate change on crop productivity partially depends on possible changes in the timing of crop development (Craufurd and Wheeler 2009).

A relatively broad overview of phenological trends for perennial and agricultural crops in Germany was given by Estrella et al. (2007). They concluded that perennial crops exhibited a significantly higher temperature response to mean spring temperature than the annual crops. Comprehensive investigations of oat phenology in Germany by Siebert and Ewert (2012) confirmed the earlier onset of all phenological stages for spring cereals (up to −17 days for yellow ripeness) and a shortening of most phenological phases (up to 14 days between sowing and yellow ripeness of oat). For UK, Williams and Abberton (2004) found an advanced beginning of first flowering of white clover of about 7.5 days per decade since 1978. These dates were significantly, negatively correlated with air temperatures during February and March and soil temperatures between January and April. An earlier flowering of winter wheat since 1950 by 0.8–1.8 days per decade was reported for the U.S. Great Plains by Hu et al. (2005). Finally, for China regional changes in temperature have shifted crop phenology and affected crop yields (i.e. rice, wheat, and maize) during 1981–2000 (Tao et al. 2006).

As already mentioned above, not only annual crops showed changes in the timing of phenological events. These trends are also manifested for perennial crops (Chmielewski et al. 2004, 2009a). For instance the beginning of fruit tree blossom in Germany has significantly advanced in the last years, mainly due to milder winters and significantly warmer spring temperatures (Fig. 29.2).

In Germany the annual mean air temperature between 1961 and 2005 increased by 1.4 °C. The strongest temperature rise was observed in winter (2.3 °C) followed by spring (1.8 °C) and summer (1.6 °C). Only in autumn no significant temperature change was registered. Figure 29.2 shows that mainly the very early blossoming fruit crops such as apricot (DOY 108) and peach (DOY 111) show the strongest trend of more than 15 days in 45 years.

Even on an international scale, there are some evidences for a shift of phenological phases of perennial crops. Already, rising spring temperatures in north-eastern USA between 1965 and 2001 advanced spring phenology of apples and grapes (Wolfe et al. 2005). Changes in the growth intervals for wine grapes were also found by Jones and Davis (2000) and Duchêne and Schneider (2005). The longest annual record of cherry blossom in the world exists for Kyoto/Japan. This unique time series shows that cherries are currently flowering earlier than they have at any time during the last 1,200 years (Primack et al. 2009). For the same region Doi (2007) reports on an earlier blossom of apricots and Fujisawa and Kobayashi (2010) on an advanced timing of apple phenology.

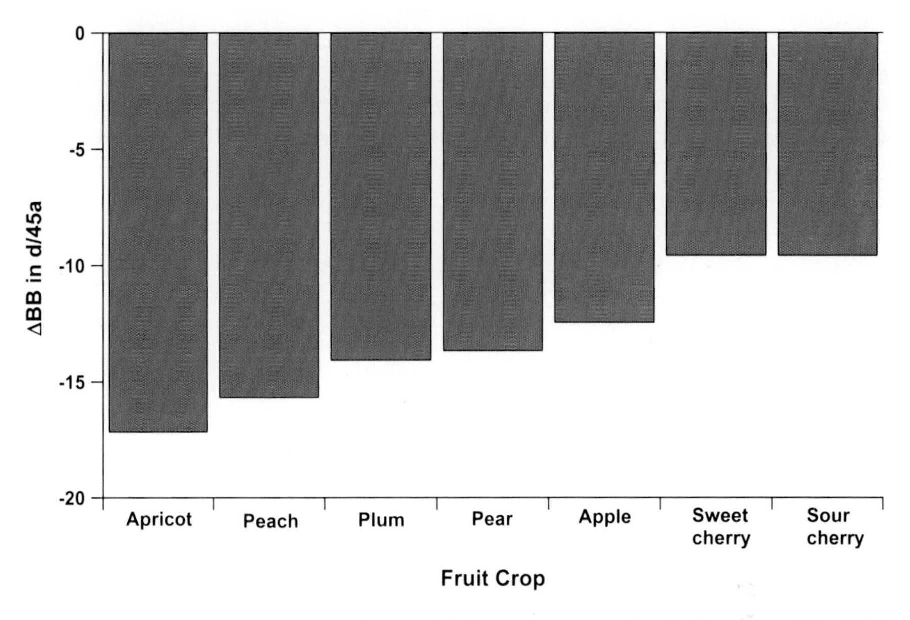

Fig. 29.2 Average changes in the beginning of fruit tree blossom (ΔBB) in Germany, 1961–2005

29.3 Applications of Phenology in Agriculture

Phenological observations in agriculture and horticulture provide basic information for farmers. Knowing the valuable phenological information helps decision-making for farmers, i.e. to correctly time operations such as *planting, fertilizing, irrigating, crop protection* and to *predict phenophases*. Phenological data are also helpful for scientific applications, such as investigations of crop-weather relationships and to describe the microclimate of crop stands. The following paragraphs give some examples.

29.3.1 Definitions of Growing Season

The average length of the growing season in a region sets the environmental limits for crop production. Each crop needs a certain time for growth, development, and yield formation.

Since most biological processes are bound to water, growth starts above the freezing point, usually at approximately 3–5 °C for most C_3-plants. With increasing temperature the biochemical processes are accelerated (rapidly above 10 °C) up to the temperature where enzyme systems are destroyed and cells die (Hörmann and Chmielewski 2001).

There are a lot of ways to define the general length of growing season in mid- and higher latitudes. The *climatic growing season* is defined in terms of climatic variables. For instance, thresholds of air temperature can define it, with common values between 0 and 10 °C. According to Chmielewski and Köhn (2000), the beginning of growing season was defined as the first day of the year $j = i_l$ on which the mean daily air temperature was $T_j \geq 5.0$ °C with the assumption that on the following 30 days the sum of differences S_D remains positive (Eq. 29.1). If S_D becomes negative the calculation has to stop and the next day j in the year on which $T_j \geq 5.0$ °C must to be selected as starting day i_l for the summation and has to be tested to give a positive S_D.

$$S_D = \left[\sum_{j=i_1+1}^{i+30} (T_j - 5^{\circ}C) \right] \geq 0, \quad i_1 \in \{1, \ldots, 365(366)\} \qquad (29.1)$$

Correspondingly, the end of growing season was defined as the day of the year $j = k_l$ $(k_l > i_l)$ on which the average daily temperature was $T_j < 5.0$ °C, under the condition that:

$$S_D = \left[\sum_{j=k_1+1}^{365(366)} (T_j - 5^{\circ}C) \right] < 0, \quad k_1 \in \{i_1 + 1, \ldots, 365(366)\} \qquad (29.2)$$

The threshold of 5 °C is often used to fix the general growing season for plants in temperate zones. Sometimes for winter cereals such as winter rye, or winter barley a threshold of 3 °C is recommended (Reiner et al. 1979). In Germany, the climatic growing season (Eqs. 29.1 and 29.2) starts on average on 6 March and ends on 2 November (1971–2000). Thus, it lasts altogether 241 days (s = 18.9 days) and covers relatively well the period of agricultural activities (Chmielewski et al. 2009a), because the fields are usually ploughed in March, so that the sowing of spring cereals can start in the end of March or beginning of April. The field works end with the sowing of winter cereals and the harvest of grain maize and sugar beet until to the end of October (see Table 29.1).

A similar definition was used by Mitchell and Hulme (2002), who suggested introducing the length of growing season as an indicator of climatic changes. They defined the beginning of growing season as the start of a period when the daily mean air temperature is greater than 5 °C for five consecutive days. The period ends on the day prior to the first subsequent period when daily mean air temperature is less than 5 °C for five consecutive days. For crops the frost-free season (Goodrich 1984), or the time between the last killing frosts in spring and the first killing frost in autumn (Critchfield 1966; Brown 1976), is also used. For Germany, this period comprises 185 days (1971–2000). On average, the last frost is here observed on 24 April and the first frost starts on 26 October.

Table 29.1 Average growing time of selected field crops in Germany, Experimental Station Berlin-Dahlem, 1953–2010

Crop	Seeding (day/month)	Harvest (day/month)	Growing time (days)
Winter rye	28/09	30/07	306[a]
Spring barley	03/04	29/07	117
Oats	03/04	05/08	124
Maize	03/05	17/10	168
Potato	25/04	18/09	147
Sugar beet	16/04	23/10	190

[a]193 days, if winter time is removed

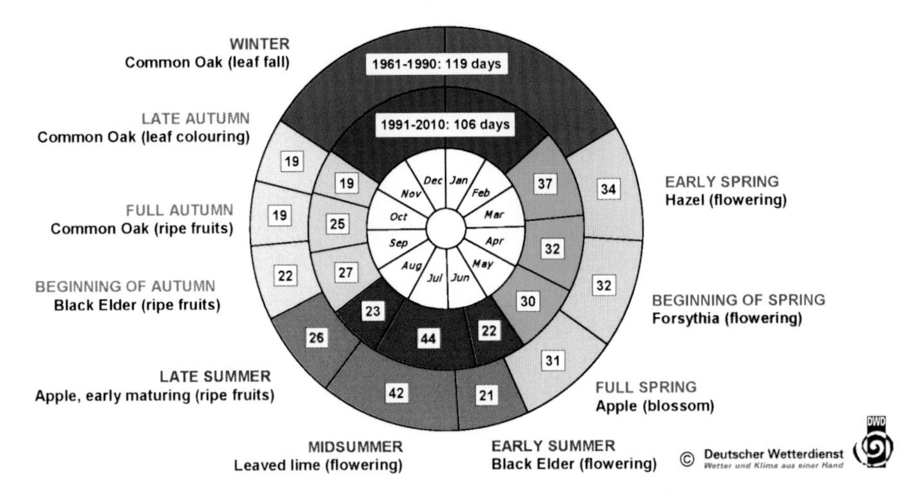

Fig. 29.3 Average duration of phenological seasons in Germany, 1961–1990 (*outer circle*) and 1991–2010 (*inner circle*) (Source: DWD)

The length of growing season can also be calculated using phenological events. Schnelle (1955, 1961) defined this period as the number of days between sowing of spring cereals and winter wheat. For the beginning of growing season, the timing of phenological events of the natural vegetation like bud burst, leafing, and flowering is often used. Accordingly, the end of growing season is then defined by autumn colouring and leaf fall of trees and shrubs. For instance, the German Weather Service (DWD) defines the beginning of growing season (early-spring) with the beginning of hazel flowering (18 February, 1991–2010) and the end of growing season with the leaf fall of common oak (4 November). The whole year is divided into nine different phenological seasons, which are defined by phenological indicator stages (Fig. 29.3). Additionally, Fig. 29.3 shows the average changes in the length of the individual seasons between 1961–1990 and 1991–2010, indicating an advancement of all phases in the most recent period.

The individual *growing time* for agricultural crops is the time from sowing to ripeness or harvest and lasts between about 3 and 6 months. Some differences exist among varieties. This means that within the climatic or general growing season different crops can grow within a specified *crop rotation*.

In warmer regions with a long growing season it is always possible to choose the optimal cropping sequence and crop rotation, including catch crops. In colder regions with a limited climatic growing season, one can grow in extreme cases only one crop within the very short growing time. For instance, sub-arctic adapted spring barley cultivars have a growing time of less than 80 days. In contrast to this, sugar beets have a relatively long growing time, of more than 6 months (Table 29.1), which is why sugar beet cropping is limited to regions with a relatively long growing season. The length of growing season also determines how much time is available to cultivate crops in spring or autumn as well as to harvest them at the end of summer. This influences the work sequence and the labour load of farms.

29.3.2 Selection of Growing Zones

Phenological observations can help to select favourable and unfavourable areas for agricultural and horticultural production. Figure 29.4 shows the spring frost hazard for sweet cherry growing in Bavaria (Rötzer et al. 1997). This map is based on the beginning of cherry blossom and minimum temperatures below −2 °C in April and May.

Phenological data can further be used for site classifications and to find crops and varieties that grow well under the given climatic conditions. Mainly, fruit and vegetable growing is very sensitive to the site selection. Therefore, phenological observations in horticulture have a great value to define optimal cropping areas, to evaluate the cultivability of fruit crops and vegetable in a given region, and to select individual species and varieties. Poorly adapted varieties show higher yield fluctuations and even yield depressions. For outdoor vegetable growing in mid-latitudes, mainly the warmest areas with an early spring development are recommended. These areas are easier to find with phenological maps. Phenological observations are therefore helpful to sketch regional and local cropping plans. For instance, maps of the beginning of growing season in Germany clearly show the favourable areas for fruit, vine, and vegetable growing. The regions with an early beginning of growing season are usually the warmest areas in spring and they preferably used for fruit growing. Most of them are located along the river valleys of Rhine and Elbe. In these regions, plant development is clearly advanced, compared with the spatial average and the frost risk is reduced.

29.3.3 Phenology and Frost Risk

Fruit growers need phenological information mainly in the flowering period and during ripeness. For these individuals it is important to know the date of latest frosts

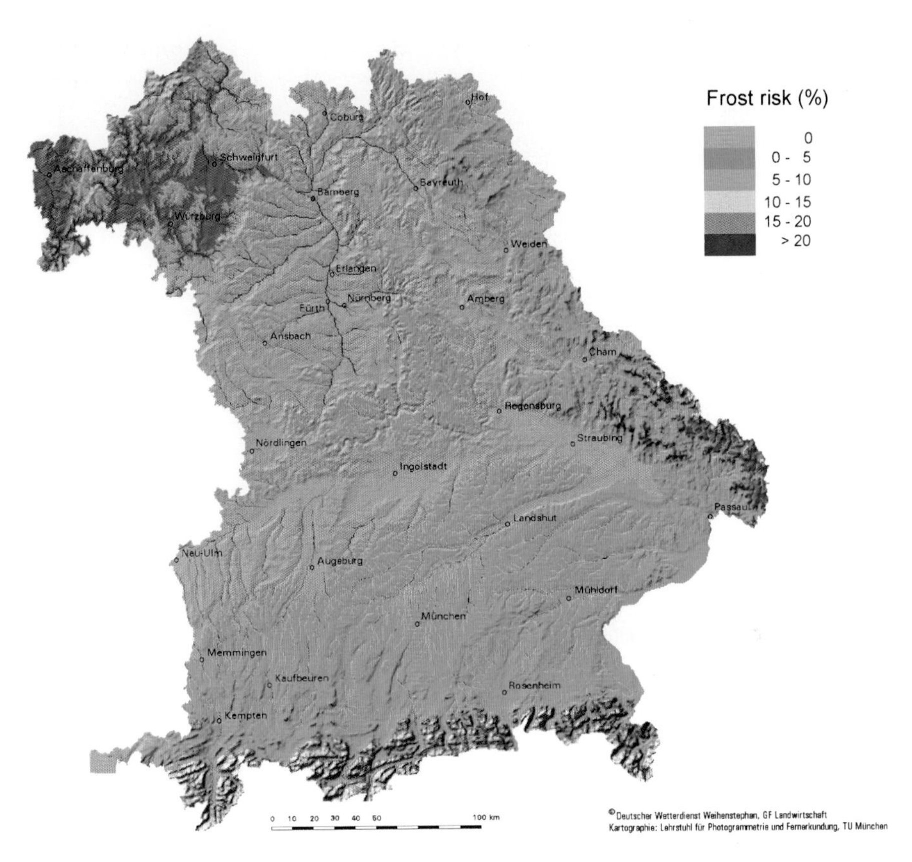

Fig. 29.4 Areas with frost hazard for sweet cherry growing in Bavaria/Germany (Rötzer et al. 1997, used with permission)

in a region, especially those that occur after first bloom. Frost damages on fruit crops generally depend on the developmental stage of the flower bud and on the strength and duration of the frost event. There are also differences between the fruit crops and varieties. The buds are most hardy during winter dormancy. If they start to swell and to expand, frost hardiness rapidly decreases. Table 29.2 gives some critical temperatures for damages on the flowers of fruit species between the beginning (BBCH 60) and the end of blossom (BBCH 69). These data are important for frost control and to evaluate the frost hazard in an orchard.

Early frosts before the beginning of blossom may cause masked injuries on flower buds, but the damage is not as great as it would be in the period of blossom or after the beginning of fruit set. Frost during the blossoming period can harm the blossoms, so that noticeable or total crop failures can occur. This happens if the ovary is frozen. For instance, frosts in the beginning of May 2011 resulted in significant damages, partially of more than 70 % in many orchards and vineyards

Table 29.2 Critical temperatures for light (10 %) and strong damages (90 %) after 30 min exposure on fruit crops for first bloom (BBCH 60), full bloom (BBCH 65), post bloom (BBCH 69), (Murray 2011)

Fruit crop (Damage in %)	First bloom (10 ... 90)	Full bloom (10 ... 90)	Post bloom (10 ... 90)
Apple	−2.2 ... −3.9	−2.2 ... −3.9	−2.2 ... −3.9
Apricot	−3.9 ... −7.2	−2.8 ... −5.6	−2.2 ... −3.9
Peach	−3.3 ... −6.1	−2.8 ... −4.4	−2.2 ... −3.9
Pear	−2.8 ... −5.0	−2.2 ... −4.4	−2.2 ... −4.4
Plume	−2.8 ... −5.0	−2.2 ... −5.0	−2.2 ... −5.0
Sour cherry	−2.2 ... −4.4	−2.2 ... −3.9	–
Sweet cherry	−2.2 ... −3.9	−2.2 ... −3.9	−2.2 ... −3.9

in Germany. Dry and cold wind exacerbated the situation. In order to investigate possible changes in the frost hazard due to climate change, phenological modelling is very important (see Sect. 29.3.6).

29.3.4 Crop Management and Timing of Field Work

For sustainable crop management, phenological data are essential to meet the right dates for *irrigation*, *fertilizing*, and *crop protection*. For example, maize can grow well in regions where the mean air temperature from May to September is above 15 °C, but in some regions the rainfall is a limiting factor for growth, so that irrigation becomes necessary. The highest water demand for cereal is between the beginning of stem elongation and flowering (BBCH 31-BBCH 61). Irrigation at the right time is a prerequisite for a sufficient grain yield. At the same time, the demand on nutrients is also very high. In just 5 weeks around the period of heading, 75 % of nutrients are taken up. Other cereals as wheat, barley, and rye have similar demands on water and nutrient supply. According to the nutrient requirement, N-fertilization of cereals is recommended at the beginning of the growing season (to promote the process of tillering as well as the formation of spikelets), in the period of stem elongation (to moderate the reduction of tillers and spikelets as well as florets), and if necessary shortly before heading (to encourage the grain size and protein content in the grain).

The application of herbicides for weed control is possible after emergence, for example at BBCH 13 (three-leaf-stage). With the appearance of the last leaf, the flag leaf (BBCH 37), growth regulators can be applied to avoid stalk breakage and the risk of grain lodging. These examples show how phenological information in agriculture is important. Phenological data are prerequisites for adapted and sustainable crop management and to obtain sufficient crop yield. They help farmers to trace the plant development, to monitor the yield formation processes and to find out the optimal time for cultural practices.

Table 29.3 Three relevant phenological phases for the yield formation of spring cereals

Period	Yield forming processes
1: BBCH 12 (25 April) – BBCH 31 (18 May) (2-leaf-stage – Beginning of stem elongation)	– Formation of side-shoots – Tillering Beginning of growth apex development: – Formation of spikelets (D^*: 27/04 – 01/05) – Differentiation of spikes (A^*: 03/05 – 10/05) – Formation of florets (B^*: 11/05 – 16/05)
2: BBCH 31 (18 May) – BBCH 65 (15 June) (Beginning of stem elongation – Anthesis)	– Reduction of tillers and spikelets – Differentiation and reduction of florets – Flowering
3: BBCH 65 (15 June) – BBCH 87 (18 July) (Anthesis – Yellow ripeness)	– Formation of kernels – Growth and ripeness of kernels

The dates are given for spring barley at the Experimental Station Berlin-Dahlem, 1955–2010, data for the development at the growth apex 1966–2005
D* double ridge stage, A* development of spikelets, B* formation of floret primordial

29.3.5 Phenological Data to Study Yield Formation and Microclimate

Crop yields are influenced by the variability of weather in many ways. For cereals the yield formation is very complex. Phenological observations can help to divide the growing time of crops into different periods which are important for the *yield formation* (Table 29.3). Thus, it is possible to find out relationships between weather and crop yield for different periods in which the individual yield component is affected. The first period includes the transition from the pure vegetative (BBCH12-D*) to the generative development, which starts with the development of the growth apex (D*: double ridge stage). The other two periods include the stem elongation – flowering phase and the grain filling – ripeness phase.

Investigations by Chmielewski and Köhn (1999b) showed that for spring barley in the first period mainly the number of kernels per ear, in the second period the crop density, and in the last period the kernel weight are influenced by weather. For winter rye, it was also possible to find out the relevant phenological phases with regard to the yield formation (Chmielewski and Köhn 2000). Thus, phenological observations are absolutely necessary to describe and to understand the yield formation of crops in detail. Crop models use phenological information as well, to steer physiological processes in the model. Therefore, phenological models are always subroutines of mechanistic crop models (see Sect. 29.3.6).

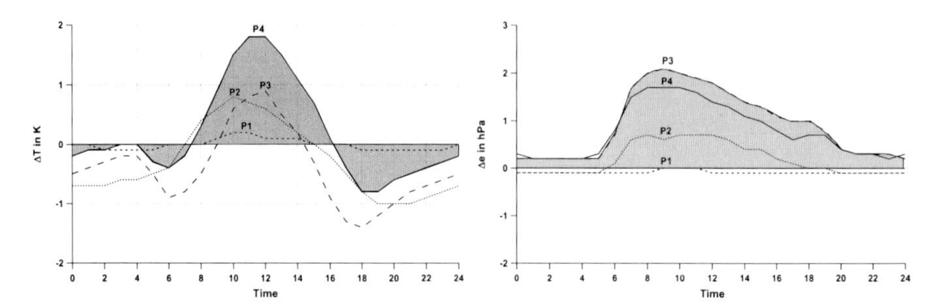

Fig. 29.5 Average anomalies in air temperature (ΔT) and in water vapour pressure (Δe) between a winter rye stand and a bare surface in 0.2 m height during different developmental periods. P1: leaf development and tillering period (BBCH 10–29), P2: stem elongation and heading period (BBCH 30–49), P3: flowering period (BBCH 51–69), P4: grain filling and ripeness period (BBCH 71–89), at the Experimental station in Berlin-Dahlem, 1981–1999 (positive anomalies mean the stand is warmer and more wet)

Crop stands have their own climate that can differ tremendously from the climatic conditions at a meteorological station. The *microclimate* of crop stands (phytoclimate) depend on various meteorological (solar radiation, air temperature, precipitation, wind speed) and plant-morphological factors (structure of plant cover, plant height, density of stand, etc.) and thus also on the plant development. Therefore, phenological observations are essential to analyse the microclimate and to separate periods with different climatic conditions. During the day the air within a crop stand is warmer and wetter than the air above a bare soil. The phytoclimate varies during the daytime and with the growth and development of the crop stand (Fig. 29.5).

The largest air temperatures differences occur in the last developmental period (P4) at noon. Then the winter rye stand in 0.20 m height is significantly warmer, compared with the air above a bare soil at the same level. During the flowering phase (P3) the differences in water vapour are the largest. They are already reduced in the grain filling and ripeness period, because then the stand becomes increasingly dryer (Wittchen and Chmielewski 2005).

These specific phytoclimatic conditions have effects on the phytopathological situation within the crop stand, because the development of fungi and insects is determined by certain temperatures and humidity levels, typical of each species. In the case of insects, the air temperature in the breeding places of insects is important for the eggs to hatch and for the larvae to be able to perform their various transformations.

For fungi, beside air temperature mainly the air moisture or the water on plant organs plays important roles. Since the phytoclimate depends on the plant development, the infection risk also changes with plant development. For example, for winter rye the infestation frequency is very high in the last two phases (P3, P4), because of the higher air humidity in these two periods, compared with the climatic conditions outside the stand. As a measure for the infection risk in phytopatological models often the relative humidity is used. Here mainly the time periods with values of at least 75 % are relevant.

29.3.6 Phenological Modelling of Perennial and Field Crops

In agriculture and horticulture, phenological models can be used by extension services to forecast phenological stages of fruit trees, nuts, strawberries, and field crops. Generally, phenological models use different approaches, such as pure forcing or combined chilling and forcing models to predict the timing of phenological stages. A detailed survey about these different approaches is given in Chap. 15. Here, only some special features in modelling phenological stages of field and fruit crops are added.

For field crops of mid- or higher latitudes very often simple *Growing Degree Day* (GDD) models, also referred as thermal time or spring warming models, are used to calculate the timing of the consecutive phenological stages. Here, in the simplest way T_i is the daily mean air temperature and T_{BF} the base temperature, which represents the lower cardinal temperature for the beginning of crop development.

Pure GDD models are mainly efficient for crops which have no winter rest such as *spring cereals*, *maize*, *potato* or *sugar beet*. Equation 29.3 describes the 'state of forcing' $S_f(t)$, were $R_f(T_i)$ is the 'forcing rate function', which is expressed in our case by GDD (Eq. 29.4). The forcing requirement F^* is fulfilled at time t_2 if $S_f(t)$ for $t = t_2$ is equal of greater F^* for the first time.

$$S_f(t) = \sum_{i=t_1}^{t} R_f(T_i), \text{ where} : S_f(t_2) \geq F^* \tag{29.3}$$

$$R_f(T_i) = \max(0., T_i - T_{BF}) \tag{29.4}$$

GDD are calculated from the beginning of sowing (t_1) until maturity, representing the accelerating influence of higher temperatures on plant development. F^* are the GDD, which are necessary to reach a certain stage of development.

Figure 29.6 shows a thermal time model for spring barley. The average sowing date was 7 April at mean temperatures of 9.2 °C. Between sowing and emergency 116 GDD are necessary (MAE = 1.8 days). Flowering and grain filling of spring barley starts if 874 or 989 GDD are reached, respectively. The mean absolute error (MAE) of the model lies between 1 and 5 days with the tendency to increasing values at the two last ripening stages (BBCH 87, 89). At the end of the growing time also deficits in soil water can affect the ripeness of grain. This effect is not considered in the simple GDD models, but could be relevant at least in some extreme years (McMaster and Wilhelm 2003). Additionally, Lawn et al. (1993) suggested the consideration of an upper cardinal temperature T_o in warm such as Mediterranean or subtropical climates, above which the rate of development linearly decreases.

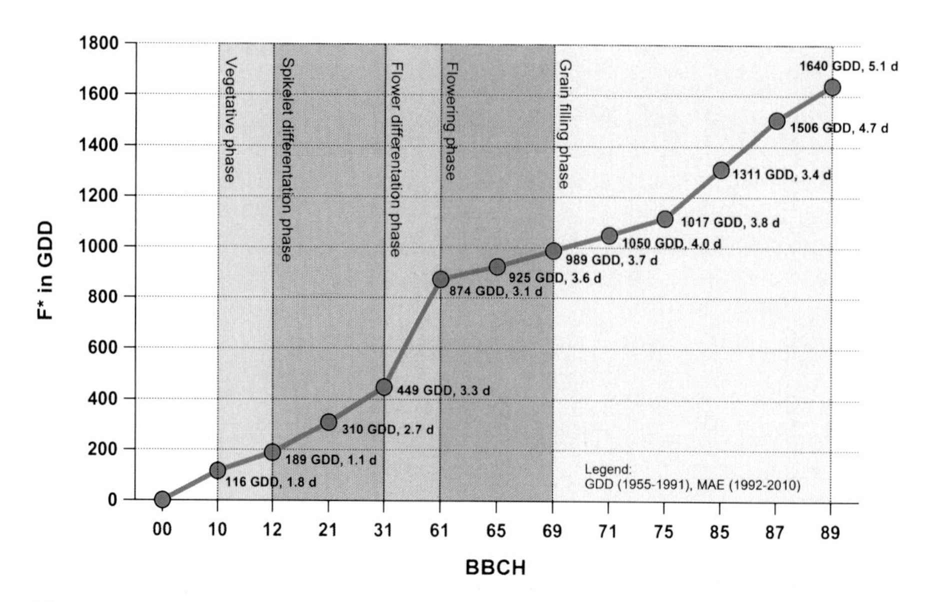

Fig. 29.6 Thermal time model for the development of spring barley, data from the Experimental Station Berlin-Dahlem, model optimization with data from 1955–1991, model validation (MAE: mean absolute error) with data from 1992 to 2010

For *winter cereals* such as winter wheat, -barley, -rye, and -rape similar approaches can be used to describe the plant development, but here the effects of *vernalisation* and *photoperiod* have to be considered, additionally. This makes the phenological models for these crops much more complex.

Vernalisation is induced by a decreasing daylength in autumn. During this process the crop produces substances which inhibit the generative development. Thus, it is protected not to start the generative development (D*) and not to shoot before winter. Lower temperatures in winter, mainly between about 0 and 10 °C (Wareing and Phillips 1981; Weir et al. 1984), over a certain time (40–50 days) reduces the growth-inhibiting substances, so that after the end of vernalisation the generative development begins (see Table 29.3). If in spring the days are longer than the nights, the plant can start to shoot (photoperiodic reaction of long-day plants). For example winter wheat needs daylengths between >13 and 15 h, depending on the cultivar.

For this reason at least two different model approaches must be combined to describe the development of winter crops. Between sowing and emergence and again from shooting to maturity thermal time models can be applied (Eqs. 29.3 and 29.4). However, between emergency or 2-leave stage in autumn and the beginning of shooting in spring combined chilling/forcing (CF) models must be used, which consider both the process of vernalisation, by calculating chilling days (CD), and the effect of daylength to trigger the beginning of shooting. A survey of existing phenological models for winter wheat is given by Mirschel and Kretschmer (1990) and Mirschel et al. (1990). Investigations by Mirschel et al. (2005) to winter rye and winter barley showed that the most difficult phase to model is indeed the beginning of shooting. For this stage the highest root mean square error between observed and calculated data was found. For agricultural extension, Roßberg et al. (2005) developed an ontogenetic model SIMONTO for winter cereals and winter rape, which combines the model approaches of ONTO (Wernecke and Claus 1992, 1996) and CERES-WHEAT (Ritchie et al. 1988). The model includes the process of vernalisation between BBCH 10 and BBCH 30 and the photoperiodic effect. Although only two input variables are required (temperature and latitude), the model provides good results.

For perennial crops similar processes as mentioned above have to be considered to model the beginning of blossom (Chmielewski et al. 2011). They also need a certain period with lower temperatures to be released from winter rest. This period is probably also induced by shorter days and sinking temperatures in autumn. The period of vernalisation, here referred to *winter dormancy*, also protects the fruit tree from blossoming too early. Instead of chilling days, *chilling hours* (*units/portions*) can be used to describe the release winter dormancy, because these data are available in the literature for different fruit crops. Winter chill demand strongly depends on the fruit crop and cultivar. For each fruit crop a wide range of varieties with different chilling requirements (500–1,800 chilling hours) exist.

According to Weinberger (1950), temperatures between 0 and 7.2 °C are effective to release winter rest. Other chilling model approaches, the Utah model (Richardson et al. 1974), the positive Utah model (Linsley-Noakes et al. 1995) or the North Carolina Model (Shaltout and Unrath 1983) calculate chilling units (CU: temperature-weighted chilling hours). The probably most physiological based model approach was presented by Fishman et al. (1987a, b). Here, the reaction of the tree to cold and warm spells during winter is modelled relatively realistic by accumulated chill portions (CP).

The release of winter dormancy for an individual site strongly depends on the chilling requirement of the crop/variety and the hourly temperature course – in Germany, mainly between October and February. Temperature departures during this period can strongly influence the accumulation of chilling hours. For example, a very cold winter (December-January: $\Delta T = -5.2$ °C) after a mild autumn (October–November: $\Delta T = +1.2$ °C) lead to a very low chilling hour accumulation in Berlin-Dahlem until the end of January (only 644 CH in 1969/1970). Conversely, a mild winter ($\Delta T = +4.8$ °C) which followed a cold autumn ($\Delta T = -0.9$ °C) increased the chilling hour accumulation enormous (2,132 CH until 31 January in 1974/1975). The difference between these two extreme cases is 1,488 CH.

Detailed investigations for Germany have shown that for low chill cultivars up to about 1,000 CH, dormancy is released relatively uniform across Germany. For these varieties the chilling requirement is almost fulfilled to the end of the year. However, for high chill varieties ($\geq$1,500 CH) the chilling fulfilment can last until the end January and February. For these cultivars the chilling requirement is fulfilled somewhat later in the colder continental, south-eastern areas than in the maritime north-western regions, so that there is a NW-SE gradient across Germany (Chmielewski et al. 2012).

In order to model the phenological stages for fruit crops in spring, chilling models can be used to calculate the end of winter dormancy. The simplest way would be the CH approach. For these calculations hourly temperatures (T_{ih}) are required which can be approximately generated from the daily extremes (Linsley-Noakes et al. 1995). In Eq. 29.5 dormancy is released if the state of chilling $S_c(t)$ reaches the chilling requirement C^* at the time $t = t_1$. The chilling rate function $R_c(T_{ih})$ according to Weinberger (1950) only considers hourly temperatures between 0 and 7.2 °C (Eq. 29.6).

$$S_c(t) = \sum_{i=t_0}^{t} \sum_{h=1}^{24} R_c(T_{ih}), \text{where} : S_c(t_1) \geq C^* \tag{29.5}$$

$$R_c(T_{ih}) = \begin{cases} 0\text{CH, if } T_{ih} \leq 0^\circ\text{C or } T_{ih} \geq 7.2\,^\circ\text{C} \\ 1\text{CH, if } 0^\circ\text{C} < \text{T}_{ih} < 7.2\,^\circ\text{C} \end{cases} \tag{29.6}$$

If the dormancy is released at the time t_1 (endodormancy), the buds are theoretically able to react to favourable environmental conditions. This can be proofed if the branches of a tree are cut and placed into a warm environment. They will start to flower after several weeks. In most cases the temperatures in the orchard are still too cold to stimulate any development, so that the buds stay dormant (ecodormancy). Thermal time models (Eq. 29.3), which accumulate GDD above a lower temperature threshold (T_{BF}), are suitable to describe the second phase, after dormancy released. If the daily temperatures exceed the base temperature, the buds of the tree can grow and develop.

The effect of daylength on the beginning of leafing is still controversially discussed, but increasingly investigated (Basler and Körner 2012). Some species such as cherries, plums (Heide and Prestrud 2004) and peaches (Erez et al. 1966) probably show a photoperiodic reaction. For other species (e.g. apple, pear) it is assumed that there is no clear response to the daylength (Heide 2008). Recent studies by Blümel and Chmielewski (2012a) as well as Matzneller et al. (2013) showed that the inclusion of a daylength-term in the original GDD approach (Eq. 29.4) improved the performance of the combined CH/CP-GDD models, remarkably (Eq. 29.7).

$$R_f(T_i) = \max(0., T_i - T_{BF}).\left(\frac{DL}{10\text{h}}\right)^{EXPO} \tag{29.7}$$

Here, *DL* is the time between sunrise and sunset in hours (h) which depends on the geographic location and the Julian Day. *EXPO* is an additional model parameter, which weights the importance of *DL* on the fruit crop. The constant in the denominator (10 h) is a normalization parameter to make the magnitude of the calculated F^* values comparable to the original GDD approach. The accumulated forcing units are here given in *photo-thermal units* (PTU). The advantages of these combined sequential CH/CP-GDD models which additionally incorporate daylength can be summarized as follows:

1. The models had meaningful model parameter estimations,
2. The differences between modelled and observed data, even in the case of an external verification, were relatively low (see Table 8.4),
3. The optimized model can be used to calculate the beginning of blossom not only for the region for which it was developed,
4. These models could be used to project possible shifts in the timing of fruit tree blossom relatively realistic, if CP are calculated.

The best results were always achieved for the combined sequential CP-GDD models (Blümel and Chmielewski 2012b; Matzneller et al. 2013).

29.4 Application of Phenological Models in Relation to Climate Change

In order to study the impact of climate change on natural or managed ecosystems, phenological models are of great importance. They can act as separated models to investigate possible shifts in the timing of phenological events, or they could be integrated as subroutines in comprehensive ecosystem, water budget, or yield models. These complex models can be used to evaluate possible changes in late frost hazard, in yield formation, irrigation demand, pest and disease infestation, etc. A prerequisite for such studies is the availability of phenological models which do not only work for current climates, but also for changed conditions. For the optimization and verification of these models high-quality, standardized phenological observations are advantages. The *International Phenological Gardens* in Europe (IPG) and the *Global Phenological Monitoring Programme* (GPM) are great resources for those applications (see Chap. 8).

Climate projections indicate that air temperature will rise in many parts of the world and in all seasons. For Germany, the mean annual air temperature could increase up to about 3 °C (REMO-UBA, A1B). A temperature rise of almost 4 °C is expected for the winter months (Chmielewski et al. 2009b).

Possible impacts of climate change on field crops in Germany are discussed and summarised in Chmielewski (2007, 2009, 2011). This field of research is widely investigated on the global, regional and local level and the results are summarised in several textbooks and scientific articles (e.g. Parry 2007; Reynolds 2010). However, potential impacts of climate variability and climate change on perennial crops have not been as widely studied as compared to annual crops and these studies are in some points much more complicated (Lobell and Field 2011).

Rising winter temperatures can influence the vernalisation or winter dormancy of field and perennial crops. A comprehensive review about this topic is given by Campoy et al. (2011). Investigations for Germany have shown that the date of

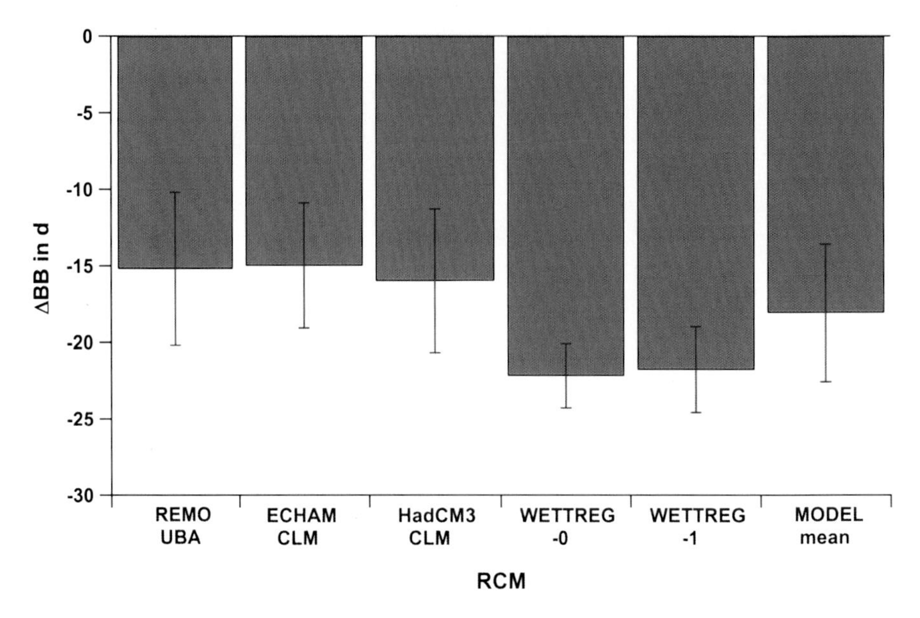

Fig. 29.7 Possible shifts in the beginning of apple blossom (early maturing variety) due to climate change in Hessen/Germany (2071–2100 vs. 1971–2000) on the basis of 5 different RCMs (GHG emission scenario A1B) and average of all 5 models (MODEL mean). 95 % confidence intervals are given. Model parameters: $C^* = 70.0$ CP, $T_{BF} = 2.40$ °C, $F^* = 425.4$ PTU, $EXPO = 1.280$, $RMSE_{val} = 3.76$ days (Blümel and Chmielewski 2012b)

dormancy release in the future can change, depending on the chilling demand of the crop/cultivar and on the region which is investigated (Chmielewski et al. 2012). Additionally, differences among the chilling hour models exist. This already was shown by Luedeling and Brown (2011) in a global analysis. For cultivars with a chilling requirement up to about 1,000 CH,CU,CP/18 all chilling models calculated a later release of dormancy towards the end of this century for Germany (REMO-UBA, GHG emission scenario A1B). However, for crops with a chilling requirement of more than 1,200 CU,CP/18 the Utah, Positive Utah and Dynamic model projected a slightly advanced release of dormancy with increasing chilling requirement of the crops. For this reason combined chilling/forcing models are necessary to calculate possible changes in the timing of spring events for perennials and winter crops due to climate change.

Figure 29.7 shows possible shifts in the beginning of apple blossom in Germany (2071–2100) on the basis of a combined CP-GDD model which additionally considers the daylength (Eq. 29.7). The chilling requirement (C^*) was calculated by the Dynamic model. According to these calculations the beginning of apple blossom in Hessen could advance by 18 days (± 4.5 days). There are some differences between the RCMs of maximum 7 days. These results were subsequently used to evaluate future changes in late frost hazard (Blümel and Chmielewski 2012b).

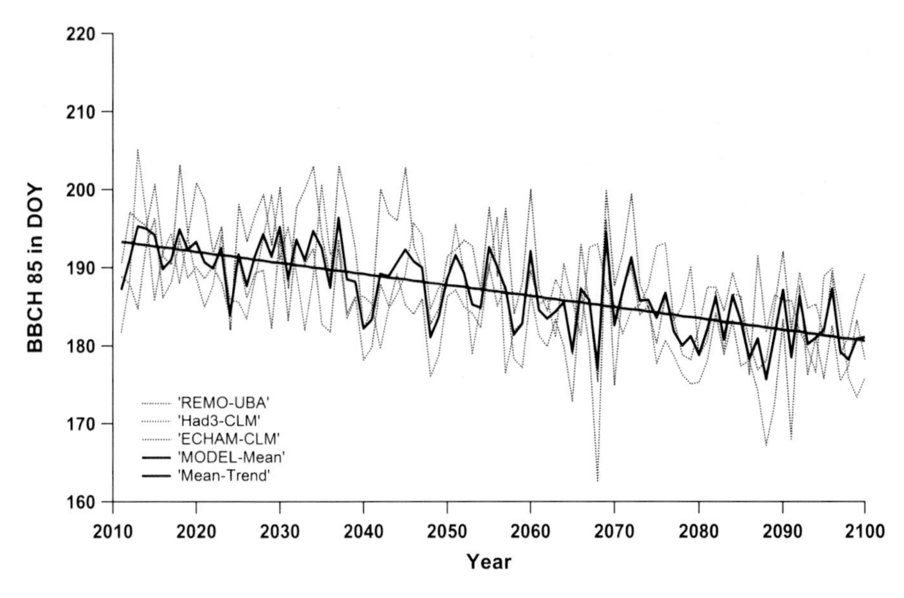

Fig. 29.8 Possible shifts in dough ripeness of spring barley (BBCH 85) at Berlin-Dahlem due to climate change 2010–2100 on the basis of 5 different RCMs (GHG emission scenario A1B) and average of all 5 models (MODEL mean) (Model parameter according to Fig. 29.6: $T_{BF} = 0\,°C$, $F^* = 1{,}311$ GDD)

For spring cereals the use of simple GDD models is an option to calculate possible shifts in the timing of phenological stages due to climate change. Figure 29.8 shows possible changes in the ripeness for spring barley (BBCH 85), calculated for the Experimental Station in Berlin-Dahlem, according to different RCMs. On average the ripening time could advance by about 10 ± 3 days, if the currently optimal sowing date is not changed. A future increasing frequency of droughts during the grain filling period can additionally advance the ripening date.

These results are generally important for adaptation strategies. Small changes in the length of growing season can already influence the *choice of varieties*. Fast ripening varieties can be replaced by slower ripening ones, if the growing season extends. This measure could have positive effects on the yield variability and on the yield level. More distinct changes in climatic growing season length by several weeks, as expected for Germany (>3 months, according to REMO-UBA, GHG emission scenario A1B, Chmielewski et al. 2009b) can influence the possibilities for cultivar selection, *catch cropping*, and *crop rotation*.

Catch cropping depends on the time available after harvest in the late summer or early autumn, before another crop is cultivated. Thus, catch cropping is only possible in regions where the growing season in autumn is long enough and the air temperatures and precipitation are still favourable for plant growing. For example an advanced harvest date of spring cereals (Fig. 29.8) would improve the conditions for catch cropping, so that it is possible to grow legumes, before winter cereal is sown.

Crop rotation is a system in which the crops on a certain plot are followed by other crops according to a predefined plan. Normally the crops are changed annually, but in some areas where the growing season is sufficiently long, multiple cropping is possible. An extension in the length of growing season can improve the scope for multiple cropping and crop rotation. For example, if the end of growing season is extended the sowing time of winter cereals can be shifted to the end of the year, so that in the space before crops with a relatively long growing time (such as sugar beets) can grow. In the field, vegetable multiple cropping is common. Here, also an extension of growing season can improve the crop rotation and number of harvests within a year.

Similar model approaches (effective temperature sums) are common to project or to assess the occurrence of pests and diseases. Annual monitoring of codling moth (*Cydia pomonella*) in Germany's northernmost apple growing region (Lower Elbe valley) point to the existence of a partial second generation associated with warm years such as 2006, 2008, and 2010. A recent climate-impact study has indicated a future increase in the annual number of codling moth generations in all German fruit growing regions (Chmielewski et al. 2009a). Further, *Diplodia seriata*, cause of black rot of apples, is a new pathogen in the Lower Elbe valley whose appearance is probably associated with higher temperatures and more frequent rainstorm events during the vegetation period (Weber 2009). There is no doubt that further research is necessary on this very complex topic.

References

Basler D, Körner C (2012) Photoperiod sensitivity of bud burst in 14 temperate forest tree species. Agr For Meteorol 165:73–81

Blümel K, Chmielewski FM (2012a) Shortcomings of classical phenological forcing models and a way to overcome them. Agr Forest Meteorol 164:10–15. doi:10.1016/j.agrformet.2012.05.001

Blümel K, Chmielewski FM (2012b) Climate change in Hessen – chances, risks, and costs for fruit growing and viniculture. Annual Report (in German), Hessen State Office for Environment and Geology (HLUG). http://www.agrar.hu-berlin.de/agrarmet/forschung/fp/CHARIKO_html. Accessed 2012

Broekhuizen S, Zadoks JC (1967) Proposal for a decimal code of growth stages in cereals. Stichting Netherlands Graan-Centrum, Wageningen

Brown JA (1976) Shortening of growing season in the U.S. corn belt. Nature 260:420–421

Bruns E (2001) Phänologie im Deutschen Wetterdienst. Mitteilungen der DMG 1–2

Campoy JA, Ruiz D, Egea J (2011) Dormancy in temperate fruit trees in a global warming context. A Rev Scientia Horticul 130:357–372

Cevik E (2011) Einfluss von Klima und Witterung auf phänologische Merkmale der Sommergerste und ihre Beziehung zum Ertrag. BSc thesis, HU Berlin

Chmielewski FM (2007) Folgen des Klimawandels für die Land- und Forstwirtschaft. In: Endlicher W, Gerstengarbe FW (eds) Der Klimawandel – Einblicke, Rückblicke und Ausblicke. Eigenverlag, Potsdam

Chmielewski FM (2009) Landwirtschaft und Klimawandel: In: Geographischen Rundschau 9: 61. Klimawandel im Industriezeitalter, pp 28–35

Chmielewski FM (2011) Klimawandel und Landwirtschaft in der Metropolregion Hamburg. In: Storch H, Claußen M (eds) Klimabericht der Metropolregion Hamburg. Springer, Berlin/Heidelberg, pp 211–227

Chmielewski FM, Köhn W (1999a) The long-term agrometeorological field experiment at Berlin-Dahlem. Agr For Meteorol 96:39–48

Chmielewski FM, Köhn W (1999b) Impact of weather on yield components of spring cereals over 30 years. Agr For Meteorol 96:49–58

Chmielewski FM, Köhn W (2000) Impact of weather on yield and yield components of winter rye. Agr For Meteorol 102:253–261

Chmielewski FM, Müller A, Bruns E (2004) Climate changes and trends in phenology of fruit trees and field crops in Germany 1961–2000. Agr For Meteorol 121(1–2):69–78

Chmielewski FM, Blümel K, Henniges Y, Müller A (2009a) Klimawandel und Obstbau in Deutschland. Endbericht des BMBF-Verbundprojekts KliO. Eigenverlag Humboldt-Universität zu Berlin 237 S. http://www.agrar.hu-berlin.de/agrarmet/forschung/fp//KliO_html. Accessed 2012

Chmielewski FM, Blümel K, Henniges Y, Müller A, Weber RWS (2009b) Klimawandel: Chancen, Risiken und Kosten für den deutschen Obstbau. In: Mahammadzadeh M, Biebeler H, Bardt H (Hrsg) Klimaschutz und Anpassung an die Klimafolgen – Strategien, Maßnahmen und Anwendungsbeispiele. Institut der deutschen Wirtschaft Köln Medien GmbH, pp 279–286

Chmielewski FM, Blümel K, Henniges Y, Blanke M, Weber RWS, Zoth M (2011) Phenological models for the beginning of apple blossom in Germany. Meteorol Z 20:487–496

Chmielewski FM, Blümel K, Páleošvá I (2012) Climate change and shifts of dormancy release for deciduous fruit crops in Germany. Clim Res 54:209–219. doi:10.3354/cr01115

Clive-James W (1971) Growth stages key for cereals. Can Pl Dis Surv 51:42–43

Craufurd PQ, Wheeler TR (2009) Climate change and the flowering time of annual crops. J Exp Bot 60(9):2529–2539

Critchfield HJ (1966) General climatology. Prentice-Hall , Englewood Cliffs

Doi H (2007) Winter flowering phenology of Japanese apricot *Prunus mume* reflects climate change across Japan. Climat Res 34:99–104

Duchêne E, Schneider C (2005) Grapevine and climatic changes: a glance at the situation in Alsace. Agron Sustain Dev 25:93–99

Erez A, Samish RM, Lavee S (1966) The role of light in leaf and flower bud break of the peach (*Prunus persica*). Physiol Plantarum 19:650–659

Estrella N, Sparks T, Menzel A (2007) Trends and temperature response in the phenology of crops in Germany. Glob Change Biol 13:1737–1747

Feekes W (1941) De tarwe en haar milieu. Versl techn Techn Tarwe Commissie 12:523–888 and 17:560–561

Fishman S, Erez A, Couvillon GA (1987a) The temperature-dependence of dormancy breaking in plants-mathematical analysis of a 2-step model involving a cooperative transition. J Theor Biol 124:473–483

Fishman S, Erez A, Couvillon GA (1987b) The temperature-dependence of dormancy breaking in plants-computer simulation of processes studied under controlled temperatures. J Theor Biol 126:309–321

Fujisawa M, Kobayashi K (2010) Apple (*Malus pumila* var. domestica) phenology is advancing due to rising air temperature in northern Japan. Glob Change Biol 16:2651–2660. doi:10.1111/j.1365-2486.2009.02126.x

Goodrich S (1984) Checklist of vascular plants of the Canyon and Church Mountain (Utah, USA). Great Basin Nat 44:277–295

Heide OM (2008) Interaction of photoperiod and temperature in the control of growth and dormancy of *Prunus* species. Sci Horticul 115:309–314

Heide OM, Prestrud AK (2004) Low temperatures, but not photoperiod, controls growth cessation and dormancy induction and release in apple and pear. Tree Physiol 25:109–114

Hörmann G, Chmielewski FM (2001) Consequences for agriculture and forestry. In: Lozan JL, Hupfer P, Graßl H (eds) The climate of the 21st century. Wissenschaftliche Auswertungen, Hamburg

Hu Q, Weiss A, Feng S, Baenziger P (2005) Earlier winter wheat heading dates and warmer spring in the U.S. Great Plains. Agr For Meteorol 135:284–290

Jones GV, Davis RE (2000) Climate influences on grapevine phenology, grape composition, and wine production and quality for Bordeaux, France. Am J Enol Viticult 51:249–261

Keller C, Baggiolini M (1954) Les stades repéres dans la vegetation du blé. Revue Romande d' Agricult 10(3):17–30

Körschens M (1997) Die wichtigsten Dauerfeldversuche der Welt. Übersicht, Bedeutung, Ergebnisse. Arch Agron Soil Sci 42(3–4):157–168

Large EC (1954) Growth stages in cereals. Illustration of the Feekes Scale Plant Pathol 3:128–129

Lawn RJ, Summerfield RJ, Ellis RH, Roberts EH, Chay PM, Brouwer JB, Rose JL, Yeates SJ (1993) Towards the reliable prediction of time to flowering in six annual crops. VI. Applications in crop improvement. Exp Agric 31:89–108

Linsley-Noakes GC, Louw M, Allan P (1995) Estimating daily positive Utah Chill units using daily minimum and maximum temperatures. J SA Soc Hort Sci 5:19–23

Lobell DB, Field CB (2011) California perennial crops in a changing climate. Clim Chang 109 (suppl 1):317–333. doi:10.1007/s10584-011-0303-6

Luedeling E, Brown PH (2011) A global analysis of the comparability of winter chill models for fruit and nut trees. Int J Biometeorol 55:411–421

Matzneller P, Blümel K, Chmielewski FM (2013) Models for the beginning of sour cherry blossom. Int J Biometeorol. doi:10.1007/s00484-013-0651-1

McMaster GS, Wilhelm WW (2003) Phenological response of wheat and barley to water and temperature: improving simulation models. J Agric Sci 141:129–147

Meier U (1997) Growth stages of mono- and dicotyledonous plants. BBCH-Monograph Blackwell, Berlin

Mirschel W, Kretschmer H (1990) Vergleich existierender Ontogenesemodelle für Winterweizen Arch Acker- Pflanzenbau Bodenkd 34:683–690

Mirschel W, Kretschmer H, Matthäus E (1990) Dynamisches Modell zur Abschätzung der Ontogenese von Winterweizen unter Berücksichtigung des Wasser- und Stickstoff-versorgungszustandes. Arch Acker- Pflanzenbau Bodenkd 34:691–699

Mirschel W, Wenkel KO, Schultz A, Pommerening J, Verch G (2005) Dynamic phenological model for winter rye and winter barley. Euro J Agron 23:123–135

Mitchell TD, Hulme M (2002) Length of the growing season. Weather 5(57):196–198

MLUV (2009) Dauerfeldversuche in Brandenburg und Berlin. Beiträge für eine nachhaltige landwirtschaftliche Bodennutzung. Heidelberg, Berlin.

Murray M (2011) Critical temperatures for frost damage on fruit trees. Utah State Univ Extension and Utah Plant Pest Diagnostic Lab. IPM-012-11

Parry M (2007) The implications of climate change for crop yields, global food supply and risk of hunger. SAT eJournal ejournal icrisat org 4(1):44

Petr J (1966) A precise phenological scale for grain cereals. Rostl Vyroba 12:207–212

Primack RB, Higuchi H, Miller-Rushing AJ (2009) The impact of climate change on cherry trees and other species in Japan. Biol Conserv 142:1943–1949

Reiner L, Mangstl A, Straß F, Teuteberg W, Panse E, Kürten PW, Meier B, Grosskopf W, Deecke U, Kühne P, Schwerdtle JG (1979) Winterroggen aktuell. DLG Verlag, Frankfurt

Reynolds MP (ed.) (2010) Climate change and crop production. CABI climate change series v. 1. CAB International, Oxfordshire

Richardson EA, Seeley SD, Walker DR (1974) A model for estimating the completion of rest for "Redhaven" and "Elberta" peach trees. Hortscience 1:331–332

Ritchie TJ, Godwin DC, Otter-Nacke S (1988) CERES-Wheat. A simulation model of wheat growth and development. Texas A&M Univ Press, College Station

Roßberg D, Jörg E, Falke K (2005) SIMONTO – ein neues Ontogenesemodell für Wintergetreide und Winterraps. Nachrichtenbl Deut Pflanzenschutzd 57:74–80

Rötzer T, Würländer W, Häckel H (1997) Agrar- und Umweltklimatologischer Atlas von Bayern. Selbstverlag Deutscher Wetterdienst, Weihenstephan

Schnelle F (1955) Pflanzenphänologie. Akademische Verlagsgesellschaft Geest & Portig K-G, Leipzig, p 299

Schnelle F (1961) Agro-phenological annual course of the German and European agricultural regions. German Geographic Meeting, Wiesbaden

Shaltout AD, Unrath CR (1983) Rest completion prediction model for Starkrimson delicious apples. J Am Soc HortSci 108:957–961

Siebert S, Ewert F (2012) Spatio-temporal patterns of phenological development in Germany in relation to temperature and day length. Agr For Meteorol 152:44–57

Strauß R, Bleiholder H, van den Bomm T, Buhr L, Hack H, Heb M, Klose R, Meier U, Weber E (1994) Einheitliche Codierung der phänologischen Entwicklungsstadien mono- und dikotyler Pflanzen. Erweiterte BBCH-Skala, Basel

Tao F, Yokozawa M, Xu Y, Hayashi Y, Zhang Z (2006) Climate changes and trends in phenology and yields of field crops in China, 1981–2000. Agr For Meteorol 138:82–92

Wareing PF, Phillips IDJ (1981) Growth and differentiation in plants, 3rd edn. Pergamon Press, Oxford

Weber RWS (2009) Possible impacts of climate change on harmful fungi in orchards. The examples of fruit rot pathogens on apples (in German). Erwerbs-Obstbau 51:115–120

Weinberger JH (1950) Chilling requirements of peach varieties. Proc Am Soc Hort Sci 56:122–128

Weir AH, Bragg PL, Porter JR, Rayner JH (1984) A winter wheat crop simulation model without water nutrient limitations. J Agric Sci Cam 102:371–382

Wernecke P, Claus S (1992) Extension and improvement of descriptive models for the ontogenesis of wheat plants. Modelling Geo-Biosphere 1. Catena Verlag, Cremlingen-Destedt

Wernecke P, Claus S (1996) Modelle der Ontogenese für die Kulturarten Winterweizen, Wintergerste und Winterraps. In: Mühle H, Claus S (Hrsg.) Reaktionsverhalten von agrarischen Ökosytemen homogener Areale, pp 105–120

Williams T, Abberton M (2004) Earlier flowering between 1962 and 2002 in agricultural varieties of white clover. Oecologia 138:122–126

Wittchen U, Chmielewski FM (2005) Phytoclimate of winter rye stands. Meteorol Z 14 (2):183–189

Wolfe DW, Schwartz MD, Lakso AN, Otsuki Y, Pool RM, Shaulis NJ (2005) Climate change and shifts in spring phenology of three horticultural woody perennials in northeastern USA. Int J Biometeorol 49:303–309

Zadoks JC, Chang TT, Konzak CF (1974) A decimal code for the growth stages of cereals. Weed Res 14:415–421

Chapter 30
Winegrape Phenology

Gregory V. Jones

Abstract Globally wine production has become an increasingly important economic activity for over 50 countries and numerous regions within countries. The annual growth cycle of winegrapes begins in the spring with bud break, culminating with leaf fall in autumn followed by winter dormancy. The growth cycle is strongly tied to climate with narrow geographic zones providing the conditions by which the plant can produce quality fruit that can be made into quality wine. Therefore the knowledge of the timing of winegrape phenology is important for understanding the suitability of different varieties to certain climatic zones. As a result, winegrape phenology has become an important industry and scientific tool to better understand how climate variability and change impacts viticulture and wine production. Observations across numerous varieties, in many regions, and over many years has shown that that winegrape phenology has trended earlier with a general shortening of the interphases and are a sensitive indicator for both short term variability and long term trends in climate.

30.1 Introduction

Grapevines are a woody, herbaceous tree-climbing plant/shrub of the *Vitaceae* family with a largely uncertain origin. Fossilized remains have been found in Paleocene and Eocene deposits indicating that vines have been around for at least 37 Ma (Galet 1979). While approximately 24,000 varieties of vines have been named, it is thought that one fifth or less are probably genuinely distinct varieties and less than 150 are cultivated to any degree (Coombe and Dry 1988). However, within Europe Lacombe et al. (2011) catalogued 1,902 grape varieties (both scions

G.V. Jones (✉)
Department of Environmental Studies, Southern Oregon University, Ashland, OR 97520, USA
e-mail: gjones@sou.edu

M.D. Schwartz (ed.), *Phenology: An Integrative Environmental Science*,
DOI 10.1007/978-94-007-6925-0_30,

and rootstocks) that are officially authorized for cultivation in one or more European countries. All grapevine varieties belong to the genus *Vitis*, including the *Euvitis* (true grapes with both European and North American species) and *Muscadinia* (whose fruit is more properly called muscadine) subgenera, with most found mainly in the temperate zones in the Northern Hemisphere. Of the main species, *vitis vinifera*, which is a Eurasian native, is responsible for most of the table grapes, raisin grapes, grape juice, wine, and vinegar produced today.

Most grapevines evolved with adaptive features such as tendrils for climbing (there are a few shrub-like varieties) and a very high hydraulic conductivity that enabled them to survive in a forest habitat. As part of the forest adaptation, the light-driven formation of flowers at the forest canopy replaced tendrils (homologous structures) at the terminal nodes. Today, grape growers attempt to maximize this characteristic by managing vines such that vigor is constrained and a perennial structure that promotes initiation and differentiation of fruitful buds is produced. In addition, it is thought that the early grapevines were dioecious but that over time hermaphroditic forms appeared. All species in the *Vitis* genus hybridize easily (highly heterozygous) and will mutate under different environmental conditions. This trait, while producing a large diversity within the genus, has lead to difficulties in identifying the parentage of some of today's most prized varieties.

The cultivation of grapevines predates written history (McGovern 2003). Archeological findings in the Caucasian Mountains, near the town of Shiraz in ancient Persia, indicate that viticulture (the cultivation of grapes) existed as early as 3,500 B.C. (Penning-Roswell 1989). *Vitis vinifera* ("wine-bearing vine") was first domesticated in this region and soon spread to Assyria, Babylon and to the shores of the Black Sea. The Assyrians, the Phoenicians, the Greeks, and the Romans furthered viticulture and viniculture (the process of fermenting grapes into wine) and spread its knowledge to Palestine, Egypt, North Africa, the Iberian Peninsula, and throughout Europe to as far north as the British Isles (Unwin 1991). During the Dark Ages grape growing declined throughout most of Europe and would have died out had it not been for the Christian monks who preserved the methods of viticulture and made vast improvements in cellaring techniques (Loubere 1990). With new world explorations, Europeans carried the vine with them and helped establish the industry in regions that were well suited for cultivation.

Today viticulture is an ever-growing agribusiness, even more so as more is learned about the health properties of drinking wine (Mansson 2001). Better understanding of the biology of the vine and climatic constraints on the individual varietals of *V. vinifera* has opened up new regions to viticulture (Schultz and Jones 2010). In general, mechanization of viticulture and viniculture has furthered the grower's ability to produce at a greater profit and new technologies in genetics and traditional plant breeding (clones and rootstocks) have reduced the susceptibility of *V. vinifera* to many parasites and diseases (Bisson et al. 2002).

30.2 Phenological Cycle and Characteristics

Like all agricultural and natural plant systems, grape producers and viticulture scientists need crop developmental scales that are easy to use, universally accepted, and accurate. Grapevine growth stage identification has been carried out to better provide standards by which growers worldwide can communicate information (see Coombe 1995 for a review). In addition, the study of grapevine phenology has allowed growers to better understand whether or not a given variety can produce a crop within a given climate regime (e.g., Tesic et al. 2002a, b; Jones et al. 2005b; Jones 2006); has facilitated husbandry by providing the structure by which cultural and chemical practices can be applied at optimum periods during a plant's growth (Matthews et al. 1987); and has been useful in estimating crop yields (Clingeleffer et al. 1997).

Four generally accepted grapevine growth descriptive systems, each based upon principal growth stages with varying levels of secondary growth stages, have been established. (1) Baggiolini (1952), later amended by Baillod and Baggiolini (1993), was the first system proposed for grapevines and consisted of 16 stages from bud break to leaf fall that was widely accepted to aid pesticide applications. (2) Eichhorn and Lorenz (1977) followed with a more comprehensive system that entails 22 stages from "winter bud" to "end of leaf fall" that has been accepted by many in the industry. (3) The BBCH system was developed by the European Union to standardize the many scales that were in use for single plant species or group of related species and has been proposed as a prototype of a universal scale–the Extended BBCH Scale for all monocot and dicot crops (Lorenz et al. 1995; Meier 1997). The adaptation of the BBCH system to grapevines resulted in seven macro-stages of growth with numerous micro-stages in between. (4) In a thorough assessment of descriptive schemes for the grapevine, Coombe (1995) found that the BBCH system included terms that were not universally understood, had a zigzag rather than continuous developmental sequence, and had minor errors in the descriptions. From this assessment and detailed data analysis from phenology trials in Australia, the author suggests the Modified Eichhorn and Lorenz system (Modified E-L) as meeting universal requirements for grapevines. The Modified E-L provides a 47-stage system,[1] which includes a simple listing of eight major growth stages along with detailed intermediate stages (Fig. 30.1).

The growth cycle of grapevines starts in the early spring with the shoot and inflorescence development stage. This stage commences with the breaking of the dormant winter stage and a rise in sap or bleeding of the vine. Four to six weeks after the sap starts to rise, the vine starts to produce foliage (tendrils and leaves) from the latent primary buds formed at the end of the previous year. This stage, termed budburst or bud break (debourrement by the French), is the budding out of the vine before the floral parts develop later in the spring. Once shoot growth reaches roughly 10 cm (Modified E-L stage 12, Fig. 30.1), numerous leaves are

[1] Conversions from one system to another can be found in Coombe (1995).

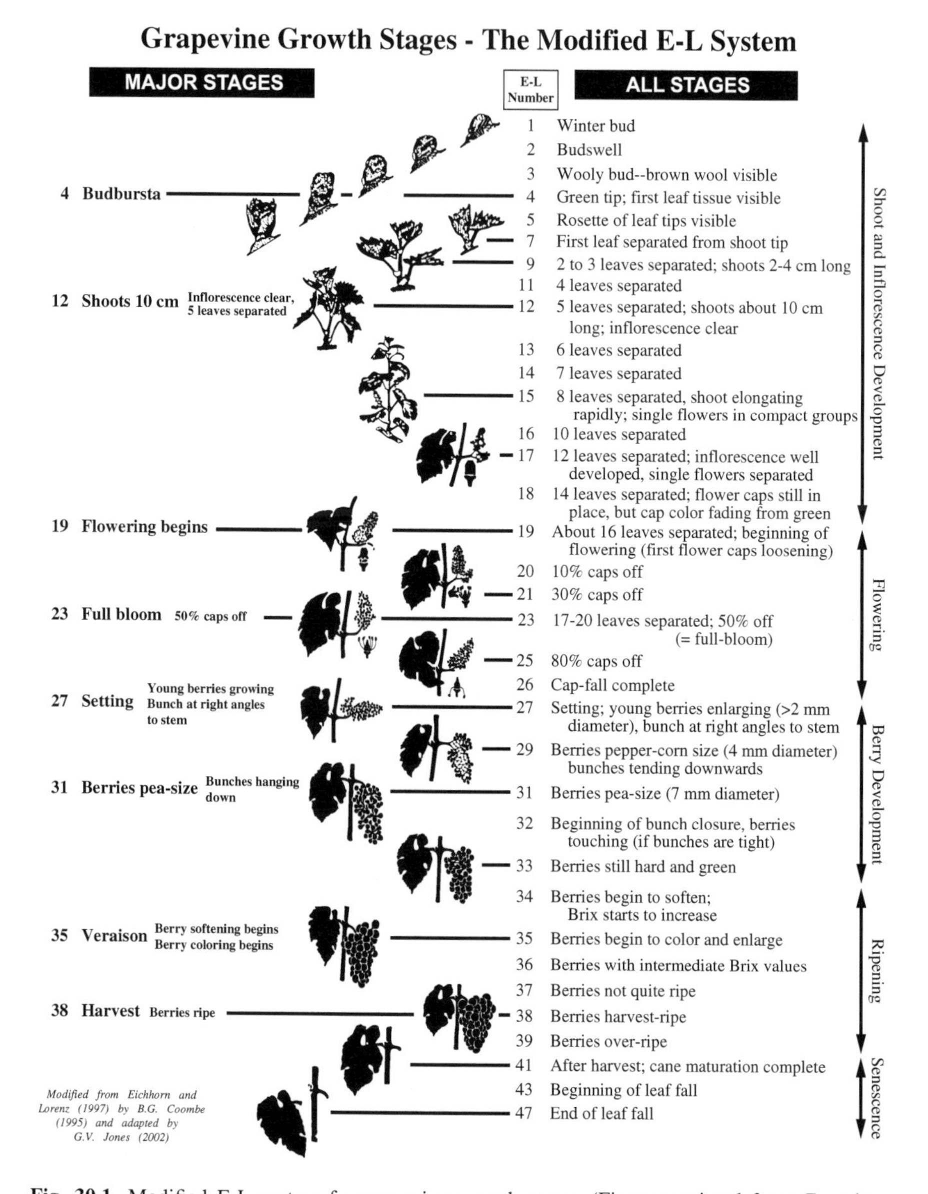

Fig. 30.1 Modified E-L system for grapevine growth stages (Figure reprinted from Coombe (1995), with permission from the Australian Society of Viticulture and Oenology)

separated and the inflorescence becomes clearly visible. The flowering (floraison) stage starts in late spring or early summer when the young shoots differentiate their meristems and put forth flower clusters at nodes along the vine. Each flower in the cluster is covered by a cap that, through further growth, breaks away (anthesis) revealing the flower parts of the vine. Full bloom is considered when 50 % of the caps have fallen off. After flowering, the green berries set (nouaison) on the clusters (only 20–60 % of the potential berries set) and develop at right angles to the stem. The berries grow in bulk over the next 2 months and when they reach pea size the clusters start to hang downward. During this stage, there is very little chemical differentiation in the fruit, except for a slight change in acidity. Nearing the end of berry development, however, the berries go through a period of rapid physical and chemical change called véraison, which initiates the ripening stage. During this stage, the grapes soften, enlarge, start accumulating sugar, and change color to translucent greenish-white (white varietals) or to a red-purple hue (red varietals). In managed systems, the berries are considered harvest-ripe when the sugar/acid ratio and flavor profile is optimized for a particular variety or desired wine style. After harvest the vines begin their senescence stage with the canes maturing (turning hard and woody) and the leaves falling. While the onset and duration of each of the main phenological stages of V. vinifera grapevines varies spatially and for individual varieties, they are very consistent for the physiology of the main varietals in a given region and can be approximated by:

- Stage I: Shoot and Inflorescence Development – commencing around the end of March or first week of April (end of September to early October in the Southern Hemisphere).
- Stage II: Flowering – generally occurring in the first few weeks of June (late November to early December in the Southern Hemisphere).
- Stage III: Berry Development – from the end of flowering in mid-June (mid-December in the Southern Hemisphere) to the ripening stage.
- Stage IV: Ripening – starts with véraison near the end of July or the first week of August (late January to early February in the Southern Hemisphere).
- Stage V: Senescence – from harvest, late September through early November (late March through late April in the Southern Hemisphere) and leaf fall, over the winter months leading back to bud break.

30.3 Phenology, Weather, and Climate

It has been said, “viticulture is perhaps the most geographically expressive of all agricultural industries” (de Blij 1983, p. 112). Along with this geographical signature comes the concomitant climate of a region, which influences and controls the characteristics and quality of the grapes and wine produced there (van Leeuwen et al. 2004; Jones 2006). The Mediterranean and marine west-coast climate regimes are synonymous with the majority of the viticulture regions of the world (however, both table and winegrape growing has become more viable in other areas as knowledge and varietal development for those regions has grown) as the mild wet-winter, dry-summer climate found there is ideal for the cultivation of *V. vinifera* grapes (Jones et al. 2012). The main regions are found along the Mediterranean basin of Europe, the *fynbos* of South Africa, the *mallee* of southern Australia, the *mattoral* of Chile and Argentina, and the *chaparral* and coastal valleys of western North America.

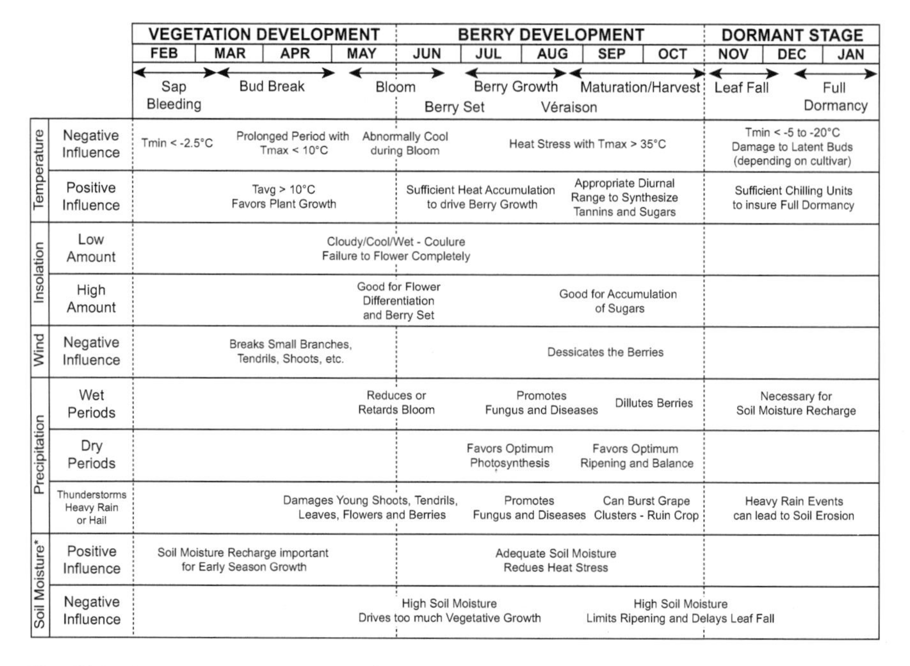

Fig. 30.2 Weather and climate influences on grapevine development and phenological growth stages (After Crespin 1987; Jones 1997; Jones et al. 2012)

30.3.1 Weather and Climate Influences

Broad weather and climate influences on growing winegrapes include winter severity, spring and fall frost severity and timing, precipitation frequency and timing, and the characteristics of growing period temperatures. In general, the growing season of winegrapes varies from region to region but averages approximately 170–190 days (Mullins et al. 1992) spanning typically from March to September (warmer climates) or April to October (cooler climates). Research has also shown that during the winter there is a minimum winter temperature that grapevines can withstand. This minimum ranges from −5 to −20 °C and is chiefly controlled by micro-variations in location and topography (Winkler et al. 1974; Amerine et al. 1980). Temperatures below these thresholds will damage plant tissue by the rupturing of cells, the denaturing of enzymes by dehydration, and the disruption of membrane function (Mullins et al. 1992). Prescott (1965) also notes that an area is suitable for grape production if the mean temperature of the warmest month is more than 18.9 °C and that of the coldest month exceeds −1.1 °C. Furthermore, the length of the frost-free season is important to the onset of bud break, flowering, and the timing of harvest.

Each of the major phenological stages of *V. vinifera* grapevines are governed by critical climatic influences (Fig. 30.2). Temperature effects are evident in the spring where vegetative growth is initiated by prolonged average daytime temperatures above 10 °C (Amerine and Winkler 1944). During this stage, temperatures below −2.5 °C can adversely affect the growth of the herbaceous parts of the plant and hard freezes can reduce the yield significantly (Riou 1994). During floraison (period during bloom) and throughout the growth of the berries, extremes of heat can be detrimental to the vines. While a few days of temperatures greater than 30 °C can be beneficial to ripening potential, prolonged periods can induce heat stress in the plant and lead to premature véraison, the elimination of the berries through abscission, permanent enzyme inactivation, and partial or total failure of flavor ripening. During the maturation stage, a pronounced diurnal temperature range effectively synthesizes the tannins and sugars in the grapes. Early frost or freezes during maturation can lead to the rupture of the grapes along with significant loss of volume. During the dormant stage, a temperature minimum or effective chilling unit (hours below a certain temperature) is generally needed to effectively set the latent primary buds (Galet 1979).

Throughout the phenological stages of the grapevine, the amount of insolation is critical in maintaining the proper levels of photosynthetic activity for the production of assimilate. The most critical stages come during the development of the berries starting at floraison and continuing through the harvest (Fig. 30.2). During floraison, high amounts of insolation result in effective meristem differentiation into flowers. During this stage, prolonged periods of cloudiness, cold temperatures, and excessive rain results in *coulure* or the failure to fully flower and set the berries. During maturation, insolation mainly acts to control the amount of sugar accumulation in the grapes, and therefore, the potential wine alcohol content.

The role that the wind plays in the growth of the grapevine and the production of fruit is mainly through the effects on vine health and yield. This is manifested in both a physical nature through direct contact with the vines and through physiological effects of stomata closure and reduced disease infestations. During the early stages of vegetative growth, high winds can break off the new shoots, delaying and even reducing the amount of flowering (Fig. 30.2). As the berries proceed through véraison and into the maturation stage, high winds can desiccate the fruit resulting in lower volume and quality. However, drying winds that occur at night and early morning can help reduce the occurrence of fungus borne diseases by limiting the formation of dew on the leaves and berries. Nighttime winds can also be beneficial in that they can help limit the occurrence of radiation frosts.

During the growth cycle, the weather conditions that can most severely afflict the vines and berries are associated with moisture. Atmospheric moisture, in the form of humidity and rainfall, hastens the occurrence of fungal diseases (i.e., powdery mildew, downy mildew, botrytis bunch rot). In extreme cases, water stress resulting from high evaporative demand can manifest itself in leaf loss, severe reductions in vine metabolism, and fruit damage or loss. Even mild periods of moisture stress can substantially reduce the relative level of photosynthesis, resulting in lower fruit yields and quality. Fungal diseases can be problematic in the morning when high temperatures and condensation combine to hasten the disease. The formation of fungal diseases can cause defoliation, reduced sugar accumulation, and a reduction in winter hardiness (Amerine et al. 1980). The occurrence of rain during critical growth stages can lead to devastating effects. While ample precipitation during the vegetative stage is beneficial, during flowering rainfall can reduce or retard inflorescence differentiation and during berry growth rainfall can enhance the likelihood of fungal diseases (Fig. 30.2). As the berries mature rainfall can further fungal maladies, yellow and dilute the berries, leading to a reduction in sugar and flavor levels, and severely limit yield and quality. Examination of the world's viticulture regions suggests that there is no upper limit on the amount of precipitation needed for optimum grapevine growth and production (Gladstones 1992; Jones et al. 2012), but grapevine viability seems to be limited in some hot climates by rainfall amounts less than 600–750 mm, although this can be overcome by regular irrigation. Extreme meteorological events, such as thunderstorms and hail, while rare in most viticultural regions, can be extremely detrimental to the crop by damaging the leaves, tendrils, and berries and if they occur during maturation can split the grapes, causing oxidation, premature fermentation, and a severe reduction in volume and quality of the yield (Winkler et al. 1974).

While grapevines are planted in a wide diversity of landscapes, the sites that produce the best quality winegrapes are generally planted on moderate slopes with good sunlight exposure (aspects). South-facing slopes (Northern Hemisphere) provide for more insolation, and therefore photosynthesis, with a south-facing slope of 8° providing a 77° noon sun angle at 44°N latitude (a 12 % insolation increase over a flat site). Depending on numerous other factors such as obstructions (e.g., trees, buildings, other hills, etc.), a properly situated slope can enhance growth and maturation or limit disease problems (Jackson and Schuster 1987). The slope's aspect influences phenological development and canopy characteristics through the amount and timing of insolation received. In general, northwest, north, and northeast aspects will experience delayed grape growth stages, less sunlight, and lower evapotranspiration rates from the soil and canopy (Northern Hemisphere). Southeast, south, southwest, and west aspects, on the other hand, will tend to exhibit earlier grape growth stages and show varying increases in insolation and evapotranspiration (Northern Hemisphere). In addition, cold air drainage problems from the pooling of cold air in low spots or via obstructions on sloping land can delay a plant's phenology or even reduce its viability (Wolf 1997).

Soil effects on grapevine phenology are mostly through water retention and plant water relationships (Fig. 30.2). Gravely to rocky soils provide good drainage and higher heat storage, which accelerates phenological development, while heavy clay soils, which retain moisture, can slow down growth, and inhibit productivity. The temperature of the soil can also have a strong influence on vine growth and fruitfulness with warmer spring soils hastening early season growth (Robinson 1994). In addition to the relative amount of sunlight, the composition and color of the soil, the local topography, and drainage capabilities are all factors affecting canopy temperatures, especially at night.

30.3.2 Bioclimatic Indices

Some of the first viticulture-climate studies were conducted by A. P. de Candolle in France during the mid nineteenth century. It was de Candolle's observation that vine growth generally started when the mean daily temperature reached 10 °C that led to the notion of a heat summation above a base temperature that could define vine growth stages and grape maturation. Amerine and Winkler (1944) furthered this research in California, developing an index based on growing degree-days [summed over April through October (Northern Hemisphere) using a base temperature of 10 °C] that is widely used as a criterion for determining a given region's ability to grow different varieties and produce different wine styles. Others have since refined this index to be applicable across a wider range of wine regions (Jones et al. 2009, 2010; Hall and Jones 2010; Anderson et al. 2012). Additional bioclimatic indices have been used to characterize a region's potential for viticulture and are mostly developed on the basis of heat accumulation. Various forms of a heliothermal index have been used (Branas 1974; Huglin 1978) along with a latitude-temperature index (Jackson and Cherry 1988) to help define the suitability of a region to the planting of certain varietals. Smart and Dry (1980) developed a simple classification of viticultural climates that uses five dimensions of mean temperatures, continentality, sunlight hours, aridity, and relative humidity. Gladstones (1992) developed a 'Biologically Effective Degree-Day' index (BEDD) by imposing an upper limit on mean temperatures (19 °C), a correction factor for latitude, and a correction for each month's temperature range.

The majority of the bioclimatic indices developed for viticulture described above were indirectly based on growth characteristics, productivity, and potential wine styles and not phenology. To better capture phenological effects Jones (2006) used grapevine growth and ripening for many of the world's most recognizable and planted varieties to show that high quality wine production is limited to 13–21 °C average growing season temperatures (GST; Fig. 30.3). The climate-maturity zoning in Fig. 30.3 was developed based upon both climate and plant growth for many varieties grown in cool to hot regions throughout the world's benchmark areas for those winegrapes. While many of these varieties are grown and produce wines outside of their individual bounds depicted in Fig. 30.3, these are more bulk wine (high yielding) for the lower end market and do not typically attain the typicity or quality for those same cultivars in their ideal climate. Furthermore, GSTs below 13 °C are typically limited to hybrids or very early ripening cultivars that do not necessarily have large-scale commercial appeal. At the upper limits of suitability, some production can also be found with GSTs greater than 21 °C, although these climates are mostly limited to fortified wines, table grapes and raisins (Fig. 30.3). The climate-maturity relationships shown in Fig. 30.3 provide information on the known limits of any individual variety. For example, Pinot Noir is a relatively cool climate variety with one of the smallest climate niches (14–16 °C) of the top planted varieties worldwide. Across this 2 °C niche the variety produces lighter, more elegant wines in the cooler areas (e.g., Tamar Valley of Tasmania) to full-bodied, fruit-driven wines in the warmer areas (e.g., Russian River Valley of California).

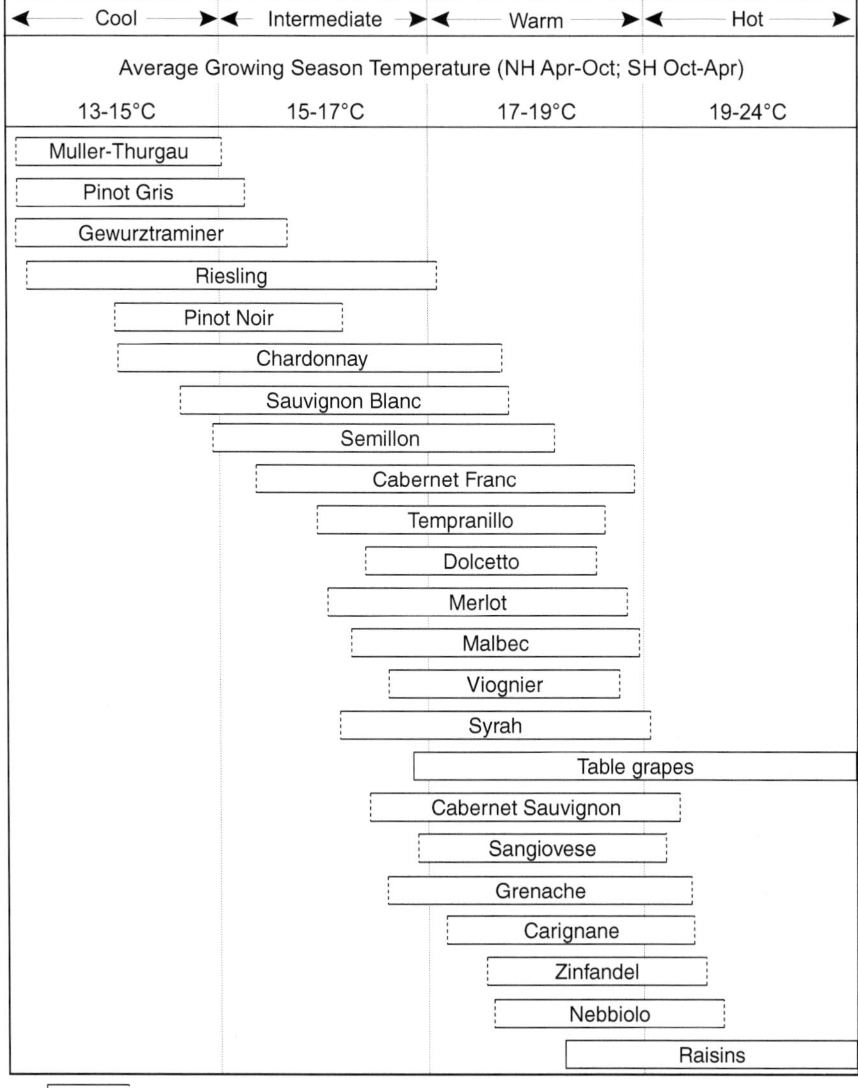

Fig. 30.3 Climate-maturity groupings based on relationships between phenological requirements and growing season average temperatures for high to premium quality wine production in the world's benchmark regions for many of the world's most common cultivars. The *dashed line* at the end of the *bars indicates* that some adjustments may occur as more data become available, but changes of more than ±0.2–0.5 °C are highly unlikely (Jones 2006)

30.4 Phenological Observations, Trends, and Modeling

30.4.1 Phenological Observations and Trends

Long-term historical records of European viticulture have been maintained for nearly a 1,000 years with harvest information being observed and recorded initially by the monks during the Middle Ages and later by the merchants and the prominent producers (Le Roy Ladurie 1971; Le Roy Ladurie and Baulant 1980). From these historic records grape harvest starting dates (GHD) are the most available with some records dating to the fourteenth century (Chuine et al. 2004). Although not true phenological events due to cultural and economic influences on when to harvest, GHD still provided a general measure of plant growth and fruit ripening due to climate. As such GHDs have been used as proxy data detailing spring or growing season temperatures during the past eight centuries (Daux et al. 2012; Krieger et al. 2010). These include GHD records for Burgundy (Chuine et al. 2004) and Besançon (Garnier et al. 2011) in France, Austria (Maurer et al. 2009), Switzerland (Meier et al. 2007), and Hungary (Kiss et al. 2011). Daux et al. (2012) detailed these GHD datasets and numerous others that together comprise 378 series across France (>90 % of the data) and numerous other European countries. These GHDs correlated well with other long-term records of climate as evidenced in glacial advancements and retreats, ice core analyses, palynological studies, varve chronologies, and dendrochronologies (Krieger et al. 2010). The GHD records showed moderate to strong correlations with spring through summer temperatures and showed that the European region has experienced wide fluctuations of climate and viticultural productivity (Le Roy Ladurie 1971; Penning-Roswell 1989) since the fourteenth century. However, the GHD also reveal that the last few decades have seen some of the warmest years in the data record, with 2003 likely being the warmest (Chuine et al. 2004).

Contemporary phenological observations by individual growers, through collective networks, or research-focused studies have been used to assess regional differences in plant maturity potential, to determine the timing of cultural practices, and provide some measure of a given vintage's development. While no worldwide single method of observing phenological events exists, it is common to indicate that an event (e.g., bud break, flowering, etc.) has occurred when 40–60 % of the plants in a given area (an entire vineyard or within a vineyard block containing a specific variety) are exhibiting the event (Coombe and Dry 1988). In general, bud break, flowering, véraison, and harvest dates are the most observed events with very few growers noting any of the more detailed micro-stages in the Modified E-L or BBCH systems. Of these, McIntyre et al. (1982) and Jimenez and Ruiz (1995), showed that the onset of bud break and flowering was very consistent from variety to variety while véraison and maturity (date of harvest) was less predictable due to greater vine management variability between growers.

The rate of development between the growing season phenological stages varies with variety, climate, and topography. In regions with cool climates, early ripening varieties produce well and in hot climates, late ripening varieties are better suited. In addition, differences in vine management affect harvest dates, yield, and quality. Growers who want higher yields generally will harvest lower quality fruit and produce moderate to low quality wine, whereas higher quality grapes are associated with lower yields. The timing of these developmental stages is also related to the ability of the vine to produce, with early and fully expressed phenological events usually resulting in larger yields (Mullins et al. 1992; Jones 1997). Additionally, the pace by which vines go through their stages has been related to vintage quality with shorter intervals and earlier harvests generally resulting in higher quality (Jones and Davis 2000b). During flowering, the weather is crucial and can ultimately determine the vine yield. The failure to flower properly may mean that the vines will develop clusters with few or no berries. During maturation, the crop can be greatly affected by rainfall and high humidity, which can induce fungal diseases. These diseases rob moisture from the berries, which reduces the yield and can render much of the harvest useless. Jackson and Lombard (1993) noted that excess precipitation had a negative effect on fruit quality by increasing vigor, delaying phenological events, reducing berry set, and increasing disease pressure.

Barbeau et al. (1998a, b) established a connection between soil texture, soil temperature, and vine phenology in Cabernet Franc vines in the Loire Valley of France. The authors found that sites with good drainage had earlier phenological events, whereas heavy clay soils and soils with perched water tables had later phenological events. In addition, the sites that produced earlier phenological events showed increases in accumulated sugar, achieved better anthocyanin levels (color pigments), and retained optimum acidity. In examining soil fertility, Costantini et al. (1996) conducted complex soil analyses in Montepulciano, Italy and found that fertile soils increased yield and berry weights, while infertile soils were detrimental to yield and berry weights. Both situations produced lower quality wine while intermediate soil types provided optimum yields and much better quality. Tesic et al. (2002a, b) also found that increased vegetative growth, mostly attributed to fertile soils, was associated with late phenology in Cabernet Sauvignon grapevines in New Zealand. The authors also used indices of vine precocity (flowering) for Cabernet Sauvignon to characterize viticultural environments and develop a site index in New Zealand, which appears to have potential use in vineyard zoning, site assessment, and site selection. In addition, Souriau and Yiou (2001), using grape harvest dates from northeastern France and Switzerland, showed significant correlations between harvest dates and the North Atlantic Oscillation (NAO) and suggested using the record "as an interesting proxy" to reconstruct the NAO back in time.

Long-term trend analyses of multiple phenological events for winegrapes have been more numerous over the past decade as the interest in better understanding the plant system and its relationship to climate becomes more important. Jones and Davis (2000b) analyzed one of the longest, scientifically controlled winegrape phenology observation networks in Bordeaux, France. Using bud break, flowering, véraison, and harvest dates from 1952 to 1997 for Merlot and Cabernet Sauvignon the authors found that a high positive correlation existed between each successive event. The results showed that predicting maturity dates was very possible as the season progressed from one phenological event to the next. The growing season averaged 193 days (from bud break to harvest) and declined by 13 days over the period (each of the interphases declined in length from 4 to 10 days). Climate parameters influencing the timing of the events indicated that precipitation and general cloudiness tended to delay events, while increased insolation and the relative number of days over 25 and 30 °C hastened the events. In an integrated synoptic air mass and circulation analysis, Jones and Davis (2000a) related the mean climate characteristics, phenology, yield, and wine quality in Bordeaux to variations in air mass and circulation frequencies. The results indicated that increased frequencies of cold- and moisture-producing events (e.g., low pressure and frontal passages) during bud break, flowering, and berry set both delayed vine phenology and reduced yield and quality. The delayed phenology in Bordeaux also showed strong correlations with sugar and acid levels (which largely determine quality) as delayed events resulted in higher acid levels, while early events resulted in greater sugar levels. Likewise, nearly half of the variation in wine quality (as told by vintage ratings) was related to earlier phenological timing and shorter stages between phenological events.

Many other analyses show similar results as to those found in Bordeaux. Work by Braslavska (2000), studying grapevine phenology of the Müller-Thurgau variety in Slovakia from 1971 to 2000, found no trend in the dates of bud break and first leaf, however flowering, véraison, and harvest dates were earlier by 8, 11, and 15 days, respectively during the time period. In addition, the length of time from bud break to harvest in Dolné Plachtince, Slovakia declined by 15 days over the period and was related to an increase in degree-day accumulation and the relative number days with Tmax >25 °C. In California's premier wine producing areas of Napa and Sonoma, Nemani et al. (2001) found that that higher yields and quality were related to an asymmetric warming (greatest warming at night and in the spring), a reduction in frost occurrence, advanced spring growth, and an increase in the growing season length. Duchêne and Schneider (2005) found similar phenological trends for Riesling in Alsace, France but also noted that potential alcohol levels at harvest have increased by 2.5 % (by volume) over the last 30 years and was highly correlated to significantly warmer ripening periods and earlier phenology. Jones et al. (2005b) examined characteristics and trends in phenology for nine locations and 15 varieties in France, Italy, Spain, Germany, and Slovakia. Collectively the results revealed significantly earlier events (6–18 days) with shorter intervals between events (4–14 days) that were strongly influenced by changes in spring through summer temperatures and/or heat accumulation indices. Both Petrie and Sadras (2008) and Webb et al. (2011) found similar results with trends to earlier harvests across numerous regions and varieties in Australia. Examining viticulture, wine and climate relationships for 1952–2006 in the Alt Penedès, Priorat, and Segrià regions of NE Spain Ramos et al. (2008) found strong correlations between regional warming and earlier phenological events, higher wine quality, but reduced production due to decreases in precipitation.

Recent work by Bock et al. (2011) examined phenological events and intervals, and composition (acid and sugar content at harvest) of white grape cultivars (Müller-Thurgau, Riesling and Silvaner) and their relationship with climate in Lower Franconia, Germany. Over 1949–2010 the research found that the phenology of grapevines in the region tended towards earlier occurrences with a shortening of phenological intervals. The research also found that the relative amounts of sugar in the grapes at harvest tended to increase, while acid levels decreased. Over 1951–2005 Urhausen et al. (2011) found that bud break and flowering for seven white varieties trended earlier by 2 weeks along the upper Moselle River between Luxembourg and Germany. Similar to other studies the authors found that sugar levels at harvest increased while acid levels decreased. Following on the work of Barbeau et al. (1998a, b) in the Loire Valley of France, Neethling et al. (2012) found that warming temperatures in the region during 1960–2010 have led to earlier harvests with higher sugar concentrations and lower titratable acidity for the main varieties grown there. Examining 18 varieties at a long-term (1964–2009) research trial in the Veneto region of Italy, Tomasi et al. (2011) found that the budbreak to harvest period covered mid-April to late September, averaging 156 days but varying 55 days across varieties. The main phenological events and intervals between events exhibited a 25–45 day variation between the earliest and latest years, with the bloom to véraison growth interval showing the lowest year to year variation. During the time period, trends of 13–19 days earlier were found for bloom, véraison, and harvest dates, while budbreak exhibited high interannual variation and no trend. Tomasi et al. (2011) also note that similar characteristics and trends for the main phenological events were found for early, middle, and late maturing varieties. The research also identified significant breakpoints in the phenology time series that averaged 1990–1991 across all varieties, with early and middle ripening varieties shifting sooner than late ripening varieties. During 1964–2009, the growing season climate differences were 2.0 °C between the years with the shortest and those with the longest budbreak to harvest intervals.

30.4.2 Phenological Modeling

While previous research has suggested that refinements to climate indices that help predict grapevine phenology are needed (Winkler et al. 1974; McIntyre et al. 1987), definitive, universally applicable formulations have not been fully developed. This is likely due to the variation in vine responses across climates and growth characteristics of early versus late maturing varieties (Winkler et al. 1974; van Leeuwen et al. 2008). Furthermore, Giomo et al. (1996) discussing the usefulness of various climate indices in grapevine growth analyses, indicated that most were developed to study global to regional climates and were not useful at the sub-region level. McIntyre et al. (1987) also showed that a simple summation of the number of days in an interphase, averaged over a long period, is a better predictor of phenological events than the single sine degree-day method (Zalom et al. 1983) used in California.

One prominent area of interest in grapevine phenological modeling has been optimizing the lower and upper temperature thresholds for plant growth or shutdown in degree-day formulations. While these thresholds have typically been defined by their influence on phenological event timing, they are commonly derived from photosynthetic activity limits. For example, it has been shown that very little photosynthesis occurs in grapevine leaves when temperatures are <5 °C (Kriedemann 1968). However, 10 °C is the most commonly recognized and used base temperature for winegrapes (Jones et al. 2010). Yet Jimenez and Ruiz (1995) noted that using accumulated degree-days above 0 °C or the number of days between events is a better predictor of phenology on average than degree-days above a 10 °C threshold. Furthermore, while there is strong evidence for a 4 °C base temperature for numerous varieties of grapevines in Australia (Moncur et al. 1989) and a 5 °C base using the BRIN model across numerous regions and varieties in France (García de Cortázar-Atauri et al. 2009), there has been little confirmation of these thresholds across other wine regions and for a wider range of varieties. There is also some evidence that grapevines have a maximum temperature threshold of ~32–35 °C (Jackson 2000), although others have found that optimum net photosynthesis occurs over a wide range of temperature (25–35 °C), making it difficult to pinpoint a universal upper temperature threshold (Kriedemann 1968). The application of an upper threshold of 19 °C (average temperature) to heat accumulation in the BEDD formulation attempted to quantify this issue (Gladstones 1992), but was identified by trial and error and recent examination has not been able to quantify its usefulness across a range of regions (Jones et al. 2009, 2010; Hall and Jones 2010; Anderson et al. 2012).

Numerous studies based on climate parameters (mostly temperature) have been used to try to predict the dates of the individual phenological events. Swanepoel et al. (1990) developed a bud break model using cultivar specific constants that explains over 80 % of the variability in the events in the warm climate region of Stellenbosch, South Africa. Nendel (2010) found that predicting grapevine budbreak to within ±2.5 days is possible, but only with an appropriately fixed start date for heat accumulation and site specific temperature data. Due et al. (1993) revealed that average temperatures, summed from specific dates and not one event to another, were good predictors of events. Calò et al. (1994) found that the length of the interval from bud break to flowering and the flowering date were best modeled using daily average maximum temperatures rather than averages or summations of temperature. Jones and Davis (2000b) also found that degree-days do not readily predict the majority of phenological timing, and yield or quality in Bordeaux, France. Tesic et al. (2002b) revealed an interesting relationship between the relative coldness of the winter on bud break with warm winters inducing an early, slow, and heterogeneous event and cold winters a later, more rapid, and homogeneous bud break. Jones et al. (2005b) found that maximum temperatures were better predictors of early season events (bud break and bloom) and véraison for some regions and varieties, while heat accumulation indices were better predictors of véraison in other regions and harvest across all varieties and regions. Parker et al. (2011) have attempted to define a single, process-based phenological model to predict flowering and véraison for grapevines. The research finds that a base temperature of 0 °C calculated from the 60th day of the year (for the Northern hemisphere) was found to be the most optimum model. From this work the authors promote a general phenological model (GFV) to characterize the timing of flowering and véraison for the grapevine. However, this model was developed for European conditions and has not been tested across a wide range of climate types over which grapevines are grown in the new world. While some success has been achieved, all of the authors agree that more detail is needed to increase the understanding of the processes and the accuracy of these predictive models.

One of the most tested and integrated grapevine phenological models is VineLOGIC (Godwin et al. 2002). VineLOGIC is designed to simulate how different varieties respond to different environmental conditions and management practices. The system incorporates numerous plant and site characteristics that influence vine growth and productivity. These include the grape variety, rootstock, weather, soil conditions, irrigation timing and amounts, water salinity, and how the vine is pruned and trellised. Output from the model includes information on bud break, véraison, time of ripening, the potential yield and estimates of the timing and severity of stress events. VineLOGIC has been widely used by the wine industry in Australia and to model the effects of climate change (Webb et al. 2007; see below).

30.4.3 Assessment of Climate Change Impacts

Similar to many other natural and managed plant systems, grapevine phenology has been mostly trending earlier with shorter interphases between events (Jones et al. 2005b; Tomasi et al. 2011). The trends have been shown to be driven strongly by climate with a 3–8 day response per 1 °C of warming over the last 30–50 years across numerous regions and varieties (Jones 2007). Early growth events (e.g., bud break) tend to show the lower response, while later season events (e.g., véraison and harvest) exhibit the greatest response due to an integrated effect over the season. Projections of further changes in climate in wine regions globally and the potential effects to both wine production and quality (Jones et al. 2005a) have heightened the interest in using grapevine phenology to model the impacts.

Early work by Bindi et al. (1996) compared different models of future climate change for Italy and found a composite 23 day reduction in the interval from bud break to harvest for Cabernet Sauvignon and Sangiovese grapes and attributed the changes to elevated CO_2 levels and temperature increases. Modeling future impacts from climate change on grapevine phenology in Australia, Webb et al. (2007) showed that the impacts varied by wine region and variety. The results projected that bud break for Cabernet Sauvignon will likely be earlier by 4–8 days by 2030 and by 6–11 days by 2050. Along with projections for earlier harvest for each variety studied, the growing season duration (from bud break to harvest) was projected to be compressed across all regions. The authors also noted that some regions may be adversely affected by the chilling requirement not being met in future warmer climates and that harvest was projected to occur both earlier in the year and in a warmer climate, or producing a 'dual warming impact.' Examining the effects of climate change on grapevine phenological stages in Alsace, France Duchêne et al. (2010) used a degree-day model to simulate budbreak, flowering and véraison for Riesling and Gewürztraminer. The research found that, compared with its timing in 1976–2008, véraison is predicted to advance by up to 23 days and mean temperatures during the 35 days following véraison are projected to increase by more than 7 °C by the end of the twenty-first century for both varieties. The authors further note that such changes will likely have a significant impact on grape and wine quality. A further examination of the genetic variability of phenology found that even the latest maturing varieties would experience warmer ripening periods than observed today (Duchêne et al. 2010).

Others have used grapevine phenology to examine future impacts due to pathogens and pests that greatly affect vines and their fruit. Salinari et al. (2006) used an empirical disease model to examine primary infections of downy mildew (*Plasmopara viticola*) on grapevines in numerous wine regions worldwide. The authors then ran the disease model under outputs from GCMs for 2030, 2050, and 2080. The results point to a likely advance of first disease outbreaks in the spring, which could lead to more severe infections, due to the polycyclic nature of the pathogen. The disease scenarios point to probable grower changes in the timing and frequency of fungicide applications, which would likely increase the overall cost of disease management. Caffarra et al. (2012) examined both plant disease and insect development using detailed phenological models in order to improve the understanding of the host-pest/pathogen system and their interactions. Combining grapevine phenological models with climate change scenarios the authors showed that the European grapevine moth (*Lobesia botrana*) will likely have greater impacts on crop yields due to an increased asynchrony between the larvae-resistant growth stages of grapevine and larvae of the pest. For powdery mildew (*Uncinula necator*) Caffarra et al. (2012) show that future warming will likely reduce disease pressure in lower elevation sites due to a decrease in the time of optimum temperatures for infections. However, the authors noted that cooler, higher elevation sites will likely see greater infections due to more time during the optimum temperature window for powdery mildew. Both results point to a strong effect of host-plant development on pest and disease pressure and suggest that current interactions between host-pest and host-disease may be altered by climate change.

30.5 Conclusions

The phenology of grapevines is extremely sensitive to climate and has become an increasingly important tool to better understand the plant system, its relationship to pest and disease timing and impacts, and its ability to produce quality fruit and wine at economically sustainable yields. However, the major drawback to grapevine phenological research (many other systems as well) has always been the spatial and temporal resolution of the observed events. On an annual basis, phenological research has been hampered by incomplete records caused by the death or translocation of the observer, lack of funding, and/or lack of interest. On a seasonal basis, 1–3 macro-stages are generally the most observed events leaving the vast majority of the plant system's physiology unaccounted for. In spite of these issues, grapevine observations from many wine regions have led to the refinement of models that have improved climate index development and use. Furthermore, the climate sensitivity of grapevine phenology has contributed to our knowledge of climate change impacts, revealing a strong coupling between temperature, earlier events, and a shortening of the intervals between phases. Understanding climate change impacts has also led to the development of numerous research groups and modeling efforts either focusing on or utilizing grapevine phenology to assess future conditions. These include the adaptation of the STICS model to the grapevine (García de Cortázar-Atauri 2006), LACCAVE (Long term impacts and adaptations to climate change in viticulture and enology) in France, FENOVITIS and ENVIROCHANGE in Italy, and SIAMVITI (Climate change in Viticulture: Scenarios, Impacts and Adaptation Measures) in Portugal. As these projects unfold more will become known about the characteristics and timing of grapevine phenology, its relationship to climate, and provide assessments of the best measures of adaptation in the face of a changing climate.

References

Amerine MA, Winkler AJ (1944) Composition and quality of musts and wines of California grapes. Hilgardia 15:493–675

Amerine MA, Berg HW, Kunkee RE, Ough CS, Singleton VL, Webb AD (1980) The technology of wine making, 4th edn. AVI Publishing Company, Westport

Anderson JD, Jones GV, Tait A, Hall A, Trought MTC (2012) Analysis of viticulture region climate structure and suitability in New Zealand. Journal International des Sciences de la Vigne and du Vin 46(3):149–165

Baggiolini M (1952) Les stades repères dans le développement annuel de la vigne el leur utilisation practique. Revue Romande d'Agriculture de Viticulture et d'Arboriculture 8:4–6

Baillod M, Baggiolini M (1993) Les stades repères de la vigne, Revue Suisse de Viticulture. Arboiculture et Horticulture 25:7–9

Barbeau G, Asselin C, Morlat R (1998a) Estimate of the viticultural potential of the Loire valley "terroirs" according to a vine's cycle precocity index. Bulletin de L'OIV 805–806:247–262

Barbeau G, Morlat R, Asselin C, Jacquet A, Pinard C (1998b) Behaviour of the cabernet franc grapevine variety in various terroirs of the Loire Valley: Influence of the precocity on the composition of the harvested grapes for a normal climatic year (example of the year 1988). Journal International des Sciences de la Vigne and du Vin 32(2):69–81

Bindi M, Fibbi L, Gozzini B, Orlandini S, Miglietta F (1996) Modeling the impact of future climate scenarios on yield and yield variability of grapevine. Clim Res 7:213–224

Bisson LF, Waterhouse AL, Ebeler SE, Walker MA, Lapsley JT (2002) The present and future of the international wine industry. Nature 418:696–699

Bock A, Sparks T, Estrella N, Menzel A (2011) Changes in the phenology and composition of wine from Franconia, Germany. Clim Res 50:69–81

Branas J (1974) Viticulture. Dehan, Montpellier

Braslavska O (2000) Tendencies and trends in the grapevine growing season at the locality dolne plachtince in 1971–2000. Národný Klimatický Program SR 8:69–78
Caffarra A, Rinaldi M, Eccel E, Rossi V, Pertot I (2012) Modelling the impact of climate change on the interaction between grapevine and its pests and pathogens: European grapevine moth and powdery mildew. Agric Ecosyst Environ 148:89–101
Calò A, Tomasi D, Costacurta A, Boscaro S, Aldighieri R (1994) The effect of temperature thresholds on grapevine (Vitis sp.) bloom: an interpretive model. Rivesta Viticultura Enologia 1:3–14
Chuine I, Yiou P, Viovy N, Seguin B, Daux V, Le Roy LE (2004) Grape ripening as a past climate indicator. Nature 432:289–290
Clingeleffer PR, Sommer KJ, Krstic M, Small G, Welsh M (1997) Winegrape crop prediction and management. Australian New Zealand Wine Ind J 12(4):354–359
Coombe BG (1995) Adoption of a system for identifying grapevine growth stages. Aust J Grape Wine Res 1:104–110
Coombe BG, Dry PR (1988) Viticulture, volume 1 – resources and volume 2 – practices. Winetitles, Adelaide
Costantini EAC, Arcara PG, Cherubini P, Campostrini F, Storchi P, Pierucci M (1996) Soil and climate functional characters for grape ripening and wine quality of "vino nobile di montepulciano". Acta Hort 427:45–56
Crespin Y, LeBerre M, Uvietta P (1987) The system climate-vine-vinegrower: example of a dynamic model. Bulletin de l'OIV 60:5–26
Daux V, Garcia de Cortazar-Atauri I, Yiou P, Chuine I, Garnier E, Le Roy LE, Mestre O, Tardaguila J (2012) An open-database of grape harvest dates for climate research: data description and quality assessment. Clim Past Discuss 7:3823–3858
de Blij HJ (1983) Geography of viticulture: rationale and resource. J Geogr 82(3):112–121
Duchêne E, Schneider C (2005) Grapevine and climatic changes: a glance at the situation in Alsace. Agron Sustain Dev 25:93–99
Duchêne E, Huard F, Dumas V, Schneider C, Merdinoglu D (2010) The challenge of adapting grapevine varieties to climate change. Clim Res 41:193–204
Due G, Morris M, Pattison S, Coombe BG (1993) Modeling grapevine phenology against weather: considerations based a large data set. Agr For Meteorol 56:91–106
Eichhorn KW, Lorenz DH (1977) Phönologische entwicklungsstadien der rebe, Nachrichtenblatt des Deutschen Pflanzenschutzdienstes (Braunschweig) 29:119–120
Galet P (1979) A practical ampelography. Comstock Publishing, London
García de Cortázar-Atauri I (2006) Adaptation du modèle STICS à la vigne (Vitis vinifera L.). Utilisation dans le cadre d'une étude du changement climatique à l'échelle de la France. SupAgro
García de Cortázar-Atauri I, Brisson N, Gaudilliere J-P (2009) Performance of several models for predicting budburst date of grapevine (Vitis vinifera L.). Int J Biometeorol 53:317–326
Garnier E, Daux V, Yiou P, Garcia de Cortazar I (2011) Grapevine harvest dates in Besançon (France) between 1525 and 1847: social outcomes or climatic evidence? Clim Change 104:783–801
Giomo A, Borsetta P, Zironi R (1996) Grape quality: research on the relationships between grape composition and climatic variables. In: Poni S, Peterlunger E, Iacono F, Intrieri C (eds) Proceedings of the workshop strategies to optimize wine grape quality, Acta Hort
Gladstones J (1992) Viticulture and environment. Winetitles, Adelaide
Godwin DC, White RJG, Sommer KJ, Walker RR, Goodwin I, Clingeleffer PR (2002) VineLOGIC – a model of grapevine growth, development and water use. In: Dundon C, Hamilton R, Johnstone R, Partridge S (eds) Managing water. Australian Society of Viticulture and Oenology Inc., Adelaide
Hall A, Jones GV (2010) Spatial analysis of climate in winegrape growing regions in Australia. Aust J Grape Wine Res 16:389–404

Huglin P (1978) Nouveau mode d'evaluation des possibilities heliothermiques d'un milieu viticole. CR Academy of Agriculture in France (111726)

Jackson RS (2000) Wine science: principles, practice, perception. Academic Press, San Diego

Jackson DI, Cherry NJ (1988) Prediction of a district's grape-ripening capacity using a latitude-temperature index. Am J Enol Viticult 39(1):19–28

Jackson DI, Lombard PB (1993) Environmental and management practices affecting grape composition and wine quality – a review. Am J Enol Viticult 44(4):409–430

Jackson D, Schuster D (1987) The production of grapes and wine in cool climates. Buttersworths Horticultural Books, Wellington

Jimenez J, Ruiz V (1995) Phenological development of vitis vinifera L. In castilla – La Mancha (Spain), a study of 21 cultivars (10 red and 11 white cultivars). Acta Hort 388:105–110

Jones GV (1997) A synoptic climatological assessment of viticultural phenology. Dissertation, Department of Environmental Sciences, University of Virginia

Jones GV (2006) Climate and terroir: impacts of climate variability and change on wine. In fine wine and terroir – the geoscience perspective. In: Macqueen RW, Meinert LD (eds) Geoscience Canada reprint series number 9. Geological Association of Canada, St. John's, Newfoundland

Jones GV (2007) Climate change and the global wine industry. In: Proceedings from the 13 Australian Wine Industry Technical Conference, Adelaide

Jones GV, Davis RE (2000a) Using a synoptic climatological approach to understand climate/ viticulture relationships. Int J Climatol 20:813–837

Jones GV, Davis RE (2000b) Climate influences on grapevine phenology, grape composition, and wine production and quality for Bordeaux, France. Am J Enol Viticult 51(3):249–261

Jones GV, Duchêne E, Tomasi D, Yuste J, Braslavksa O, Schultz H, Martinez C, Boso S, Langellier F, Perruchot C, Guimberteau G (2005a) Changes in European winegrape phenology and relationships with climate. Proceedings of GESCO 2005. Geisenheim, Germany

Jones GV, White MA, Cooper OR, Storchmann K (2005b) Climate change and global wine quality. Clim Change 73(3):319–343

Jones GV, Moriondo M, Bois B, Hall A, Duff A (2009) Analysis of the spatial climate structure in viticulture regions worldwide. Le Bulletin de l'OIV 82(944,945,946):507–518

Jones GV, Duff AA, Hall A, Myers J (2010) Spatial analysis of climate in winegrape growing regions in the western United States. Am J Enol Viticult 61:313–326

Jones GV, Reid R, Vilks A (2012) Climate, grapes, and wine: structure and suitability in a variable and changing climate. In: Dougherty P (ed) The geography of wine: regions, terrior, and techniques. Springer Press, Dordrecht/New York

Kiss A, Wilson R, Bariska I (2011) An experimental 392-year documentary- based multi-proxy (vine and grain) reconstruction for May-July temperatures for Közseg, West Hungary. Int J Biometeorol 55:595–611

Kriedemann PE (1968) Photosynthesis in vine leaves as a function of light intensity, temperature, and leaf age. Vitis 7:213–220

Krieger M, Lohmann G, Laepple T (2010) Climate signatures of grape harvest dates. Clim Past Discuss 6:1525–1550

Lacombe T et al (2011) Grapevine European catalogue: towards a comprehensive list. Vitis 50 (2):65–68

Le Roy LE (1971) Times of feast, times of famine: a history of climate since the year 1000. Doubleday & Company, New York

Le Roy LE, Baulant M (1980) Grape harvests from the fifteenth through the nineteenth centuries. J Interdiscipl Hist 10:839–849

Lorenz DH, Eichhorn KW, Bleiholder H, Klose R, Meier U, Weber E (1995) Phenological growth stages of the grapevine (Vitis vinifera L. sp. vinifera) – codes and descriptions according to the extended BBCH scale. Australian J Grape Wine Res 1:91–103

Loubere LA (1990) The wine revolution in France. Princeton University Press, Princeton

Mansson P-H (2001) "Wine and Health", "The Case for Red Wine", and "Wine and Health Trailblazers". The Wine Spectator, pp 32–65, Dec 15, 2001

Matthews MA, Anderson MM, Schultz HR (1987) Phenologic and growth responses to early and late season water deficits in Cabernet Franc. Vitis 26:147–160

Maurer C, Koch E, Hammerl C, Hammerl T, Pokorny E (2009) BACCUS temperature reconstruction for the period 16th to 18th centuries from Viennese and Klosterneuburg grape harvest dates. J Geophys Res 114, D22106. doi:10.1029/2009JD011730

McGovern PE (2003) Ancient wine: the search for the origins of viniculture. Princeton University Press, Princeton

McIntyre GN, Lider LA, Ferrari NL (1982) The chronological classification of grapevine phenology. Am J Enol Viticult 33(2):80–85

McIntyre GN, Kliewer WM, Lider LA (1987) Some limitations of the degree day system as used in viticulture in California. Am J Enol Viticult 38(2):128–132

Meier U (1997) Growth stages of mono- and dicotyledonous plants. BBCH-Monograph. Blackwell Wissenschafts-Verlag, Berlin

Meier N, Rutishauser T, Pfister C, Wanner H, Luterbacher J (2007) Grape harvest dates as a proxy for Swiss April to August temperature reconstructions back to AD 1480. Geophys Res Lett 34, L20705. doi:10.1029/2007GL031381

Moncur MW, Rattigan K, MacKenzie DH, McIntyre GN (1989) Base temperatures for budbreak and leaf appearance of grapevines. Am J Enol Viticult 40(1):21–26

Mullins MG, Bouquet A, Williams LE (1992) Biology of the grapevine. Cambridge University Press, Great Britain

Neethling E, Barbeau G, Bonnefoy C, Quénol H (2012) Change in climate and berry composition for grapevine varieties cultivated in the Loire Valley. Clim Res 53:89–101

Nemani RR, White MA, Cayan DR, Jones GV, Running SW, Coughlan JC (2001) Asymmetric climatic warming improves California vintages. Clim Res 19(1):25–34

Nendel C (2010) Grapevine bud break prediction for cool winter climates. Int J Biometeorol 54:231–241

Parker AK, García de Cortázar-Atauri I, van Leeuwen C, Chuine I (2011) General phenological model to characterise the timing of flowering and véraison of Vitis vinifera L. Aust J Grape Wine Res 17:206–216

Penning-Roswell E (1989) Wines of bordeaux, 6th edn. Penguin Books, London

Petrie PR, Sadras VO (2008) Advancement of grapevine maturity in Australia between 1993 and 2006: putative causes, magnitude of trends and viticultural consequences. Aust J Grape Wine Res 14:33–45

Prescott JA (1965) The climatology of the vine (Vitis vinifera L.) the cool limits of cultivation. Transcr R Soc S A 89:5–23

Ramos MC, Jones GV, Martínez-Casasnovas JA (2008) Structure and trends in climate parameters affecting winegrape production in northeast Spain. Clim Res 38:1–15

Riou C (1994) The effect of climate on grape ripening: application to the zoning of sugar content in the European community. Commission Europeenne, Luxembourg

Robinson J (1994) The oxford companion to wine, 1st edn. Oxford University Press, New York

Salinari F, Giosue S, Tubiello FN, Rettori A, Rossi V, Spanna F, Rosenzweig C, Gullino ML (2006) Downy mildew (Plasmopara viticola) epidemics on grapevine under climate change. Glob Change Biol 12(7):1299–1307

Schultz HR, Jones GV (2010) Climate induced historic and future changes in viticulture. J Wine Res 21(2):137–145

Smart RE, Dry PR (1980) A climatic classification for Australian viticultural regions. Aust Grapegrow Winemaker 196:8–16

Souriau A, Yiou P (2001) Grape harvest dates for checking NAO paleoreconstructions. Geophys Res Lett 28(20):3895–3896

Swanepoel JJ, de Villiers FS, Pouget R (1990) Predicting the date of bud burst in grapevines. S A J Enol Viticult 11(1):46–49

Tesic D, Woolley DJ, Hewett EW, Martin DJ (2002a) Environmental effect on cv Cabernet Sauvignon (Vitis Vinifera L.) grown in Hawkes Bay, New Zealand, 1. Phenology and characterization of viticultural environments. Aust J Grape Wine Res 8:15–26

Tesic D, Woolley DJ, Hewett EW, Martin DJ (2002b) Environmental effect on cv Cabernet Sauvignon (Vitis Vinifera L.) grown in Hawkes Bay, New Zealand, 2. Development of a site index. Aust J Grape Wine Res 8:27–35

Tomasi D, Jones GV, Giust M, Lovat L, Gaiotti F (2011) Grapevine phenology and climate change: relationships and trends in the Veneto region of Italy for 1964–2009. Am J Enol Viticult 62(3):329–339

Unwin T (1991) Wine and the vine: a historical geography of viticulture and the wine trade. Routledge, London

Urhausen S, Brienen S, Kapala A, Simmer C (2011) Climatic conditions and their impact on viticulture in the Upper Moselle region. Clim Change 109(3–4):349–373

van Leeuwen C, Friant P, Chone X, Tregoat O, Koundouras S, Dubourdieu D (2004) Influence of climate, soil, and cultivar on terroir. Am J Enol Vitic 55:207–217

van Leeuwen C et al (2008) Heat requirements for grapevine varieties are essential information to adapt plant material in a changing climate. In: Proceedings of the 7th International Terroir Congress, Changins, Switzerland. Agroscope Changins-Wädenswil: Switzerland

Webb LB, Whetton PH, Barlow EWR (2007) Modelled impact of future climate change on the phenology of winegrapes in Australia. Aust J Grape Wine Res 13:165–175

Webb LB, Whetton PH, Barlow EWR (2011) Observed tends in winegrape maturity in Australia. Glob Change Biol 17:2707–2719

Winkler AJ, Cook JA, Kliewere WM, Lider LA (1974) General viticulture, 4th edn. University of California Press, Berkley

Wolf TK (1997) Site Selection for Commercial Vineyards. Virginia Agricultural Experiment Station, Winchester, Virginia, Publication Number 463–016

Zalom FG, Goodell PB, Wilson WW, Bentley WJ (1983) Degree-days: the calculation and the use of heat units in pest management, Leaflet no. 21373. Division of Agriculture and Natural Resources, University of California, Davis

Chapter 31
Phenology in Higher Education: Ground-Based and Spatial Analysis Tools

Kirsten M. de Beurs, Robert B. Cook, Susan Mazer, Brian Haggerty, Alisa Hove, Geoffrey M. Henebry, LoriAnne Barnett, Carolyn L. Thomas, and Bob R. Pohlad

Abstract New spatial analysis methods and an increasing amount of remote sensing data are the necessary tools for scaling from ground-based phenological measurements to larger ecosystem, continental, and global processes. However, since remote sensing data and tools are not straightforward to master, training at the higher education level is often necessary. Curricula and training programs linking these integral components of phenological research are sorely needed because the number of people with requisite skills in the use of a growing array of sophisticated analytical tools and collected remote sensing data is still quite small. In this chapter we provide a series of examples of field-based approaches to college- and

K.M. de Beurs (✉)
Department of Geography and Environmental Sustainability, The University of Oklahoma, Norman, OK 73019, USA
e-mail: kdebeurs@ou.edu

R.B. Cook
Environmental Sciences Division, Oak Ridge National Laboratory, Oak Ridge, TN 37831, USA

S. Mazer • B. Haggerty
Department of Ecology, Evolution, and Marine Biology, University of California, Santa Barbara, CA 93106, USA

A. Hove
Biology Department, Warren Wilson College, Asheville, NC 28815, USA

G.M. Henebry
Geographic Information Science Center of Excellence, South Dakota State University, Brookings, SD 57007, USA

L. Barnett
Education Program, USA National Phenology Network, National Coordinating Office, Tucson, AZ 85721, USA

C.L. Thomas • B.R. Pohlad
Ferrum College, School of Natural Sciences and Mathematics, Ferrum, VA 24088, USA

M.D. Schwartz (ed.), *Phenology: An Integrative Environmental Science*,
DOI 10.1007/978-94-007-6925-0_31,

university-level phenological education. We then guide the reader through the resources that are available for the integration of remote sensing with land-based phenological monitoring and suggest potential ways of using these resources.

31.1 Introduction

In 1974, Forest Stearns wrote a chapter on "Phenology and Environmental Education" for the book "Phenology and Seasonality Modeling" edited by Helmut Lieth. Stearns remarked on the utility of phenology as a theme for every level of education because it helps to make the complexity of interactions between organisms and their environment relevant to students (Stearns 1974). However, he found that there were no articles on phenology in *American Biology Teacher* and only one in *Science Teacher*, the two leading science education journals of the day. Almost 40 years later, the interest in phenology among educators, ecologists, botanists, zoologists, geographers, climatologists, and amateur naturalists has increased greatly. As a result, many more phenological education resources are now available online, including indoor and outdoor activities for audiences ranging from K-12 to college students to the general public. In the United States, for example, some agencies and institutions tasked with educating students and the public about climate change have been developing phenology-themed programs, many of which make use of the growing number of internet-based public participation programs. Moreover, phenological education provides a unique opportunity for participants to explore the relationship between science, nature, and themselves. These educational programs help students develop critical thinking skills and an understanding of how science and the natural world affect their daily lives. The program in the U.S. that facilitates the collection, reporting, and interpretation of phenological data among public participation programs and across federal, non-governmental, and academic institutions is the USA National Phenology Network (USA-NPN; www.usanpn.org).

Educational objectives of the USA-NPN include recruiting and training scientists, students, teachers, outdoor educators, land stewards, and community members to contribute accurate phenological data to the growing national database and to participate in its quantitative analysis and interpretation (USA-NPN National Coordinating Office 2012). Other outcomes for participants include understanding the importance of phenology as an indicator of the health of our environment, spending more time engaged in nature, and becoming scientifically literate by engaging in scientific data collection, analysis, and interpretation. A variety of education and training materials are available on the USA-NPN's Education website (www.usanpn.org/education), both specific to its on-line data collection interface, *Nature's Notebook*, as well as to other phenology programs, including a suite of "phenological and climate change literacy" resources designed specifically for K-12, college, and public audiences, created by the Phenology Stewardship Program at the University of California, Santa Barbara (Mazer lab, Department of Ecology, Evolution and Marine Biology). The USA-NPN Education Program is also creating a suite of site-based engagement materials for facilitating community-based phenological education programs. Additional examples of well-known and successful public participation programs include the international Global Learning and Observations to Benefit the Environment (GLOBE) program, Canada's PlantWatch, Project Budburst (USA), and the California Phenology Project (described later in this chapter). Each of these programs offers materials for educators to incorporate the monitoring of plant phenology into the activities of traditional and non-traditional classrooms.

Since phenology is an integrative science drawing from numerous scientific disciplines, there are many topics in environmental science with which to engage students and the public in learning phenology. The interdisciplinary nature of phenological research challenges current and new generations of scientists, educators, and students to learn a variety of Science, Technology, Engineering, and Mathematics (STEM) skills across several disciplines. However, resources for educating certain audiences (e.g., college, university) and disciplines (e.g., remote sensing, spatial analysis) are currently limited. This is especially unfortunate because, while there has been considerable attention paid to promoting the recording of phenological observations by younger students (grades K-12) and citizen scientists, the number of people with requisite skills to use the growing array of sophisticated analytical tools and collected data is still quite small (Dickinson et al. 2010).

New spatial analysis methods and an increasing amount of satellite data are the necessary tools for using ground-based measurements to make inferences about larger ecosystem, continental, and global processes. However, these analytical tools and data sets are not straightforward to master and training is often necessary (Dickinson et al. 2010). Although a small suite of educational and training materials has been developed to address the gaps for undergraduate and post-graduate audiences (described in the field based section below: Haggerty and Mazer 2009; Haggerty et al. 2012a, b; Hove et al. 2012a, b), these resources do not directly link ground-level phenological observations with remotely sensed measurements. Both curricula and training programs to link these integral components of phenological research are sorely needed.

In this chapter we provide examples of field-based, phenology-themed college- and university-level education and then guide the reader through resources available for the integration of remote sensing with land-based phenological monitoring. We also suggest potential ways of using the resources. As there are many resources available, it is difficult to be comprehensive. The resources discussed here focus primarily on ongoing initiatives in North America that have English language web links. While we are aware of some interesting initiatives going on in other parts of the world, many of their websites are not available in English.

31.2 Field-Based Activities and Examples

To engage all components of the public in nationwide (and international) phenological monitoring efforts, it is essential to: expand educational efforts at the college and university level; provide training in the interpretation of phenological data and its link to climate change; expand public outreach efforts; demonstrate real-world applications of phenology and its relevance to career training and choices; and include phenological training for pre-service and active teachers enrolled in credential or professional development programs. Undergraduate training programs should provide greater breadth and depth particularly in the STEM skills necessary to train the next generation of scientists interested in the intersection of biology, ecology, geography, and climatology. To demonstrate the relationships among these disciplines, undergraduate interdisciplinary programs should promote and illustrate the use of long-term phenological data in scientific research. Overall, there are far fewer phenological educational settings aimed at training young scientists at the undergraduate level than K-12 phenology programs. However, over the past few years, college-based phenology programs have started to develop (Chen 2003; Long and Wyse 2012); both the USA-NPN and the CPP now offer materials appropriate for college settings (http://www.usanpn.org/cpp/education).

Three examples of classroom-based phenology training materials recently designed for undergraduate and post-graduate audiences include: the *Phenology Handbook* (Haggerty and Mazer 2009); the *Primer on Herbarium-Based Phenological Research* for tracking historical trends in flowering phenology (Haggerty et al. 2012a), which includes a sample data set and a guided spreadsheet-based analytical exercise (Haggerty et al. 2012b); and a suite of annotated undergraduate lectures and seminar modules designed to guide discussions of the primary research literature (Hove et al. 2012a, b). All of these materials, and others for K-12 and public audiences, are described and freely available online at the California Phenology Project's education website (www.usanpn.org/cpp/education). These materials may help educators provide a useful entry point into phenological research for their students by introducing the motivation for phenological monitoring, its history in the U.S. and elsewhere, the ecological interpretation of long-term phenological shifts, the links between phenology and climate change, the protocols of the USA-NPN, and the botanical observational skills necessary for the accurate reporting of the phenological status of individual plants.

31.2.1 California Phenology Project

The California Phenology Project (www.usanpn.org/cpp) is the first statewide effort to assess the effects of climate change on California's diverse landscapes. Established in 2010 with funding from the National Park Service Climate Change Response Program, the project includes a citizen science program that contributes data directly to USA-NPN's *Nature's Notebook*. The CPP takes a "train-the-trainer" approach, where training workshops are delivered primarily for participants who will themselves deliver continued training sessions for their staff, for place-based volunteers dedicated to monitoring geo-referenced and labeled plants at particular national parks, and for the public. The long-term aims of the CPP include: testing and refining USA-NPN protocols for ground-based phenological monitoring; creating tools and infrastructure to establish long-term phenological monitoring in California national parks, University of California Natural Reserve System, and public lands; informing decisions routinely made by land managers that depend on the timing of phenological events; and developing materials to help park staff and educators to communicate to the public how the seasonal cycles of natural resources in the parks are affected by inter-annual variation in climate and by climate change.

The CPP was funded initially to establish phenological monitoring programs in seven national parks and recreation areas in California. These pilot parks range from those with high visitation from nearby urban populations (e.g., Santa Monica Mountains National Recreation Area [NRA], Golden Gate NRA, and John Muir National Historic Site) to remote parks with relatively little visitation and a small community of volunteers (e.g., Lassen Volcanic National Park). The pilot parks were also selected to represent a wide range of ecosystems and biogeographic regions. Coastal region chaparral, prairie, and forests are represented by Santa Monica Mountains NRA, Golden Gate NRA, John Muir National Historic Site, and Redwood National and State Parks. Montane plant communities are represented by Lassen Volcanic National Park and Sequoia & Kings Canyon National Parks. Desert communities are represented by Joshua Tree National Park. Several of these parks work actively with outdoor schools and/or local teachers (e.g., NatureBridge, the National Park Service "Parks as Classrooms" and Teacher-Ranger-Teacher programs- www.nps.gov/learn/trt/, and regional school districts) to incorporate phenological monitoring into residential or single-visit programs.

The CPP offers materials and phenology-themed lesson plans to guide educators using a variety of approaches, including seminar modules (appropriate for advanced undergraduate or graduate education, and including guided discussions of the primary literature), practical instructions for the use of herbarium specimens to detect historical changes in phenology (targeted towards undergraduates and adult citizen scientists), annotated lectures (to introduce university students and citizen scientists to the study of phenology and its link to climate and climate change), hands-on interactive activities (appropriate for middle-school through adult education), outdoor activities (for middle school through adult education), and a step-by-step guide for the construction and use of native plant gardens designed for phenological monitoring.

Plant species adapted to Mediterranean, semi-arid, and arid environments, such as the California chaparral, high- and low-elevation deserts, and the southwestern U.S. present a number of challenges for phenological monitoring whether on-the-ground or remotely sensed. For example, compared to the highly seasonal temperate zone, where the onset of an individual plant's growing season can be clearly defined by the opening of large, dormant, well-protected winter buds and the subsequent synchronous emergence of young leaves, the growing season for many species adapted to Mediterranean, frost-free, and desert environments is poorly defined. Many species of long-lived perennials, shrubs, and trees adapted to these habitats are semi-deciduous and do not produce visible vegetative buds that remain dormant until the growing season is initiated. Rather, they produce leaves opportunistically from meristematic tissue (typically located in the axils of leaves or at stem tips) whenever conditions allow, usually after precipitation events. Consequently, a "growing season" of such plants is an episodic process that can recur multiple times from spring through autumn. The total length of the growing season may be determined as much by the temporal distribution of rainfall events as by the total amount of rain occurring during the wet season. Compared to the temperate zone, where the end of the growing season is determined by the onset of cool temperatures and the timing of the first frost, the termination of annual growth is not well-defined in frost-free environments of the western and southwestern U.S.

Consequently, participants in phenological monitoring networks who observe plants in these environments must be counseled that the onset of vegetative growth may occur multiple times throughout the year, such that capturing all of these events requires an extended period of vigilance that is not essential in more seasonal environments. When the California Phenology Project offers training events focusing on these species (e.g., *Adenostoma fasciculatum*, *Baccharis pilularis*, *Coleogyne ramosissima*, *Eriogonum fasciculatum*, *Larrea tridentata*, and *Mimulus auriantiacus*), instructors take care to point out the locations on plants (apical meristems) where new growth (both vegetative and reproductive) occurs. This training helps participants to seek evidence for newly occurring leaf growth in the absence of large, conspicuous buds.

31.2.2 Phenology Gardens and Trails

Phenology gardens are planned landscapes established for the purpose of monitoring plant and animal phenology. Phenology gardens have been planted and monitored by agricultural climatologists since the 1950s to study and better predict growing seasons across Europe (International Phenology Gardens) and across the U.S. (Lilac Phenology Network). Although established with short-term interests in mind, accumulated data from these projects have helped to form the foundation of knowledge about long-term phenological responses to climate change in Europe (Menzel 2000; Chmielewski and Rotzer 2001) and the U.S. (Cayan et al. 2001; Primack and Miller-Rushing 2009). With careful planning such gardens are able to distinguish environmentally-induced from genetically-based variation in phenology and to detect genetic variation in the phenological response to climate change.

More recently, the educational values of phenology gardens are being identified and developed. In *Phenology Gardens: a practical guide for integrating phenology into garden planning and education* (Haggerty et al. 2012c), the authors provide conceptual background and suggestions for planning and establishing phenology gardens to maximize scientific, educational, ecological, and societal goals (including aligning gardens with USA National Phenology Network programs). They also provide case studies from four native plant phenology gardens established in southern coastal California in collaboration with the USA-NPN, US Geological

Survey, and US Fish & Wildlife Service. A network of more than 15 phenology gardens, each containing a similar suite of native plant species, has been established at schools, community centers, and universities in the region, largely due to continued efforts by the Ventura office of the US Fish & Wildlife Service. Phenological data collected in the gardens have been reported to the USA-NPN's *Nature's Notebook*.

The USA-NPN also has developed an implementation guide for creating Phenology Trails, *The Phenology Trail Guide: An experiential education tool for site-based community engagement* (USA-NPN National Coordinating Office 2012). Phenology Trails are networks of *Nature's Notebook* observation sites linked together to provide participants places to visit, enjoy nature, collect data, and learn about supporting organizations and their efforts related to phenological research. Such trails further serve the purpose of collective engagement, meaningful learning, and development of community.

Phenology gardens and trails can be established relatively easily in K-12, college, or public settings, thereby creating an outdoor classroom that can be revisited easily over time for phenological monitoring and basic instruction in plant biology. With some planning, phenology gardens can be integrated into many curricular goals and topics for all ages. Thus, phenology gardens provide a valuable setting for developing a broad range of concepts and skills spanning STEM, humanities, and fine arts topics. As such, phenology gardens can provide a single unifying platform for integrative environmental education. For example, Haggerty, Hove, and Mazer developed several activities that use USA-NPN protocols, including *Flight of the Pollinators* (Haggerty et al. 2012d), *Ethnophenology* (Haggerty et al. 2012e), *Phenology Relay Race* (Haggerty 2012a), and standards-aligned lesson plans (Haggerty 2012b). These resources are described and available on the California Phenology Project's education website (www.usanpn.org/cpp/education).

31.2.3 *Phenology at Ferrum College*

As an illustration we discuss how the study of phenology is integrated into science instructions at Ferrum College (www.ferrum.edu), a small undergraduate college in Virginia. Faculty at Ferrum College have introduced phenology into horticulture and environmental sciences courses by using the Project BudBurst website to teach phenology and climate change. Here are two examples of class assignments.

31.2.3.1 Making Phenological Observations of Campus Species

This module was used in Introductory Horticulture (HOR/AGR 219) at Ferrum College in the spring semester of 2010. This introductory class targets both horticulture and agriculture majors. The purpose of the exercise is to increase student awareness of plant cycles by using the plants in the Ferrum Community Arboretum for observation. Horticulture and agriculture students must make decisions on plant management based upon predicted climate change. These decisions may, for example, include planting dates, germination rate, and pest management. By completing this exercise and monitoring plant phenological schedules, students can begin to understand the complexities of plant cycles as they relate to climate as well as the growth and management of plants in controlled settings. By entering their data into the national database, students in subsequent courses can use these data to discuss phenological trends that may be related to climate change.

In lab, students were told they would be doing an exercise on the phenology of plants. Each student was given a pre-test to determine their familiarity with phenology, plant identification, and plant growth as it relates to climate change. Students were then given a handout obtained from the Project Budburst site and asked to review the information. Following a discussion which focused on phenology of the plant species growing on campus and plant life cycles, the class went to the computer laboratory to log onto the Project BudBurst website (neoninc.org/budburst). After reading the introductory information, they were instructed to enter the online BudBurst guide.

31.2.3.2 Using Project Budburst to Study the Effects of Climate Change on Plant Phenology

The goal of the module developed for this project was to acquaint students with the impacts of climate change on the seasonal biological cycles by having them contribute to a worldwide phenological study. The students recorded and reported their phenological observations using the Project BudBurst website.

Student teams (two students) in *Environmental Sciences and Issues in Appalachia* (ESC 110) participated in this project by choosing a plant (tree, shrub, or annual plant) from the Project BudBurst list of plants of interest. Students made twice-weekly observations and took photos with a digital camera or their cell phones for 7 weeks during the spring semester of 2010 at Ferrum College. Student observations and photos were uploaded to ANGEL, the learning management system at Ferrum College. At the end of the observation period (end of April), each student team had approximately 14 observations and photos. Some student teams did not observe first leaf and most did not observe first bud because of the timing of the spring semester relative to the onset of spring in Ferrum, Virginia (37°59′N, 79°59′W, elevation = 437 m) which starts at the very end of the spring semester. The student collected observations and photos were collated and summarized at the end of the class project by the student project assistant. The data were summarized by plant species and phenophase and then entered into the Project BudBurst database by the student project assistant. Assessment of these class activities was accomplished by administering a pre-test and a post-test and making note of number of observations and photos.

31.3 Remote Sensing

There is a place in college-level classes for the analysis of satellite imagery without the need to provide more than basic knowledge in remote sensing. Students can be introduced to basic remote sensing principles and the tools to download remotely sensed data. In addition, they can learn how to interpret image time series with respect to phenological monitoring. Several resources are available for downloading and processing of satellite data. In this section we limit our discussion to resources that focus directly on links between remote sensing and phenological data.

Land surface phenology (LSP) explores how quasi-periodic events in terrestrial vegetation (e.g., budburst, leaf out, flowering, senescence) appear when observed by remote sensing technologies. LSP can be studied by means of vegetation indices calculated from optical sensors, such as the Advanced Very High Resolution Radiometer (AVHRR) polar orbiting sensors as well as the newer Moderate Resolution Imaging Spectroradiometer (MODIS) sensors on the Terra and Aqua satellites which provide higher spatial resolutions (250–1000m) than the older but still operational AVHRR series (1–8 km). Land surface phenological metrics are primarily based on image time series of vegetation indices (VI). These phenological metrics aim to retrieve onset of greening, timing of the peak of the growing season, senescence, and the growing season length based on analysis of the VI curve (Reed et al. 1994; Zhang et al. 2003, 2004; White et al. 2009; de Beurs and Henebry 2010a). Vegetation indices such as the Normalized Difference Vegetation Index (NDVI) and the Enhanced Vegetation Index (EVI) are surrogate measures for aboveground net primary production, and in recent years, considerable effort has been made to link the global NDVI variability to temperature, precipitation, and atmospheric CO_2 (Myneni et al. 1997; Tucker et al. 2001; Lee et al. 2002; Dye and Tucker 2003; Zhou et al. 2003).

As such, LSP provides an important method for detecting responses to climate change in terrestrial ecosystems. Changes in LSP (sometimes erroneously called "greenness") have often been detected as trends in NDVI products over multiple years (e.g. Dye and Tucker 2003; Beck et al. 2006; Julien et al. 2006; Bradley et al. 2007; Potter et al. 2007; de Beurs et al. 2009; White et al. 2009; de Beurs and Henebry 2010b). These changes in the timing or intensity of the phenological signal over multiple years are frequently interpreted as resulting directly from climate change, particularly warming or droughts (Potter and Brooks 2001; de Beurs et al. 2009; Brown et al. 2010). Thus, there is great potential for college and graduate training in LSP and remote sensing to contribute to new generations of climate change scientists.

Although college-level geography classes generally provide students with an introduction to remote sensing and training using remote-sensing software packages such as ERDAS Imagine, PCI Geomatics, or Exelis ENVI, remote sensing classes do not necessarily discuss land surface phenology and how to link these observations with ground level observations of individual plants, populations, or communities. This omission is especially unfortunate given the number of resources that are available to facilitate the use and analysis of long-term image time series. Moreover, even courses that are not specifically focused on remote sensing may now introduce the analysis of satellite imagery in their programs. The tools and data sources discussed in the last section of this chapter are provided as a resource for instructors aiming to enhance their college-level classes by training students not only to collect and to interpret ground level observations, but also to link these observations with satellite observations (Table 31.1).

31.3.1 USGS Remote Sensing Phenology

The USGS website Remote Sensing Phenology provides an overview of how remote sensing can be used to monitor phenology (http://phenology.cr.usgs.gov/index.php). This resource also provides a short explanation of data, sensors, and methods available, and offers the opportunity to download remotely sensed phenological indicators at 1km spatial resolution over the Continental United States (CONUS) derived from AVHRR data. The data are available from 1989 to 2010. The downloaded files are available as png images, which can be viewed in any image viewing program, and as flat binary files in band sequential format that can be opened with specialized remote sensing programs such as Exelis ENVI, ERDAS Imagine, or PCI Geomatica, as well as ESRI ArcMap. While there are no specific educational modules available, the data could be incorporated easily into college classes.

31.3.2 ORNL DAAC (Distributed Active Archive Center for Biogeochemical Dynamics)

The National Laboratory Distributed Active Archive Center (ORNL DAAC; daac.ornl.gov/MODIS/modis.html) provides users with easy-to-use data files in text or GeoTIFF format of MODIS phenology products for any site on land, user-selected area from one pixel up to 200 $\times$ 200 km, and time period between 2000 and the present (Table 31.2; SanthanaVannan et al. 2009). In addition, the subsets are online in interactive time series plots that reveal the timing of vegetation events. The products are thus especially well-suited to be used in a college course on the monitoring of land surface phenology. Students are able to select a defined research area using a Google Map interface and then to download subsets of satellite data up to 4,000 km^2. The ORNL DAAC offers the download of most basic MODIS products including vegetation indices (MOD13Q1), leaf area index (LAI), fraction of photosynthetically active radiation (FPAR) (MOD15A2), and gross primary productivity (MOD17A2). The selected data can be downloaded as GeoTIFF files in their original sinusoidal projection. These images can be incorporated into any remote sensing program and ESRI's ArcMap. In addition

Table 31.1 Available resources for land surface phenology in education

	Websites
Remote sensing data sources	
USGS Remote sensing phenology	http://phenology.cr.usgs.gov/index.php
ORNL DAAC Land product subsets	http://daac.ornl.gov/MODIS/modis.shtml
Web-enabled Landsat data	http://weld.cr.usgs.gov/
Vegetation index and phenology	http://measures.arizona.edu/MODIS_Project.php
Phenocam network	http://phenocam.unh.edu/webcam/
Real time phenology monitoring	http://tethys.dges.ou.edu/Twofiles/ and http://tethys.dges.ou.edu/EVI/
Software	
Season	http://tethys.dges.ou.edu/
Timesat	http://www.nateko.lu.se/timesat/timesat.asp

Table 31.2 Spatial and temporal resolution for MODIS products that provide vegetation phenology

Products	*Short name*	*Nominal spatial resolution (meters)*	*Temporal resolution (days)*
Land cover dynamics	MOD12Q2	500	8
Vegetation indices	MOD13Q1	250	16
Leaf area index and fPAR[a]	MOD15A2	1,000	8
Gross primary productivity	MOD17A2	1,000	8

From https://lpdaac.usgs.gov/products/modis_products_table (Accessed on November 12, 2012)
[a]*fPAR*: fraction of photosynthetically active radiation

to imagery, the ORNL DAAC tool provides a good overview of the selected data within an easily navigable interface. The website also provides a general overview of the land surface phenology (using the MOD12 land surface dynamic product) for the selected area.

The *Remote Sensing and Phenology* class at the University of Oklahoma, co-taught by de Beurs and Hobson, teaches students how to use the ORNL DAAC data. After learning about the MODIS sensors, the quality of the information that they provide, and the different products that are available, the students spend several weeks investigating land surface phenology around the globe using data from the ORNL DAAC. In one exercise, the students are asked to select satellite data from one year for three different areas of the same biome around the world. They use the Season program (described below) to explore the land surface phenology visible in the satellite data and they learn how the land surface phenology can differ within biomes, depending on its location. The students summarize their findings in PowerPoint slides that are discussed collectively in class. Another exercise has the students compare the land surface phenology from different biomes. The students investigate the changes in land surface phenology for these biomes over multiple years. The students in the *Remote Sensing and Phenology* class also conduct ground level phenological monitoring in a small forest on the campus of The University of Oklahoma. This university forest, called Oliver's

Woods, contains a trail with more than 100 geo-referenced trees representing more than 15 different species (oudaily.com/news/2012/may/01/class-collects-information-plant-growing-seasons). Students link these ground-based observations with the MODIS data retrieved from the ORNL DAAC.

31.3.3 Web-Enabled Landsat Data (WELD)

The NASA-funded WELD project provides Landsat 7 ETM + terrain corrected mosaics at several temporal resolutions for the conterminous United States and Alaska. The Landsat 7 images offer 30-m resolution, which is finer than the MODIS (250–500 m) or AVHRR (1–8 km) data that are frequently used for phenological monitoring. In addition to top of atmosphere (TOA) reflectance data, the project delivers NDVI data and several quality flags. The 'what you see is what you get' ordering system allows the ordering of custom size mosaics that are delivered as geotiff data and can easily be incorporated into remote sensing software or ESRI ArcMap. In addition to the ordering of image mosaics, the systems allows for the ordering of time series from a specific 30 m by 30 m area. These data are delivered as text files that can be analyzed easily in spreadsheet software. This last option is particularly interesting for those who do not need entire image mosaics. One interesting exercise for students is to investigate the land surface phenology over a given land cover type across a north–south gradient, for example by selecting one pixel per state from North Dakota to Texas. The data can be plotted in spreadsheet software to investigate the effect of latitude on the development of land surface phenology. Another potential exercise is to compare the land surface phenology of different crop types (e.g., spring wheat, soybeans, corn, and winter wheat), within and across latitudes or elevations.

31.3.4 PhenoCam Network

The PhenoCam network is a continent-wide monitoring effort in North America (focused in the U.S.) that provides automated, near-surface remote sensing through the use of high-resolution webcams. The images are uploaded to the PhenoCam website every half hour (Richardson et al. 2007; www.oeb.harvard.edu/faculty/richardson/phenocam.html) and are available for educational and scientific use after registration (see Chap. 22 in this volume). Although the PhenoCam network does not offer any specific educational opportunities, it is not difficult to incorporate these images into college-level education. For example, when learning how to interpret land surface phenology signals, it is often beneficial for students to observe how each vegetation index corresponds to the phenological progress of vegetation on the ground. While a ground-based investigation close to campus is often ideal, students may supplement these data with observations from other ecosystems, such as those monitored by the phenocams.

31.3.5 Real Time Phenological Monitoring Application

If we are interested in the analysis of land surface phenology and the coupling of these remotely sensed observations with ground observations, it may not be necessary to learn specialized (and relatively expensive) remote sensing software packages. A freely available online application (tethys.dges.ou.edu/Twofiles/) was developed by de Beurs to enable the monitoring of land surface phenology over North America. The application allows for the selection of one 0.05° by 0.05° area based on a Google Earth map of any location in North America. Upon selection of an area, the application then reveals the average land surface phenology of all available years (currently 2000–2011) based on phenology derived from the MODIS Nadir BRDF-Adjusted Reflectance (NBAR; MOD43C3). The user can request standard errors for the NDVI values based on all years to better understand temporal variability, and then compare the phenological cycle of the current year (now 2012) to that of previous years (Fig. 31.1). The user may calculate the start and end of the growing season based on satellite data by experimenting with different thresholds for the midpoint NDVI method (White et al. 1997). This method was found to be one of the most accurate methods in a large land surface phenology model evaluation study (White et al. 2009). The users can explore the effect of the different thresholds for the estimation of the start and the end of the growing season. Data can be exported as jpeg files and the time series can be transferred to spreadsheets for subsequent analysis. A similar application is available to compare the standard vegetation indices NDVI and EVI.

31.3.6 Season Software

The Season program is a freely available custom developed software package developed by de Beurs (Fig. 31.2) that allows colleges and universities to adopt remote sensing in classes that address phenology. It can ingest and process data from the ORNL DAAC and it provides a range of widely used methods to estimate the start and the end of the growing season. This flexibility enables students to experiment with a range of different methods to enhance their understanding of the differences among methods. Students can investigate individual grid cells and output the information into basic spreadsheet programs or as jpg files that can be incorporated into written reports (Fig. 31.2). They are also able to investigate a range of MODIS products including the basic vegetation indices NDVI and EVI, but also LAI, FPAR, and MODIS land surface temperature. This software was developed for de Beurs' *Remote Sensing and Phenology* classes at Virginia Tech and at The University of Oklahoma. The program also has been successfully used in professional workshops on land surface phenology at conferences of the US Chapter of the International Association for Landscape Ecology (US-IALE) in 2008, 2010, and 2012.

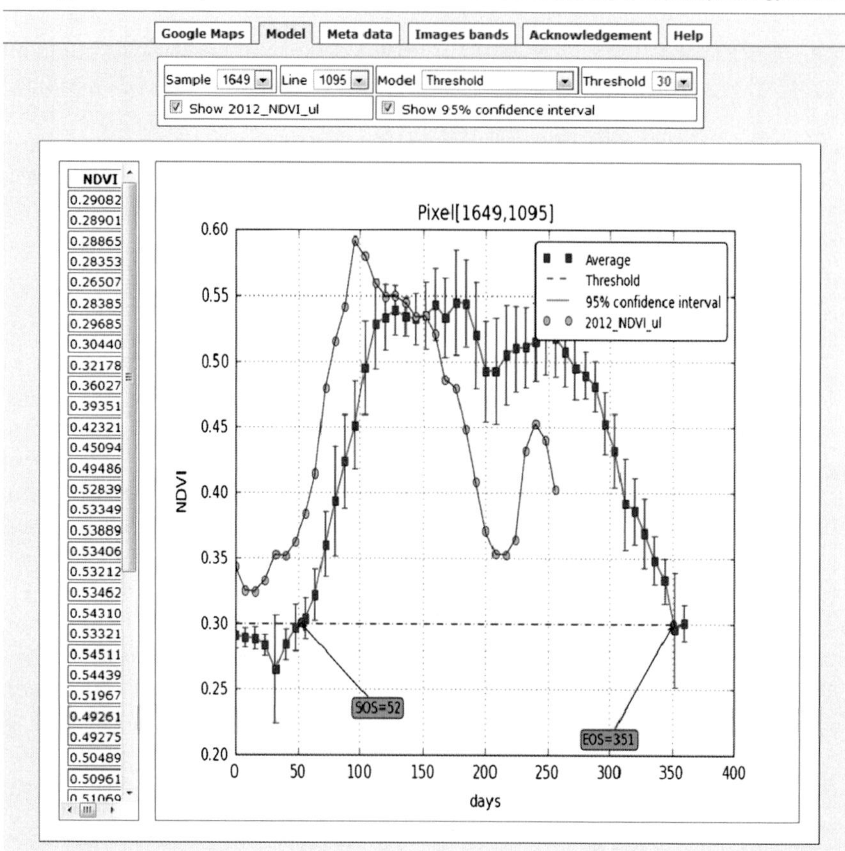

Fig. 31.1 Web application to monitor the phenology based on MODIS BRDF (MOD43C3) data. The figure shows the average NDVI between 2000 and 2011 for The University of Oklahoma in Norman (OK) in *blue*. The *yellow dots* give the current ongoing vegetation development for 2012. It is easy to see that the spring in 2012 has developed more quickly relative to the average of the previous 12 years (The data were downloaded from USGS's Land Processes Distributed Active Archive Center: https://lpdaac.usgs.gov/)

31.3.7 Timesat Software

Timesat is another freely available software package that is specifically developed for the analysis of image time series to investigate phenology. Timesat allows for the smoothing of satellite image time series using Savitzky-Golay filtering, asymmetrical Gaussian, or double logisitic functions (Jonnson and Eklundh 2002, 2004). The package has been used in a range of scientific studies (Verbesselt et al. 2006; Huemann et al. 2007; Gao et al. 2008), including the development of the MODIS North American Carbon Program phenology product (accweb.nascom.nasa.gov/)

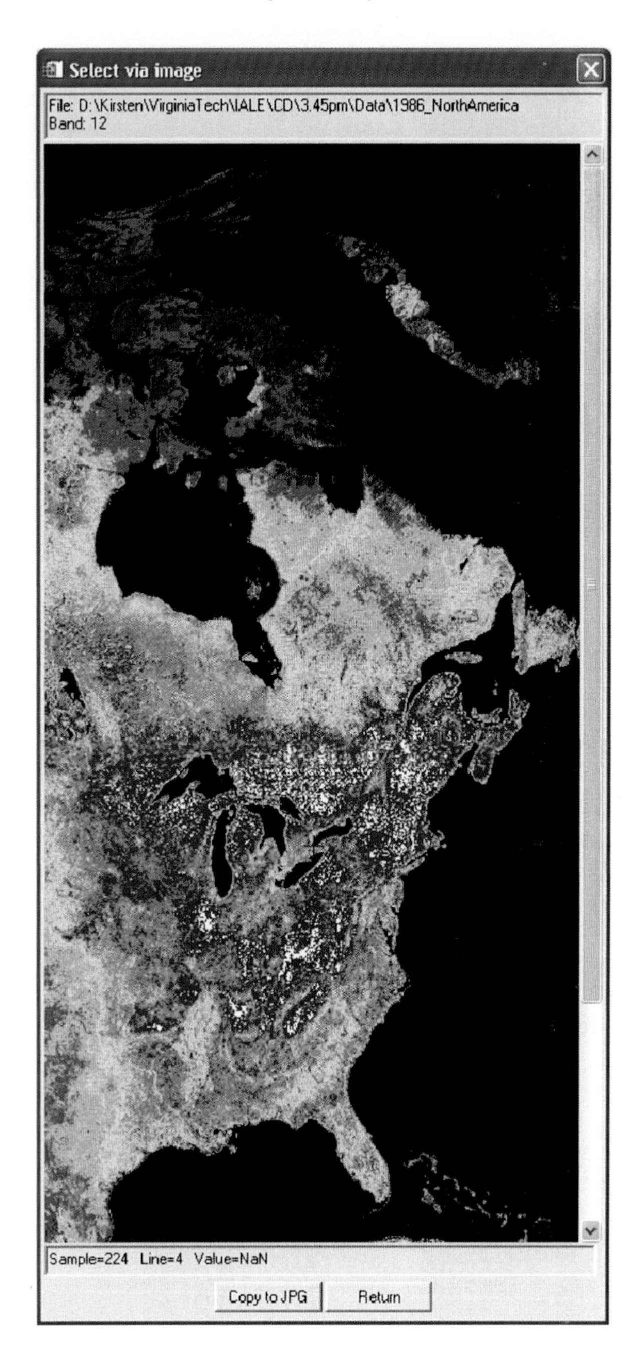

Fig. 31.2 Software to determine land surface phenology variables such as start and end of season

that is currently only available through 2008. Output of the program includes phenological metrics such as the beginning and end of the growing season and smoothed versions of the original input data. While the TIMESAT software is freely available, it is not appropriate for introductory level college courses since there is a relatively steep learning curve. In addition, specialized remote sensing software, such as Exelis ENVI or ERDAS Imagine is necessary to create the input data for TIMESAT. More advanced students, however, may enjoy experimenting with the program.

31.4 Conclusions

The interdisciplinary nature of phenological research challenges current and new generations of scientists, educators, and students to learn a variety of Science, Technology, Engineering, and Mathematics (STEM) skills across several disciplines with the goal of developing the critical thinking skills necessary for understanding human relationships to the natural world. There is an increasing interest in training college and university students in remote sensing and spatial analytical methods. Likewise, working with satellite data and spatial analyses are essential skills for researchers wishing to extrapolate landscape-level processes from ground-based phenological measurements. Curricula and training programs linking these vital aspects of phenological research are just beginning to develop but are sorely needed and can provide post-secondary students engagement in real-world projects, thus informing career choice.

This chapter has discussed a number of resources for training students to take field measurements, such as those provided by Project Budburst and the USA National Phenology Network, as well as tools focused on the analysis of land surface phenology data.

Acknowledgements We would like to acknowledge Elisabeth Beaubien for her substantial contributions to an earlier version of this chapter.

References

Beck PSA, Atzberger C, Hogda KA, Johansen B, Skidmore AK (2006) Improved monitoring of vegetation dynamics at very high latitudes: a new method using MODIS NDVI. Remote Sens Environ 100(3):321–334

Bradley BA, Jacob RW, Hermance JF, Mustard JF (2007) A curve fitting procedure to derive intern-annual phenologies from time series of noisy satellite NDVI data. Remote Sens Environ 106:137–145

Brown ME, de Beurs K, Vrieling A (2010) The response of African land surface phenology to large scale climate oscillations. Remote Sens Environ 114(10):2286–2296. doi:10.1016/j.rse.2010.05.005

Cayan DR, Kammerdiener SA, Dettinger MA, Caprio JM, Peterson DH (2001) Changes in the onset of spring in the western United States. Bull Am Meteorol Soc 82:399–415

Chen X (2003) East Asia. In: Schwartz MD (ed) Phenology: an integrative environmental science. Tasks for vegetation science, vol 39. Kluwer, Dordrecht

Chmielewski FM, Rotzer T (2001) Response of tree phenology to climate change across Europe. Agr For Meteorol 108:101–112 de Beurs KM, Henebry GM (2010a) Spatio-temporal statistical methods for modelling land surface phenology. In: Hudson IL, Keatley MR (eds) Phenological research: methods for environmental and climate change analysis. Springer, Dordrecht. doi:10.1007/978-90-481-3335-2_9

de Beurs KM, Henebry GM (2010b) A land surface phenology assessment of the northern polar regions using MODIS reflectance time series. Can J Rem Sens 36:S87–S110

de Beurs KM, Wright CK, Henebry GM (2009) Dual scale trend analysis distinguishes climatic from anthropogenic effects on the vegetated land surface. Environ Res Lett 4:045012

Dickinson JL, Zuckerberg B, Bonter DN (2010) Citizen science as an ecological research tool: challenges and benefits. In: Futuyma DJ, Shafer HB, Simberloff D (eds). Annu Rev Ecol Evol System 41:149–172. Annual Reviews, Palo Alto. doi:10.1146/annurev-ecolsys-102209-144636

Dye DG, Tucker CJ (2003) Seasonality and trends of snow-cover, vegetation index, and temperature in northern Eurasia. Geophys Res Lett 30(7):1405. doi:10.1029/2002GL016384

Gao F, Morisette JT, Wolfe RE, Ederer G, Pedelty J, Masuoka E, Myneni R, Tan B, Nightingale JM (2008) An algorithm to produce temporally and spatially continuous MODIS-LAI time series. IEEE Trans Geosci Remote Sens 5(1):1–5

Haggerty BP (2012a) Phenology relay race. Activity guide for K-16 and public audiences reinforcing phenological monitoring protocols of the USA National Phenology Network. Available online at the California Phenology Project website (www.usanpn.org/cpp/education)

Haggerty BP (2012b) Phenology gardens: lesson plans I & II. Standards-aligned lesson plans for K-12 audiences to learn USA National Phenology Network monitoring protocols. Available online at the California Phenology Project website (www.usanpn.org/cpp/education)

Haggerty BP, Mazer SJ (2009) The phenology handbook: a guide to phenological monitoring for students, teachers, families, and nature enthusiasts. Extended conceptual and historical background, practical advice, field guides, and educational activities for introducing phenology to a variety of audiences. Available online at the USA National Phenology Network website www.usanpn.org/education

Haggerty BP, Hove AA, Mazer SJ (2012a) Primer on herbarium-based phenological research. Guide for college and public audiences for understanding the use of preserved plants in climate change research. Available online at the California Phenology Project website www.usanpn.org/cpp/education

Haggerty BP, Hove AA, Mazer SJ (2012b) Skeletons in the closet: preserved plants reveal phenological responses to climate change. Guide and practice data set for college and public audiences for analyzing long-term phenological data. Available online at the California Phenology Project website www.usanpn.org/cpp/education

Haggerty BP, Hove AA, Mazer SJ (2012c) Phenology gardens: a practical guide for integrating phenology into garden planning and education. Available online at the California Phenology Project website www.usanpn.org/cpp/education

Haggerty BP, Hove AA, Mazer SJ (2012d) Flight of the pollinators. Activity guide merging phenology and pollination biology for K-16 and public audiences. Available online at the California Phenology Project website www.usanpn.org/cpp/education

Haggerty BP, Hove AA, Mazer SJ (2012e) Ethnophenology. Activity guide merging phenology and ethnobotany for K-16 and public audiences. Available online at the California Phenology Project website www.usanpn.org/cpp/education

Hove AA, Mazer SJ, Haggerty BP (2012a) Phenology: the science of the seasons. Three-part annotated lecture series for undergraduates. Available online at the California Phenology Project website www.usanpn.org/cpp/education

Hove AA, Mazer SJ, Haggerty BP (2012b) Advanced topics in phenological and climate change research. Nine-part graduate seminar series with peer-reviewed literature. Available online at the California Phenology Project website www.usanpn.org/cpp/education

Huemann BW, Seaquist JW, Eklundh L, Jönsson P (2007) AVHRR derived phenological change in the Sahel and Soudan, Africa, 1982–2005. Remote Sens Environ 108:385–392

Jönsson P, Eklundh L (2002) Seasonality extraction by function fitting to time-series of satellite sensor data. IEEE Trans Geosci Remote Sens 40(8):1824–1831

Jönsson P, Eklundh L (2004) TIMESAT - a program for analyzing time-series of satellite sensor data. Comput Geosci 30:833–845

Julien Y, Sobrino JA, Verhoef W (2006) Changes in land surface temperatures and NDVI values over Europe between 1982 and 1999. Remote Sens Environ 103:43–55

Lee R, Yu F, Price KP, Ellis J, Shi P (2002) Evaluating vegetation phenological patterns in Inner Mongolia using NDVI time-series analysis. Int J Remote Sens 23(12):2505–2512

Long T, Wyse S (2012) A season for inquiry: investigating phenology in local campus trees. Science 335(6071):932–933. doi:10.1126/science.1213528

Menzel A (2000) Trends in phenological phases in Europe between 1951 and 1996. Int J Biometeorol 44:76–81

Myneni RB, Keeling CD, Tucker CJ, Asrar G, Nemani RR (1997) Increased plant growth in the northern high latitudes from 1981 to 1991. Nature 386(6626):698–702

USA-NPN National Coordinating Office (2012) The phenology trail guide: an experiential education tool for site-based community engagement. USA-NPN Programmatic/Technical Series 2012–001

Potter CS, Brooks V (2001) Global analysis of empirical relations between annual climate and seasonality of NDVI. Int J Remote Sens 19(15):2921–2948

Potter C, Kumar V, Klooster S, Nemani RR (2007) Recent history of trends in vegetation greenness and large-scale ecosystem disturbances in Eurasia. Tellus 59B:260–272

Primack RB, Miller-rushing AJ (2009) The role of botanical gardens in climate change research. New Phytol 182:303–313

Reed BC, Brown JF, VanderZee D, Loveland TR, Merchant JW, Ohlen DO (1994) Measuring phenological variability from satellite imagery. J Veg Sci 5:703–714

Richardson A, Jenkins J, Braswell B, Hollinger D, Ollinger S, Smith M-L (2007) Use of digital webcam images to track spring green-up in a deciduous broadleaf forest. Oecologia 152(2):323–334. doi:10.1007/s00442-006-0657-z

SanthanaVannan SK, Cook RB, Holladay SK, Olsen LM, Dadi U, Wilson BE (2009) A web-based subsetting service for regional scale MODIS land products. IEEE J Sel Top Appl 2 (4):319–328. doi:10.1109/JSTARS.2009.2036585

Stearns FW (1974) Phenology and environmental education. In: Lieth H (ed) Phenology and seasonality modeling. Springer – Verlag, New York

Tucker CJ, Slayback DA, Pinzon JE, Los SO, Myneni RB, Taylor MG (2001) Higher northern latitude normalized difference vegetation index and growing season trends from 1982 to 1999. Int J Biometeorol 45(4):184–190

Verbesselt J, Jönsson P, Lhermitte S, van Aardt J, Coppin P (2006) Evaluating satellite and climate data derived indices as fire risk indicators in Savanna ecosystems. IEEE Trans Geosci Remote Sens 44(6):1622–1632

White MA, Thornton PE, Running SW (1997) A continental phenology model for monitoring vegetation responses to interannual climatic variability. Glob Biogeochem Cycles 11 (2):217–234

White MA, de Beurs KM, Didan K, Inouye DW, Richardson AD, Jensen OP, O'Keefe J, Zhang G, Nemani RR, van Leeuwen WJD, Brown JF, de Wit A, Schaepman M, Lin X, Dettinger M, Bailey AS, Kimball JS, Schwartz MD, Baldocchi DD, Lee JT, Lauenroth WK (2009) Intercomparison, interpretation, and assessment of spring phenology in North America estimated from remote sensing for 1982–2006. Glob Change Biol 15(10):2335–2359

Zhang X, Friedl MA, Schaaf CB, Strahler AH, Hodges JCF, Gao F, Reed BC, Huete A (2003) Monitoring vegetation phenology using MODIS. Remote Sens Environ 84(3):471–475

Zhang X, Friedl MA, Schaaf CB, Strahler AH (2004) Climate controls on vegetation phenological patterns in northern mid- and high latitudes inferred from MODIS data. Glob Change Biol 10 (7):1133–1145

Zhou L, Kaufmann RK, Tian Y, Myneni RB, Tucker CJ (2003) Relation between interannual variations in satellite measures of northern forest greenness and climate between 1982 and 1999. J Geophys Res 108(D1):ACL 3-1–3-16. doi:10.1029/2002JD002510

Index

M.D. Schwartz (ed.), *Phenology: An Integrative Environmental Science*, DOI 10.1007/978-94-007-6925-0, © Springer Science+Business Media B.V. 2013

W

Y

Printed by Publishers' Graphics LLC